HANDBOOK
OF
MULTISENSOR
DATA FUSION

THE ELECTRICAL ENGINEERING
AND APPLIED SIGNAL PROCESSING SERIES
Edited by Alexander Poularikas

The Advanced Signal Processing Handbook:
Theory and Implementation for Radar, Sonar,
and Medical Imaging Real-Time Systems
Stergios Stergiopoulos

The Transform and Data Compression Handbook
K.R. Rao and P.C. Yip

Handbook of Multisensor Data Fusion
David Hall and James Llinas

Handbook of Antennas in Wireless Communications
Lal Chand Godara

Forthcoming Titles

Propagation Data Handbook for Wireless Communications
Robert Crane

The Digital Color Imaging Handbook
Guarav Sharma

Handbook of Neural Network Signal Processing
Yu Hen Hu and Jeng-Neng Hwang

Applications in Time Frequency Signal Processing
Antonia Papandreou-Suppappola

Noise Reduction in Speech Applications
Gillian Davis

Signal Processing in Noise
Vyacheslav Tuzlukov

Electromagnetic Radiation and the Human Body:
Effects, Diagnosis and Therapeutic Technologies
Nikolaos Uzunoglu and Konstantina S. Nikita

Digital Signal Processing with Examples in MATLAB®
Samuel Stearns

HANDBOOK OF MULTISENSOR DATA FUSION

Edited by
DAVID L. HALL, PH.D.
JAMES LLINAS, PH.D.

CRC Press
Boca Raton   London   New York   Washington, D.C.

Library of Congress Cataloging-in-Publication Data

Hall, David L.
    Handbook of multisensor data fusion / David L. Hall and James Llinas.
        p.   cm. -- (Electrical engineering and applied signal processing)
    Includes bibliographical references and index.
    ISBN 0-8493-2379-7 (alk. paper)
        1. Multisensor data fusion--Handbooks, manuals, etc. I. Llinas, James. II. Title. III.
    Series.

TK5102.9 .H355 2001
681'.2--dc21   2001025085

Visit the CRC Press Web site at www.crcpress.com

© 2001 by CRC Press LLC

International Standard Book Number 0-8493-2379-7
Library of Congress Card Number 2001025085
Printed in the United States of America   2 3 4 5 6 7 8 9 0
Printed on acid-free paper

PREFACE

Multisensor data fusion is an emerging technology applied to Department of Defense (DoD) areas such as automated target recognition (ATR), identification-friend-foe-neutral (IFFN) recognition systems, battlefield surveillance, and guidance and control of autonomous vehicles. Non-DoD applications include monitoring of complex machinery, environmental surveillance and monitoring systems, medical diagnosis, and smart buildings. Techniques for data fusion are drawn from a wide variety of disciplines, including signal processing, pattern recognition, statistical estimation, artificial intelligence, and control theory. The rapid evolution of computers, proliferation of micro-mechanical/electrical systems (MEMS) sensors, and the maturation of data fusion technology provide a basis for utilization of data fusion in everyday applications.

This book is intended to be a comprehensive resource for data fusion system designers and researchers, providing information on terminology, models, algorithms, systems engineering issues, and examples of applications. The book is divided into four main parts. Part I introduces data fusion terminology and models. Chapter 1 provides a general introduction to data fusion and terminology. Chapter 2 introduces the Joint Directors of Laboratories (JDL) data fusion process model, widely used to assist in understanding DoD applications. In Chapter 3, Jeffrey Uhlmann discusses the problem of multitarget, multisensor tracking and introduces the challenges of data association and correlation. Chapter 4, by Ed Waltz, introduces concepts of image and spatial data fusion, and in Chapter 5 Richard Brooks and Lynne Grewe describe issues of data registration for image fusion. Chapter 6, written by Richard Antony, discusses issues of data fusion focused on situation assessment and database management. Finally, in Chapter 7, Joseph Carl contrasts some approaches to combining evidence using probability and fuzzy set theory.

A perennial problem in multisensor fusion involves combining data from multiple sensors to track moving targets. Gauss originally addressed this problem for estimating the orbits of asteroids by developing the method of least squares. In its most general form, this problem is not tractable. In general, we do not know *a priori* how many targets exist or how to assign observations to potential targets. Hence, we must simultaneously estimate the state (e.g., position and velocity) of N targets based on M sensor reports and also determine which of the M reports belong to (or should be assigned to) each of the N targets. This problem may be complicated by closely spaced, maneuvering targets with potential observational clutter and false alarms.

Part II of this book presents alternative views of this multisensor, multitarget tracking problem. In Chapter 8, T. Kirubarajan and Yaakov Bar-Shalom present an overview of their approach for probabilistic data association (PDA) and the joint PDA (JPDA) methods. These have been useful in dense target tracking environments. In Chapter 9, Jeffrey Uhlmann describes another approach using an approximate method for addressing the data association combination problem. A classical Bayesian approach to target tracking and identification is described by Lawrence D. Stone in Chapter 10. This has been applied to problems in target identification and tracking for undersea vehicles. Recent research by Aubrey B. Poore, Suihua Lu, and Brian J. Suchomel is summarized in Chapter 11. Poore's approach combines the problem of estimation and data association by generalizing the optimization problem, followed by development of efficient computational methods. In Chapter 12, Simon Julier and Jeffrey K. Uhlmann discuss issues related to the estimation of target error and how to treat the codependence between sensors. They extend this work to nonlinear systems in Chapter 13. Finally, in Chapter 14, Ronald Mahler provides a very extensive discussion of multitarget, multisensor tracking using an approach based on random set theory.

Part III of this book addresses issues of the design and development of data fusion systems. It begins with Chapter 15 by Ed Waltz and David L. Hall, and describes a systemic approach for deriving data fusion system requirements. Chapter 16 by Christopher Bowman and Alan Steinberg provides a general discussion of the systems engineering process for data fusion systems including the selection of appropriate architectures. In Chapter 17, David L. Hall, James Llinas, Christopher L. Bowman, Lori McConnel, and Paul Applegate provide engineering guidelines for the selection of data fusion algorithms. In Chapter 18, Richard Antony presents a discussion of database management support, with applications to tactical data fusion. New concepts for designing human-computer interfaces (HCI) for data fusion systems are summarized in Chapter 19 by Mary Jane Hall, Sonya Hall, and Timothy Tate. Performance assessment issues are described by James Llinas in Chapter 20. Finally, in Chapter 21, David L. Hall and Alan N. Steinberg present the *dirty secrets* of data fusion. The experience of implementing data fusion systems described in this section was primarily gained on DoD applications; however, the lessons learned should be of value to system designers for any application.

Part IV of this book provides a taste of the breadth of applications to which data fusion technology can be applied. Mary L. Nichols, in Chapter 22, presents a limited survey of some DoD fusion systems. In Chapter 23, Carl S. Byington and Amulya K. Garga describe the use of data fusion to improve the ability to monitor complex mechanical systems. Robert J. Hansen, Daniel Cooke, Kenneth Ford, and Steven Zornetzer provide an overview of data fusion applications at the National Aeronautics and Space Administration (NASA) in Chapter 24. In Chapter 25, Richard R. Brooks describes an application of data fusion funded by DARPA. Finally, in Chapter 26, Hans Keithley describes how to determine the utility of data fusion for C4ISR. This fourth part of the book is not by any means intended to be a comprehensive survey of data fusion applications. Instead, it is included to provide the reader with a sense of different types of applications. Finally, Part V of this book provides a list of Internet Web sites and news groups related to multisensor data fusion.

The editors hope that this handbook will be a valuable addition to the bookshelves of data fusion researchers and system designers. We remind the reader that data fusion remains an evolving discipline. Even for classic problems, such as multisensor, multitarget tracking, competing approaches exist. The book has sought to identify and provide a representation of the leading methods in data fusion. The reader should be advised, however, that there are disagreements in the data fusion community (especially by some of the contributors to this book) concerning which method is *best*. It is interesting to read the descriptions that the authors in this book present concerning the relationship between their own techniques and those of the other authors. Many of this book's contributors have written recent texts that advocate a particular method. These authors have condensed or summarized that information as a chapter here.

We take the view that each competing method must be considered in the context of a specific application. We believe that there is no such thing as a generic data fusion system. Instead, there are numerous applications to which data fusion techniques can be applied. In our view, there is no such thing as a magic approach or technique. Even very sophisticated algorithms may be corrupted by a lack of *a priori* information or incorrect information concerning sensor performance. Thus, we advise the reader to become a knowledgeable and demanding consumer of fusion algorithms.

We hope that this text will become a companion to other texts on data fusion methods and techniques, and that it assists the data fusion community in its continuing maturation process.

Acknowledgment

The editors acknowledge the support and dedication of Ms. Natalie Nodianos, who performed extensive work to coordinate with the contributing authors. In addition, she assisted the contributing authors in clarifying and improving their manuscripts. Her attention to detail and her insights have greatly assisted in developing this handbook. In addition, the editors acknowledge the extensive work done by Mary Jane Hall. She provided support in editing, developed many graphics, and assisted in coordinating the final review process. She also provided continuous encouragement and moral support throughout this project. Finally, the editors would like to express their appreciation for the assistance provided by Barbara L. Davies.

Editors

David L. Hall, Ph.D., is the associate dean of research and graduate studies for The Pennsylvania State University School of Information Sciences and Technology. He has conducted research in data fusion and related technical areas for more than 20 years and has lectured internationally on data fusion and artificial intelligence. In addition, he has participated in the implementation of real-time data fusion systems for several military applications. He is the author of three textbooks (including *Mathematical Techniques in Multisensor Data Fusion*, published by Artech House, 1992) and more than 180 technical papers. Prior to joining the Pennsylvania State University, Dr. Hall worked at HRB Systems (a division of Raytheon, E-Systems), at the Computer Sciences Corporation, and at the MIT Lincoln Laboratory. He is a senior member of the IEEE. Dr. Hall earned a master's and doctorate degrees in astrophysics and an undergraduate degree in physics and mathematics.

James Llinas, Ph.D., is an adjunct research professor at the State University of New York at Buffalo. An expert in data fusion, he coauthored the first integrated book on the subject (*Multisensor Data Fusion*, published by Artech House, 1990) and has lectured internationally on the subject for over 15 years. For the past decade, he has been a technical advisor to the Defense Department's Joint Directors of Laboratories Data Fusion Panel. His experience in applying data fusion technology to different problem areas ranges from complex defense and intelligence-system applications to nondefense diagnosis. His current projects include basic and applied research in automated reasoning, distributed, cooperative problem solving, avionics information fusion architectures, and the scientific foundations of data correlation. He earned a doctorate degree in industrial engineering.

Contributors

Richard Antony
VGS Inc.
Fairfax, Virginia

Paul Applegate
Consultant
Buffalo, New York

Yaakov Bar-Shalom
University of Connecticut
Storrs, Connecticut

Christopher L. Bowman
Consultant
Broomfield, Colorado

Richard R. Brooks
The Pennsylvania State University
University Park, Pennsylvania

Carl S. Byington
The Pennsylvania State University
University Park, Pennsylvania

Joseph W. Carl
Harris Corporation
Annapolis, Maryland

Daniel Cooke
NASA Ames Research Center
Moffett Field, California

Kenneth Ford
Institute for Human and Machine
    Cognition
Pensacola, Florida

Amulya K. Garga
The Pennsylvania State University
University Park, Pennsylvania

Lynne Grewe
California State University
Hayward, California

David L. Hall
The Pennsylvania State University
University Park, Pennsylvania

Mary Jane M. Hall
TECH REACH Inc.
State College, Pennsylvania

Capt. Sonya A. Hall
Minot AFB
Minot, North Dakota

Robert J. Hansen
University of West Florida
Pensacola, Florida

Simon Julier
IDAK Industries
Jefferson City, Missouri

Hans Keithley
Office of the Secretary of Defense
    Decision Support Center
Arlington, Virginia

T. Kirubarajan
University of Connecticut
Storrs, Connecticut

James Llinas
State University of New York
Buffalo, New York

Suihua Lu
Colorado State University
Fort Collins, Colorado

Ronald Mahler
Lockheed Martin
Eagan, Minnesota

Capt. Lori McConnel
USAF/Space Warfare Center
Denver, Colorado

Mary L. Nichols
The Aerospace Corporation
El Segundo, California

Aubrey B. Poore
Colorado State University
Fort Collins, Colorado

Alan N. Steinberg
Utah State University
Logan, Utah

Lawrence D. Stone
Metron, Inc.
Reston, Virginia

Brian J. Suchomel
Numerica, Inc.
Fort Collins, Colorado

Timothy Tate
Naval Training Command
Arlington, Virginia

Jeffrey K. Uhlmann
University of Missouri
Columbia, Missouri

Ed Waltz
Veridian Systems
Ann Arbor, Michigan

Steven Zornetzer
NASA Ames Research Center
Moffett Field, California

Contents

Part I   Introduction to Multisensor Data Fusion

1   Multisensor Data Fusion   David L. Hall and James Llinas
1.1   Introduction 1-1
1.2   Multisensor Advantages.. 1-2
1.3   Military Applications 1-3
1.4   Nonmilitary Applications... 1-4
1.5   Three Processing Architectures 1-5
1.6   A Data Fusion Process Model.. 1-6
1.7   Assessment of the State of the Art 1-8
1.8   Additional Information 1-10
Reference 1-10

**2   Revisions to the JDL Data Fusion Model   Alan N. Steinberg and
Christopher L. Bowman**
2.1   Introduction 2-1
2.2   What Is Data Fusion? What Isn't?... 2-1
2.3   Models and Architectures.. 2-4
2.4   Beyond the Physical 2-12
2.5   Comparison with Other Models 2-15
2.6   Summary 2-17
References 2-18

**3   Introduction to the Algorithmics of Data Association in Multiple-Target
Tracking   Jeffrey K. Uhlmann**
3.1   Introduction 3-1
3.2   Ternary Trees 3-10
3.3   Priority *kd*-Trees 3-13
3.4   Conclusion 3-17
Acknowledgments 3-17
References 3-17

4 The Principles and Practice of Image and Spatial Data Fusion
Ed Waltz

4.1	Introduction	4-1
4.2	Motivations for Combining Image and Spatial Data	4-2
4.3	Defining Image and Spatial Data Fusion	4-3
4.4	Three Classic Levels of Combination for Multisensor Automatic Target Recognition Data Fusion	4-5
4.5	Image Data Fusion for Enhancement of Imagery Data	4-10
4.6	Spatial Data Fusion Applications	4-11
4.7	Summary	4-15
	References	4-15

5 Data Registration *Richard R. Brooks and Lynne Grewe*

5.1	Introduction	5-1
5.2	Registration Problem	5-2
5.3	Review of Existing Research	5-3
5.4	Registration Using Meta-Heuristics	5-5
5.5	Wavelet-Based Registration of Range Images	5-7
5.6	Registration Assistance/Preprocessing	5-9
5.7	Conclusion	5-10
	Acknowledgments	5-10
	References	5-11

6 Data Fusion Automation: A Top-Down Perspective *Richard Antony*

6.1	Introduction	6-1
6.2	Biologically Motivated Fusion Process Model	6-8
6.3	Fusion Process Model Extensions	6-14
6.4	Observations	6-22
	Acknowledgments	6-25
	References	6-25

7 Contrasting Approaches to Combine Evidence *Joseph W. Carl*

7.1	Introduction	7-1
7.2	Alternative Approaches to Combine Evidence	7-2
7.3	An Example Data Fusion System	7-18
7.4	Contrasts and Conclusion	7-31
	Appendix 7.A The Axiomatic Definition of Probability	7-31
	References	7-31

Part II Advanced Tracking and Association Methods

8 Target Tracking Using Probabilistic Data Association-Based Techniques with Applications to Sonar, Radar, and EO Sensors *T. Kirubarajan and Yaakov Bar-Shalom*

    8.1   Introduction **8-1**
    8.2   Probabilistic Data Association... **8-2**
    8.3   Low Observable TMA Using the ML-PDA Approach with Features................... **8-8**
    8.4   The IMMPDAF for Tracking Maneuvering Targets **8-17**
    8.5   A Flexible-Window ML-PDA Estimator for Tracking Low Observable (LO) Targets **8-27**
    8.6   Summary **8-37**
    References **8-37**

9 An Introduction to the Combinatorics of Optimal and Approximate Data Association *Jeffrey K. Uhlmann*

    9.1   Introduction **9-1**
    9.2   Background **9-2**
    9.3   Most Probable Assignments **9-4**
    9.4   Optimal Approach **9-5**
    9.5   Computational Considerations... **9-7**
    9.6   Efficient Computation of the JAM... **9-8**
    9.7   Crude Permanent Approximations.. **9-8**
    9.8   Approximations Based on Permanent Inequalities............................ **9-10**
    9.9   Comparisons of Different Approaches **9-12**
    9.10 Large-Scale Data Associations... **9-15**
    9.11 Generalizations... **9-17**
    9.12 Conclusions **9-17**
    Acknowledgments **9-18**
    Appendix 9.A Algorithm for Data Association Experiment **9-18**
    References **9-19**

10 A Bayesian Approach to Multiple-Target Tracking *Lawrence D. Stone*

    10.1 Introduction **10-1**
    10.2 Bayesian Formulation of the Single-Target Tracking Problem....................... **10-3**
    10.3 Multiple-Target Tracking without Contacts or Association (Unified Tracking)............ **10-8**
    10.4 Multiple-Hypothesis Tracking (MHT)... **10-12**
    10.5 Relationship of Unified Tracking to MHT and Other Tracking Approaches **10-22**
    10.6 Likelihood Ratio Detection and Tracking **10-23**
    References **10-30**

11 Data Association Using Multiple Frame Assignments *Aubrey B. Poore, Suihua Lu, and Brian J. Suchomel*

11.1 Introduction	11-1
11.2 Problem Background	11-2
11.3 Assignment Formulation of Some General Data Association Problems	11-3
11.4 Multiple Frame Track Initiation and Track Maintenance	11-8
11.5 Algorithms	11-10
11.6 Future Directions	11-14
Acknowledgments	11-16
References	11-16

12 General Decentralized Data Fusion with Covariance Intersection (CI) *Simon Julier and Jeffrey K. Uhlmann*

12.1 Introduction	12-1
12.2 Decentralized Data Fusion	12-2
12.3 Covariance Intersection	12-5
12.4 Using Covariance Intersection for Distributed Data Fusion	12-8
12.5 Extended Example	12-10
12.6 Incorporating Known Independent Information	12-13
12.7 Conclusions	12-19
Appendix 12.A The Consistency of CI	12-21
Appendix 12.B MATLAB Source Code (Conventional CI and Split CI)	12-23
Acknowledgments	12-21
References	12-24

13 Data Fusion in Nonlinear Systems *Simon Julier and Jeffrey K. Uhlmann*

13.1 Introduction	13-1
13.2 Estimation in Nonlinear Systems	13-2
13.3 The Unscented Transformation (UT)	13-5
13.4 Uses of the Transformation	13-8
13.5 The Unscented Filter (UF)	13-12
13.6 Case Study: Using the UF with Linearization Errors	13-13
13.7 Case Study: Using the UF with a High-Order Nonlinear System	13-15
13.8 Multilevel Sensor Fusion	13-18
13.9 Conclusions	13-20
Acknowledgments	13-21
References	13-21

14 Random Set Theory for Target Tracking and Identification *Ronald Mahler*

14.1 Introduction	14-3
14.2 Basic Statistics for Tracking and Identification	14-10
14.3 Multitarget Sensor Models	14-12

14.4 Multitarget Motion Models **14**-14
14.5 The FISST Multisource-Multitarget Calculus **14**-15
14.6 FISST Multisource-Multitarget Statistics **14**-19
14.7 Optimal-Bayes Fusion, Tracking, ID **14**-23
14.8 Robust-Bayes Fusion, Tracking, ID **14**-26
14.9 Summary and Conclusions **14**-30
Acknowledgments **14**-30
References **14**-30

Part III Systems Engineering and Implementation

15 Requirements Derivation for Data Fusion Systems *Ed Waltz and David L. Hall*
    15.1 Introduction **15**-1
    15.2 Requirements Analysis Process **15**-2
    15.3 Engineering Flow-Down Approach **15**-3
    15.4 Enterprise Architecture Approach **15**-5
    15.5 Comparison of Approaches **15**-6
    References **15**-8

16 A Systems Engineering Approach for Implementing Data Fusion Systems *Christopher L. Bowman and Alan N. Steinberg*
    16.1 Scope **16**-1
    16.2 Architecture for Data Fusion **16**-2
    16.3 Data Fusion System Engineering Process **16**-7
    16.4 Fusion System Role Optimization **16**-17
    References **16**-38

17 Studies and Analyses with Project Correlation: An In-Depth Assessment of Correlation Problems and Solution Techniques *James Llinas, Lori McConnel, Christopher L. Bowman, David L. Hall, and Paul Applegate*
    17.1 Introduction **17**-1
    17.2 A Description of the Data Correlation (DC) Problem **17**-3
    17.3 Hypothesis Generation **17**-4
    17.4 Hypothesis Evaluation **17**-8
    17.5 Hypothesis Selection **17**-9
    17.6 Summary **17**-17
    References **17**-18

18 Data Management Support to Tactical Data Fusion *Richard Antony*
    18.1 Introduction **18**-1
    18.2 Database Management Systems **18**-2

18.3 Spatial, Temporal, and Hierarchical Reasoning **18**-3
18.4 Database Design Criteria **18**-6
18.5 Object Representation of Space **18**-14
18.6 Integrated Spatial/Nonspatial Data Representation **18**-16
18.7 Sample Application **18**-17
18.8 Summary and Conclusions **18**-25
Acknowledgments **18**-25
References **18**-25

19 Removing the HCI Bottleneck: How the Human-Computer
Interface (HCI) Affects the Performance of Data Fusion Systems
Mary Jane M. Hall, Sonya A. Hall, and Timothy Tate
19.1 Introduction **19**-1
19.2 A Multimedia Experiment **19**-3
19.3 Summary of Results **19**-5
19.4 Implications for Data Fusion Systems **19**-9
Acknowledgment **19**-10
References **19**-11

20 Assessing the Performance of Multisensor Fusion Processes
James Llinas
20.1 Introduction **20**-1
20.2 Test and Evaluation of the Data Fusion Process **20**-3
20.3 Tools for Evaluation: Testbeds, Simulations, and Standard Data Sets **20**-7
20.4 Relating Fusion Performance to Military Effectiveness — Measures of Merit **20**-11
20.5 Summary **20**-17
References **20**-17

21 Dirty Secrets in Multisensor Data Fusion   *David L. Hall and Alan N.
Steinberg*
21.1 Introduction **21**-1
21.2 The JDL Data Fusion Process Model **21**-2
21.3 Current Practices and Limitations in Data Fusion **21**-2
21.4 Research Needs **21**-7
21.5 Pitfalls in Data Fusion **21**-9
21.6 Summary **21**-10
References **21**-10

Part IV   Sample Applications

22 A Survey of Multisensor Data Fusion Systems   *Mary L. Nichols*
22.1 Introduction **22**-1
22.2 Recent Survey of Data Fusion Activities **22**-1
22.3 Assessment of System Capabilities **22**-2

References 22-7

23 Data Fusion for Developing Predictive Diagnostics for
   Electromechanical Systems   *Carl S. Byington and Amulya K. Garga*
   23.1 Introduction 23-1
   23.2 Aspects of a CBM System 23-3
   23.3 The Diagnosis Problem 23-4
   23.4 Multisensor Fusion Toolkit 23-7
   23.5 Application Examples 23-8
   23.6 Concluding Remarks 23-29
   Acknowledgments 23-30
   References 23-30

24 Information Technology for NASA in the 21st Century   *Robert J.*
   Hansen, Daniel Cooke, Kenneth Ford, and Steven Zornetzer
   24.1 Introduction 24-1
   24.2 NASA Applications 24-2
   24.3 Critical Research Investment Areas for NASA 24-3
   24.4 High-Performance Computing and Networking 24-5
   24.5 Conclusions 24-6

25 Data Fusion for a Distributed Ground-Based Sensing System
   Richard R. Brooks
   25.1 Introduction 25-1
   25.2 Problem Domain 25-2
   25.3 Existing Systems 25-3
   25.4 Prototype Sensors for SenseIT 25-4
   25.5 Software Architecture 25-5
   25.6 Declarative Language Front-End 25-6
   25.7 Subscriptions 25-6
   25.8 Mobile Code 25-7
   25.9 Diffusion Network Routing 25-7
   25.10 Collaborative Signal Processing 25-7
   25.11 Information Security 25-8
   25.12 Summary 25-8
   Acknowledgments and Disclaimers 25-8
   References 25-8

26 An Evaluation Methodology for Fusion Processes Based on Information
   Needs   *Hans Keithley*
   26.1 Introduction 26-1
   26.2 Information Needs 26-2
   26.3 Key Concept 26-6

26.4   Evaluation Methodology 26-6
References 26-9

Part V   Resources

Web Sites and News Groups Related to Data Fusion

Data Fusion Web Sites..A-1
News Groups..A-3
Other World Wide Web Information A-4
Government Laboratories and Agencies A-4

Index

Index I-1

I

Introduction to Multisensor Data Fusion

1 **Multisensor Data Fusion** *David L. Hall and James Llinas* 1-1
Introduction • Multisensor Advantages • Military Applications • Nonmilitary
Applications • Three Processing Architectures • A Data Fusion Process Model •
Assessment of the State of the Art • Additional Information

2 **Revisions to the JDL Data Fusion Model** *Alan N. Steinberg and
Christopher L. Bowman* 2-1
Introduction • What Is Data Fusion? What Isn't? • Models and Architectures • Beyond
the Physical • Comparison with Other Models • Summary

3 **Introduction to the Algorithmics of Data Association in Multiple-Target
Tracking** *Jeffrey K. Uhlmann* 3-1
Introduction • Ternary Trees • Priority *kd*-Trees • Conclusion • Acknowledgments

4 **The Principles and Practice of Image and Spatial Data Fusion** *Ed Waltz* 4-1
Introduction • Motivations for Combining Image and Spatial Data • Defining Image and
Spatial Data Fusion • Three Classic Levels of Combination for Multisensor Automatic Target
Recognition Data Fusion • Image Data Fusion for Enhancement of Imagery Data • Spatial
Data Fusion Applications • Summary

5 **Data Registration** *Richard R. Brooks and Lynne Grewe* 5-1
Introduction • Registration Problem • Review of Existing Research • Registration Using
Meta-Heuristics • Wavelet-Based Registration of Range Images • Registration
Assistance/Preprocessing • Conclusion • Acknowledgments

6 **Data Fusion Automation: A Top-Down Perspective** *Richard Antony* 6-1
Introduction • Biologically Motivated Fusion Process Model • Fusion Process Model
Extensions • Observations • Acknowledgments

7 **Contrasting Approaches to Combine Evidence** *Joseph W. Carl* 7-1
Introduction • Alternative Approaches to Combine Evidence • An Example Data Fusion
System • Contrasts and Conclusion • Appendix 7.A The Axiomatic Definition of
Probability

1

Multisensor Data Fusion

David L. Hall
The Pennsylvania State University

James Llinas
State University of New York

1.1   Introduction 1-1
1.2   Multisensor Advantages.. 1-2
1.3   Military Applications 1-3
1.4   Nonmilitary Applications 1-4
1.5   Three Processing Architectures 1-5
1.6   A Data Fusion Process Model 1-6
1.7   Assessment of the State of the Art...................................... 1-8
1.8   Additional Information 1-10
Reference 1-10

Integration or fusion of data from multiple sensors improves the accuracy of applications ranging from target tracking and battlefield surveillance to nondefense applications such as industrial process monitoring and medical diagnosis.

1.1 Introduction

In recent years, significant attention has focused on multisensor data fusion for both military and nonmilitary applications. Data fusion techniques combine data from multiple sensors and related information to achieve more specific inferences than could be achieved by using a single, independent sensor.

The concept of multisensor data fusion is hardly new. As humans and animals have evolved, they have developed the ability to use multiple senses to help them survive. For example, assessing the quality of an edible substance may not be possible using only the sense of vision; the combination of sight, touch, smell, and taste is far more effective. Similarly, when vision is limited by structures and vegetation, the sense of hearing can provide advanced warning of impending dangers. Thus, multisensory data fusion is naturally performed by animals and humans to assess more accurately the surrounding environment and to identify threats, thereby improving their chances of survival.

While the concept of data fusion is not new, the emergence of new sensors, advanced processing techniques, and improved processing hardware have made real-time fusion of data increasingly viable. Just as the advent of symbolic processing computers (e.g., the SYMBOLICs computer and the Lambda machine) in the early 1970s provided an impetus to artificial intelligence, recent advances in computing and sensing have provided the capability to emulate, in hardware and software, the natural data fusion capabilities of humans and animals. Currently, data fusion systems are used extensively for target tracking, automated identification of targets, and limited automated reasoning applications. Data fusion technology has rapidly advanced from a loose collection of related techniques to an emerging true engineering discipline with standardized terminology, collections of robust mathematical techniques, and established system design principles.

Applications for multisensor data fusion are widespread. Military applications include automated target recognition (e.g., for smart weapons), guidance for autonomous vehicles, remote sensing, battlefield surveillance, and automated threat recognition systems, such as identification-friend-foe-neutral (IFFN) systems. Nonmilitary applications include monitoring of manufacturing processes, condition-based maintenance of complex machinery, robotics, and medical applications.

Techniques to combine or fuse data are drawn from a diverse set of more traditional disciplines, including digital signal processing, statistical estimation, control theory, artificial intelligence, and classic numerical methods. Historically, data fusion methods were developed primarily for military applications. However, in recent years, these methods have been applied to civilian applications and a bidirectional transfer of technology has begun.

1.2 Multisensor Advantages

Fused data from multiple sensors provides several advantages over data from a single sensor. First, if several identical sensors are used (e.g., identical radars tracking a moving object), combining the observations will result in an improved estimate of the target position and velocity. A statistical advantage is gained by adding the N independent observations (e.g., the estimate of the target location or velocity is improved by a factor proportional to $N^{\frac{1}{2}}$), assuming the data are combined in an optimal manner. This same result could also be obtained by combining N observations from an individual sensor.

A second advantage involves using the relative placement or motion of multiple sensors to improve the observation process. For example, two sensors that measure angular directions to an object can be coordinated to determine the position of an object by triangulation. This technique is used in surveying and for commercial navigation. Similarly, the use of two sensors, one moving in a known way with respect to another, can be used to measure instantaneously an object's position and velocity with respect to the observing sensors.

A third advantage gained by using multiple sensors is improved observability. Broadening the baseline of physical observables can result in significant improvements. Figure 1.1 provides a simple example of a moving object, such as an aircraft, that is observed by both a pulsed radar and a forward-looking infrared (FLIR) imaging sensor. The radar can accurately determine the aircraft's range but has a limited ability to determine the angular direction of the aircraft. By contrast, the infrared imaging sensor can accurately determine the aircraft's angular direction but cannot measure range. If these two observations are correctly associated (as shown in Figure 1.1), the combination of the two sensors provides a better

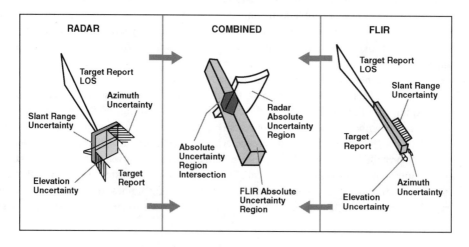

FIGURE 1.1   A moving object observed by both a pulsed radar and an infrared imaging sensor.

TABLE 1.1 Representative Data Fusion Applications for Defense Systems

Specific Applications	Inferences Sought by Data Fusion Process	Primary Observable Data	Surveillance Volume	Sensor Platforms
Ocean surveillance	Detection, tracking, identification of targets and events	EM signals Acoustic signals Nuclear-related Derived observations	Hundreds of nautical miles Air/surface/sub- surface	Ships Aircraft Submarines Ground-based Ocean-based
Air-to-air and surface-to-air defense	Detection, tracking, identification of aircraft	EM radiation	Hundreds of miles (strategic) Miles (tactical)	Ground-based Aircraft
Battlefield intelligence, surveillance, and target acquisition	Detection and identification of potential ground targets	EM radiation	Tens of hundreds of miles about a battlefield	Ground-based Aircraft
Strategic warning and defense	Detection of indications of impending strategic actions Detection and tracking of ballistic missiles and warheads	EM radiation Nuclear-related	Global	Satellites Aircraft

determination of location than could be obtained by either of the two independent sensors. This results in a reduced error region, as shown in the fused or combined location estimate. A similar effect may be obtained in determining the identity of an object based on observations of an object's attributes. For example, there is evidence that bats identify their prey by a combination of factors, including size, texture (based on acoustic signature), and kinematic behavior.

1.3 Military Applications

The Department of Defense (DoD) community focuses on problems involving the location, characterization, and identification of dynamic entities such as emitters, platforms, weapons, and military units. These dynamic data are often termed an order-of-battle database or order-of-battle display (if superimposed on a map display). Beyond achieving an order-of-battle database, DoD users seek higher-level inferences about the enemy situation (e.g., the relationships among entities and their relationships with the environment and higher level enemy organizations). Examples of DoD-related applications include ocean surveillance, air-to-air defense, battlefield intelligence, surveillance and target acquisition, and strategic warning and defense. Each of these military applications involves a particular focus, a sensor suite, a desired set of inferences, and a unique set of challenges, as shown in Table 1.1.

Ocean surveillance systems are designed to detect, track, and identify ocean-based targets and events. Examples include antisubmarine warfare systems to support Navy tactical fleet operations and automated systems to guide autonomous vehicles. Sensor suites can include radar, sonar, electronic intelligence (ELINT), observation of communications traffic, infrared, and synthetic aperture radar (SAR) observations. The surveillance volume for ocean surveillance may encompass hundreds of nautical miles and focus on air, surface, and subsurface targets. Multiple surveillance platforms can be involved and numerous targets can be tracked. Challenges to ocean surveillance involve the large surveillance volume, the combination of targets and sensors, and the complex signal propagation environment — especially for underwater sonar sensing. An example of an ocean surveillance system is shown in Figure 1.2.

Air-to-air and surface-to-air defense systems have been developed by the military to detect, track, and identify aircraft and anti-aircraft weapons and sensors. These defense systems use sensors such as radar, passive electronic support measures (ESM), infrared identification-friend-foe (IFF) sensors, electro-optic

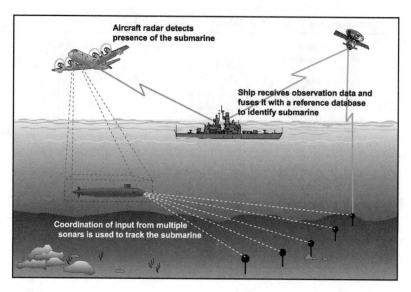

FIGURE 1.2   An example of an ocean surveillance system.

image sensors, and visual (human) sightings. These systems support counter-air, order-of-battle aggregation, assignment of aircraft to raids, target prioritization, route planning, and other activities. Challenges to these data fusion systems include enemy countermeasures, the need for rapid decision making, and potentially large combinations of target-sensor pairings. A special challenge for IFF systems is the need to confidently and non-cooperatively identify enemy aircraft. The proliferation of weapon systems throughout the world has resulted in little correlation between the national origin of a weapon and the combatants who use the weapon.

Battlefield intelligence, surveillance, and target acquisition systems attempt to detect and identify potential ground targets. Examples include the location of land mines and automatic target recognition. Sensors include airborne surveillance via SAR, passive electronic support measures, photo reconnaissance, ground-based acoustic sensors, remotely piloted vehicles, electro-optic sensors, and infrared sensors. Key inferences sought are information to support battlefield situation assessment and threat assessment.

1.4   Nonmilitary Applications

A second broad group addressing data fusion problems are the academic, commercial, and industrial communities. They address problems such as the implementation of robotics, automated control of industrial manufacturing systems, development of smart buildings, and medical applications. As with military applications, each of these applications has a particular set of challenges and sensor suites, and a specific implementation environment (see Table 1.2).

Remote sensing systems have been developed to identify and locate entities and objects. Examples include systems to monitor agricultural resources (e.g., to monitor the productivity and health of crops), locate natural resources, and monitor weather and natural disasters. These systems rely primarily on image systems using multispectral sensors. Such processing systems are dominated by automatic image processing. Multispectral imagery — such as the Landsat satellite system and the SPOT system — is used. A technique frequently used for multisensor image fusion involves adaptive neural networks. Multi-image data are processed on a pixel-by-pixel basis and input to a neural network to classify automatically the contents of the image. False colors are usually associated with types of crops, vegetation, or classes of objects. Human analysts can readily interpret the resulting false color synthetic image.

A key challenge in multi-image data fusion is coregistration. This problem requires the alignment of two or more photos so that the images are overlaid in such a way that corresponding picture elements

TABLE 1.2 Representative Nondefense Data Fusion Applications

Specific Applications	Inferences Sought by Data Fusion Process	Primary Observable Data	Surveillance Volume	Sensor Platforms
Condition-based maintenance	Detection, characterization of system faults Recommendations for maintenance/ corrections	EM signals Acoustic signals Magnetic Temperatures X-rays	Microscopic to hundreds of feet	Ships Aircraft Ground-based (e.g., factories)
Robotics	Object location/recognition Guide the locomotion of robot (e.g., "hands" and "feet")	Television Acoustic signals EM signals X-rays	Microscopic to tens of feet about the robot	Robot body
Medical diagnoses	Location/identification of tumors, abnormalities, and disease	X-rays NMR Temperature IR Visual inspection Chemical and biological data	Human body volume	Laboratory
Environmental monitoring	Identification/location of natural phenomena (e.g., earthquakes, weather)	SAR Seismic EM radiation Core samples Chemical and biological data	Hundreds of miles Miles (site monitoring)	Satellites Aircraft Ground-based Underground samples

(pixels) on each picture represent the same location on earth (i.e., each pixel represents the same direction from an observer's point of view). This coregistration problem is exacerbated by the fact that image sensors are nonlinear and perform a complex transformation between the observed three-dimensional space and a two-dimensional image.

A second application area, which spans both military and nonmilitary users, is the monitoring of complex mechanical equipment such as turbo machinery, helicopter gear trains, or industrial manufacturing equipment. For a drivetrain application, for example, sensor data can be obtained from accelerometers, temperature gauges, oil debris monitors, acoustic sensors, and infrared measurements. An online condition-monitoring system would seek to combine these observations in order to identify precursors to failure, such as abnormal gear wear, shaft misalignment, or bearing failure. The use of such condition-based monitoring is expected to reduce maintenance costs and improve safety and reliability. Such systems are beginning to be developed for helicopters and other platforms (see Figure 1.3).

1.5   Three Processing Architectures

Three basic alternatives can be used for multisensor data: (1) direct fusion of sensor data, (2) representation of sensor data via *feature vectors*, with subsequent fusion of the feature vectors, or (3) processing of each sensor to achieve high-level inferences or decisions, which are subsequently combined. Each of these approaches utilizes different fusion techniques as described and shown in Figures 1.4a, 1.4b, and 1.4c.

If the multisensor data are commensurate (i.e., if the sensors are measuring the same physical phenomena, such as two visual image sensors or two acoustic sensors), then the raw sensor data can be directly combined. Techniques for raw data fusion typically involve classic estimation methods, such as Kalman filtering. Conversely, if the sensor data are noncommensurate, then the data must be fused at the feature/state vector level or decision level.

FIGURE 1.3   Mechanical diagnostic testbed used by The Pennsylvania State University to perform condition-based maintenance research.

Feature-level fusion involves the extraction of representative features from sensor data. An example of feature extraction is the cartoonist's use of key facial characteristics to represent the human face. This technique — which is popular among political satirists — uses key features to evoke recognition of famous figures. Evidence confirms that humans utilize a feature-based cognitive function to recognize objects. In the case of multisensor feature-level fusion, features are extracted from multiple sensor observations and combined into a single concatenated feature vector that is input to pattern recognition techniques such as neural networks, clustering algorithms, or template methods.

Decision-level fusion combines sensor information after each sensor has made a preliminary determination of an entity's location, attributes, and identity. Examples of decision-level fusion methods include weighted decision methods (voting techniques), classical inference, Bayesian inference, and Dempster-Shafer's method.

1.6   A Data Fusion Process Model

One of the historical barriers to technology transfer in data fusion has been the lack of a unifying terminology that crosses application-specific boundaries. Even within military applications, related but distinct applications — such as IFF, battlefield surveillance, and automatic target recognition — used different definitions for fundamental terms, such as correlation and data fusion. To improve communications among military researchers and system developers, the Joint Directors of Laboratories (JDL) Data Fusion Working Group, established in 1986, began an effort to codify the terminology related to data fusion. The result of that effort was the creation of a process model for data fusion and a data fusion lexicon, shown in Figure 1.5. The JDL process model, which is intended to be very general and useful across multiple application areas, identifies the processes, functions, categories of techniques, and specific techniques applicable to data fusion. The model is a two-layer hierarchy. At the top level, shown in Figure 1.5, the data fusion process is conceptualized by sensor inputs, human-computer interaction, database management, source preprocessing, and four key subprocesses:

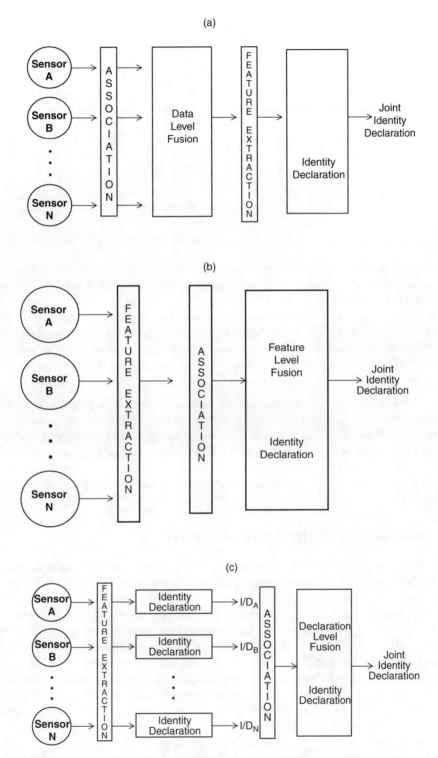

FIGURE 1.4 (a) Direct fusion of sensor data. (b) Representation of sensor data via feature vectors and subsequent fusion of the feature vectors. (c) Processing of each sensor to achieve high-level inferences or decisions that are subsequently combined.

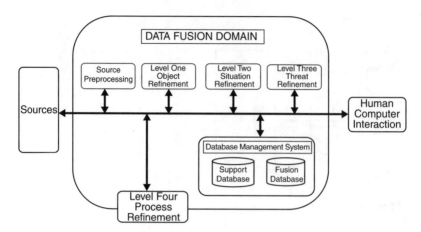

FIGURE 1.5   Joint Directors of Laboratories (JDL) process model for data fusion.

Level 1 processing (Object Refinement) is aimed at combining sensor data to obtain the most reliable
   and accurate estimate of an entity's position, velocity, attributes, and identity;
Level 2 processing (Situation Refinement) dynamically attempts to develop a description of current
   relationships among entities and events in the context of their environment;
Level 3 processing (Threat Refinement) projects the current situation into the future to draw inferences
   about enemy threats, friend and foe vulnerabilities, and opportunities for operations;
Level 4 processing (Process Refinement) is a meta-process that monitors the overall data fusion process
   to assess and improve real-time system performance.

For each of these subprocesses, the hierarchical JDL model identifies specific functions and categories of
techniques (in the model's second layer) and specific techniques (in the model's lowest layer). Implementation of data fusion systems integrates and interleaves these functions into an overall processing flow.

The data fusion process model is augmented by a hierarchical taxonomy that identifies categories of
techniques and algorithms for performing the identified functions. An associated lexicon has been
developed to provide a consistent definition of data fusion terminology. The JDL model is described in
more detail in Chapter 2.

1.7   Assessment of the State of the Art

The technology of multisensor data fusion is rapidly evolving. There is much concurrent ongoing research
to develop new algorithms, to improve existing algorithms, and to assemble these techniques into an
overall architecture capable of addressing diverse data fusion applications.

The most mature area of data fusion process is Level 1 processing — using multisensor data to
determine the position, velocity, attributes, and identity of individual objects or entities. Determining
the position and velocity of an object based on multiple sensor observations is a relatively old problem.
Gauss and Legendre developed the method of least squares for determining the orbits of asteroids.[1]
Numerous mathematical techniques exist for performing coordinate transformations, associating observations to observations or to tracks, and estimating the position and velocity of a target. Multisensor
target tracking is dominated by sequential estimation techniques such as the Kalman filter. Challenges in
this area involve circumstances in which there is a dense target environment, rapidly maneuvering targets,
or complex signal propagation environments (e.g., involving multipath propagation, cochannel interference, or clutter). However, single-target tracking in excellent signal-to-noise environments for dynamically well-behaved (i.e., dynamically predictable) targets is a straightforward, easily resolved problem.

Current research focuses on solving the assignment and maneuvering target problem. Techniques such as multiple-hypothesis tracking (MHT), probabilistic data association methods, random set theory, and multiple criteria optimization theory are being used to resolve these issues. Some researchers are utilizing multiple techniques simultaneously, guided by a knowledge-based system capable of selecting the appropriate solution based on algorithm performance.

A special problem in Level 1 processing involves the automatic identification of targets based on observed characteristics or attributes. To date, object recognition has been dominated by feature-based methods in which a feature vector (i.e., a representation of the sensor data) is mapped into feature space with the hope of identifying the target based on the location of the feature vector relative to *a priori* determined decision boundaries. Popular pattern recognition techniques include neural networks and statistical classifiers. Although numerous techniques are available, the ultimate success of these methods relies on the selection of *good* features. (Good features provide excellent class separability in feature space; bad features result in greatly overlapping feature space areas for several classes of target.) More research is needed in this area to guide the selection of features and to incorporate explicit knowledge about target classes. For example, syntactic methods provide additional information about the makeup of a target. In addition, some limited research is proceeding to incorporate contextual information — such as target mobility with respect to terrain — to assist in target identification.

Level 2 and Level 3 fusion (situation refinement and threat refinement) are currently dominated by knowledge-based methods such as rule-based blackboard systems. These areas are relatively immature and have numerous prototypes, but few robust, operational systems. The main challenge in this area is to establish a viable knowledge base of rules, frames, scripts, or other methods to represent knowledge about situation assessment or threat assessment. Unfortunately, only very primitive cognitive models exist to replicate the human performance of these functions. Much research is needed before reliable and large-scale knowledge-based systems can be developed for automated situation assessment and threat assessment. New approaches that offer promise are the use of fuzzy logic and hybrid architectures, which extend the concept of blackboard systems to hierarchical and multitime scale orientations.

Finally, Level 4 processing, which assesses and improves the performance and operation of an ongoing data fusion process, has a mixed maturity. For single sensor operations, techniques from operations research and control theory have been applied to develop effective systems, even for complex single sensors such as phased array radars. In contrast, situations that involve multiple sensors, external mission constraints, dynamic observing environments, and multiple targets are more challenging. To date, considerable difficulty has been encountered in attempting to model and incorporate mission objectives and constraints to balance optimized performance with limited resources, such as computing power and communication bandwidth (e.g., between sensors and processors), and other effects. Methods from utility theory are being applied to develop measures of system performance and measures of effectiveness. Knowledge-based systems are being developed for context-based approximate reasoning. Significant improvements will result from the advent of smart, self-calibrating sensors, which can accurately and dynamically assess their own performance.

Data fusion has suffered from a lack of rigor with regard to the test and evaluation of algorithms and the means of transitioning research findings from theory to application. The data fusion community must insist on high standards for algorithm development, test, and evaluation; creation of standard test cases; and systematic evolution of the technology to meet realistic applications. On a positive note, the introduction of the JDL process model and emerging nonmilitary applications are expected to result in increased cross discipline communication and research. The nonmilitary research in robotics, condition-based maintenance, industrial process control, transportation, and intelligent buildings will produce innovations that will cross-fertilize the entire field of data fusion technology. The many challenges and opportunities related to data fusion establish it as an exciting research field with numerous applications.

1.8   Additional Information

Additional information about multisensor data fusion may be found in the following references:

- D. L. Hall, *Mathematical Techniques in Multisensor Data Fusion*, Artech House, Inc. (1992) — provides details on the mathematical and heuristic techniques for data fusion
- E. Waltz and J. Llinas, *Multisensor Data Fusion*, Artech House, Inc. (1990) — presents an excellent overview of data fusion especially for military applications
- L. A. Klein, *Sensor and Data Fusion Concepts and Applications*, SPIE Optical Engineering Press, Volume TT 14 (1993) — presents an abbreviated introduction to data fusion
- R. Antony, *Principles of Data Fusion Automation*, Artech House, Inc. (1995) — provides a discussion of data fusion processes with special focus on database issues to achieve computational efficiency
- A multimedia computer-based training package, "Introduction to Data Fusion, A multimedia computer-based training package" — available from Artech House, Inc., Boston, MA, 1995.
- A data fusion lexicon is available from TECH REACH Inc. at http://www.techreachinc.com.

Reference

1. Sorenson, H.W., Least-squares estimation: from Gauss to Kalman, *IEEE SPECTRUM*, July 1970, 63–68.

2

Revisions to the JDL Data Fusion Model

2.1 Introduction 2-1
2.2 What Is Data Fusion? What Isn't?.......................... 2-1
The Role of Data Fusion • Definition of Data Fusion
2.3 Models and Architectures................................. 2-4
Data Fusion "Levels" • Association and Estimation • Context
Sensitivity and Situation Awareness • Attributive and
Relational Functions
2.4 Beyond the Physical.. 2-12
2.5 Comparison with Other Models....................... 2-15
Dasarathy's Functional Model • Bedworth and O'Brien's
Comparison among Models and Omnibus
2.6 Summary.. 2-17
References 2-18

Alan N. Steinberg
Utah State University

Christopher L. Bowman
Consultant

2.1 Introduction

The data fusion model, developed in 1985 by the U.S. Joint Directors of Laboratories (JDL) Data Fusion Group*, with subsequent revisions, is the most widely used system for categorizing data fusion-related functions. The goal of the JDL Data Fusion Model is to facilitate understanding and communication among acquisition managers, theoreticians, designers, evaluators, and users of data fusion techniques to permit cost-effect system design, development, and operation.[1,2]

This chapter discusses the most recent model revision (1998): its purpose, content, application, and relation to other models.[3]

2.2 What Is Data Fusion? What Isn't?

2.2.1 The Role of Data Fusion

Often, the role of data fusion has been unduly restricted to a subset of the relevant processes. Unfortunately, the universality of data fusion has engendered a profusion of overlapping research and development in many applications. A jumble of confusing terminology (illustrated in Figure 2.1) and ad hoc methods in a variety of scientific, engineering, management, and educational disciplines obscures the fact that the same ground has been plowed repeatedly.

---

*Now recharted as the Data and Information Fusion Group within the Deputy Director for Research and Engineering's Information System Technology Panel at the U.S. Department of Defense.

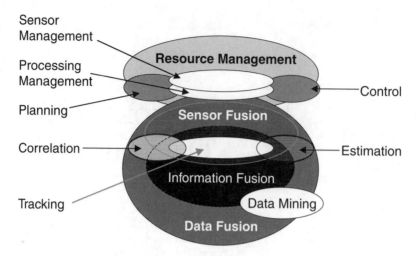

FIGURE 2.1   (Con)fusion of terminology.

Often, the role of data fusion has been unduly restricted to a subset of processes and its relevancy has been limited to particular state estimation problems. For example, in military applications, such as targeting or tactical intelligence, the focus is on estimating and predicting the state of specific types of entities in the external environment (e.g., targets, threats, or military formations). In this context, the applicable sensors/sources that the system designer considers are often restricted to sensors that directly collect data from targets of interest.

Ultimately, however, such problems are inseparable from other aspects of the system's assessment of the world. In a tactical system, this will involve estimation of one's own state in relation to the relevant external entities: friends, foes, neutrals, and background. Estimation of the state of targets and threats cannot be separated from the problems of estimating one's own location and motion, of calibrating one's sensor performance and alignment, and of validating one's library of target sensor and environment models. The data fusion problem, then, becomes that of achieving a consistent, comprehensive estimate and prediction of some relevant portion of the world state. In such a view, data fusion involves exploiting all sources of data to solve all relevant state estimation/prediction problems, where relevance is determined by utility in forming plans of action.

The data fusion problem, therefore, encompasses a number of interrelated problems: estimation and prediction of states of entities both external and internal to the acting system, and the interrelations among such entities. Evaluating the system's models of the characteristics and behavior of all of these external and organic entities is, likewise, a component of the overall problem of estimating the actual world state.

Making the nontrivial assumption that the universe of discourse for a given system can be partitioned into an unknown but finite number of entities of interest, the problem of consistently estimating a multi-object world state can be defined as shown in Figure 2.2.[4] Here, $x_1...,x_k$ are entity states, so the global state estimation problem becomes one of finding the finite set of entity states X with maximum *a posteriori* likelihood.

The complexity of the data fusion system engineering process is characterized by difficulties in

- representing the uncertainty in observations and in models of the phenomena that generate observations;
- combining noncommensurate information (e.g., the distinctive attributes in imagery, text, and signals);
- maintaining and manipulating the enormous number of alternative ways of associating and interpreting large numbers of observations of multiple entities.

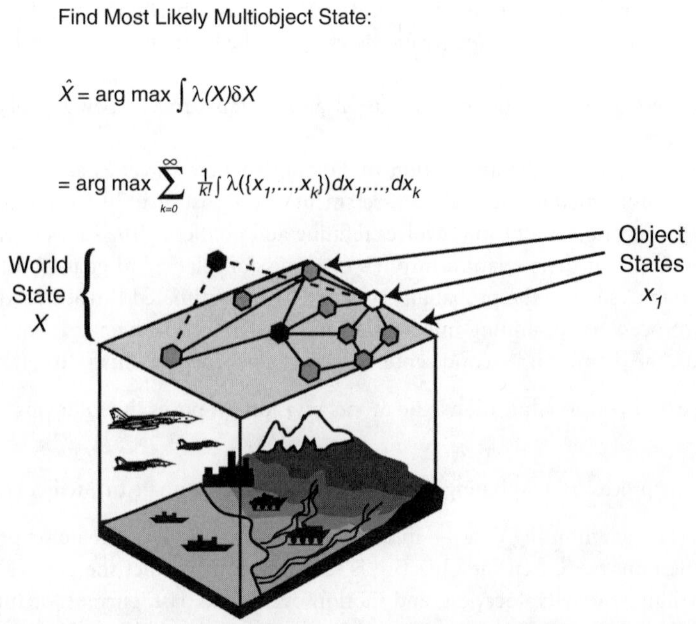

Find Most Likely Multiobject State:

$$\hat{X} = \arg\max \int \lambda(X)\delta X$$

$$= \arg\max \sum_{k=0}^{\infty} \frac{1}{k!} \int \lambda(\{x_1,...,x_k\})dx_1,...,dx_k$$

FIGURE 2.2 Global state estimation problem.

Deriving general principles for developing and evaluating data fusion processes — whether automatic or manual — will help to take advantage of the similarity in the underlying problems of data association and combination that span engineering, analysis, and cognitive situations. Furthermore, recognizing the common elements of diverse data fusion problems can provide extensive opportunities for synergistic development. Such synergy — enabling the development of information systems that are cost-effective and trustworthy — requires common performance evaluation measures, system engineering methodologies, architecture paradigms, and multispectral models of targets and data collection systems.

2.2.2 Definition of Data Fusion

The initial JDL Data Fusion Lexicon defined data fusion as:

A process dealing with the association, correlation, and combination of data and information from single and multiple sources to achieve refined position and identity estimates, and complete and timely assessments of situations and threats, and their significance. The process is characterized by continuous refinements of its estimates and assessments, and the evaluation of the need for additional sources, or modification of the process itself, to achieve improved results.[1]

As the above discussion suggests, this initial definition is rather too restrictive. A definition is needed that can capture the fact that similar underlying problems of data association and combination occur in a very wide range of engineering, analysis, and cognitive situations. In response, the initial definition requires a number of modifications:

1. Although the concept *combination of data* encompasses the broad range of problems of interest, *correlation* does not. Statistical correlation is merely one method for generating and evaluating hypothesized associations among data.
2. Association is not an essential ingredient in combining multiple pieces of data. Recent work in random set models of data fusion provides generalizations that allow state estimation of multiple targets without explicit report-to-target association.[4-6]

3. *Single or multiple sources* is comprehensive; therefore, it is superfluous in a definition.
4. The reference to *position and identity estimates* should be broadened to cover all varieties of state estimation.
5. *Complete* assessments are not required in all applications; *timely*, being application-relative, is superfluous.
6. *Threat assessment* limits the application to situations where threat is a factor. This description must also be broadened to include any assessment of the cost or utility implications of estimated situations. In general, data fusion involves refining and predicting the states of entities and aggregates of entities and their relation to one's own mission plans and goals. Cost assessments can include variables such as the probability of surviving an estimated threat situation.
7. Not every process of combining information involves collection management or process refinement. Thus, the definition's second sentence is best construed as illustrative, not definitional.

Pruning these extraneous qualifications, the model revision proposes the following concise definition for data fusion:[3]

Data fusion is the process of combining data or information to estimate or predict entity states.

Data fusion involves combining data — in the broadest sense — to estimate or predict the state of some aspect of the universe. Often the objective is to estimate or predict the physical state of entities: their identity, attributes, activity, location, and motion over some past, current, or future time period. If the job is to estimate the state of people (or any other sentient beings), it may be important to estimate or predict the individuals' and groups' informational and perceptual states and the interaction of these with physical states (this point is discussed in Section 2.5).

Arguments about whether *data fusion* or some other label best describes this very broad concept are pointless. Some people have adopted terms such as *information integration* in an attempt to generalize earlier, narrower definitions of data fusion (and, perhaps, to distance themselves from old data fusion approaches and programs). However, relevant research should not be neglected simply because of shifting terminological fashion. Although no body of common and accepted usage currently exists, this broad concept is an important topic for a unified theoretical approach and, therefore, deserves its own label.

2.3   Models and Architectures

The use of the JDL Data Fusion Model in system engineering can best be explained by considering the role of models in system architectures in general. According to the IEEE definition,[7] an *architecture* is a "structure of components, their relationships, and the principles and guidelines governing their design and evolution over time." Architectures serve to coordinate capabilities to achieve interoperability and affordability. As such, general requirements for an architecture are that it must

1. Identify a focused purpose,
2. Facilitate user understanding/communication,
3. Permit comparison and integration,
4. Promote expandability, modularity, and reusability,
5. Promote cost-effective system development,
6. Apply to the required range of situations.

The JDL Model has been used to develop an architecture paradigm for data fusion[8-10] (as discussed in Chapter 18); however, in reality, the JDL Model is merely an element of an architecture. A model is an abstract description of a set of functions or processes that may be components of a system of a particular type, without indication of software or physical implementation. That being the case, the previous list of architectural virtues applies, with the exception of item (1), which is relevant only to specific system architectures.

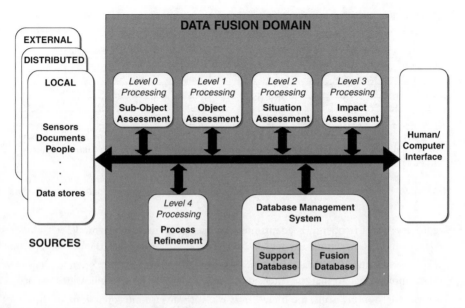

FIGURE 2.3 Revised JDL data fusion model (1998).[3]

The JDL Model was designed to be a *functional* model — a set of definitions of the functions that could comprise any data fusion system. Distinguishing functional models from *process* models and other kinds of models is important. Process models specify the interaction among functions within a system. Examples of process models include Boyd's Observe, Orient, Decide and Act (OODA) loop, the Predict, Extract, Match and Search (PEMS) loop, and the UK Intelligence cycle and waterfall process models cited by Bedworth and O'Brien.[11]

Another type of model is a *formal* model, constituting a set of axioms and rules for manipulating entities. Examples are probabilistic, possibilistic, and evidential reasoning frameworks.*

A model should clarify the elements of problems and solutions to facilitate recognition of commonalities in problems and in solutions. Among questions that a model should help answer are the following:

- Has the problem been solved before?
- Has the same problem appeared in a different form and is there an existing solution?
- Is there a related problem with similar constraints?
- Is there a related problem with the same unknowns?
- Can the problem be subdivided into parts that are easier to solve?
- Can the constraints be relaxed to transform the problem into a familiar one?[12]

2.3.1 Data Fusion "Levels"

Of the many ways to differentiate types of data fusion functions, the JDL model has gained the widest usage. The JDL model's differentiation of functions into fusion levels (depicted in Figure 2.3) provides a useful distinction among data fusion processes that relate to the refinement of "objects," "situations," "threats," and "processes."[2]

---

* This is seen as equivalent to the concept of *framework* as used in Reference 11.

TABLE 2.1 Characterization of the Revised Data Fusion Levels

Data Fusion Level	Association Process	Estimation Process	Entity Estimated
L.0 — Sub-Object Assessment	Assignment	Detection	Signal
L.1 — Object Assessment		Attribution	Individual Object
L.2 — Situation Assessment	Aggregation	Relation	Aggregation (Situation)
L.3 — Impact Assessment		Plan Interaction	Effect (situation, given plans)
L.4 — Process Refinement	Planning	(Control)	(Action)*

* Process Refinement does not involve estimation, but rather control. Therefore, its product is a control sequence, which — by the duality of estimation and control — relates to a controlled entity's actions as an estimate relates to an actual state.[15]

Nonetheless, several concerns must be raised with regard to the ways in which these JDL data fusion levels have been used in practice:

- The JDL levels have frequently been misinterpreted as specifying a *process* model (i.e., as a canonical guide for process flow within a system — "perform Level 1 fusion first, then Levels 2, 3, and 4...).
- The original JDL model names and definitions (e.g., "threat refinement") seem to focus on tactical military applications, so that the extension of the concepts to other applications is not obvious.
- For these and other reasons, the literature is rife with diverse interpretations of the data fusion levels. The levels have been interpreted as distinguishing any of the following: (a) the kinds of association and/or characterization processing involved, (b) the kinds of entities being characterized, and (c) the degree to which the data used in the characterization has already been processed.

The objectives in the 1998 revision of the definitions for the levels are (a) to provide a useful categorization representing logically different types of problems, which are generally (though not necessarily) solved by different techniques and (b) to maintain a degree of consistency with regard to terminology. The former is a matter of engineering; the latter is a language issue.

Figure 2.3 shows the suggested revised model. The proposed new definitions are as follows:

- Level 0 — Sub-Object Data Assessment: estimation and prediction of signal- or object-observable states on the basis of pixel/signal-level data association and characterization.
- Level 1 — Object Assessment: estimation and prediction of entity states on the basis of inferences from observations.
- Level 2 — Situation Assessment: estimation and prediction of entity states on the basis of inferred relations among entities.
- Level 3 — Impact Assessment: estimation and prediction of effects on situations of planned or estimated/predicted actions by the participants (e.g., assessing susceptibilities and vulnerabilities to estimated/predicted threat actions, given one's own planned actions).
- Level 4 — Process Refinement (an element of Resource Management): adaptive data acquisition and processing to support mission objectives.

Table 2.1 provides a general characterization of these concepts. Note that the levels are differentiated first on the basis of types of estimation process, which roughly correspond to the types of entity for which state is estimated.

2.3.2 Association and Estimation

In the common cases where the fusion process involves explicit association in performing state estimates, a corresponding distinction is made among the types of association processes. Figure 2.4 depicts assignment matrices that are typically formed in each of these processing levels. The examples have the form of two-dimensional matrices, as commonly used in associating reports to tracks.

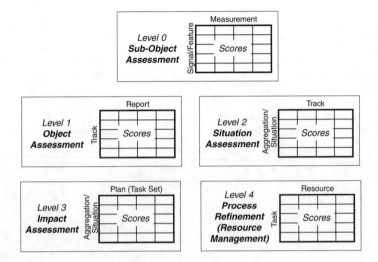

FIGURE 2.4   Assignment matrices for various data fusion "levels."

Level 0 association involves hypothesizing the presence of a signal (i.e., of a common source of sensed energy) and estimating its state. Level 0 associations can include (a) signal detection obtained by integrating a time series of data (e.g., the output of an analog-to-digital converter) and (b) feature extraction from a region in imagery. In this case, a region could correspond to a cluster of closely spaced objects, or to part of an object, or simply to a differentiable spatio-temporal region.

Level 1 association involves selecting observation reports (or tracks from prior fusion nodes in a processing sequence) for inclusion in a track. Such a track is a hypothesis that a certain set of reports is the total set of reports available to the system referencing some individual entity. Global Level 1 hypotheses map the set of observations available to the system to tracks. For systems in which observations are assumed to be associated with only one track, this is a set-partitioning problem; more generally, it is a set-covering problem.

Level 2 association involves associating tracks (i.e., hypothesized entities) into aggregations. The state of the aggregate entity is represented as a network of relations among aggregation elements. Any variety of relations — physical, organizational, informational, and perceptual — can be considered, as appropriate to the given information system's mission. As the class of estimated relationships and the numbers of interrelated entities broaden, the term *situation* is used to refer to an aggregate object of estimation. A model for such development is presented by Steinberg and Washburn.[14]

Level 3 association is usually implemented as a prediction, drawing particular kinds of inferences from Level 2 associations. Level 3 fusion estimates the impact of an assessed situation (i.e., the outcome of various plans as they interact with one another and with the environment). The impact estimate can include likelihood and cost/utility measures associated with potential outcomes of a player's planned actions.

Because Level 2 has been defined so broadly, Level 3 is actually a subset of Level 2. Whereas Level 2 involves estimating or predicting all types of relational states, Level 3 involves predicting some of the relationships between a specific player and his environment, including interaction with other players' actions, given the player's action plan and that of every other player. More succinctly, Level 2 concerns relations in general: paradigmatically third-person, objective relations. Level 3 concerns first-person relations — involving the system or its user — with an attendant sense of subjective utility.

Level 4 processing involves planning and control, not estimation. As discussed by Bowman,[15] just as a formal duality exists between estimation and control, there is a similar duality between association and planning. Therefore, Level 4 association involves assigning resources to tasks.

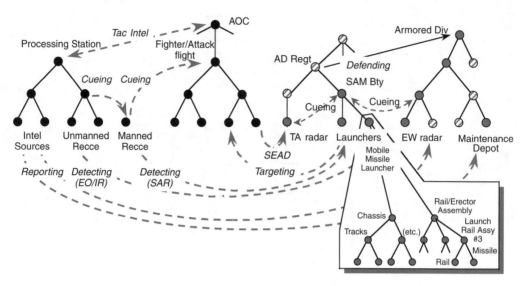

FIGURE 2.5 A Level 2 hypothesis with imbedded Level 1 hypotheses.

2.3.3   Context Sensitivity and Situation Awareness

Once again, the JDL model is a functional model, not a process model. Therefore, it would be a mistake to assume that the information flow in data fusion must proceed strictly from Level 1 to Level 2 to Level 3. Such a mistake has, unfortunately, been common with system designers. A "bottom-up" fusion process is justified only under the following conditions:

- Sensor observations can be partitioned into measurements, each of which originates from, at most, one real entity.
- All information relevant to the estimation of an entity state is contained in the measurement of the individual entity.

Neither of these conditions is necessarily true, and the second is usually false.

The value of estimating entity states on the basis of context is becoming increasingly apparent. A system that integrates data association and estimation processes of all "levels" will permit entities to be understood as parts of complex situations. A relational analysis, as illustrated in Figure 2.5, permits evidence applicable to a local estimation problem to be propagated through a complex relational network.

Note that inferencing based on hypothesized relationships among entities can occur within and between all of the data fusion levels. Figure 2.6 depicts typical information flow across the data fusion levels. Level 0 functions combine measurements to generate estimates of signals or features. At Level 1, signal/feature reports are combined to estimate the states of objects. These are combined, in turn, at Level 2 to estimate situations (i.e., states of aggregate entities). Level 3, according to this logical relationship, seems to be out of numerical sequence. It is a "higher" function than the planning function of Level 4. Indeed, Process Refinement (Level 4) processes can interact with association/estimation data fusion processes in a variety of ways, managing the operation of individual fusion nodes or that of larger ensembles of such nodes. The figure reinforces the point that the data fusion levels are not to be taken as a prescription for the sequencing of a system's process flow. Processing partitioning and flow must be designed in terms of the individual system requirements, as discussed in Chapter 16.

2.3.4   Attributive and Relational Functions

Table 2.1 shows that association within Levels 0 and 1 involves assignment, while Levels 2 and 3 association involves aggregation. This can be modeled as the distinction between

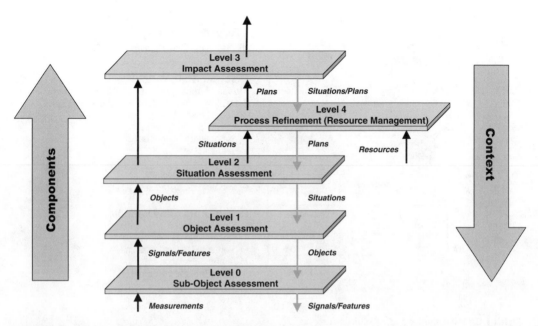

FIGURE 2.6 Characteristic data flow among the "levels."

- estimation on the basis of observations: $(x|Z)$ or $(X|Z)$ for entity or world states, given a set of observations, Z, and
- estimation on the basis of inferred relations among entities: $(x|R)$ or $(X|R)$, where R is a set of ordered n-tuples $<x_1,...,x_{n-1},r>$, the x_i being entity states and r a relational state

Figure 2.5 provides an example of the relationship of Level 1 and 2 hypotheses. A Level 2 hypothesis can be modeled as a directed graph, the nodes of which may correspond to entity tracks and, therefore, to Level 1 hypotheses. More precisely, a node in a Level 2 hypothesis corresponds to a perceived entity. The set of observations associated directly with that node can be considered to be a Level 1 hypothesis imbedded in the Level 2 structure. Of course, entities can be inferred from their context alone, without having been observed directly. For example, in the SA-6 battery of Figure 2.6, the estimation of the presence of launchers at three corners of a diamond pattern may support the inference of a fourth launcher in the remaining corner. The figure further illustrates the point that hypotheses regarding physical objects (e.g., the mobile missile launcher at the lower right of Figure 2.5) may themselves be Level 2 relational constructs.

2.3.4.1 Types of Relationships

Assembling an exhaustive list of relationships of interest is impossible, which is one reason that Level 2 fusion (Situation Assessment) is generally more difficult than Level 1 fusion. Level 2 problems are generally more difficult than Level 1 problems. The process model for aggregate entities — particularly those involving human activity — is often poorly understood, being less directly inferable from underlying physics than Level 1 observable attributes. For this reason, automation of Situational Awareness has relied on so-called cognitive techniques that are intended to copy the inference process of human analysts. However, knowledge extraction is a notoriously difficult undertaking. Furthermore, Level 2 problems often involve a much higher dimensionality, corresponding to the relations that may be part of an inference. Finally, no general metric exists for assessing the relevance of data in these unspecified, high-dimension spaces, unlike the simple distance metrics commonly used for Level 1 validation gating. Relationships of interest to particular context exploitation or situation awareness concerns can include:

- Spatio-temporal relationships;
- Part/whole relationships;

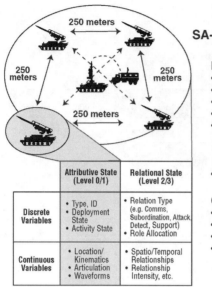

Aggregate Entity State
SA-6 Battery Combat Formation

Discrete State Variable:
- Type of Aggregate Entity: SA-6 Bty
- Deployment State: Combat Formation
- Readiness State: Operational
- Activity State: Target Search
- Subordinate Elements:
  – 4 launchers
  – 1FC Radar
  – 1Command Vehicle
- Subordination: Iraqi 4th AD Rgt

Continuous State Variable:
- Cluster Dimensions
- Cluster Orientation
- Element Relative Locations
- Temporal Relations
  (State Transition Criteria)

	Attributive State (Level 0/1)	Relational State (Level 2/3)
Discrete Variables	• Type, ID • Deployment State • Activity State	• Relation Type (e.g. Comms, Subordination, Attack, Detect, Support) • Role Allocation
Continuous Variables	• Location/Kinematics • Articulation • Waveforms	• Spatio/Temporal Relationships • Relationship Intensity, etc.

FIGURE 2.7   Attributive and relational state example.

- Organizational relationships (e.g., X is a subordinate unit to Y) and roles (e.g., X is the unit commander, company clerk, CEO, king, or court jester of Y);
- Various causal relations, whereby X changes the state of Y:
  - Physical state (damaging, destroying, moving, invading, repairing)
  - Informational state (communicating, informing, revealing)
  - Perceptual or other mental state (persuading, deceiving, intimidating)
  - Financial or legal state (paying, fining, authorizing, forbidding, sentencing)
  - Intentional relationships, whereby X *wishes* to change the state of Y (targeting, jamming, cajoling, lying to);
- Semantic relationships (X is of the same type as Y);
- Similarity relationships (X is taller than Y);
- Legal relationships (X owns Y, X leases Y to Z);
- Emotional relationships (love, hate, fear);
- Biological relationships (kinship, ethnicity).

2.3.4.2   Attributive and Relational Inferencing Example

Figure 2.7 provides an example of the attributive and relational states within and among the elements of an aggregate entity. Steinberg and Washburn[14] discuss formal methods for inferring relational states to refine entity-level and aggregate-level state estimates. A Bayesian network technique is used to combine

- the estimate of an entity state, X_i, based on a set of observations, Z_i, in a Level 1 hypothesis (track) and
- the estimate of an entity state, X_i, based on a set of relations, R_i, among nodes (tracks) in a Level 2 hypothesis (aggregation).

The distribution of discrete states, x_d, for X, *given its assignment to the given node in a Level 2 hypothesis*, ζ, will be determined by this "evidence" from each of these sources:

$$P_{L2}(x_d, \varsigma) = \frac{p_{L1}(x_d)\Lambda_\varsigma(x_d)}{\displaystyle\sum_{x_d} p_{L1}(x_d)\Lambda_\varsigma(x_d)} \tag{2.1}$$

where $p_{L1}(x_d)$ is the probability currently assigned to discrete state, x_d, by Level 1 data fusion of observations associated with node X, and $\Lambda(x_d)$ is the evidence communicated to X from the tracks related to Y in a Level 2 association hypothesis.

The evidence from the nodes communicating with X will be the product of evidence from each such node Y:

$$\Lambda_\varsigma(x_d) = \prod_{\langle X,Y\rangle \in \varsigma} \Lambda_Y(x_d) \tag{2.2}$$

The factors $\Lambda_Y(x_d)$ are interpreted in terms of relational states among entities as follows. Ordered pairs of entities are hypothesized as having *relational states*, $r_i(X,Y)$. A given track, Y, may be involved in several competing relations relative to X with probability distributions $p[r_i(X,Y)]$.[*]

Updating a track, Y, contributes information for evaluating the probability of each state, x, of a possible related entity, X. As with attributive states, relational states, r, can be decomposed into discrete and continuous components, r_d and r_c (as exemplified in Figure 2.6). Then this *contextual evidence* is given by

$$\Lambda_Y(x_d) = \sum_{y_d} p_{L1}(y_d) p[y_d | x_d] = \int \sum_{y_d} p_{L1}(y_d) p[y_d | rx_d] p[r | x_d] dr$$

$$= \sum_{y_d} p_{L1}(y_d) \sum_{r_d} p[y_d | r_d, x_d] p[r_d | x_d] \int p[r_c | r_d, x_d] dr \tag{2.3}$$

Inferences can be drawn about a hypothesized entity denoted by track X_i, given the Level 2 hypothesis that the entity corresponding to X_i stands in a particular relationship to another hypothesized entity corresponding to a track X_j. In the example shown in Figure 2.8 (based on sets of relationships as illustrated in Figure 2.7), it is assumed that an entity — elliptically referred to as X_1 — has been estimated to have probabilities $p(x_1)$ of being an entity of types and activity states x_1 on the basis of Level 1 association of sensor reports z_1 and z_2. Then, if X_1 and X_2 meet the criteria of particular relationships for any states x_1 and x_2 of X_1 and X_2, respectively, inferences can be drawn regarding the probabilities as to the type and activity of X_2.

For example, given the estimate that X_1 and X_2 stand in certain spatio-temporal and other relationships, as listed in Figure 2.7, there is a mutual reinforcement of pairs of Level 1 state estimates $<x_1,x_2>$ that are consistent with this relationship (e.g., that X_1 is a Straight Flush radar and X_2 is an SA-6 surface-to-air missile battery) and suppression of nonconsistent state pairs. Conditioned on this association, the estimate of the likelihood of track X_2 can be refined (i.e., the hypothesis that the associated observations — z_3 in Figure 2.8 — relate to the same entity). Furthermore, likelihood and state estimates to other nodes adjoining X_2 can be further propagated (e.g., to infer the battery-association and the type and activity of a missile launcher, X_3, hypothesized on the basis of observations z_4 and z_5). As noted above, the presence, identity, and activity state of entities that have not been observed can be inferred (e.g., the presence of

---

[*] For simplicity, the present discussion is limited to binary relations. In cases where more complex relations are relevant, a second order can be employed, whereby entities can have binary links to nodes representing n-ary relations.[16]

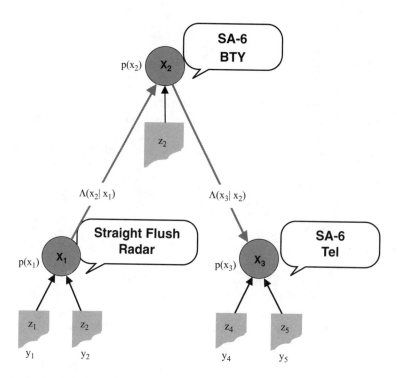

FIGURE 2.8 Attributive and relational inferencing example.

a full complement of launchers and other associated equipment can be inferred, conditioned on the assessed presence of an SA-6 battery).

Each node in a Level 2 hypothesis combines the effects of evidence from all adjacent nodes and propagates the updated probability distributions and likelihood (i.e., association confidence) regarding an entity state to the other nodes. Loops in the inference flow occur; however, methods have been defined to deal with them.

2.3.4.3 A Generalization about the Levels

Level 1 data fusion involves estimating and predicting the state of inferred entities based on observed features. Level 2 data fusion involves estimating and predicting the state of inferred entities on the basis of relationships to other inferred entities. Because of their reliance on these inference mechanisms, Levels 0 and 3 are seen as special cases of Levels 1 and 2, respectively (as illustrated in Figure 2.9):

- Level 0 is a special case of Level 1, where entities are signals/features.
- Level 3 is a special case of Level 2, where relations are first-person relations.

Earlier, this chapter asserted that Level 4 fusion is not fusion at all, but a species of Resource Management; therefore, only two super-levels of fusion remain, and these are partitioned by type of data association. A secondary partitioning by type of entity characterized distinguishes within these super-levels. Section 2.5 presents the case for an even finer partitioning within the JDL levels.

2.4 Beyond the Physical

In general, then, the job of data fusion is that of estimating or predicting the state of some aspect of the world. When that aspect includes people (or any other information systems, for that matter), it can be relevant to include a consideration of informational and perceptual states and their relations to physical states. *Informational state* refers to the data available to the target. *Perceptual state* refers to the target's own estimate of the world state.[17] (See Chapter 15.)

The JDL Data Fusion Model (1998 revision) distinguishes data fusion processes in terms of "levels" based on the types of processes involved:

– Level 1 fusion involves *attribution-based state estimation*:

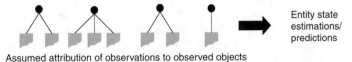

Assumed attribution of observations to observed objects

– Level 2 fusion involves *relation-based state estimation*:

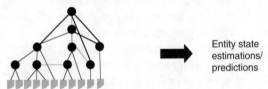

Assumed relationships among observed objects

FIGURE 2.9   Attributive and relational inferencing.

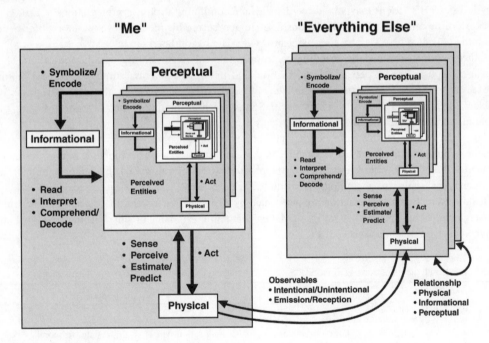

FIGURE 2.10   Entity states: three aspects.

A person or other information system (represented by the box at the left of Figure 2.10) senses physical stimuli as a function of his physical state in relation to that of the stimulating physical world. These include both stimuli originating outside the person's body and those originating from within.

The person can combine multiple sensory reports to develop and refine estimates of perceived entities (i.e., tracks), aggregations, and impacts on his plans and goals (Levels 1–3 fusion). This ensemble of perceived entities and their interrelationships is part of the person's *perceptual state*. As depicted in the figure, his perceptual state can include an estimation of physical, informational, and perceptual states and relations of things in the world. The person's perceptions can be encoded symbolically for manipulation, communication, or storage. The set of symbolic representations available to the person is his *informational state*. Informational state can encompass available data stores such as databases and documents. The notion of informational state is probably more applicable to a closed system (e.g., a nonnetworked computer) than to a person, for whom the availability of information is generally a matter of degree. The tripartite view of reality developed by Waltz[17] extends the work of philosopher Karl Popper. The status of information as a separable aspect of reality is certainly subject to discussion. Symbols can have both a physical and a perceptual aspect: they can be expressed by physical marks or sounds, but their interpretation (i.e., recognizing them orthographically as well as semantically) is a matter of perception.

As seen in this example, symbol recognition (e.g., reading) is clearly a perceptual process. It is a form of context-sensitive model-based processing. The converse process, that of representing perceptions symbolically for purpose of recording or communicating them, produces a physical product — text, sounds, etc. Such physical products must be interpreted as symbols before their informational content can be accessed. Whether there is more to information than these physical and perceptual aspects remains to be demonstrated. Furthermore, the distinction between information and perception is not the difference between what a person *knows* and what he *thinks* (cf. Plato's *Theatetus*, in which knowledge is shown to involve true opinion plus some sense of understanding). Nonetheless, the notion of informational state is useful as a topic for estimation because knowing what information is available to an entity (e.g., an enemy commander's sources of information) is an important element in estimating (and influencing) his perceptual state and, therefore, in predicting (and influencing) changes.

The person acts in response to his perceptual state, thereby affecting his and the rest of the world's physical state. His actions may include comparing and combining various representations of reality: his network of perceived entities and relationships. He may search his memory or seek more information from the outside. These are processes associated with data fusion Level 4.

Other responses can include encoding perceptions in symbols for storage or communication. These can be incorporated in the person's physical actions and, in turn, are potential stimuli to people (including the stimulator himself) and other entities in the physical world (as depicted at the right of Figure 2.10). Table 2.2 describes the elements of state estimation for each of the three aspects shown in Figure 2.10. Note the recursive reference in the bottom right cell.

Figure 2.11 illustrates this recursive character of perception. Each decision maker interacts with every other one on the basis of an estimate of current, past, and future states. These include not only estimates of who is doing what, where, and when in the physical world, but also what their informational states and perceptual states are (including, "What do they think of *me*?").

If state estimation and prediction are performed by an automated system, that system may be said to possess physical and perceptual states, the latter containing estimates of physical, informational, and perceptual states of some aspects of the world.

TABLE 2.2   Elements of State Estimation

	Attributive State		Relational State	
Object Aspect	Discrete	Continuous	Discrete	Continuous
Physical	Type, ID Activity state	Location/kinematics Waveform parameters	Causal relation type Role allocation	Spatio-temporal relationships
Informational	Available data types Available data records and quantities	Available data values Accuracies Uncertainties	Informational relation type Info source/ recipient role allocation	Source data quality, quantity, timeliness Output quality, quantity, timeliness
Perceptual	Goals Priorities	Cost assignments Confidence Plans/schedules	Influence relation type Influence source/recipient role allocation	Source confidence World state estimates (per this table)

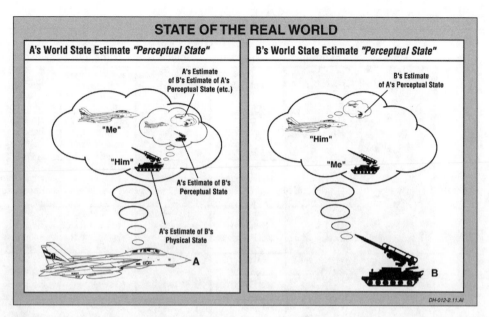

FIGURE 2.11 World states and nested state estimates.

2.5 Comparison with Other Models

2.5.1 Dasarathy's Functional Model

Dasarathy[18] has defined a very useful categorization of data fusion functions in terms of the types of data/information that are processed and the types that result from the process. Table 2.3 illustrates the types of inputs/outputs considered. Processes corresponding to the cells in the highlighted diagonal X region are described by Dasarathy, using the abbreviations *DAI-DAO, DAI-FEO, FEI-FEO, FEI-DEO,* and *DEI-DEO*. A striking benefit of this categorization is the natural manner in which technique types can be mapped into it.

TABLE 2.3 Interpretation of Dasarathy's Data Fusion I/O Model

INPUT	OUTPUT		
	Data	**Features**	**Objects**
Data	Signal Detection DAI-DAO	Feature Extraction DAI-FEO	Gestalt-Based Object Characterization DAI-DEO
Features	Model-Based Detection/ Feature Extraction FEI-DAO	Feature Refinement FEI-FEO	(Feature-Based) Object Characterization FEI-DEO
Objects	Model-Based Detection/ Estimation DEI-DAO	Model-Based Feature Extract DEI-FEO	Object Refinement DEI-DEO

Level 0   *Level 1*

TABLE 2.4　Expansion of Dasarathy's Model to Data Fusion Levels 0–4

INPUT \ OUTPUT	Data	Features	Objects	Relations	Impacts	Responses
Data	Signal Detection *DAI-DAO*	Feature Extraction *DAI-FEO*	Gestalt-Based Object Extract *DAI-DEO*	Gestalt-Based Situation Assessment *DAI-RLO*	Gestalt-Based Impact Assessment *DAI-IMO*	Reflexive Response *DAI-RSO*
Features	Model-Based Detection/ Feature Extraction *FEI-DAO*	Feature Refinement *FEI-FEO*	Object Characterization *FEI-DEO*	Feature-Based Situation Assessment *FEI-RLO*	Feature-Based Impact Assessment *FEI-IMO*	Feature-Based Response *FEI-RSO*
Objects	Model-Based Detection/ Estimation *DEI-DAO*	Model-Based Feature Extraction *DEI-FEO*	Object Refinement *DEI-DEO*	Entity-Relational Situation Assessment *DEI-RLO*	Entity-Based Impact Assessment *DEL-IMO*	Entity- Relation Based Response *DEI-RSO*
Relations	Context- Sensitive Detection/Est *RLI-DAO*	Context- Sensitive Feature Extraction *RLI-FEO*	Context- Sensitive Object Refinement *RLI-DEO*	Micro/Macro Situation Assessment *RLI-RLO*	Context- Sensitive Impact Assessment *RLI-IMO*	Context- Sensitive Response *RLI-RSO*
Impacts	Cost-Sensitive Detection/Est *IMI-DAO*	Cost-Sensitive Feature Extraction *IMI-FEO*	Cost- Sensitive Object Refinement *IMI-DEO*	Cost-Sensitive Situation Assessment *IMI-RLO*	Cost-Sensitive Impact Assessment *IMI-IMO*	Cost- Sensitive Response *IMI-RSO*
Responses	Reaction- Sensitive Detection/Est *RSI-DAO*	Reaction- Sensitive Feature Extraction *RSI-FEO*	Reaction- Sensitive Object Refinement *RSI-DEO*	Reaction- Sensitive Sit Assessment *RSI-RLO*	Reaction- Sensitive Impact Assessment *RSI-RLO*	Reaction- Sensitive Response *RSI-RSO*
	Level 0		Level 1	Level 2	Level 3	Level 4

We have augmented the categorization as shown in the remaining matrix cells by adding labels to these cells, relating input/output (I/O) types to process types, and filling in the unoccupied cells in the original matrix.

Note that Dasarathy's original categories represent constructive, or data-driven, processes in which organized information is extracted from relatively unorganized data. Additional processes — FEI-DAO, DEI-DAO, and DEI-FEO — can be defined that are analytic, or model-driven, such that organized information (a model) is analyzed to estimate lower-level data (features or measurements) as they relate to the model. Examples include predetection tracking (an FEI-DAO process), model-based feature-extraction (DEI-FEO), and model-based classification (DEI-DAO). The remaining cell in Table 2.3 — DAO-DEO — has not been addressed in a significant way (to the authors' knowledge) but could involve the direct estimation of entity states without the intermediate step of feature extraction.

Dasarathy's categorization can readily be expanded to encompass Level 2, 3, and 4 processes, as shown in Table 2.4. Here, rows and columns have been added to correspond to the object types listed in Figure 2.4.

Dasarathy's categories represent a useful refinement of the JDL levels. Not only can each of the levels (0–4) be subdivided on the basis of input data types, but our Level 0 can also be subdivided into detection processes and feature-extraction processes.*

Of course, much of Table 2.4 remains virgin territory; researchers have seriously explored only its northwest quadrant, with tentative forays southeast. Most likely, little utility will be found in either the northeast or the southwest. However, there may be gold buried somewhere in those remote stretches.

---

* A Level 0 remains a relatively new concept in data fusion (although quite mature in the detection and signal processing communities); therefore, it hasn't been studied to a great degree. The extension of formal data fusion methods into this area must evolve before the community will be ready to begin partitioning it. Encouragingly, Bedworth and O'Brien[11] describe a similar partitioning of Level 1-related functions in the Boyd and UK Intelligence Cycle models.

TABLE 2.5   Bedworth and O'Brien's Comparison of Data Fusion-related Models[11]

Activity being undertaken	Waterfall model	JDL Model	Boyd Loop	Intelligence Cycle
Command execution			Act	Disseminate
Decision making process	Decision making	Level 4	Decide	
Threat assessment		Level 3		Evaluate
Situation assessment	Situation assessment	Level 2	Orient	
Information processing	Pattern processing	Level 1		Collate
	Feature extraction			
Signal processing	Signal Processing	Level 0		
Source/sensor acquisition	Sensing		Observe	Collect

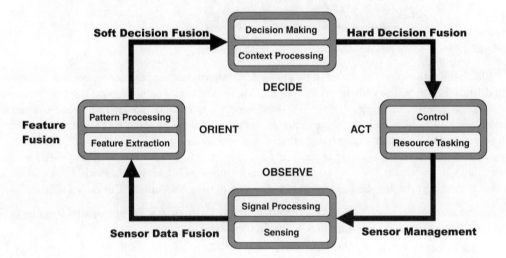

FIGURE 2.12   The "Omnibus" process model.[11]

2.5.2   Bedworth and O'Brien's Comparison among Models and Omnibus

Bedworth and O'Brien[11] provide a commendable comparison and attempted synthesis of data fusion models. That comparison is summarized in Table 2.5. By comparing the discrimination capabilities of the various process models listed — and of the JDL and Dasarathy's *functional* models — Bedworth and O'Brien suggest a comprehensive "Omnibus" *process* model as represented in Figure 2.12.

As noted by Bedworth and O'Brien, an information system's interaction with its environment need not be the single cyclic process depicted in Figure 2.12. Rather, the OODA process is often hierarchical and recursive, with analysis/decision loops supporting detection, estimation, evaluation, and response decisions at several levels (illustrated in Figure 2.13).

2.6   Summary

The goal of the JDL Data Fusion Model is to serve as a functional model for use by diverse elements of the data fusion community, to the extent that such a community exists, and to encourage coordination and collaboration among diverse communities. A model should clarify the elements of problems and solutions to facilitate recognition of commonalties in problems and in solutions. The virtues listed in Section 2.3 are significant criteria by which any functional model should be judged.[12]

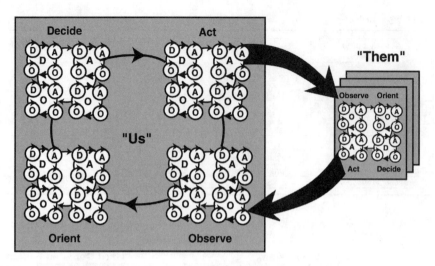

FIGURE 2.13   System interaction via interacting fractal OODA loops.

Additionally, a functional model must be amenable to implementation in process models. A functional model must be compatible with diverse instantiations in architectures and allow foundation in theoretical frameworks. Once again, the goal of the functional model is to facilitate understanding and communication among acquisition managers, theoreticians, designers, evaluators, and users of data fusion systems to permit cost-effect system design, development, and operation.

The revised JDL model is aimed at providing a useful tool of this sort. If used appropriately as part of a coordinated system engineering methodology (as discussed in Chapter 16), the model should facilitate research, development, test, and operation of systems employing data fusion. This model should

- Facilitate communications and coordination among theoreticians, developers, and users by providing a common framework to describe problems and solutions.
- Facilitate research by representing underlying principles of a subject. This should enable researchers to coordinate their attack on a problem and to integrate results from diverse researchers. By the same token, the ability to deconstruct a problem into its functional elements can reveal the limits of our understanding.
- Facilitate system acquisition and development by enabling developers to see their engineering problems as instances of general classes of problems. Therefore, diverse development activities can be coordinated and designs can be reused. Furthermore, such problem abstraction should enable the development of more cost-effective engineering methods.
- Facilitate integration and test by allowing the application of performance models and test data obtained with other applications of similar designs.
- Facilitate system operation by permitting a better sense of performance expectations, derived from experiences with entire classes of systems. Therefore, a system user will be able to predict his system's performance with greater confidence.

References

1. White, Jr., F.E., *Data Fusion Lexicon*, Joint Directors of Laboratories, Technical Panel for C³, Data Fusion Sub-Panel, Naval Ocean Systems Center, San Diego, 1987.
2. White, Jr., F.E., "A model for data fusion," *Proc. 1st Natl. Symp. Sensor Fusion*, vol. 2, 1988.
3. Steinberg, A.N., Bowman, C.L., and White, Jr., F.E., "Revisions to the JDL Data Fusion Model," *Proc. 3rd NATO/IRIS Conf.*, Quebec City, Canada, 1998.
4. Mahler, R.. "A Unified Foundation for Data Fusion," *Proc. 1994 Data Fusion Sys. Conf.*, 1994.

5. Goodman, I.R., Nguyen, H.T., and Mahler, R., *New Mathematical Tools for Data Fusion*, Artech House, Inc., Boston, 1997.

6. Mori, S., "Random sets in data fusion: multi-object state-estimation as a foundation of data fusion theory," *Proc. Workshop on Applications and Theory of Random Sets,* Springer-Verlag, 1997.

7. *C4ISR Architecture Framework, Version 1.0*, C4ISR ITF Integrated Architecture Panel, CISA-0000-104-96, June 7, 1996.

8. *Data Fusion System Engineering Guidelines*, SWC Talon-Command Operations Support Technical Report 96-11/4, vol. 2, 1997.

9. *Engineering Guidelines for Data Correlation Algorithm Characterization*, TENCAP SEDI Contractor Report, SEDI-96-00233, 1997.

10. Steinberg, A.N. and Bowman, C.L., "Development and application of data fusion engineering guidelines," *Proc. 10th Natl. Symp. Sensor Fusion*, 1997.

11. Bedworth, M. and O'Brien, J., "The Omnibus model: a new model of data fusion?", *Proc. 2nd Intl. Conf. Information Fusion*, 1999.

12. Polya, G., *How To Solve It*, Princeton University Press, Princeton, NJ, 1945.

13. Antony, R., *Principles of Data Fusion Automation,* Artech House, Inc., Boston, 1995.

14. Steinberg, A.N. and Washburn, R.B., "Multi-level fusion for Warbreaker intelligence correlation," *Proc. 8th Natl. Symp. Sensor Fusion*, 1995.

15. Bowman, C.L., "The data fusion tree paradigm and its dual," *Proc. 7th Natl. Symp. Sensor Fusion*, 1994.

16. Curry, H.B. and Feys, R., *Combinatory Logic*, North-Holland, Amsterdam, 1974.

17. Waltz, E., *Information Warfare: Principles and Operations*, Artech House, Inc., Boston, 1998.

18. Dasarathy, B., *Decision Fusion*, IEEE Computer Society Press, 1994.

3

Introduction to the Algorithmics of Data Association in Multiple-Target Tracking

3.1 Introduction 3-1
  Keeping Track • Nearest Neighbors • Track Splitting and
  Multiple Hypotheses • Gating • Binary Search and *kd*-Trees
3.2 Ternary Trees 3-10
3.3 Priority *kd*-Trees 3-13
  Applying the Results
3.4 Conclusion 3-17
Acknowledgments 3-17
References 3-17

Jeffrey K. Uhlmann
University of Missouri

3.1 Introduction

When a major-league outfielder runs down a long fly ball, the tracking of a moving object looks easy. Over a distance of a few hundred feet, the fielder calculates the ball's trajectory to within an inch or two and times its fall to within milliseconds. But what if an outfielder were asked to track 100 fly balls at once? Even 100 fielders trying to track 100 balls simultaneously would likely find the task an impossible challenge.

Problems of this kind do not arise in baseball, but they have considerable practical importance in other realms. The impetus for the studies described in this chapter was the Strategic Defense Initiative (SDI), the plan conceived in the early 1980s for defending the U.S. against a large-scale nuclear attack. According to the terms of the original proposal, an SDI system would be required to track tens or even hundreds of thousands of objects — including missiles, warheads, decoys, and debris — all moving at speeds of up to 8 kilometers per second. Another application of multiple-target tracking is air-traffic control, which attempts to maintain safe separations among hundreds of aircraft operating near busy airports. In particle physics, multiple-target tracking is needed to make sense of the hundreds or thousands of particle tracks emanating from the site of a high-energy collision. Molecular dynamics has similar requirements.

The task of following a large number of targets is surprisingly difficult. If tracking a single baseball, warhead, or aircraft requires a certain measurable level of effort, then it might seem that tracking 10 similar objects would require at most 10 times as much effort. Actually, for the most obvious methods of solving the problem, the difficulty is proportional to the square of the number of objects; thus, 10 objects demand 100 times the effort, and 10,000 objects increase the difficulty by a factor of 100 million. This combinatorial explosion is a first hurdle to solving the multiple-target tracking problem. In fact, exploiting all information to solve the problem optimally requires exponentially scaling effort. This chapter, however, considers computational issues that arise for any proposed multiple-target tracking system.*

Consider how the motion of a single object might be tracked, based on a series of position reports from a sensor such as a radar system. To reconstruct the object's trajectory, plot the successive positions in sequence and then draw a line through them (as shown on the left-hand side of Figure 3.1). Extending this line yields a prediction of the object's future position. Now, suppose you are tracking 10 targets simultaneously. At regular time intervals 10 new position reports are received, but the reports do not have labels indicating the targets to which they correspond. When the 10 new positions are plotted, each report could, in principle, be associated with any of the 10 existing trajectories (as illustrated on the right-hand side of Figure 3.1). This need to consider every possible combination of reports and tracks makes the difficulty of all n-target problem proportional to — or on the order of — n^2, which is denoted as $O(n^2)$.

Over the years, many attempts have been made to devise an algorithm for multiple-target tracking with better than $O(n^2)$ performance. Some of the proposals offered significant improvements in special circumstances or for certain instances of the multiple-target tracking problem, but they retained their $O(n^2)$ worst-case behavior. However, recent results in the theory of spatial data structures have made possible a new class of algorithms for associating reports with tracks — algorithms that scale better than quadratically in most realistic environments. In degenerate cases, in which all of the targets are so densely clustered that they cannot be individually resolved, there is no way to avoid comparing each report with each track. When each report can be feasibly associated only with a constant number of tracks on average, subquadratic scaling is achievable. This will become clear later in the chapter. Even with the new methods, multiple-target tracking remains a complex task that strains the capacity of the largest and fastest supercomputers. However, the new methods have brought important problem instances within reach.

3.1.1   Keeping Track

The modern need for tracking algorithms began with the development of radar during World War II. By the 1950s, radar was a relatively mature technology. Systems were installed aboard military ships and aircraft and at airports. The tracking of radar targets, however, was still performed manually by drawing lines through blips on a display screen. The first attempts to automate the tracking process were modeled closely on human performance. For the single-target case, the resulting algorithm was straightforward — the computer accumulated a series of positions from radar reports and estimated the velocity of the target to predict its future position.

Even single-target tracking presented certain challenges related to the uncertainty inherent in position measurements. A first problem involves deciding how to represent this uncertainty. A crude approach is to define an error radius surrounding the position estimate. This practice implies that the probability of finding the target is uniformly distributed throughout the volume of a three-dimensional sphere. Unfortunately, this simple approach is far from optimal. The error region associated with many sensors is highly nonspherical; radar, for example, tends to provide accurate range information but has relatively poorer radial resolution. Furthermore, one would expect the actual position of the target to be closer on average to the mean position estimate than to the perimeter of the error volume, which suggests, in turn, that the probability density should be greater near the center.

A second difficulty in handling uncertainty is determining how to interpolate the actual trajectory of the target from multiple measurements, each with its own error allowance. For targets known to have constant velocity (e.g., they travel in a straight line at constant speed), there are methods for calculating tile straight-line path that best fits, by some measure, the series of past positions. A desirable property of this approach is that it should always converge on the correct path — as the number of reports increases, the difference between the estimated velocity and the actual velocity should approach zero. On the other hand, retaining all past reports of a target and recalculating the entire trajectory every time a new report

---

* The material in this chapter updates and supplements material that first appeared in *American Scientist.*[1]

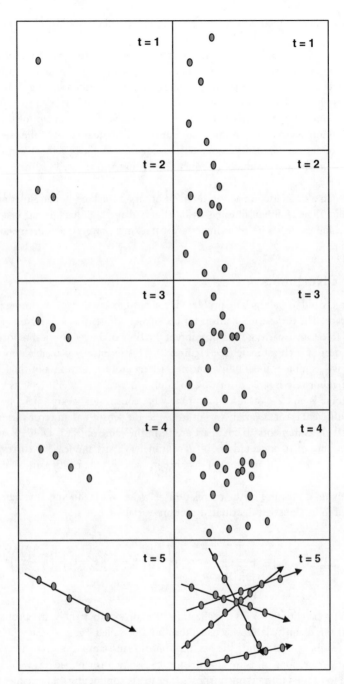

FIGURE 3.1 The information available for plotting a track consists of position reports (shown as dots) from a sensor such as a radar system. In tracking a single target (left), one can accumulate a series of reports and then fit a line or curve corresponding to those data points to estimate the object's trajectory. With multiple targets (right), there is no obvious way to determine which object has generated each report. Here, five reports appear initially at timestep $t = 1$, then five more are received at $t = 2$. Neither the human eye nor a computer can easily distinguish which of the later dots goes with which of the earlier ones. (In fact, the problem is even more difficult given that the reports at $t = 2$ could be newly detected targets that are not correlated with the previous five reports.) As additional reports arrive, coherent tracks begin to emerge. The tracks from which these reports were derived are shown in the lower panels at $t = 5$. Here and in subsequent figures, all targets are assumed to have constant velocity in two dimensions. The problem is considerably more difficult for ballistic or maneuvering trajectories in three dimensions.

arrives is impractical. Such a method would eventually exceed all constraints on computation time and storage space.

A near-optimal method for addressing a large class of tracking problems was developed in 1960 by R.E. Kalman.[2] His approach, referred to as *Kalman filtering*, involves the recursive fusion of noisy measurements to produce an accurate estimate of the state of a system of interest. A key feature of the Kalman filter is its representation of state estimates in terms of mean vectors and error covariance matrices, where a covariance matrix provides an estimate (usually a conservative over-estimate) of the second moment of the error distribution associated with the mean estimate. The square root of the estimated covariance gives an estimate of the standard deviation. If the sequence of measurement errors are statistically independent, the Kalman filter produces a sequence of conservative fused estimates with diminishing error covariances.

Kalman's work had a dramatic impact on the field of target tracking in particular and data fusion in general. By the mid-1960s, Kalman filtering was a standard methodology. It has become as central to multiple-target tracking as it has been to single-target tracking; however, it addresses only one aspect of the overall problem.

3.1.2 Nearest Neighbors

What multiple targets add to the tracking problem is the need to assign each incoming position report to a specific target track. The earliest mechanism for classifying reports was the nearest-neighbor rule. The idea of the rule is to estimate each object's position at the time of a new position report, and then assign the report to the nearest such estimate (see Figure 3.2). This intuitively plausible approach is especially attractive because it decomposes the multiple-target tracking problem into a set of single-target problems.

The nearest-neighbor rule is straightforward to apply when all tracks and reports are represented as points; however, there is no clear means for defining what constitutes "nearest neighbors" among tracks and reports with different error covariances. For example, if a sensor has an error variance of 1 cm, then the probability that measurements 10 cm apart are from the same object is $O(10^{-20})$, whereas measurements having a variance of 10 cm could be 20–30 centimeters apart and feasibly correspond to the same object. Therefore, the appropriate measure of distance must reflect the relative uncertainties in the mean estimates.

The most widely used measure of the correlation between two mean and covariance pairs $\{x1, P1\}$, which are assumed to be Gaussian-distributed random variables, is[3,4]

$$P_{association}\left(\mathbf{x}_1, \mathbf{x}_2\right) = \frac{1}{\sqrt{2\pi\left|\left(\mathbf{P}_2 + \mathbf{P}_2\right)\right|}} \exp\left(-\frac{1}{2}\left(\mathbf{x}_1 - \mathbf{x}_2\right)\left(\mathbf{P}_1 + \mathbf{P}_2\right)^{-1}\left(\mathbf{x}_1 - \mathbf{x}_2\right)^T\right) \tag{3.1}$$

which reflects the probability that $\mathbf{x}_1$ is a realization of $\mathbf{x}_2$ or, symmetrically, the probability that $\mathbf{x}_2$ is a realization of $\mathbf{x}_1$. If this quantity is above a given threshold — called a gate — then the two estimates are considered to be feasibly correlated. If the assumption of Gaussianity does not hold exactly — and it generally does not — then this measure is heuristically assumed (or hoped) to yield results that are at least good enough to be used for ranking purposes (i.e., to say confidently that one measurement is more likely than another measurement to be associated with a given track). If this assumption approximately holds, then the gate will tend to discriminate high- and low-probability associations. Accordingly, the nearest-neighbor rule can be redefined to state that a report should be assigned to the track with which it has the highest association ranking. In this way, a multiple-target problem can still be decomposed into a set of single-target problems.

The nearest-neighbor rule has strong intuitive appeal, but doubts and difficulties connected with it soon emerged. For example, early implementers of the method discovered problems in creating initial tracks for multiple targets. In the case of a single target, two reports can be accumulated to derive a velocity estimate, from which a track can be created. For multiple targets, however, there is no obvious

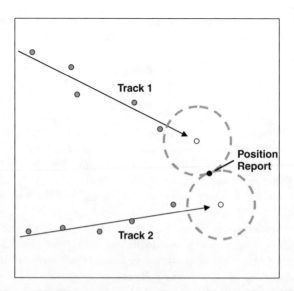

FIGURE 3.2   The nearest-neighbor rule is perhaps the simplest approach for determining which tracked object produced a given sensor report. When a new position report arrives, all existing tracks are projected forward to the time of the new measurement. (In this diagram, earlier target positions are indicated by dots and projected positions by circles; the new position report is labeled.) Then, the distance from the report to each projected position is calculated, and the report is associated with the nearest track. More generally, the distance calculation is computed to reflect the relative uncertainties (covariances) associated with each track and report. In the situation depicted above, the report would be assigned to Track 1, based purely on its Euclidean proximity to the report. If this assignment is erroneous, the subsequent tracking process will be adversely affected.

way to deduce such initial velocities. The first two reports received could represent successive positions of a single object or the initial detection of two distinct objects. Every subsequent report could be the continuation of a known track or the start of a new one. To make matters worse, almost every sensor produces some background rate of spurious reports, which give rise to spurious tracks. Thus, the tracking system needs an additional mechanism to recognize and delete tracks that do not receive any subsequent confirming reports.

Another difficulty with the nearest-neighbor rule becomes apparent when reports are misclassified, as will inevitably happen from time to time if the tracked objects are close together. A misassignment can cause the Kalman-filtering process to converge very slowly, or fail to converge altogether, in which case the track cannot be predicted. Moreover, tracks updated with misassigned reports (or not updated at all) will tend to correlate poorly with subsequent reports and may, therefore, be mistaken as spurious by the track-deletion mechanism. Mistakenly deleted tracks then necessitate subsequent track initiations and a possible repetition of the process.

3.1.3   Track Splitting and Multiple Hypotheses

A robust solution to the problem of assignment ambiguities is to create multiple *hypothesis tracks*. Under this scheme, the tracking system does not have to commit immediately or irrevocably to a single assignment of each report. If a report is highly correlated with more than one track, an updated copy of each track can be created; subsequent reports can be used to determine which assignment is correct. As more reports come in, the track associated with the correct assignment will rapidly converge on the true target trajectory, whereas the falsely updated tracks are less likely to be correlated with subsequent reports.

This basic technique is called track splitting.[3] One of its worrisome consequences is a proliferation in the number of tracks upon which a program must keep tabs. The proliferation can be controlled with the same track deletion mechanism used in the nearest-neighbor algorithm, which scans through all the tracks from time to time and eliminates those that have a low probability of association with recent reports. A more sophisticated approach to track splitting, called multiple-hypothesis tracking, maintains a history of track branchings, so that as soon as one branch is confirmed, the alternative branches can be pruned away.

Track splitting in its various forms[5] is a widely applied strategy for handling the ambiguities inherent in correlating tracks with reports from multiple targets. It is also used to minimize the effects of spurious reports when tracking a single target. Nevertheless, some serious difficulties remain. First, track splitting does not completely decompose a multiple-target tracking problem into independent single-target problems, the way the nearest-neighbor strategy was intended to function. For example, two hypothesis tracks may lock onto the trajectory of a single object. Because both tracks are valid, the standard track-deletion mechanism cannot eliminate either of them. The deletion procedure has to be modified to detect redundant tracks and, therefore, cannot look at just one track at a time. This coupling between multiple tracks is theoretically troubling; however, experience has shown that it can be managed in practice at low computational cost.

A second problem is the difficulty of deciding when a position report and a projected track are correlated closely enough to justify creating a new hypothesis track. If the correlation threshold is set too high, correct assignments may be missed so often as to prevent convergence of the Kalman filter. If the threshold is too low, the number of hypotheses could grow exponentially. The usual practice is to set the threshold low enough to ensure convergence, and then add another mechanism to limit the rate of hypothesis generation. A simple strategy is to select the n hypothesis candidates with the highest probabilities of association, where n is the maximum number of hypotheses that computational resource constraints will allow. This "greedy" method often yields good performance.

Even with these enhancements, the tracking algorithm makes such prodigious demands on computing resources that large problems remain beyond practical reach. Monitoring the computation to see how much time is spent in various subtasks shows that calculating probabilities of association is, by far, the biggest expense. The program gets bogged down projecting target tracks to the time of a position report and calculating association probabilities. Because this is the critical section of the algorithm, further effort has focused on improving performance in this area.

3.1.4   Gating

The various calculations involved in estimating a probability of association are numerically intensive and inherently time consuming. Thus, one approach to speeding up the tracking procedure is to streamline or fine-tune these calculations — to encode them more efficiently without changing their fundamental nature. An obvious example is to calculate

$$\text{dist}^2\left(\mathbf{x}_1, \mathbf{x}_2\right) = \left(\mathbf{x}_1 - \mathbf{x}_2\right)\left(\mathbf{P}_1 + \mathbf{P}_2\right)^{-1}\left(\mathbf{x}_1 - \mathbf{x}_2\right)^T \qquad (3.2)$$

rather than the full probability of association. This measure is proportional to the logarithm of the probability of association and is commonly referred to as the Mahalanobis distance or log-likelihood measure.[4] Applying a suitably chosen threshold to this quantity yields a method for obtaining the same set of feasible pairs, while avoiding a large number of numerically intensive calculations.

An approach for further reducing the number of computations is to minimize the number of log-likelihood calculations by performing a simpler preliminary screening of tracks and sensor reports. Only if a track report pair passes this computationally inexpensive feasibility check is there a need to complete the log-likelihood calculation. Multiple gating tests also can be created for successively weeding out infeasible pairs, so that each gate involves more calculations but is applied to considerably fewer pairs than the previous gate.

Several geometric tests could serve as gating criteria. For example, if each track is updated, on average, every five seconds, and the targets are known to have a maximum speed of 10 kilometers per second, a track and report more than 50 kilometers apart are not likely to be correlated. A larger distance may be required to take into account the uncertainty measures associated with both the tracks and the reports.

Simple gating strategies can successfully reduce the numerical overhead of the correlation process and increase the number of targets that can be tracked in real time. Unfortunately, the benefits of simple gating diminish as the number of targets increases. Specifically, implementers of gating algorithms have found that increasing the number of targets by a factor of 20 often increases the computational burden by a factor of more than 100. Moreover, the largest percentage of computation time is still spent in the correlation process, although now the bulk of the demand is for simple distance calculations within the gating algorithm. This implies that the quadratic growth in the number of gating tests is more critical than the constant numerical overhead associated with the individual tests. In other words, simple gating can reduce the average cost of each comparison, but what is really needed is a method to reduce the sheer number of comparisons. Some structure must be imposed on the set of tracks that will allow correlated track-report pairs to be identified without requiring every report to be compared with every track.

The gating problem is difficult conceptually because it demands that most pairs of tracks and reports be excluded from consideration *without ever being examined*. At the same time, no track-report pair whose probability of association exceeds the correlation threshold can be disregarded. Until the 1980s, the consensus in the tracking literature was that these constraints were impossible to satisfy simultaneously. Consequently, the latter constraint was often sacrificed by the use of methods that did allow some, but *hopefully* few, track-report pairs to be missed even though their probabilities of association exceeded the threshold. This seemingly reasonable compromise, however, has led to numerous ad hoc schemes that either fail to adequately limit the number of comparisons or fail to adequately limit the number of missed correlations. Some approaches are susceptible to both problems.

Most of the ad hoc strategies depend heavily on the distribution of the targets. A common approach is to identify clusters of targets that are sufficiently separated that reports from targets in one cluster will never have a significant probability of association with tracks from another cluster.[6] This allows the correlation process to determine from which cluster a particular report could have originated and then compare the report only to the tracks in that cluster. The problem with this approach is that the number of properly separated clusters depends on the distribution of the targets and, therefore, cannot be controlled by the clustering algorithm (Figure 3.3). If $O(n)$ tracks are partitioned into $O(n)$ clusters, each consisting of a constant number of tracks, or into a constant number of clusters of $O(n)$ tracks, the method still results in a computational cost that is proportional to the comparison of every report to every track. Unfortunately, most real-world tracking problems tend to be close to one of these extremes.

A gating strategy that avoids some of the distribution problems associated with clustering involves partitioning the space in which the targets reside into grid cells. Each track can then be assigned to a cell according to its mean projected position. In this way, the tracks that might be associated with a given report can be found by examining only those tracks in cells within close proximity to the report's cell. The problem with this approach is that its performance depends heavily on the size of the grid cells, as well as on the distribution of the targets (Figure 3.4). If the grid cells are large and the targets are densely distributed in a small region, every track will be within a nearby cell. Conversely, if the grid cells are small, the algorithm may spend as much time examining cells (most of which may be empty) as would be required to simply examine each track.

3.1.5   Binary Search and *kd*-Trees

The deficiencies of grid methods suggest the need for a more flexible data structure. The main requirement imposed on the data structure has already been mentioned — it must allow all proximate track-report pairs to be identified without having to compare every report with every track (unless every track is within the prescribed proximity to every report).

A clue to how real-time gating might be accomplished comes from one of the best-known algorithms in computer science: binary search. Suppose one is given a sorted list of *n* numbers and asked to find

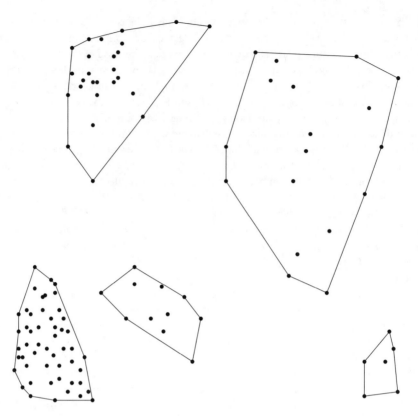

FIGURE 3.3 Clustering algorithms may produce spatially large clusters with few points and spatially small ones with many points.

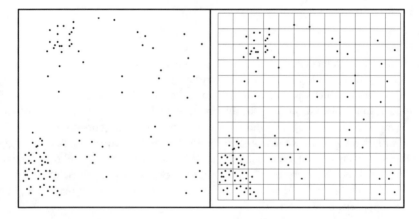

FIGURE 3.4 Grids may have a few cells with many points, while the remaining cells contain few or no points.

out whether or not a specific number, q, is included in the list. The most obvious search method is simply to compare q with each number in sequence; in the worst case (when q is the last number or is not present at all), the search requires n comparisons. There is a much better way. Because the list is sorted, if q is found to be greater than a particular element of the list, one can exclude from further consideration not only that element but all those that precede it in the list. This principle is applied optimally in binary search. The algorithm is recursive — first compare q to the median value in the list of numbers (by definition, the median will be found in the middle of a sorted list). If q is equal to the median value, then stop, and report that the search was successful. If q is greater than the median value, then apply the

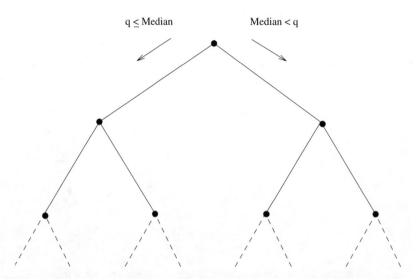

FIGURE 3.5 Each node in a binary search tree stores the median value of the elements in its subtree. Searching the tree requires a comparison at each node to determine whether the left or right subtree should be searched.

same procedure recursively to the sublist greater than the median; otherwise apply it to the sublist less than the median (Figure 3.5). Eventually either q will be found — it will be equal to the median of some sublist — or a sublist will turn out to be empty, at which point the procedure terminates and reports that q is not present in the list.

The efficiency of this process can be analyzed as follows. At every step, half of the remaining elements in the list are eliminated from consideration. Thus, the total number of comparisons is equal to the number of halvings, which in turn is $O(\log n)$. For example, if n is 1,000,000, then only 20 comparisons are needed to determine if a given number is in the list.

Binary search can also be used to find all elements of the list that are within a specified range of values (*min, max*). Specifically, it can be applied to find the position in the list of the largest element less than *min* and the position of the smallest element greater than *max*. The elements between these two positions then represent the desired set. Finding the positions associated with *min* and *max* requires $O(\log n)$ comparisons. Assuming that some operation will be carried out on each of the m elements of the solution set, the overall computation time for satisfying a range query scales as $O(\log n + m)$.

Extending binary search to multiple dimensions yields a *kd*-tree.[7] This data structure permits the fast retrieval of all 3-D points; for example, in a data set whose x coordinate is in the range (x_{min}, x_{max}), whose y coordinate is in the range (y_{min}, y_{max}) and whose z coordinate is in the range (z_{min}, z_{max}). The *kd*-tree for $k = 3$ is constructed as follows: The first step is to list the x coordinates of the points and choose the median value, then partition the volume by drawing a plane perpendicular to the x-axis through this point. The result is to create two subvolumes, one containing all the points whose x coordinates are less than the median and the other containing the points whose x coordinates are greater than the median. The same procedure is then applied recursively to the two subvolumes, except that now the partitioning planes are drawn perpendicular to the y-axis and they pass through points that have median values of the y coordinate. The next round uses the z coordinate, and then the procedure returns cyclically to the x coordinate. The recursion continues until the subvolumes are empty.*

---

* An alternative generalization of binary search to multiple dimensions is to partition the dataset at each stage according to its distance from a selected set of points;[8-14] those that are less than the median distance comprise one branch of the tree, and those that are greater comprise the other. These data structures are very flexible because they offer the freedom to use an appropriate application-specific metric to partition the dataset; however, they are also much more computationally intensive because of the number of distance calculations that must be performed.

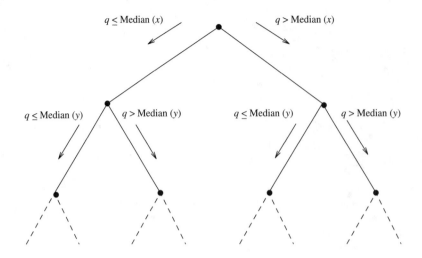

A *kd*-tree partitions on a different coordinate at each level in the tree.

FIGURE 3.6   A *kd*-tree is analogous to an ordinary binary search tree, except that each node stores the median of the multidimensional elements in its subtree projected onto one of the coordinate axes.

Searching the subdivided volume for the presence of a specific point with given x, y, and z coordinates is a straightforward extension of standard binary search. As in the one-dimensional case, the search proceeds as a series of comparisons with median values, but now attention alternates among the three coordinates. First the x coordinates are compared, then the y, then the z, and so on (Figure 3.6). In the end, either the chosen point will be found to lie on one of the median planes, or the procedure will come to an empty subvolume.

Searching for all of the points that fall within a specified interval is somewhat more complicated. The search proceeds as follows: If x_{min} is less than the median-value x coordinate, the left subvolume must be examined. If x_{max} is greater than the median value of x, the right subvolume must be examined. At the next level of recursion, the comparison is done using y_{min} and y_{max}, then z_{min} and z_{max}.

A detailed analysis[15-17] of the algorithm reveals that for k dimensions (provided that k is greater than 1), the number of comparisons performed during the search can be as high as $O(n^{1-1/k} + m)$; thus in three dimensions the search time is proportional to $O(n^{2/3} + m)$. In the task of matching n reports with n tracks, the range query must be repeated n times, so the search time scales as $O(n * n^{2/3} + m)$ or $O(n^{5/3} + m)$. This scaling is better than quadratic, but not nearly as good as the logarithmic scaling observed in the one-dimensional case, which works out for n range queries to be $O(n \log n + m)$. The reason for the penalty in searching a multidimensional tree is the possibility at each step that both subtrees will have to be searched without necessarily finding an element that satisfies the query. (In one dimension, a search of both subtrees implies that the median value satisfies the query.) In practice, however, this seldom happens, and the worst-case scaling is rarely seen. Moreover, for query ranges that are small relative to the extent of the dataset — as they typically are in gating applications — the observed query time for *kd*-trees is consistent with $O(\log^{1+\varepsilon} + n)$, where $\varepsilon \gg 0$.

3.2   Ternary Trees

The *kd*-tree is provably optimal for satisfying multidimensional range queries if one is constrained to using only linear (i.e., $O(n)$) storage.[16,17] Unfortunately, it is inadequate for gating purposes because the track estimates have spatial extent due to uncertainty in their exact position. In other words, a *kd*-tree would be able to identify all track points that fall within the observation uncertainty bounds. It would fail, however, to return any imprecisely localized map item whose uncertainty region intersects the

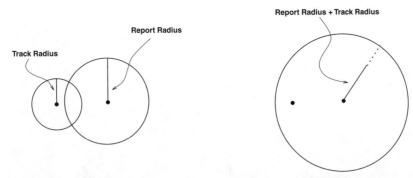

If the position uncertainties are thresholded, then gating requires intersection detection.

If the largest track radius is added to all the report radii, then the tracks can be treated as points.

FIGURE 3.7   Transferring uncertainty from tracks to reports reduces intersection queries to range queries.

observation region, but whose mean position does not. Thus, the gating problem requires a data structure that stores sized objects and is able to retrieve those objects that intersect a given query region associated with an observation.

One approach for solving this problem is to shift all of the uncertainty associated with the tracks onto the reports.[18,19] The nature of this transfer is easy to understand in the simple case of a track and a report whose error ellipsoids are spherical and just touching. Reducing the radius of the track error sphere to zero, while increasing the radius of the report error sphere by an equal amount, leaves the enlarged report sphere just touching the point representing the track, so the track still falls within the gate of the report (Figure 3.7). Unfortunately, when this idea is applied to multiple tracks and reports, the query region for every report must be enlarged in all directions by an amount large enough to accommodate the largest error radius associated with any track. Techniques have been devised to find the minimum enlargement necessary to guarantee that every track correlated with a given report will be found;[19] however, many tracks with large error covariances can result in such large query regions that an intolerable number of uncorrelated tracks will also be found.

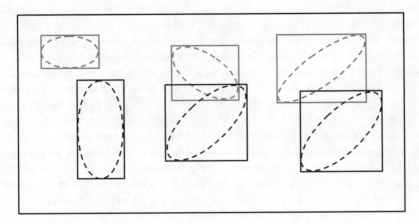

FIGURE 3.8   The intersection of error boxes offers a preliminary indication that a track and a report probably correspond to the same object. A more definitive test of correlation requires a computation to determine the extent to which the error ellipses (or their higher-dimensional analogs) overlap, but such computations can be too time consuming when applied to many thousands of track/report pairs. Comparing bounding boxes is more computationally efficient; if they do not intersect, an assumption can be made that the track and report do not correspond to the same object. However, intersection does not necessarily imply that they do correspond to the same object. False positives must be weeded out in subsequent processing.

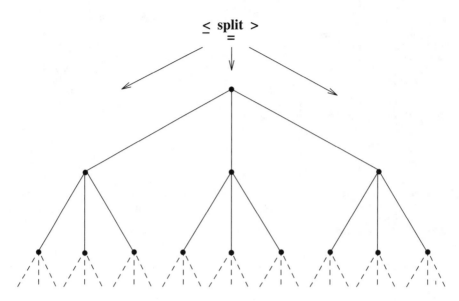

FIGURE 3.9 Structure of a ternary tree. In a ternary tree, the boxes in the left subtree fall on one side of the partitioning (split) plane; the boxes in the right subtree fall to the other side of the plane; and the boxes in the middle subtree are strictly cut by the plane.

A solution that avoids the need to inflate the search volumes is to use a data structure that can satisfy ellipsoid intersection queries instead of range queries. One such data structure that has been applied in large scale tracking applications is an enhanced form of *kd*-tree that stores coordinate-aligned boxes.[1,20] A box is defined as the smallest rectilinear shape, with sides parallel to the coordinate axes, that can entirely surround a given error ellipsoid (see Figure 3.8). Because the axes of the ellipse may not correspond to those of the coordinate system, the box may differ significantly in size and shape from the ellipse it encloses. The problem of determining optimal approximating boxes is presented in Reference 21.

An enhanced form of the *kd*-tree is needed for searches in which one range of coordinate values is compared with another range, rather than the simpler case in which a range is compared with a single point. A binary tree will not serve this purpose because it is not possible to say that one interval is entirely greater than or less than another when they intersect. What is needed is a ternary tree, with three descendants per node (Figure 3.9). At each stage in a search of the tree, the maximum value of one interval is compared with the minimum of the other, and vice versa. These comparisons can potentially eliminate either the left subtree or the right subtree. In either case, examining the middle subtree — the one made up of nodes representing boxes that might intersect the query interval — is necessary. Because all of the boxes in a middle subtree intersect the plane defined by the split value, however, the dimensionality of the subtree can be reduced by one, causing subsequent searches to be more efficient.

The middle subtree represents obligatory search effort; therefore, one goal is to minimize the number of boxes that straddle the split value. However, if most of the nodes fall to the left or right of the split value, then few nodes will be eliminated from the search, and query performance will be degraded. Thus, a tradeoff must be made between the effects of unbalance and of large middle subtrees. Techniques have been developed for adapting ternary trees to exploit distribution features of a given set of boxes,[20] but they cannot easily be applied when boxes are inserted and deleted dynamically. The ability to dynamically update the search structure can be very important in some applications; this topic is addressed in subsequent sections of this chapter.

3.3   Priority *kd*-Trees

The ternary tree represents a very intuitive approach to extending the *kd*-tree for the storage of boxes. The idea is that, in one dimension, if a balanced tree is constructed from the minimum values of each interval, then the only problematic cases are those intervals whose *min* endpoints are less than a split value while their *max* endpoints are greater. Thus, if these cases can be handled separately (i.e., in separate subtrees), then the rest of the tree can be searched the same way as an ordinary binary search tree. This approach fails because it is not possible to ensure simultaneously that all subtrees are balanced and that the extra subtrees are sufficiently small. As a result, an entirely different strategy is required to bound the worst-case performance.

A technique is known for extending binary search to the problem of finding intersections among one-dimensional intervals.[22,23] The priority search tree is constructed by sorting the intervals according to the first coordinate as in an ordinary one-dimensional binary search tree. Then down every possible search path, the intervals are ordered by the second endpoint. Thus, the intervals encountered by always searching the left subarray will all have values for their first endpoint that are less than those of intervals with larger indices (i.e., to their right). At the same time, though, the second endpoints in the sequence of intervals will be in ascending order. Because any interval whose second endpoint is less than the first endpoint of the query interval cannot possibly produce an intersection, an additional stopping criterion is added to the ordinary binary search algorithm.

The priority search tree avoids the problems associated with middle subtrees in a ternary tree by storing the *min* endpoints in an ordinary balanced binary search tree, while storing the *max* endpoints in priority queues stored along each path in the tree. This combination of data structures permits the storage of n intervals, such that intersection queries can be satisfied in worst-case $O(\log n + m)$ time, and insertions and deletions of intervals can be performed in worst-case $O(\log n)$ time. Thus, the priority search tree generalizes binary search on points to the case of intervals, without any penalty in terms of errors. Unfortunately, the priority search tree is defined purely for intervals in one dimension.

Whereas the *kd*-tree can store multidimensional points, but not multidimensional ranges, the priority search tree can store one-dimensional ranges, but not multiple dimensions. The question that arises is whether the *kd*-tree can be extended to store boxes efficiently, or whether the priority search tree can be extended to accommodate the analogue of intervals in higher dimensions (i.e., boxes). The answer to the question is "yes" for both data structures, and the solution is, in fact, a combination of the two.

A priority *kd*-tree[24] is defined as follows: given a set S of k-dimensional box intervals (lo_i, hi_i), $1 < i < k$, a priority *kd*-tree consists of a *kd*-tree constructed from the *lo* endpoints of the intervals with a priority set containing up to k items stored at each node (Figure 3.10).* The items stored at each node are the minimum set so that the union of the *hi* endpoints in each coordinate includes a value greater than the corresponding *hi* endpoint of any interval of any item in the subtree. Searching the tree proceeds exactly as for all ordinary priority search trees, except that the intervals compared at each level in the tree cycle through the k dimensions as in a search of a *kd*-tree.

The priority *kd*-tree can be used to efficiently satisfy box intersection queries. Just as important, however, is the fact that it can be adapted to accommodate the dynamic insertion and deletion of boxes in optimal $O(\log n)$ time by replacing the *kd*-tree structure with a divided *kd*-tree structure.[25] The difference between the divided *kd*-tree and an ordinary *kd*-tree is that the divided variant constructs a d-layered tree in which each layer partitions the data structure according to only one of the d coordinates. In three dimensions, for example, the first layer would partition on the x coordinate, the next layer on y, and the last layer on z. The number of levels per layer/coordinate is determined so as to minimize query

---

* Other data structures have been independently called "priority *kd*-trees" in the literature, but they are designed for different purposes.

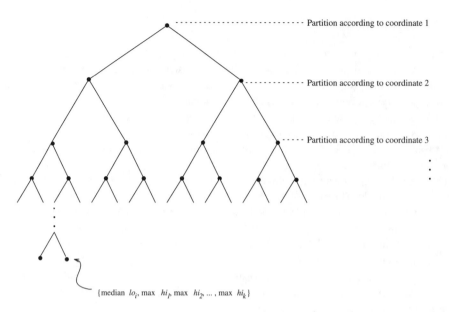

FIGURE 3.10 Structure of a priority *kd*-tree. The priority *kd*-tree stores multidimensional boxes, instead of vectors. A box is defined by an interval (lo_i, hi_i) for each coordinate i. The partitioning is applied to the *lo* coordinates analogously to an ordinary *kd*-tree. The principal difference is that the maximum *hi* value for each coordinate is stored at each node. These *hi* values function analogously to the priority fields of a priority search tree. In searching a priority *kd*-tree, the query box is compared to each of the stored values at each visited node. If the node partitions on coordinate i, then the search proceeds to the left subtree if lo_i is less than the median lo_i associated with the node. If hi_i is greater than the median lo_i, then the right subtree must be searched. The search can be terminated, however, if for any j, lo_j of the query box is greater than the hi_j stored at the node.

time complexity. The reason for stratifying the tree into layers for the different coordinates is to allow updates within the different layers to be treated just like updates in ordinary one-dimensional binary trees.

Associating priority fields with the different layers results in a dynamic variant of the priority *kd*-tree, which is referred to as a *Layered Box Tree*. Note that the *i* priority fields, for coordinates l,...,*i*, need to be maintained at level *i*. This data structure has been proven[26] to be maintainable at a cost of $O(\log n)$ time per insertion or deletion and can satisfy box intersection queries $O(n^{1-1/k} \log^{1/k} n + m)$, where *m* is the number of boxes in *S* that intersect a given query box b. A relatively straightforward variant[27] of the data structure improves the query complexity to $O(n^{1-1/k} + m)$, which is optimal.

The priority *kd*-tree is optimal among the class of linear-sized data structures, i.e., ones using only $O(n)$ storage, but asymptotically better $O(\log^k n + m)$ query complexity is possible if $O(n \log^{k-1} n)$ storage is used.[16,17] However, the extremely complex structure, called a range-segment tree, requires $O(\log^k n)$ update time, and the query performance is $O(\log^k n + m)$. Unfortunately, this query complexity holds in the *average case*, as well as in the worst case, so it can be expected to provide superior query performance in practice only when *n* is extremely large. For realistic distributions of objects, however, it may never provide better query performance practice. Whether or not that is the case, the range-segment tree is almost never used in practice because the values of $n^{1-1/k}$ and $\log^k n$ are comparable even for *n* as large as 1,000,000, and for datasets of that size the storage for the range-segment tree is multiplied by a factor of $\log^2(1,000,000) = 400$.

3.3.1 Applying the Results

The method in which multidimensional search structures are applied in a tracking algorithm can be summarized as follows: tracks are recorded by storing the information — such as current positions, velocities, and accelerations — that a Kalman filter needs to estimate the future position of each candidate target. When a new batch of position reports arrives, the existing tracks are projected forward to the time of the reports. An error ellipsoid is calculated for each track and each report, and a box is constructed around each ellipsoid. The boxes representing the track projections are organized into a multidimensional tree. Each box representing a report becomes the subject of a complete tree search; the result of the search is the set of all track boxes that intersect the given report box. Track-report pairs whose boxes do not intersect are excluded from all further consideration. Next the set of track-report pairs whose boxes do overlap is examined more closely to see whether the inscribed error ellipsoids also overlap. Whenever this calculation indicates a correlation, the track is projected to the time of the new report. Tracks that consistently fail to be associated with any reports are eventually deleted; reports that cannot be associated with any existing track initiate new tracks.

The approach for multiple-target tracking described above ignores a plethora of intricate theoretical and practical details. Unfortunately, such details must eventually be addressed, and the SDI forced a generation of tracking, data fusion, and sensor system researchers to face all of the thorny issues and constraints of a real-world problem of immense scale. The goal was to develop a space-based system to defend against a full-scale missile attack against the U.S. Two of the most critical problems were the design and deployment of sensors to detect the launch of missiles at the earliest moment possible in their 20-minute mid-course flight, and the design and deployment of weapons systems capable of destroying the detected missiles. Although an automatic tracking facility would clearly be an integral component of any SDI system, it was not generally considered a "high risk" technology. Tracking, especially of aircraft, had been widely studied for more than 30 years, so the tracking of nonmaneuvering ballistic missiles seemed to be a relatively simple engineering exercise. The principal constraint imposed by SDI was that the tracking be precise enough to predict a missile's future position to within a few meters, so that it could be destroyed by a high-energy laser or a particle-beam weapon.

The high-precision tracking requirement led to the development of highly detailed models of ballistic motion that took into account the effects of atmospheric drag and various gravitational perturbations over the earth. By far the most significant source of error in the tracking process, however, resulted from the limited resolution of existing sensors. This fact reinforced the widely held belief that the main obstacle to effective tracking was the relatively poor quality of sensor reports. The impact of large numbers of targets seemed manageable; just build larger, faster computers. Although many in the research community thought otherwise, the prevailing attitude among funding agencies was that if 100 objects could be tracked in real time, then little difficulty would be involved in building a machine that was 100 times faster — or simply having 100 machines run in parallel — to handle 10,000 objects.

Among the challenges facing the SDI program, multiple-target tracking seemed far simpler than what would be required to further improve sensor resolution. This belief led to the awarding of contracts to build tracking systems in which the emphasis was placed on high precision at any cost in terms of computational efficiency. These systems did prove valuable for determining bounds on how accurately a single cluster of three to seven missiles could be tracked in an SDI environment, but ultimately pressures mounted to scale up to more realistic numbers. In one case, a tracker that had been tested on five missiles was scaled up to track 100, causing the processing time to increase from a couple of hours to almost a month of nonstop computation for a simulated 20-minute scenario. The bulk of the computations was later determined to have involved the correlation step, where reports were compared against hypothesis tracks.

In response to a heightened interest in scaling issues, some researchers began to develop and study prototype systems based on efficient search structures. One of these systems demonstrated that 65 to 100 missiles could be tracked in real time on a late-1980s personal workstation. These results were based on the assumption that a good-resolution radar report would be received every five seconds for every missile, which is unrealistic in the context of SDI; nevertheless, the demonstration did provide convincing evidence that SDI trackers could be adapted to avoid quadratic scaling. A tracker that had been installed at the SDI National Testbed in Colorado Springs achieved significant performance improvements after a tree-based search structure was installed in its correlation routine; the new algorithm was superior for as few as 40 missiles. Stand-alone tests showed that the search component could process 5,000 to 10,000 range queries in real time on a modest computer workstation of the time. These results suggested that the problem of correlating vast numbers of tracks and reports had been solved. Unfortunately, a new difficulty was soon discovered.

The academic formulation of the problem adopts the simplifying assumption that all position reports arrive in batches, with all the reports in a batch corresponding to measurements taken at the same instant of all of the targets. A real distributed sensor system would not work this way; reports would arrive in a continuing stream and would be distributed over time. In order to determine the probability that a given track and report correspond to the same object, the track must be projected to the measurement time of the report. If every track has to be projected to the measurement time of every report, the combinatorial advantages of the tree-search algorithm is lost.

A simple way to avoid the projection of each track to the time of every report is to increase the search radius in the gating algorithm to account for the maximum distance an object could travel during the maximum time difference between any track and report. For example, if the maximum speed of a missile is 10 kilometers per second, and the maximum time difference between any report and track is five seconds, then 50 kilometers would have to be added to each search radius to ensure that no correlations are missed. For boxes used to approximate ellipsoids, this means that each side of the box must be increased by 100 kilometers.

As estimates of what constitutes a realistic SDI scenario became more accurate, members of the tracking community learned that successive reports of a particular target often would be separated by as much as 30 to 40 seconds. To account for such large time differences would require boxes so immense that the number of spurious returns would negate the benefits of efficient search. Demands for a sensor configuration that would report on every target at intervals of 5 to 10 seconds were considered unreasonable for a variety of practical reasons. The use of sophisticated correlation algorithms seemed to have finally reached its limit. Several heuristic "fixes" were considered, but none solved the problem.

A detailed scaling analysis of the problem ultimately pointed the way to a solution. Simply accumulate sensor reports until the difference between the measurement time of the current report and the earliest report exceeds a threshold. A search structure is then constructed from this set of reports, the tracks are projected to the mean time of the reports, and the correlation process is performed with the maximum time difference being no more than half of the chosen time-difference threshold. The subtle aspect of this deceptively simple approach is the selection of the threshold. If it is too small, every track will be projected to the measurement time of every report. If it is too large, every report will fall within the search volume of every track. A formula has been derived that, with only modest assumptions about the distribution of targets, ensures the optimal trade-off between these two extremes.

Although empirical results confirm that the track file projection approach essentially solves the time difference problem in most practical applications, significant improvements are possible. For example, the fact that different tracks are updated at different times suggests that projecting all of the tracks at the same points in time may be wasteful. An alternative approach might take a track updated with a report at time t_i and construct a search volume sufficiently large to guarantee that the track gates with any report of the target arriving during the subsequent s seconds, where s is a parameter similar to the threshold used for triggering track file projections. This is accomplished by determining the region of space the target could conceivably traverse based on its kinematic state and error covariance. The box circumscribing this search volume can then be maintained in the search structure until time $t_i + s$, at which point it becomes stale and must be replaced with a search volume that is valid from time $t_i + s$ to time $t_i + 2s$. However, if before becoming stale it is updated with a report at time t_j, $t_i < t_j < t_i + s$, then it must be replaced with a search volume that is valid from time t_j to time $t_j + s$.

The benefit of the enhanced approach is that each track is projected only at the times when it is updated or when all extended period has passed without an update (which could possibly signal the need to delete the track). In order to apply the approach, however, two conditions must be satisfied. First, there must be a mechanism for identifying when a track volume has become stale and needs to be recomputed. It is, of course, not possible to examine every track upon the receipt of each report because the scaling of the algorithm would be undermined. The solution is to maintain a priority queue of the times at which the different track volumes will become invalid. A priority queue is a data structure that can be updated efficiently and supports the retrieval of the minimum of n values in $O(\log n)$ time. At the time a report is received, the priority queue is queried to determine which, if any, of the track volumes have become stale. New search volumes are constructed for the identified tracks, and the times at which they will become invalid are updated in the priority queue.

The second condition that must be satisfied for the enhanced approach is a capability to incrementally update the search structure as tracks are added, updated, recomputed, or deleted. The need for such a capability was hinted at in the discussion of dynamic search structures. Because the layered box tree supports insertions and deletions in $O(\log n)$ time, the update of a track's search volume can be efficiently accommodated. The track's associated box is deleted from the tree, an updated box is computed, and then the result is inserted back into the tree. In summary, the cost for processing each report involves updates of the search structure and the priority queue, at $O(\log n)$ cost, plus the cost of determining the set of tracks with which the report could be feasibly associated.

3.4   Conclusion

The correlation of reports with tracks numbering in the thousands can now be performed in real time on a personal computer. More research on large-scale correlation is needed, but work has already begun on implementing efficient correlation modules that can be incorporated into existing tracking systems. Ironically, by hiding the intricate details and complexities of the correlation process, these modules give the appearance that multiple-target tracking involves little more than the concurrent processing of several single-target problems. Thus, a paradigm with deep historical roots in the field of target tracking is at least partially preserved.

Note that the techniques described in this chapter are applicable only to a very restricted class of tracking problems. Other problems, such as the tracking of military forces, demand more sophisticated approaches. Not only does the mean position of a military force change, its shape also changes. Moreover, reports of its position are really only reports of the positions of its parts, and various parts may be moving in different directions at any given instant. Filtering out the local deviations in motion to determine the net motion of the whole is beyond the capabilities of a simple Kalman filter. Other difficult tracking problems include the tracking of weather phenomena and soil erosion. The history of multiple-target tracking suggests that, in addition to new mathematical techniques, new algorithmic techniques will certainly be required for any practical solution to these problems.

Acknowledgments

The author gratefully acknowledges support from the Naval Research Laboratory, Washington, DC.

References

1. Uhlmann, J.K., Algorithms for multiple-target tracking, *American Scientist*, 80(2), 1992.
2. Kalman, R.E., A new approach to linear filtering and prediction problems, *ASME, Basic Eng.*, 82:34–45, 1960.
3. Blackman, S., *Multiple-Target Tracking with Radar Applications*, Artech House, Inc., Norwood, MA, 1986.
4. Bar-Shalom, Y. and Fortmann, T.E., *Tracking and Data Association*, Academic Press, 1988.
5. Bar-Shalom, Y. and Li, X.R., *Multitarget-Multisensor Tracking: Principles and Techniques*, YBS Press, 1995.
6. Uhlmann J.K., Zuniga M.R., and Picone, J.M., Efficient approaches for report/cluster correlation in multitarget tracking systems, *NRL Report 9281*, 1990.
7. Bentley, J., Multidimensional binary search trees for associative searching, *Communications of the ACM*, 18, 1975.

8. Yianilos, P.N., Data structures and algorithms for nearest neighbor search in general metric spaces, in *SODA,* 1993.

9. Ramasubramanian, V. and Paliwal, K., An efficient approximation-elimination algorithm for fast nearest-neighbour search on a spherical distance coordinate formulation, *Pattern Recogntion Letters,* 13, 1992.

10. Vidal, E., An algorithm for finding nearest neighbours in (approximately) constant average time complexity, *Pattern Recognition Letters,* 4, 1986.

11. Vidal, E., Rulot, H., Casacuberta, F., and Benedi, J., On the use of a metric-space search algorithm (aesa) for fast dtw-based recognition of isolated words, *Trans. Acoust. Speech Signal Process.,* 36, 1988.

12. Uhlmann, J.K., Metric trees. *Applied Math. Letters,* 4, 1991.

13. Uhlmann, J.K., Satisfying general proximity/similarity queries with metric trees, *Info. Proc. Letters,* 2, 1991.

14. Uhlmann, J.K., Implementing metric trees to satisfy general proximity/similarity queries, *NRL Code 5570 Technical Report,* 9192, 1992.

15. Lee, D.T. and Wong, C.K., Worst-case analysis for region and partial region searches in multidimensional binary search trees and quad trees, *Acta Informatica,* 9(1), 1997.

16. Preparata, F. and Shamos, M., *Computational Geometry,* Springer-Verlag, 1985.

17. Mehlhorn, Kurt, *Multi-dimensional Searching and Computational Geometry,* Vol. 3, Springer-Verlag, Berlin, 1984.

18. Uhlmann, J.K. and Zuniga, M.R., Results of an efficient gating algorithm for large-scale tracking scenarios, *Naval Research Reviews,* 1:24–29, 1991.

19. Zuniga, M.R., Picone, J.M., and Uhlmann, J.K., Efficient algorithm for unproved gating combinatorics in multiple-target tracking, Submitted to *IEEE Transactions on Acrospace and Electronic Systems,* 1990.

20. Uhlmann, J.K., Adaptive partitioning strategies for ternary tree structures, *Pattern Recognition Letters,* 12:537–541, 1991.

21. Collins, J.B. and Uhlmann, J.K., Efficient gating in data association for multivariate Gaussian distributions, *IEEE Trans. Aerospace and Electronic Systems,* 28, 1990.

22. McCreight, E.M., Priority search trees, *SIAM J. Comput.,* 14(2):257–276, May 1985.

23. Wood, D., *Data, Structures, Algorithms, and Performance,* Addison-Wesley Publishing Company, 1993.

24. Uhlmann, J.K., Dynamic map building and localization for autonomous vehicles, *Engineering Sciences Report,* Oxford University, 1994.

25. van Kreveld, M. and Mvermars, M., Divided *kd*-trees, *Algorithmica,* 6:840–858, 1991.

26. Boroujerdi, A. and Uhlmann, J.K., Large-scale intersection detection using layered box trees, AIT-DSS Report, 1998.

27. Uhlmann, J.K. and Kuo, E., Achieving optimal query time in layered trees, 2001 (in preparation).

4

The Principles and Practice of Image and Spatial Data Fusion*

4.1 Introduction 4-1
4.2 Motivations for Combining Image and Spatial Data 4-2
4.3 Defining Image and Spatial Data Fusion 4-3
4.4 Three Classic Levels of Combination for Multisensor
     Automatic Target Recognition Data Fusion 4-5
     Pixel-Level Fusion • Feature-Level Fusion • Decision-Level
     Fusion • Multiple-Level Fusion
4.5 Image Data Fusion for Enhancement of Imagery
     Data 4-10
     Multiresolution Imagery • Dynamic Imagery • Three-
     Dimensional Imagery
4.6 Spatial Data Fusion Applications 4-11
     Spatial Data Fusion: Combining Image and Non-Image Data
     to Create Spatial Information Systems • Mapping, Charting
     and Geodesy (MC&G) Applications
4.7 Summary 4-15
References 4-15

Ed Waltz
Veridian Systems

4.1 Introduction

The joint use of imagery and spatial data from different imaging, mapping, or other spatial sensors has the potential to provide significant performance improvements over single sensor detection, classification, and situation assessment functions. The terms *imagery fusion* and *spatial data fusion* have been applied to describe a variety of combining operations for a wide range of image enhancement and understanding applications. Surveillance, robotic machine vision, and automatic target cueing are among the application areas that have explored the potential benefits of multiple sensor imagery. This chapter provides a framework for defining and describing the functions of image data fusion in the context of the Joint Directors of Laboratories (JDL) data fusion model. The chapter also describes representative methods and applications.

Sensor fusion and data fusion have become the de facto terms to describe the general abductive or deductive combination processes by which diverse sets of related data are joined or merged to produce

---

*Adapted from the principles and practice of image and spatial data fusion, in *Proceedings of the 8th National Data Fusion Conference,* Dallas, Texas, March 15–17, 1995, pp. 257–278.

a product that is greater than the individual parts. A range of mathematical operators has been applied to perform this process for a wide range of applications. Two areas that have received increasing research attention over the past decade are the processing of imagery (two-dimensional information) and spatial data (three-dimensional representations of real-world surfaces and objects that are imaged). These processes combine multiple data views into a composite set that incorporates the best attributes of all contributors. The most common product is a spatial (three-dimensional) model, or virtual world, which represents the best estimate of the real world as derived from all sensors.

4.2   Motivations for Combining Image and Spatial Data

A diverse range of applications has employed image data fusion to improve imaging and automatic detection/classification performance over that of single imaging sensors. Table 4.1 summarizes representative and recent research and development in six key application areas.

Satellite and airborne imagery used for military intelligence, photogrammetric, earth resources, and environmental assessments can be enhanced by combining registered data from different sensors to refine the spatial or spectral resolution of a composite image product. Registered imagery from different passes (multitemporal) and different sensors (multispectral and multiresolution) can be combined to produce composite imagery with spectral and spatial characteristics equal to or better than that of the individual contributors.

Composite SPOT™ and LANDSAT satellite imagery and 3-D terrain relief composites of military regions demonstrate current military applications of such data for mission planning purposes.[1-3] The Joint National Intelligence Development Staff (JNIDS) pioneered the development of workstation-based systems to combine a variety of image and nonimage sources for intelligence analysts[4] who perform

TABLE 4.1   Representative Range of Activities Applying Spatial and Imagery Fusion

	Activities	Sponsors
	Satellite/Airborne Imaging	
Multiresolution image sharpening	Multiple algorithms, tools in commercial packages	U.S., commercial vendors
Terrain visualization	Battlefield visualization, mission planning	Army, Air Force
Planetary visualization-exploration	Planetary mapping missions	NASA
	Mapping, Charting and Geodesy	
Geographic information system (GIS) generation from multiple sources	Terrain feature extraction, rapid map generation	DARPA, Army, Air Force
Earth environment information system	Earth observing system, data integration system	NASA
	Military Automatic Target Recognition ATR	
Battlefield surveillance	Various MMW/LADAR/FLIR	Army
Battlefield seekers	Millimeter wave (MMW)/forward looking IR (FLIR)	Army, Air Force
IMINT correlation	Single Intel IMINT correlation	DARPA
IMINT-SIGINT/MTI correlation	Dynamic database	DARPA
	Industrial Robotics	
3-D multisensor inspection	Product line inspection	Commercial
Non-destructive inspection	Image fusion analysis	Air Force, commercial
	Medical Imaging	
Human body visualization, diagnosis	Tomography, magnetic resonance imaging, 3-D fusion	Various R&D hospitals

- registration — spatial alignment of overlapping images and maps to a common coordinate system;
- mosaicking — registration of nonoverlapping, adjacent image sections to create a composite of a larger area;
- 3-D mensuration-estimation — calibrated measurement of the spatial dimensions of objects within in-image data.

Similar image functions have been incorporated into a variety of image processing systems, from tactical image systems such as the premier Joint Service Image Processing System (JSIPS) to Unix- and PC-based commercial image processing systems. Military services and the National Imagery and Mapping Agency (NIMA) are performing cross intelligence (i.e., IMINT and other intelligence source) data fusion research to link signals and human reports to spatial data.[5]

When the fusion process extends beyond imagery to include other spatial data sets, such as digital terrain data, demographic data, and complete geographic information system (GIS) data layers, numerous mapping applications may benefit. Military intelligence preparation of the battlefield (IPB) functions (e.g., area delimitation and transportation network identification), as well as wide area terrain database generation (e.g., precision GIS mapping), are complex mapping problems that require fusion to automate processes that are largely manual. One area of ambitious research in this area of spatial data fusion is the U.S. Army Topographic Engineering Center's (TEC) efforts to develop automatic terrain feature generation techniques based on a wide range of source data, including imagery, map data, and remotely sensed terrain data.[6] On the broadest scale, NIMA's Global Geospatial Information and Services (GGIS) vision includes spatial data fusion as a core functional element.[7] NIMA's Mapping, Charting and Geodesy Utility Software package (MUSE), for example, combines vector and raster data to display base maps with overlays of a variety of data to support geographic analysis and mission planning.

Real-time automatic target cueing/recognition (ATC/ATR) for military applications has turned to multiple sensor solutions to expand spectral diversity and target feature dimensionality, seeking to achieve high probabilities of correct detection/identification at acceptable false alarm rates. Forward-looking infrared (FLIR), imaging millimeter wave (MMW), and light amplification for detection and ranging (LADAR) sensors are the most promising suite capable of providing the diversity needed for reliable discrimination in battlefield applications. In addition, some applications seek to combine the real-time imagery to present an enhanced image to the human operator for driving, control, and warning, as well as manual target recognition.

Industrial robotic applications for fusion include the use of 3-D imaging and tactile sensors to provide sufficient image understanding to permit robotic manipulation of objects. These applications emphasize automatic object position understanding rather than recognition (e.g., the target recognition) that is, by nature, noncooperative).[8]

Transportation applications combine millimeter wave and electro-optical imaging sensors to provide collision avoidance warning by sensing vehicles whose relative rates and locations pose a collision threat.

Medical applications fuse information from a variety of imaging sensors to provide a complete 3-D model or enhanced 2-D image of the human body for diagnostic purposes. The United Medical and Dental Schools of Guy's and St. Thomas' Hospital (London, U.K.) have demonstrated methods for registering and combining magnetic resonance (MR), positron emission tomography (PET), and computer tomography (CT) into composites to aid surgery.[9]

4.3 Defining Image and Spatial Data Fusion

In this chapter, image and spatial data fusion are distinguished as subsets of the more general data fusion problem that is typically aimed at associating and combining 3-D data about *sparse point-objects located in space*. Targets on a battlefield, aircraft in airspace, ships on the ocean surface, or submarines in the 3-D ocean volume are common examples of targets represented as point objects in a three-dimensional space model.

Image data fusion, on the other hand, is involved with associating and combining complete, spatially filled sets of data in 2-D (images) or 3-D (terrain or high resolution spatial representations of real objects).

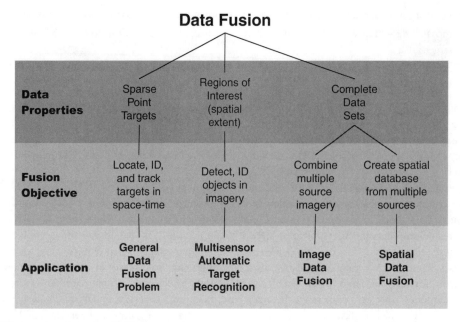

FIGURE 4.1   Data fusion application taxonomy.

Herein lies the distinction: image and spatial data fusion requires data representing every point on a surface or in space to be fused, rather than selected points of interest.

The more general problem is described in detail in introductory texts by Waltz and Llinas[10] and Hall,[11] while the progress in image and spatial data fusion is reported over a wide range of the technical literature, as cited in this chapter.

The taxonomy in Figure 4.1 distinguishes the data properties and objectives that distinguish four categories of fusion applications.

In all of the image and spatial applications cited above, the common thread of the fusion function is its emphasis on the following distinguishing functions:

- **Registration** involves spatial and temporal alignment of physical items within imagery or spatial data sets and is a prerequisite for further operations. It can occur at the raw image level (i.e., any pixel in one image may be referenced with known accuracy to a pixel or pixels in another image, or to a coordinate in a map) or at higher levels, relating objects rather than individual pixels. Of importance to every approach to combining spatial data is the accuracy with which the data layers have been spatially aligned relative to each other or to a common coordinate system (e.g., geo-location or geo-coding of earth imagery to an earth projection). Registration can be performed by traditional *internal* image-to-image correlation techniques (when the images are from sensors with similar phenomena and are highly correlated)[12] or by *external* techniques.[13] External methods apply in-image control knowledge or as-sensed information that permits accurate modeling and estimation of the true location of each pixel in two- or three-dimensional space.

- The **combination** function operates on multiple, registered "layers" of data to derive composite products using mathematical operators to perform integration; mosaicking; spatial or spectral refinement; spatial, spectral or temporal (change) detection; or classification.

- **Reasoning** is the process by which intelligent, often iterative search operations are performed *between* the layers of data to assess the meaning of the entire scene at the highest level of abstraction and of individual items, events, and data contained in the layers.

The image and spatial data fusion functions can be placed in the JDL data fusion model context to describe the architecture of a system that employs imagery data from multiple sensors and spatial data

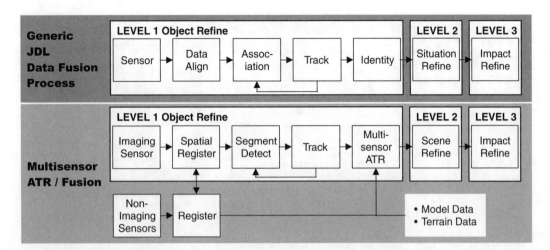

FIGURE 4.2 Image of a data fusion functional flow can be directly compared to the joint directors of labs (JDL) data fusion subpanel model of data fusion.

(e.g., maps and solid models) to perform detection, classification, and assessment of the meaning of information contained in the scenery of interest.

Figure 4.2 compares the JDL general model[14] with a specific multisensor ATR image data fusion functional flow to show how the more abstract model can be related to a specific imagery fusion application. The Level 1 processing steps can be directly related to image counterparts:

- *Alignment* — The alignment of data into a common time, space, and spectral reference frame involves spatial transformations to warp image data to a common coordinate system (e.g., projection to an earth reference model or three-dimensional space). At this point, nonimaging data that can be spatially referenced (perhaps not to a point, but often to a region with a specified uncertainty) can then be associated with the image data.

- *Association* — New data can be correlated with previous data to detect and segment (select) targets on the basis of motion (temporal change) or behavior (spatial change). In time-sequenced data sets, target objects at time t are associated with target objects at time $t-1$ to discriminate newly appearing targets, moved targets, and disappearing targets.

- *Tracking* — When objects are tracked in dynamic imagery, the dynamics of target motion are modeled and used to predict the future location of targets (at time $t+1$) for comparison with new sensor observations.

- *Identification* — The data for segmented targets are combined from multiple sensors (at any one of several levels) to provide an assignment of the target to one or more of several target classes.

Level 2 and 3 processing deals with the aggregate of targets in the scene and other characteristics of the scene to derive an assessment of the "meaning" of data in the scene or spatial data set.

In the following sections, the primary image and spatial data fusion application areas are described to demonstrate the basic principles of fusion and the state of the practice in each area.

4.4 Three Classic Levels of Combination for Multisensor Automatic Target Recognition Data Fusion

Since the late 1970s, the ATR literature has adopted three levels of image data fusion as the basic design alternatives offered to the system designer. The terminology was adopted to describe the point in the traditional ATR processing chain at which registration and combination of different sensor data occurred. These functions can occur at multiple levels, as described later in this chapter. First, a brief overview of

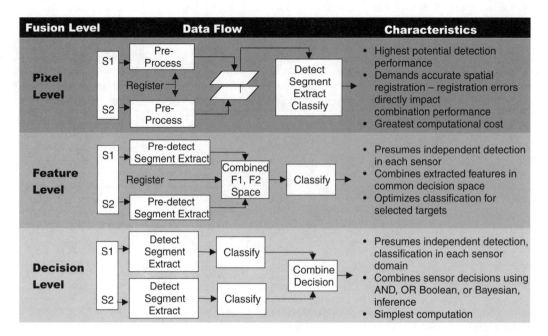

FIGURE 4.3   Three basic levels of fusion are provided to the multisensor ATR designer as the most logical alternative points in the data chain for combining data.

TABLE 4.2   Most Common Decision-Level Combination Alternatives

Decision Type	Method	Description
Hard Decision	Boolean	Apply logical AND, OR to combine independent decisions.
	Weighted Sum Score	Weight sensors by inverse of covariance and sum to derive score function.
	M-of-N	Confirm decision based on m-out-of-n sensors that agree.
Soft Decision	Bayesian	Apply Bayes rule to combine sensor independent conditional probabilities.
	Dempster-Shafer	Apply Dempster's rule of combination to combine sensor belief functions.
	Fuzzy Variable	Combine fuzzy variables using fuzzy logic (AND, OR) to derive combined membership function.

the basic alternatives and representative research and development results is presented. (Broad overviews of the developments in ATR in general, with specific comments on data fusion, are available in other literature.[15-17])

4.4.1   Pixel-Level Fusion

At the lowest level, *pixel-level fusion* uses the registered pixel data from all image sets to perform detection and discrimination functions. This level has the potential to achieve the greatest signal detection performance (if registration errors can be contained) at the highest computational expense. At this level, detection decisions (pertaining to the presence or absence of a target object) are based on the information from all sensors by evaluating the spatial and spectral data from all layers of the registered image data. A subset of this level of fusion is *segment-level fusion*, in which basic detection decisions are made independently in each sensor domain, but the segmentation of image regions is performed by evaluation of the registered data layers.

   Fusion at the pixel level involves accurate registration of the different sensor images before applying a combination operator to each set of registered pixels (which correspond to associated measurements in each sensor domain at the highest spatial resolution of the sensors.) Spatial registration accuracies should be subpixel to avoid combination of unrelated data, making this approach the most sensitive to registration errors. Because image data may not be sampled at the same spacing, resampling and warping of images is generally required to achieve the necessary level of registration prior to combining pixel data.

In the most direct 2-D image applications of this approach, coregistered pixel data may be classified on a pixel-by-pixel basis using approaches that have long been applied to multispectral data classification.[18] Typical ATR applications, however, pose a more complex problem when dissimilar sensors, such as FLIR and LADAR, image in different planes. In such cases, the sensor data must be projected into a common 2-D or 3-D space for combination. Gonzalez and Williams, for example, have described a process for using 3-D LADAR data to infer FLIR pixel locations in 3-D to estimate target pose prior to feature extraction.[19] Schwickerath and Beveridge present a thorough analysis of this problem, developing an eight-degree of freedom model to estimate both the target pose and relative sensor registration (*coregistration*) based on a 2-D and 3-D sensor.[20]

Delanoy et al. demonstrated pixel-level combination of *spatial interest images* using Boolean and fuzzy logic operators.[21] This process applies a spatial feature extractor to develop multiple interest images (representing the relative presence of spatial features in each pixel), before combining the interest images into a single detection image. Similarly, Hamilton and Kipp describe a *probe-based technique* that uses spatial templates to transform the direct image into probed images that enhance target features for comparison with reference templates.[22,23] Using a limited set of television and FLIR imagery, Duane compared pixel-level and feature-level fusion to quantify the relative improvement attributable to the pixel-level approach with well-registered imagery sets.[24]

4.4.2 Feature-Level Fusion

At the intermediate level, *feature-level fusion* combines the features of objects that are detected and segmented in the individual sensor domains. This level presumes independent detectability of objects in all of the sensor domains. The features for each object are independently extracted in each domain; these features crate a common feature space for object classification.

Such feature-level fusion reduces the demand on registration, allowing each sensor channel to segment the target region and extract features without regard to the other sensor's choice of target boundary. The features are merged into a common decision space only after a spatial association is made to determine that the features were extracted from objects whose centroids were spatially associated.

During the early 1990s, the Army evaluated a wide range of feature-level fusion algorithms for combining FLIR, MMW, and LADAR data for detecting battlefield targets under the Multi-Sensor Feature Level Fusion (MSFLF) Program of the OSD Multi-Sensor Aided Targeting Initiative. Early results demonstrated marginal gains over single sensor performance and reinforced the importance of careful selection of complementary features to specifically reduce single sensor ambiguities.[25]

At the feature level of fusion, researchers have developed model-based (or model-driven) alternatives to the traditional statistical methods, which are inherently data driven. Model-based approaches maintain target and sensing models that predict all possible views (and target configurations) for comparison with extracted features rather than using a more limited set of real signature data for comparison.[26] The application of model-based approaches to multiple-sensor ATR offers several alternative implementations, two of which are described in Figure 4.4. The Adaptive Model Matching approach performs feature extraction (FE) and comparison (match) with predicted features for the estimated target pose. The process iteratively searches to find the best model match for the extracted features.

4.4.2.1 Discrete Model Matching Approach

A multisensor model-based matching approach described by Hamilton and Kipp[27] develops a relational tree structure (hierarchy) of 2-D silhouette templates. These templates capture the spatial structure of the most basic all-aspect target "blob" (at the top or *root* node), down to individual target hypotheses at specific poses and configurations. This predefined search tree is developed on the basis of model data

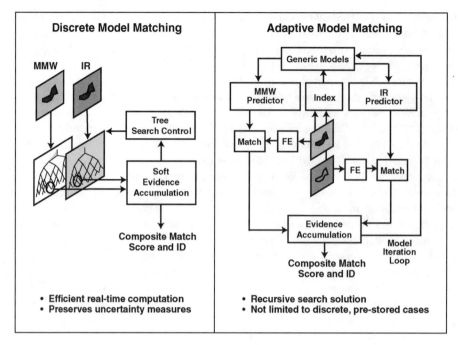

FIGURE 4.4   Two model-based sensor alternatives demonstrate the use of a prestored hierarchy of model-based templates or an online, iterative model that predicts features based upon estimated target pose.

for each sensor, and the ATR process compares segmented data to the tree, computing a composite score at each node to determine the path to the most likely hypotheses. At each node, the evidence is accumulated by applying an operator (e.g., weighted sum, Bayesian combination, etc.) to combine the score for each sensor domain.

4.4.2.2   Adaptive Model Matching Approach

Rather than using prestored templates, this approach implements the sensor/target modeling capability within the ATR algorithm to dynamically predict features for direct comparison. Figure 4.4 illustrates a two-sensor extension of the one-sensor, model-based ATR paradigm (e.g., ARAGTAP[28] or MSTAR[29] approaches) in which independent sensor features are predicted and compared *iteratively*, and evidence from the sensors is accumulated to derive a composite score for each target hypothesis.

Larson et al. describe a model-based IR/LADAR fusion algorithm that performs extensive pixel-level registration and feature extraction before performing the model-based classification at the extracted feature level.[30] Similarly, Corbett et al. describe a model-based feature-level classifier that uses IR and MMW models to predict features for military vehicles.[31] Both of these follow the adaptive generation approach.

4.4.3   Decision-Level Fusion

Fusion at the *decision level* (also called *post-decision* or *post-detection* fusion) combines the decisions of independent sensor detection/classification paths by Boolean (AND, OR) operators or by a heuristic score (e.g., M-of-N, maximum vote, or weighted sum). Two methods of making classification decisions exist: hard decisions (single, optimum choice) and soft decisions, in which decision uncertainty in each sensor chain is maintained and combined with a composite measure of uncertainty.

The relative performance of alternative combination rules and independent sensor thresholds can be optimally selected using distribution data for the features used by each sensor.[32] In decision-level fusion, each path must independently detect the presence of a candidate target and perform a classification on the candidate. These detections and/or classifications (the sensor *decisions*) are combined into a fused decision. This approach inherently assumes that the signals and signatures in each independent sensor chain are sufficient to perform independent detection before the sensor decisions are combined. This approach is much less sensitive to spatial misregistration than all others and permits accurate association of detected targets to occur with registration errors over an order of magnitude larger than for pixel-level fusion. Lee and Vleet have shown procedures for estimating the registration error between sensors to minimize the mean square registration error and optimize the association of objects in dissimilar images for decision-level fusion.[33]

Decision-level fusion of MMW and IR sensors has long been considered a prime candidate for achieving the level of detection performance required for autonomous precision-guided munitions.[34] Results of an independent two-sensor (MMW and IR) analysis on military targets demonstrated the relative improvement of two-sensor decision-level fusion over either independent sensor.[35-37] A summary of ATR comparison methods was compiled by Diehl, Shields, and Hauter.[38] These studies demonstrated the critical sensitivity of performance gains to the relative performance of each contributing sensor and the independence of the sensed phenomena.

4.4.4 Multiple-Level Fusion

In addition to the three classic levels of fusion, other alternatives or combinations have been advanced. At a level even higher than the decision level, some researchers have defined *scene-level* methods in which target detections from a low-resolution sensor are used to cue a search-and-confirm action by a higher resolution sensor. Menon and Kolodzy described such a system, which uses FLIR detections to cue the analysis of high spatial resolution laser radar data using a nearest neighbor neural network classifier.[39] Maren describes a *scene structure* method that combines information from hierarchical structures developed independently by each sensor by decomposing the scene into element representations.[40] Others have developed hybrid, multilevel techniques that partition the detection problem to a high level (e.g., decision level) and the classification to a lower level. Aboutalib et al. described a hybrid algorithm that performs decision-level combination for detection (with detection threshold feedback) and feature-level classification for air target identification in IR and TV imagery.[41]

Other researchers have proposed multi-level ATR architectures, which perform fusion at all levels, carrying out an appropriate degree of combination at each level based on the ability of the combined information to contribute to an overall fusion objective. Chu and Aggarwal describe such a system that integrates pixel-level to scene-level algorithms.[42] Eggleston has long promoted such a knowledge-based ATR approach that combines data at three levels, using many partially redundant combination stages to reduce the errors of any single unreliable rule.[43,44] The three levels in this approach are

- Low level — Pixel-level combinations are performed when image enhancement can aid higher-level combinations. The higher levels adaptively control this fine grain combination.
- Intermediate symbolic level — Symbolic representations (*tokens*) of attributes or features for segmented regions (*image events*) are combined using a symbolic level of description.
- High level — The scene or context level of information is evaluated to determine the *meaning* of the overall scene, by considering all intermediate-level representations to derive a situation assessment. For example, this level may determine that a scene contains a brigade-sized military unit forming for attack. The derived situation can be used to adapt lower levels of processing to refine the high-level hypotheses.

Bowman and DeYoung described an architecture that uses neural networks at all levels of the conventional ATR processing chain to achieve pixel-level performances of up to 0.99 probability of correct identification for battlefield targets using pixel-level neural network fusion of UV, visible, and MMW imagery.[45]

Pixel, feature, and decision-level fusion designs have focused on combining imagery for the purposes of *detecting and classifying* specific targets. The emphasis is on limiting processing by combining only the most likely regions of target data content and combining at the minimum necessary level to achieve the desired detection/classification performance. This differs significantly from the next category of image fusion designs, in which all data must be combined to form a new spatial data product that contains the best composite properties of all contributing sources of information.

4.5 Image Data Fusion for Enhancement of Imagery Data

Both still and moving image data can be combined from multiple sources to enhance desired features, combine multiresolution or differing sensor look geometries, mosaic multiple views, and reduce uncorrelated noise.

4.5.1 Multiresolution Imagery

One area of enhancement has been in the application of *band sharpening* or *multiresolution image fusion* algorithms to combine differing resolution satellite imagery. The result is a composite product that enhances the spatial boundaries in lower resolution multispectral data using higher resolution panchromatic or Synthetic Aperture Radar (SAR) data.

Veridian-ERIM International has applied its Sparkle algorithm to the band sharpening problem, demonstrating the enhancement of lower-resolution SPOT™ multispectral imagery (20-meter ground sample distance or GSD) with higher resolution airborne SAR (3-meter GSD) and panchromatic photography (1-meter) to sharpen the multispectral data. Radar backscatter features are overlayed on the composite to reveal important characteristics of the ground features and materials. The composite image preserves the spatial resolution of the pancromatic data, the spectral content of the multispectral layers, and the radar reflectivity of the SAR.

Vrabel has reported the relative performance of a variety of band sharpening algorithms, concluding that Veridian ERIM International's Sparkle algorithm and a color normalization (CN) technique provided the greatest GSD enhancement and overall utility.[46] Additional comparisons and applications of band sharpening techniques have been published in the literature.[47-50]

Imagery can also be mosaicked by combining overlapping images into a common block, using classical photogrammetric techniques (bundle adjustment) that use absolute ground control points and tie points (common points in overlapped regions) to derive mapping polynomials. The data may then be *forward resampled* from the input images to the output projection or *backward resampled* by projecting the location of each output pixel onto each source image to extract pixels for resampling.[51] The latter approach permits spatial deconvolution functions to be applied in the resampling process. Radiometric *feathering* of the data in transition regions may also be necessary to provide a gradual transition after overall balancing of the radiometric dynamic range of the mosaicked image is performed.[52] Such mosaicking fusion processes have also been applied to three-dimensional data to create composite digital elevation models (DEMs) of terrain.[53]

4.5.2 Dynamic Imagery

In some applications, the goal is to combine different types of real-time video imagery to provide the clearest possible composite video image for a human operator. The David Sarnoff Research Center has applied wavelet encoding methods to selectively combine IR and visible video data into a composite video image that preserves the most desired characteristics (e.g., edges, lines, and boundaries) from each data set.[54] The Center later extended the technique to combine multitemporal and moving images into composite mosaic scenes that preserve the "best" data to create a current scene at the best possible resolution at any point in the scene.[55,56]

4.5.3 Three-Dimensional Imagery

Three-dimensional perspectives of the earth's surface are a special class of image data fusion products that have been developed by *draping* orthorectified images of the earth's surface over digital terrain models. The 3-D model can be viewed from arbitrary static perspectives, or a dynamic *fly-through*, which provides a visualization of the area for mission planners, pilots, or land planners.

TABLE 4.3   Basic Image Data Fusion Functions Provided in Several Commercial Image Processing Software Packages

	Function	Description
Registration	Sensor-platform modeling	Model sensor-imaging geometry; derive correction transforms (e.g., polynomials) from collection parameters (e.g., ephemeris, pointing, and earth model)
	Ground Control Point (GCP) calibration	Locate known GCPs and derive correction transforms
	Warp to polynomial	Spatially transform (warp) imagery to register pixels to regular grid or to a digital terrain model
	Orthorectify to digital terrain model	
	Resample imagery	Resample warped imagery to create fixed pixel-sized image
Combination	Mosaic imagery	Register adjacent and overlapped imagery; resample to common pixel grid
	Edge feathering	Combine overlapping imagery data to create smooth (feathered) magnitude transitions between two image components
	Band sharpening	Enhance spatial boundaries (high-frequency content) in lower resolution band data using higher resolution registered imagery data in a different band

Off-nadir regions of aerial or spaceborne imagery include a horizontal displacement error that is a function of the elevation of the terrain. A digital elevation model (DEM) is used to correct for these displacements in order to accurately overlay each image pixel on the corresponding post (i.e., terrain grid coordinate). Photogrammetric orthorectification functions[57] include the following steps to combine the data:

- DEM preparation — the digital elevation model is transformed to the desired map projection for the final composite product.
- Transform derivation — platform, sensor, and the DEM are used to derive mapping polynomials that will remove the horizontal displacements caused by to terrain relief, placing each input image pixel at the proper location on the DEM grid.
- Resampling — The input imagery is resampled into the desired output map grid.
- Output file creation — The resampled image data (x, y, and pixel values) and DEM (x, y, and z) are merged into a file with other geo-referenced data, if available.
- Output product creation — Two-dimensional image maps may be created with map grid lines, or three-dimensional visualization perspectives can be created for viewing the terrain data from arbitrary viewing angles.

The basic functions necessary to perform registration and combination are provided in an increasing number of commercial image processing software packages (see Table 4.3), permitting users to fuse static image data for a variety of applications.

4.6   Spatial Data Fusion Applications

Robotic and transportation applications include a wide range of applications similar to military applications. Robotics applications include relatively short-range, high-resolution imaging of cooperative target objects (e.g., an assembly component to be picked up and accurately placed) with the primary objectives of position determination and inspection. Transportation applications include longer-range sensing of vehicles for highway control and multiple sensor situation awareness within a vehicle to provide semi-autonomous navigation, collision avoidance, and control.

The results of research in these areas are chronicled in a variety sources, beginning with the 1987 Workshop on Spatial Reasoning and MultiSensor Fusion,[58] and many subsequent SPIE conferences.[59-63]

4.6.1   Spatial Data Fusion: Combining Image and Non-Image Data to Create Spatial Information Systems

One of the most sophisticated image fusion applications combines diverse sets of imagery (2-D), spatially referenced nonimage data sets, and 3-D spatial data sets into a composite spatial data information system. The most active area of research and development in this category of fusion problems is the development of geographic information systems (GIS) by combining earth imagery, maps, demographic and infrastructure or facilities mapping (geospatial) data into a common spatially referenced database.

Applications for such capabilities exist in three areas. In civil government, the need for land and resource management has prompted intense interest in establishing GISs at all levels of government. The U.S. Federal Geographic Data Committee is tasked with the development of a National Spatial Data Infrastructure (NSDI), which establishes standards for organizing the vast amount of geospatial data currently available at the national level and coordinating the integration of future data.[64]

Commercial applications for geospatial data include land management, resources exploration, civil engineering, transportation network management, and automated mapping/facilities management for utilities.

The military application of such spatial databases is the intelligence preparation of the battlefield (IPB),[65] which consists of developing a spatial database containing all terrain, transportation, ground-cover, manmade structures, and other features available for use in real-time situation assessment for command and control. The Defense Advanced Research Projects Agency (DARPA) Terrain Feature Generator is one example of a major spatial database and fusion function defined to automate the functions of IPB and geospatial database creation from diverse sensor sources and maps.[66]

To realize efficient, affordable systems capable of accommodating the volume of spatial data required for large regions and performing reasoning that produces accurate and insightful information depends on two critical technology areas:

- *Spatial Data Structure* — Efficient, linked data structures are required to handle the wide variety of vector, raster, and nonspatial data sources. Hundreds of point, lineal, and areal features must be accommodated. Data volumes are measured in terabytes and short access times are demanded for even broad searches.
- *Spatial Reasoning* — The ability to reason in the context of dynamically changing spatial data is required to assess the "meaning" of the data. The reasoning process must perform the following kinds of operations to make assessments about the data:
  - Spatial measurements (e.g., geometric, topological, proximity, and statistics)
  - Spatial modeling
  - Spatial combination and inference operations, in uncertainty
  - Spatial aggregation of related entities
  - Multivariate spatial queries

Antony surveyed the alternatives for representing spatial and spatially referenced semantic knowledge[67] and published the first comprehensive data fusion text[68] that specifically focused on spatial reasoning for combining spatial data.

4.6.2   Mapping, Charting and Geodesy (MC&G) Applications

The use of remotely sensed image data to create image maps and generate GIS base maps has long been recognized as a means of automating map generation and updating to achieve currency as well as accuracy.[69-71] The following features characterize integrated geospatial systems:

- *Currency* — Remote sensing inputs enable continuous update with change detection and monitoring of the information in the database.
- *Integration* — Spatial data in a variety of formats (e.g., raster and vector data) is integrated with meta data and other spatially referenced data, such as text, numerical, tabular, and hypertext

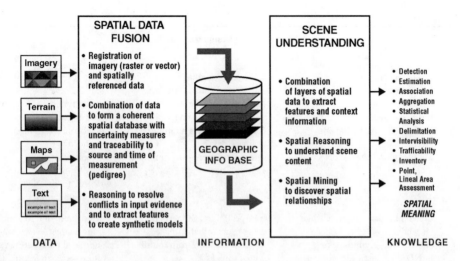

FIGURE 4.5 The spatial data fusion process flow includes the generation of a spatial database and the assessment of spatial information in the database by multiple users.

formats. Multiresolution and multiscale spatial data coexist, are linked, and share a common reference (i.e., map projection).

- *Access* — The database permits spatial query access for multiple user disciplines. All data is traceable and the data accuracy, uncertainty, and entry time are annotated.
- *Display* — Spatial visualization and query tools provide maximum human insight into the data content using display overlays and 3-D capability.

Ambitious examples of such geospatial systems include the DARPA Terrain Feature Generator, the European ESPRIT II MultiSource Image Processing System (MuSIP),[72,73] and NASA's Earth Observing Systems Data and Information System (EOSDIS).[74]

Figure 4.5 illustrates the most basic functional flow of such a system, partitioning the data integration (i.e., database generation) function from the scene assessment function. The integration functions spatially registers and links all data to a common spatial reference and also combines some data sets by mosaicking, creating composite layers, and extracting features to create feature layers. During the integration step, higher-level spatial reasoning is required to resolve conflicting data and to create derivative layers from extracted features. The output of this step is a registered, refined, and traceable spatial database.

The next step is scene assessment, which can be performed for a variety of application functions (e.g., further feature extraction, target detection, quantitative assessment, or creation of vector layers) by a variety of user disciplines. This stage extracts information in the context of the scene, and is generally query driven.

Table 4.4 summarizes the major kinds of registration, combination, and reasoning functions that are performed, illustrating the increasing levels of complexity in each level of spatial processing. Faust described the general principles for building such a geospatial database, the hierarchy of functions, and the concept for a blackboard architecture expert system to implement the functions described above.[75]

4.6.2.1 A Representative Example

The spatial reasoning process can be illustrated by a hypothetical military example that follows the process an image or intelligence analyst might follow in search of critical mobile targets (CMTs). Consider the layers of a spatial database illustrated in Figure 4.6, in which recent unmanned air vehicle (UAV) SAR data (the top data layer) has been registered to all other layers, and the following process is performed (process steps correspond to path numbers on the figure):

TABLE 4.4   Spatial Data Fusion Functions

	Increasing Complexity and Processing		
	Registration	Combination	Reasoning
Data Fusion Functions	Image registration Image-to-terrain registration Orthorectification Image mosaicking, including radiometric balancing and feathering Multitemporal change detection	Multiresolution image sharpening Multispectral classification of registered imagery Image-to-image cueing Spatial detection via multiple layers of image data Feature extraction using multilayer data	Image-to-image cross layer searches Feature finding: extraction by roaming across layers to increase detection, recognition, and confidence Context evaluation Image-to-nonimage cueing (e.g., IMINT to SIGINT) Area delimitation
Examples	Coherent radar imagery change detection SPOT™ imagery mosaicking LANDSAT magnitude change detection	Multispectral image sharpening using panchromatic image 3-D scene creation from multiple spatial sources	Area delimitation to search for critical target Automated map feature extraction Automated map feature updating

Note: Spatial data fusion functions include a wide variety of registration, combination, and reasoning processes and algorithms.

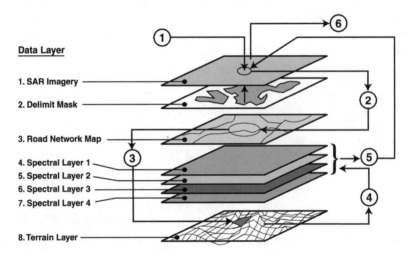

FIGURE 4.6   Target search example uses multiple layers of spatial data and applies iterative spatial reasoning to evaluate alternative hypotheses while accumulating evidence for each candidate target.

1. A target cueing algorithm searches the SAR imagery for candidate CMT targets, identifying potential targets in areas within the allowable area of a predefined delimitation mask (Data Layer 2).*
2. Location of a candidate target is used to determine the distance to transportation networks (which are located in the map Data Layer 3) and to hypothesize feasible paths from the network to the hide site.
3. The terrain model (Data Layer 8) is inspected along all paths to determine the feasibility that the CMT could traverse the path. Infeasible path hypotheses are pruned.
4. Remaining feasible paths (on the basis of slope) are then inspected using the multispectral data (Data Layers 4, 5, 6, and 7). A multispectral classification algorithm is scanned over the feasible

---

*This mask is a *derived* layer produced, by a spatial reasoning process in the scene generation stage, to delimit the entire search region to only those allowable regions in which a target may reside.

paths to assess ground load-bearing strength, vegetation cover, and other factors. Evidence is accumulated for slope and these factors (for each feasible path) to determine a composite path likelihood. Evidence is combined into a likelihood value and unlikely paths are pruned.

5. Remaining paths are inspected in the recent SAR data (Data Layer 1) for other significant evidence (e.g., support vehicles along the path, recent clear cut) that can support the hypothesis. Supportive evidence is accumulated to increase likelihood values.

6. Composite evidence (target likelihood plus likelihood of feasible paths to candidate target hide location) is then used to make a final target detection decision.

In the example presented in Figure 4.6, the reasoning process followed a spatial search to accumulate (or discount) evidence about a candidate target. In addition to target detection, similar processes can be used to

- Insert data in the database (e.g., resolve conflicts between input sources),
- Refine accuracy using data from multiple sources, etc.,
- Monitor subtle changes between existing data and new measurements, and
- Evaluate hypotheses about future actions (e.g., trafficability of paths, likelihood of flooding given rainfall conditions, and economy of construction alternatives).

4.7 Summary

The fusion of image and spatial data is an important process that promises to achieve new levels of performance and integration in a variety of application areas. By combining registered data from multiple sensors or views, and performing intelligent reasoning on the integrated data sets, fusion systems are beginning to significantly improve the performance of current generation automatic target recognition, single-sensor imaging, and geospatial data systems.

References

1. Composite photo of Kuwait City in *Aerospace and Defense Science,* Spring 1991.
2. *Aviation Week and Space Technology,* May 2, 1994, 62.
3. Composite multispectral and 3-D terrain view of Haiti in *Aviation Week and Space Technology,* October 17, 1994, 49.
4. Robert Ropelewski, Team Helps Cope with Data Flood, *Signal,* August 1993, 40–45.
5. Intelligence and Imagery Exploitation, Solicitation BAA 94-09-KXPX, *Commerce Business Daily,* April 12, 1994.
6. Terrain Feature Generation Testbed for War Breaker Intelligence and Planning, Solicitation BAA 94-03, *Commerce Business Daily,* July 28, 1994; Terrain Visualization and Feature Extraction, Solicitation BAA 94-01, *Commerce Business Daily,* July 25, 1994.
7. *Global Geospace Information and Services (GGIS),* Defense Mapping Agency, Version 1.0, August 1994, 36–42.
8. M.A. Abidi and R.C. Gonzales, Eds., *Data Fusion in Robotics and Machine Intelligence,* Academic Press, Boston, 1993.
9. Derek L.G. et al., *Accurate Frameless Registration of MR and CT Images of the Head: Applications in Surgery and Radiotherapy Planning,* Dept. of Neurology, United Medical and Dental Schools of Guy's and St. Thomas's Hospitals, London, SE1 9R, U.K., 1994.
10. Edward L. Waltz and James Llinas, *Multisensor Data Fusion,* Norwood, MA: Artech House, 1990.
11. David L. Hall, *Mathematical Techniques in Multisensor Data Fusion,* Norwood, MA: Artech House, 1992.
12. W.K. Pratt, Correlation Techniques of Image Registration, *IEEE Trans. AES,* May 1974, 353–358.

13. L. Gottsfield Brown, A Survey of Image Registration Techniques, *Computing Surveys,* 1992, Vol. 29, 325–376.

14. Franklin E. White, Jr., Data Fusion Subpanel Report, *Proc. Fifth Joint Service Data Fusion Symp.,* October 1991, Vol. I, 335–361.

15. Bir Bhanu, Automatic Target Recognition: State-of-the-Art Survey, *IEEE Trans. AES,* Vol. 22, No. 4, July 1986, 364–379.

16. Bir Bhanu and Terry L. Jones, Image Understanding Research for Automatic Target Recognition, *IEEE AES,* October 1993, 15–23.

17. Wade G. Pemberton, Mark S. Dotterweich, and Leigh B. Hawkins, An Overview of ATR Fusion Techniques, *Proc. Tri-Service Data Fusion Symp.,* June 1987, 115–123.

18. Laurence Lazofson and Thomas Kuzma, Scene Classification and Segmentation Using Multispectral Sensor Fusion Implemented with Neural Networks, *Proc. 6th Nat'l. Sensor Symp.,* August 1993, Vol. I, 135–142.

19. Victor M. Gonzales and Paul K. Williams, Summary of Progress in FLIR/LADAR Fusion for Target Identification at Rockwell, *Proc. Image Understanding Workshop,* ARPA, November 1994, Vol. I, 495–499.

20. Anthony N.A. Schwickerath and J. Ross Beveridge, Object to Multisensor Coregistration with Eight Degrees of Freedom, *Proc. Image Understanding Workshop,* ARPA, November 1994, Vol. I, 481–490.

21. Richard Delanoy, Jacques Verly, and Dan Dudgeon, Pixel-Level Fusion Using "Interest" Images, *Proc. 4th National Sensor Symp.,* August 1991, Vol. I, 29.

22. Mark K. Hamilton and Theresa A. Kipp, Model-based Multi-Sensor Fusion, *Proc. IEEE Asilomar Circuits and Systems Conf.,* November 1993.

23. Theresa A. Kipp and Mark K. Hamilton, Model-based Automatic Target Recognition, *4th Joint Automatic Target Recognition Systems and Technology Conf.,* November 1994.

24. Greg Duane, Pixel-Level Sensor Fusion for Improved Object Recognition, *Proc. SPIE Sensor Fusion,* 1988, Vol. 931, 180–185.

25. D. Reago, et al., Multi-Sensor Feature Level Fusion, *4th Nat'l. Sensor Symp.,* August 1991, Vol. I, 230.

26. Eric Keydel, Model-Based ATR, Tutorial Briefing, Environmental Research Institute of Michigan, February 1995.

27. M.K. Hamilton and T.A. Kipp, ARTM: Model-Based Mutisensor Fusion, *Proc. Joint NATO AC/243 Symp. on Multisensors and Sensor Data Fusion,* November 1993.

28. D.A. Analt, S.D. Raney, and B. Severson, An Angle and Distance Constrained Matcher with Parallel Implementations for Model Based Vision, *Proc. SPIE Conf. on Robotics and Automation,* Boston, MA, October 1991.

29. Model-Driven Automatic Target Recognition Report, ARPA/SAIC System Architecture Study Group, October 14, 1994.

30. James Larson, Larry Hung, and Paul Williams, FLIR/Laser Radar Fused Model-based Target Recognition, *4th Nat'l. Sensor Symp.,* August 1991, Vol. I, 139–154.

31. Francis Corbett et al., Fused ATR Algorithm Development for Ground to Ground Engagement, *Proc. 6th Nat'l. Sensor Symp.,* August 1993, Vol. I, 143–155.

32. James D. Silk, Jeffrey Nicholl, David Sparrow, Modeling the Performance of Fused Sensor ATRs, *Proc. 4th Nat'l. Sensor Symp.,* August 1991, Vol. I, 323–335.

33. Rae H. Lee and W.B. Van Vleet, Registration Error Between Dissimilar Sensors, *Proc. SPIE Sensor Fusion,* 1988, Vol. 931, 109–114.

34. J.A. Hoschette and C.R. Seashore, IR and MMW Sensor Fusion for Precision Guided Munitions, *Proc. SPIE Sensor Fusion,* 1988, Vol. 931, 124–130.

35. David Lai and Richard McCoy, A Radar-IR Target Recognizer, *Proc. 4th Nat'l. Sensor Symp.,* August 1991, Vol. I, 137.

36. Michael C. Roggemann et al., An Approach to Multiple Sensor Target Detection, *Sensor Fusion II, Proc. SPIE* Vol. 1100, March 1989, 42–50.

37. Kris Siejko et al., Dual Mode Sensor Fusion Performance Optimization, *Proc. 6th Nat'l. Sensor Symp.*, August 1993, Vol. I, 71–89.

38. Vince Diehl, Frank Shields, and Andy Hauter, Testing of Multi-Sensor Automatic Target Recognition and Fusion Systems, *Proc. 6th Nat'l. Sensor Fusion Symp.*, August 1993, Vol. I, 45–69.

39. Murali Menon and Paul Kolodzy, Active/passive IR Scene Enhancement by Markov Random Field Sensor Fusion, *Proc. 4th Nat'l. Sensor Symp.*, August 1991, Vol. I, 155.

40. Alianna J. Maren, A Hierarchical Data Structure Representation for Fusing Multisensor Information, Sensor Fusion II, *Proc. SPIE* Vol. 1100, March 1989, 162–178.

41. A. Omar Aboutalib, Lubong Tran, and Cheng-Yen Hu, Fusion of Passive Imaging Sensors for Target Acquisition and Identification, *Proc. 5th Nat'l. Sensor Symp.*, June 1992, Vol. I, 151.

42. Chen-Chau Chu and J.K. Aggarwal, Image Interpretation Using Multiple Sensing Modalities, *IEEE Trans. on Pattern Analysis and Machine Intelligence*, August 1992, Vol. 14, No. 8, 840–847.

43. Peter A. Eggleston and Harles A. Kohl, Symbolic Fusion of MMW and IR Imagery, *Proc. SPIE Sensor Fusion*, 1988, Vol. 931, 20–27.

44. Peter A. Eggleston, *Algorithm Development Support Tools for Machine Vision*, Amerinex Artificial Intelligence, Inc., (n.d., received February 1995).

45. Christopher Bowman and Mark DeYoung, Multispectral Neural Network Camouflaged Vehicle Detection Using Flight Test Images, *Proc. World Conf. on Neural Networks*, June 1994.

46. Jim Vrabel, MSI Band Sharpening Design Trade Study, *Presented at 7th Joint Service Data Fusion Symp.*, October 1994.

47. P.S. Chavez, Jr. et al., Comparison of Three Different Methods to Merge Multiresolution and Multispectral Data: LANDSAT™ and SPOT Panchromatic, *Photogrammetric Engineering and Remote Sensing*, March 1991, Vol. 57, No. 3, 295–303.

48. Kathleen Edwards and Philip A. Davis, The Use of Intensity-Hue-Saturation Transformation for Producing Color-Shaded Relief Images, *Photogrammetric Engineering and Remote Sensing*, November 1994, Vol. 60, No. 11, 1379–1374.

49. Robert Tenney and Alan Willsky, Multiresolution Image Fusion, DTIC Report AD-B162322L, January 31, 1992.

50. Barry N. Haack and E. Terrance Slonecker, Merging Spaceborne Radar and Thematic Mapper Digital Data for Locating Villages in Sudan, *Photogrammetric Engineering and Remote Sensing*, October 1994, Vol. 60, No. 10, 1253–1257.

51. Christopher C. Chesa, Richard L. Stephenson, and William A. Tyler, Precision Mapping of Spaceborne Remotely Sensed Imagery, *Geodetical Info Magazine*, March 1994, Vol. 8, No. 3, 64–67.

52. Roger M. Reinhold, Arc Digital Raster Imagery (ADRI) Program, Air Force Spatial data Technology Workshop, Environmental Research Institute of Michigan, July 1991.

53. In So Kweon and Takeo Kanade, High Resolution Terrain Map from Muliple Sensor Data, *IEEE Trans. Pattern Analysis and Machine Intelligence*, Vol. 14, No. 2, February 1992.

54. Peter J. Burt, Pattern Selective Fusion of IR and Visible Images Using Pyramid Transforms, *Proc. 5th Nat'l. Sensor Symp.*, June 1992, Vol. I, 313–325.

55. M. Hansen et al., Real-time Scene Stabilization and Mosaic Construction, *Proc. Image Understanding Workshop*, ARPA, November 1994, Vol. I, 457–465.

56. Peter J. Burt and P. Anandan, Image Stabilization by Registration to a Reference Mosaic, *Proc. Image Understanding Workshop*, ARPA, November 1994, Vol. I, 425–434.

57. Christopher C. Chiesa and William A. Tyler, Data Fusion of Off-Nadir SPOT Panchromatic Images with Other Digital Data Sources, *Proc. 1990 ACSM-ASPRS Annual Convention*, Denver, March 1990, 86–98.

58. Avi Kak and Su-shing Chen (Eds.), *Proc. Spatial Reasoning and Multi-Sensor Fusion Workshop*, AAAI, October 1987.

59. Paul S. Shenker (Ed.), Sensor Fusion: Spatial Reasoning and Scene Interpretation, *SPIE* Vol. 1003, November 1988.

60. Paul S. Shenker, Sensor Fusion III: 3-D Perception and Recognition, *SPIE* Vol. 1383, November 1990.

61. Paul S. Shenker, Sensor Fusion IV: Control Paradigms and Data Structures, *SPIE* Vol. 1611, November 1991.

62. Paul S. Shenker, Sensor Fusion V, *SPIE* Vol. 1828, November 1992.

63. Paul S. Shenker, Sensor Fusion VI, *SPIE* Vol. 2059, November 1993.

64. *Content Standards for Digital Geographic Metadata,* Federal Geographic Data Committee, Washington D.C., June 8, 1994.

65. *Intelligence Preparation of the Battlefield,* FM-34-130, HQ Dept. of the Army, May 1989.

66. *Development and Integration of the Terrain Feature Generator (TFG),* Solicitation DACA76-94-R-0009, Commerce Business Daily Issue PSA-1087, May 3, 1994.

67. Richard T. Antony, Eight Canonical Forms of Fusion: A Proposed Model of the Data Fusion Process, *Proc. of 1991 Joint Service Data Fusion Symp.,* Vol. III, October 1991.

68. Richard T. Antony, *Principles of Data Fusion Automation,* Norwood, MA: Artech House Inc., 1995.

69. R.L. Shelton and J.E. Estes, Integration of Remote Sensing and Geographic Information Systems, *Proc. 13th Int'l. Symp. Remote Sensing of the Environment,* Environmental Research Institute of Michigan, April 1979, 463–483.

70. John E. Estes and Jeffrey L. Star, Remote Sensing and GIS Integration: Towards a Prioritized Research Agenda, *Proc. 25th Int'l Symp. on Remote Sensing and Change,* Graz Austria, April 1993, Vol. I, 448–464.

71. J.L. Star (Ed.), *The Integration of Remote Sensing and Geographic Information Systems,* American Society for Photogrammetry and Remote Sensing, 1991.

72. G. Sawyer et al., MuSIP Multi-Sensor Image Processing System, *Image and Vision Computing,* Vol. 11, No. 1, January-February 1993, 25–34.

73. D.C. Mason et al., Spatial Database Manager for a Multi-source Image Understanding System, *Image and Vision Computing,* Vol. 10, No. 9, November 1992, 589–609.

74. Nahum D. Gershon and C. Grant Miller, Dealing with the Data Deluge, *IEEE Spectrum,* July 1993, 28–32.

75. Nickolas L. Faust, *Design Concept for Database Building,* Project 2851 Newsletter, May 1989, 17–25.

5

Data Registration

5.1	Introduction	5-1
5.2	Registration Problem	5-2
5.3	Review of Existing Research	5-3
5.4	Registration Using Meta-Heuristics	5-5
5.5	Wavelet-Based Registration of Range Images	5-7
5.6	Registration Assistance/Preprocessing	5-9
5.7	Conclusion	5-10
	Acknowledgments	5-10
	References	5-11

Richard R. Brooks
The Pennsylvania State University

Lynne Grewe
California State University

5.1 Introduction

Sensor fusion refers to the use of multiple sensor readings to infer a single piece of information. Inputs may be received from a single sensor over a period of time. They may be received from multiple sensors of the same or different types. Inputs may be raw data, extracted features, or higher-level decisions. This process provides increased robustness and accuracy in machine perception. This is conceptually similar to the use of repeated experiments to establish parameter values using statistics.[1] Several reference books have been published on sensor fusion.[2-4]

One decomposition of the sensor fusion process is shown in Figure 5.1. Sensor readings are gathered, preprocessed, compared, and combined, and a final result is derived. An essential preprocessing step for comparing readings from independent physical sensors is transforming all input data into a common coordinate system. This is referred to as *data registration*. In this chapter, we describe data registration, provide a review of existing methods, and discuss some recent results.

Data registration transformation is often assumed to be known *a priori*, partially because the problem is not trivial. Traditional methods are based on methods developed by cartographers. These methods have a number of drawbacks and often make invalid assumptions concerning the input data.

Although data input includes raw sensor readings, features extracted from sensor data, and higher-level information, registration is a preprocessing stage and, therefore, is usually applied only to either raw data or extracted features. Sensor readings can have one to n dimensions. The number of dimensions will not necessarily be an integer. Most techniques deal with data of two or three dimensions; however, same approaches can be trivially applied to one-dimensional readings. Depending on the sensing modalities used, occlusion may be a problem with data in more than two dimensions, causing data in the environment to be obscured by the relative position of objects in the environment. The specific case studies presented in this chapter use image data in two dimensions and range data in $2\frac{1}{2}$ dimensions.

This chapter is organized as follows. Section 5.2 gives a formal definition of image registration. Section 5.3 provides a brief survey of existing methods. Section 5.4 discusses meta-heuristic techniques that have been used for image registration. This includes objective functions for sensor readings with various types

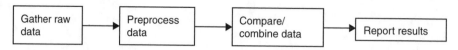

FIGURE 5.1   Decomposition of sensor fusion process.

of noise. Section 5.5 discusses a multiresolution implementation of image registration. Section 5.6 provides a brief summary discussion.

5.2   Registration Problem

Competitive multiple sensor networks consist of a large number of physical sensors providing readings that are at least partially redundant. The first step in fusing multiple sensor readings is registering them to a common frame of reference.[5] "Registration" refers to finding the correct mapping of one image onto another. When an inaccurate estimate of the registration is known, finding the exact registration is referred to as *refined registration*. Another survey of image registration can be found in Brown.[6]

As shown in Figure 5.2, the general image registration problem is, given two N-dimensional sensor readings, find the function F which best maps the reading from sensor two, $S_2(x_1,...,x_n)$ onto the reading from sensor one, $S_1(x_1,...,x_n)$. Ideally, $F(S_2(x_1,...,x_n)) = S_1(x_1,...,x_n)$. Because all sensor readings contain some amount of measurement error or noise, the ideal case rarely occurs.

Many processes require that data from one image, called the *observed image*, be compared with or mapped to another image, called the *reference image*. As a result, a wide range of critical applications depends on image registration.

Perhaps the largest amount of image registration research is focused on medical imaging. One application is sensor fusion to combine outputs from several medical imaging technologies, such as PET and MRI, to form a more complete image of internal organs.[7] Registered images are then used for medical diagnosis of illness[8] and automated control of radiation therapy.[9] Similar applications of registered and fused images are common[11] in military applications (e.g. terrain "footprints"),[10] remote sensing applications, and robotics. A novel application is registering portions of images to estimate motion. Descriptions of motion can then be used to construct intermediate images in television transmissions. Jain and Jain describe the applications of this to bandwidth reduction in video communications.[12] These are some of the more recent applications that rely on accurate image registration. Methods of image registration have been studied since the beginning of the field of cartography.

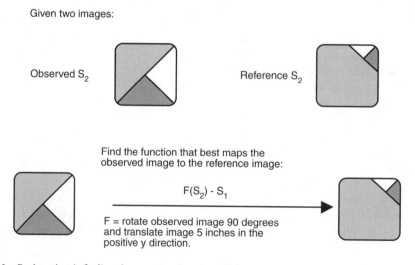

FIGURE 5.2   Registration is finding the mapping function $F(S_2)$.

TABLE 5.1   Image Registration Methods

Algorithm	Image Type	Matching Method	Interpolation Function	Transforms Supported	Comments
Andrus	Boundary maps	Correlation	None	Gruence	Noise intolerant, small rotations
Barnea	No restriction	Improved correlation	None	Translation	No rotation, scaling noise, rubber sheet
Barrow	No restriction	Hill climbing	Parametric chamfer	Gruence	Noise tolerant, small displacement
Brooks Iyengar	No restriction	Elitist gen. Alg.	None	Gruence	Noise tolerant, tolerates periodicity
Cox	Line segments	Hill climbing	None	Gruence	Matches using small number of features
Davis	Specific shapes	Relaxation	None	Affine	Matches shapes
Goshtasby 1986	Control points	Various	Piecewise linear	Rubber sheet	Fits images using mapped points
Goshtasby 1987	Control points	Various	Piecewise cubic	Rubber sheet	Fits images using mapped points
Goshtasby 1988	Control points	Various	Lease squares	Rubber sheet	Fits images using mapped points
Jain	Sub-images	Hill climbing	None	Translation	Small translations, no rotation, no noise
Mandara	Control points	Classic G.A.S.A.	Bi-linear	Rubber sheet	Fits 4 fixed points using error fitness
Mitiche	Control points	Least squares	None	Affine	Uses control points
Oghabian	Control points	Sequential search	Least squares	Rubber sheet	Assumes small displacement
Pinz	Control points	Tree search	None	Affine	Difficulty with local minima
Stockman	Control points	Cluster	None	Affine	Assumes landmarks, periodicity problem
Wong	Intensity differences	Exhaustive search	None	Affine	Uses edges, intense computation

5.3   Review of Existing Research

This section discusses the current state of research concerning image registration. Image registration is a basic problem in image processing, and a large number of methods have been proposed.

Table 5.1 summarizes the features of representative image registration methods discussed in this section. The discussion is followed by a detailed discussion of the established methodologies, and algorithms currently in use. Each is explored in more detail in the remainder of the section.

The traditional method of registering two images is an extension of methods used in cartography. A number of control points are found in both images. The control points are matched, and this match is used to deduce equations that interpolate all points in the new image to corresponding points in the reference image.[13,14]

Several algorithms exist for each phase of this process. Control points must be unique and easily identified in both images. Control points have been explicitly placed in the image by the experimenter[9] and edges have been defined by intensity changes,[15] specific points peculiar to a given image,[16] line intersections, center of gravity of closed regions, or points of high curvature.[13] The type of control point that should be used primarily depends on the application and contents of the image. For example, in medical image processing, the contents of the image and approximate poses are generally known *a priori*.

Similarly, many methods have been proposed for matching control points in the observed image to the control points in the reference image. The obvious method is to correlate a template of the observed image.[17,18] Another widely used approach is to calculate the transformation matrix, which describes the mapping with the least square error.[11,16,19] Other standard computational methods, such as relaxation and hill-climbing, have also been used.[12,20,21] Pinz et al. use a hill-climbing algorithm to match images and note the difficulty posed by local minima in the search space; to overcome this, they run a number of attempts in parallel with different initial conditions.[22]

Some interesting methods have been implemented that consider all possible transformations. Stockman et al. construct vectors between all pairs of control points in an image.[10] For each vector in each image, an affine transformation matrix is computed which converts the vector from the observed image to one of the vectors from the reference image. These transformations are then plotted, and the region containing the largest number of correspondences is assumed to contain the correct transformation.[10] This method is computationally expensive because it considers the power set of control points in each image. Wong and Hall match scenes by extracting edges or intensity differences and constructing a tree of all possible matches that fall below a given error threshold.[15] They reduce the amount of computation needed by stopping all computation concerning a potential matching once the error threshold is exceeded; however, this method remains computationally intensive. Dai and Khorram extract affine transform invariant features based on the central moments of regions found in remote sensing images.[23] Regions are defined by zero-crossing points. Similarly, Yang and Cohen describe a moments-based method for registering images using affine transformations given sets of control points.[24]

Registration of multisensor data to a three-dimensional scene, given a knowledge of the contents of the scene, is discussed by Chellappa.[25] The use of an extended Kalman filter (EKF) to register moving sensors in a sensor fusion problem is discussed by Zhou.[26] Mandara and Fitzpatrick have implemented a very interesting approach[8] using simulated annealing and genetic algorithm heuristics to find good matches between two images. They find a rubber sheet transformation, which fits two images by using linear interpolation around four control points, and assume that the images match approximately at the beginning. A similar approach has been espoused by Matsopoulos.[27]

A number of researchers have used multiresolution methods to prune the search space considered by their algorithms. Mandara and Fitzpatrick[8] use a multiresolution approach to reduce the size of their initial search space for registering medical images using simulated annealing and genetic algorithms. This work influenced Oghabian and Todd-Prokopek, who similarly reduced their search space when registering brain images with small displacements.[7] Pinz adjusted both multiresolution scale space and step size in order to reduce the computational complexity of a hill-climbing registration method.[22] These researchers believe that by starting with low-resolution images, they can reject large numbers of possible matches and find the correct match by progressively increasing the resolution. Note that in images with a strong periodic component, a number of low-resolution matches may be feasible. In such cases, the multiresolution approach will be unable to prune the search space and, instead, will increase the computational load. Another problem with a common multiresolution approach, the wavelet transform, is its sensitivity to translation.[28]

A number of methods have been proposed for fitting the entire image around the control points once an appropriate match has been found. Simple linear interpolation is computationally straightforward.[8] Goshtasby has explored using a weighted least-squares approach,[19] constructing piecewise linear interpolation functions within triangles defined by the control points,[13] and developing piecewise cubic interpolation functions.[29] These methods create nonaffine rubber sheet transformation functions to attempt to reduce the image distortion caused by either errors in control point matching, or differences in the sensors that constructed the image.

Several algorithms exist for image registration. The algorithms described have some common drawbacks. The matching algorithms assume that a small number of distinct features can be matched,[10,16,30] that specific shapes are to be matched,[31] that no rotation exists, or that the relative displacement is small.[7,8,12,16,17,21] Refer to Table 5.1 for a summary of many of these points.

Choosing a small number of control points is not a trivial problem and has a number of inherent drawbacks. For example, the control point found may be a product of measurement noise. When two readings have more than a trivial relative displacement, control points in one image may not exist in the other image. This requires considering the power set of the control points. When an image contains periodic components, control points may not define a unique mapping of the observed image to the reference image. Additional problems exist. The use of multiresolution cannot always trim the search space and, if the image is dominated by periodic elements, it will only increase the computational complexity of an algorithm.[7,8,22]

Many algorithms attempt to minimize the square error over the image; however, this does not consider the influence of noise in the image.[7,8] Most of the existing methods are sensitive to noise.[7,16,17] Section 5.4 discusses meta-heuristics based methods, which try to overcome these drawbacks. Section 5.5 discusses a multiresolution approach.

5.4   Registration Using Meta-Heuristics

This section discusses research on automatically finding a *gruence* (i.e., translation and rotation) registering two overlapping images. Results from this research have previously been presented in a number of sources.[2,32-35] This approach attempts to correctly calibrate two two-dimensional sensor readings with identical geometries. These assumptions about the sensors can be made without a loss of generality because

- A method that works for two readings can be extended to register any number of readings sequentially.
- The majority of sensors work in one or two dimensions. Extensions of calibration methods to more dimensions is desirable, but not imperative.
- Calibration of two sensors presupposes known sensor geometry. If geometries are known, a function can be derived that maps the readings as if the geometries were identical when a registration is given.

This approach finds gruences because these functions best represent the most common class of problems. The approach used can be directly extended to include the class of all affine transformations by adding scaling transformations.[36] It does not consider "rubber sheet" transformations that warp the contents of the image because these transformations mainly correct local effects after use of an affine transformation correctly matches the images.[14] It assumes that any rubber sheet deformations of the sensor image are known and corrected before the mapping function is applied, or that their effects over the image intersections are negligible.

The computational examples used pertain to two sensors returning two-dimensional gray scale data from the same environment. The amount of noise and the relative positions of the two sensors are not known. Sensor two is translated and rotated by an unknown amount with relation to sensor one.

If the size or content of the overlapping areas is known, a correlation using the contents of the overlap on the two images could find the point where they overlap directly. Use of central moments could also find relative rotation of the readings. When the size or content of the areas is unavailable, this approach is impossible.

In this work, the two sensors have identical geometric characteristics. They return readings covering a circular region, and these readings overlap. Both sensors' readings contain noise. What is not known, however, is the relative positions of the two sensors. Sensor two is translated and rotated by an unknown amount with relation to sensor one.

The best way to solve this problem depends on the nature of the terrain being observed. If unique landmarks can be identified in both images, those points can be used as control points. Depending on the number of landmarks available, minor adjustments may be needed to fit the readings exactly. Goshtasby's methods could be used at that point.[13,19,29]

Thus, the problem to be solved is, given noisy gray scale data readings from sensor one and sensor two, find the optimal set of parameters (x-displacement, y-displacement, and angle of rotation) that defines the center of the sensor two image relative to the center of the sensor one image. These parameters would provide the optimal mapping of sensor two readings to the readings from sensor one. This can be done using meta-heuristics for optimization. Brooks describes implementations of genetic algorithms, simulated annealing, and tabu search for this problem.[2] Chen applies TRUST, a subenergy tunneling approach from Oak Ridge National Laboratories.[35]

To measure optimality, a fitness function can be used. The fitness function provides a numerical measure of the goodness of a proposed answer to the registration problem. Brooks derives a fitness function for sensor readings corrupted with Gaussian noise:[2]

$$\frac{\sum\left(read_1\left(x,y\right)read_2\left(x',y'\right)\right)^2}{K\left(W\right)^2} = \frac{\sum\left(gray_1\left(x,y\right)gray_2\left(x',y'\right)\right)^2 - \sum\left(noise_1\left(x,y\right)noise_2\left(x',y'\right)\right)^2}{K\left(W\right)^2}$$

(5.1)

where

w	is a point in the search space
$K(W)$	is the number of pixels in the overlap for w
(x',y')	is the point corresponding to (x,y)
$read_1(x,y)read_2(x',y')$	is the pixel value returned by sensor 1 (2) at point (x,y) (x',y')
$gray_1(x,y)gray_2(x',y')$	is the noiseless value for sensor 1 (2) at (x,y) (x',y')
$noise_1(x,y)noise_2(x',y')$	is the noise in the sensor 1 (2) reading at (x,y) (x',y')

The equation is derived by separating the sensor reading into information and additive noise components. This means the fitness function is made up of two components: (a) lack of fit, and (b) stochastic noise. The lack of fit component has a unique minimum when the two images have the same gray scale values in the overlap (i.e., when they are correctly registered). The noise component follows a Chi-squared distribution, whose expected value is proportional to the number of pixels in the region where the two sensor readings intersect. Dividing the difference squared by the cardinality of the overlap, makes the expected value of the noise factor constant. Dividing by the cardinality squared favors large intersections. For a more detailed explanation of this derivation, see Brooks.[2]

Other noise models simply modify the fitness function. Another common noise model addresses salt-and-pepper noise typically caused by either malfunctioning pixels in electronic cameras or dust in optical systems. In this model, the correct gray-scale value in a picture is replaced by a value of 0 (255) with an unknown probability $p(q)$. An appropriate fitness function for this type of noise is Equation 5.2.

$$\begin{matrix} read_1\left(x,y\right)\neq 0 \\ read_1\left(x,y\right)\neq 255 \\ read_2\left(x',y'\right)\neq 0 \\ read_2\left(x',y'\right)\neq 255 \end{matrix} \qquad \sum \frac{\left(read_1\left(x,y\right)-read_2\left(x',y'\right)\right)^2}{K\left(W\right)}$$

(5.2)

A similar function can be derived for uniform noise by using the expected value $E[(U_1 - U_2)^2]$ of the squared difference of two uniform variables U_1 and U_2. An appropriate fitness function is then given by

$$\sum \frac{\left(read_1\left(x,y\right)-read_2\left(x',y'\right)\right)^2}{E\left[\left(U_1-U_2\right)^2\right]K\left(W\right)}$$

(5.3)

Figure 5.3 shows the best fitness function value found by simulated annealing, elitist genetic algorithms, classic genetic algorithms, and tabu search versus the number of iterations performed. In Brooks, elitist genetic algorithms out-perform the other methods attempted. Further work by Chen indicates that

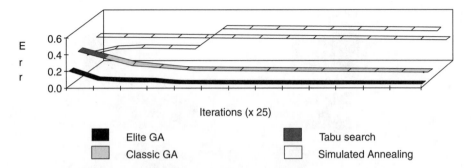

FIGURE 5.3 Fitness function results variance 1.

TRUST is more efficient than the elitist genetic algorithms.[35] These studies show that optimization techniques can work well on the problem, even in the presence of large amounts of noise. This is surprising because the fitness functions take the difference of noise-corrupted data — essentially a derivative. Derivatives are sensitive to noise. Further inspection of the fitness functions explains this surprising result. Summing over the area of intersection is equivalent to integrating over the area of intersection. Implicitly, integrating counteracts the derivative's magnification of noise.

Matsopoulos uses affine, b-linear, and projective transformations to register medical images of the retina.[27] The techniques tested include genetic algorithms, simulated annealing, and the downhill simplex method. They use image correlation as a fitness function. For their application, much preprocessing is necessary, which removes sensor noise. Their results indicate the superiority of genetic algorithms for automated image registration. This is consistent with Brooks' results.[2,34]

5.5 Wavelet-Based Registration of Range Images

This section uses range sensor readings. More details are provided by Grewe.[39] Range images consist of pixels with values corresponding to range or depth rather than photometric information. The range image represents a perspective of a three-dimensional world. The registration approach described herein can be trivially applied to other kinds of images, including one-dimensional readings. If desired, the approaches described by Brooks[2,34] and Chen[35] can be directly extended to include the class of all affine transformations by adding scaling transformations. This section discusses an approach for finding these transformations.

The approach uses a multiresolution technique, the wavelet transform, to extract features used to register images. Other researchers have also applied wavelets to this problem, including using locally maximum wavelet coefficient values as features from two images.[37] The centroids of these features are used to compute the translation offset between the two images. A principle components analysis is then performed and the eigenvectors of the covariance matrix provide an orthogonal reference system for computing the rotation between the two images. (This use of a simple centroid difference is subject to difficulties when the scenes only partially overlap and, hence, contain many other features.)

In another example, the wavelet transform is used to obtain a complexity index for two images.[38] The complexity measure is used to determine the amount of compression appropriate for the image. Compression is then performed, yielding a small number of control points. The images, made up of control points for rotations, are tested to determine the best fit.

The system described in Grewe[39] is similar to some of the previous work discussed. Similar to DeVore,[38] Grewe uses wavelets to compress the amount of data used in registration. Unlike previous wavelet-based systems prescribed by Sharman[37] and DeVore,[38] Grewe's[39] capitalizes on the hierarchical nature of the wavelet domain to further reduce the amount of data used in registration. Options exist to perform a hierarchical search or simply to perform registration inside one wavelet decomposition level. Other system options include specifying an initial registration estimate, if known, and the choice of the wavelet

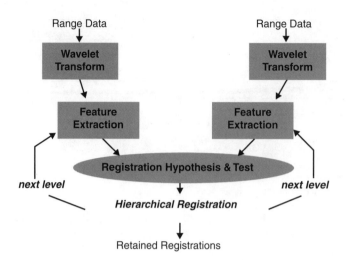

FIGURE 5.4   Block diagram of WaveReg system.

decomposition level in which to perform or start registration. At higher decomposition levels, the amount of data is significantly reduced, but the resulting registration will be approximate. At lower decomposition levels, the amount of data is reduced to a lesser extent, but the resulting registration is more exact. This allows the user to choose between accuracy and speed as necessary.

Figure 5.4 shows a block diagram of the system. It consists of a number of phases, beginning with the transformation of the range image data to the wavelet domain. Registration can be performed on only one decomposition level of this space to reduce registration complexity. Alternately, a hierarchical registration across multiple levels will extract features from a wavelet decomposition level as a function of a number of user-selected parameters, which determine the amount of compression desired in the level. Matching features from the two range images are used to hypothesize the transformation between the two images and are evaluated. The "best" transformations are retained. This process is explained in the following paragraphs.

First, a Daubechies-4 wavelet transform is applied to each range image. The wavelet data is compressed by thresholding the data to eliminate low magnitude wavelet coefficients. The wavelet transform produces a series of 3-D edge maps at different resolutions. A maximal wavelet value indicates a relatively sharp change in depth.

Features, special points of interest in the wavelet domain, are simply points of maximum value in the current wavelet decomposition level under examination. These points are selected so that no two points are close to each other. The minimum distance is scaled with the changing wavelet level under examination. Figure 5.5 shows features detected for different range scenes at different wavelet levels. Notice how these correspond to points of sharp change in depth.

Using a small number of feature points allows this approach to overcome the wavelets transform's sensitivity to translation. Stone[28] proposed another method for overcoming the sensitivity to translation. Stone noted that the low-pass portions of the wavelet transform are less sensitive to translation and that coarse to fine registration of images using the wavelet transform should be robust.

The next stage involves hypothesizing correspondences between features extracted from the two unregistered range images. Each hypothesis represents a possible registration and is subsequently evaluated for its goodness. Registrations are compared and the best retained.

Hypothesis formation begins at a default wavelet decomposition level. Registrations retained at this level are further "refined" at the next lower level, L-1. This process continues until the lowest level in the wavelet space is reached.

For each hypothesis, the corresponding geometric transformation relating the matched features is calculated, and the remaining features from one range image are transformed into the other's space. This

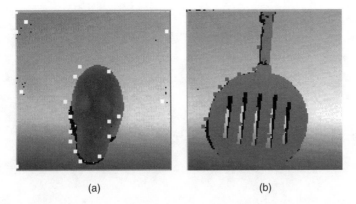

(a)   (b)

FIGURE 5.5 Features detected, approximate location indicated by white squares: (a) for wavelet Level 2 and (b) for wavelet Level 1.

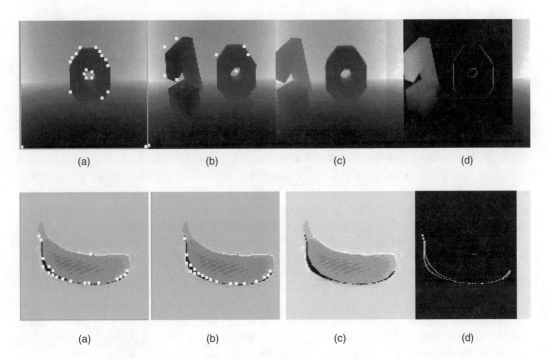

(a)   (b)   (c)   (d)

(a)   (b)   (c)   (d)

FIGURE 5.6 (a) Features extracted Level 1, Image 1, (b) Features extracted Level 1, Image 2, (c) Merged via averaging registered images, (d) Merged via subtraction of registered images.

greatly reduces the computation involved in hypothesis evaluation in comparison to those systems that perform non-feature-based registration. Next, features not part of the hypothesis are compared. Two features match if they are close in value and location. Hypotheses are ranked by the number of features matched and how closely the features match. Examples are given in Figure 5.6.

5.6 Registration Assistance/Preprocessing

All of the registration techniques discussed herein operate on the basic premise that there is identical content in the data sets being compared. However, the difficulty in registration pertains to the fact that

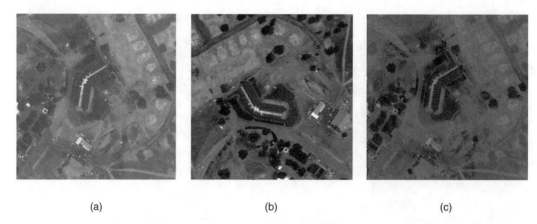

      (a)   (b)   (c)

FIGURE 5.7 (a) Image 1, (b) Image 2, (c) New Image 1 corrected to appear more like Image 2 in photometric content.

the content is the same semantically, but often not numerically. For example, sensor readings taken at different times of the day can lead to lighting changes that can significantly alter the underlying data values. Also, weather changes can lead to significant changes in data sets. Registration of these kinds of data sets can be improved by first preprocessing the data. Figure 5.7 shows some preliminary work by Grewe[40] on the process of altering one image to appear more like another image in terms of photometric values. Such systems may improve registration systems of the future.

5.7 Conclusion

Addressing the data registration problem is an essential preprocessing step in multisensor fusion. Data from multiple sensors must be transformed onto a common coordinate system. This chapter provided a survey of existing methods, including methods for finding registrations and applying registrations to data after they have been found. In addition, example approaches were described in detail.

Brooks[2] and Chen[35] detail meta-heuristic-based optimization methods that can be applied to raw data. Of these methods, TRUST, a new meta-heuristic from Oak Ridge National Laboratories, is the most promising. Fitness functions have been given for readings corrupted with Gaussian, uniform, and salt-and-pepper noise. Because these methods use raw data, they are computationally intensive.

Grewe[39] presents a wavelet-based approach to registering range data. Features are extracted from the wavelet domain. A feedback approach is then applied to search for good registrations. Use of the wavelet domain compresses the amount of data that must be considered, providing for increased computational efficiency. Drawbacks to using feature-based methods have also been discussed in the chapter.

Acknowledgments

Efforts were sponsored by the Defense Advance Research Projects Agency (DARPA) and Air Force Research Laboratory, Air Force Materiel Command, USAF, under agreement number F30602-99-2-0520 (Reactive Sensor Network). The U.S. Government is authorized to reproduce and distribute reprints for Governmental purposes notwithstanding any copyright annotation thereon. The views and conclusions contained herein are those of the authors and should not be interpreted as necessarily representing the official policies or endorsements, either expressed or implied, of the Defense Advanced Research Projects Agency (DARPA), the Air Force Research Laboratory, or the U.S. Government.

References

1. D. C. Montgomery, *Design and Analysis of Experiments, 4th Edition*, John Wiley & Sons, NY, 1997.
2. R. R. Brooks and S. S. Iyengar, *Multi-Sensor Fusion: Fundamentals and Applications with Software*, Prentice Hall, PTR, Upper Saddle River, NJ, 1998.
3. B. V. Dasarathy, *Decision Fusion*, IEEE Computer Society Press, Los Alamitos, CA, 1994.
4. D. L. Hall, *Mathematical Techniques in Multi-sensor Fusion*, Artech House, Norwood, MA, 1992.
5. R. C. Gonzalez and R. E. Woods, *Digital Image Processing*, Addison-Wesley, Reading, PA, 302, 1993.
6. L. G. Brown, A Survey of Image Registration Techniques, *ACM Computing Surveys*, 24(4), 325–376, 1992.
7. M. A. Oghabian and A. Todd-Pokropek, Registration of brain images bay a multiresolution sequential method, *Information Processing in Medical Imaging*, 165–174, Springer, Berlin, 1991.
8. V. R. Mandara and J. M. Fitzpatrick, Adaptive search space scaling in digital image registration, *IEEE Transactions on Medical Imaging*, 8(3), 251–262, 1989.
9. C. A. Palazzini, K. K. Tan, and D. N. Levin, Interactive 3-D patient-image registration, *Information Processing in Medical Imaging*, 132–141, Springer, Berlin, 1991.
10. G. Stockman, S. Kopstein, and S. Benett, Matching images to models for registration and object detection via clustering, *IEEE Transactions on Pattern Analysis and Machine Intelligence*, 4(3), 229–241, 1982.
11. P. Van Wie and M. Stein, A LANDSAT digital image rectification system, *IEEE Transactions on Geosci. Electron.*, vol. GE-15, July 1977.
12. J. R. Jain and A. K. Jain, Displacement measurement and its application in interface image coding, *IEEE Transactions on Communications*, vol. COM-29 No. 12, 1799–1808, 1981.
13. A. Goshtasby, Piecewise linear mapping functions for image registration, *Pattern Recognition*, 19(6), 459–466, 1986.
14. G. Wolberg, *Digital Image Warping*, IEEE Computer Society Press, Los Alamos, 1964.
15. R. Y. Wong and E. L. Hall, Performance Comparison of Scene Matching Techniques, *IEEE Transaction on Pattern Analysis and Machine Intelligence*, vol. PAMI-1 No. 3, 325–330, 1979.
16. A. Mitiche and K. Aggarwal, Contour registration by shape specific points for shape matching comparison, *Vision, Graphics and Image Processing*, vol. 22, 396–408, 1983.
17. J. Andrus and C. Campbell, Digital Image Registration Using Boundary Maps, *IEEE Trans. Comput.*, vol. 19, 935–940, 1975.
18. D. Barnea and H. Silverman, A Class of Algorithms for Fast Digital Image Registration, *IEEE Trans. Comput.*, vol. C-21, No. 2, 179–186, 1972.
19. A. Goshtasby, Image Registration by Local Approximation Methods, *Image and Vision Computing*, 6(4), 255–261, 1988.
20. B. R. Horn and B. L. Bachman, Using Synthetic Images with Surface Models, *Commun. ACM*, vol. 21, 914–924, 1977.
21. H. G. Barrow, J. M. Tennenbaum, R. C. Bolles, and H. C. Wolf, Parametric Correspondence and Chamfer Matching: Two New Techniques for Image Matching, *Proc. Int. Joint Conf. Artificial Intelligence*, 659–663, 1977.
22. A. Pinz, M. Prontl, and H. Ganster, Affine Matching of Intermediate Symbolic Presentations, *CAIP '95 Proc. LNCS* 970, Hlavac and Sara (Eds.), Springer-Verlag, 359–367, 1995.
23. X. Dai and S. Khorram, A Feature-Based Image Registration Algorithm Using Improved Chain-Code Representation Combined with Invariant Moments, *IEEE Trans. GeoScience and Remote Sensing*, 37(5), 2351–2362, 1999.
24. Z. Yang and F. S. Cohen, Cross-weighted Moments and Affine Invariants for Image Registration and Matching, *IEEE Trans. Pattern Analysis and Machine Intelligence*, 21(8), 804–814, 1999.
25. R. Chellappa, Q. Zheng, P. Burlina, C. Shekhar, and K. B. Eom, On the Positioning of Multisensory Imagery for Exploitation and Target Recognition, *Proc. IEEE.*, 85(1), 120–138, 1997.

26. Y. Zhou, H. Leung, and E. Bosse, Registration of mobile sensors using the parallelized extended Kalman filter, *Optical Engineering*, 36(3), 780–788, 1997.

27. G. K. Matsopoulos, N. A. Mouravliansky, K. K. Delibasis, and K. S. Nikita, Automatic Retinal Image Registration Scheme Using Global Optimization Techniques, *IEEE Trans. Inform. Technol. Biomed.*, 3(1), 47–68, 1999.

28. H. S. Stone, J. Lemoigne, and M. McGuire, The Translation Sensitivity of Wavelet-Based Registration, *IEEE Trans. Pattern Analysis and Machine Intelligence*, 21(10), 1074–1081, 1999.

29. A. Goshtasby, Piecewise cubic mapping functions for image registration, *Pattern Recognition*, 20(5), 525–535, 1987.

30. H. S. Baird, *Model-based Image Matching Using Location*, MIT Press, Boston, MA, 1985.

31. L. S. Davis, Shape Matching Using Relaxation Techniques, *IEEE Trans. Pattern Analysis and Machine Intelligence*, 1(1), 60–72, 1979.

32. R. R. Brooks and S. S. Iyengar, Self-Calibration of A Noisy Multiple Sensor System with Genetic Algorithms, Self-Calibrated Intelligent Optical Sensors and Systems, SPIE, Bellingham, WA, *Proc. SPIE Internat. Symp. Intelligent Systems and Advanced Manufacturing*, 1995.

33. R. R. Brooks, S. S. Iyengar, and J. Chen, Automatic Correlation and Calibration of Noisy Sensor Readings using Elite Genetic Algorithms, *Artificial Intelligence*, 1996.

34. R. R. Brooks, *Robust Sensor Fusion Algorithms: Calibration and Cost Minimization*, Ph.D. Dissertation in Computer Science, Louisiana State University, Baton Rouge, LA, 1996.

35. Y. Chen, R. R. Brooks, S. S. Iengar, N. S. V. Rao, and J. Barhen, Efficient Global Optimization for Image Registration, *IEEE Trans. Knowledge and Data Engineering*, in press.

36. F. S. Hill, *Computer Graphics*, Prentice Hall, Englewood Cliffs, NJ, 1990.

37. R. Sharman, J. M. Tyler, and O. S. Pianykh, Wavelet-Based Registration and Compression of Sets of Images, *SPIE Proc.*, vol. 3078, 497–505, 1997.

38. R. A. DeVore, W. Shao, J. F. Pierce, E. Kaymaz, B. T. Lerner, and W. J. Campbell, Using nonlinear wavelet compression to enhance image registration, *SPIE Proc.*, vol. 3078, 539–551.

39. L. Grewe and R. R. Brooks, Efficient Registration in the Compressed Domain, *Wavelet Applications VI, SPIE Proc.*, H. Szu, Ed., vol. 3723, Aerosense 1999.

40. L. Grewe, Image Correction, Technical Report, CSUMB, 2000.

6

Data Fusion Automation: A Top-Down Perspective

6.1	Introduction	6-1
	Biological Fusion Metaphor • Command and Control Metaphor • Puzzle-Solving Metaphor • Evidence Combination • Information Requirements • Problem Dimensionality • Commensurate and Noncommensurate Data	
6.2	Biologically Motivated Fusion Process Model	6-8
6.3	Fusion Process Model Extensions	6-14
	Short-, Medium-, and Long-Term Knowledge • Fusion Classes • Fusion Classes and Canonical Problem-Solving Forms	
6.4	Observations	6-22
	Observation 1 • Observation 2 • Observation 3 • Observation 4 • Observation 5	

Richard Antony
VGS Inc.

Acknowledgment	6-25
References	6-25

6.1 Introduction

This chapter offers a conceptual-level view of the data fusion process and discusses key principles associated with both data analysis and information combination. The discussion begins with a high-level view of data fusion requirements and analysis options. Although the discussion focuses on tactical situation awareness development, a much wider range of applications exists for this technology.

After motivating the concepts behind effective information combination and decision making through a series of easily understood metaphors, the chapter

- Presents a top-down view of the data fusion process,
- Discusses the inherent complexities of combining uncertain, erroneous, and fragmentary information,
- Offers a taxonomic approach for distinguishing classes of fusion algorithms, and
- Identifies key algorithm requirements for practical and effective machine-based reasoning.

6.1.1 Biological Fusion Metaphor

Multiple sensory fusion in biological systems provides a natural metaphor for studying artificial data fusion systems. As with any good metaphor, consideration of a simpler or more familiar phenomenon can provide valuable insight into the study of a more complex or less familiar process.

Even the most primitive animals sense their environment, develop some level of situation awareness, and react to the acquired information. Situation awareness directly supports survival of the species by assisting in the acquisition of food and the avoidance of animals of prey. A barn owl, for instance, fuses visual and auditory information to help accurately locate mice under very low light conditions, while a mouse responds to threatening visual and auditory cues to attempt to avoid being caught by an owl.

In general, natural selection has tended to favor the development of more capable senses (sensors) and more effective utilization of the derived information (exploitation and fusion). Color vision in humans, for instance, is believed to have been a natural adaptation that permitted apes to more easily locate ripe fruit among vegetation. Situation awareness in animals can rely on a single, highly developed sense, or on multiple, often less capable senses. A hawk depends principally on a highly acute visual search and tracking capability, while a shark primarily relies on its sense of smell when hunting. Sexual attraction can depend primarily on sight (plumage), smell (pheromones), or sound (mating call). For humans, sight is arguably the most vital sense, with hearing a close second. Dogs, on the other hand, rely most heavily on the senses of smell and hearing, with vision typically acting as a secondary information source.

Sensory input in biological organisms typically supports both sensory cueing and situation awareness development. Sounds cue the visual sense to the presence and the general direction of an important event. Information gained by the aural sense (i.e., direction, speed, and tentative object classification) is then combined (fused) with the information gathered by the visual system to produce more complete, higher confidence, or higher level situation awareness. In many cases, multiple sensory fusion can be critical to successful decision making. Food that looks appetizing (sight) might be extremely salty (taste), spoiled (smell), or too hot (touch). At the other extreme, fusion of multiple sensory input might be unnecessary if the various senses provide highly redundant information. Bacon frying in a pan need not be seen, smelled, and tasted to be positively identified; each sense, taken separately, could perform such a function.

Although discarding apparently redundant information may seem to be prudent, such information can aid in sorting out conflicts, both intentional (deception) and unintentional (confusion). While single-source deception is reasonably straightforward to perpetrate, deception across multiple senses (sensor modalities) is considerably more difficult. For example, successful hunting and fishing depend, to a large degree, on effective multisource deception. Duck hunters use both visual decoys and mating calls to simultaneously provide deceptive visual and auditory information. Because deer can sense danger through the sense of smell, sound, and sight, the shrewd hunter must mask his scent (or stay down-wind), make little or no noise, and remain motionless if the deer looks in his direction. Even in nonadversarial applications, data fusion requires resolution of unintentional conflicts among supporting data sources in order to deal effectively with the inherent uncertainty in both the measurement and decision spaces.

Multiple sensory fusion need not be restricted to the familiar five senses of sight, sound, smell, taste, and touch. Internal signals, such as acidity of the stomach, coupled with visual and/or olfactory cues, can trigger hunger pains. The fusion of vision, inner-ear balance information, and muscle feedback signals facilitate motor control. In a similar manner, measurement and signature intelligence (MASINT) in a tactical application focuses on the collection and analysis of a wide range of nontraditional information classes.

6.1.2 Command and Control Metaphor

The game of chess provides a literal metaphor for military command and control (C²), as well as an abstract metaphor for any system that senses and reacts to its environment. Both chess players and battlefield commanders require a clear picture of the "playing field" to properly evaluate the options available to them and their opponents. In both chess and C², opposing players command numerous individual resources (i.e., pieces or units) that possess a range of characteristics and capabilities. Resources and strategies vary over time. Groups of chess pieces are analogous to higher-level organizations on the battlefield. The chessboard represents domain constraints to movement that are similar to constraints posed by terrain, weather, logistics, and other features of the military problem domain. Player-specific strategies are analogous to tactics, while legal moves represent established doctrine. In both domains, the overall objective of an opponent may be known, while specific tactics and subgoals must be deduced.

Despite a chess player's complete knowledge of the chess board (*all domain constraints*), the location of all pieces (*own and opponent-force locations*), and all legal moves (*own and opponent-force doctrine*), and his ability to exercise direct control over all of his own assets, chess remains a highly challenging game. Metaphorically similar to chess, tactical situation development has numerous domain characteristics that make it an even more challenging problem.

First, battlefield commanders normally possess neither a complete nor fully accurate picture of their own forces or those of their adversaries. Forced to deal with incomplete and inaccurate force structure knowledge, as well as location uncertainty, chess players would be reduced to guessing the location and composition of an adversary's pieces, somewhat akin to playing "Battleship," the popular children's game.

Second, individual sensors provide only limited observables, coverage, resolution, and accuracy. Thus, the analysis of individual sensor reports tend to lead to ambiguous and rather local interpretations. Third, domain constraints in tactical situation awareness are considerably more complex than the well-structured (and level) playing field in chess. Fourth, doctrinal knowledge in the tactical domain tends to be more difficult to exploit effectively and far less reliable that its counterpart in chess.

A wide range of other application-motivated metaphors can also be useful for studying specific fusion applications. Data fusion, for example, seems destined to play a significant role in the development of future "smart highway" control systems where a simple car driving metaphor can be applied to study sensor requirements and fusion opportunities. The underpinning of such a system is a sophisticated control capability that optimally resolves a range of conflicting requirements, such as (1) expedite the movement of both local and long distance traffic, (2) ensure maximum safety for all vehicles, and (3) create the minimum environmental impact. The actors in the metaphor are drivers (or automated vehicle control systems), the rules of the game are the "rules of the road," and domain constraints are the road network and traffic control means. Individual players possess individualized objectives and tactics; road characteristics and vehicle performance capabilities provide physical constraints on the problem solution.

6.1.3 Puzzle-Solving Metaphor

Situation awareness development requires the production and maintenance of an adequate multiple level-of-abstraction picture of a (dynamic) situation; therefore, the data fusion process can be compared to assembling a complex jigsaw puzzle for which no picture of the completed scene exists. While assembling puzzles that contain hundreds of pieces (*information fragments*) can challenge an individual's skill and patience, the production of a comprehensive situational picture, created by fusing disparate and fragmentary sensor-derived information, represents an even more challenging task. Although a completed jigsaw puzzle represents a fixed scene, the process of collecting and integrating the numerous information fragments clearly evolves over time. Time, on the other hand, represents a key dimension in highly dynamic tactical situation awareness applications.

The partially completed puzzle (*fused situation awareness product*) illustrated in Figure 6.1 contains numerous aggregate objects (i.e., forest and meadow), each composed of simpler objects (i.e., trees and ground cover). Each of these objects, in turn, have been assembled from multiple puzzle pieces, some representing a section of bark on a single tree trunk, others a grassy area associated with a meadow. In terms of the metaphor then, sensor-derived information can be associated with individual puzzle pieces, providing little more information than color and texture, as well as pieces that depict higher level of abstraction objects.

At the beginning of the reconstruction process, problem solving necessarily relies on general analysis strategies (e.g., locate border pieces). Because little context exists to direct either puzzle piece selection or puzzle piece placement, at the early stages of the process, rather simple, brute-force pattern matching strategies are needed. A predominately blue-colored piece, for example, might represent either sky or water with little basis for distinguishing between the two interpretations. Unless they came from an unopened box, there may be no assurance that the scattered pieces on the table all belong in the puzzle

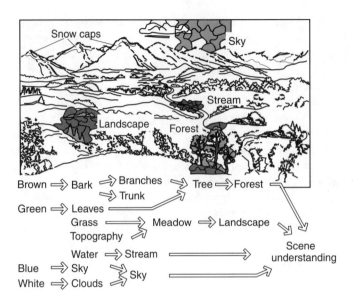

FIGURE 6.1 Puzzle-solving metaphor example.

under construction. However, once certain sections of the puzzle have been filled in, the assembly process (*fusion*) tends to become much more goal-directed.

Fitting a single puzzle piece supports both scene entropy reduction as well as higher level-of-abstraction scene interpretation. As regions of the puzzle begin to take form, identifiable features in the scene emerge (e.g., trees, grass, and cliffs) and higher-level interpretations can be developed (e.g., forest, meadows, and mountains). By supporting the placement of the individual pieces, as well as the goal-driven search (*sensor resource management*) for specific pieces, the *context* provided by the developing multiple level-of-abstraction picture of the scene (*situation awareness product*) helps further focus the reconstruction process (*fusion process optimization*).

Just as duplicate or erroneous pieces can significantly complicate puzzle assembly, redundant and irrelevant sensor-derived information similarly burdens machine-based situation development. Therefore, goal-directed information collection offers a two-fold benefit: critical information requirements are satisfied and the collection (and subsequent analysis) of unnecessary information is minimized. Although numerous puzzle pieces may be yet unplaced (*undetected objects*) and perhaps some pieces are actually missing (*information not collectible by the available sensor suite*), a reasonably comprehensive, multiple level-of-abstraction understanding of the overall scene (*situation awareness*) gradually emerges.

Three broad classes of knowledge are apparent in the puzzle reconstruction metaphor:

- Individual puzzle pieces — collected information fragments, i.e., *sensor-derived knowledge,*
- Puzzle-solving strategies, such as edge detection and pattern matching — a priori *reasoning knowledge*
- World knowledge, such as the relationship between meadows and grass — *domain context knowledge.*

To investigate the critical role that each knowledge form plays in fusion product development, recast the analysis in terms of a building construction metaphor. Puzzle pieces (*sensor input*) are clearly the building blocks required to assemble the scene (*fused situation awareness product*). A priori reasoning knowledge represents construction knowledge and skills, and context provides the nails and mortar that "glue" the sensor input together to form a coherent whole. When too many puzzle pieces (or building blocks) are missing (*inadequate sensor-derived information*), scene reconstruction (or building construction) becomes difficult or impossible.

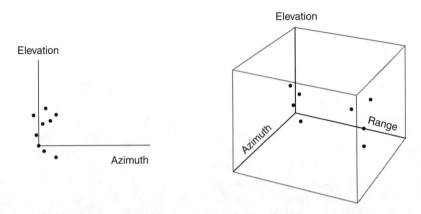

FIGURE 6.2 (a) Two-dimensional measurements and (b) the corresponding three-dimensional measurement space.

A simple example demonstrates how both the complexity of the fusion process and the quality of the resultant product are sensitive to the availability of adequate information. Figure 6.2(a) illustrates a cluster of azimuth and elevation measurements associated with two separate groups of air targets. Given the spatial overlap between the data sets, reliable target-to-group assignment may not be possible, regardless of the selected analysis paradigm or the extent of algorithm training. However, with the addition of range measurements (*increased measurement space dimensionality*), two easily separable clusters become readily apparent (Figure 6.2(b)). Because the information content of the original 2-D data set was fundamentally inadequate, even sophisticated clustering algorithms would be unable to discriminate between the two target groups. However, with the addition of the third measurement dimension, a simple clustering algorithm easily handles the decision task.

Reasoning knowledge can be implemented using a spectrum of problem solving paradigms (e.g., rules, procedures, and statistical-based algorithms), evidence combination strategies (e.g., Bayes, Dempster-Shafer, and fuzzy set theory), and decision-making approaches (e.g., rule instantiation and parametric algorithms). In general, the process of solving a complex puzzle (or performing automated situation awareness) benefits from **both** bottom-up (deductive-based) and top-down (goal-directed) reasoning that exploits relationships among the hierarchy of domain entities (i.e., primitive, composite, aggregate, and organizational).

In the puzzle-solving metaphor, *context knowledge* refers to relevant domain knowledge **not** explicitly contained within a puzzle piece (*non-sensor-derived knowledge*). Humans routinely apply a wide range of contextual knowledge during analysis and decision making.* For example, context-sensitive evaluation of Figure 6.1 permits the determination that the picture is a summer scene in the western U.S. The season and location are deduced from the presence of deciduous trees in full leaf (summer) in the foreground and jagged snow-capped mountain peaks in the distance (western U.S.). In a similar fashion, the exploitation of context knowledge in automated fusion systems can promote much more effective and comprehensive interpretations of sensor-derived information.

In both puzzle assembly and automated situation development, determining when an *adequate* situation representation has been achieved can be difficult. In the puzzle reconstruction problem, although the general landscape characteristics might be evident, missing puzzle pieces could depict denizens of the woodland community that can be hypothesized, but for which no compelling evidence yet exists. On the other hand, individual puzzle pieces might contain partial or ambiguous information. For

---

* This fact partially accounts for the disparity in performance between manual and automated approaches to data fusion.

example, the presence of a section of log wall in the evolving scene suggests the possibility of a log cabin. However, additional evidence is required to validate such a hypothesis.

6.1.4   Evidence Combination

Reliance on a single information source can lead to ambiguous, uncertain, and inaccurate situation awareness. Data fusion seeks to overcome such limitations by synergistically combining all relevant (and available) information sources leading to the generation of consistent, accurate, comprehensive, and global situation awareness. A famous poem by John Godfrey Saxe,* written more than a century ago, aptly demonstrates both the need for and challenge of effectively combining fragmentary information.

The poem describes an attempt by six blind men to gain a first-hand understanding of an elephant. The first man happens to approach the elephant from the side and surmises that an elephant must be something like a wall. The second man touches the tusk and imagines an elephant to be like a spear. The third man approaches the trunk and decides an elephant is similar to a snake. The fourth man reaches out and touches a leg and determines an elephant to be much like a tree. The fifth man chances to touch an ear and imagines an elephant must be like a fan. The sixth man grabs the tail and concludes an elephant is similar to a rope. While each man's assessment is entirely consistent within his own limited sensory space and myopic frame of reference, unless the six observations are effectively integrated (*fused*), a true picture of an elephant fails to emerge.

Among other insights, the puzzle-solving metaphor illustrated that (1) complex dependencies can exist among and between information fragments and the completed situation description, and (2) determining whether an individual puzzle piece actually belongs to the scene being assembled can be difficult. Even when the collected information is known to be relevant, based strictly on local interpretations, determining whether a given blue-colored piece represents sky, water, or some other feature class may not be possible. Much like assembling observations, hunting for clues, and evaluating motives required during criminal investigations, a similar approach to information combination is required by general situation awareness systems. Just as at the outset of a criminal investigation, a single strand of hair might appear insignificant, but it could later prove to be the key piece of evidence that discriminates among several suspects. Similarly, a seemingly irrelevant piece of sensor-derived information might ultimately link observations with motives, or provide other significant situational awareness benefits. Thus, not only is the information content (*information measure*) associated with a given piece of data important; its relationship to the overall fusion task is also vital to achieving successful information fusion. As a direct consequence of this observation, the development of a comprehensive information theoretical framework for the data fusion process appears to be problematic. Only through a top-down, holistic treatment of the analysis task can the content of a single information fragment be properly assessed and its true value to the overall fusion process be fully realized.

6.1.5   Information Requirements

Because no widely accepted formal theory exists for determining when adequate information has been assembled to support a given fusion task, empirical measures of performance generally must be relied upon to evaluate the effectiveness of both individual fusion algorithms and an overall fusion system. In general, data fusion performance can be enhanced by

- Technical improvements in sensor measurements (i.e., longer range, higher resolution, improved signal-to-noise ratio, better accuracy, higher reliability);
- Increased measurement space dimensionality afforded by heterogeneous sensors that provide at least partially independent information;

---

* Saxe, J. G., "The Blind Man and the Elephant," *The Poetical Works of John Godfrey Saxe*, Boston, MA: Houghton, Mifflin and Company, 1882.

- Spatially distributed sensors providing improved coverage, perspective, and measurement reliability;
- Relevant non-sensor-derived domain knowledge to constrain the information combination and decision-making process.

In general, effective data fusion automation requires the development of robust, context-sensitive algorithms that are practical to implement. The first two requirements reflect the "quality of performance" of the algorithm, while the latter reflects cost/benefit tradeoffs associated with meeting a wide range of implicit and explicit performance objectives. In general, *robust* performance argues for the use of all potentially relevant sensor-derived information sources and reasoning knowledge. Achieving *context-sensitive* performance argues for maximal utilization of relevant non-sensor-derived information. On the other hand, to be *practical* to implement and efficient enough to employ in an operational setting, the algorithms may need to compromise some fusion performance quality. Consequently, system developers must quantify or otherwise assess the value of these various information sources in light of system requirements, moderated by programmatic, budgetary, and performance constraints (e.g., decision time-line and hardware capability). The interplay between achieving optimal algorithm robustness and context-sensitivity, on the one hand, and a practical implementation, on the other, is a fundamental tension associated with virtually any form of machine-based reasoning directed at solving complex, real-world problems.

6.1.6 Problem Dimensionality

Effective situational awareness, with or without intentional deception, generally benefits from the collection and analysis of a wide range of observables. As a result of the dynamic nature of many problem domains, observables can change with time and, in some cases, may require continuous monitoring. In a tactical application, objects of interest can be stationary (fixed or currently nonmoving), quasistationary (highly localized motion), or moving. Individual objects possess characteristics that constrain their behavior. Objects emit different forms of electromagnetic energy that vary with time and can indicate the state of the object. Object emissions include *intentional* or active emissions, such as radar, communications, and data link signals, as well as *unintentional* or passive emissions, such as acoustic, magnetic, or thermal signatures generated by internal heat sources or environmental loading. Patterns of physical objects and their behavior provide indications of organization, tactics, and intent. Patterns of emissions, both active and passive, can reveal the same. For example, a sequence of signals emitted from a surface-to-air missile radar over time representing search, lock-on, launch, and hand-over clearly indicates hostile intent.

A single sensor modality is incapable of measuring all relevant information dimensions; therefore, multiple sensor classes often must be relied upon to detect, track, classify, and infer the likely intent of a host of objects, from submarines and surface vessels, to land, air, and space-based objects. Certain sensor classes lend themselves to surveillance applications, providing both wide-area and long-range coverage, and readily automated target detection capability. Examples of such sensor classes include signals intelligence (SIGINT) for collecting active emissions, moving target indication (MTI) radar for detecting and tracking moving targets against a high clutter background, and synthetic aperture radar (SAR) for detecting stationary targets. Appropriately cued, other sensor classes that possess narrower fields of view and that typically operate at much shorter ranges may be capable of providing higher fidelity measurement to support refined analysis. Geospatial and other intelligence databases can provide the static domain context within which the target-sensed data must be interpreted, while environmental sensors generate dynamic context estimates, such as weather and current atmospheric conditions.

6.1.7 Commensurate and Noncommensurate Data

Although the fusion of *similar* (commensurate) information would seem to be more straightforward than the fusion of *dissimilar* (noncommensurate) information, that is not always the case. Three examples are offered to highlight the varying degrees of difficulty associated with the combination of multiple-source data. First, consider the relative simplicity of fusing registered electronic intelligence (ELINT) data and real-time synthetic aperture radar (SAR) imagery. Although these sensors measure dramatically different information dimensions, both sources provide reasonably wide area coverage, relatively good geolocation, and highly complementary information. As a consequence, the fusion process tends to be straightforward. Even when an ELINT sensor provides little more than target line-of-bearing, the ELINT and SAR measurements can potentially be combined by simply overlaying the two data sets. If the line-of-bearing intercepts a single piece of equipment in the SAR image, the radar system class, as well as its precise location, would be known. This information, in turn, can support the identification of other nearby objects in the image (e.g., missile launchers normally associated with track-while-scan radar).

At the other end of the spectrum, the fusion of information from two or more identical sensors can present a significant challenge. Consider, for example, fusing data sets obtained from spatially separated forward-looking infrared (FLIR) radars. Although FLIR imagery provides good azimuth and elevation resolution, it does not directly measure range. Because the range and view angles to targets will be different for multiple sensors, combining such data sets demands sophisticated registration and normalization.

Finally, consider the fusion of two bore-sited sensors: light-intensified and forward-looking infrared (FLIR). The former device amplifies low intensity optical images to enhance night vision. When coupled with the human's natural ability to separate moving objects from the relatively stationary background, such devices permit visualization of the environment and detection of both stationary and moving objects. However, such devices offer limited capability for the detection of stationary personnel and equipment located in deep shadows or under extremely low ambient light levels (e.g., heavy cloud cover, no moon, or inside buildings). Rather than detecting reflected energy, FLIR devices detect thermal radiation from objects. Consequently, these devices support the detection of humans, vehicles, and operating equipment based on their higher temperature relative to the background. Consequently, with bore-sighted sensors, pixel-by-pixel combination of the two separate images may be feasible, providing a highly effective night vision capability.

6.2 Biologically Motivated Fusion Process Model

A hierarchically organized *functional-level* model of data fusion is presented in Chapter 2. In contrast, this section focuses on a *process-level* model. While the functional model describes **what** analysis functions or processes need to be performed, a process-level model describes at a high level of abstraction **how** this analysis is accomplished.

The goal of data fusion, as well as most other forms of data processing, is to turn data into useful information. In perhaps the simplest possible view, all of the required information is assumed to be present within a set of sensor measurements. Thus, the role of data fusion is extraction of information embedded in a data set (separating the wheat from the chaff). In this case, fusion algorithms can be characterized as a function of

- Observables
- Current situation description (e.g., target track files and current situation description)
- *A priori* declarative knowledge (e.g., distribution functions, templates, constraint sets, filters, and decision threshold values).

As shown in Figure 6.3(a), the fusion process output provides updates to the *situation description*, as well as feedback to the *reasoning knowledge base* to support knowledge refinement (learning).

Signal processing, statistical hypothesis testing, target localization performed by intersecting two independently derived error ellipses, and target identification based on correlation of an image with a set of rigid templates are simple examples of such a fusion model. In general, this "information extraction" view of data fusion makes a number of unstated, simplifying assumptions including the existence of

- Adequate information content in the sensor observables
- Adequate sensor update rates
- Homogeneous sensor data

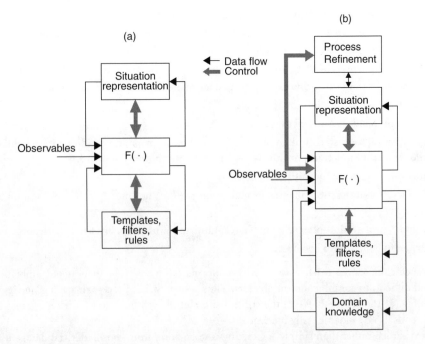

FIGURE 6.3 (a) Basic fusion process model and (b) generalized process model.

- Relatively small number of readily distinguishable targets
- Relatively high resolution sensors
- High reliability sensors
- Full sensor coverage of the area of interest
- Stationary, Gaussian random interference.

When such assumptions are appropriate, data analysis tends to be straightforward and an "information extraction" fusion model is adequate. Rigid template-match paradigms typically perform well when a set of observables closely matches a single template and are uncorrelated with the balance of the templates. Track association algorithms perform well against a small number of moving, widely spaced targets provided the radar generates relatively high update rates. The combination of similar features is often more straightforward than the combination of disparate features. When the sensor data possesses adequate information content, high confidence analysis is possible. High signal-to-noise ratios tend to enhance signal detection. High resolution sensors reduce ambiguity and uncertainty with respect to feature measurements (e.g., location and frequency). High reliability sensors maximize sensor availability. Adequate sensor coverage provides a "complete" view of the areas of interest. Statistical-based reasoning is generally simplified when signal interference can be modeled as a Gaussian random process.

Typical applications where such assumptions are realistic, include

- Track assignment in low target-density environments or for ballistic targets that obey well-established physical laws of motion
- Classification of military organizations based on associated radio types
- Detection of signals and targets exhibiting high signal-to-background ratio.

However, numerous real-world data fusion tasks exhibit one or more of the following complexities:

- Large number of target and nontarget entities (e.g., garbage trucks may be nearly indistinguishable from armored personnel carriers);
- Within-class variability of individual targets (e.g., hatch open vs. hatch closed);

- Low data rates (exacerbating track association problems);
- Multiple sensor classes (disparate numeric and symbolic observables can be difficult to combine);
- Inadequate sensor coverage of areas of interest (i.e., inadequate number of sensors, obscuration due to terrain and foliage, radio frequency interference, weather, or counter measures);
- Inadequate set of sensor observables (e.g., inadequate input space dimensionality);
- Inadequate sensor resolution;
- Registration and measurement errors;
- Inadequate *a priori* statistical knowledge (e.g., unknown prior and conditional probabilities, multimodal density functions, or non-Gaussian and nonstationary statistics);
- Processing and communication latencies;
- High level-of-abstraction analysis product required (i.e., not merely platform location and identification);
- Complex propagation phenomenon (i.e., multipath, diffraction, or atmospheric attenuation);
- Purposefully deceptive behavior.

When such complexities exist, sensor-derived information tends to be incomplete, ambiguous, erroneous, and difficult to combine and/or abstract. Thus, a data fusion process that relies on rigid composition among (1) the observables, (2) the current situation description, and (3) a set of rigid templates or filters, tends to be fundamentally inadequate.

As stated earlier, rather than simply "extracting" information from sensor-derived data, effective data fusion requires the combination, consolidation, organization, and abstraction of information. Such analysis can enhance the fusion product, its confidence, and its ultimate utility in at least four ways:

1. Existing sensors can be improved to provide better resolution, accuracy, sensitivity, and reliability.
2. Additional similar sensors can be employed to improve the coverage and/or confidence in the domain observables.
3. Dissimilar sensors can be used to increase the dimensionality of the observation space, permitting the measurement of at least partially independent target attributes (a radar can offer excellent range and azimuth resolution, while an ELINT sensor can provide target identification).
4. Additional domain knowledge and context constraints can be utilized.

While the first three recommendations effectively *increase* the information content and/or dimensionality of the observables, the latter effectively *reduces* the decision space dimensionality by constraining the possible decision states.

Observables can be treated as *explicit* knowledge (i.e., knowledge that is explicitly provided by the sensors). Context knowledge, on the other hand, represents *implicit* (or non-sensor-derived) knowledge. Although human analysts routinely use both forms in performing fusion tasks, automated approaches have traditionally relied almost exclusively on the former.

As an example of the utility of implicit domain knowledge, consider the extrapolation of the track of a ground-based vehicle that has been observed moving along the relatively straight-line path shown in Figure 6.4. Although the target is a wheeled vehicle traveling along a road with a hairpin curve just beyond the last detection point, a purely statistical-based tracker will likely attempt to extend the track through the hill (the reason for the curve in the road) and into the lake on the other side.

Although tracking aircraft, ballistic projectiles, and naval vessels using statistical-based motion models has been highly successful, adapting such algorithms to tracking ground vehicles has proved to be a considerable challenge. Tracked and wheeled vehicles typically exhibit many more degrees of freedom than a high performance aircraft or naval vessel because they can stop and move in an unpredictable manner. Additional complications include the potentially large numbers of ground vehicles, nonresolvable individual vehicles, terrain and vegetation masking, and infrequent target update rates. However, through the application of relevant domain constraints (e.g., mobility, observability, vehicle class behavior, and vehicle group behavior), the expectation-based analysis process can be effectively constrained,

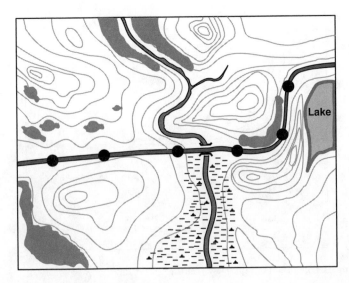

FIGURE 6.4 Road-following target tracking model.

thus helping to manage the additional degrees of freedom. In much the same way that a system of equations with too many unknowns does not produce a unique solution, "missing" domain knowledge can lead to an "underdamped" Kalman filter solution to ground target tracking. In recognition of the benefits of context-sensitive analysis, domain-sensitive ground target tracking models have received considerable interest in recent years.

In addition to the importance of reasoning in context, the road-following target tracking problem also dramatically illustrates the critical role of *paradigm selection* in the algorithm development process. Rather than demonstrating the failure of a statistical-based tracker, the above example illustrates its misapplication. Applying a purely statistical approach to this problem assumes (perhaps unwittingly) that domain constraints are either irrelevant or insignificant. However, in this application, domain constraints tend to be stronger than the relatively weak constraints on platform motion provided by a strictly statistical-based motion model.

Paradigm selection, in fact, must be viewed as a key component of successful data fusion automation. Consequently, algorithm developers must ensure that both the capability and limitations of a selected problem-solving paradigm are appropriately matched to the requirements of the fusion task they are attempting to automate.

To illustrate the importance of both context-sensitive reasoning and paradigm selection, consider the problem of analyzing the time-stamped radar detections from multiple closely spaced targets, some with potentially crossing trajectories, as illustrated in Figure 6.5. A traditional statistical tracking algorithm typically associates the "closest" (with respect to a specified evaluation metric) new detection to an existing track. A human analyst, on the other hand, would quite naturally invoke a context-sensitive model of vehicle behavior. By employing multiple behavior models, alternative interpretations of the observations can be made. False hypotheses can be eliminated once adequate information is obtained to resolve the associated ambiguity.

Emulating such an analysis strategy requires the time-stamped detections to be associated with local cultural and topographic features. In addition, the analysis model(s) must accommodate individual vehicle-class capabilities, as well as *a priori* class-specific behavioral knowledge. By doing so, it can be inferred that tracks 1–3 would be highly consistent with a road-following behavior, tracks 4 and 5 would be determined to be most consistent with a minimum terrain-gradient following behavior, while track 6 would be found to be inconsistent with any ground-based vehicle behavior model. By evaluating track updates from targets 1–3 with respect to road association, estimated vehicle speed, and observed inter-target spacing (assuming individual targets are resolvable), it can be deduced that targets 1–3 are wheeled

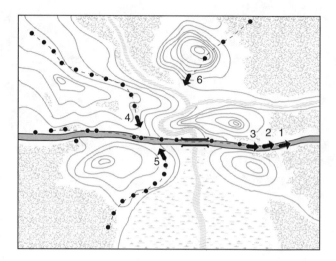

FIGURE 6.5   Example of the fusion of multiple-target tracks over time.

vehicles traveling in a convoy along a secondary road. Based on the maximum observed vehicle speeds and the associated surface conditions along their trajectories, tracks 4 and 5 can be deduced to be tracked vehicles. Finally, because of its relatively high speed and the rugged terrain in the vicinity, track 6 would be determined to be most consistent with a low-flying airborne target. Because the velocity of target 6 is too low to be a fixed-wing aircraft, the target can be inferred to be a helicopter.

Targets may be moving at one instant of time and stationary at another and communicating during one interval and silent during another, resulting in four mutually exclusive target states: (1) moving, nonemitting, (2) moving emitting, (3) nonmoving, nonemitting, and (4) nonmoving, emitting. Over time, many entities in the domain may change between two or more of these four states. Thus, if the situation awareness product is to be continuously maintained, data fusion inherently involves a recursive analysis. Table 6.1 provides a mapping between these four target states and a wide range of sensor classes. As shown, the ability to track entities through these state changes effectively requires multiple source sensor data.

In general, individual targets exhibit complex patterns of behavior that can help discriminate object classes and identify activities of interest. Consider the scenario depicted in Figure 6.6, showing the movement of a tactical erectable missile launcher (TEL) between time t_0 and time t_6. At t_0, the vehicle is in a location that makes it difficult to detect. At t_1, the vehicle is moving along a dirt road at velocity v_1. At time t_2, the vehicle continues along the road and begins communicating with its support elements. At time t_3, the vehicle is traveling off road at velocity v_3 along a minimum terrain gradient path. At time t_4, the target has stopped moving and begins to erect its launcher. At time t_5, just prior to launch, radar emissions begin. At time t_6, the vehicle is traveling to a new hide location at velocity v_6.

TABLE 6.1   Mapping between Sensor Classes and Target States

	Sensor Classes								
Target Classes	MTI Radar	SAR	Laser Radar	COMINT	ELINT	FLIR	Optical	Acoustic	Measurement and Signature
Moving/emitting	•		•	•	•	•		•	•
Moving/nonemitting	•		•			•		•	
Nonmoving/emitting		•		•	•	•	•		•
Nonmoving/nonemitting		•				•	•		

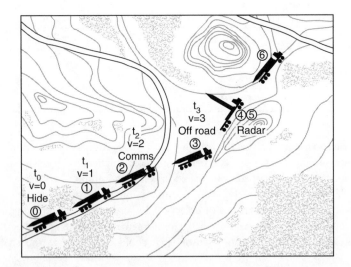

FIGURE 6.6 Dynamic target scenario showing sensor snapshots over time.

TABLE 6.2 Interpretation of Scenario Depicted in Figure 6.6

1	2	3	4	5	6	7	8
State	State Class	Velocity	Emission	Potentially Contributing Sensors	Local Interpretation	High-level Interpretation	Global Interpretation
0	Nonmoving/ nonemitting	0		SAR FLIR Imagery Video	Light foliage	Concealment	Hide
1	Moving/ nonemitting	v_1		MTI FLIR Video	Road association	High-speed mobility	Move to launch
2	Moving/ emitting	v_2	Comm type 1	MTI FLIR Video COMINT	Road association C2 network active	High-speed mobility Coordination status	
3	Moving/ nonemitting	v_3		MTI FLIR Video	Off road Good mobility	Minimum terrain gradient path Local goal seeking	
4	Nonmoving/ nonemitting	v_4		MTI FLIR Imagery	Open, flat Good mobility Good visibility	Tactical activity or staging area	Launch preparation and launch
5	Nonmoving/ emitting	v_5	Comm type 1 and 2; radar	SAR FLIR Imagery Video SIGINT	Coordination status Prelaunch transmission	Launch indication	
6	Moving/ nonemitting	v_6		MTI FLIR Video	High-speed travel	Rapid movement Road seeking	Move to hide

Table 6.2 identifies sensor classes that could contribute to the detection and identification of the various target states. Opportunities for effective sensor cross cueing for the TEL scenario discussed earlier are shown in the "Potentially Contributing Sensors" column. At the lowest level of abstraction, observed

TABLE 6.3   Mapping between Sensor Classes and Activities for a Bridging Operation

State	MTI Radar	SAR	COMINT	ELINT	FLIR	Optical	Acoustic
Engineers move to river bank	•		•			•	•
Construction activity		•	•	•	•	•	•
Forces move toward river bank	•		•	•		•	•
Forces move from opposite side of river	•		•			•	•

behavior can be interpreted with respect to a highly local perspective, as indicated in column 6, "Local Interpretation." By assuming that the object is performing some higher level behavior, progressively more global interpretations can be developed as indicated in columns 7 and 8.

Individual battle space objects are typically organized into operational or functional-level units, enabling observed behavior among groups of objects to be analyzed to generate higher level situation awareness products. Table 6.3 categorizes the behavioral fragments of an engineer battalion engaged in a bridge-building operation and identifies sensors that could contribute to the recognition of each fragment.

Situation awareness development involves the recursive refinement of a composite multiple level-of-abstraction scene description. Consequently, the generalized fusion process model shown in Figure 6.3(b) supports the effective combination of (1) domain observables, (2) *a priori* reasoning knowledge, and (3) the multiple level-of-abstraction/multiple-perspective fusion product. The process refinement loop controls both effective information combination and collection management. Each element of the process model is potentially sensitive to implicit (non-sensor-derived) domain knowledge.

6.3   Fusion Process Model Extensions

Recasting the generalized fusion process model within a biologically motivated framework establishes its relationship to the more familiar manual analysis paradigm. With suitable extensions, this biological framework leads to the development of a problem-solving taxonomy that categorizes the spectrum of machine-based approaches to reasoning. Drawing on this taxonomy of problem solving approaches helps to

- Reveal underlying similarities and differences between apparently disparate data analysis paradigms,
- Explore fundamental shortcomings of classes of machine-based reasoning approaches,
- Demonstrate the critical role of a database management system in terms of its support to both algorithm development and algorithm performance,
- Identify opportunities for developing more powerful approaches to machine-based reasoning.

6.3.1   Short-, Medium-, and Long-Term Knowledge

The various knowledge forms involved in the fusion process model can be compared with short-term, medium-term and long-term memory. *Short-term memory* retains highly transient short-term knowledge; *medium-term memory* retains dynamic, but somewhat less transient medium-term knowledge;* and *long-term memory* retains relatively static long-term knowledge. Thus, just as short-, medium-, and long-term memory suggest the durability of the information in biological systems, short-, medium-, and long-term knowledge relate to the durability of the information in machine-based reasoning applications.

---

* In humans, medium-term memory appears to be stored in the hippocampus in a midprocessing state between short-term and long-term memory, helping to explain why, after a trauma, a person often loses all memory from a few minutes to a few days.

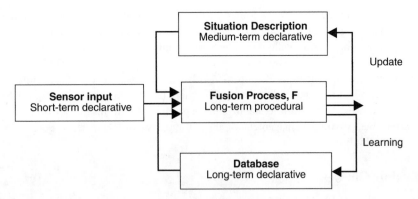

FIGURE 6.7 Biologically motivated metaphor for the data fusion process.

Within this metaphor, sensor data relates to the short-term knowledge, while long-term knowledge relates to relatively static factual and procedural knowledge. Because the goal of both biological and artificial situation awareness systems is the development and maintenance of the *current relevant perception* of the environment, the dynamic *situation description* represents medium-term memory. In both biological and tactical data fusion systems, *current* emphasizes the character of the dynamically changing scene under observation, as well as the potentially time-evolving analysis process that could involve interactions among a network of distributed fusion processes. Memory limitations and the critical role medium-term memory plays in both biological and artificial situation awareness systems enables only **relevant** states to be maintained. Because sensor measurements are inherently information-limited, real-world events are often nondeterministic, and uncertainties often exist in the reasoning process, a disparity between *perception* and *reality* must be expected.

As illustrated in Figure 6.7, sensor observables represent short-term declarative knowledge and the situation description represents medium-term declarative knowledge. Templates, filters, and the like are static declarative knowledge; domain knowledge includes both static (long-term) and dynamic (medium- and short-term) declarative context knowledge; and F represents the fusion process reasoning (long-term procedural) knowledge. Thus, as in biological situation awareness development, machine-based approaches require the interaction among short-, medium-, and long-term declarative knowledge, as well as long-term procedural knowledge. Medium-term knowledge tends to be highly perishable, while long-term declarative and procedural knowledge is both learned and forgotten much more slowly. With the exception of the difference in the time constants, learning of long-term knowledge and update of the situation description are fully analogous operations.

In general, short-, medium-, and long-term knowledge can be either *context-sensitive* or *context-insensitive*. In this chapter, context is treated as a conditional dependency among objects, attributes, or functions (e.g., $f(x_1,x_2|x_3 = a)$). Thus, context represents both explicit and implicit dependencies or conditioning that exist as a result of the state of the current situation representation or constraints imposed by the domain and/or the environment.

Short-term knowledge is dynamic, perishable, and highly context sensitive. Medium-term knowledge is less perishable and is learned and forgotten at a slower rate than short-term knowledge. Medium-term knowledge maintains the context-sensitive situation description at all levels of abstraction. The inherent context-sensitivity of short- and medium-term knowledge indicates that effective interpretation can be achieved only through consideration of the broadest possible context.

Long-term knowledge is relatively nonperishable information that may or may not be context-sensitive. *Context-insensitive* long-term knowledge is either generic knowledge, such as terrain/elevation, soil type, vegetation, waterways, cultural features, system performance characteristics, and coefficients of fixed-parameter signal filters, or context-free knowledge that simply ignores any domain sensitivity. *Context-sensitive* long-term knowledge is specialized knowledge, such as enemy Tables of Equipment, context-conditioned rule sets, doctrinal knowledge, and special-purpose two-dimensional map overlays (e.g., mobility maps or field-of-view maps). The specialization of long-term knowledge can be either fixed (*context-specific*) or conditionally dependent on dynamic or static domain knowledge (*context-general*).

Attempts at overcoming limitations of context-free algorithms often relied on fixed context algorithms that lack both generality and extensibility. The development of algorithms that are implicitly *sensitive* to relevant domain knowledge, on the other hand, tends to produce algorithms that are both more powerful and more extensible. Separate management of these four classes of knowledge potentially enhances database maintainability.

6.3.2　Fusion Classes

The fusion model depicted in Figure 6.3(b) views the process as the composition among (1) short-term declarative, (2) medium-term declarative, (3) long-term declarative, and (4) long-term procedural knowledge. Based on such a characterization, 15 distinct data fusion classes can be defined as illustrated by Table 6.4, representing all combinations of the four classes of knowledge.

Fusion classes provide a simple characterization of fusion algorithms, permitting a number of straightforward observations to be made. For example, only algorithms that employ short-term knowledge are sensitive to a dynamic input space, while only algorithms that employ medium-term knowledge are sensitive to the existing situation awareness product. Only algorithms that depend on long-term declarative knowledge are sensitive to static domain constraints.

While data fusion algorithms can rely on any possible combination of short-term, medium-term, and long-term declarative knowledge, every algorithm employs some form of procedural knowledge. Such knowledge may be either explicit or implicit. *Implicit* procedural knowledge is implied knowledge, while *explicit* procedural knowledge is formally represented knowledge. In general, implicit procedural knowledge tends to be associated with rigid analysis paradigms (i.e., cross correlation of two signals), whereas explicit procedural knowledge supports more flexible and potentially more powerful reasoning forms (e.g., model-based reasoning).

All fusion algorithms rely on some form of procedural knowledge; therefore, the development of a procedural knowledge taxonomy provides a natural basis for distinguishing approaches to machine-based reasoning. For our purposes, *procedural* knowledge will be considered to be long-term *declarative knowledge* and its associated *control knowledge*. Long-term declarative knowledge, in turn, is either *specific* or

TABLE 6.4　Fusion Classes

Fusion Class	Declarative Knowledge Class			Procedural Knowledge
	Short-Term Knowledge	Medium-Term Knowledge	Long-Term Knowledge	
1	•			
2		•		
3			•	
4	•	•		
5		•	•	
6	•		•	
7	•	•	•	
8				•
9	•			•
10		•		•
11			•	•
12	•	•		•
13		•	•	•
14	•		•	•
15	•	•	•	•

general. Specific declarative knowledge represents fixed (static) facts, transformations, or templates, such as filter transfer functions, decision trees, sets of explicit relations, object attributes, exemplars, or univariate density functions. *General* declarative knowledge, on the other hand, characterizes not just the value of individual attributes, but the relationships among attributes. Thus, object models, production-rule condition sets, parametric models, joint probability density functions, and semantic constraint sets are examples of *general* long-term declarative knowledge. Consequently, *specific* long-term declarative knowledge supports relatively fixed and rigid reasoning, while *general* long-term declarative knowledge supports more flexible approaches to reasoning.

Fusion algorithms that rely on specific long-term declarative knowledge are common when these three conditions **all** hold true:

- The decision process has relatively few degrees of freedom (attributes, parameters, dimensions).
- The problem attributes are relatively independent (no complex interdependencies among attributes).
- Relevant reasoning knowledge is static.

Thus, static problems characterized by moderate-sized state spaces and static domain constraints tend to be well served by algorithms that rely on specific long-term declarative knowledge.

At the other end of the spectrum are problems that possess high dimensionality and complex dependencies and are inherently dynamic. For such problems, reliance on algorithms that employ specific long-term declarative knowledge inherently limits the robustness of their performance. While such algorithms might yield acceptable performance for highly constrained problem sets, their performance tends to degrade rapidly as conditions deviate from nominal or as the problem set is generalized. In addition, dependence on specific declarative knowledge often leads to computation and/or search requirements exponentially related to the problem size. Thus, algorithms based on general long-term declarative knowledge can offer significant benefits when one or more of the following hold:

- The decision process has a relatively large number of degrees of freedom.
- The relationships among attributes are significant (attribute dependency).
- Reasoning is temporally sensitive.

Control knowledge can be grouped into two broad classes: *rigid* and *flexible*. *Rigid* control knowledge is appropriate for simple, routine tasks that are static and relatively context-insensitive. The computation of the correlation coefficient between an input data set and a set of stored exemplar patterns is an example of a simple rigid control strategy. *Flexible* control knowledge, on the other hand, supports more complex strategies, such as multiple-hypothesis, opportunistic, and mixed-initiative approaches to reasoning. In addition to being flexible, such knowledge can be characterized as either *single level-of-abstraction* or *multiple level-of-abstraction*. The former implies a relatively local control strategy, while the latter supports more global reasoning strategies. Based on these definitions, four distinct classes of control knowledge exist:

- Rigid, single level-of-abstraction;
- Flexible, single level-of-abstraction;
- Rigid, multiple level-of-abstraction;
- Flexible, multiple level-of abstraction.

Given the two classes of declarative knowledge and the four classes of control knowledge, there exist eight distinct forms of procedural knowledge.

In general, there are two fundamental approaches to reasoning: *generation-based* and *hypothesis-based*. Viewing analysis as a "black box" process with only its inputs and outputs available enables a simple distinction to be made between the two reasoning modalities. Generation-based problem-solving approaches "transform" a set of input states into output states; hypothesis-based approaches begin with output states and hypothesize and, ultimately, validate input states. Numerous reasoning paradigms such as filtering, neural networks, template match approaches, and forward-chained expert systems rely on

TABLE 6.5   Biologically Motivated Problem-Solving Form Taxonomy

Canonical Form #	Procedural Knowledge						Gen/ Hyp
	Declarative		Control				
	Specific	General	Rigid	Flexible	Single Level of Abstraction	Multiple Levels of Abstraction	
I	•		•		•		Gen
II	•		•		•		Hyp
III	•		•			•	Gen
IV	•		•			•	Hyp
V	•			•	•		Gen
VI	•			•	•		Hyp
VII	•			•		•	Gen
VIII	•			•		•	Hyp
IX		•	•		•		Gen
X		•	•		•		Hyp
XI		•	•			•	Gen
XII		•	•			•	Hyp
XIII		•		•	•		Gen
XIV		•		•	•		Hyp
XV		•		•		•	Gen
XVI		•		•		•	Hyp

generation-based reasoning. Other paradigms, such as backward-chained expert systems and certain graph-based and model-based reasoning approaches, rely on the hypothesis-based paradigm. Hybrid approaches utilize both reasoning modalities.

In terms of object-oriented reasoning, generation-based approaches tend to emphasize bottom-up analysis, while hypothesis-based reasoning often relies on top-down reasoning. Because both generation-based and hypothesis-based approaches can utilize any of the eight forms of procedural knowledge,[16] canonical problem solving (or paradigm) forms can be defined, as shown in Table 6.5.

Existing problem-solving taxonomies are typically constructed in a bottom-up fashion, by clustering similar problem-solving techniques and then grouping the clusters into more general categories. The categorization depicted in Table 6.5, on the other hand, being both *hierarchical* and *complete*, represents a true taxonomy. In addition to a convenient organizational framework, this taxonomy forms the basis of a "capability-based" paradigm classification scheme.

6.3.3   Fusion Classes and Canonical Problem-Solving Forms

Whereas a *fusion class* characterization categorizes the classes of data utilized by a fusion algorithm, the *canonical problem solving form* taxonomy can help characterize the potential robustness, context-sensitivity, and efficiency of a given algorithm. Thus, the two taxonomies serve different, yet fully complementary purposes.

6.3.3.1   The Lower-Order Canonical Forms

6.3.3.1.1   Canonical Forms I and II

Canonical forms I and II represent the simplest generation-based and hypothesis-based analysis approaches, respectively. Both of these canonical forms employ specific declarative knowledge and simple, rigid, single level-of-abstraction control. Algorithms based on these canonical form approaches generally

- Perform rather fixed data-independent operations,
- Support only implicit temporal reasoning (time series analysis),
- Rely on explicit inputs,
- Treat problems at a single level-of-abstraction.

Signal processing, correlation-based analysis, rigid template match, and artificial neural systems are typical examples of these two canonical forms. Such approaches are straightforward to implement; therefore, examples of these two forms abound.

Early speech recognition systems employed relatively simple canonical form I class algorithms. In these approaches, an audio waveform of individual spoken words was correlated with a set of prestored exemplars of all words in the recognition system's vocabulary. The exemplar achieving the highest correlation above some threshold was declared the most likely candidate. Because the exemplars were obtained during a training phase from the individual used to test its performance, these systems were highly speaker-dependent. The algorithm clearly relied on specific declarative knowledge (specific exemplars) and rigid, single level-of-abstraction control (exhaustive correlation followed by rank ordering of candidates). Although easy to implement and adequate in certain idealized environments (speaker-dependent, high signal-to-noise ratio, nonconnected word-speech applications), the associated exhaustive generation-and-test operation made the approach too inefficient for large vocabulary systems, and too brittle for noisy, speaker-independent, and connected-speech applications.

Although artificial neural systems are motivated by their biological counterpart, current capabilities of undifferentiated artificial neural systems (ANS) generally fall short of the performance of even simple biological organisms. Whereas humans are capable of complex, context-sensitive, multiple level-of-abstraction reasoning based on robust world models, ANS effectively filter or classify a set of input states. While humans can learn as they perform tasks, the ANS weight matrix is typically frozen (except in certain forms of clustering) during the state-transition process.

Regardless of the type of training, the nature of the nonlinearity imposed by the algorithm, or the specific details of the connection network, pretrained ANS represent static, specific long-term declarative knowledge; the associated control element is clearly static, rigid, and single level-of-abstraction. Most neural networks are used in generation-based processing applications and therefore possess all the key characteristics of all canonical form I problem-solving forms. Typical of canonical form I approaches, neural network performance tends to be brittle for problems of general complexity (because they are not model based) and non-context-sensitive (because they rely on either a context-free or highly context-specific weight matrix). Widely claimed properties of neural networks, such as robustness and ability to generalize, tend to be dependent on the data set and on the nature and extent of data set preprocessing.

Although the computational requirements of most canonical form I problem-solving approaches increase dramatically with problem complexity, artificial neural systems can be implemented using high concurrency hardware realizations to effectively overcome this limitation. Performance issues are not necessarily eliminated, however, because before committing a network to hardware (and during any evolutionary enhancements), extensive retraining and testing may be required.

6.3.3.1.2 Canonical Forms III-VIII

Canonical form III and IV algorithms utilize specific declarative knowledge and rigid, multiple level-of-abstraction control knowledge. Although such algorithms possess most of the limitations of the lowest order problem solving approaches, canonical form III and IV algorithms, by virtue of their support to multiple level-of-abstraction control, tend to be somewhat more efficient than canonical forms I and II. Simple recursive, multiple resolution, scale-space, and relaxation-based algorithms are examples of these forms.

As with the previous four problem-solving forms, canonical form V and VI algorithms rely on specific declarative knowledge. However, rather than rigid control, these algorithms possess a flexible, single level-of-abstraction control element that can support multiple hypothesis approaches, dynamic reasoning, and limited context-sensitivity.

Canonical form VII and VIII approaches employ specific declarative and flexible, multiple level-of-abstraction control knowledge. Although fundamentally non-model-based reasoning forms, these forms support flexible, mixed top-down/bottom-up reasoning.

6.3.3.2 The Higher-Order Canonical Forms

As a result of their reliance on specific declarative knowledge, the eight lower-order canonical form approaches represent the core of most numeric-based approaches to reasoning. In general, these lower-order form approaches are unable to effectively mimic the high-level semantic and cognitive processes employed by human decision makers. The eight higher-level canonical forms, on the other hand, provide significantly better support to semantic and symbolic-based reasoning.

6.3.3.2.1 Canonical Forms IX and X

Canonical forms IX and X rely on general declarative knowledge and rigid, single level-of-abstraction control, representing simple model-based transformation and model-based constraint set evaluation approaches, respectively. General declarative knowledge supports more dynamic and more context-sensitive reasoning than specific declarative knowledge. However, because these two canonical forms rely on rigid, single level-of-abstraction control, canonical form IX and X algorithms tend to be inefficient.

The motivation behind expert system development was to emulate the human reasoning process in a restricted problem domain. An expert system rule-set generally contains both formal knowledge (e.g., physical laws and relationships), as well as heuristics and "rules-of-thumb" gleaned from practical experience. Although expert systems can accommodate rather general rule condition and action sets, the associated control structure is typically quite rigid (i.e., sequential condition set evaluation, followed by straightforward resolution of which instantiated rules should be allowed to fire). In fact, the separation of procedural knowledge into modular IF/THEN rule-sets (general declarative knowledge) that are evaluated using a rigid, single level-of-abstraction control structure (rigid control knowledge) represents the hallmark of the pure production-rule paradigm. Thus, demanding rule modularity and a uniform control structure effectively relegates conventional expert system approaches to the two lowest-order, model-based, problem-solving forms.

6.3.3.2.2 Canonical Forms XI through XIV

Problem solving associated with canonical forms XI and XII relies on a general declarative element and rigid, multiple level-of-abstraction control. Consequently, these forms support both top-down and bottom-up reasoning. Production rule paradigms that utilize a hierarchical rule-set are an example of such an approach.

Canonical forms XIII and XIV employ procedural knowledge that possesses a general declarative element and flexible, single level-of-abstraction control. As a result, these canonical forms can support sophisticated single level-of-abstraction, model-based reasoning.

6.3.3.2.3 Canonical Forms XV and XVI

Canonical form XV and XVI paradigms employ general declarative knowledge and flexible, multiple level-of-abstraction control; therefore, they represent the most powerful generation-based and hypothesis-based problem-solving forms, respectively. Although few canonical form XV and XVI fusion algorithms have achieved operational status, efficient algorithms that perform sophisticated, model-based reasoning, while meeting rather global optimality criteria, can be reasonably straightforward to develop.[1]

The HEARSAY speech understanding system[2] was an early attempt at building a higher-order reasoning system. This system, developed in the early 1980s, treated speech recognition as both inherently context-sensitive and multiple level-of-abstraction. HEARSAY employed a hierarchy of models appropriate at the various levels-of-abstraction within the problem domain, from signal processing to perform formant tracking and spectral analysis for phoneme extraction, to symbolic reasoning for meaning extraction. Higher-level processes, with their broader perspective and higher-level knowledge, provided some level of control over the lower-level processes. Importantly, HEARSAY viewed speech understanding in a holistic fashion with each level of the processing hierarchy treated as a critical component of the fully integrated analysis process.

6.3.3.3 Characteristics of the Higher-Order Canonical Forms

Five key algorithm issues have surfaced during the preceding discussion:

- Robustness
- Context-sensitivity
- Extensibility

- Maintainability
- Efficiency

Each of these issues is discussed briefly below.

6.3.3.3.1   Robustness

Robustness measures the fragility of a problem-solving approach to changes in the input space. Algorithm robustness depends, quite naturally, on both the quality and efficacy of the models employed. The development of an "adequate" model depends, in turn, on the complexity of the process being modeled. A problem that intrinsically exhibits few critical degrees of freedom would logically require a simpler model than one that possesses many highly correlated features.

As a simple illustration, consider the handwritten character recognition problem. Although handwritten characters possess a large number of degrees-of-freedom (e.g., line thickness, character orientation, style, location, size, color, darkness, and contrast ratio), a simple model can capture the salient attributes of the character "H" (i.e., two parallel lines connected at their approximate centers by a third line segment). Thus, although the handwritten character intrinsically possesses many degrees-of-freedom, most are not relevant for distinguishing the letter "H" from other handwritten characters. Conversely, in a non-model-based approach, each character must be compared with a complete set of exemplar patterns for all possible characters. Viewed from this perspective, a non-model-based approach can require consideration of all combinations of both relevant and nonrelevant problem attributes.

6.3.3.3.2   Context Sensitivity

Context refers to both the static domain constraints (natural and cultural features, physical laws) and dynamic domain constraints (current location of all air defense batteries) relevant to the problem-solving process. Dynamic short-term and medium-term knowledge are generally context-sensitive, while *a priori* long-term reasoning knowledge may or may not be sensitive to context.

Context-sensitive long-term knowledge (both declarative and procedural) is conditional knowledge that must be specialized by static or dynamic domain knowledge (e.g., mobility map or current dynamic Order of Battle). Context-insensitive knowledge is generic, absolute, relatively immutable knowledge that is effectively domain independent (e.g., terrain obscuring radar coverage or wide rivers acting as obstacles to ground-based vehicles). Such knowledge is fundamentally unaffected by the underlying context. Context-specific knowledge is long-term knowledge that has been specialized for a given, fixed context. Context-free knowledge simply ignores any effects related to the underlying context.

In summary, *context-sensitivity* is a measure of a problem's dependency on implicit domain knowledge and constraints. As such, canonical forms I–IV are most appropriate for tasks that require either *context-insensitive* or *context-specific* knowledge. Because canonical forms V–VIII possess flexible control, all are potentially sensitive to problem context. General declarative knowledge can be sensitive to non-sensor-derived domain knowledge (e.g., a mobility map, the weather, the current ambient light level, or the distance to the nearest river); therefore, all higher order canonical forms are potentially context-sensitive. Canonical forms XIII–XVI support both *context-sensitive declarative* and *context-sensitive control* knowledge and, therefore, are the only *fully context-sensitive* problem-solving forms.

6.3.3.3.3   Extensibility and Maintainability

Extensibility and maintainability are two closely related concepts. *Extensibility* measures the "degree of difficulty" of extending the knowledge base to accommodate domain changes or to support related applications. *Maintainability* measures the "cost" of storing and updating knowledge. Because canonical forms I–VIII rely on a specific declarative knowledge, significant modifications to the algorithm can be required for even relatively minor domain changes. Alternatively, because they employ general declarative knowledge, canonical forms IX–XVI tend to be much more extensible.

The domain sensitivity of the various canonical form approaches varies considerably. The lower-order canonical form paradigms typically rely on context-free and context-specific knowledge, leading to relatively nonextensible algorithms. Because context-specific knowledge may be of little value when the problem context changes (e.g., a mobility map that is based on dry conditions cannot be used to support

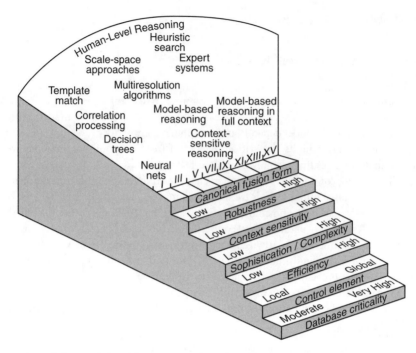

FIGURE 6.8   General characteristics of the sixteen canonical fusion forms and associated problem-solving paradigms.

analysis during a period of flooding), canonical form I–IV approaches tend to exhibit brittle performance as the problem context changes. Attempting to support context-sensitive reasoning using context-specific knowledge can lead to significant database maintainability problems.

Conversely, context-insensitive knowledge (e.g., road, bridge, or terrain-elevation databases) is unaffected by context changes. Context-insensitive knowledge remains valid when the context changes; however, context-sensitive knowledge may need to be redeveloped. Therefore, database maintainability benefits from the separation of these two knowledge bases. *Algorithm extensibility* is enhanced by model-based approaches and *knowledge base maintainability* is enhanced by the logical separation of context-sensitive and context-insensitive knowledge.

6.3.3.3.4   Efficiency
Algorithm efficiency measures the relative performance of algorithms with respect to computational and/or search requirements. Although exceptions exist, for complex, real-world problem solving, the following generalizations often apply:

- *Model-based reasoning* tends to be more efficient than *non-model-based reasoning*.
- *Multiple level-of-abstraction reasoning* tends to be more efficient than *single level-of-abstraction reasoning*.

The general characteristics of the 16 canonical forms are summarized in Figure 6.8.

6.4   Observations

This chapter concludes with five general observations pertaining to data fusion automation.

6.4.1   Observation 1

Attempts to automate many complex, real-world fusion tasks face a considerable challenge. One obvious explanation relates to the disparity between manual and algorithmic approaches to data fusion. For example, humans

- Are adept at model-based reasoning (which supports robustness and extensibility),
- Naturally employ domain knowledge to augment formally supplied information (which supports context-sensitivity),
- Update or modify existing beliefs to accommodate new information as it becomes available (which supports dynamic reasoning),
- Intuitively differentiate between context-sensitive and context-insensitive knowledge (which supports maintainability),
- Control the analysis process in a highly focused, often top-down fashion (which enhances efficiency).

As a consequence, manual approaches to data fusion tend to be inherently dynamic, robust, context-sensitive, and efficient. Conversely, traditional paradigms used to implement data fusion algorithms have tended to be inherently static, nonrobust, non-context-sensitive, and inefficient. Many data fusion problems exhibit complex, and possibly dynamic, dependencies among relevant features, advocating the practice of

- Relying more on the higher order problem solving forms,
- Applying a broader range of supporting databases and reasoning knowledge,
- Utilizing more powerful, global control strategies.

6.4.2   Observation 2

Although global phenomena naturally require global analysis, local phenomena can benefit from both a local and a global analysis perspective. As a simple example, consider the target track assignment process typically treated as a strictly local analysis task. With a conventional canonical form I approach to target tracking, track assignment is based on recent, highly local behavior (often assuming a Markoff process). For ground-based objects, a vehicle's historical trajectory and its maximum performance capabilities provide rather weak constraints on future target motion. A "road-constrained target extrapolation strategy," for example, provides much stronger constraints on ground-vehicle motion than a purely statistical-based approach. As a result, the latter tends to generate highly under-constrained solutions.

Although applying nearby domain constraints could adequately explain the local behavior of an object (e.g., constant velocity travel along a relatively straight, level road), a more global viewpoint is required to interpret global behavior. Figure 6.9 demonstrates local (i.e., concealment, minimum terrain gradient, and road seeking), medium-level (i.e., river-crossing and road-following), and global (i.e., reinforce at unit) interpretations of a target's trajectory over space and time. The development and maintenance of such a multiple level-of-abstraction perspective is a critical underlying requirement for automating the situation awareness development process.

6.4.3   Observation 3

Production systems have historically performed better against static, well-behaved, finite-state diagnostic-like problems than against problems that possess complex dependencies and exhibit dynamic, time-varying behavior. These shortcomings occur because such systems rely on rigid, single level-of-abstraction control that is often insensitive to domain context. Despite this fact, during the early 1990s, expert systems were routinely applied to dynamic, highly context-sensitive problem domains, often with disappointing results.

The lesson to be learned is that both the strengths and limitations of a selected problem-solving paradigm must be fully understood by the algorithm developer from the outset. When an appropriately constrained task was successfully automated using an expert system approach, developers often found that the now well-understood problem could be more efficiently implemented using another paradigm. In such cases, better results were obtained by using either an alternative canonical form IX or X problem-solving approach or a lower-order, non-model-based approach.

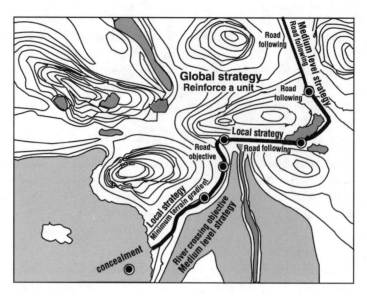

FIGURE 6.9   Multiple level-of-abstraction situation understanding.

When an expert system proved to be inadequate for handling a given problem, artificial neural systems were often seen as an alternative or preferred approach. Neural networks require no programming; therefore, the paradigm appeared ideal for handling ill-defined or poorly understood problems. While expert systems could have real-time performance problems, artificial neural systems promised high performance hardware implementations. In addition, the adaptive nature of the neural net learning process often seemed to match real-world, dynamically evolving problem-solving requirements. However, most artificial neural systems operate more like a statistical or fuzzy pattern recognizer than as a sophisticated reasoning system capable of generalization, reasoning by analogy, and abstract inference. As indicated by the reasoning class taxonomy, while expert systems represent a lower-order *model-based reasoning* approach, a neural network represents the lowest-order *non-model-based reasoning* approach.

6.4.4   Observation 4

Radar systems typically employ a single statistical-based algorithm for tracking air targets, regardless of whether an aircraft is flying at an altitude of 20 kilometers or just above tree-top level. Likewise, such algorithms are generally insensitive as to whether the target is a high performance fighter aircraft or a relatively low speed helicopter. Suppose a nonfriendly high-performance reconnaissance aircraft is flying just above a river as it snakes through a mountainous region. There exist a wide range of problems associated with tracking such a target, including dealing with high clutter return, terrain masking, and multipath effects. In addition, an airborne radar system may have difficulty tracking the target as a result of high acceleration turns associated with an aircraft following a highly irregular surface feature. The inevitable track loss and subsequent track fragmentation errors typically would require intervention by a radar analyst. Tracking helicopters can be equally problematic. Although they fly more slowly, such targets can hover, fly below tree-top level, and execute rapid directional changes.

Tracking performance can potentially be improved by making the tracking analysis sensitive to target class-specific behavior, as well as to constraints posed by the domain. For example, the recognition that the aircraft is flying just above the terrain suggests that surface features are likely to influence the target's trajectory. When evaluated with respect to "terrain feature-following models," the trajectory would be discovered to be highly consistent with a "river-following flight path." Rather than relying on past behavior to predict future target positions, a tracking algorithm could anticipate that the target is likely to continue to follow the river.

In addition to potentially improving tracking performance, the interpretation of sensor-derived data within context also permits more abstract interpretations. If the aircraft were attempting to avoid radar detection by one or more nearby surface-to-air missile batteries, a nap of the earth flight profile could indicate hostile intent. Even more global interpretations can be hypothesized. Suppose a broader view of the "situation picture" reveals another unidentified aircraft operating in the vicinity of the river-following target. By evaluating the apparent coordination between the two aircraft, the organization and mission of the target group can be conjectured. For example, if the second aircraft begins jamming friendly communication channels just as the first aircraft reaches friendly airspace, the second aircraft's role can be inferred to be "standoff protection for the primary collection or weapon delivery aircraft." The effective utilization of relevant domain knowledge and physical domain constraints offers the potential for developing both more effective and higher level-of-abstraction interpretations of sensor-derived information.

6.4.5   Observation 5

Indications and warnings, as well as many other forms of expectation-based analysis have traditionally relied on relatively rigid doctrinal and tactical knowledge. However, contemporary data fusion applications often must support intelligence applications where flexible, ill-defined, and highly creative tactics and doctrine are employed. Consequently, the credibility of any analysis that relies on rigid expectation-based behavior needs to be carefully scrutinized. Although the lack of strong, reliable *a priori* knowledge handicaps all forms of expectation-based reasoning, the use of relevant logical, physical, and logistical context at least partially compensates for the lack of more traditional problem domain constraints.

Acknowledgment

The preparation of this chapter was funded by CECOM I2WD, Fort Monmouth, NJ.

References

1. Antony, R. T., *Principles of Data Fusion Automation,* Artech House Inc., Boston, 1995.
2. Erman, L. D. et al., The HEARSAY II Speech Understanding System: Integrating Knowledge to Resolve Uncertainty, *Computing Surveys*, 12(2), 1980.

7

Contrasting Approaches to Combine Evidence

7.1   Introduction **7-1**
7.2   Alternative Approaches to Combine Evidence **7-2**
       The Probability Theory Approach • The Possibility Theory
       Approach • The Belief Theory Approach • Methods of
       Combining Evidence
7.3   An Example Data Fusion System **7-18**
       System Context • Collections of Spaces • The System in
       Operation • Summary
7.4   Contrasts and Conclusion **7-31**
       Appendix 7.A The Axiomatic Definition of Probability **7-31**
       References **7-31**

Joseph W. Carl
Harris Corporation

7.1 Introduction

A broad consensus holds that a probabilistic approach to evidence accumulation is appropriate because it enjoys a powerful theoretical foundation and proven guiding principles. Nevertheless, many would argue that probability theory is not suitable for practical implementation on complex real-world problems. Further debate arises when considering people's subjective opinions regarding events of interest. Such debate has resulted in the development of several alternative approaches to combining evidence.[1-3] Two of these alternatives, possibility theory (or fuzzy logic)[4-6] and belief theory (or Dempster-Shafer theory),[7-10] have each achieved a level of maturity and a measure of success to warrant their comparison with the historically older probability theory.

This chapter first provides some background on each of the three approaches to combining evidence in order to establish notation and to collect summary results about the approaches. Then an example system that accumulates evidence about the identity of an aircraft target is introduced. The three methods of combining evidence are applied to the example system, and the results are contrasted. At this point, possibility theory is dropped from further consideration in the rest of the chapter because it does not seem well suited to the sequential combination of information that the example system requires. Finally, an example data fusion system is constructed that determines the presence and location of mobile missile batteries. The evidence is derived from multiple sensors and is introduced into the system in temporal sequence, and a software component approach is adopted for its implementation. Probability and belief theories are contrasted within the context of the example system.

One key idea that emerges for simplifying the solution of complex, real-world problems involves collections of spaces. This is in contradistinction to collections of events in a common space. Although the spaces are all related to each other, considering each space individually proves clearer and more manageable. The relationships among the spaces become explicit by considering some as fundamental representations of what is known about the physical setting of a problem, and others as arising from observation processes defined at various knowledge levels.

The data and processes employed in the example system can be encapsulated in a component-based approach to software design, regardless of the method adopted to combine evidence. This leads naturally to an implementation within a modern distributed processing environment.

Contrasts and conclusions are stated in Section 7.4.

7.2   Alternative Approaches to Combine Evidence

Probability is much more than simply a relative frequency. Rather, there is an axiomatic definition[11] of probability that places it in the general setting of measure theory. As a particular measure, it has been crafted to possess certain properties that make it useful as the basis for modeling the occurrence of events in various real-world settings. Some critics (fuzzy logicians among them) have asserted that probability theory is too weak to include graded membership in a set; others have asserted that probability cannot handle non-monotonic logic. In this chapter, both of these assertions are demonstrated by example to be unfounded. This leads to the conclusion that fuzzy logic and probability theory have much in common, and that they differ primarily in their methods for dealing with unions and intersections of events (characterized as sets). Other critics have asserted that probability theory cannot account for imprecise, incomplete, or inconsistent information. Evidence is reviewed in this chapter to show that interval probabilities can deal with imprecise and incomplete information in a natural way that explicitly keeps track of what is known and what is not known. The collection of spaces concept (developed in Section 7.3) provides an explicit means that can be used with any of the approaches to combine evidence to address the inconsistencies.

7.2.1   The Probability Theory Approach

The definition of a probability space tells what properties an assignment of probabilities must possess, but it does not indicate what assignment should be made in a specific setting. The specific assignment must come from our understanding of the physical situation being modeled, as shown in Figure 7.1. The definition tells us how to construct probabilities for events that are mutually exclusive (i.e., their set

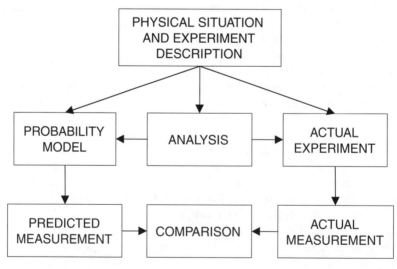

FIGURE 7.1   The comparison of predictions with measurements places probability models on firm scientific ground.

representations are disjoint). Generally speaking, when collections of events are not mutually exclusive, a new collection of mutually exclusive events (i.e., disjoint sets) must first be constructed.

Consider the desirable properties for measuring the plausibility of statements about some specific experimental setting. Given that

1. The degree of plausibility can be expressed by a real number,
2. The extremes of the plausibility scale must be compatible with the truth values of logic,
3. An infinitesimal increase in the plausibility of statement A implies an infinitesimal decrease in the plausibility of the statement not-A,
4. The plausibility of a statement must be independent of the order in which the terms of the statement are evaluated,
5. All available evidence must be used to evaluate plausibility, and
6. Equivalent statements must have the same plausibility,

then the definition of a probability space follows as a logical consequence.[12] Further, the definition implies that the probability measure has properties (1) through (6). Hence, any formalism for measuring the plausibility of statements must necessarily be equivalent to the probability measure, or it must abandon one or more of the properties listed.

7.2.1.1 Apparent Paradoxes and the Failure of Intuition

Some apparent paradoxes about probability theory reappear from time to time in various forms. Two will be discussed — Bertrand's paradox and Hughes' paradox. A dice game that cannot be lost is then described. This will help to make the point that human intuition can fail with regard to the outcome of probability-space models. A failure of intuition is probably the underlying reason for the frequent underestimation of the power of the theory.

7.2.1.1.1 *Bertrand's Paradox*

Bertrand's paradox[13] begins by imagining that lines are drawn at random to intersect a circle to form chords. Suppose that the coordinates of the center of the circle and the circle's radius are known. The length of each chord can then be determined from the coordinates of the midpoint of the chord, which might be assumed to be uniformly distributed within the circle. The length of each chord can also be determined from the distance from the center of the chord to the center of the circle, which might be assumed to be uniformly distributed between zero and the radius of the circle. The length of each chord can also be determined from the angle subtended by the chord, which might be assumed to be uniformly distributed between 0 and 180 degrees. The length of each chord is certainly the same, regardless of the method used to compute it.

Bertrand asked, "What is the probability that the length of a chord will be longer than the side of an inscribed equilateral triangle?" Three different answers to the question appear possible depending on which of the three assumptions is made. How can that be if the lengths must be the same? A little reflection reveals that the lengths may indeed be the same when determined by each method, but that assumptions have been made about three different related quantities, none of which is directly the length. In fact, the three quantities cannot simultaneously be distributed in the same way. Which one is correct? Jaynes[14] has shown that only the assumption that chord centers are uniformly distributed within the circle provides an answer that is invariant under infinitesimal translations and rotations.

Bertrand's paradox touches on the *principle of indifference*: if no reason exists for believing that any one of n mutually exclusive events is more likely than any other, a probability of $1/n$ is assigned to each event. This is a valid principle, but it must be applied with caution to avoid pitfalls. Suppose, for instance, four cards — two black and two red — are shuffled and placed face down on a table. Two cards are picked at random. What is the probability they are the same color? One person reasons, "They are either both black, or they are both red, or they are different; in two cases the colors are the same, so the answer is $2/3$." A second person reasons, "No, the cards are either the same or they are different; the answer is $1/2$." They are both wrong, as shown in Figure 7.2. There is simply no substitute for careful analysis.

	C_1	C_2	C_3	C_4
	R	R	B	B
	R	B	R	B
	R	B	B	R
	B	R	R	B
	B	R	B	R
	B	B	R	R

There are 6 equally likely ways to place 2 red and 2 black cards

FIGURE 7.2 No matter which two cards one picks, P(same color) = 1/3.

7.2.1.1.2 Hughes' Paradox

The Hughes paradox arose in the context of pattern recognition studies during the late 1960s and early 1970s. Patterns were characterized as vectors, and rules to decide a pattern's class membership were studied using a collection of samples of the patterns. The collection size was held constant. The performance of a decision rule was observed experimentally to often improve as the dimension of the pattern vectors increased — up to a point. The performance of the decision rule decreased beyond that point. This led some investigators to conclude that there was an optimal dimension for pattern vectors. However, most researchers believed that the performance of a Bayes-optimal classifier never decreases as the dimension of the pattern vectors increases. This can be attributed to the fact that a Bayes-optimal decision rule, if given irrelevant information, will just throw the information away. (See, for example, the "theorem of irrelevance."[15]). The confusion was compounded by the publication of Hughes' paper,[16] which seemed to prove that an optimal dimension existed for a Bayes classifier. As a basis for his proof, Hughes constructed a monotonic sequence of data quantizers that provided the Bayes classifier with a finer quantization of the data at each step. Thus, the classifier dealt with more data at each step of the sequence. Hughes thought that he had constructed a sequence of events in a common probability space. However, he had not; he had constructed a sequence of probability spaces.[17] Because the probability-space definition was changing at each step of the sequence, the performance of a Bayes classifier in one space was not simply related to the performance of a Bayes classifier in another space. There was no reason to expect that the performances would be monotonically related in the same manner as the sequence of classifiers. This experience sheds light on how to construct rules to accumulate evidence in data fusion systems: accumulating evidence can change the underlying probability-space model in subtle ways for which researchers must account.

7.2.1.1.3 A Game That Can't Be Lost

This next example demonstrates that people do not have well-developed intuition about what can happen in probability spaces. Given the four nonstandard, fair, six-sided dice shown in Figure 7.3, play the following game. First, pick one of the dice. Then have someone else pick one of the remaining three. Both of you roll the die that you have selected; the one with the highest number face up wins. You have the advantage, right? Wrong! No matter which die you pick, one of the remaining three will win at this game two times out of three. Call the dice A, B, C, and D. A beats B with probability 2/3, B beats C with probability 2/3, C beats D with probability 2/3, and D beats A with probability 2/3 — much like the childhood game rock-scissors-paper, this game involves nontransitive relationships. People typically think about "greater than" as inducing a transitive relation among ordinary numbers. Their intuition fails when operating in a domain with nontransitive relations.[18] In this sense, probability-space models can deal with non-monotonic logic.

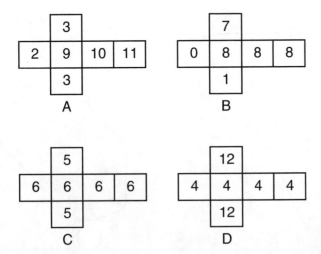

FIGURE 7.3 These dice (reported in 1970 by Martin Gardner to be designed by Bradley Efron at Stanford University) form the basis of a game one cannot lose.

The point of this section is to emphasize that the physical situation at hand is critically important. Considerable work may be required to construct an accurate probability model, but the effort can be very rewarding. The power of probability theory is that it tells us how to organize and quantify what is known in terms that lead to minimizing the expected cost of making decisions.

7.2.1.2 Observation Processes and Random Variables

In many physical settings, an item of interest cannot be directly accessed. Instead, it can only be indirectly observed. For example, the receiver in a communication system must observe a noise-corrupted modulated signal for some interval of time to decide which message was sent. Based on the sampling theorem, a received signal can be characterized completely by a vector of its samples taken at an appropriate rate. The sample vectors are random vectors; their components are joint random variables. The random variables of interest arise from some well-defined observation processes implemented as modules in a data fusion system. It is important to be precise about random variables that can characterize observation processes.

Formally, a random variable is a measurable function defined on a sample space (e.g., $(f:S \rightarrow R)$ or $(f:S \rightarrow R_n)$, indicating scalar or vector random variables taking values on the real line or its extension to n-dimensional space). The probability distribution on the random variable is induced by assigning to each subset of $R(R_n)$, termed *events*, the same probability as the subset of S that corresponds to the inverse mapping from the event-subset to S. This is the formal definition of a measurable function. In Figure 7.4, the event, B, occurs when the random variable takes on values in the indicated interval on the real line. The image of B under the inverse mapping is a subset of S, called B'. This results in $P(B) = P(B')$, even though B and B' are in different spaces.

The meaning of this notation when observation processes are involved should be emphasized. If the set, A, in Ω represents an event defined on the sample space, and if the set, B, in R represents an event defined on the real line through a random variable, then one set must be mapped into a common space with the other. This enables a meaningful discussion about the set $\{f(A) \& B\}$, or about the set $\{A \& f^{-1}(B)\}$. The joint events $[A \& B]$ can similarly be discussed, taking into consideration the meaning in terms of the set representations of those events. In other words, $P[A \& B] = P[\{f(A) \& B\}] = P[\{A \& f^{-1}(B)\}]$. Note that even when a collection of sets, A_i, for $i = 1,2,...,n$, partitions some original sample space, the images of those sets under the observation mapping, $f(A_i)$, will not, in general, partition the new sample space. In this way, probability theory clearly accommodates concepts of measurement vectors belonging to a set (representing a cause) with graded membership.

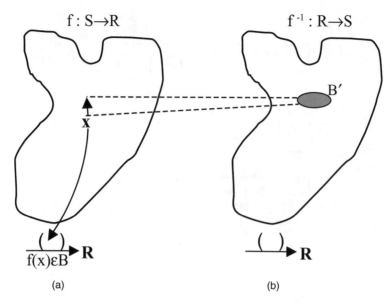

FIGURE 7.4   (a) Forward mappings and (b) inverse mappings relate the sample space to an observation space.

7.2.1.3   Bayes' Theorem

There may be modules in a data fusion system that observe the values of random variables (or vectors) and compute the probability that the observed values have some particular cause. The causes partition their sample space. Bayes' theorem is employed to compute the probability of each possible cause, given some observation event. Suppose $A_1, A_2, \ldots A_n$ form a collection of subsets of S (representing causes) that partition S. Then for any observation event, B, with $P(B) > 0$,

$$P\left(A_i \middle| B\right) = \frac{P\left(B \middle| A_i\right) P\left(A_i\right)}{P\left(B\right)} \tag{7.1}$$

and

$$P\left(B\right) = \sum_{i=1}^{n} P\left(B \middle| A_i\right) P\left(A_i\right) \tag{7.2}$$

The quantities $P(A_i|B)$ and $P(B|A_i)$ are termed *conditional probabilities*; the quantities $P(A_i)$ and $P(B)$ are termed *marginal probabilities*. The quantities $P(B|A_i)$ and $P(A_i)$ are termed *a priori probabilities* because they represent statements that can be made prior to knowing the value of any observation. Again, note that Bayes theorem remains true for events represented by elements of Ω, as well as for random events defined through an observation process. This can cause some confusion. The original sample space and the observation space are clearly related, but they are separate probability spaces. Knowing which space you are operating in is important.

Note that Bayes' theorem assumes that some event is *given* (i.e., it has unequivocally occurred). Often this is not the case in a data fusion system. Suppose, for example, that an event, E, is observed with confidence 0.9. This could be interpreted to mean that E has occurred with probability 0.9, and that its alternatives occur with a combined probability of 0.1. Assuming two alternatives, $A1$ and $A2$, interval probabilities can be employed to conclude that E occurred with probability 0.9, A1 occurred with probability x, and A2 occurred with probability $0.1 - x$, where $0 \le x \le 0.1$. Finally, assuming that one of the possible causes of the observed events is C, and noting that a true conditioning event does not yet exist, a superposition of probability states can be defined. Thus, combining the results from using Bayes' theorem on each of the possible observed events and weighting them together gives

$$P(C) = (0.9) * P(C|E) + (x) * P(C|A1) + (0.1 - x) * P(C|A2) \tag{7.3}$$

where $0 \le x \le 0.1$.

This particular form of motivation for the resulting probability interval does not seem to appear in the substantial literature on interval probabilities. Yet, it has a nice physical feel to it. To compensate for the uncertainty of not knowing the current state, an interval probability is created from a superposition of possible event states. As addressed in the next section of this chapter, enough evidence may later be accumulated to "pop" the superposition and declare with acceptable risk that a true conditioning event has occurred.

7.2.1.4 Bayes-Optimal Data Fusion

The term *Bayes-optimal* means minimizing risk, where risk is defined to be the expected cost associated with decision-making. There are costs associated with correct decisions, as well as with incorrect decisions, and typically some types of errors are more costly than others. Those who must live with the decisions made by the system must decide the cost structure associated with any particular problem. Once decided, the cost structure influences the optimal design through an equation that defines expected cost. To simplify notation, just the binary-hypotheses case will be presented; the extension to multiple-hypotheses is straightforward.

Suppose there are just two underlying causes of some observations, C1 or C2. Then there are four elements to the cost structure:

1. C_{11}, the cost of deciding C1 when really C1 (a correct decision);
2. C_{22}, the cost of deciding C2 when really C2 (another correct decision);
3. C_{21}, the cost of deciding C2 when really C1 (an error; sometimes a miss); and
4. C_{12}, the cost of deciding C1 when really C2 (an error; sometimes a false alarm).

The expected cost is simply Risk $= E\{\text{cost}\} = C_{11}P_{11} + C_{22}P_{22} + C_{21}P_{21} + C_{12}P_{12}$, where the indicated probabilities have the obvious meaning. Suppose the observation process produces a measurement vector, X, and define two regions in the associated vector space: $R_1 = \{X|\text{decide } C1\}$, and $R_2 = \{X|\text{decide } C2\}$. Let $p(X|C1)$ denote the conditional probability density function of a specific value of the measurement vector given C1. Let $p(X|C2)$ denote the conditional probability density function of a specific value of the measurement vector given C2. Let $p(X)$ denote the marginal probability density function on the measurement vector. Then, as shown elsewhere,[19] minimize risk by forming the *likelihood ratio* and comparing it to a threshold:

Decide C1 if

$$l(X) = \frac{p(X|C1)}{p(X|C2)} > \frac{(C_{12} - C_{22})P(C2)}{(C_{21} - C_{11})P(C1)} \tag{7.4}$$

otherwise, decide C2. Because applying the same monotonic function to both sides preserves the inequality, an equivalent test is (for example) to decide C1 if

$$\log\left[l(X)\right] > \log\left[\frac{\left(C_{12} - C_{22}\right)P(C2)}{\left(C_{21} - C_{11}\right)P(C1)}\right] \tag{7.5}$$

and decide *C2* otherwise.

An equivalent test that minimizes risk is realized by comparing $d(X) = (C_{21} - C_{22})P(C_2)p(X|C_2) - (C_{12} - C_{11})P(C_1)p(X|C_1)$ to 0. That is, decide *C1* if $d(X) < 0$, and decide *C2* otherwise. In some literature,[20,21] $d(X)$ is called a *discriminant* function; it has been used together with nonparametric estimators (e.g., potential functions or Parzen estimators) of the conditional probabilities as the basis for pattern recognition systems, including neural networks.

An important property of this test in any of its equivalent forms is its ability to optimally combine prior information with measurement information. It is perhaps most obvious in the likelihood ratio form that relative probability is what is important — how much greater p(X|C1) is than $p(X|C2)$ — rather than the specific values of the two conditional probabilities. When one is sufficiently greater than the other, there may be acceptable risk in "popping" a superposition of probability states to declare a true conditioning event has occurred. Finally, note that in any version of the test, knowledge of the form of the optimal decision rule is a focal point and guide to understanding a particular problem domain.

7.2.1.5 Exploiting Lattice Structure

Many researchers likely under-appreciate the fact that the lattice structure induced by the event relationships within a probability space can be exploited to determine the probability of events, perhaps in interval form, from partial information about some of the probabilities. To be precise, consider $S = \{x_1, x_2, \ldots, x_N\}$ to be an exhaustive collection of *N* mutually exclusive (simple, or atomic) events. The set 2^S is the set of all possible subsets of *S*. Suppose unnormalized probabilities (e.g., as odds) are assigned to *M* events in 2^S, say E_k for $k = 1, 2, \ldots, M$, where *M* may be less than, equal to, or greater than *N*. The next section of this chapter partially addresses the question: under what conditions can the probabilities of x_i be inferred?

7.2.1.5.1 A Characteristic Matrix

Consider only the case $M = N$. Define **C**, an $N \times N$ matrix with elements $c_{i,j} = 1$ if $\{x_j\} \subset E_i$, and 0 otherwise. **C** can be called the characteristic matrix for the E_ks. Also, define **P**, an $N \times 1$ vector with elements p_k ($k = 1, 2, \ldots, N$) that are the assigned unnormalized probabilities of E_k. From the rule for combining probabilities of mutually exclusive events, $\mathbf{P} = \mathbf{C}\,\mathbf{X}$, where **X** is an $N \times 1$ vector with elements $P[\{x_i\}]$, some or all of which are unknown. Clearly, $\mathbf{X} = \mathbf{C}^{-1}\,\mathbf{P}$. For this last equation to be solvable, the determinant of **C** must be nonzero, which means the rows/columns of **C** are linearly independent. Put another way, the collection $\{E_k \mid k=1,2,\ldots,N\}$ must "span" the simple events.

7.2.1.5.2 Applicability

The characteristic matrix defined above provides a mathematically sound, intuitively clear method of determining the probabilities of simple events from the probabilities of compound events derived from them, including combining evidence across knowledge sources. Bayes' theorem is not used to obtain any of these results, and the question of how to assign prior probabilities does not arise. The lattice structure implicit in the definition of probabilities is simply exploited. This concept and its use are discussed later in this chapter, where methods of combining evidence are considered.

7.2.2 The Possibility Theory Approach

Possibility theory considers a body of knowledge represented as subsets of some established reference set, *S*. (Most often in literature on possibility theory the domain of discourse is denoted Ω. This discussion uses *S* to minimize the introduction of new notation for each approach. It will remain clear that the syntax and semantics of possibility theory differs from those of probability theory.) Denote the collection of all subsets of *S* as $\Omega = 2^S$. In the case that *S* has an infinite number of elements, Ω denotes a sigma-algebra (the definition of Ω given for probability in Appendix 7.A defines a sigma-algebra). Most of the time this chapter restricts S to finite cardinality for reasons of simplicity. This is not a severe restriction, since in practical systems the representable body of knowledge will be finite.

There are two distinguished subsets of Ω, the empty set, ϕ, and the set S itself. Let C denote a confidence function that maps the elements of Ω into the interval $[0, 1]$, $C:\Omega \rightarrow [0, 1]$. It is required that $C(\varphi) = 0$ and that $C(S) = 1$. φ can be called the "impossible" or "never true" event, and S can be called the "sure" or "always true" event. Note that $C(A) = 0$ does not imply $A = \varphi$ and $C(A) = 1$ does not imply $A = S$, where $A \in \Omega$.

In order to have a minimum of coherence, any confidence function should be monotonic with respect to inclusion, which requires that $A \subseteq B$ implies $C(A) \leq C(B)$. This is interpreted to mean that if a first event is a restriction of (or implies) a second event, then there is at least as much confidence in the occurrence of the second as in the occurrence of the first. Immediate consequences of this monotonicity are that $C(A\cup B) \geq \max[C(A), C(B)]$ and $C(A\cap B) \leq \min[C(A), C(B)]$.

The limiting case, $C(A\cup B) = \max[C(A), C(B)]$, can be taken as an axiom that defines a *possibility measure*.[22] (Zadeh was the first to use the term *possibility measures* to describe confidence measures that obey this axiom. He denoted them $\Pi(\cdot)$, the convention that is followed in this chapter.) The term "possibility" for this limiting case can be motivated, even justified, by the following observations (this motivation follows a similar treatment in Dubois and Prade.)[5]

Suppose $E \in \Omega$ is such that $C(E) = 1$. Define a particular possibility measure as $\Pi_1(A) = 1$ if $A \cap E \neq \varphi$ and 0 otherwise. Then interpret $\Pi_1(A) = 1$ to mean A is possible. Also, since $\Pi_1(A \cup \text{not-}A) = \Pi_1(S) = 1$, $\max[\Pi_1(A), \Pi_1(\text{not-}A)] = 1$. Interpret this to mean that of two contradictory events, at least one is possible. However, one being possible does not prevent the other from being possible, too. This is consistent with the semantics of judged possibilities, which invokes little commitment. Finally, $\Pi_1(A \cup B) = \max[\Pi_1(A), \Pi_1(B)]$ seems consistent with notions of physical possibility: to realize $A \cup B$ requires only the easiest (i.e., the most possible) of the two to be realized.

Because "max" is a reflexive, associative, and transitive operator, any possibility measure can be represented in terms of the (atomic) elements of S: $\Pi_1(A) = \sup\{\pi_1(a)|a\in A\}$, where "sup" stands for *supremum* (that is, for least upper bound), $A \in \Omega$, $a \in S$, and $\pi_1(a) = \Pi_1(\{a\})$. Call $\pi_1(a)$ a *possibility distribution* (defined on S). Consider a possibility distribution to be *normalized* if there exists at least one $a \in S$ such that $\pi_1(a) = 1$. If S is infinite, a possibility distribution exists only if the axiom is extended to include infinite unions of events.[23]

Now take the limiting case $C(A \cap B) = \min[C(A), C(B)]$ as a second axiom of possibility theory, and call set functions that satisfy this axiom *necessity measures*. The term "necessity" for this limiting case can be motivated, even justified, by the following observations.

Suppose $E\in \Omega$ is such that $C(E) = 1$. Define a particular necessity measure as $N_1(A) = 1$ if $E \subseteq A$, and 0 otherwise. $N_1(A) = 1$ clearly means that A is necessarily true. This is easy to verify from the definitions: if $\Pi_1(A) = 1$ then $N_1(\text{not-}A)] = 0$, and if $\Pi_1(A) = 0$ then $N_1(\text{not-}A)] = 1$. Thus, $\Pi_1(A) = 1 - N_1(\text{not-}A)]$. This is interpreted to mean that if an event is necessary, its contrary is impossible, or, conversely, if an event is possible its contrary is absolutely not necessary. This last equation expresses a duality between the possible and the necessary, at least for the particular possibility and necessity functions used here.

Because "min" is a reflexive, associative, transitive operator, this duality implies it is always appropriate to construct a necessity measure from a possibility distribution:

$$N_1\left(A\right) = \inf\left\{1 - \pi_1\left(a\right)\middle|a \notin A\right\} \tag{7.6}$$

where "inf" stands for *infemum* (or greatest lower bound).

Several additional possibility and necessity relationships can be quickly derived from the definitions. For example:

1. $\min[N_1(A), N_1(\text{not-}A)] = 0$ (if an event is necessary its complement is not the least bit necessary).
2. $\Pi_1(A) \geq N_1(A)$ for all $A \in \Omega$ (an event becomes possible before it becomes necessary).
3. $\Pi_1(A) + \Pi_1(\text{not-}A) \geq 1$.
4. $N_1(A) + N_1(\text{not-}A) \leq 1$.

Thus, the relationship between the possibility (or the necessity) of an event and the possibility (or necessity) of its contrary is weaker than in probability theory, and both possibility and necessity numbers are needed to characterize the uncertainty of an event. However, both probability and possibility can be characterized in terms of a distribution function defined on the atomic members of the reference set.

Now adopt this motivation and justification to call arbitrary functions, $\Pi(A)$ and $N(A)$, possibility and necessity functions, respectively, if they satisfy the two axioms given above and can be constructed from a distribution, $\pi(a)$, as $\Pi(A) = \sup\{\pi(a) | a \in A\}$ and $N(A) = \inf\{1 - \pi(a) \mid a \notin A\}$ for all $a \in A$. It is straightforward to show that all the properties defined here for $\Pi_1(A)$ and $N_1(A)$ hold for these arbitrary possibility and necessity functions provided that $0 \leq \pi(a) \leq 1$ for all $a \in S$ and provided $\pi(a)$ is normalized (i.e., there exists at least one $a \in S$ such that $\pi(a) = 1$). The properties would have to be modified if the distribution function is not normalized. In the sequel it is assumed that possibility distribution functions are normalized.

A relationship exists between possibility theory and fuzzy sets. To understand this relationship, some background on fuzzy sets is also needed.

L. A. Zadeh introduced fuzzy sets in 1965.[24] Zadeh noted that there is no unambiguous way to determine whether or not a particular real number is much greater than one. Likewise, no unambiguous way exists of determining whether or not a particular person is in the set of tall people. Ambiguous sets like these arise naturally in our everyday life. The aim of fuzzy set theory is to deal with such situations wherein sharply defined criteria for set membership are absent.

Perhaps the most fundamental aspect of fuzzy sets that differentiates them from ordinary sets is the domain on which they are defined. A fuzzy set is a function defined on some (ordinary) set of interest, S, termed the domain of discourse. As discussed earlier in this chapter, probability is defined on a *collection* of ordinary sets, 2^S. This is a profound difference. Measure theory and other topics within the broad area of real analysis employ collections of subsets of some given set (such as the natural numbers or the real line) in order to avoid logical problems that can otherwise arise.[25]

Another difference between fuzzy sets and probability theory is that fuzzy sets leave vague the meaning of membership functions and the operations on membership functions beyond a generalization of the characteristic functions of ordinary sets (note that the terms *fuzzy set* and *fuzzy membership function* refer to the same thing). To understand this, let $\{x\}$ be the domain from which the elements of an ordinary set are drawn. The characteristic function of the ordinary "crisp" set is defined to have value 1 if and only if x is a member of the set, and to have the value 0 otherwise. A fuzzy set is defined to have a membership function that satisfies $0 \leq f(x) \leq 1$. In this sense, the characteristic function of the ordinary set is included as a special case. However, the interpretation of the fuzzy membership function is subjective, rather than precise; some researchers have asserted that it does not correspond to a probability interpretation[26] (although that assertion is subject to debate). This suggests that fuzzy membership functions will prove useful in possibility theory as possibility distribution functions, but not directly as possibility measures.

Operations on fuzzy sets are similarly motivated by properties of characteristic functions. Table 7.1 summarizes the definitions of fuzzy sets, including those that result from operations on one or more other fuzzy sets. There, $f(x)$ denotes a general fuzzy set, and $f_A(x)$ denotes a particular fuzzy set, A. "Max" and "min" played a role in the initial definition of fuzzy sets. Thus, fuzzy intersection suggests a possibility measure, fuzzy intersection suggests a necessity measure, and if a fuzzy set is thought of as a possibility distribution, the connection that $f(x)$ can equal $\Pi(\{x\})$ for $x \in S$ is established.

Assigning numerical values as the range of a membership function is no longer essential. One generalization of Zadeh's original definition that now falls within possibility theory is the accommodation of word labels in the range for a fuzzy membership function. This naturally extends fuzzy sets to include

TABLE 7.1 Summary Definition of Fuzzy Membership Functions

Operation	Definition
Empty Set	f is empty iff $f(x) = 0 \; \forall \; x$
Complement	$f^C = 1 - f(x) \; \forall \; x$
Equality	$f_A = f_B$ iff $f_A(x) = f_B(x) \; \forall \; x$
Inclusion	$f_A \subseteq f_B$ iff $f_A(x) \leq f_B(x) \; \forall \; x$
Disjunction	$f_{A \cup B}(x) = \max \, [f_A(x), f_B(x)]$
Conjunction	$f_{A \cap B}(x) = \min \, [f_A(x), f_B(x)]$
Convexity	A is a convex fuzzy set $\Leftrightarrow f_A[kx_1 + (1 - k)x_2] > \min \, [f_A(x_1), f_A(x_2)]$ for all x_1 and x_2 in X, and for any constant, k, in the interval $[0,1]$.
Algebraic Product	$\lvert f_A(x) \, f_B(x) \rvert$
Algebraic Sum	$\lvert f_A(x) + f_B(x) \rvert \leq 1$
Absolute Difference	$\lvert f_A(x) - f_B(x) \rvert$
Entropy Ratio	$\dfrac{\displaystyle\int dx \min\left[f_A(x), 1 - f_A(x)\right]}{\displaystyle\int dy \max\left[f_A(y), 1 - f_A(y)\right]}$

language, creating an efficient interface with rule-based expert systems. Architectures created using this approach are often referred to as fuzzy controllers.[27] Except for this difference in range, the fuzzy sets in a fuzzy controller continue to be combined as indicated in Table 7.1.

7.2.3 The Belief Theory Approach

Dempster[7] and Shafer[8] start with an exhaustive set of mutually exclusive outcomes of some experiment of interest, S, and call it the *frame of discernment*. (In much of the literature on Dempster-Shafer theory, the frame of discernment is denoted Θ. This discussion uses S to minimize the introduction of new notation for each approach. It will remain clear that the syntax and semantics of belief theory differ from those of probability theory.) Dempster-Shafer then form $\Omega = 2^S$, and assign a belief, $B(A)$, to any set $A \subset \Omega$. (In some literature on belief theory, the set formed is $2^S - \varphi$, but this can cause confusion and makes no difference, as the axioms will show.) The elements of S can be called atomic events; the elements of Ω can be called molecular if they are not atomic. The interpretation of a molecular event is that any one of its atomic elements is "in it," but not in a constructive sense. The evidence assigned to a molecular event cannot be titrated; it applies to the molecular event as a whole. The mass of evidence is also sometimes called the basic probability assignment; it satisfies the following axioms:

1. $m \, (\phi) = 0$
2. $m \, (A) > 0$ for all $A \in \Omega$
3. $\displaystyle\sum_{A \subseteq \Omega} m(A) = 1$

Note that although these axioms bear some similarity to the axioms for probability, they are not the same. Belief and probability are not identical. The crucial difference is that axiom 3 equates unity with the total accumulated evidence assigned to all elements of Ω, whereas an axiom of probability equates unity with $S \in \Omega$. Belief theorists interpret S to mean a state of maximal ignorance, and the evidence for S is transferable to other elements of Ω as knowledge becomes manifest, that is, as ignorance diminishes. Hence, in the absence of any evidence, in a state of total ignorance, assign m(S) =1 and to all other elements of Ω assign a mass of 0. In time, as knowledge increases, some other elements of Ω will have assigned nonzero masses of evidence. Then, if $m(A) > 0$ for some $A \in \Omega$, $m(S) < 1$ in accord with the reduction of ignorance. This ability of belief theory to explicitly deal with ignorance is often cited as a useful property of the approach. However, this property is not unique to belief theory.[28]

Belief theory further defines a belief function in terms of the mass of evidence. The mass of evidence assigned to a particular set is committed exactly to the set, and not to any of the constituent elements of the set. Therefore, to obtain a measure of total belief committed to the set, add the masses of evidence associated with all the sets that are subsets of the given set. For all sets A and B in Ω, define

$$Bel(A) = \sum_{B \subseteq A} m(B) \qquad (7.7)$$

Given $Bel(A)$, $m(B)$ can be recovered as follows:

$$m(B) = \sum_{A \subseteq B} (-1)^{|B|-|A|} Bel(A) \qquad (7.8)$$

Thus, given either representation, the other can be recovered. This transform pair is known as the Möbius transformation.[29]

$B \in \Omega$ is a *focal element* of the belief system if $m(B) > 0$. (Confusingly, some authors seem to equate the focal elements of a belief system with the atomic events. That definition would not be sufficient to obtain the results cited here.) The union of all the focal elements of a belief system is called the *core* of the belief system, denoted C. It should be apparent that $Bel(A) = 1$ if and only if $C \subseteq A$. It should also be apparent that if all the focal elements are atomic events, then $Bel(A)$ is the classical probability measure defined on S. It is this last property that leads some authors to assert that belief theory (or Dempster-Shafer theory) is a generalization of probability theory. However, a generalization should also be expected to do something the other cannot do, and this has not been demonstrated. Indeed, Dempster explicitly acknowledges that there are stronger constraints on belief theory than on probability theory.[5,23,30] Dempster was well aware that his rule of combination (still to be discussed) leads to more constrained results than probability theory, but he preferred it because it allows an artificial intelligence system to get started with zero initial information about priors.

This belief function has been called the *credibility function*, denoted $Cr(A)$, and also the *support* for A, denoted $Su(A)$. In the sequel, $Su(A)$ will be used in keeping with the majority of the engineering literature. By duality, a *plausibility function*, denoted $Pl(A)$, can be defined in terms of the support function:

$$Pl(A) = 1 - Su(\text{not-}A)$$

$$= 1 - \sum_{B \subseteq (\Omega - A)} m(B) \qquad (7.9)$$

$$= 1 - \sum_{A \cap B = \phi} m(B) = \sum_{A \cap B \neq \phi} m(B)$$

Thus, the plausibility of A is 1 minus the sum of the mass of evidence assigned to all the subsets of Ω that have an *empty* intersection with A. Equivalently, it is the sum of the mass of evidence assigned to all the subsets of Ω that have a *nonempty* intersection with A. An example should help to solidify these definitions. Suppose $S = \{x, y, z\}$. Then $\Omega = \{\phi, \{x\}, \{y\}, \{z\}, \{x,y\}, \{x,z\}, \{y,z\}, S\}$. The credibility and the plausibility of all the elements of Ω can be computed by assigning a mass of evidence to some of the elements of Ω as shown in Table 7.2.

For any set $A \in \Omega$ $Su(A) \leq Pl(A)$, $Su(A) + Su(\text{not-}A) \leq 1$, and $Pl(A) + Pl(\text{not-}A) \geq 1$.

The relationship between support and plausibility leads to the definition of an interval, $[Su(A), Pl(A)]$. What is the significance of this interval? The support of a proposition can be interpreted as the total

TABLE 7.2 An Example Clarifying Belief System Definitions

Event A	Mass of Evidence m(A)	Support $\displaystyle\sum_{B \subseteq A} m(B)$	Plausibility $\displaystyle 1 - \sum_{A \cap B = \phi} m(B)$
φ	0	0	0
$\{x\}$	m_x	m_x	$1 - m_y - m_z - m_{yz}$
$\{y\}$	m_y	m_y	$1 - m_x - m_z - m_{xz}$
$\{z\}$	m_z	m_z	$1 - m_x - m_y - m_{xy}$
$\{x,y\}$	m_{xy}	$m_x + m_y + m_{xy}$	$1 - m_z$
$\{x,z\}$	m_{xz}	$m_x + m_z + m_{xz}$	$1 - m_y$
$\{y,z\}$	m_{yz}	$m_y + m_z + m_{yz}$	$1 - m_x$
S	$1 - \Sigma$ (all other masses)	1	1

mass of evidence that has been transferred to the proposition, whereas the plausibility of the proposition can be interpreted as the total mass of evidence that has either already been transferred to the proposition or is still free to transfer to it. Thus, the interval spans a spectrum of belief from that which is already available to that which may yet become available given the information at hand.

7.2.4 Methods of Combining Evidence

Each of the three theories just reviewed has its own method of combining evidence. This section provides an example problem as a basis of comparison (this example follows Blackman[10]). Suppose there are four possible targets operating in some area, which are called t_1, t_2, t_3, and t_4. Suppose t_1 is a friendly interceptor (fighter aircraft), t_2 is a friendly bomber, t_3 is a hostile interceptor, and t_4 is a hostile bomber.

7.2.4.1 Getting Started

This is enough information to begin to define a probability space. Define $S = \{t_1, t_2, t_3, t_4\}$ and form $\Omega = 2^S$. Clearly, $\varphi \in \Omega$ and $P[\varphi] = 0$. Also, $S \in \Omega$ and $P[S] = 1$ (i.e., one of the targets will be observed because S is exhaustive).

This provides enough information for a possibility system to establish its universe of discourse, $S = \{t_1, t_2, t_3, t_4\}$. However, there is no clearly defined way to characterize the initial ignorance of which target may be encountered. Note that there is *not* a constraint of the form $f_0(S) = 1$. A possible choice is $f_0(x) = 1$ if $x \in S$, and 0 otherwise, corresponding to an assignment of membership equal to nonmembership for each of the possible targets about which no information is initially available. Another possible choice is $f_0(x) = 0$, corresponding to an assignment of the empty set to characterize that no target is present prior to the receipt of evidence. As noted above, {Low, Medium, High} could also be chosen as the range, and $f_0(x) = $ Low could be assigned for all x in S. In order to be concrete, choose $f_0(x) = 0$.

This is also enough information to begin to construct a belief system. Accepting this knowledge at face value and storing it as a single information string, $\{t_1 \cup t_2 \cup t_3 \cup t_4\}$, with unity belief (which implies $P(\Omega) = 1$, as required), minimizes the required storage and computational resources of the system.

7.2.4.2 Receipt of First Report

Suppose a first report comes in from a knowledge source that states, "I am 60 percent certain the target is an interceptor." All three systems map the attribute "interceptor" to the set $\{t_1, t_3\}$.

7.2.4.2.1 Probability Response

Based on the first report, $P[\{t_1,t_3\}|1^{st} \text{ report}\}] = 0.6$. A probability approach requires that $P[\text{not}\{t_1,t_3\}|1^{st} \text{ report}\}] = 1 - P[\{t_1,t_3\}|1^{st} \text{ report}\}]$. The set complement is with respect to S, so $P[\{t_2,t_4\}|1^{st} \text{ report}\}] = 0.4$. The status of knowledge at this point is summarized on the lattice structure based on subsets of S, as shown in Figure 7.5.

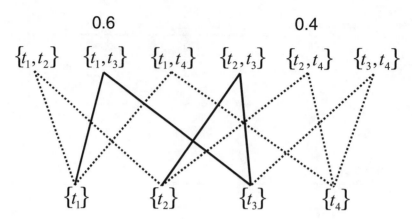

FIGURE 7.5   The lattice imposes exploitable constraints.

With $p_i = P[\{t_i\}]$ the constraints from this lattice structure lead to the conclusion that:

$$0 \le p_1 \le 0.6 \tag{7.10}$$

$$0 \le p_2 \le 0.4 \tag{7.11}$$

$$p_3 = 0.6 - p_1 \left(0 \le p_3 \le 0.6\right) \tag{7.12}$$

$$p_4 = 0.4 - p_2 \left(0 \le p_4 \le 0.4\right) \tag{7.13}$$

7.2.4.2.2   Possibility Response
The possibility system interprets the message as saying that $\Pi(\{t_1\} \cup \{t_3\}) = 0.6. = \max[\Pi(\{t_1\}), \Pi(\{t_3\}]$. This implies only that $0 \le \Pi(\{t_1\} \le 0.6$, and $0 \le \Pi(\{t_3\}) \le 0.6$, and does not express any constraining relationship between $\Pi\{t_1\}$ and $\Pi\{t_3\}$. Because $\{S\} = \{t_1,t_3\} \cup \{t_2,t_4\}$ and $\Pi(\{S\}) = 1$, $\max[\Pi\{t_1,t_3\}, \Pi\{t_2,t_4\}] = \max[0.6, \Pi\{t_2,t_4\}] = 1$, which implies that $\Pi\{t_2,t_4\}] = 1$. From this, the conclusion is reached that $0 \le \Pi\{t_2\} \le 1$, and $0 \le \Pi\{t_4\} \le 1$, again without any constraint between $\Pi\{t_2\}$ and $\Pi\{t_4\}$. This contributes little.

7.2.4.2.3   Belief Response
Since the reported information is not certain regarding whether or not the target is an interceptor, the belief system transfers some of the mass of evidence from S as follows: $m_1(S) = 0.4$ and $m_1(\{t_1, t_3\}) = 0.6$ (all the others are assigned zero). From these the support and the plausibility for these two propositions can be computed, as shown in Table 7.3. This is all that can be inferred; no other conclusions can be drawn at this point.

7.2.4.3   Receipt of Second Report

Next, a second report comes in from a knowledge source that states, "I am 70 percent sure the target is hostile." All three systems map the attribute "hostile" to the set $\{t_3,t_4\}$.

TABLE 7.3   Belief Support and Plausibility

Event	Support	Plausibility
$\{t_1, t_3\}$	0.6	1.0
S	1.0	1.0

7.2.4.3.1 *Probability Combination*

At this point, $P[\{t_3,t_4\}|2^{nd} \text{ report}\}] = 0.7$ and $P[\{t_1,t_2\}|2^{nd} \text{ report}] = 0.3$. Thus, the following probabilities for six possible pairings of four targets result from the second report:

$$P\left[\left\{t_1,t_2\right\}\right]=0.3 \tag{7.14}$$

$$P\left[\left\{t_1,t_3\right\}\right]=0.6 \tag{7.15}$$

$$P\left[\left\{t_1,t_4\right\}\right]=1-x \tag{7.16}$$

$$P\left[\left\{t_2,t_3\right\}\right]=x \tag{7.17}$$

$$P\left[\left\{t_2,t_4\right\}\right]=0.4 \tag{7.18}$$

$$P\left[\left\{t_3,t_4\right\}\right]=0.7 \tag{7.19}$$

Now, using Equations 7.13, 7.14, and 7.16 (because this choice involves only three unknowns; any such choice necessarily provides the same answers), together with a characteristic matrix, gives

$$\begin{bmatrix} 1 & 1 & 0 \\ 1 & 0 & 1 \\ 0 & 1 & 1 \end{bmatrix} \begin{bmatrix} P\left[\left\{t_1\right\}\right] \\ P\left[\left\{t_2\right\}\right] \\ P\left[\left\{t_3\right\}\right] \end{bmatrix} = \begin{bmatrix} 0.3 \\ 0.6 \\ x \end{bmatrix} \tag{7.20}$$

and any standard technique for matrix inversion can be used to obtain

$$0.5\begin{bmatrix} 1 & 1 & -1 \\ 1 & -1 & 1 \\ -1 & 1 & 1 \end{bmatrix} \begin{bmatrix} 0.3 \\ 0.6 \\ x \end{bmatrix} = \begin{bmatrix} P\left[\left\{t_1\right\}\right] \\ P\left[\left\{t_2\right\}\right] \\ P\left[\left\{t_3\right\}\right] \end{bmatrix} \tag{7.21}$$

This leads to the conclusions that

$$P\left[\left\{t_1\right\}\right]=0.45-\left(x/2\right) \tag{7.22}$$

$$P\left[\left\{t_2\right\}\right]=\left(x/2\right)-0.15 \tag{7.23}$$

$$P\left[\left\{t_3\right\}\right]=0.15+\left(x/2\right) \tag{7.24}$$

$$P\left[\left\{t_4\right\}\right]=0.55-\left(x/2\right) \tag{7.25}$$

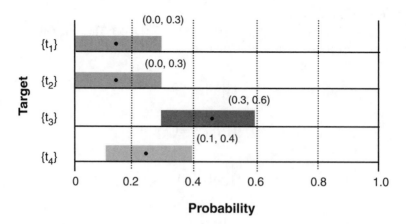

FIGURE 7.6   Even partial information supports a preliminary decision ("•" marks the mid-range of x).

The fact that these are all probabilities constrains x to the range $0.3 \leq x \leq 0.9$, which, in turn, implies

$$0.0 \leq P\big[\big\{t_1\big\}\big] \leq 0.3 \qquad\qquad (7.26)$$

$$0.0 \leq P\big[\big\{t_2\big\}\big] \leq 0.3 \qquad\qquad (7.27)$$

$$0.3 \leq P\big[\big\{t_3\big\}\big] \leq 0.6 \qquad\qquad (7.28)$$

$$0.1 \leq P\big[\big\{t_4\big\}\big] \leq 0.4 \qquad\qquad (7.29)$$

This is all that can be deduced until at least one additional piece of independent information arrives. However, even this partial information supports a preliminary decision, as shown in Figure 7.6.

7.2.4.3.2   Possibility Combination
Possibility theory combines information using min and max beginning with the following observed facts: (1) $\Pi[\{t_1,t_3\}] = 0.6$, and (2) $\Pi[\{t_3,t_4\}] = 0.7$. Since $\{t_3\} = \{t_1,t_3\} \cap \{t_3,t_4\}$, $N(\{t_3\}) = \min[0.6, 0.7] = 0.6$. However, since there are no defined constraints among the individual possibilities aside from the min-max rules of combination, little else can be deduced. Therefore, possibility theory does not seem well suited to a sequential *combination* of information even though it may be an effective way to assign measures of belief in other settings (e.g., in fuzzy controllers[27]). For this reason, there will be no further consideration of possibility theory in this chapter.

7.2.4.3.3   Belief Combination
The belief system represents the information in the second report as an assignment of masses of evidence from a second source: $m_2(S) = 0.3$ and $m_2(\{t_3,t_4\}) = 0.7$. Dempster's rule is then used to combine the masses of evidence from the two independent sources. In this case the calculations are particularly simple, as shown in Table 7.4. Again, the support and the plausibility of each of these events can be computed, as shown in Table 7.5.

Obviously Dempster's rule does not directly inform belief in the presence of individual targets, yet the physical situation often presents only one target. This and related phenomena have led some authors to point out that belief theory is weak in its ability to result in a decision.[28]

TABLE 7.4   Application of Dempster's Rule

	$m_2(\{t_3,t_4\}) = 0.7$	$m_2(S) = 0.3$
$m_1(\{t_1, t_3\}) = 0.6$	$m\{t_3\}) = 0.42$	$m(\{t_1, t_3\}) = 0.18$
$m_1(S) = 0.4$	$m(\{t_3,t_4\}) = 0.28$	$m(S) = 0.12$

TABLE 7.5   Combined Support and Plausibility

Event	Support	Plausibility
$\{t_3\}$	0.42	1.0
$\{t_1,t_3\}$	0.60	1.0
$\{t_3,t_4\}$	0.70	1.0
S	1.0	1.0

7.2.4.4 Inconsistent Evidence

Inconsistency is said to occur when one knowledge source assigns a mass of evidence to one event (set), a second knowledge source assigns a mass of evidence to a different event (set), and the two events have nothing in common — the intersection of their set representations is null. This situation has not so far arisen in the target identification example considered above and will now be introduced.

Suppose a third report comes in that states "I am 80 percent sure one of the known targets is present: $\{t_2\}$ is twice as likely as $\{t_1\}$; $\{t_3\}$ is three times as likely as $\{t_1\}$; and $\{t_4\}$ is twice as likely as $\{t_1\}$."

7.2.4.4.1 Probability Resolution

This third report calls into question whether or not the probability space so far constructed to model the situation is accurate. Do the four targets represent an exhaustive set of outcomes, or don't they? One possibility is that other target types are possible; another possibility is that there really is no target present. So the probability space must be modified to consider the possibility that something else can happen and guarantee that the atomic events really exhaustively span all possible outcomes. Therefore, define an additional atomic event, O, called "other" to denote the set of whatever other undifferentiated possibilities there might be. The probability system then represents the third report as stating that $P[\{t_1\}] = 0.1$; $P[\{t_2\}] = 0.2$; $P[\{t_3\}] = 0.3$; $P[\{t_4\}] = 0.2$; and $P[O] = 0.2$. Now in order to combine this report with the earlier reports, the earlier probability results must be mapped to the new probability space just defined, otherwise they are simply incommensurate.

For example, based on the first report, $P[\{t_1,t_3\}|1^{st}\text{ report}\}] = 0.6$. A probability approach requires that $P[\text{not}\{t_1,t_3\}|1^{st}\text{ report}\}] = 1 - P[\{t_1,t_3\}|1^{st}\text{ report}\}]$. The set complement is with respect to S, so if S includes O, $P[\{t_2,t_4,O\}|1^{st}\text{ report}\}] = 0.4$. Similarly, when $P[\{t_3,t_4\}|2^{nd}\text{ report}\}] = 0.7$ and S includes O, $P[\{t_1,t_2,O\}|2^{nd}$ report] $= 0.3$. This requires a complete new analysis that obviates the analysis reported in Sections 7.2.4.2.1 and 7.2.4.3.1, above, and there is not sufficient space in this chapter to do it again. Suffice it to say that the results from the two messages agree qualitatively with the third message, but there are quantitative disparities. Utility theory has been developed to address such situations. A rational person is expected to choose an alternative that has the greatest utility. So, how can utility be assigned to these two alternatives? One way is to equate utility with the number of corroborating reports; this is appropriate if all data sources have equal veracity. Since two reports are consistent for the first alternative, and only one report is self-consistent for the second alternative, the data fusion system would prefer the first alternative if this utility function is adopted. Utility theory also offers the means to create a linear combination of the alternatives, and the number of corroborating reports can again be used to form the weights. The computations are omitted.

7.2.4.4.2 Belief Resolution

The belief system represents the third report as stating that $m[\{t_1\}] = 0.1$; $m[\{t_2\}] = 0.2$; $m[\{t_3\}] = 0.3$; $m[\{t_4\}] = 0.2$; and $m[S] = 0.2$. The assignment of a mass of evidence to S accounts for the uncertainty

TABLE 7.6   The Dempster-Shafer Rule Applied to Inconsistent Evidence

	$m_{1,2}[\{t_3\}] = 0.42$	$m_{1,2}[\{t_1, t_3\}] = 0.18$	$m_{1,2}[\{t_3, t_4\}] = 0.28$	$m_{1,2}[S] = 0.12$
$m_3[\{S\}] = 0.2$	$m[\{t_3\}] = 0.084$	$m[\{t_1, t_3\}] = 0.036$	$m[\{t_3, t_4\}] = 0.056$	$m[S] = 0.024$
$m_3[\{t_1\}] = 0.1$	$k = 0.042$	$m[\{t_1\}] = 0.018$	$k = 0.028$	$m[\{t_1\}] = 0.012$
$m_3[\{t_2\}] = 0.2$	$k = 0.084$	$k = 0.036$	$k = 0.056$	$m[\{t_2\}] = 0.024$
$m_3[\{t_3\}] = 0.3$	$m[\{t_3\}] = 0.126$	$m[\{t_3\}] = 0.054$	$m[\{t_3\}] = 0.084$	$m[\{t_3\}] = 0.036$
$m_3[\{t_4\}] = 0.2$	$k = 0.084$	$k = 0.036$	$m_3[\{t_4\}] = 0.056$	$m_3[\{t_4\}] = 0.024$

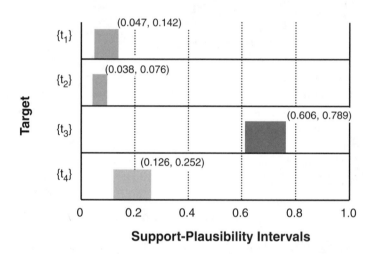

FIGURE 7.7   Support-plausibility intervals result from combining three bodies of evidence.

as to whether or not one of the known targets is actually present. The Dempster-Shafer rule of combination applies as before, but with one modification. When the evidence is inconsistent, their products of masses of evidence are assigned to a measure of inconsistency, termed k. The results from this first part of the procedure are shown in Table 7.6.

The next step is to sum all the corresponding elements of the matrix. Thus, for example, the total mass of evidence assigned to inconsistency, k, is $0.042 + 0.028 + 0.084 + 0.036 + 0.056 + 0.084 + 0.036 = 0.366$. Finally, divide the summed masses of evidence by the normalizing factor $(1 - k)$, which has the value 0.634 in this example. The results for individual targets follow: $m[\{t_1\}] = (0.018 + 0.012)/0.634 = 0.047$; $m[\{t_2\}] = 0.022/0.634 = 0.038$; $m[\{t_3\}] = (0.084 + 0.084 + 0.126 + 0.054 + 0.036)/0.634 = 0.606$; and $m[\{t_4\}] = (0.056 + 0.024)/0.634 = 0.126$. The resulting Support-Plausibility intervals are diagrammed in Figure 7.7.

7.3   An Example Data Fusion System

The characterization of components needed in a data fusion system begins with standard techniques, such as structured, object-oriented, or component-based analysis. A complete analysis is beyond the scope of this chapter; however, the following example should help clarify and demonstrate the concepts discussed herein. The first step in any method of analyzing system requirements is to establish the system context. The system context is summarized in a context diagram that represents a jumping-off point for the abstract decomposition that follows.

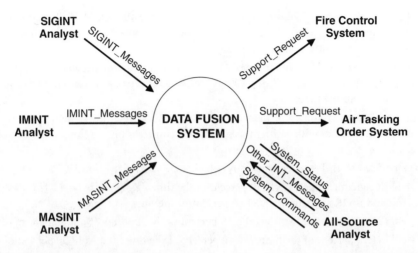

FIGURE 7.8 The Level 0 diagram establishes the system boundary and clarifies what is considered to be inside or outside the system.

7.3.1 System Context

Suppose that an adversarial ground force armed with mobile ground-to-ground missiles has been deployed to harass a friendly force in a fixed location within a contested region. The friendly force is supported by an all-source intelligence center that provides target location data to a fire control system and to an air tasking order system. The fire control system directs artillery, while the air tasking order system automatically requests air support. The data fusion system resides within the all-source intelligence center, and is required to

- Interface with other elements within the center that provide signals intelligence (SIGINT) messages, measurements and analysis intelligence (MASINT) messages, and image intelligence (IMINT) reports;
- Analyze the messages received from those elements to determine the presence and location of the mobile missile launchers;
- Report those locations and any other available information about the status of the located launchers to a human analyst who will determine the optimal response to the threat posed by the launchers.

The human controls the follow-on flow of location and status information to either the fire control system or the air tasking order system.

The system context is summarized in the context diagram, which in structured analysis is known as the Level 0 diagram and is shown in Figure 7.8. Level 0 establishes the system boundary and clarifies what information is regarded as being internal or external to the system.

7.3.1.1 Intelligence Preparation of the Battlefield

Suppose intelligence preparation of the battlefield (IPB) has estimated the following composition of mobile missile batteries operating in the contested region:

- 12 batteries, each with 1 vehicle of type 1 (V1)

  10 with 3 vehicles of type 2 (V2)

  2 with 2 V2

  8 with 3 vehicles of type 3 (V3)

  3 with 2 V3

  1 with 1 V3

- 11 of the *V*1 have SIGINT emitter type 1 (*E*1) and 6 of the *V*1 have SIGINT emitter type 2 (*E*2); all 12 *V*1 have at least one of these two types of emitters. When *V*1 has both emitter types, only one emitter is on at a time, and it is used half the time.
- 24 of the *V*2 have SIGINT emitter type 3 (*E*3) and 17 of them have *E*2; all 34 *V*2 have at least one of these two types of emitters. When *V*2 has both emitter types, only one emitter is on at a time, and it is used half the time.
- 22 of the *V*3 have *E*1 and 19 of them have *E*3; all 31 *V*3 have at least one of these two types of emitters. When *V*3 has both emitter types, only one emitter is on at a time, and it is used half the time.
- Image reports (IMINT) correctly identify vehicle type 98% of the time.
- *V*1 yield IR signature type 1 (*IR*1) 10 percent of the time; IR signature type 2 (*IR*2) 60 percent of the time; and no IR signature (*NoIR*) 30 percent of the time.
- *V*2 yield *IR*1 80 percent of the time, *IR*2 10 percent of the time, and *NoIR* 10 percent of the time.
- *V*3 yield *IR*1 10 percent of the time, *IR*2 40 percent of the time, and *NoIR* 50 percent of the time.
- Batteries are composed of vehicles arrayed within a radius of 1 kilometer centered on *V*1.

7.3.1.2   Initial Estimates

The example experiment involves receiving an intelligence report (i.e., a report that an emitter or an IR signature has been detected) and then determining the vehicle type. There are 77 vehicles. The IPB estimates given in Section 7.3.1.1 indicate that there are nine configurations of vehicle/emitter, as shown in Table 7.7. Furthermore, there are nine configurations of vehicle/IR-signature, as listed in Table 7.8.

TABLE 7.7   Nine Vehicle/Emitter Configurations

Config. No.	Vehicle/Emitter Configuration	Quantity
1	*V*1 with *E*1	6
2*	*V*1 with *E*1 and *E*2	5
3	*V*1 with *E*2	1
4	*V*2 with *E*2	10
5*	*V*2 with *E*2 and *E*3	7
6	*V*2 with *E*3	17
7	*V*3 with *E*1	12
8*	*V*3 with *E*1 and *E*3	10
9	*V*3 with *E*3	9
Total		77

* Note: Each emitter is on half the time, one at a time.

TABLE 7.8   Nine Vehicle/IR-Signature Configurations

Config. No.	Vehicle/IR Signature-Configuration	Quantity
1	*V*1 with *IR*1	1.2
2	*V*1 with *IR*2	7.2
3	*V*1 with *NoIR*	3.6
4	*V*2 with *IR*1	27.2
5	*V*2 with *IR*2	3.4
6	*V*2 with *NoIR*	3.4
7	*V*3 with *IR*1	3.1
8	*V*3 with *IR*2	12.4
9	*V*3 with *NoIR*	15.5
Total		77

7.3.1.2.1 *Initial Probability Estimates*

These considerations (and others) lead to the following prior probabilities:

$$P\left[E1\middle|V1\right] = 8.5/12 \tag{7.30}$$

$$P\left[E1\middle|V2\right] = 0 \tag{7.31}$$

$$P\left[E1\middle|V3\right] = 17/31 \tag{7.32}$$

$$P\left[E2\middle|V1\right] = 3.5/12 \tag{7.33}$$

$$P\left[E2\middle|V2\right] = 13.5/34 \tag{7.34}$$

$$P\left[E2\middle|V3\right] = 0 \tag{7.35}$$

$$P\left[E3\middle|V1\right] = 0 \tag{7.36}$$

$$P\left[E3\middle|V2\right] = 20.5/34 \tag{7.37}$$

$$P\left[E3\middle|V3\right] = 14/31 \tag{7.38}$$

$$P\left[IR1\middle|V1\right] = 0.1 \tag{7.39}$$

$$P\left[IR1\middle|V2\right] = 0.8 \tag{7.40}$$

$$P\left[IR1\middle|V3\right] = 0.1 \tag{7.41}$$

$$P\left[IR2\middle|V1\right] = 0.6 \tag{7.42}$$

$$P\left[IR2\middle|V2\right] = 0.1 \tag{7.43}$$

$$P\left[IR2\middle|V3\right] = 0.4 \tag{7.44}$$

$$P\left[NoIR\middle|V1\right] = 0.3 \tag{7.45}$$

$$P\left[NoIR\middle|V2\right] = 0.1 \tag{7.46}$$

$$P\left[NoIR\middle|V3\right] = 0.5 \tag{7.47}$$

$$P[V1|IMINT] = 0.98 \tag{7.48}$$

$$P[V2|IMINT] = 0.98 \tag{7.49}$$

$$P[V3|IMINT] = 0.98 \tag{7.50}$$

$$P[IR1] = 31.5/77 \tag{7.51}$$

$$P[IR2] = 23/77 \tag{7.52}$$

$$P[NoIR] = 22.5/77 \tag{7.53}$$

$$P[V1] = 12/77 \tag{7.54}$$

$$P[V2] = 34/77 \tag{7.55}$$

$$P[V3] = 31/77 \tag{7.56}$$

$$P[E1] = 25.5/77 \tag{7.57}$$

$$P[E2] = 17/77 \tag{7.58}$$

$$P[E3] = 34.5/77 \tag{7.59}$$

From these, the initial values of the posterior probabilities can be computed using Bayes' rule. In anticipation of an example that follows, examine $P[V1|E1]$, $P[V2|E1]$, and $P[V3|E1]$ (other initial posterior probabilities can, of course, be computed in a similar manner):

$$P[V1|E1] = P[E1|V1] * P[V1]/P[E1] = (8.5/12)*(12/77)*(77/25.5) = 0.333 \tag{7.60}$$

$$P[V1|E2] = P[E2|V1] * P[V1]/P[E2] = (3.5/12)*(12/77)*(77/17) = 0.206 \tag{7.61}$$

$$P[V1|E3] = P[E3|V1] * P[V1]/P[E3] = 0 \tag{7.62}$$

7.3.1.2.2   Initial Belief Estimates

The representation of uncertainty within the Dempster-Shafer approach is an assignment of mass based either on observations reported by knowledge sources or on defined rules. Some rules typically come from an understanding of the problem domain, such as from IPB. To be concrete, consider the prior probabilities to also define a mass of evidence distribution, and from them compute the initial support and plausibility values for each event of interest (these computations are omitted to conserve space in this chapter). Note, however, that a belief system could express the conditional probability information in the IPB in the form of rules. For example, $P[V1|IMINT] = 0.98$ could be expressed as the rule, "If IMINT reports $V1$, then $V1$ occurs with mass of evidence 0.98." Or, "If $E2$ is reported, then the mass of evidence for $V3$ is zero." Additional examples are presented in Section 7.3.2.2 below.

7.3.2 Collections of Spaces

This section considers how this example data fusion system can most efficiently be constructed. This will prove useful when examining the system in operation. This section summarizes some of the aspects of human decision making that motivate the use of collections of spaces. It also characterizes the modules that can implement the collection-of-spaces approach.

7.3.2.1 Motivation

Pearl[31] provided a summary of human performance in decision-making tasks that contrasts the brute force applications of the Bayesian theory. This section is based on his ideas.

The enumeration of all propositions of interest, and all combinations in which they can occur, is exponentially complex. This means that practical systems that attempt to define a joint probability function in a brute force way — by listing arguments in a table and trying to manipulate the table to compute marginal and conditional probabilities — are doomed to fail. In practice, many of the entries in such a matrix will be zero — most combinations of evidence never occur in nature. Pearl noted that humans seem to counter this complexity by only dealing with a small number of propositions at a time. Although humans make probabilistic judgments quickly and reliably when making pair-wise conditional statements (such as the likelihood of finding a target based on observing a certain feature), they estimate joint probabilities of many propositions poorly, hesitantly, and only with difficulty. Further, humans may be reluctant to estimate even pair-wise conditional statements in numerical terms, but they usually state with confidence whether or not two propositions are independent (that is, whether or not one statement influences the truth of the other). Even three-way dependency statements (e.g., measurement M implies target presence given condition C) are handled with confidence and consistency.

This suggests that the fundamental building blocks of human knowledge are not exhaustive entries in a table to estimate joint probabilities. Instead, human knowledge builds on low-order marginal and conditional probabilities defined over small clusters of propositions. Notions of dependence within clusters and of independence between clusters seem basic to human reasoning. Our limited short-term memory and narrow focus of attention seem to imply that "… we reason over fairly local domains incrementally along parallel pathways whose structure implicitly codes information at the knowledge level itself."[31] This apparent manner in which humans manage the complexity of decision making in real world settings can be captured in an approach that unites the concept of collections of probability spaces with Bayesian methods.

7.3.2.2 Component-Based Implementation

A software component is a unit of software with the following characteristics: (1) it is discrete and functionally well defined; (2) it has standardized, clear, and usable interfaces to its methods; and (3) it runs in a container, either with other components or as a stand-alone entity.[32-34] A component may contain object classes, methods, and data that can be reused in a manner similar to the reuse of hardware components of a system (although a component need not be object-oriented). The conceived collection-of-spaces component constitutes such a reusable component.

A collection of related spaces imparts a common nature to components of a data fusion system — a system that may be distributed. Although the spaces exist at various levels of modeling abstraction and observation representation, the common nature provides the foundation for component definition and integration, as indicated in Figure 7.9. Each component comprises knowledge and evidence in a local domain. A local space and its associated observation processes model the domain. This means that the spaces in which the mutually exclusive causes lie are explicitly modeled, and the observation processes are analyzed in physical terms to explicitly characterize the evidence that can be measured. This results in C, a computation component with

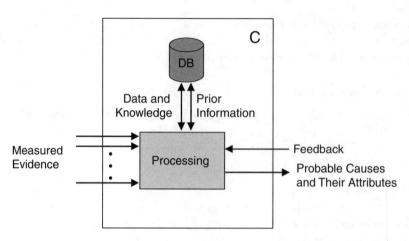

FIGURE 7.9   Components in the system represent local domains, but they share a common nature.

- Input — measured evidence in a common observation space (vector or scalar variables or events and/or logical assertions in a defined domain);
- Internal data — prior information (e.g., prior marginal and conditional probabilities), if available, and data and knowledge about attributes of each cause;
- Processing:
  - Calculate posterior values for occurrence of each cause (e.g., Bayes' theorem and/or exploit lattice structure, Dempster-Shafer combining of evidence mass, fuzzy logic min-max, and/or rule-based calculation),
  - Calculate a figure of merit for each cause (e.g., probability interval, support-plausibility interval, or rule-based assertion),
  - Determine most likely causes from figures of merit,
  - Associate attributes with likely causes (e.g., from local database or from the properties of an observation process),
  - Determine routing of output data,
  - Accept and process feedback;
- Output — likely or plausible causes with figures of merit (e.g., attributes of causes and routing information);
- Feedback — from other components in the data fusion system (e.g., data and knowledge updates and updates affecting prior information).

7.3.2.3   Component Examples

Consider as examples two components that share the task of determining the vehicle type based on message traffic that flows into the system.

Both probability and belief systems are built on the foundation of an exhaustive set of possible outcomes, S, and go on to consider subsets of S. Therefore, the space defined for any component consists of S, Ω (the collection of subsets of S), and the assignment of either an initial probability or an initial mass of evidence distribution (and its associated support and plausibility) for as many possible outcomes as is feasible. Two examples follow.

1. $S_1 = \{V1, V2, V3\}$ because these comprise the totality of observable outcomes with respect to vehicle type. There are $2^3 = 8$ elements in Ω_1. The initial (prior) probabilities are computed from the data stored in the database:

$$P\big[\{V1\}\big] = 12/77 \tag{7.63}$$

$$P[\{V2\}] = 34/77 \qquad (7.64)$$

$$P[\{V3\}] = 31/77 \qquad (7.65)$$

These same values could be used to initialize a mass of evidence distribution, and the resulting support and plausibility could be calculated.

2. $S_2 = \{E1, E2, E3\}$ because these comprise the totality of observable outcomes with respect to emitter type. There are $2^3 = 8$ elements in Ω_2. The initial (prior) probabilities are computed from the data stored in the database:

$$P[\{E1\}] = 25.5/77 \qquad (7.66)$$

$$P[\{E2\}] = 17/77 \qquad (7.67)$$

$$P[\{E3\}] = 34.5/77 \qquad (7.68)$$

These same values could be used to initialize a mass of evidence distribution, and the resulting support and plausibility could be calculated.

7.3.3 The System in Operation

Now suppose a first message arrives from the SIGINT analyst that states $E1$ has been identified at location (x_1, y_1) at time t_1 with confidence 0.9. Then, suppose a later message arrives from the MASINT analyst that states NoIR is detected at location (x_1, y_1) at time t_2 and that t_2 is shortly after t_1. The MASINT analyst appends a note that states the IR detector has a miss rate of 0.05.

7.3.3.1 The Probability System Response

The emitter probability component employs interval probabilities to update the elements of its probability space lattice: state $E1$ with probability 0.9; state $E2$ with probability x; state $E3$ with probability $0.1 - x$, where $0 \le x \le 0.1$; and the others based on the subset relationships. The vehicle component interprets this to mean a superposition of probability states because no true conditioning event yet exists. The vehicle component responds by weighting the three states together:

$$(0.9) * P[V1|E1] + (x) * P[V1|E2] + (0.1 - x) * P[V1|E3] = (0.9) * (0.333) + (x) * (0.206) \qquad (7.69)$$

This leads to a range of values that expresses an updated P[V1] at time t_1 and location (x_1, y_1), which is denoted as $P_{1,1}[V1]$: $0.300 \le P_{1,1}[V1] \le 0.321$. Similar ranges are computed for V2 and V3, respectively: $0.059 \le P_{1,1}[V2] \le 0.079$ and $0.600 \le P_{1,1}[V3] \le 0.641$. The vehicle component represents these as interval probabilities assigned to the elements of its probability space lattice, as shown in Figure 7.10. Note that $\Sigma P[V_i]$ must equal 1, and that simultaneous choices exist within these three intervals that satisfy this constraint.

Even the first message results in a preliminary identification of vehicle type. Indeed, if the maximum possible values are assigned to $P_{1,1}[V1]$ (0.321) and $P_{1,1}[V2]$ (0.079), then $P_{1,1}[V3] = 0.600$. The ratio $P_{1,1}[V3]/P_{1,1}[V1]$ would equal $0.600/0.321 = 1.87$, which can be tested against the threshold defined in Section 7.2.4, above:

$$\text{Decide } V3 \text{ if } 1.87 > \frac{(C_{3,1} - C_{1,1})P[V1]}{(C_{1,3} - C_{3,3})P[V3]} \qquad (7.70)$$

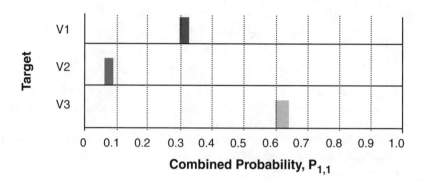

FIGURE 7.10   The first message results in a preliminary identification of vehicle type.

Assuming that the costs of mistakes are equal, the threshold becomes $P[V1]/P[V3] = 12/31 = 0.387$. Clearly, $1.87 > 0.387$, which leads to a decision with (assumed) acceptable risk that $V3$ has been detected at (x_1, y_1). However, let us postpone this decision to examine the effect that the receipt of additional evidence has on the probability intervals.

The IR component employs interval probabilities to update the elements of its probability space lattice: state *NoIR* with probability 0.95; state *IR*1 with probability x; state *IR*2 with probability $0.05 - x$, where $0 \leq x \leq 0.05$; and the others based on the subset relationships. The vehicle component could interpret this to mean a superposition of probability states because no true conditioning event yet exists. The vehicle component would then respond by weighting the three states together. As before, three ranges of values would express the updated probabilities of $V1$, $V2$, and $V3$:

$$0.154 \leq P_{1,2}\left[V1\right] \leq 0.168 \tag{7.71}$$

$$0.146 \leq P_{1,2}\left[V2\right] \leq 0.181 \tag{7.72}$$

$$0.661 \leq P_{1,2}\left[V3\right] \leq 0.683 \tag{7.73}$$

However, in this case, the vehicle component has already made an initial assessment of vehicle type at location (x_1, y_1), and a determination must be made about how the probability intervals evolve in this setting where observations arrive in temporal sequence. Computing $P[V_i|E_j, IR_k]$ from $P[V_i, E_j, IR_k]$ and $P[E_j, IR_k]$ using the intelligence preparation of the battlefield information presented in Section 7.3.1.1, yields

$$P\left[V_i, E_j, IR_k\right] = P\left[E_j|V_i\right]P\left[IR_k|V_i\right]P\left[V_i\right] \tag{7.74}$$

$$P\left[E_j, IR_k\right] = \sum P\left[V_i, E_j, IR_k\right] \tag{7.75}$$

$$P\left[V_i|E_j, IR_k\right] = \frac{P\left[V_i, E_j, IR_k\right]}{P\left[E_j, IR_k\right]} \tag{7.76}$$

Next, each $P[V_i|E_j, IR_k]$ must be weighted by the *reported* values of $P[E_j]$ and $P[IR_k]$, and then summed over j and k to obtain $P_c[V_i]$, which denotes the combined probability interval for V_i. The required computations are shown in Tables 7.9, 7.10, and 7.11(a–c).

TABLE 7.9 Computation of $P[V_i, E_j, IR_k]$ from the IPB

i	j	k	$P[E_j|V_i] \, P[IR_k|V_i] \, P[V_i]$
1	1	1	(8.5/12) (0.3) (12/77)
1	1	2	(8.5/12) (0.1) (12/77)
1	1	3	(8.5/12) (0.6) (12/77)
1	2	1	(3.5/12) (0.3) (12/77)
1	2	2	(3.5/12) (0.1) (12/77)
1	2	3	(3.5/12) (0.6) (12/77)
2	2	1	(13.5/34) (0.1) (34/77)
2	2	2	(13.5/34) (0.8) (34/77)
2	2	3	(13.5/34) (0.1) (34/77)
2	3	1	(20.5/34) (0.1) (34/77)
2	3	2	(20.5/34) (0.8) (34/77)
2	3	3	(20.5/34) (0.1) (34/77)
3	1	1	(17/31) (0.5) (31/77)
3	1	2	(17/31) (0.1) (31/77)
3	1	3	(17/31) (0.4) (31/77)
3	3	1	(14/31) (0.5) (31/77)
3	3	2	(14/31) (0.1) (31/77)
3	3	3	(14/31) (0.4) (31/77)

TABLE 7.10 Computation of $P[E_j, IR_k]$ from Information in Table 7.9

j	k	$P[E_j, IR_k]$
1	1	(8.5/12) (0.3) (12/77) + (17/31) (0.5) (31/77)
1	2	(8.5/12) (0.1) (12/77) + (17/31) (0.1) (31/77)
1	3	(8.5/12) (0.6) (12/77) + (17/31) (0.4) (31/77)
2	1	(3.5/12) (0.3) (12/77) + (13.5/34) (0.1) (34/77)
2	2	(3.5/12) (0.1) (12/77) + (13.5/34) (0.8) (34/77)
2	3	(3.5/12) (0.6) (12/77) + (13.5/34) (0.1) (34/77)
3	1	(20.5/34) (0.1) (34/77) + (14/31) (0.5) (31/77)
3	2	(20.5/34) (0.8) (34/77) + (14/31) (0.1) (31/77)
3	3	(20.5/34) (0.1) (34/77) + (14/31) (0.4) (31/77)

TABLE 7.11a Computation of $P[V_1|E_j, IR_k]$

j	k	$P[V_1|E_j, IR_k]$	Reported $P[E_j]$	Reported $P[IR_k]$	$P[V_1, E_j, IR_k]$
1	1	0.231	0.9	0.95	0.178
1	2	0.333	0.9	x	[0, 0.015]
1	3	0.429	0.9	$0.05 - x$	[0, 0.019]
2	1	0.438	y	0.95	[0, 0.042]
2	2	0.032	y	x	[0, 0.000]
2	3	0.609	y	$0.05 - x$	[0, 0.005]

Note: These values are then weighted by reported information and summed to yield $P_c[V1] = [0.178, 0.259]$. Note that $0 \le x \le 0.05$ and $0 \le y \le 0.10$.

These computations result in the updated probability intervals depicted in Figure 7.11:

$$0.178 \le P_c\big[V1\big] \le 0.259 \qquad (7.77)$$

$$0.000 \le P_c\big[V2\big] \le 0.088 \qquad (7.78)$$

TABLE 7.11b Computation of $P[V_2|E_j,IR_k]$

j	k	$P[V_2|E_j,IR_k]$	Reported $P[E_j]$	Reported $P[IR_k]$	$P[V_2,E_j,IR_k]$
2	1	0.563	y	0.95	[0, 0.053]
2	2	0.967	y	x	[0, 0.005]
2	3	0.391	y	$0.05 - x$	[0, 0.002]
3	1	0.227	$0.1 - y$	0.95	[0, 0.022]
3	2	0.921	$0.1 - y$	x	[0, 0.005]
3	3	0.268	$0.1 - y$	$0.05 - x$	[0, 0.001]

Note: These values are then weighted by reported information and summed to yield $P_c[V2] = [0.000, 0.088]$. Note that $0 \le x \le 0.05$ and $0 \le y \le 0.10$.

TABLE 7.11c Computation of $P[V_3|E_j,IR_k]$

j	k	$P[V_3|E_j,IR_k]$	Reported $P[E_j]$	Reported $P[IR_k]$	$P[V_3,E_j,IR_k]$
1	1	0.939	0.9	0.95	0.803
1	2	0.095	0.9	x	[0, 0.004]
1	3	0.889	0.9	$0.05 - x$	[0, 0.040]
3	1	0.773	y	0.95	[0, 0.073]
3	2	0.079	y	x	[0, 0.000]
3	3	0.732	y	$0.05 - x$	[0, 0.004]

Note: These values are then weighted by reported information and summed to yield $P_c[V3] = [0.803, 0.924]$. Note that $0 \le x \le 0.05$ and $0 \le y \le 0.10$.

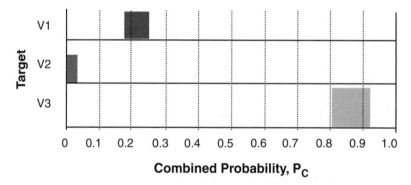

FIGURE 7.11 The combined interval estimates result from the probability-evolution computations (compare with Figure 7.10).

$$0.803 \le P_c\left[V3\right] \le 0.924 \tag{7.79}$$

Note again the simultaneous choices within these three intervals that satisfy the constraint $\sum P[V_i] = 1$. Assigning the minimum possible values to $P_c[V3]$ (0.803) and to $P_c[V2]$ (0) gives a maximum possible value for $P_c[V1]$ of 0.197. The ratio $P_c[V3]/P_c[V1]$ then equals $0.803/0.197 = 4.08$, which provides even greater confidence (i.e., reduced risk) than before in a decision that $V3$ is present at (x_1, y_1). At this point, the superposition of probability states could be "popped" and $V3$ could be declared to be present at location (x_1, y_1) as a true conditioning event.

Also note that an engineering approximation is available to simplify the calculations. Product terms that include x and y contribute little to the summations when x and y are on the order of 0.1 or less. Ignoring such terms would enable analysts to arrive more quickly at the estimates $P_c[V1] \approx 0.178$ and $P_c[V3] \approx 0.803$, and then $P_c[V2] \approx 0.019$ in order to satisfy the known constraint. Note that these approximations comprise one set of choices that simultaneously lie within the three probability intervals.

TABLE 7.12 Combination of Rule-Provided Information with Emitter Information

	$m_e(E1) = 0.9$	$m_e(S) = 0.1$
$m_{r1}(V1) = 0.333$	$m_1(V1) = (0.333)(0.9) = 0.3$	—
$m_{r1}(V2) = 0$	$m_1(V2) = (0)(0.9) = 0$	—
$m_{r1}(V3) = 0.667$	$m_1(V3) = (0.667)(0.9) = 0.6$	—
$m_{r1}(S) = 0$	—	$m_1(S) = 0.1$

TABLE 7.13 Computation of Support-Plausibility Intervals

Vehicle	Support	Plausibility
V1	0.3	0.4
V2	0	0.1
V3	0.6	0.7

In addition, other components can be defined. For example, a vehicle-tracking component could use standard probability methods to keep track of detected vehicles' trajectories over time. Another component could identify batteries by using a distance metric to group detected vehicles into candidate batteries and track the batteries as entities over time.

7.3.3.2 The Belief System Response

After the first message arrives, the emitter belief component assigns a mass of evidence value to state $E1$ of 0.9 and a mass of evidence value to state S of 0.1. The vehicle component uses a rule base that is constructed by essentially duplicating the calculation of initial posterior conditional probabilities from the information in the IPB. Three examples follow.

Rule 1: If emitter $E1$ is reported, then $m(V1) = 0.333$.
Rule 2: If emitter $E1$ is reported, then $m(V2) = 0$.
Rule 3: If emitter $E1$ is reported, then $m(V3) = 0.667$.

The vehicle component combines the information from the emitter component with the information in its rule base by multiplying the rule-provided masses of evidence by an appropriate mass of evidence from the emitter component, and by transferring any mass of evidence assigned by the emitter component to S as shown in Table 7.12. Note that the combined masses of evidence sum to one, as required; it is easy to show that this is always the outcome when transferring the ignorance this way.

The vehicle component then computes the Support-Plausibility interval for each vehicle, as shown in Table 7.13. A comparison of these intervals with those produced by the probability system after receiving the first message (see Figure 7.10) shows that they are qualitatively the same. Unlike probability, however, belief theory is not equipped with a clear-cut decision-making rule.

After the second message arrives, the IR belief component assigns a mass of evidence value to state *NoIR* of 0.95 and a mass of evidence value to state S of 0.05. The vehicle component also contains rules that relate MASINT reports to vehicles. Three examples follow.

Rule 1: If *NoIR* is reported, then $m(V1) = 0.160$.
Rule 2: If *NoIR* is reported, then $m(V2) = 0.151$.
Rule 3: If *NoIR* is reported, then $m(V3) = 0.689$.

The vehicle component combines the information from the second message with the information in its rule base by multiplying the rule-provided masses of evidence by an appropriate mass of evidence from the message, and by transferring the mass of evidence assigned by the IR component to S as shown in Table 7.14.

The vehicle component now combines the masses of evidence from Tables 7.12 and 7.14 using the standard Dempster-Shafer combination rule as shown in Table 7.15. The inconsistency values from

TABLE 7.14   Combination of Rule-Provided Information
with IR Information

	$m_{ir}(NoIR) = 0.95$	$m_{ir}(S) = 0.05$
$m_{r2}(V1) = 0.160$	$m_2 = (0.160)(0.95) = 0.152$	—
$m_{r2}(V2) = 0.151$	$m_2 = (0.151)(0.95) = 0.143$	—
$m_{r2}(V3) = 0.689$	$m_2 = (0.689)(0.95) = 0.655$	—
$m_{r2}(S) = 0$	—	$m_2 = 0.05$

TABLE 7.15   Combination of Masses of Evidence Derived from Messages 1 and 2

	$m_1(V1) = 0.3$	$m_1(V2) = 0$	$m_1(V3) = 0.6$	$m_1(S) = 0.1$
$m_2(V1) = 0.152$	$m_c(V1) = 0.0456$	$k = 0$	$k = 0.0912$	$m_c(V1) = 0.0152$
$m_2(V2) = 0.143$	$k = 0.0429$	$m_c(V2) = 0$	$k = 0.0858$	$m_c(V2) = 0.0143$
$m_2(V3) = 0.655$	$k = 0.1965$	$k = 0$	$m_c(V3) = 0.3930$	$m_c(V3) = 0.0655$
$m_2(S) = 0.05$	$m_c(V1) = 0.0150$	$m_c(V2) = 0$	$m_c(V3) = 0.0300$	$m_c(S) = 0.0050$

TABLE 7.16   Calculation of Normalized Combined
Masses of Evidence

$m_c(V1)$	$(0.0456 + 0.0152 + .0150)/0.5836 = 0.1299$
$m_c(V2)$	$(0.0143)/0.5836 = 0.0245$
$m_c(V3)$	$(0.3930 + 0.0300 + 0.0655)/0.5836 = 0.8370$
$m_c(S)$	$(0.0050)/0.5836 = 0.0086$

TABLE 7.17   Computation of New Support-Plausibility Intervals

Vehicle	Support	Plausibility
V1	0.1299	0.1385
V2	0.0245	0.0331
V3	0.8370	0.8456

Table 7.15 are summed to calculate the normalization factor $(1 - k) = (1 - 0.4164) = 0.5836$. Then the new combined masses of evidence are calculated as shown in Table 7.16. Finally, the Support-Plausibility interval for each vehicle is calculated as shown in Table 7.17.

A comparison of these intervals with those produced by the probability system, after receiving the second message (see Figure 7.10), shows that they are again qualitatively the same. Even though belief theory is not equipped with a clear-cut decision-making rule, the situation that presents itself here clearly justifies a decision that vehicle 3 has been detected.

7.3.4   Summary

This chapter has introduced all of the key concepts and has provided a vehicle-identification example to show that components can characterize local domains. These components can communicate with each other to accumulate evidence up an abstraction hierarchy. New spaces can be formed in ways that relate to the earlier spaces but that account for the differences in the level of abstraction, as well as for the amount and kinds of evidence available. Partial information can be collected in local domains, and the domains can eventually result in a situation level. However, capturing all of the information in a single space is unnecessary and impractical. The changes in the spaces track the changes in the levels of our human understanding.

7.4   Contrasts and Conclusion

Other authors have described related ideas. Peter Cheeseman has argued strongly for adoption of Bayesian techniques in favor of alternative methods of combining evidence,[35] and Judea Pearl[31] has developed techniques for implementing Bayesian methods in a distributed network environment. Here, the idea of a collection of spaces has been proposed as the idea that both clarifies the theoretical underpinnings of data fusion methods and makes their implementation practical.

The highly modular approach that is described herein is well suited to a modern component-based software design pattern. Multiple processes, each matched to the computations of individual components, could lead naturally to real-time systems that solve real-world data fusion problems.

Appendix 7.A   The Axiomatic Definition of Probability

Formally, a *probability space* is a three-tuple, (S, Ω, P), where

- S is a set of observable outcomes from some experiment of interest (the totality of outcomes).
- Ω is a collection of subsets of S with the following properties:
    1. If A is an element of Ω then, the complement of A (with respect to S) is also an element of Ω.
    2. If both A and B are elements of Ω, then the union of A and B is also an element of Ω.
    3. If A_i are elements of Ω for $i = 1, 2, \ldots$, then any countable union of A_i is also an element of Ω.
- P is a set-function defined on the elements of Ω (termed events) that has the following properties:
    1. To each event A in Ω there is assigned a nonnegative real number, $P(A)$ (that is, $0 \, \Omega \, P(A)$).
    2. $P(S) = 1$.
    3. For A and B both in Ω, if the intersection of A and B is empty, then $P(A \text{ or } B) = P(A) + P(B)$.

and, if the intersection of A_i and A_j is empty when $i \neq j$, then $P(\cup A_i) = \sum P(A_i)$.

The axioms presented here are in essentially the same form as proposed first (in 1933) by the Russian mathematician Andrei Nikolaevich Kolmogorov.[36]

References

1. F. G. Cozman, Introduction to the Theory of Sets of Probabilities, http://www.cs.cmu.edu/~fgcozman/QuasiBayesianInformation/.
2. H. E. Kyburg, Jr., Interval Valued Probabilities, http://ippserv.rug.ac.be.
3. *Proceedings of the First International Symposium on Imprecise Probabilities and Their Applications* (ISIPTA), Ghent, Belgium, June/July 1999.
4. L. A. Zadeh, *Fuzzy Sets and Applications: Selected Papers by L. A. Zadeh*, NY: John Wiley & Sons, Inc., 1987.
5. D. Dubois and H. Prade, *Possibility Theory: An Approach to Computerized Processing of Uncertainty*, NY: Plenum Press, 1988.
6. M. A. Abidi and R. C. Gonzalez, Eds., *Data Fusion in Robotics and Machine Intelligence*, San Diego, CA: Academic Press, 1992, 481–505.
7. A. P. Dempster, A generalization of Bayesian inference, *J. Royal Statistical Soc., Series B*, vol. 30, 205–247, 1968.
8. G. Shafer, *A Mathematical Theory of Evidence*, Princeton, NJ: Princeton University Press, 1976.
9. D. L. Hall, *Mathematical Techniques in Multisensor Data Fusion*, Boston: Artech House, Inc., 1992, 179–187.
10. S. S. Blackman, *Multiple Target Tracking with Radar Applications*, Norwood, MA: Artech House, Inc., 1986, 380–387. (The example in section 2.4 of this chapter follows Blackman's example.)
11. W. B. Davenport, *Probability and Random Processes*, NY: McGraw-Hill Book Co., 1970, 46–49.

12. R. T. Cox, Of inference and inquiry — an essay in inductive logic, *The Maximum Entropy Formalism*, Levine and Tribus, Eds., Cambridge, MA: MIT Press, 1979.

13. A. Papoulis, *Probability, Random Variables, and Stochastic Processes*, NY: McGraw-Hill Book Co., 1965, 11–12.

14. E. T. Jaynes, The well-posed problem, *Foundations of Physics*, vol. 3, 1973, 477–493.

15. Wozencraft and Jacobs, *Principles of Communication Engineering*, 2nd Edition, NY: John Wiley & Sons, 1967, 220.

16. G. F. Hughes, On the mean accuracy of statistical pattern recognition accuracy, *IEEE Trans. Info. Th.*, vol. IT-14, 55–63, January 1968.

17. J. M. Van Campenhout, On the peaking of the Hughes mean recognition accuracy: the resolution of an apparent paradox, *IEEE Trans. Sys., Man, & Cybernetics*, vol. SMC-8, 390–395, May 1978.

18. M. Gardner, The paradox of the non-transitive dice and the principle of indifference, *Scientific American*, 110–114, December 1970.

19. H. L. Van Trees, *Detection, Estimation, and Modulation Theory, Part 1*, NY: John Wiley & Sons, 1968, 23–36.

20. W. S. Meisel, *Computer-Oriented Approaches to Pattern Recognition*, NY and London: Academic Press, 1972, 98–118.

21. R. O. Duda and P. E. Hart, *Pattern Classification and Scene Analysis*, NY: John Wiley & Sons, 1973, 88–94.

22. L. A. Zadeh, Fuzzy sets as a basis for a theory of possibility, *Fuzzy Sets and Systems*, vol. 1, 3–28, 1978.

23. H. T. Nguyen, Some mathematical tools for linguistic probabilities, *Fuzzy Sets and Systems*, vol. 2, 53–65, 1979.

24. L. A. Zadeh, Fuzzy sets, *Information and Control*, 338–353, 1965.

25. P. R. Halmos, *Naïve Set Theory*, NY: Van Nostrand, 1960.

26. J. C. Bezdek and S. K. Pal, Eds., Background, significance, and key points, *Fuzzy Models for Pattern Recognition*, NY: IEEE Press, 1992, chap. 1.

27. J. M. Mendel, Fuzzy logic systems for engineering: a tutorial, *Proc. IEEE*, 83(3), 345–377, 1995.

28. H. E. Kyburg, Jr., Bayesian and non-Bayesian evidential updating, *Artificial Intelligence*, vol. 31, 271–293, 1987.

29. P. Smets, What is Dempster-Shafer's model? *Advances in Dempster-Shafer Theory of Evidence*, R. R. Yager, M. Fedrizzi, and J. Kasprzyk, Eds., NY: John Wiley & Sons, 1994, 5–34.

30. A. P. Dempster, Upper and lower probabilities induced by a multi-valued mapping, *Annals of Math. Stat.*, vol. 38, 325–339, 1967.

31. J. Pearl, Fusion, propagation, and structuring in belief networks, *Int'l. J. AI*, 241–288, 1986.

32. S. Garone, Managing Component-Based Development: SELECT Software Tools, an IDC paper, www.selectst.com/downloads/IDC/IDC.asp.

33. C. Szyperski, *Component Software: Beyond Object-Oriented Programming*, NY: ACM Press, 1998.

34. A. W. Brown, Ed., *Component-Based Software Engineering: Selected Papers from the Software Engineering Institute*, Los Alamos, CA: IEEE Computer Society Press, 1996.

35. P. Cheeseman, In defense of probability, *Int. J. Comput. Artificial Intelligence*, 1002–1009, 1985.

36. A. N. Kolmogorov, *Foundations of the Theory of Probability*, NY: Chelsea Publishing Co., 1956.

II

Advanced Tracking and Association Methods

8 **Target Tracking Using Probabilistic Data Association-Based Techniques with Applications to Sonar, Radar, and EO Sensors**   *T. Kirubarajan and Yaakov Bar-Shalom* 8-1
Introduction • Probabilistic Data Association • Low Observable TMA Using the ML-PDA Approach with Features • The IMMPDAF for Tracking Maneuvering Targets • A Flexible-Window ML-PDA Estimator for Tracking Low Observable (LO) Targets • Summary

9 **An Introduction to the Combinatorics of Optimal and Approximate Data Association**   *Jeffrey K. Uhlmann* 9-1
Introduction • Background • Most Probable Assignments • Optimal Approach • Computational Considerations • Efficient Computation of the JAM • Crude Permanent Approximations • Approximations Based on Permanent Inequalities • Comparisons of Different Approaches • Large-Scale Data Associations • Generalizations • Conclusions • Acknowledgments • Appendix 9.A Algorithm for Data Association Experiment

10 **A Bayesian Approach to Multiple-Target Tracking**   *Lawrence D. Stone* 10-1
Introduction • Bayesian Formulation of the Single-Target Tracking Problem • Multiple-Target Tracking without Contacts or Association (Unified Tracking) • Multiple-Hypothesis Tracking (MHT) • Relationship of Unified Tracking to MHT and Other Tracking Approaches • Likelihood Ratio Detection and Tracking

11 **Data Association Using Multiple Frame Assignments**   *Aubrey B. Poore, Suihua Lu, and Brian J. Suchomel* 11-1
Introduction • Problem Background • Assignment Formulation of Some General Data Association Problems • Multiple Frame Track Initiation and Track Maintenance • Algorithms • Future Directions

12 **General Decentralized Data Fusion with Covariance Intersection (CI)**   *Simon Julier and Jeffrey K. Uhlmann* 12-1
Introduction • Decentralized Data Fusion • Covariance Intersection • Using Covariance Intersection for Distributed Data Fusion • Extended Example • Incorporating Known Independent Information • Conclusions • Appendix 12.A   The Consistency of CI • Appendix 12.B   MATLAB Source Code (Conventional CI and Split CI)

13 Data Fusion in Nonlinear Systems *Simon Julier and Jeffrey K. Uhlmann* **13**-1
Introduction • Estimation in Nonlinear Systems • The Unscented Transformation • Uses of the Transformation • The Unscented Filter • Case Study: Using the UF with Linearization Errors • Case Study: Using the UF with a High-Order Nonlinear System • Multilevel Sensor Fusion • Conclusions

14 Random Set Theory for Target Tracking and Identification *Ronald Mahler* **14**-1
Introduction • Basic Statistics for Tracking and Identification • Multitarget Sensor Models • Multitarget Motion Models • The FISST Multisource-Multitarget Calculus • FISST Multisource-Multitarget Statistics • Optimal-Bayes Fusion, Tracking, ID • Robust-Bayes Fusion, Tracking, ID • Summary and Conclusions

8

Target Tracking Using Probabilistic Data Association-Based Techniques with Applications to Sonar, Radar, and EO Sensors

T. Kirubarajan
University of Connecticut

Yaakov Bar-Shalom
University of Connecticut

8.1 Introduction **8**-1
8.2 Probabilistic Data Association... **8**-2
    Assumptions • The PDAF Approach • Measurement
    Validation • The State Estimation • The State and
    Covariance Update • The Prediction Equations • The
    Probabilistic Data Association • The Parametric PDA • The
    Nonparametric PDA
8.3 Low Observable TMA Using the ML-PDA Approach
    with Features **8**-8
    Amplitude Information Feature • Target Models • Maximum
    Likelihood Estimator Combined with PDA — The ML-
    PDA • Cramér-Rao Lower Bound for the Estimate • Results
8.4 The IMMPDAF for Tracking Maneuvering Targets **8**-17
    Coordinate Selection • Track Formation • Track
    Maintenance • Track Termination • Simulation Results
8.5 A Flexible-Window ML-PDA Estimator for Tracking
    Low Observable (LO) Targets **8**-27
    The Scenario • Formulation of the ML-PDA Estimator •
    Adaptive ML-PDA • Results
8.6 Summary... **8**-37
References **8**-37

8.1 Introduction

In tracking targets with less-than-unity probability of detection in the presence of false alarms (clutter), data association — deciding *which* of the received multiple measurements to use to update each track — is crucial. A number of algorithms have been developed to solve this problem.[1-4] Two simple solutions are the Strongest Neighbor Filter (SNF) and the Nearest Neighbor Filter (NNF). In the SNF, the signal with the highest intensity among the validated measurements (in a gate) is used for track update and the others are discarded. In the NNF, the measurement closest to the predicted measurement is used. While these simple techniques work reasonably well with benign targets in sparse scenarios, they begin to fail as the false alarm rate increases or with low observable (low probability of target detection) maneuvering targets.[5,6] Instead of using only one measurement among the received ones and discarding the others, an alternative approach is to use all of the validated measurements with different weights (probabilities), known as Probabilistic Data Association (PDA).[3] The standard PDA and its numerous improved versions have been shown to be very effective in tracking a single target in clutter.[6,7]

Data association becomes more difficult with multiple targets where the tracks compete for measurements. Here, in addition to a track validating multiple measurements as in the single target case, a measurement itself can be validated by multiple tracks (i.e., contention occurs among tracks for measurements). Many algorithms exist to handle this contention. The Joint Probabilistic Data Association (JPDA) algorithm is used to track multiple targets by evaluating the measurement-to-track association probabilities and combining them to find the state estimate.[3] The Multiple-Hypothesis Tracking (MHT) is a more powerful (but much more complex) algorithm that handles the multitarget tracking problem by evaluating the likelihood that there is a target given a sequence of measurements.[4] In the tracking benchmark problem[8] designed to compare the performance of different algorithms for tracking highly maneuvering targets in the presence of electronic countermeasures, the PDA-based estimator, in conjunction with the Interacting Multiple Model (IMM) estimator, yielded one of the best solutions. Its performance was comparable to that of the MHT algorithm.[6,9]

This chapter presents an overview of the PDA technique and its application for different target-tracking scenarios. Section 8.2 summarizes the PDA technique. Section 8.3 describes the use of the PDA technique for tracking low observable targets with passive sonar measurements. This target motion analysis (TMA) is an application of the PDA technique, in conjunction with the maximum likelihood (ML) approach for target motion parameter estimation via a batch procedure. Section 8.4 presents the use of the PDA technique for tracking highly maneuvering targets and for radar resource management. It illustrates the application of the PDA technique for recursive state estimation using the IMMPDAF. Section 8.5 presents a state-of-the-art sliding-window (which can also expand and contract) parameter estimator using the PDA approach for tracking the state of a maneuvering target using measurements from an electro-optical sensor. This, while still a batch procedure, offers the flexibility of varying the batches depending on the estimation results.

8.2   Probabilistic Data Association

The PDA algorithm calculates in real-time the probability that each validated measurement is attributable to the target of interest. This probabilistic (Bayesian) information is used in a tracking filter, the PDA filter (PDAF), which accounts for the measurement origin uncertainty.

8.2.1   Assumptions

The following assumptions are made to obtain the recursive PDAF state estimator (tracker):

- There is only one target of interest whose state evolves according to a dynamic equation driven by process noise.
- The track has been initialized.
- The past information about the target is summarized approximately by

$$p\left[x(k)\big|Z^{k-1}\right] = N\left[x(k); \hat{x}(k|k-1), P(k|k-1)\right] \tag{8.1}$$

where $N[x(k); \hat{x}(k|k-1)]$ denotes the normal probability density function (pdf) with argument $x(k)$, mean $\hat{x}(k|k-1)$, and covariance matrix $P(k|k-1)$. This assumption of the PDAF is similar to the GPB1 (Generalized Pseudo-Bayesian) approach,[10] where a single "lumped" state estimate is a quasi-sufficient statistic.

- At each time, a validation region as in Reference 3 is set up (see Equation 8.4).
- Among the possibly several validated measurements, at most one of them can be target-originated — if the target was detected and the corresponding measurement fell into the validation region.
- The remaining measurements are assumed to be false alarms or clutter and are modeled as independent identically distributed (iid) measurements with uniform spatial distribution.
- The target detections occur independently over time with known probability PD.

These assumptions enable a state estimation scheme to be obtained, which is almost as simple as the Kalman filter, but much more effective in clutter.

8.2.2 The PDAF Approach

The PDAF uses a decomposition of the estimation with respect to the origin of each element of the latest set of validated measurements, denoted as

$$Z(k) = \left\{ z_i(k) \right\}_{i=1}^{m(k)} \tag{8.2}$$

where $z_i(k)$ is the i-th validated measurement and $m(k)$ is the number of measurements in the validation region at time k.

The cumulative set (sequence) of measurements* is

$$Z^k = \left\{ Z(j) \right\}_{j=1}^{k} \tag{8.3}$$

8.2.3 Measurement Validation

From the Gaussian assumption (Equation 8.1), the validation region is the elliptical region

$$V(k, \gamma) = \left\{ Z : \left[z - \hat{z}(k|k-1) \right]' S(k)^{-1} \left[z - \hat{z}(k|k-1) \right] \le \gamma \right\} \tag{8.4}$$

where γ is the gate threshold and

$$S(k) = H(k) P(k|k-1) H(k)' + R(k) \tag{8.5}$$

is the covariance of the innovation corresponding to the true measurement. The volume of the validation region (Equation 8.4) is

$$V(k) = c_{n_z} \left| \gamma S(k) \right|^{1/2} = c_{n_z} \gamma^{\frac{n_z}{2}} \left| S(k) \right|^{1/2} \tag{8.6}$$

---

* When the running index is a time argument, a sequence exists; otherwise it is a set where the order is not relevant. The context should indicate which is the case.

where the coefficient c_{n_z} depends on the dimension of the measurement (it is the volume of the n_z-dimensional unit hypersphere: $c_1 = 2$, $c_2 = \pi$, $c_3 = 4\pi/3$, etc.).

8.2.4   The State Estimation

In view of the assumptions listed, the association events

$$\theta_i(k) = \begin{cases} \{z_i(k) \text{ is the target originated measurement}\} & i = 1,\dots m(k) \\ \{\text{none of the measurements is target originated}\} & i = 0 \end{cases} \tag{8.7}$$

are mutually exclusive and exhaustive for $m(k) \geq 1$.

Using the total probability theorem[10] with regard to the above events, the conditional mean of the state at time k can be written as

$$\hat{x}(k|k) = E\left[x(k)\big|Z^k\right]$$

$$= \sum_{i=0}^{m(k)} E\left[x(k)\big|\theta_i(k), Z^k\right] P\left\{\theta_i(k)\big|Z^k\right\} \tag{8.8}$$

$$= \sum_{i=0}^{m(k)} \hat{x}_i(k|k)\beta_i(k)$$

where $\hat{x}_i(k|k)$ is the updated state conditioned on the event that the i-th validated measurement is correct, and

$$\beta_i(k) \triangleq P\left\{\theta_i(k)\big|Z_k\right\} \tag{8.9}$$

is the conditional probability of this event — the association probability, obtained from the PDA procedure presented in the next subsection.

The estimate conditioned on measurement i being correct is

$$\hat{x}_i(k|k) = \hat{x}(k|k-1) + W(k)v_i(k) \qquad i = 1,\dots, m(k) \tag{8.10}$$

where the corresponding innovation is

$$v_i(k) = z_i(k) - \hat{z}(k|k-1) \tag{8.11}$$

The gain $W(k)$ is the same as in the standard filter

$$W(k) = P(k|k-1)H(k)'S(k)^{-1} \tag{8.12}$$

since, conditioned on $\theta_i(k)$, there is no measurement origin uncertainty.

For $i = 0$ (i.e., if none of the measurements is correct) or $m(k) = 0$ (i.e., there is no validated measurement)

$$\hat{x}_0\big(k|k\big) = \hat{x}\big(k|k-1\big) \tag{8.13}$$

8.2.5 The State and Covariance Update

Combining Equations 8.10 and 8.13 into Equation 8.8 yields the state update equation of the PDAF

$$\hat{x}\big(k|k\big) = \hat{x}\big(k|k-1\big) + W\big(k\big)v\big(k\big) \tag{8.14}$$

where the combined innovation is

$$v\big(k\big) = \sum_{i-1}^{m(k)} \beta_i\big(k\big)v_i\big(k\big) \tag{8.15}$$

The covariance associated with the updated state is

$$P\big(k|k\big) = \beta_0\big(k\big)P\big(k|k-1\big) + \big[1 - \beta_0\big(k\big)\big]P^c\big(k|k\big) + \tilde{P}\big(k\big) \tag{8.16}$$

where the covariance of the state updated with the correct measurement is[3]

$$P^c\big(k|k\big) = P\big(k|k-1\big) - W\big(k\big)S\big(k\big)W\big(k\big)' \tag{8.17}$$

and the spread of the innovations term (similar to the spread of the means term in a mixture[10]) is

$$\tilde{P}\big(k\big) \overset{\Delta}{=} W\big(k\big)\left[\sum_{i=1}^{m(k)} \beta_i\big(k\big)v_i\big(k\big)v_i\big(k\big)' - v\big(k\big)v\big(k\big)'\right]W\big(k\big)' \tag{8.18}$$

8.2.6 The Prediction Equations

The prediction of the state and measurement to $k + 1$ is done as in the standard filter, i.e.,

$$\hat{x}\big(k+1|k\big) = F\big(k\big)\hat{x}\big(k|k\big) \tag{8.19}$$

$$\hat{z}\big(k+1|k\big) = H\big(k+1\big)\hat{x}\big(k+1|k\big) \tag{8.20}$$

The covariance of the predicted state is, similarly,

$$P\big(k+1|k\big) = F\big(k\big)P\big(k|k\big)F\big(k\big)' + Q\big(k\big) \tag{8.21}$$

where $P(k|k)$ is given by Equation 8.16.

The innovation covariance (for the correct measurement) is, again, as in the standard filter

$$S\big(k+1\big) = H\big(k+1\big)P\big(k+1|k\big)H\big(k+1\big)' + R\big(k+1\big) \tag{8.22}$$

8.2.7   The Probabilistic Data Association

To evaluate the association probabilities, the conditioning is broken down into the past data Z^{k-1} and the latest data $Z(k)$. A probabilistic inference can be made on both the number of measurements in the validation region (from the clutter density, if known) and on their location, expressed as:

$$\beta_i(k)=P\left\{\theta_i(k)\middle|Z^k\right\}=P\left\{\theta_i(k)\middle|Z(k),m(k),Z^{k-1}\right\} \tag{8.23}$$

Using Bayes' formula, the above is rewritten as

$$\beta_i(k)=\frac{1}{c}p\left[Z(k)\middle|\theta_i(k),m(k),Z^{k-1}\right]p\left\{\theta_i(k)\middle|m(k),Z^{k-1}\right\}\qquad i=0,\dots,m(k) \tag{8.24}$$

The joint density of the validated measurements conditioned on $\theta_i(k)$, $i \neq 0$, is the product of

- The (assumed) Gaussian pdf of the correct (target-originated) measurements
- The pdf of the incorrect measurements, which are assumed to be uniform in the validation region whose volume $V(k)$ is given in Equation 8.6.

The pdf of the correct measurement (with the P_G factor that accounts for restricting the normal density to the validation gate) is

$$p\left[z_i(k)\middle|\theta_i(k),m(k),Z^{k-1}\right]=p_G^{-1}N\left[z_i(k);z(k|k-1),S(k)\right]=p_G^{-1}N\left[v_i(k);0,S(k)\right] \tag{8.25}$$

The pdf from Equation 8.24 is then

$$p\left[Z(k)\middle|\theta_i(k),m(k),Z^{k-1}\right]=\left\{V(k)^{-m(k)+1}P_G^{-1}N\left[\ \right]\right. \tag{8.26}$$

The probabilities of the association events conditioned only on the number of validated measurements are

$$p\left[Z(k)\middle|\theta_i(k),m(k),Z^{k-1}\right]=\begin{cases}V(k)^{-m(k)+1}P_G^{-1}N\left[v_i(k);0,S(k)\right]&i=1,\dots,m(k)\\V(k)^{-m(k)}&i=0\end{cases} \tag{8.27}$$

where $\mu_F(m)$ is the probability mass function (pmf) of the number of false measurements (false alarms or clutter) in the validation region.

Two models can be used for the pmf $\mu_F(m)$ in a volume of interest V:

1. A Poisson model with a certain spacial density λ

$$\mu_F(m)=e^{-\lambda V}\frac{(\lambda V)^m}{m!} \tag{8.28}$$

2. A diffuse prior model[3]

$$\mu_F(m)=\mu_F(m-1)=\delta \tag{8.29}$$

where the constant δ is irrelevant since it cancels out.

Using the (parametric) Poisson model in Equation 8.27 yields

$$
\gamma_i\big[m(k)\big]=\begin{cases}P_D P_G\Big[P_D P_G m(k)+\big(1-P_D P_G\big)\lambda V(k)\Big]^{-1} & i=1,\dots,m(k)\\[2ex] \big(1-P_D P_F\big)\lambda V(k)\Big[P_D P_G m(k)+\big(1-P_D P_G\big)\lambda V(k)\Big]^{-1} & i=0\end{cases}
\tag{8.30}
$$

The (nonparametric) diffuse prior (Equation 8.29) yields

$$
\gamma_i\big[m(k)\big]=\begin{cases}\dfrac{1}{m(k)}P_D P_G & i=1,\dots,m(k)\\[2ex] \big(1-P_D P_G\big) & i=0\end{cases}
\tag{8.31}
$$

The nonparametric model (Equation 8.31) can be obtained from Equation 8.30 by setting

$$
\lambda=\frac{m(k)}{V(k)}
\tag{8.32}
$$

i.e., replacing the Poisson parameter with the sample spatial density of the validated measurements. The volume $V(k)$ of the elliptical (i.e., Gaussian-based) validation region is given in Equation 8.6.

8.2.8 The Parametric PDA

Using Equations 8.30 and 8.26 with the explicit expression of the Gaussian pdf in Equation 8.24 yields, after some cancellations, the final equations of the parametric PDA with the Poisson clutter model

$$
\beta_i(k)=\begin{cases}\dfrac{e_i}{b+\sum_{j=1}^{m(k)}e_j} & i=1,\dots,m(k)\\[3ex] \dfrac{b}{b+\sum_{j=1}^{m(k)}e_j} & i=0\end{cases}
\tag{8.33}
$$

where

$$
e_i\overset{\Delta}{=}e^{-\frac{1}{2}v_i(k)'S(k)^{-1}v_i(k)}
\tag{8.34}
$$

$$
b\overset{\Delta}{=}\lambda\big|2\pi S(k)\big|^{1/2}\frac{1-P_D P_G}{P_D}
\tag{8.35}
$$

The last expression above can be rewritten as

$$
b=\left(\frac{2\pi}{\lambda}\right)^{\frac{n_z}{2}}\lambda V(k)c_{n_z}^{-1}\frac{1-P_D P_G}{P_D}
\tag{8.36}
$$

8.2.9 The Nonparametric PDA

The nonparametric PDA is the same as above except for replacing $\lambda V(k)$ in Equation 8.36 by $m(k)$ — this obviates the need to know λ.

8.3 Low Observable TMA Using the ML-PDA Approach with Features

This section considers the problem of target motion analysis (TMA) — estimation of the trajectory parameters of a constant velocity target — with a passive sonar, which does not provide full target position measurements. The methodology presented here applies equally to any *target motion characterized by a deterministic equation*, in which case the initial conditions (a finite dimensional parameter vector) characterize in full *the entire motion*. In this case the (batch) maximum likelihood (ML) parameter estimation can be used; this method is more powerful than state estimation when the target motion is deterministic (it does not have to be linear). Furthermore, the ML-PDA approach makes no approximation, unlike the PDAF in Equation 8.1.

8.3.1 Amplitude Information Feature

The standard TMA consists of estimating the target's position and its constant velocity from bearings-only (wideband sonar) measurements corrupted by noise.[10] Narrowband passive sonar tracking, where frequency measurements are also available, has been studied.[11] The advantages of narrowband sonar are that it does not require a maneuver of the platform for observability, and it greatly enhances the accuracy of the estimates. However, not all passive sonars have frequency information available. In both cases, the intensity of the signal at the output of the signal processor, which is referred to as *measurement amplitude* or *amplitude information* (AI), is used implicitly to determine whether there is a valid measurement. This is usually done by comparing it with the detection threshold, which is a design parameter.

This section shows that the measurement amplitude carries valuable information and that its use in the estimation process increases the observability even though the amplitude information cannot be correlated to the target state directly. Also superior global convergence properties are obtained.

The pdf of the envelope detector output (i.e., the AI) a when the signal is due to noise only is denoted as $p_0(a)$ and the corresponding pdf when the signal originated from the target is $p_1(a)$. If the signal-to-noise ratio (SNR — this is the SNR in a resolution cell, to be denoted later as SNR_c) is d, the density functions of noise only and target-originated measurements can be written as

$$p_0(a) = a \exp\left(-\frac{a^2}{2}\right) \quad a \geq 0 \tag{8.37}$$

$$p_1(a) = \frac{a}{1+d} \exp\left(-\frac{a^2}{2(1+d)}\right) \quad a \geq 0 \tag{8.38}$$

respectively. This is a Rayleigh fading amplitude (Swerling I) model believed to be the most appropriate for shallow water passive sonar.

A suitable threshold, denoted by τ, is used to declare a detection. The probability of detection and the probability of false alarm are denoted by P_D and P_{FA}, respectively. Both P_D and P_{FA} can be evaluated from the probability density functions of the measurements. Clearly, in order to increase P_D, the threshold τ must be lowered. However, this also increases P_{FA}. Therefore, depending on the SNR, τ must be selected to satisfy two conflicting requirements.*

The density functions given above correspond to the signal at the envelope detector output. Those corresponding to the output of the threshold detector are

---

*For other probabilistic models of the detection process, different SNR values correspond to the same P_D, P_{FA} pair. Compared to the Rician model receiver operating characteristic (ROC) curve, the Rayleigh model ROC curve requires a higher SNR for the same pair (P_D, P_{FA}), i.e., the Rayleigh model considered here is pessimistic.

$$\rho_0^\tau(a) = \frac{1}{P_{FA}} p_0(a) = \frac{1}{P_{FA}} a \exp\left(-\frac{a^2}{2}\right) \quad a > \tau \tag{8.39}$$

$$\rho_1^\tau(a) = \frac{1}{P_D} p_1(a) = \frac{1}{P_D} \frac{a}{1+d} a \exp\left(-\frac{a^2}{2(1+d)}\right) \quad a > \tau \tag{8.40}$$

where $\rho_0^\tau(a)$ is the pdf of the validated measurements that are caused by noise only, and $\rho_1^\tau(a)$ is the pdf of those that originated from the target. In the following, a is the amplitude of the candidate measurements. The amplitude likelihood ratio, ρ, is defined as

$$\rho = \frac{p_1^\tau(a)}{p_0^\tau(a)} \tag{8.41}$$

8.3.2 Target Models

Assume that n sets of measurements, made at times $t = t_1, t_2,\dots, t_n$, are available.

For bearings-only estimation, the target motion is defined by the four-dimensional parameter vector

$$x \triangleq \left[\xi(t_0), \ \eta(t_0), \ \dot\xi, \ \dot\eta\right] \tag{8.42}$$

where $\xi(t_0)$ and $\eta(t_0)$ are the distances of the target in the east and north directions, respectively, from the origin at the reference time t_0. The corresponding velocities, assumed constant, are $\dot\xi$ and $\dot\eta$, respectively. This assumes deterministic target motion (i.e., no process noise[10]). Any other deterministic motion (e.g., constant acceleration) can be handled within the same framework.

The state of the platform at t_i ($i = 1,\dots, n$) is defined by

$$x_p(t_i) \triangleq \left[\xi_p(t_i), \ \eta_p(t_i), \ \dot\xi_p(t_i), \ \dot\eta_p(t_i)\right]' \tag{8.43}$$

The relative position components in the east and north directions of the target with respect to the platform at t_i are defined by $r_\xi(t_i, x)$ and $r_\eta(t_i, x)$, respectively. Similarly, $v_\xi(t_i, x)$ and $v_\eta(t_i, x)$ define the relative velocity components. The true bearing of the target from the platform at t_i is given by

$$\theta_i(x) \triangleq \tan^{-1}\left[r_\xi(t_i, x)/r_\eta(t_i, x)\right] \tag{8.44}$$

The range of possible bearing measurements is

$$U_\theta \triangleq \left[\theta_1, \theta_2\right] \subset \left[0, 2\pi\right] \tag{8.45}$$

The set of measurements at t_i is denoted by

$$Z(i) \triangleq \left\{z_j(i)\right\}_{j=1}^{m_i} \tag{8.46}$$

where m_i is the number of measurements at t_i, and the pair of bearing and amplitude measurements $z_j(i)$, is defined by

$$z_j(i) \triangleq \begin{bmatrix} \beta_{ij} & a_{ij} \end{bmatrix}' \tag{8.47}$$

The cumulative set of measurements during the entire period is

$$Z^n \triangleq \left\{ Z(i) \right\}_{i=1}^n \tag{8.48}$$

The following additional assumptions about the statistical characteristics of the measurements are also made:[11]

1. The measurements at two different sampling instants are conditionally independent, i.e.,

$$p\Big[Z(i_1), Z(i_2)\big|x\Big] = p\Big[Z(i_1)\big|x\Big] p\Big[Z(i_2)\big|x\Big] \quad \forall i_1 \neq i_2 \tag{8.49}$$

   where $p[\cdot]$ is the probability density function.
2. A measurement that originated from the target at a particular sampling instant is received by the sensor only once during the corresponding scan with probability P_D and is corrupted by zero-mean Gaussian noise of known variance. That is

$$\beta_{ij} = \theta_i(x) + \in_{ij} \tag{8.50}$$

   where $\in_{ij} \sim \mathcal{N}\Big[0, \sigma_\theta^2\Big]$ is the bearing measurement noise. Due to the presence of false measurements, the index of the true measurement is not known.
3. The false bearing measurements are distributed uniformly in the surveillance region, i.e.,

$$\beta_{ij} \sim \mathcal{U}\Big[\theta_1, \theta_2\Big] \tag{8.51}$$

4. The number of false measurements at a sampling instant is generated according to a Poisson law with a known expected number of false measurements in the surveillance region. This is determined by the detection threshold at the sensor (exact equations are given in Section 8.3.5).

For narrowband sonar (with frequency measurements) the target motion model is defined by the five-dimensional vector

$$x \triangleq \begin{bmatrix} \xi(t_1), & \eta(t_1), & \dot{\xi}, & \dot{\eta}, & \gamma \end{bmatrix} \tag{8.52}$$

where γ is the unknown emitted frequency assumed constant. Due to the relative motion between the target and platform at t_i, this frequency will be Doppler shifted at the platform. The (noise-free) shifted frequency, denoted by $\gamma_i(x)$, is given by

$$\gamma_i(x) = \gamma\left[1 - \frac{v_\xi(t_i, x)\sin\theta_i(x) + v_\eta(t_i, x)\cos\theta_i(x)}{c}\right] \tag{8.53}$$

where c is the velocity of sound in the medium. If the bandwidth of the signal processor in the sonar is $[\Omega_1, \Omega_2]$, the measurements can lie anywhere within this range. As in the case of bearing measurements, we assume that an operator is able to select a frequency subregion $[\Gamma_1, \Gamma_2]$ for scanning. In addition to the bearing surveillance region given in Equation 8.45, the region for frequency is defined as

$$U_\gamma \overset{\Delta}{=} [\Gamma_1, \Gamma_2] \subset [\Omega_1, \Omega_2] \tag{8.54}$$

The noisy frequency measurements are denoted by f_{ij} and the measurement vector is

$$z_j(i) \overset{\Delta}{=} [\beta_{ij}, \ f_{ij}, \ a_{ij}]' \tag{8.55}$$

As for the statistical assumptions, those related to the conditional independence of measurements (assumption 1) and the number of false measurements (assumption 4) are still valid. The equations relating the number of false alarms in the surveillance region to detection threshold are given in Section 8.3.5.

The noisy bearing measurements satisfy Equation 8.50 and the noisy frequency measurements f_{ij} satisfy

$$f_{ij} = \gamma_i(x) + v_{ij} \tag{8.56}$$

where $v_{ij} \sim \mathcal{N}[0, \sigma_\gamma^2]$ is the frequency measurement noise.

It is also assumed that these two measurement noise components are conditionally independent. That is,

$$p(\epsilon_{ij}, v_{ij} | x) = p(\epsilon_{ij} | x) p(v_{ij} | x) \tag{8.57}$$

The measurements resulting from noise only are assumed to be uniformly distributed in the entire surveillance region.

8.3.3 Maximum Likelihood Estimator Combined with PDA — The ML-PDA

In this section we present the derivation and implementation of the maximum likelihood estimator combined with the PDA technique for both bearings-only tracking and narrowband sonar tracking. If there are m_i detections at t_i, one has the following mutually exclusive and exhaustive events:[3]

$$\epsilon_j(i) \overset{\Delta}{=} \begin{cases} \{\text{measurement } z_j(i) \text{ is from the target}\} & j = 1, \ldots, m_i \\ \{\text{all measurements are false}\} & j = 0 \end{cases} \tag{8.58}$$

The pdf of the measurements corresponding to the above events can be written as

$$p(Z(i) | e_j(i), x) = \begin{cases} u^{1-m_i} p(\beta_{ij}) \rho_{ij} \prod_{j=1}^{m_i} p_0^\tau(a_{ij}) & j = 1, \ldots, m_i \\ u^{-m_i} \prod_{j=1}^{m_i} p_0^\tau(a_{ij}) & j = 0 \end{cases} \tag{8.59}$$

where $u = U_\theta$ is the area of the surveillance region.

Using the total probability theorem, the likelihood function of the set of measurements at t_i can be expressed as

$$p\big[Z(i)\big|x\big]=u^{-m_i}\big(1-P_D\big)\prod_{j=1}^{m_i}p_0^{\tau}\big(a_{ij}\big)\mu_f\big(m_i\big)+\frac{u^{1-m_i}P_D\mu_f\big(m_i-1\big)}{m_i}\prod_{j=1}^{m_i}p_0^{\tau}\big(a_{ij}\big)\sum_{j=1}^{m_i}p\big(b_{ij}\big)\rho_{ij}$$

$$=u^{-m_i}\big(1-P_D\big)\prod_{j=1}^{m_i}p_0^{\tau}\big(a_{ij}\big)\mu_f\big(m_i\big)+\frac{u^{1-m_i}P_D\mu_f\big(m_i-1\big)}{m_i}\prod_{j=1}^{m_i}p_0^{\tau}\big(a_{ij}\big) \qquad (8.60)$$

$$\cdot\sum_{j=1}^{m_i}\frac{1}{\sqrt{2\pi}\sigma_{\theta}}\exp\left(-\frac{1}{2}\left[\frac{\beta_{ij}-\theta_i(x)}{\sigma_{\theta}}\right]^2\right)\rho_{ij}$$

where $\mu_f(m_i)$ is the Poisson probability mass function of the number of false measurements at t_i. Dividing the above by $p[Z(I)|\varepsilon_0(I), x]$ yields the *dimensionless* likelihood ratio $\Phi_i[Z(I), x]$ at t_i. Then

$$\Phi_i\big[Z(i),x\big]=\frac{p\big[Z(i),x\big]}{p\big[Z(i)\big|\varepsilon_0(i),x\big]}$$

$$(8.61)$$

$$=\big(1-P_D\big)+\frac{P_D}{\lambda}\sum_{j=1}^{m_i}\frac{1}{\sqrt{2\pi}\sigma_{\theta}}\rho_{ij}\exp\left(-\frac{1}{2}\left[\frac{\beta_{ij}-\theta_i(x)}{\sigma_{\theta}}\right]^2\right)$$

where λ is the expected number of false alarms per unit area. Alternately, the log-likelihood ratio at t_i can be defined as

$$\phi_i\big[Z(i),x\big]=\ln\left[\big(1-P_D\big)+\frac{P_D}{\lambda}\sum_{j=1}^{m_i}\frac{1}{\sqrt{2\pi}\sigma_{\theta}}\rho_{ij}\exp\left(-\frac{1}{2}\left[\frac{\beta_{ij}-\theta_i(x)}{\sigma_{\theta}}\right]^2\right)\right] \qquad (8.62)$$

Using conditional independence of measurements, the likelihood function of the entire set of measurements can be written in terms of the individual likelihood functions as

$$p\big[Z^n\big|x\big]=\prod_{i=1}^{n}p\big[Z(i)\big|x\big] \qquad (8.63)$$

Then the dimensionless likelihood ratio for the entire data is given by

$$\Phi\big[Z^n,x\big]=\prod_{i=1}^{n}\Phi_i\big[Z(i),x\big] \qquad (8.64)$$

From the above, the total log-likelihood ratio $\Phi_i[Z(i), x]t_i$ can be expressed as

$$\Phi\big[Z^n,x\big]=\sum_{i=1}^{n}\ln\left[\big(1-P_D\big)+\frac{P_D}{\lambda}\sum_{j=1}^{m_i}\frac{1}{\sqrt{2\pi}\sigma_{\theta}}\rho_{ij}\exp\left(-\frac{1}{2}\left[\frac{\beta_{ij}-\theta_i(x)}{\sigma_{\theta}}\right]^2\right)\right] \qquad (8.65)$$

The maximum likelihood estimate (MLE) is obtained by finding the state $x = \hat{x}$ that maximizes the total log-likelihood function. In deriving the likelihood function, the gate probability mass, which is the probability that a target-originated measurement falls within the surveillance region, is assumed to be one. The operator selects the appropriate region.

Arguments similar to those given earlier can be used to derive the MLE when frequency measurements are also available. Defining $\varepsilon_j(i)$ as in Equation 8.58, the pdf of the measurements is

$$p\left[Z(i)\middle|\varepsilon_j(i),x\right] = \begin{cases} u^{1-m_i} p\left(\beta_{ij}\right) p\left(f_{ij}\right) p_{ij} \prod_{j=1}^{m_i} p_0^{\tau}\left(a_{ij}\right) & j=1,\ldots,m_i \\ u^{-m_i} \prod_{j=1}^{m_i} p_0^{\tau}\left(a_{ij}\right) & j=0 \end{cases} \tag{8.66}$$

where $u = U_\theta U_\gamma$ is the volume of the surveillance region.

After some lengthy manipulations, the total log-likelihood function is obtained as

$$\phi\left[Z^n,x\right] = \sum \ln\left[\left(1-P_D\right) + \frac{P_D}{\lambda}\sum_{j=1}^{m_i} \frac{\rho_{ij}}{2\pi\sigma_\theta\sigma_\gamma} \exp\left(-\frac{1}{2}\left[\frac{\beta_{ij}-\theta_i(x)}{\sigma_\theta}\right]^2 - \frac{1}{2}\left[\frac{f_{ij}-\gamma_i(x)}{\sigma_\gamma}\right]^2\right)\right] \tag{8.67}$$

For narrowband sonar, the MLE is found by maximizing Equation 8.67.

This section demonstrated the essence of the use of the PDA — all the measurements are accounted for and the likelihood function is evaluated using the total probability theorem, similar to Equation 8.8. However, since Equation 8.67 is *exact* (for the parameter estimation formulation), there is no need for the approximation in Equation 8.1, which is necessary in the PDAF for state estimation.

The same ML-PDA approach is applicable to the estimation of the trajectory of an exoatmospheric ballistic missile.[12,13] The modification of this fixed-batch ML-PDA estimator to a flexible (sliding/expanding/contracting) procedure is discussed in Section 8.5 and demonstrated with an actual electro-optics (EO) data example.

8.3.4 Cramér-Rao Lower Bound for the Estimate

For an unbiased estimate, the Cramér-Rao lower bound (CRLB) is given by

$$E\left\{\left(x-\hat{x}\right)\left(x-\hat{x}\right)'\right\} \geq J^{-1} \tag{8.68}$$

where J is the Fisher information matrix (FIM) given by

$$J = E\left\{\left[\nabla_x \ln p\left(Z^n\middle|x\right)\right]\left[\nabla_x \ln p\left(Z^n\middle|x\right)\right]'\right\}\Big|_{x=x_{true}} \tag{8.69}$$

Only in simulations will the true value of the state parameter be available. In practice CRLB is evaluated at the estimate.

As expounded in Reference 14, the FIM J is given in the present ML-PDA approach for the bearings-only case — *wideband sonar* — by

$$J = q_2\left(P_D, \lambda v_g, g\right) \sum_{i=1}^{n} \frac{1}{\sigma_\theta^2}\left[\nabla_x \theta_i(x)\right]\left[\nabla x \theta i(x)\right]' \tag{8.70}$$

where $q_2(P_D, \lambda v_g, g)$ is the *information reduction factor* that accounts for the loss of information resulting from the presence of false measurements and less-than-unity probability of detection,[3] and the expected number of false alarms per unit volume is denoted by λ.

In deriving Equation 8.70, only the bearing measurements that fall within the validation region

$$V_g^i(x) \triangleq \left\{ \beta_{ij} : \frac{\left| \beta_{ij} - \theta_i(x) \right|}{\sigma_\theta} \leq g \right\} \tag{8.71}$$

at t_i were considered. The validation region volume (*g*-sigma region), v_g, is given by

$$v_g = 2\sigma_\theta g \tag{8.72}$$

The information reduction factor $q_2(P_D, \lambda v_g, g)$ for the present two-dimensional measurement situation (bearing *and* amplitude) is given by

$$q_2\left(P_D, \lambda v_g, g\right) = \frac{1}{1+d} \sqrt{\frac{2}{\pi}} \sum_{m=1}^{\infty} \frac{\mu_f(m-1)}{\left(g P_{FA}\right)^{m-1}} I_2\left(m, P_D, g\right) \tag{8.73}$$

where $I_2(m, P_D, g)$ is a 2*m*-fold integral given in Reference 14 where numerical values of $q_2(P_D, \lambda v_g, g)$ for different combinations of P_D and λv_g are also presented. The derivation of the integral is based on Bar-Shalom and Li.[3] In this implementation, $g = 5$ was selected. Knowing P_D and λv_g, P_{FA} can be determined by using

$$\lambda v_g = P_{FA} \frac{v_g}{V_c} \tag{8.74}$$

where V_c is the resolution cell volume of the signal processor (discussed in more detail in Section 8.3.5). Finally, d, the SNR, can be calculated from P_D and λv_g.

The rationale for the term *information reduction factor* follows from the fact that the FIM for zero false alarm probability and unity target detection probability, J_0, is given by Reference 10

$$J_0 = \sum_{i=1}^{n} \frac{1}{\sigma_\theta^2} \left[\nabla_x \theta_i(x) \right] \left[\nabla_x \theta_i(x) \right]' \tag{8.75}$$

Equations 8.70 and 8.75 clearly show that $q_2(P_D, \lambda v_g, g)$, which is always less than or equal to unity, represents the loss of information due to clutter.

For *narrowband sonar* (bearing and frequency measurements), the FIM is given by

$$J = q_2\left(P_D, \lambda v_g, g\right) \sum_{i=1}^{n} \left\{ \frac{1}{\sigma_\theta^2} \left[\nabla_x \theta_i(x) \right] \left[\nabla_x \theta_i(x) \right]' + \frac{1}{\sigma_\theta^2} \left[\nabla_x \gamma_i(x) \right] \left[\nabla_x \gamma_i(x) \right]' \right\} \tag{8.76}$$

where $q_2(P_D, \lambda v_g, g)$ for this three-dimensional measurement (bearing, frequency, and amplitude) case is evaluated[14] using

$$q_2\left(P_D, \lambda v_g, g\right) = \frac{1}{1+d} \sum \frac{2^{m-1}\mu_f(m-1)}{\left(g^2 P_{FA}\right)^{m-1}} I_2\left(m, P_D, g\right) \tag{8.77}$$

The expression for $I_2(m, P_D, g)$ and the numerical values for $q_2(P_D, \lambda v_g, g)$ are also given by Kirubarajan and Bar-Shalom.[14]

For narrowband sonar, the validation region is defined by

$$V_g^i(x) \triangleq \left\{ (\beta_{ij}, f_{ij}) : \left[\frac{\beta_{ij} - \theta_i(x)}{\sigma_\theta} \right]^2 + \left[\frac{f_{ij} - \theta_i(x)}{\sigma \gamma} \right]^2 \right\} \le g^2 \tag{8.78}$$

and the volume of the validation region, v_g, is

$$v_g = \sigma^\theta \sigma^\gamma g^2 \tag{8.79}$$

8.3.5 Results

Both the bearings-only and narrowband sonar problems with amplitude information were implemented to track a target moving at constant velocity. The results for the narrowband case are given below, accompanied by a discussion of the advantages of using amplitude information by comparing the performances of the estimators with and without amplitude information.

In narrowband signal processing, different bands in the frequency domain are defined by an appropriate cell resolution and a center frequency about which these bands are located. The received signal is sampled and filtered in these bands before applying FFT and beamforming. Then the angle of arrival is estimated using a suitable algorithm.[15] As explained earlier, the received signal is registered as a valid measurement only if it exceeds the threshold τ. The threshold value, together with the SNR, determines the probability of detection and the probability of false alarm.

The signal processor was assumed to consist of the frequency band [500Hz, 1000Hz] with a 2048-point FFT. This results in a frequency cell whose size is given by

$$C_\gamma = 500/2048 \approx 0.25\text{Hz} \tag{8.80}$$

Regarding azimuth measurements, the sonar is assumed to have 60 equal beams, resulting in an azimuth cell C_θ with size

$$C_\theta = 180°/60 = 3.0° \tag{8.81}$$

Assuming uniform distribution in a cell, the frequency and azimuth measurement standard deviations are given by*

$$\sigma_\gamma = 0.25/\sqrt{12} = 0.0722\text{Hz} \tag{8.82}$$

$$\sigma_\theta = 3.0/\sqrt{12} = 0.866° \tag{8.83}$$

The SNR_C in a cell** was taken as 6.1dB and $P_D = 0.5$. The estimator is not very sensitive to an incorrect P_D. This is verified by running the estimator with an incorrect P_D on the data generated with a different

---

* The "uniform" factor $\sqrt{12}$ corresponds to the worst case. In practice, σ_θ and σ_γ are functions of the 3dB-bandwidth and of the SNR.

** The commonly used SNR, designated here as SNR_1, is signal strength divided by the noise power in a 1-Hz bandwidth. SNR_C is signal strength divided by the noise power in a resolution cell. The relationship between them, for $C_\gamma = 0.25$Hz is $SNR_C = SNR_1 - 6$dB. SNR_C is believed to be the more meaningful SNR because it determines the ROC curve.

P_D. Differences up to 0.15 are tolerated by the estimator. The corresponding SNR in a 1-Hz bandwidth SNR_1 is 0.1dB. These values give

$$\tau = 2.64 \tag{8.84}$$

$$P_{FA} = 0.306 \tag{8.85}$$

From P_{FA}, the expected number of false alarms per unit volume, denoted by λ, can be calculated using

$$P_{FA} = \lambda C_\theta C_\gamma \tag{8.86}$$

Substituting the values for C_θ and λ gives

$$\lambda = \frac{0.0306}{3.0 \times 0.25} = 0.0407/\text{deg} \cdot \text{Hz} \tag{8.87}$$

The surveillance regions for azimuth and frequency, denoted by U_θ and U_γ, respectively, are taken as

$$U_\theta = \left[-20°, 20°\right] \tag{8.88}$$

$$U_\gamma = \left[747\text{Hz}, 753\text{Hz}\right] \tag{8.89}$$

The expected number of false alarms in the entire surveillance region and that in the validation gate V_g can be calculated. These values are 9.8 and 0.2, respectively, where the validation gate is restricted to $g = 5$. These values mean that, for every true measurement that originated from the target, there are about 10 false alarms that exceed the threshold.

The estimated tracks were validated using the hypothesis testing procedure described in Reference 14. The track acceptance test was carried out with a miss probability of five percent.

To check the performance of the estimator, simulations were carried out with clutter only (i.e., without a target) and also with a target present; measurements were generated accordingly. Simulations were done in batches of 100 runs.

When there was no target, irrespective of the initial guess, the estimated track was always rejected. This corroborates the accuracy of the validation algorithm given by Kirubarajan and Bar-Shalom.[14]

For the set of simulations with a target, the following scenario was selected: the target moves at a speed of 10 m/s heading west and 5 m/s heading north starting from (5000 m, 35,000 m). The signal frequency is 750 Hz. The target parameter is $x = $ [5000 m, 35,000 m, −10 m/s, 5 m/s, 750 Hz]. The motion of the platform consisted of two velocity legs in the northwest direction during the first half, and in the northeast direction during the second half of the simulation period with a constant speed of 7:1 m/s. Measurements were taken at regular intervals of 30 s. The observation period was 900 s. Figure 8.1 shows the scenario including the target true trajectory (solid line), platform trajectory (dashed line), and the 95% probability regions of the position estimates at the initial and final sampling instants based on the CRLB (Equation 8.76). The initial and the final positions of the trajectories are marked by I and F, respectively. The purpose of the probability region is to verify the validity of the CRLB as the actual parameter estimate covariance matrix from a number of Monte Carlo runs.[4]

Figure 8.1 shows the 100 tracks formed from the estimates. Note that in all but six runs (i.e., 94 runs) the estimated trajectory endpoints fall in the corresponding 95% uncertainty ellipses.

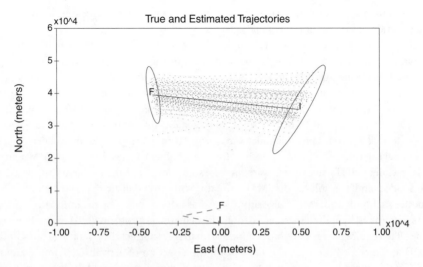

FIGURE 8.1 Estimated tracks from 100 runs for narrowband sonar with AI.

TABLE 8.1 Results of 100 Monte Carlo Runs
for Narrowband Sonar with AI (SNRC = 6:1 dB)

Unit	x_{true}	x_{init}	$\bar{x}$	σ_{CRLB}	$\hat{\sigma}$
m	5000	−12,000 to 12,000	4991	667	821
m	35,000	49,000 to 50,000	35,423	5576	5588
m/s	−10	−16 to 5	−9.96	0.85	0.96
m/s	5	−4 to 9	4.87	4.89	4.99
Hz	750	747 to 751	749.52	2.371	2.531

Table 8.1 gives the numerical results from 100 runs. Here $\bar{x}$ is the average of the estimates, $\hat{\sigma}$ the variance of the estimates evaluated from 100 runs, and σ_{CRLB} the theoretical CRLB derived in Section 8.3.4. The range of initial guesses found by rough grid search to start off the estimator are given by x_{init}.

The efficiency of the estimator was verified using the normalized estimation error squared (NEES)[10] defined by

$$\in_x \overset{\Delta}{=} \left(x - \hat{x}\right)' J\left(x - \hat{x}\right) \qquad (8.90)$$

where $\bar{x}$ is the estimate, and J is the FIM (Equation 8.76). Assuming approximately Gaussian estimation error, the NEES is chi-square distributed with n degrees-of-freedom where n is the number of estimated parameters. For the 94 accepted tracks the NEES obtained was 5.46, which lies within the 95% confidence region [4:39; 5:65]. Also note that each component of $\bar{x}$ is within $2\hat{\sigma}/\sqrt{100}$ of the corresponding component of x_{true}.

8.4 The IMMPDAF for Tracking Maneuvering Targets

Target tracking is a problem that has been well studied and documented. Some specific problems of interest in the single-target, single-sensor case are tracking maneuvering targets,[10] tracking in the presence of clutter,[3] and electronic countermeasures (ECM). In addition to these tracking issues, a complete tracking system for a sophisticated electronically steered antenna radar has to consider radar scheduling, waveform selection, and detection threshold selection.

Although many researchers have worked on these issues and many algorithms are available, there had been no standard problem comparing the performances of the various algorithms. Rectifying this, the first benchmark problem[16] was developed, focusing only on tracking a maneuvering target and pointing/scheduling a phased array radar. Of all the algorithms considered for this problem, the interacting multiple model (IMM) estimator yielded the best performance.[17] The second benchmark problem[9] included false alarms (FA) and ECM — specifically, a stand-off jammer (SOJ) and range gate pull off (RGPO) — as well as several possible radar waveforms (from which the resource allocator has to select one at every revisit time). Preliminary results for this problem showed that the IMM and multiple-hypothesis tracking (MHT) algorithms were the best solutions.[6,9] For the problem considered, the MHT algorithm yielded similar results as the IMM estimator with probabilistic data association filter (IMMP-DAF) modules,[3] although the MHT algorithm was one to two orders of magnitude costlier computationally (as many as 40 hypotheses were needed*). The benchmark problem of Reference 18 was upgraded in Reference 8 to require the radar resource allocator/manager to select the operating constant false alarm rate (CFAR) and included the effects of the SOJ on the direction of arrival (DOA) measurements; also the SOJ power was increased to present a more challenging benchmark problem. While, in Reference 18, the primary performance criterion for the tracking algorithm was minimization of radar energy, the primary performance was changed in Reference 8 to minimization of a weighted combination of radar time and energy.

This section presents the IMMPDAF technique for automatic track formation, maintenance, and termination. The coordinate selection for tracking, radar scheduling/pointing and the models used for mode-matched filtering (the modules inside the IMM estimator) are also discussed. These cover the target tracking aspects of the solution to the benchmark problem. These are based on the benchmark problem tracking and sensor resource management.[6,8]

8.4.1   Coordinate Selection

For target tracking in track-dwell mode of the radar, the number of detections at scan k (time t_k) is denoted by m_k. The m-th detection report $\bar{\zeta}_m$ (t_k) $(m = 1,2,\ldots,m_k)$ consists of a time stamp t_k, range r_m, bearing b_m, elevation e_m, amplitude information (AI) ρ_m given by the SNR, and the standard deviations of bearing and elevation measurements, σ_m^b and σ_m^e, respectively. Thus,

$$\bar{\zeta}_m\left(t_k\right)=\left[t_k, \ \ r_m, \ \ b_m, \ \ e_m, \ \ \rho_m, \ \ \sigma_m^b, \ \ \sigma_m^e\right]' \qquad (8.91)$$

where the overbar indicates that this is in the radar's spherical coordinate system.

The AI is used only to declare detections and select the radar waveform for the next scan. Since the use of AI, for example, as in Reference 17, can be counterproductive in discounting RGPO measurements, which generally have higher SNR than target-originated measurements, AI is not utilized in the estimation process itself. Using the AI would require a separate model for the RGPO intensity, which cannot be estimated in real time due to its short duration and variability.[17]

For target tracking, the measurements are converted from spherical coordinates to Cartesian coordinates, and then the IMMPDAF is used on these converted measurements. This conversion avoids the use of extended Kalman filters and makes the problem linear.[4] The converted measurement report ζ_m (t_k) corresponding to $\bar{\zeta}_m(t_k)$ is given by[6]

$$\zeta_m\left(t_k\right)=\left[t_k, \ \ x_m, \ \ y_m, \ \ z_m, \ \ \rho_m, \ \ R_m\right] \qquad (8.92)$$

---

* The more recent IMM-MHT (as opposed to Kalman filter-based MHT) requires six to eight hypotheses.

where x_m, y_m, z_m, and R_m are the three position measurements in the Cartesian frame and their covariance matrix, respectively. The converted values are

$$x_m = r_m \cos(b_m) \cos(e_m) \tag{8.93}$$

$$y_m = r_m \sin(b_m) \cos(e_m) \tag{8.94}$$

$$z_m = r_m \sin(e_m) \tag{8.95}$$

$$R_m = T_m \cdot diag\left[(\sigma_k^r)^2, (\sigma_m^b)^2, (\sigma_m^e)^2 \right] \cdot T_m' \tag{8.96}$$

where σ_k^r is the standard deviation of range measurements at scan k and T_m is the spherical-to-Cartesian transformation matrix given by

$$T_m = \begin{bmatrix} \cos(b_m)\cos(e_m) & -r_m \sin(b_m)\cos(e_m) & -r_m \cos(b_m)\cos(e_m) \\ \sin(b_m)\cos(e_m) & r_m \cos(b_m)\cos(e_m) & -r_m \sin(b_m)\cos(e_m) \\ \sin(e_m) & 0 & r_m \cos(e_m) \end{bmatrix} \tag{8.97}$$

For the scenarios considered here, this transformation is practically unbiased and there is no need for the debiasing procedure of Reference 4.

8.4.2 Track Formation

In the presence of false alarms, track formation is crucial. Incorrect track initiation will result in target loss. In Reference 3, an automatic track formation/deletion algorithm in the presence of clutter is presented based on the IMM algorithm. In the present benchmark problem, a noisy measurement corresponding to the target of interest is given in the first scan.* Forming new tracks for each validated measurement (based on a velocity gate) at subsequent scans, as suggested in Reference 3 and as implemented in Reference 6, is expensive in terms of both radar energy and computational load. In this implementation, track formation is simplified and handled as follows:

Scan 1 (t = 0s) — As defined by the benchmark problem, there is only one (target-originated, noisy) measurement. The position component of this measurement is used as the starting point for the estimated track.

Scan 2 (t = 0.1s) — The beam is pointed at the location of the first measurement. This yields, possibly, more than one measurement and these measurements are gated using the maximum possible velocity of the targets to avoid the formation of impossible tracks. This validation region volume, which is centered on the initial measurement, is given by

$$V_{xyz} = \left[2\left(\dot{x}_{max}\delta_2 + 3\sqrt{R_{m_2}^x} \right) \right]\left[2\left(\dot{y}_{max}\delta_2 + 3\sqrt{R_{m_2}^y} \right) \right]\left[2\left(\dot{z}_{max}\delta_2 + 3\sqrt{R_{m_2}^z} \right) \right] \tag{8.98}$$

where $\delta_2 = 0{:}1\text{s}$ is the sampling interval and $\dot{x}_{max}\delta_2, \dot{y}_{max}\delta_2$, and $z_{max}\delta_2$ are the maximum speeds in the X, Y, and Z directions respectively; $R_{m_2}^x, R_{m_2}^y$, and $R_{m_2}^z$ are the variances of position measurements in these directions obtained from the diagonal components of Equation 8.96. The maximum speed in each direction is assumed to be 500 m/s.

---

*Assuming that this is a search pulse without (monopulse) split-beam processing, the angular errors are uniformly distributed in the beam.

The measurement in the first scan and the measurement with the highest SNR in the second scan are used to form a track with the two-point initialization technique.[10] The track splitting used in References 3 and 6 was found unnecessary — the strongest validated measurement was adequate. This technique yields the position and velocity estimates and the associated covariance matrices in all three coordinates.

Scan 3 (t = 0.2s) — The pointing direction for the radar is given by the predicted position at $t = 0.2\ s$ using the estimates at scan 2. An IMMPDA filter with three models discussed in the sequel is initialized with the estimates and covariance matrices obtained at the second scan. The acceleration component for the third order model is assumed zero with variance $(a_{max})^2$, where $a_{max} = 70\ m/s^2$ is the maximum expected acceleration of the target.

From scan 3 on, the track is maintained using the IMMPDAF as described in Section 8.4.3. In order to maintain a high SNR for the target-originated measurement during track formation, a high-energy waveform is used. Also, scan 3 dwells are used to ensure target detection. This simplified approach cannot be used if the target-originated measurement is not given at the first scan. In that case, the track formation technique in Reference 3 can be used.

Immediate revisit with sampling interval 0.1s is carried out during track formation because the initial velocity of the target is not known — in the first scan only the position is measured and there is no *a priori* velocity. This means that in the second scan the radar must be pointed at the first scan position, assuming zero velocity. Waiting longer to obtain the second measurement could result in the loss of the target-originated measurement due to incorrect pointing. Also, in order to make the IMM mode probabilities converge to the correct values as quickly as possible, the target is revisited at a high rate.

8.4.3   Track Maintenance

The true state of the target at t_k is

$$x\left(t_k\right) = \left[x\left(t_k\right)\ \dot{x}\left(t_k\right)\ \ddot{x}\left(t_k\right)\ y\left(t_k\right)\ \dot{y}\left(t_k\right)\ \ddot{y}\left(t_k\right)\ z\left(t_k\right)\ \dot{z}\left(t_k\right)\ \ddot{z}\left(t_k\right)\right]'$$

where $x(t_k)$, $y(t_k)$, and $z(t_k)$ are the positions, $\dot{x}(t_k)$, $\dot{y}(t_k)$, and $\dot{z}(t_k)$ are the velocities, and $\ddot{x}(t_k)$, $\ddot{y}(t_k)$, and $\ddot{z}(t_k)$ are the accelerations of the target in the corresponding coordinates, respectively. The measurement vector consists of the Cartesian position components at t_k and is denoted by $z(t_k)$.

Assuming that the target motion is linear in the Cartesian coordinate system, the true state of the target can be written as

$$x\left(t_k\right) = F\left(\delta_k\right)x\left(t_{k-1}\right) + \Gamma\left(\delta_k\right)v\left(t_{k-1}\right) \tag{8.99}$$

and the target-originated measurement is related to the state according to

$$z\left(t_k\right) = Hx\left(t_k\right) + w\left(t_k\right) \tag{8.100}$$

where $\delta_k = t - t_{k-1}$. The white Gaussian noise sequences $v(t_k)$ and $w(t_k)$ are independent and their covariances are $Q(\delta_k)$ and $R(t_k)$, respectively.

With the above matrices, the predicted state $\hat{x}(t_k^-)$ at time t_k is

$$\hat{x}\left(t_k^-\right) = F\left(\delta_k\right) + \hat{x}\left(t_{k-1}\right) \tag{8.101}$$

and the predicted measurement is

$$\hat{z}\left(t_k^-\right) = H\hat{x}\left(t_k^-\right) \tag{8.102}$$

with associated innovation covariance

$$S(t_k) = HP(t_k^-)H' + R(t_k)$$ (8.103)

where $P(t_k^-)$ is the predicted state covariance to be defined in Equation 8.117 and $R(t_k)$ is the (expected) measurement noise covariance.

8.4.3.1 Probabilistic Data Association

During track maintenance, each measurement at scan t_k is validated against the established track. This is achieved by setting up a validation region centered around the predicted measurement at t_k^-. The validation region is

$$\left[z(t_k) - \hat{z}(t_k^-)\right]' S(t_k)^{-1} \left[z(t_k) - \hat{z}(t_k^-)\right] \leq \gamma$$ (8.104)

where $S(t_k)$ is the expected covariance of the innovation corresponding to the correct measurement and $\gamma = 16$ (0.9989 probability mass[3]) is the gate size. The appropriate covariance matrix to be used in the above is discussed in the sequel.

The set of measurements validated for the track at t_k is

$$Z(k) = \left\{z_m(t_k), m = 1, 2, \ldots, m_k\right\}$$ (8.105)

where m_k is the number of measurements validated and associated with the track. Also, the cumulative set of validated measurements up to and including scan k is denoted by Z_1^k. All unvalidated measurements are discarded.

With these m_k validated measurements at t_k, one has the following mutually exclusive and exhaustive events:

$$\varepsilon_m(t_k) \stackrel{\Delta}{=} \begin{cases} \left\{\text{measurement } z_m(t_k) \text{ is from the target}\right\} & m = 1, \ldots, m_k \\ \left\{\text{all measurements are false}\right\} & m = 0 \end{cases}$$ (8.106)

Using the nonparametric version of the PDAF,[4] the validated measurements are associated probabilistically to the track. The combined target state estimate is obtained as

$$\hat{x}(t_k) = \sum_{m=0}^{m_k} \beta_m(t_k)\hat{x}_m(t_k)$$ (8.107)

where $\beta_m(t_k)$ is the probability that the m-th validated measurement is correct and $\hat{x}_m(t_k)$ is the updated state conditioned on that event. The conditionally updated states are given by

$$\hat{x}_m(t_k) = \hat{x}(t_k^-) + W_m(t_k)v_m(t_k) \quad m = 1, 2, \ldots m_k$$ (8.108)

where $W_m(t_k)$ is the filter gain and $v_m(t_k) = z_m(t_k) - \hat{z}_m(t_k^-)$ is the innovation associated with the m-th validated measurement. The gain, which depends on the measurement noise covariance, is

$$W_m(t_k) = P(t_k^-)H'\left[HP(t_k^-)H' + R_m(t_k)\right]^{-1} \stackrel{\Delta}{=} P(t_k^-)H'S_m(t_k)^{-1}$$ (8.109)

where $R_m(t_k)$ depends on the observed SNR for measurement m.[8]

The association event probabilities $\beta_m(t_k)$ are given by

$$\beta_m(t_k) = P\{\varepsilon_m(t_k)|z_1^k\} \tag{8.110}$$

$$= \begin{cases} \dfrac{e(m)}{b + \displaystyle\sum_{j=1}^{m_k} e(j)} & m = 1,2,\ldots,m_k \\[4ex] \dfrac{b}{b + \displaystyle\sum_{j=1}^{m_k} e(j)} & m = 0 \end{cases} \tag{8.111}$$

where

$$e(m) = N\left[v_m(t_k);0,S_m(t_k)\right] \tag{8.112}$$

$$b = m_k \frac{1-P_D}{P_D V(t_k)} \tag{8.113}$$

and P_D is the probability of detection of a target-originated measurement. The probability that a target-originated measurement, if detected, falls within the validation gate is assumed to be unity. Also, $N[v;0, S]$ denotes the normal pdf with argument v, mean zero, and covariance matrix S. The common validation volume $V(t_k)$ is the union of the validation volumes $V_m(t_k)$ used to validate the individual measurements associated with the target $V(t_k)$ and is given by

$$V_m(t_k) = \gamma^{n_z/2} V_{n_z} \left|S_m(t_k)\right|^{1/2} \tag{8.114}$$

where V_{n_z} is the volume of the unit hypersphere of dimension n_z, the dimension of the measurement $\mathbf{z}$. For the three-dimensional position measurements $V_{n_z} = \frac{4\pi}{3}$ (see Reference 3).

The state estimate is updated as

$$\hat{x}(t_k) = \hat{x}(t_k^-) + \sum_{m=1}^{m_k} \beta_m(t_k) W_m(t_k) v_m(t_k) \tag{8.115}$$

and the associated covariance matrix is updated as

$$P(t_k) = P(t_k^-) - \sum_{m=1}^{m_k} \beta_m(t_k) W_m(t_k) S_m(t_k) W_m(t_k)' + \tilde{P}(t_k) \tag{8.116}$$

where

$$P(t_k^-) = F(\delta_k) P(t_{k-1}) F(\delta_k)' + \Gamma(\delta_k) Q(\delta_k) \Gamma(\delta_k)' \tag{8.117}$$

is the predicted state covariance and the term

$$\tilde{P}(t_k) = \sum_{m=1}^{m_k} \beta_m(t_k) \left[W_m(t_k) v_m(t_k) \right] \left[W_m(t_k) v_m(t_k) \right]' -$$

$$\left[\sum_{m=1}^{m_k} \beta_m(t_k) W_m(t_k) v_m(t_k) \right] \left[\sum_{m=1}^{m_k} \beta_m(t_k) W_m(t_k) v_m(t_k) \right]'$$

(8.118)

is analogous to the *spread of the innovations* in the standard PDA.[3] Monopulse processing results in different accuracies (standard deviations) for different measurements within the same dwell. This accounts for the difference in the above equations from the standard PDA, where the measurement accuracies are assumed to be the same for all of the validated measurements.

To initialize the filter at $k = 3$, the following estimates are used:[10]

$$\hat{x}(t_2) = \begin{bmatrix} \hat{x}(t_2) \\ \dot{\hat{x}}(t_2) \\ \ddot{\hat{x}}(t_2) \\ \hat{y}(t_2) \\ \dot{\hat{y}}(t_2) \\ \ddot{\hat{y}}(t_2) \\ \hat{z}(t_2) \\ \dot{\hat{z}}(t_2) \\ \ddot{\hat{z}}(t_2) \end{bmatrix} = \begin{bmatrix} \mathbf{z}_h^x(t_2) \\ \left(\mathbf{z}_h^x(t_2) - \mathbf{z}^x(t_1) \right)/\delta_2 \\ 0 \\ \mathbf{z}_h^y(t_2) \\ \left(\mathbf{z}_h^y(t_2) - \mathbf{z}^y(t_1) \right)/\delta_2 \\ 0 \\ \mathbf{z}_h^z(t_2) \\ \left(\mathbf{z}_h^z(t_2) - \mathbf{z}^z(t_1) \right)/\delta_2 \\ 0 \end{bmatrix}$$

(8.119)

where h is the index corresponding to the validated measurement with the highest SNR in the second scan, and the superscripts x, y, and z denote the components in the corresponding directions, respectively. The associated covariance matrix can be derived[10] using the measurement covariance R_h and the maximum target acceleration a_{max}. If the two point differencing results in a velocity component that exceeds the corresponding maximum speed, it is replaced by that speed. Similarly, the covariance terms corresponding to the velocity components are upper bounded by the corresponding maximum values.

8.4.3.2 IMM Estimator Combined with the PDA Technique

In the IMM estimator it is assumed that at any time the target trajectory evolves according to one of a finite number of models, which differ in their noise levels and/or structures.[10] By probabilistically combining the estimates of the filters, typically Kalman, matched to these modes, an overall estimate is found. In the IMM-PDAF the Kalman filter is replaced with the PDA filter (given in Section 8.4.3.1 for mode-conditioned filtering of the states), which handles the data association.

Let r be the number of mode-matched filters used, $M(t_k)$ the index of the mode in effect in the semi-open interval (t_{k-1}, t_k) and $\mu_j(t_k)$ be the probability that mode j ($j = 1, 2,\ldots, r$) is in effect in the above interval. Thus,

$$\mu_j(t_k) = P\{M(t_k) = j | Z_1^k\}$$

(8.120)

The mode transition probability is defined as

$$p_{ij} = P\{M(t_k) = j | M(t_{k-1}) = i\}$$

(8.121)

The state estimates and their covariance matrix at t_k conditioned on the j-th mode are denoted by and $P_j(t_k)$, respectively.

The steps of the IMMPDAF are as follows[3]

Step 1 — Mode interaction or mixing. The mode-conditioned state estimate and the associated covariances from the previous iteration are mixed to obtain the initial condition for the mode-matched filters. The initial condition in cycle k for the PDAF matched to the j-th mode is computed using

$$\hat{x}_{0j}(t_{k-1}) = \sum_{i=1}^{r} \hat{x}_i(t_{k-1})\mu_{i|j}(t_{k-1}) \tag{8.122}$$

where

$$\mu_{i|j}(t_{k-1}) = P\left\{M(t_{k-1})=i\Big|M(t_{k-1})=j, Z_1^{k-1}\right\} = \frac{p_{ij}\mu_i(t_{k-1})}{\sum_{l=1}^{r} p_{lj}\mu_l(t_{k-1})} \qquad i,j=1,2,\ldots,r \tag{8.123}$$

are the mixing probabilities. The covariance matrix associated with Equation 8.122 is given by

$$P_{0j}(t_{k-1}) = \sum_{i=1}^{r} \mu_{i|j}(t_{k-1})\left\{P_i(t_{k-1}) + \left[\hat{x}_i(t_k)-\hat{x}_{0j}(t_{k-1})\right]\left[\hat{x}_i(t_k)-\hat{x}_{0j}(t_{k-1})\right]'\right\} \tag{8.124}$$

Step 2 — Mode-conditioned filtering. A PDAF is used for each mode to calculate the mode-conditioned state estimates and covariances. In addition, we evaluate the likelihood function $\Lambda_j(t_k)$ of each mode at t_k using the Gaussian-uniform mixture

$$\Lambda_j(t_k) \overset{\Delta}{=} p\left[Z(k)\Big|M(t_k)=j, Z_1^{k-1}\right]$$
$$= V(t_k)^{-m_l}(1-P_D) + V(t_k)^{1-m_k}\frac{P_D}{m_k}\sum_{m=1}^{m_k} e_j(m) \tag{8.125}$$

$$= \frac{P_D V(t_k)^{1-m_k}}{m_k}\left(b + \sum_{m=1}^{m_k} e_j(m)\right) \tag{8.126}$$

where $e_j(m)$ is defined in Equation 8.112 and b in Equation 8.113. Note that the likelihood function, as a pdf, has a physical dimension that depends on m_k. Since ratios of these likelihood functions are to be calculated, they all must have the same dimension, i.e., the same m_k. Thus a common validation region (Equation 8.104) is vital for all the models in the IMMPDAF. Typically the "largest" innovation covariance matrix corresponding to "noisiest" model covers the others and, therefore, this can be used in Equations 8.104 and 8.114.

Step 3 — Mode update. The mode probabilities are updated based on the likelihood of each mode using

$$\mu_j(t_k) = \frac{\Lambda_j(t_k)\sum_{l=1}^{r} p_{lj}\mu_l(t_{k-1})}{\sum_{l=1}^{r}\sum_{l=1}^{r}\Lambda_i(t_k)p_{lj}\mu_l(t_{k-1})} \tag{8.127}$$

Step 4 — State combination. The mode-conditioned estimates and covariances are combined to find the overall estimate $\hat{x}(t_k)$ and its covariance matrix $P(t_k)$, as follows:

$$\hat{x}(t_k) = \sum_{j=1}^{r} \mu_j(t_k) \hat{x}(t_k) \tag{8.128}$$

$$P(t_k) = \sum_{j=1}^{r} \mu_j(t_k) \left(P_j(t_k) + \left[\hat{x}_j(t_k) - \hat{x}(t_k)\right]\left[\hat{x}_j(t_k) - \hat{x}(t_k)\right]' \right) \tag{8.129}$$

8.4.3.3 The Models in the IMM Estimator

The selection of the model structures and their parameters is one of the critical aspects of the implementation of IMMPDAF. Designing a good set of filters requires *a priori* knowledge about the target motion, usually in the form of maximum accelerations and sojourn times in various motion modes.[10] The tracks considered in the benchmark problem span a wide variety of motion modes — from benign constant velocity motions to maneuvers up to 7g. To handle all possible motion modes and to handle automatic track formation and termination, the following models are used:

Benign motion model (M^1) — This second-order model with low noise level (to be given later) has a probability of target detection P_D given by the target's expected SNR and corresponds to the nonmaneuvering intervals of the target trajectory. For this model the process noise is, typically, assumed to model air turbulence.

Maneuver model (M^2) — This second-order model with high noise level corresponds to ongoing maneuvers. For this white noise acceleration model, the process noise standard deviation σ_{v2} is obtained using

$$\sigma_{v2} = \alpha a_{max} \tag{8.130}$$

where a_{max} is the maximum acceleration in the corresponding modes and $0.5 < \alpha \le 1$.[10]

Maneuver detection model (M^3) — This is a third-order (Wiener process acceleration) model with high level noise. For highly maneuvering targets, like military attack aircraft, this model is useful for detecting the onset and termination of maneuvers. For civilian air traffic surveillance,[19] this model is not necessary.

For a Wiener process acceleration model, the standard deviation σ_{v3} is chosen using

$$\sigma_{v3} = \min\{\beta \Delta_a \delta, a_{max}\} \tag{8.131}$$

where Δ_a is the maximum acceleration increment per unit time (jerk), δ is the sampling interval, and $0.5 < \alpha \le 1$.[3]

For the targets under consideration, $a_{max} = 70$ m/s^2 and $\Delta_a = 35$ m/s^3. Using these values, the process noise standard deviations were taken as

$$\sigma_{v1} = 3 \text{ m/s}^2 \qquad \text{(for nonmaneuvering intervals)}$$

$$\sigma_{v2} = 35 \text{ m/s}^2 \qquad \text{(for maneuvering intervals)}$$

$$\sigma_{v3} = \min\{35\delta, 70\} \quad \text{(for maneuver start/termination)}$$

In addition to the process noise levels, the elements of the Markov chain transition matrix between the modes, defined in Equation 8.121, are also design parameters. Their selection depends on the sojourn time in each motion mode. The transition probability depends on the expected sojourn time via

$$\tau_i = \frac{\delta}{1 - p_{ii}} \qquad (8.132)$$

where τ_i is the expected sojourn time of the I-th mode, p_{ii} is the probability of transition from I-th mode to the same mode and δ is the sampling interval.[10]

For the above models, p_{ii}, $I = 1,2,3$ are calculated using

$$p_{ii} = \min\left\{ u_{i,max}\left(l_i, 1 - \frac{\delta}{\tau_i} \right) \right\} \qquad (8.133)$$

where $l_i = 0{:}1$ and $u_i = 0{:}9$ are the lower and upper limits, respectively, for the I-th model transition probability.

The expected sojourn times of 15, 4, and 2s, are assumed for modes M^1, M^2, and M^3, respectively. The selection of the off-diagonal elements of the Markov transition matrix depends on the switching characteristics among the various modes and is done as follows:

$$p_{12} = 0.1(1 - p_{11}) \qquad p_{13} = 0.9(1 - p_{11})$$

$$p_{21} = 0.1(1 - p_{22}) \qquad p_{23} = 0.9(1 - p_{22})$$

$$p_{31} = 0.3(1 - p_{33}) \qquad p_{32} = 0.7(1 - p_{33})$$

The x, y, z components of target dynamics are uncoupled, and the same process noise is used in each coordinate.

8.4.4 Track Termination

According to the benchmark problem, a track is declared lost if the estimation error is greater than the two-way beam width in angles or 1.5 range gates in range. In addition to this problem-specific criterion, the IMMPDAF declares (on its own) track loss if the track is not updated for 100s. Alternatively, one can include a "no target" model,[3] which is useful for automatic track termination, in the IMM mode set. In a more general tracking problem, where the true target state is not known, the "no target" mode probability or the track update interval would serve as the criterion for track termination, and the IMMPDAF would provide a unified framework for track formation, maintenance, and termination.

8.4.5 Simulation Results

This section presents the simulation results obtained using the algorithms described earlier. The computational requirements and root-mean-square errors (RMSE) are given.

The tracking algorithm using the IMMPDAF is tested on the following six benchmark tracks (the tracking algorithm does not know the type of the target under track — the parameters are selected to handle any target):

Target 1 — A large military cargo aircraft with maneuvers up to 3g.
Target 2 — A Learjet or commercial aircraft which is smaller and more maneuverable than target 1 with maneuvers up to 4g.
Target 3 — A high-speed medium bomber with maneuvers up to 4g.
Target 4 — Another medium bomber with good maneuverability up to 6g.
Targets 5 and 6 — Fighter or attack aircraft with very high maneuverability up to 7g.

In Table 8.2, the performance measures and their averages of the IMMPDAF (in the presence of FA, RGPO, and SOJ[6,8]) are given. The averages are obtained by adding the corresponding performance metrics

TABLE 8.2 Performance of IMMPDAF in the Presence of False Alarms, Range Gate Pull-Off, and the Standoff Jammer

Target	Time Length (s)	Max. Acc. (m/s$_2$)	Man. Density (%)	Sample Period (s)	Avg. Power (W)	Pos. RMSE (m)	Vel. RMSE (m/s)	Ave. Load (kFLOPS)	Lost Tracks (%)
1	165	31	25	2.65	8.9	98.1	61.3	22.2	1
2	150	39	28.5	2.39	5.0	97.2	68.5	24.3	0
3	145	42	20	2.38	10.9	142.1	101.2	24.6	1
4	184	58	20	2.34	3.0	26.5	25.9	24.3	0
5	182	68	38	2.33	18.4	148.1	110.7	27.1	2
6	188	70	35	2.52	12.4	98.6	71.4	24.6	1
Avg.	—	—	—	**2.48**	**8.3**	—	—	**24.5**	—

of the six targets (with those of target 1 added twice) and dividing the sum by 7. In the table, the maneuver density is the percentage of the total time that the target acceleration exceeds 0.5g. The average floating point operation (FLOP) count per second was obtained by dividing the total number of floating point operations by the target track length. This is the computational requirement for target and jammer tracking, neutralizing techniques for ECM, and adaptive parameter selection for the estimator, i.e., it excludes the computational load for radar emulation.

The average FLOP requirement is 25 kFLOPS, which can be compared with the FLOP rate of 78 MFLOPS of a Pentium® processor running at 133 MHz. (The FLOP count is obtained using the built-in MATLAB function flops. Note that these counts, which are given in terms of thousands of floating point operations per second (kFLOPS) or millions of floating point operations per second (MFLOPS), are rather pessimistic — the actual FLOP requirement would be considerably lower.) Thus, the real-time implementation of the complete tracking system is possible. With the average revisit interval of 2.5s, the FLOP requirement of the IMMPDAF is 62.5 kFLOP/radar cycle. With the revisit time calculations taking about the same amount of computation as a cycle of the IMMPDAF, but running at half the rate of the Kalman filter (which runs at constant rate), the IMMPDAF with adaptive revisit time is about 10 times costlier computationally than a Kalman filter. Due to its ability to save radar resources, which are much more expensive than computational resources, the IMMPDAF is a viable alternative to the Kalman filter, which is the standard "workhorse" in many current tracking systems. (Some systems still use the α-β filter as their "work mule.")

8.5 A Flexible-Window ML-PDA Estimator for Tracking Low Observable (LO) Targets

One difficulty with the ML-PDA approach of Section 8.3, which uses a set of scans of measurements as a batch, is the incorporation of noninformative scans when the target is not present in the surveillance region for some consecutive scans. For example, if the target appears within the surveillance region of the sensor after the first few scans, the estimator can be misled by the pure clutter in those scans — the earlier scans contain no relevant information, and the incorporation of these into the estimator not only increases the amount of processing (without adding any more information), but also results in less accurate estimates or even track rejection. Also, a target could disappear from the surveillance region for a while during tracking and reappear sometime later. Again, these intervening scans contain little or no information about the target and can potentially mislead the tracker.

In addition, the standard ML-PDA estimator assumes that the target SNR, the target velocity, and the density of false alarms over the entire tracking period remain constant. In practice, this may not be the case, and then the standard ML-PDA estimator will not yield the desired results. For example, the average target SNR may vary significantly as the target gets closer to or moves away from the sensor. In addition, the target might change its course and/or speed intermittently over time. For electro-optical sensors, depending on the time of the day and weather, the number of false alarms may vary as well.

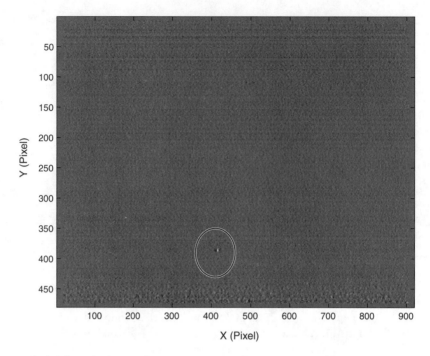

FIGURE 8.2 The last frame in the F1 Mirage sequence.

Because of these concerns, an estimator capable of handling time-varying SNR (with online adaptation), false alarm density, and slowly evolving course and speed is needed. While a recursive estimator like the IMM-PDA is a candidate, in order to operate under low SNR conditions in heavy clutter, a batch estimator is still preferred. In this section, the above problems are addressed by introducing an estimator that uses the ML-PDA with AI *adaptively* in a sliding-window fashion,[20] rather than using all the measurements in a single batch as the standard ML-PDA estimator does.[14] The initial time and the length of this sliding window are adjusted adaptively based on the information content in the measurements in the window. Thus, scans with little or no information content are eliminated and the window is moved over to scans with "informative" measurements.

This algorithm is also effective when the target is temporarily lost and reappears later. In contrast, recursive algorithms will diverge in this situation and may require an expensive track reinitialization. The standard batch estimator will be oblivious to the disappearance and may lose the whole track. This section demonstrates the performance of the adaptive sliding-window ML-PDA estimator on a real scenario with heavy clutter for tracking a fast-moving aircraft using an electro-optical (EO) sensor.

8.5.1 The Scenario

The adaptive ML-PDA algorithm was tested on an actual scenario consisting of 78 frames of Long Wave Infrared (LWIR) IR data collected during the Laptex data collection, which occurred in July, 1996 at Crete, Greece. The sequence contains a single target — a fast-moving Mirage F1 fighter jet. The 920 × 480 pixel frames, taken at a rate of 1Hz were registered to compensate for frame-to-frame line-of-sight (LOS) jitter. Figure 8.2 shows the last frame in the F1 Mirage sequence.

A sample detection list for the Mirage F1 sequence obtained at the end of preprocessing is shown in Figure 8.3. Each "x" in the figure represents a detection above the threshold.

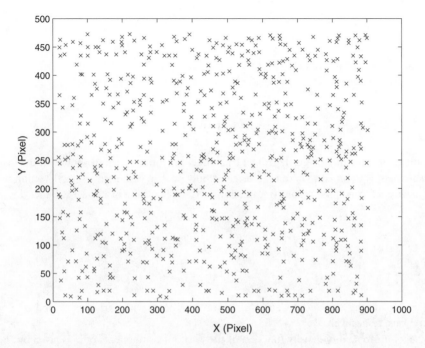

FIGURE 8.3 Detection list corresponding to the frame in Figure 8.2.

8.5.2 Formulation of the ML-PDA Estimator

This section describes the target models used by the estimator in the tracking algorithm and the statistical assumptions made by the algorithm. The ML-PDA estimator for these models is introduced, and the CRLB for the estimator and the hypothesis test used to validate the track are presented.

8.5.2.1 Target Models

The ML-PDA tracking algorithm is used on the detection lists after the data preprocessing phase. It is assumed that there are n detection lists obtained at times $t = t_1, t_2, \ldots t_n$. The i-th detection list, where $1 \le i \le n$, consists of m_i detections at pixel positions (x_{ij}, y_{ij}) along the X and Y directions. In addition to locations, the signal strength or amplitude, a_{ij}, of the j-th detection in the i-th list, where $1 \le j \le m$, is also known. Thus, assuming constant velocity over a number of scans, the problem can be formulated as a two-dimensional scenario in space with the target motion defined by the four-dimensional vector

$$\mathbf{x} \triangleq \left[\xi(t_0) \eta(t_0) \dot{\xi} \dot{\eta} \right]' \tag{8.134}$$

where $\xi(t_0)$ and $\eta(t_0)$ are the horizontal and vertical pixel positions of the target, respectively, from the origin at the reference time t_0. The corresponding velocities along these directions are assumed constant at $\dot{\xi}(t_0)$ pixel/s and $\dot{\eta}(t_0)$ pixel/s, respectively.

The set of measurements in list i at time t_i is denoted by

$$Z(i) = \left\{ z_j(i) \right\}_{j=1}^{m_i} \tag{8.135}$$

where m_i is the number of measurements at t_i. The measurement vector $z_j(i)$ is denoted by

$$z_j(i) \overset{\Delta}{=} \begin{bmatrix} x_{ij} & y_{ij} & a_{ij} \end{bmatrix}' \tag{8.136}$$

where x_{ij} and y_{ij} are observed X and Y positions, respectively.

The cumulative set of measurements made in scans t_1 through t_{-n} is given by

$$Z^n = \left\{ Z(i) \right\}_{i=1}^n \tag{8.137}$$

A measurement can either originate from a true target or from a spurious source. In the former case, each measurement is assumed to have been received only once in each scan with a detection probability P_D and to have been corrupted by zero-mean additive Gaussian noise of known variance, i.e.,

$$x_{ij} = \xi(t_i) + \epsilon_{ij} \tag{8.138}$$

$$y_{ij} = \eta(t_i) + v_{ij} \tag{8.139}$$

where ϵ_{ij} and v_{ij} are the zero-mean Gaussian noise components with variances σ_1^2 and σ_2^2 along the X and Y directions, respectively.

Thus, the joint probability density function of the position components of z_{ij} is given by

$$p(z_{ij}) = \frac{1}{2\pi\sigma_1\sigma_2} \exp\left(-\frac{1}{2}\left[\frac{x_{ij} - \xi(t_i)}{\sigma_1}\right]^2 - \frac{1}{2}\left[\frac{y_{ij} - \eta(t_i)}{\sigma_2}\right]^2 \right) \tag{8.140}$$

The false alarms are assumed to be distributed uniformly in the surveillance region and their number at any sampling instant obeys the Poisson probability mass function

$$\mu_f(m_i) = \frac{(\lambda U)^{m_i} e^{-\lambda U}}{m_i!} \tag{8.141}$$

where U is the area of surveillance and λ is the expected number of false alarms per unit of this area. Kirubarajan and Bar-Shalom[14] have shown that the performance of the ML-PDA estimator can be improved by using amplitude information (AI) of the received signal in the estimation process itself, in addition to thresholding. After the signal has been passed through the matched filter, an envelope detector can be used to obtain the amplitude of the signal. The noise at the matched filter is assumed to be narrowband Gaussian. When this is fed through the envelope detector, the output is Rayleigh distributed. Given the detection threshold, τ, the probability of detection P_D and the probability of false alarm P_{FA} are

$$P_D \overset{\Delta}{=} P\left(\text{The target-oriented measurement exceeds the threshold } \tau\right) \tag{8.142}$$

and

$$P_{FA} \overset{\Delta}{=} P\left(\text{A measurement caused by noise only exceeds the threshold } \tau\right) \tag{8.143}$$

where $P(\cdot)$ is the probability of an event.

The probability density functions at the output of the threshold detector, which corresponds to signals from the target and false alarms are denoted by $p_1^\tau(a)$ and $p_0^\tau(a)$, respectively. Then the amplitude likelihood ratio, ρ, can then be written as[3]

$$\rho = \frac{p_1^\tau(a)}{p_0^\tau(a)} = \frac{P_{FA}}{P_D(1+d)}\exp\left(\frac{a^2 d}{2(1+d)}\right) \tag{8.144}$$

where τ is the detection threshold.

8.5.2.2 The Maximum Likelihood-Probabilistic Data Association Estimator

This section focuses on the maximum likelihood estimator combined with the PDA approach. If there are m_i detections at t_i, one has the following mutually exclusive and exhaustive events[3]

$$\varepsilon_j(i) \stackrel{\Delta}{=} \begin{cases} \{\text{measurement } z_j(i) \text{ is from the target}\} & j=1,\ldots,m_i \\ \{\text{all measurements are false}\} & j=0 \end{cases} \tag{8.145}$$

The pdf of the measurements corresponding to the above events can be written as[3]

$$p\Big(Z(i)\big|\varepsilon_j(i),\mathbf{x}\Big) = \begin{cases} U^{1-m_i}\,p(z_{ij})\rho_{ij}\prod_{j=1}^{m_i}p_0^\tau(a_{ij}) & j=1,\ldots,m_i \\ U^{1-m_i}\prod_{j=1}^{m_i}p_0^\tau(a_{ij}) & j=0 \end{cases} \tag{8.146}$$

Using the total probability theorem,

$$p\big(Z(i)\big|\mathbf{x}\big) = \sum_{j=0}^{m_i} p\big(Z(i)\big|\varepsilon_j(i),\mathbf{x}\big)p\big(\varepsilon_j(i),\mathbf{x}\big)$$

$$= \sum_{j=0}^{m_i} p\big(Z(i)\big|\varepsilon_j(i),\mathbf{x}\big)p\big(\varepsilon_j(i)\big) \tag{8.147}$$

the above can be written explicitly as

$$p\big(Z(i)\big|\mathbf{x}\big) = U^{-mi}(1-P_D)\prod_{j=1}^{m_i}p_0^\tau(a_{ij})\mu_f(m_i) + \frac{U^{1-m_i}P_D\mu_f(m_i-1)}{m_i}\prod_{j=1}^{m_i}p_0^\tau(a_{ij})\sum_{j=1}^{m_i}p(z_{ij})\rho_{ij}$$

$$= U^{-mi}(1-P_D)\prod_{j=1}^{m_i}p_0^\tau(a_{ij})\mu_f(m_i) + \frac{U^{1-m_i}P_D\mu_f(m_i-1)}{2\pi\sigma_1\sigma_2 m_i}\prod_{j=1}^{m_i}p_0^\tau(a_{ij}) \tag{8.148}$$

$$\cdot \sum_{j=1}^{m_i}\rho_{ij}\exp\left(-\frac{1}{2}\left[\frac{x_{ij}-\xi(t_i)}{\sigma_1}\right]^2 - \frac{1}{2}\left[\frac{y_{ij}-\eta(t_i)}{\sigma_2}\right]^2\right)$$

To obtain the likelihood ratio, $\Phi[Z(i), \mathbf{x}]$, at t_i, divide Equation 8.148 by $p[Z(i)|\varepsilon_0(i), \mathbf{x}]$

$$\Phi\left[Z(i), \mathbf{x}\right] = \frac{p\left[Z(i), \mathbf{x}\right]}{p\left[Z(i)|\varepsilon_0(i), \mathbf{x}\right]}$$

$$= \left(1 - P_D\right) + \frac{P_D}{2\pi\lambda\sigma_1\sigma_2} \sum \rho_{ij} \exp\left(-\frac{1}{2}\left[\frac{x_{ij} - \xi(t_i)}{\sigma_1}\right]^2 - \frac{1}{2}\left[\frac{y_{ij} - \eta(t_i)}{\sigma_2}\right]^2\right)$$

(8.149)

Assuming that measurements at different sampling instants are conditionally independent, i.e.,

$$p\left[Z^n|\mathbf{x}\right] = \prod p\left[Z(i)|\mathbf{x}\right]$$

(8.150)

the total likelihood ratio[3] for the entire data set is given by

$$\Phi\left[Z^n, \mathbf{x}\right] = \prod_{i=1}^{n} \Phi_i\left[Z(i), \mathbf{x}\right]$$

(8.151)

Then, the total log-likelihood ratio, $\Phi[Z^n, \mathbf{x}]$, expressed in terms of the individual log-likelihood ratios $\phi[Z(i), \mathbf{x}]$ at sampling time instants t_i, becomes

$$\phi\left[Z^n, \mathbf{x}\right] = \sum_{i=1}^{n} \phi_i\left[Z(i), \mathbf{x}\right]$$

$$= \sum \ln\left[\left(1 - P_D\right) + \frac{P_D}{2\pi\lambda\sigma_1\sigma_2} \sum_{j=1}^{m_i} \rho_{ij} \exp\left(-\frac{1}{2}\left[\frac{x_{ij} - \xi(t_i)}{\sigma_1}\right]^2 - \frac{1}{2}\left[\frac{y_{ij} - \eta(t_i)}{\sigma_2}\right]^2\right)\right]$$

(8.152)

The maximum likelihood estimate (MLE) is obtained by finding the vector $\mathbf{x} = \hat{\mathbf{x}}$ that maximizes the total log-likelihood ratio given in Equation 8.152. This maximization is performed using a quasi-Newton (variable metric) method. This can also be accomplished by minimizing the negative log-likelihood function. In our implementation of the MLE, the Davidon-Fletcher-Powell variant of the variable metric method is used. This method is a conjugate gradient technique that finds the minimum value of the function iteratively.[21] However, the negative log-likelihood function may have several local minima; i.e., it has multiple modes. Due to this property, if the search is initiated too far away from the global minimum, the line search algorithm may converge to a local minimum. To remedy this, a multi-pass approach is used as in Reference 14.

8.5.3 Adaptive ML-PDA

Often, the measurement process begins before the target becomes visible — that is, the target enters the surveillance region of the sensor some time after the sensor started to record measurements. In addition, the target may disappear from the surveillance region for a certain period of time before reappearing.

During these periods of blackout, the received measurements are purely noise-only, and the scans of data contain no information about the target under track. Incorporating these scans into a tracker reduces its accuracy and efficiency. Thus, detecting and rejecting these scans is important to ensure the fidelity of the estimator. This subsection presents a method that uses the ML-PDA algorithm in a sliding-window fashion. In this case, the algorithm uses only a subset of the data at a time rather than all of the frames at once, to eliminate the use of scans that have no target. The initial time and the length of the sliding window are adjusted adaptively based on the information content of the data — the smallest window, and thus the fewest number of scans, required to identify the target is determined online and adapted over time.

The key steps in the adaptive ML-PDA estimator are as follows:

1. Start with a window of minimum size.
2. Run the ML-PDA estimator within this window and carry out the validation test on the estimates.
3. If the estimate is accepted (i.e., if the test is passed), and if the window is of minimum size, accept the window. The next window is the present window advanced by one scan. Go to step 2.
4. If the estimate is accepted, and if the window is greater than minimum size, try a shorter window by removing the initial scan. Go to step 2 and accept the window only if estimates are better than those from the previous window.
5. If the test fails and if the window is of minimum size, increase the window length to include one more scan of measurements and, thus, increase the information content in the window. Go to step 2.
6. If the test fails and if the window is greater than minimum size, eliminate the first scan, which could contain pure noise only. Go to step 2.
7. Stop when all scans are used.

The algorithm is described below. In order to specify the exact steps in the estimator, the following variables are defined:

W = Current window length
W_{min} = Minimum window length
$Z(t_i)$ = Scan (set) of measurements at time t_i

With these definitions, the algorithm is given below:

BEGIN PROCEDURE Adaptive ML PDA estimator(W_{min}, $Z(t_1)$, $Z(t_n)$)
    $i = 1$ — Initialize the window at the first scan.
    $W = W_{min}$ — Initially, use a window of minimum size.
    WHILE ($i + W < n$) — Repeat until the last scan at t_n.
        Do grid search for initial estimates by numerical search on $Z(t_i), Z(t_{i+1}),\ldots,Z(t_{i+w})$
        Apply ML-PDA Estimator on the measurements in $Z(t_i), Z(t_{i+1}),\ldots,Z(t_{i+w})$
        Validate the estimates
        IF the estimates are rejected
            IF ($W > W_{min}$) — Check if we can reduce the window size.
                $i = i + 1$ — Eliminate the initial scan that might be due to noise only.
            ELSEIF ($W = W_{min}$)
                $W = W + 1$ — Expand window size to include an additional scan.
            ENDIF
        ENDIF
        IF the estimates are accepted
            IF ($W > W_{min}$) — Check if we can reduce the window size.
                Try a shorter window by removing the initial scan and check if estimates are better, $i = i + 1$
            ENDIF

IF estimates for shorter window are NOT better OR ($W = W_{min}$)
Accept estimates and try next window, $i = i + 1$
ENDIF
ENDIF
END WHILE
END PROCEDURE

To illustrate the adaptive algorithm, consider a scenario where a sensor records 10 scans of measurements over a surveillance region. The target, however, appears in this region (i.e., its intensity exceeds the threshold) only after the second scan (i.e., from the third scan onward). This case is illustrated in Figure 8.4. The first two scans are useless because they contain only noise.

Consider the smallest window size required for a detection to be 5. Then the algorithm will evolve as shown in Figure 8.5. First, for the sake of illustration, assume that a single "noisy" scan present in the data set is sufficient to cause the MLE to fail the hypothesis test for track acceptance. The algorithm tries to expand the window to include an additional scan if a track detection is not made. This is done because an additional scan of data may bring enough additional information to detect the target track. The algorithm next tries to cut down the window size by removing the initial scans. This is done to check whether a better estimate can be obtained without this scan. If this initial scan is noise only, then it degrades the accuracy of the estimate. If a better estimate is found (i.e., a more accurate estimate) without this scan, the latter is eliminated. Thus, as in the example given above, the algorithm expands at the front (most recent scan used) and contracts at the rear end of the window to find the best window that produces the strongest detection, based on the validation test.

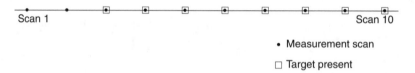

FIGURE 8.4   Scenario with a target being present for only a partial time during observation.

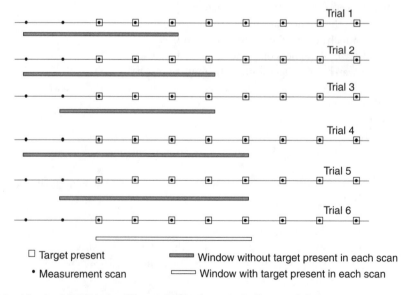

FIGURE 8.5   Adaptive ML-PDA algorithm applied to the scenario illustrated above.

TABLE 8.3 Parameters Used in the ML-PDA Algorithm for the F1 Mirage Jet

Parameter	Value
σ_1	1.25
σ_2	1.25
Min Window Size, W	10
Initial Target SNR, d_0	9.5
P_{DC}	0.70
α	0.85
π_m	5%
$\bar{v}$	5.0
$\bar{\sigma}_v$	0.15
K	4

8.5.4 Results

8.5.4.1 Estimation Results

The Mirage F1 data set consists of 78 scans or frames of LWIR IR data. The target appears late in this scenario and moves towards the sensor. There are about 600 detections per frame. In this implementation the parameters shown in Table 8.3 were chosen.

The choice of these parameters is explained below:

- σ_1 and σ_2 are, as in Equation 8.140, the standard deviations along the horizontal and vertical axes respectively. The value of 1:25 for both variables models the results of the preprocessing.
- The minimum window size, W_{min}, was chosen to be 10. The algorithm will expand this window if a target is not detected in 10 frames. Initially a shorter window was used, but the estimates appeared to be unstable. Therefore, fewer than 10 scans is assumed to be ineffective at producing an accurate estimate.
- The initial target SNR, d_0, was chosen as 9:5 dB because the average SNR of all the detections over the frames is approximately 9:0 dB. However, in most frames, random spikes were noted. In the first frame, where a target is unlikely to be present, a single spike of 15:0 dB is noted. These spikes, however, cannot and should not be modeled as the target SNR.
- A constant probability of detection (P_{DC}) of 0:7 was chosen. A value that is too high would bring down the detection threshold and increase P_{FA}.
- α is the parameter used to update the estimated target SNR with an α filter. A high value is chosen for the purpose of detecting a distant target that approaches the sensor over time and to account for the presence of occasional spikes of noise. Thus, the estimated SNR is less dependent on a detection that could originate from a noisy source and, thus, set the bar too high for future detections.
- π_m is the miss probability.
- $\bar{v}$ and $\bar{\sigma}_v$ are used in the multipass approach of the optimization algorithm.[11,14]
- The number of passes K in the multipass approach of the optimization algorithm was chosen as 4.

Figure 8.6 further clarifies the detection process by depicting the windows where the target has been detected.

From the above results, note the following:

- The first detection uses 22 scans and occurs at scan 28. This occurs because the initial scans have low-information content as the target appears late in the frame of surveillance. The IMM-MHT algorithm[22] required 38 scans for a detection, while the IMMPDA[23] required 39 scans. Some spurious detections were noticed at earlier scans, but these were rejected.

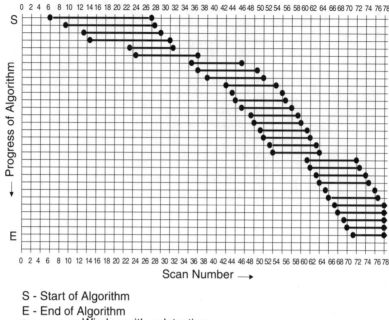

FIGURE 8.6   Progress of the algorithm showing windows with detections.

- The next few detection windows produce similar target estimates. This is because a large number of scans repeat themselves in these windows.
- After the initial detections, there is a "jump" in the scan number at which a detection is made. In addition, the estimates, particularly the velocity estimates, deteriorate. This could indicate either that the target has suddenly disappeared (became less visible) from the region of surveillance or that the target made a maneuver.
- From scan 44 onward, the algorithm stabilizes for several next windows. At scan 52, however, there is another jump in detection windows. This is also followed by a drop in the estimated target SNR, as explained above. This, however, indicates that the algorithm can adjust itself and restart after a target has become suddenly invisible. Recursive algorithms will diverge in this case.
- From scan 54 onward, the algorithm stabilizes, as indicated by the estimates. Also, a detection is made for every increasing window because the target has come closer to the sensor and, thus, is more visible. This is noted by the sharp rise in the estimated target SNR after scan 54.
- The above results provide an understanding of the target's behavior. The results suggest that the Mirage F1 fighter jet appears late in the area of surveillance and moves towards the sensor. However, initially it remains quite invisible and possibly undergoes maneuvers. As it approaches the sensor, it becomes more and more visible and, thus, easier to detect.

8.5.4.2   Computational Load

The adaptive ML-PDA tracker took 442s, including the time for data input/output, on a Pentium® III processor running at 550MHz to process the 78 scans of the Mirage F1 data. This translates into about 5.67s per frame (or 5.67s running time for one-second data), including input/output time. A more efficient implementation on a dedicated processor can easily make the algorithm real-time capable on a similar processor. Also, by parallelizing the initial grid search, which required more than 90% of the time, the adaptive ML-PDA estimator can be made even more efficient.

8.6 Summary

This chapter presented the use of the PDA technique for different tracking problems. Specifically, the PDA approach was used for parameter estimation as well as recursive state estimation. As an example of parameter estimation, track formation of a low observable target using a nonlinear maximum likelihood estimator in conjunction with the PDA technique with passive (sonar) measurements was presented. The use of the PDA technique in conjunction with the IMM estimator, resulting in the IMMPDAF, was presented as an example of recursive estimation on a radar-tracking problem in the presence of ECM. Also presented was an adaptive sliding-window PDA-based estimator that retains the advantages of the batch (parameter) estimator while being capable of tracking the motion of maneuvering targets. This was illustrated on an EO surveillance problem. These applications demonstrate the usefulness of the PDA approach for a wide variety of real tracking problems.

References

1. Bar-Shalom, Y. (Ed.), *Multitarget-Multisensor Tracking: Advanced Applications*, Vol. I, Artech House Inc., Dedham, MA, 1990. Reprinted by YBS Publishing, 1998.
2. Bar-Shalom, Y. (Ed.), *Multitarget-Multisensor Tracking: Applications and Advances*, Vol. II, Artech House Inc., Dedham, MA, 1992. Reprinted by YBS Publishing, 1998.
3. Bar-Shalom, Y. and Li, X. R., *Multitarget-Multisensor Tracking: Principles and Techniques*, Storrs, CT: YBS Publishing, 1995.
4. Blackman, S. S. and Popoli, R., *Design and Analysis of Modern Tracking Systems*, Artech House Inc., Dedham, MA, 1999.
5. Feo, M., Graziano, A., Miglioli, R., and Farina, A., IMMJPDA vs. MHT and Kalman Filter with NN Correlation: Performance Comparison, *IEE Proc. on Radar, Sonar and Navigation (Part F)*, 144(2), 49–56, 1997.
6. Kirubarajan, T., Bar-Shalom, Y., Blair, W. D., and Watson, G. A., IMMPDA Solution to Benchmark for Radar Resource Allocation and Tracking in the Presence of ECM, *IEEE Trans. Aerospace and Electronic Systems*, 34(3), 1023–1036, 1998.
7. Lerro, D., and Bar-Shalom, Y., Interacting Multiple Model Tracking with Target Amplitude Feature, *IEEE Trans. Aerospace and Electronic Systems*, AES-29, No. 2, 494–509, 1993.
8. Blair, W. D., Watson, G. A., Kirubarajan, T., and Bar-Shalom, Y., Benchmark for radar resource allocation and tracking in the presence of ECM, *IEEE Trans. Aerospace and Electronic Systems*, 34(3), 1015–1022, 1998.
9. Blackman, S. S., Dempster, R. J., Busch, M. T., and Popoli, R. F., IMM/MHT solution to radar benchmark tracking problem, *IEEE Trans. Aerospace and Electronic Systems*, Vol. 35(2), 730–738, 1999.
10. Bar-Shalom, Y. and Li, X. R., *Estimation and Tracking: Principles, Techniques and Software*, Artech House, Dedham, MA, 1993. Reprinted by YBS Publishing, 1998.
11. Jauffret, C., and Bar-Shalom, Y., Track formation with bearing and frequency measurements in clutter, *IEEE Trans. Aerospace and Electronic Systems*, AES-26, 999–1010, 1990.
12. Kirubarajan, T., Wang, Y., and Bar-Shalom, Y., Passive ranging of a low observable ballistic missile in a gravitational field using a single sensor, *Proc. 2nd International Conf. Information Fusion*, July 1999.
13. Sivananthan, S., Kirubarajan, T., and Bar-Shalom, Y., A radar power multiplier algorithm for acquisition of LO ballistic missiles using an ESA radar, *Proc. IEEE Aerospace Conf.*, March 1999.
14. Kirubarajan, T., and Bar-Shalom, Y., Target motion analysis in clutter for passive sonar using amplitude information, *IEEE Trans. Aerospace and Electronic Systems*, 32(4), 1367–1384, 1996.
15. Nielsen, R. O., *Sonar Signal Processing*, Artech House Inc., Boston, MA, 1991.
16. Blair, W. D., Watson, G. A., and Hoffman, S. A., Benchmark problem for beam pointing control of phased array radar against maneuvering targets, *Proc. Am. Control Conf.*, June 1994.

17. Blair, W. D., and Watson, G. A., IMM algorithm for solution to benchmark problem for tracking maneuvering targets, *Proc. SPIE Acquisition, Tracking and Pointing Conf.*, April, 1994.

18. Blair, W. D., Watson, G. A., Hoffman, S. A., and Gentry, G. L., Benchmark problem for beam pointing control of phased array radar against maneuvering targets in the presence of ECM and false alarms, *Proc. American Control Conf.*, June 1995.

19. Yeddanapudi, M., Bar-Shalom, Y., and Pattipati, K. R., IMM estimation for multitarget-multisensor air traffic surveillance, *Proc. IEEE*, 85(1), 80–94, 1997.

20. Chummun, M. R., Kirubarajan, T., and Bar-Shalom, Y., An adaptive early-detection ML-PDA estimator for LO targets with EO sensors, *Proc. SPIE Conf. Signal and Data Processing of Small Targets*, April 2000.

21. Press, W. H., Teukolsky, S. A., Vetterling, W. T., and, Flannery, B. P., *Numerical Recipes in C*, Cambridge University Press, Cambridge, U.K., 1992.

22. Roszkowski, S. H., Common database for tracker comparison, *Proc. SPIE Conf. Signal and Data Processing of Small Targets*, Vol. 3373, April 1998.

23. Lerro, D., and Bar-Shalom, Y., IR Target detection and clutter reduction using the interacting multiple model estimator, *Proc. SPIE Conf. on Signal and Data Processing of Small Targets*, Vol. 3373, April 1998.

9

An Introduction to the Combinatorics of Optimal and Approximate Data Association

9.1 Introduction 9-1
9.2 Background.. 9-2
9.3 Most Probable Assignments 9-4
9.4 Optimal Approach.. 9-5
9.5 Computational Considerations 9-7
9.6 Efficient Computation of the JAM 9-8
9.7 Crude Permanent Approximations.................................. 9-8
9.8 Approximations Based on Permanent Inequalities 9-10
9.9 Comparisons of Different Approaches........................... 9-12
9.10 Large-Scale Data Association 9-15
9.11 Generalizations 9-17
9.12 Conclusions 9-17
Acknowledgments... 9-18
Appendix 9.A Algorithm for Data Association Experiment.... 9-18
References.. 9-19

Jeffrey K. Uhlmann
University of Missouri

9.1 Introduction

Applying filtering algorithms to track the states of multiple targets first requires the correlation of the tracked objects with their corresponding sensor observations. A variety of probabilistic measures can be applied to each track estimate to determine independently how likely it is to have produced the current observation; however, such measures are useful only in practice for eliminating obviously infeasible candidates. Chapter 3 uses these measures to construct gates for efficiently reducing the number of feasible candidates to a number that can be accommodated within real-time computing constraints. Subsequent elimination of candidates can then be effected by measures that consider the joint relationships among the remaining track and report pairs.

After the gating step has been completed, a determination must be made concerning which feasible associations between observations and tracks are most likely to be correct. In a dense environment, however, resolving the ambiguities among various possible assignments of sensor reports to tracks may

0-8493-2379-7/01/$0.00+$1.50

be impossible. The general approach proposed in the literature for handling such ambiguities is to maintain a set of *hypothesized* associations in the hope that some will be eliminated by future observations.[1-3] The key challenge is somehow to bound the overall computational cost by limiting the proliferation of pairs under consideration, which may increase in number at a geometric rate.

If n observations are made of n targets, one can evaluate an independent estimate of the probability that a given observation is associated with a given beacon. An independently computed probability, however, may be a poor indicator of how likely a particular observation is associated with a particular track because it does not consider the extent to which other observations may also be correlated with that item. More sophisticated methods generally require a massive batch calculation (i.e., they are a function of all n beacon estimates and $O(n)$ observations of the beacons) where approximately one observation of each target is assumed.[4] Beyond the fact that a real-time tracking system must process each sensor report as soon as it is obtained, the most informative joint measures of association scale exponentially with n and are, therefore, completely useless in practice.

This chapter examines some of the combinatorial issues associated with the batch data association problem arising in tracking and correlation applications. A procedure is developed that addresses a large class of data association problems involving the calculation of permanents of submatrices of the original association matrix. This procedure yields what is termed the joint assignment matrix (JAM), which can be used to optimally rank associations for hypothesis selection. Because the computational cost of the permanent scales exponentially with the size of the matrix, improved algorithms are developed both for calculating the exact JAM and for generating approximations to it. Empirical results suggest that at least one of the approximations is suitable for real-time hypothesis generation in large-scale tracking and correlation applications. Novel theoretical results include an improved upper bound on the calculation of the JAM and new upper bound inequalities for the permanent of general nonnegative matrices. One of these inequalities is an improvement over the best previously known inequality.

9.2 Background

The batch data association problem[4-8] can be defined as follows:

> Given predictions about the states of n objects at a future time t, n measurements of that set of objects at time t, and a function to compute a probability of association between each prediction and measurement pair, calculate a new probability of association for a prediction and a measurement that is conditioned on the knowledge that the mapping from the set of predictions to the set of measurements is one-to-one.

For real-time applications, the data association problem is usually defined in terms only of estimates maintained at the current timestep of the filtering process. A more general problem can be defined that considers observations over a series of timesteps. Both problems are intractable, but the single-timestep variant appears to be more amenable to efficient approximation schemes.

As an example of the difference between these two measures of the probability of association, consider an indicator function that provides a binary measure of whether or not a given measurement is compatible with a given prediction. If the measure considers each prediction/measurement pair independently of all other predictions and measurements, then it could very well indicate that several measurements are feasible realizations of a single prediction. However, if one of the measurements has only that prediction as a feasible association, then from the constraint that the assignment is one-to-one, it and the prediction must correspond to the same object. Furthermore, it can then be concluded, based on the same constraint, that the other measurements absolutely are *not* associated with that prediction. Successive eliminations of candidate pairs will hopefully yield a set of precisely n feasible candidates that must represent the correct assignment; however, this is rare in real-world applications. The example presented in Figure 9.1 demonstrates the process.

The above association problem arises in a number of practical tracking and surveillance applications. A typical example is the following: a surveillance aircraft flies over a region of ocean and reports the

---

Assignment Example

Consider the effect of the assignment constraint on the following matrix of feasible associations:

	R_1	R_2	R_3	R_4
T_1	0	0	1	1
T_2	1	1	0	1
T_3	0	1	0	1
T_4	0	0	1	1

Every track has more than one report with which it could be feasibly assigned. The report R_1, however, can only be assigned to T_2. Given the one-to-one assignment constraint, R_1 clearly must have originated from track T_2. Making this assignment leaves us with the following options for the remaining tracks and reports:

	R_1	R_2	R_3	R_4
T_1	–	0	1	1
T_2	1	–	–	–
T_3	–	1	0	1
T_4	–	0	1	1

The possibility that R_2 originated from T_2 has been eliminated; therefore, R_2 could only have originated from the remaining candidate, T_3. This leaves:

	R_1	R_2	R_3	R_4
T_1	–	–	1	1
T_2	1	–	–	–
T_3	–	1	–	–
T_4	–	–	1	1

where two equally feasible options now exist for assigning tracks T_1 and T_4 to reports R_3 and R_4. From this ambiguity only the following can be concluded:

	R_1	R_2	R_3	R_4
T_1	–	–	0.5	0.5
T_2	1	–	–	–
T_3	–	1	–	–
T_4	–	–	0.5	0.5

FIGURE 9.1   Assignment example.

positions of various ships. Several hours later another aircraft repeats the mission. The problem then arises how to identify which reports in the second pass are associated with which from the earlier pass. The information available here includes known kinematic constraints (e.g., maximum speed) and the time difference between the various reports. If each ship is assumed to be traveling at a speed in the interval $[v_{min}, v_{max}]$, then the indicator function can identify feasible pairs simply by determining which

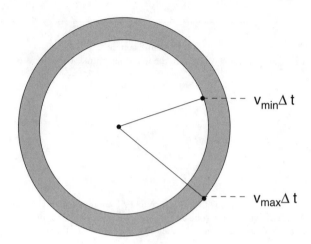

FIGURE 9.2   The inner circle represents the possible positions of the ship if it travels at minimum speed, while the outer circle represents its possible positions at maximum speed.

reports from the second pass fall within the radial interval $[v_{min}\Delta t, v_{max}\Delta t]$ about reports from the first pass (see Figure 9.2). This problem is called track initiation[4,5] because its solution provides a full position and velocity estimate — referred to as a *track* — which can then permit proper tracking.

After the tracking process has been initiated, the association problem arises again at the arrival of each batch of new reports. Thus, data association in this case is not a "one shot" problem — it is a problem of associating series of reports over a period of time in order to identify distinct trajectories. This means that attempting to remove ambiguity entirely at each step is not necessary; it is possible to retain a set of pairs in the hope that some will be eliminated at future steps. The maintenance of tentative tracks — referred to as *hypotheses* — is often termed *track splitting*. Track splitting can be implemented in several ways, ranging from methods that simply add extra tracks to the track set with no logical structure to indicate which were formed from common reports, to methods that construct a complete "family tree" of tracks so that confirmation of a single leaf can lead to the pruning of large branches. No matter what the method, the critical problem is to determine which pairs to keep and which to discard.

9.3   Most Probable Assignments

One way to deal with ambiguities arising in the joint analysis of the prior probabilities of association is to determine which of the *a priori n*! possible assignments is most probable. In this case, "most probable" means the assignment that maximizes the product of the prior probabilities of its component pairs. In other words, it is the assignment, σ_i, that maximizes

$$\prod_i a_{i\sigma_i} \qquad (9.1)$$

where a_{ij} is the matrix element giving the probability that track i is associated with report j. Unfortunately, this approach seems to require the evaluation and examination of n! products. There exists, however, a corpus of work on the closely related problem of optimal assignment (also known as maximum-weighted bipartite matching). The optimal assignment problem seeks the assignment that maximizes the *sum* of the values of its component pairs. In other words, it maximizes

$$\sum_i a_{i\sigma_i} \qquad (9.2)$$

This can be accomplished in $O(n^3)$ time.[9] Thus, the solution to the maximum product problem can be obtained with an optimal assignment algorithm simply by using the logs of the prior probabilities, with the log of zero replaced with an appropriately large negative number. The optimal assignment approach eliminates ambiguity by always assuming that the best assignment is always the correct assignment. Thus, it never maintains more than n tracks. The deficiency with this approach is that unless there are very few ambiguous pairs, many assignments will have almost the same probability. For example, if two proximate reports are almost equally correlated with each of two tracks, then swapping their indices in the optimal assignment will generate a new but only slightly less probable assignment. In fact, in most nontrivial applications, the best assignment has a very low probability of being correct.

The optimal assignment method can be viewed as the best choice of track-report pairs if the criterion is to maximize the probability of having *all* of the pairs be correct. Another reasonable optimality criterion would seek the set of n pairs that maximizes the *expected number* of pairs that are correct. To illustrate the difference between these criteria, consider the case in which two proximate reports are almost equally correlated with two tracks, and no others. The optimal assignment criterion would demand that the two reports be assigned to distinct tracks, while the other criterion would permit all four pairs to be kept as part of the n selected pairs. With all four possible pairs retained, the second criterion ensures that at least two are correct. The two pairs selected by the optimal assignment criterion, however, have almost a 0.5 probability of both being incorrect.

9.4 Optimal Approach

A strategy for generating multiple hypotheses within the context of the optimal assignment approach is to identify the k best assignments and take the union of their respective pairings.[6,10,11] The assumption is that pairs in the most likely assignments are most likely to be correct. Intuition also would suggest that a pair common to all of the k best assignments stands a far better chance of being correct than its prior probability of association might indicate. Generalizing this intuition leads to the exact characterization of the probabilities of association under the assignment constraint. Specifically, the probability that report R_i is associated with track T_j is simply the sum of the probabilities of all the assignments containing the pair, normalized by the sum of the probabilities of all assignments:[6]

$$p\left(a_{ij}\middle|\text{Assignment Constraint}\right) = \frac{1}{\sum_\sigma \prod_k a_{k\sigma_k}} \sum_{\{\sigma|\sigma_i=j\}} \prod_m a_{m\sigma_m} \tag{9.3}$$

For example, suppose the following matrix is given containing the prior probabilities of association for two tracks with two reports:

	R_1	R_2	
T_1	0.3	0.7	(9.4)
T_2	0.5	0.4	

Given the assignment constraint, the correct pair of associations must be either (T_1, R_1) and (T_2, R_2) or (T_1, R_2) and (T_2, R_1). To assess the joint probabilities of association, Equation 9.3 must be applied to each entry of the matrix to obtain:

$$\frac{1}{(.3)(.4)+(.5)(.7)} \cdot \begin{vmatrix} (.3)(.4) & (.5)(.7) \\ (.5)(.7) & (.3)(.4) \end{vmatrix} = \begin{vmatrix} 0.255 & 0.745 \\ 0.745 & 0.255 \end{vmatrix} \tag{9.5}$$

Notice that the resulting matrix is doubly stochastic (i.e., it has rows and columns all of which sum to unity), as one would expect. (The numbers in the tables have been rounded, but still sum appropriately. One can verify that the true values do as well.) Notice also that the diagonal elements are equal. This is the case for any 2 x 2 matrix because elements of either diagonal can only occur jointly in an assignment; therefore, one element of a diagonal cannot be more or less likely than the other. The matrix of probabilities generated from the assignment constraint is called the joint assignment matrix, or the JAM.

Now consider the following matrix:

$$
\begin{array}{ccc}
 & R_1 & R_2 \\
T_1 & 0.9 & 0.5 \\
T_2 & 0.4 & 0.1
\end{array}
\qquad (9.6)
$$

which, given the assignment constraint, leads to the following:

$$
\begin{array}{ccc}
 & R_1 & R_2 \\
T_1 & 0.31 & 0.69 \\
T_2 & 0.69 & 0.31
\end{array}
\qquad (9.7)
$$

This demonstrates how significant the difference can be between the prior and the joint probabilities. In particular, the pair (T_1, R_1) has a prior probability of 0.9, which is extremely high. Considered jointly with the other measurements, however, its probability drops to only 0.31.

A more extreme example is the following:

$$
\begin{array}{ccc}
 & R_1 & R_2 \\
T_1 & 0.99 & 0.01 \\
T_2 & 0.01 & 0.0
\end{array}
\qquad (9.8)
$$

which leads to:

$$
\begin{array}{ccc}
 & R_1 & R_2 \\
T_1 & 0.0 & 1.0 \\
T_2 & 1.0 & 0.0
\end{array}
\qquad (9.9)
$$

where the fact that T_2 cannot be associated with R_2 implies that there is only one feasible assignment. Examples like this show why a "greedy" selection of hypotheses based only on the independently assessed prior probabilities of association can lead to highly suboptimal results.

The examples that have been presented have considered only the ideal case where the actual number of targets is known and the number of tracks and reports equals that number. However, the association matrix can easily be augmented* to include rows and columns to account for the cases in which some reports are spurious and some targets are not detected. Specifically, a given track can have a probability of being associated with each report, as well as a probability of not being associated with any of them. Similarly, a given report has a probability of association with each of the tracks and a probability that it is a false alarm which is not associated with any target. Sometimes a report that is not associated with

---

* Remember the use of an association "matrix" is purely for notational convenience. In large applications, such a matrix would not be generated in its entirety; rather, a sparse graph representation would be created by identifying and evaluating only those entries with probabilities above a given threshold.

any tracks signifies a newly detected target. If the actual number of targets is not known, a combined probability of false alarm and probability of new target must be used. Estimates of probabilities of detection, probabilities of false alarms, and probabilities of new detections are difficult to determine because of complex dependencies on the type of sensor used, the environment, the density of targets, and a multitude of other factors whose effects are almost never known. In practice, such probabilities are often lumped together into one tunable parameter (e.g., a "fiddle factor").

9.5 Computational Considerations

A closer examination of Equation 9.3 reveals that the normalizing factor is a quantity that resembles the determinant of the matrix, but without the alternating ±1 factors. In fact, the determinant of a matrix is just the sum of products over all even permutations minus the sum of products over all odd permutations. The normalizing quantity of Equation 9.3, however, is the sum of products over all permutations. This latter quantity is called the permanent[7,12,13] of a matrix, and it is often defined as follows:

$$\mathrm{per}(A) = \sum_{\sigma} a_{1\sigma_1} a_{2\sigma_2} \cdots a_{n\sigma_n} \tag{9.10}$$

where the summation extends over all permutations Q of the integers $1,2,\ldots,n$. The Laplace expansion of the determinant also applies for the permanent:

$$\mathrm{per}(A) = \sum_{j=1}^{n} a_{ij} \cdot \mathrm{per}(A_{\bar{\imath}\bar{\jmath}})$$

$$= \sum_{i=1}^{n} a_{ij} \cdot \mathrm{per}(A_{\bar{\imath}\bar{\jmath}}) \tag{9.11}$$

where $A_{\bar{\imath}\bar{\jmath}}$ is the submatrix obtained by removing row i and column j. This formulation provides a straightforward mechanism for evaluating the permanent, but it is not efficient. As in the case of the determinant, expanding by Laplacians requires $O(n \cdot n!)$ computations. The unfortunate fact about the permanent is that while the determinant can be evaluated by other means in $O(n^3)$ time, effort exponential in n seems to be necessary to evaluate the permanent.[14]

If Equation 9.3 were evaluated without the normalizing coefficient for every element of the matrix, determining the normalizing quantity by simply computing the sum of any row or column would seem possible, since the result must be doubly stochastic. In fact, this is the case. Unfortunately, Equation 9.3 can be rewritten, using Equation 9.11, as

$$p(a_{ij}|\text{Assignment Constraint}) = a_{ij} \cdot \mathrm{per}(A_{\bar{\imath}\bar{\jmath}}) / \mathrm{per}(A) \tag{9.12}$$

where the evaluation of permanents seems unavoidable. Knowing that such evaluations are intractable, the question then is how to deal with computational issues.

The most efficient method for evaluating the permanent is attributable to Ryser:[7] Let A be an $n \times n$ matrix, let A_r denote a submatrix of A obtained by deleting r columns, let $\Pi(A_r)$ denote the product of the row sums of A_r, and let $\Sigma\Pi(A_r)$ denote the sum of the products $\Pi(A_r)$ taken over all possible A_r. Then,

$$\mathrm{per}(A) = \prod(A) - \sum\prod(A_1) + \sum\prod(A_2) - \ldots + (-1)^{n-1}\sum\prod(A_{n-1}) \tag{9.13}$$

This formulation involves only $O(2^n)$, rather than $O(n!)$, products. This may not seem like a significant improvement, but for n in the range of 10–15, the reduction in compute time is from hours to minutes on a typical workstation. For n greater than 35, however, scaling effects thwart any attempt at evaluating the permanent. Thus, the goal must be to reduce the coefficients as much as possible to permit the optimal solution for small matrices in real time, and to develop approximation schemes for handling large matrices in real time.

9.6   Efficient Computation of the JAM

Equation 9.13 is known to permit the permanent of a matrix to be computed in $O(n^2 \cdot 2^n)$ time. From Equation 9.12, then, one can conclude that the joint assignment matrix is computable in $O(n^4 \cdot 2^n)$ time (i.e., the amount of time required to compute the permanent of $A_{\bar{ij}}$ for each of the n^2 elements a_{ij}). However, this bound can be improved by showing that the joint assignment matrix can be computed in $O(n^3 \cdot 2^n)$ time, and that the time can be further reduced to $O(n^2 \cdot 2^n)$.

First, the permanent of a general matrix can be computed in $O(n \cdot 2^n)$ time. This is accomplished by eliminating the most computationally expensive step in direct implementations of Ryser's method — the $O(n^2)$ calculation of the row sums at each of the 2^n iterations. Specifically, each term of Ryser's expression of the permanent is just a product of the row sums with one of the 2^n subsets of columns removed. A direct implementation, therefore, requires the summation of the elements of each row that are not in one of the removed columns. Thus, $O(n)$ elements are summed for each of the n rows at a total cost of $O(n^2)$ arithmetic operations per term. In order to reduce the cost required to update the row sums, Nijenhuis and Wilf showed that the terms in Ryser's expression can be ordered so that only one column is changed from one term to the next.[15] At each step the algorithm updates the row sums by either adding or subtracting the column element corresponding to the change. Thus, the total update time is only $O(n)$. This change improves the computational complexity from $O(n^2 \cdot 2^n)$ to $O(n \cdot 2^n)$.

The above algorithm for evaluating the permanent of a matrix in $O(n \cdot 2^n)$ time can be adapted for the case in which a row and column of a matrix are assumed removed. This permits the evaluation of the permanents of the submatrices associated with each of the n^2 elements, as required by Equation 9.12, to calculate the joint assignment matrix in $O(n^3 \cdot 2^n)$ time. This is an improvement over the $O(n^4 \cdot 2^n)$ scaling obtained by the direct application of Ryser's formula (Equation 9.13) to calculate the permanent of each submatrix. The scaling can, however, be reduced even further. Specifically, note that the permanent of the submatrix associated with element a_{ij} is the sum of all $((n-1)2^{(n-1)})/(n-1)$ terms in Ryser's formula that do not involve row i or column j. In other words, one can eliminate a factor of $(n-1)$ by factoring the products of the $(n-1)$ row sums common to the submatrix permanents of each element in the same row. This factorization leads to an optimal JAM algorithm with complexity $O(n^2 \cdot 2^n)$.[16]

An optimized version of this algorithm can permit the evaluation of 12×12 joint assignment matrices in well under a second. Because the algorithm is highly parallelizable — the $O(2^n)$ iterative steps can be divided into k subproblems for simultaneous processing on k processors — solutions to problems of size n in the range 20–25 should be computable in real time with $O(n^2)$ processors. Although not practical, the quantities computed at each iteration could also be computed in parallel on $n \cdot 2^n$ processors to achieve an $O(n^2)$ sequential scaling. This might permit the real-time solution of somewhat larger problems, but at an exorbitant cost. Thus, the processing of $n \cdot n$ matrices, for $n > 25$, will require approximations.

Figure 9.3 provides actual empirical results (on a circa 1994 workstation) showing that the algorithm is suitable for real-time applications for $n < 10$, and it is practical for offline applications for n as large as 25. The overall scaling has terms of $n^2 2^n$ and 2^n, but the test results and the algorithm itself clearly demonstrate that the coefficient on the $n^2 2^n$ term is small relative to the 2^n term.

9.7   Crude Permanent Approximations

In the late 1980s, several researchers identified the importance of determining the *number* of feasible assignments in sparse association matrices arising in tracking applications. In this section, a recently developed approach for approximating the JAM via permanent inequalities is described which yields surprisingly good results within time roughly proportional to the number of feasible track/report pairs.

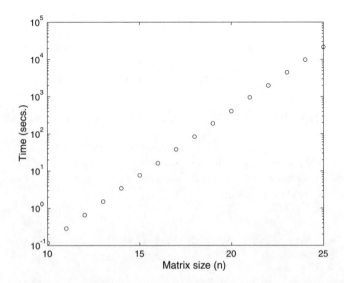

FIGURE 9.3   Tests of the new JAM algorithm reveal the expected exponentially scaling computation time.

Equation 9.12 shows that an approximation to the permanent would lead to a direct approximation of Equation 9.3. Unfortunately, research into approximating the permanent has emphasized the case in which the association matrix A has only 0–1 entries.[17] Moreover, the methods for approximating the permanent, even in this restricted case, still scale exponentially for reasonable estimates.[17-19] Even "unreasonable" estimates for the general permanent, however, may be sufficient to produce a reasonable estimate of a conditional probability matrix. This is possible because the solution matrix has additional structure that may permit the filtering of noise from poorly estimated permanents. The fact that the resulting matrix should be doubly stochastic, for example, suggests that the normalization of the rows and/or columns (i.e., dividing each row or column by the sum of its elements) should improve the estimate.

One of the most important properties of permanents relating to the doubly stochastic property of the joint assignment matrix is the following: *Multiplying a row or column by a scalar c has the effect of multiplying the permanent of the matrix by the same factor.*[7,12,13] This fact verifies that the multiplication of a row or column by some c also multiplies the permanent of any submatrix by c. This implies that the multiplication of any combination of rows and/or columns of a matrix by any values (other than zero) has no effect on the joint assignment matrix. This is because the factors applied to the various rows and columns cancel in the ratio of permanents in Equation 9.12. Therefore, the rows and columns of a matrix can be normalized in any manner before attempting to approximate the joint assignment matrix.

To see why a conditioning step could help, consider a 3×3 matrix with all elements equal to 1. If row 1 and column 1 are multiplied by 2, the following matrix is obtained:

$$\begin{vmatrix} 4 & 2 & 2 \\ 2 & 1 & 1 \\ 2 & 1 & 1 \end{vmatrix} \tag{9.14}$$

where the effect of the scaling of the first row and column has been undone.* This kind of preconditioning could be expected to improve the reliability of an estimator. For example, it seems to provide more reliable information for the greedy selection of hypotheses. This iterative process could also be useful for the post conditioning of an approximate joint assignment matrix (e.g., to ensure that the estimate is

---

*A physical interpretation of the original matrix in sensing applications is that each row and column corresponds to a set of measurements, which is scaled by some factor due to different sensor calibrations or models.

doubly stochastic). Remember that its use for preconditioning is permissible because it does nothing more than scale the rows and columns during each iteration, which does not affect the obtained joint assignment matrix. The process can be repeated for $O(n)$ iterations,* involves $O(n^2)$ arithmetic operations per iteration, and thus scales as $O(n^3)$. In absolute terms, the computations take about the same amount of compute time as required to perform an $n \times n$ matrix multiply.

9.8 Approximations Based on Permanent Inequalities

The possibility has been discussed of using crude estimators of the permanent, combined with knowledge about the structure of the joint assignment matrix, to obtain better approximations. Along these lines, four upper bound inequalities, $\varepsilon_1 - \varepsilon_4$, for the permanent are examined for use as crude estimators. The first inequality, ε_1, is a well-known result:

$$\text{per}(A) \le \prod_{i=1}^{n} r_i \tag{9.15}$$

where r_i is the sum of the elements in row i. This inequality holds for nonnegative matrices because it sums over all products that do not contain more than one element from the same row. In other words, it sums over a set larger than that of the actual permanent because it includes products with more than one element from the same column. For example, the above inequality applied to a 2×2 matrix would give $a_{11}a_{21} + a_{11}a_{22} + a_{12}a_{21} + a_{12}a_{22}$, rather than the sum over one-to-one matchings $a_{11}a_{22} + a_{12}a_{21}$. In fact, the product of the row sums is the first term in Ryser's equation. This suggests that the evaluation of the first k terms should yield a better approximation. Unfortunately, the computation required scales exponentially in the number of evaluated terms, thus making only the first two or three terms practical for approximation purposes. All of the inequalities in this section are based on the first term of Ryser's equation applied to a specially conditioned matrix. They can all, therefore, be improved by the use of additional terms, noting that an odd number of terms yields an upper bound, while an even number yields a lower bound.

This inequality also can be applied to the columns to achieve a potentially better upper bound. This would sum over products that do not contain more than one element from the same column. Thus, the following bound can be placed on the permanent:[7,12,13]

$$\text{per}(A) \le \min\left\{\prod_{i=1}^{n} r_i, \prod_{i=1}^{n} c_i\right\} \tag{9.16}$$

While taking the minimum of the product of the row sums and the product of the column sums tends to yield a better bound on the permanent, there is no indication that it is better than always computing the bound with respect to the rows. This is because the goal is to estimate the permanent of a submatrix for every element of the matrix, as required by Equation 9.12, to generate an estimate of the joint assignment matrix. If all of the estimates are of the same "quality" (i.e., are all too large by approximately the same factor), then some amount of post conditioning could yield good results. If the estimates vary considerably in their quality, however, post conditioning might not provide significant improvement.

---

*The process seems to converge rapidly if there are no nonzero elements in the matrix. Otherwise, a preprocessing step can be taken to promote more rapid convergence: When an entry must be 1 because it is the only nonzero value in its row (column), then all other values in its column (row) must become zero in the limit (assuming there exists at least one feasible assignment). Setting these values to zero and applying the same process repeatedly to the matrix seem to eliminate slower converging sequences. The author has not performed a complete theoretical analysis of the convergence behavior of the iterative renormalization, so comments on its behavior are based purely on empirical results.

The second inequality considered, ε_2, is the Jurkat-Ryser upper bound:[12,20] Given a nonnegative $n \times n$ matrix $A = [a_{ij}]$ with row sums $r_1, r_2, \ldots, r_n$ and column sums $c_1, c_2, \ldots, c_n$, where the row and column sums are indeed so that $r_k \leq r_{k+1}$ and $c_k \leq c_{k+1}$ for all k, then:

$$\text{per}(A) \leq \prod_{i=1}^{n} \min\{r_i, c_i\} \tag{9.17}$$

For cases in which there is at least one k, such that $r_k \neq c_k$, this upper bound is less than that obtained from the product of row or column sums. This has been the best of all known upper bounds since it was discovered in 1966. The next two inequalities, ε_3 and ε_4, are new results.[16]

ε_3 is defined as:

$$\text{per}(A) \leq \prod_{i=1}^{n} \sum_{j=1}^{n} \frac{a_{ij} c_i}{c_j} \tag{9.18}$$

This inequality is obtained by normalizing the columns, computing the product of the row sums of the resulting matrix, and then multiplying that product by the product of the original column sums. In other words,

$$\text{per}(A) \leq (c_1 c_2 \ldots c_n) \prod_{i=1}^{n} \sum_{j=1}^{n} a_{ij}/c_j \tag{9.19}$$

$$= \prod_{i=1}^{n} c_i \cdot \sum_{j=1}^{n} a_{ij}/c_j \tag{9.20}$$

$$= \prod_{i=1}^{n} \sum_{j=1}^{n} c_i \frac{a_{ij}}{c_j} \tag{9.21}$$

Note that the first summation is just the row sums after the columns have been normalized.

This new inequality is interesting because it seems to be the first general upper bound to be discovered that is, in some cases, superior to Jurkat-Ryser. For example, consider the following matrix:

$$\begin{vmatrix} 1 & 2 \\ 2 & 3 \end{vmatrix} \tag{9.22}$$

Jurkat-Ryser (ε_2) yields an upper bound of 15, while the estimator ε_3 yields a bound of 13.9. Although ε_3 is not generally superior, it is considered for the same reasons as the estimator ε_1.

The fourth inequality considered, ε_4, is obtained from ε_3 via ε_2:

$$\text{per}(A) \leq \prod_{i=1}^{n} \min\left\{ \sum_{j=1}^{n} \frac{a_{ij} c_i}{c_j}, \ c_i \right\} \tag{9.23}$$

The inequality is derived by first normalizing the columns. Then applying Jurkat-Ryser with all column sums equal to unity yields the following:

$$\text{per}(A) \le (c_1 c_2 \ldots c_n) \prod_{i=1}^{n} \min\left\{ \sum_{j=1}^{n} a_{ij}/c_j,\ 1 \right\} \tag{9.24}$$

$$= \prod_{i=1}^{n} c_i \cdot \min\left\{ \sum_{j=1}^{n} a_{ij}/c_j,\ 1 \right\} \tag{9.25}$$

$$= \prod_{i=1}^{n} \min\left\{ \sum_{j=1}^{n} \frac{a_{ij} c_i}{c_j},\ c_i \right\} \tag{9.26}$$

where the first summation simply represents the row sums after the columns have been normalized.

Similar to ε_1 and ε_3, this inequality can be applied with respect to the rows or columns, whichever yields the better bound. In the case of ε_4, this is critical, because one case usually provides a bound that is smaller than the other and is smaller than that obtained from Jurkat-Ryser.

In the example matrix (Equation 9.22), the ε_4 inequality yields an upper bound of 11 — an improvement over the other three estimates. Small-scale tests of the four inequalities on matrices of uniform deviates suggest that ε_3 almost always provides better bounds than ε_1; ε_2 almost always provides better bounds than ε_3; and ε_4 virtually always (more than 99% of the time) produces superior bounds to the other three inequalities. In addition to producing relatively tighter upper bound estimates in this restricted case, inequality ε_4 should be more versatile analytically than Jurkat-Ryser, because it does not involve a re-indexing of the rows and columns.

9.9   Comparisons of Different Approaches

Several of the JAM approximation methods described in this chapter have been compared on matrices containing

1. Uniformly and independently generated association probabilities
2. Independently generated binary (i.e., 0–1) indicators of feasible association
3. Probabilities of association between two three-dimensional (3D) sets of correlated objects.

The third group of matrices were generated from n tracks with uniformly distributed means and equal covariances by sampling the Gaussian defined by each track covariance to generate n reports. A probability of association was then calculated for each track/report pair.

The first two classes of matrices are examined to evaluate the generality of the various methods. These matrices have no special structure to be exploited. Matrices from the third class, however, contain structure typical of association matrices arising in tracking and correlation applications. Performance on these matrices should be indicative of performance in real-world data association problems, while performance in the first two classes should reveal the general robustness of the approximation schemes.

The approximation methods considered are

1. ε_1, the simplest of the general upper bound inequalities on the permanent. Two variants are considered:
   The upper bound taken with respect to the rows.
   The upper bound taken with respect to the rows or columns, whichever is less.
   Scaling: $O(n^3)$.
2. ε_2, the Jurkat-Ryser inequality.
   Scaling: $O(n^3 \log n)$.

3. ε_3 with two variants:

    The upper bound taken with respect to the rows.

    The upper bound taken with respect to the rows or columns, whichever is less.

    Scaling: $O(n^3)$ (see Appendix 9.A).

4. ε_4 with two variants:

    The upper bound taken with respect to the rows.

    The upper bound taken with respect to the rows or columns, whichever is less.

    Scaling: $O(n^3)$.

5. Iterative renormalization alone.

    Scaling: $O(n^3)$.

6. The standard greedy method[4] that assumes the prior association probabilities are accurate (i.e., performs no processing of the association matrix).

    Scaling: $O(n^2 \log n)$.

7. The one-sided normalization method that normalizes only the rows (or columns) of the association matrix.

    Scaling: $O(n^2 \log n)$.

The four ε estimators include the $O(n^3)$ cost of pre- and postprocessing via iterative renormalization.

The quality of hypotheses generated by the greedy method $O(n^2 \log n)$ is also compared to those generated optimally via the JAM. The extent to which the greedy method is improved by first normalizing the rows (or columns) of the association matrix is also examined. The latter method is relevant for real-time data association applications in which the set of reports (or tracks) cannot be processed in batch.

The following tables give the results of the various schemes when applied to different classes of $n \times n$ association matrices. The n best associations for each scheme are evaluated via the true JAM to determine the expected number of correct associations. The ratio of the expected number of correct associations for each approximation method and the optimal method yields the percentages in the table.* For example, an entry of 50% implies that the expected number of correct associations is half of what would be obtained from the JAM.

Table 9.1 provides a comparison of the schemes on matrices of size 20×20. Matrices of this size are near the limit of practical computability. For example, the JAM computations for a 35×35 matrix would demand more than a hundred years of nonstop computing on current high-speed workstations.

The most interesting information provided by Table 9.1 is that the inequality-based schemes are all more than 99% of optimal, with the approximation based on the Jurkat-Ryser inequality performing worst. The ε_3 and ε_4 methods performed best and always yielded identical results on the doubly stochastic matrices obtained by preprocessing. Preprocessing also improved the greedy scheme by 10 to 20%. Tables 9.2 and 9.3 show the effect of matrix size on each of the methods.

In almost all cases, the inequality-based approximations seemed to improve with matrix size. The obvious exception is the case of 5×5 0–1 matrices: The approximations are perfect because there tends to be only one feasible assignment for randomly generated 5×5 0–1 matrices. Surprisingly, the one-sided normalization approach is 80 to 90% optimal, yet scales as $O(p \log p)$, where p is the number of feasible pairings.

The one-sided approach is the only practical choice for large-scale applications that do not permit batch processing of sensor reports because the other approaches require the generation of the (preferably sparse) assignment matrix. The one-sided normalization approach requires only the set of tracks with which it gates (i.e., one row of the assignment matrix). Therefore, it permits the sequential processing of measurements. Of the methods compared here, it is the only one that satisfies online constraints in which each observation must be processed at the time it is received.

---

* The percentages are averages over enough trials to provide an accuracy of at least five decimal places in all cases except tests involving 5×5 0–1 matrices. The battery of tests for the 5×5 0–1 matrices produced some instances in which no assignments existed. The undefined results for these cases were not included in the averages, so the precision may be slightly less.

TABLE 9.1   Results of Tests of Several JAM Approximation Methods on 20×20 Association Matrices

| Method | Tests of JAM Approximations (% optimal for 20×20 Matrices) | | |
	Uniform Matrices	0-1 Matrices	3-D Spatial Matrices
ε_{1r}	99.9996186	99.9851376	99.9995588
ε_{1c}	99.9996660	99.9883342	100.0000000
Jurkat-Ryser (ε_2)	99.9232465	99.4156865	99.8275153
ε_{3r}	99.9996660	99.9992264	99.9999930
ε_{3c}	99.9997517	99.9992264	100.0000000
ε_{4r}	99.9996660	99.9992264	99.9999930
ε_{4c}	99.9997517	99.9992264	100.0000000
Iter. Normalized	99.9875369	99.9615623	99.9698123
Standard Greedy	84.7953351	55.9995049	86.3762418
One-Sided Norm.	93.6698728	82.5180206	98.0243181

Note: Each entry in the table represents the ratio of the expected number of correct associations made by the approximate method and the expected number for the optimal JAM. The subscripts r and c on the ε_{1-4} methods denote the application of the method to the rows and to the columns, respectively.

TABLE 9.2   Results of Tests of the JAM Approximation on Variously Sized Association Matrices Generated from Uniform Random Deviates

| Method | JAM Approximations on Varying Sized Uniform Matrices | | | |
	5×5	10×10	15×15	20×20
ε_{1r}	99.9465167	99.9883438	99.9996111	99.9996186
ε_{1c}	99.9465167	99.9920086	99.9996111	99.9996660
Jurkat-Ryser (ε_2)	99.8645867	99.7493972	99.8606475	99.9232465
ε_{3r}	99.9465167	99.9965856	99.9996111	99.9996660
ε_{3c}	99.9465167	99.9965856	99.9997695	99.9997517
ε_{4r}	99.9465167	99.9965856	99.9996111	99.9996660
ε_{4c}	99.9465167	99.9965856	99.9997695	99.9997517
Iter. Normalized	99.4492256	99.8650315	99.9646233	99.9875369
Standard Greedy	80.3063296	80.5927739	84.2186048	84.7953351
One-Sided Norm.	90.6688891	90.7567223	93.2058342	93.6698728

TABLE 9.3   Results of Tests of the JAM Approximation on Variously Sized Association Matrices Generated from Uniform 0–1 Random Deviates

| Method | JAM Approximations on Varying Sized 0–1 Matrices | | | |
	5×5	10×10	15×15	20×20
ε_{1r}	100.0000000	99.9063047	99.9606670	99.9851376
ε_{1c}	100.0000000	99.9137304	99.9754337	99.9883342
Jurkat-Ryser (ε_2)	100.0000000	99.7915349	99.6542955	99.4156865
ε_{3r}	100.0000000	99.9471028	99.9947939	99.9992264
ε_{3c}	100.0000000	99.9503096	99.9949549	99.9992264
ε_{4r}	100.0000000	99.9471028	99.9947939	99.9992264
ε_{4c}	100.0000000	99.9503096	99.9949549	99.9992264
Iter. Normalized	100.0000000	99.7709328	99.9256354	99.9615623
Standard Greedy	56.1658957	55.7201435	53.8121279	55.9995049
One-Sided Norm.	72.6976451	77.6314664	83.6890193	82.5180206

To summarize, the JAM approximation schemes based on the new permanent inequalities appear to yield near-optimal results. An examination of the matrices produced by these methods reveals a standard deviation of less than 3×10^{-5} from the optimal JAM computed via permanents. The comparison of expected numbers of correct assignments given in the tables, however, is the most revealing in terms of applications to multiple-target tracking. Specifically, the recursive formulation of the tracking process leads to highly nonlinear dependencies on the quality of the hypothesis generation scheme. In a dense tracking environment, a deviation of less than 1% in the expected number of correct associations can make the difference between convergence and divergence of the overall process. The next section considers applications to large-scale problems.

9.10   Large-Scale Data Association

This section examines the performance of the one-sided normalization approach. The evidence provided in the previous section indicates that the estimator ε_3 yields probabilities of association conditioned on the assignment constraint that are very near optimal. Therefore, ε_3 can be used as a baseline of comparison for the one-sided estimator for problems that are too large to apply the optimal approach. The results in the previous section demonstrates that the one-sided approach yields relatively poor estimates when compared to the optimal and near-optimal methods. However, because the latter approaches cannot be applied online to process each observation as it arrives, the one-sided approach is the only feasible alternative. The goal, therefore, is to demonstrate only that its estimates do not diminish in quality as the size of the problem increases. This is necessary to ensure that a system that is tuned and tested on problems of a given size will behave predictably when applied to larger problems.

In the best-case limit, as the amount of ambiguity goes to zero, any reasonable approach to data association should perform acceptably. In the worst-case limit, as all probabilities converge to the same value, no approach can perform any better than a simple random selection of hypotheses. The worst-case situation in which information can be exploited is when the probabilities of association appear to be uncorrelated random deviates. In such a case, only higher-order estimation of the joint probabilities of association can provide useful discriminating information. The following is an example of an association matrix that was generated from a uniform random number generator:

0.266	0.057	0.052	0.136	0.227	0.020	0.059
0.051	0.023	0.208	0.134	0.199	0.135	0.058
0.031	0.267	0.215	0.191	0.117	0.227	0.002
0.071	0.057	0.243	0.029	0.230	0.281	0.046
0.020	0.249	0.166	0.148	0.095	0.178	0.121
0.208	0.215	0.064	0.268	0.067	0.180	0.039
0.018	0.073	0.126	0.062	0.125	0.141	0.188

This example matrix was chosen from among ten that were generated because it produced the most illustrative JAM. The uniform deviate entries have been divided by n so that they are of comparable magnitude to actual association probabilities.

Applying the optimal JAM algorithm yields the following true probabilities of association conditioned on the assignment constraint:

0.502	0.049	0.037	0.130	0.191	0.014	0.077
0.075	0.026	0.244	0.167	0.243	0.136	0.109
0.034	0.310	0.182	0.193	0.093	0.186	0.003
0.088	0.055	0.244	0.027	0.237	0.278	0.071
0.022	0.285	0.139	0.144	0.076	0.144	0.191
0.259	0.200	0.043	0.279	0.048	0.124	0.048
0.021	0.074	0.111	0.060	0.113	0.119	0.501

A cursory examination of the differences between corresponding entries in the two matrices demonstrates that a significant amount of information has been extracted by considering higher order correlations. For example, the last entries in the first two rows of the association matrix are 0.059 and 0.058 — differing by less than 2% — yet their respective JAM estimates are 0.077 and 0.109, a difference of almost 30%.

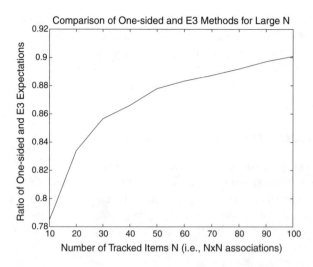

FIGURE 9.4   The performance of the one-sided approach relative to ε_3 improves with increasing N. This is somewhat misleading, however, because the expected number of correct assignments goes to zero in the limit $N \rightarrow \infty$ for both approaches.

Remarkably, despite the fact that the entries in the first matrix were generated from uniform deviates, the first entry in the first row and the last entry in the last row of the resulting JAM represents hypotheses that each have a better than 50% chance of being correct.

To determine whether the performance of the one-sided hypothesis selection approach suffers as the problem size is increased, its hypotheses were compared with those of estimator ε_3 on $n \times n$, for n in the range 10 to 100, association matrices generated from uniform random deviates. Figure 9.4 shows that the number of correct associations for the one-sided approach seems to approach the same number as ε_3 as n increases. This may be somewhat misleading, however, because the expected number of correct associations out of n hypotheses selected from $n \times n$ possible candidates will tend to decrease as n increases. More specifically, the ratio of correct associations to number of hypotheses (proportional to n) will tend to zero for all methods if the prior probabilities of association are generated at random.

A better measure of performance is how many hypotheses are necessary for a given method to ensure that a fixed number are expected to be correct. This can be determined from the JAM by summing its entries corresponding to a set of hypotheses. Because each entry contains the expectation that a particular track/report pair corresponds to the same object, the sum of the entries gives the expected number of correct assignments. To apply this measure, it is necessary to fix a number of hypotheses that must be correct, independent of n, and determine the ratio of the number of hypotheses required for the one-sided approach to that required by ε_3. Figure 9.5 also demonstrates that the one-sided method seems to approach the same performance as ε_3 as n increases in a highly ambiguous environment. In conclusion, the performance of the one-sided approach scales robustly even in highly ambiguous environments.

The good performance of the one-sided approach — which superficially appears to be little more than a crude heuristic — is rather surprising given the amount of information it fails to exploit from the association matrix. In particular, it uses information only from the rows (or columns) taken independently, thus making no use of information provided by the columns (or rows). Because the tests described above have all rows and columns of the association matrix scaled randomly, but uniformly, the worst-case performance of the one-sided approach may not have been seen. In tests in which the columns have been independently scaled by vastly different values, results generated from the one-sided approach show little improvement over those of the greedy method. (The performances of the optimal and near optimal methods, of course, are not affected.) In practice, a system in which all probabilities are scaled by the same value (e.g., as a result of using a single sensor) should not be affected by this limitation of the one-sided approach. In multisensor applications, however, association probabilities must be generated

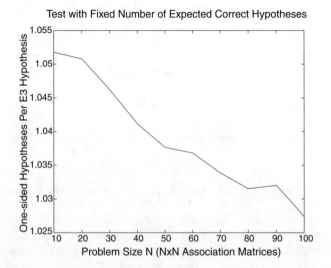

FIGURE 9.5   This test plots the number of hypotheses required by the one-sided approach to achieve some fixed expected number of correct assignments. Again, it appears that the one-sided approach performs comparably to ε_3 as N increases.

consistently. If a particular sensor is not modeled properly, and its observations produce track/report pairs with consistently low or high probabilities of association, then the hypotheses generated from these pairs by the one-sided approach will be ranked consistently low or high.

9.11   Generalizations

The combinatorial analysis of the assignment problem in previous sections has considered only the case of a single "snapshot" of sensor observations. In actual tracking applications, however, the goal is to establish tracks from a sequence of snapshots. Mathematically, the definition of a permanent can be easily generalized to apply not only to assignment matrices but also to tensor extensions. Specifically, the generalized permanent sums over all assignments of observations at timestep k, for each possible assignment of observations at timestep $k - 1$, continuing recursively down to the base case of all possible assignments of the first batch of observations to the initial set of tracks. (This multidimensional assignment problem is described more fully in Chapter 11.) Although generalizing the crude permanent approximations for application to the multidimensional assignment problem is straightforward, the computation time scales geometrically with exponent k. This is vastly better than the super-exponential scaling required for the optimal approach, but it is not practical for large values of n and k unless the gating process yields a very sparse association matrix (see Chapter 3).

9.12   Conclusions

This chapter has discussed some of the combinatorial issues arising in the batch data association problem. It described the optimal solution for a large class of data association problems involving the calculation of permanents of submatrices of the original association matrix. This procedure yields the JAM, which can be used to optimally rank associations for hypothesis selection. Because the computational cost of the permanent scales exponentially in the size of the matrix, improved algorithms have been developed both for calculating the exact JAM and for generating approximations to it. Empirical results suggest that the approximations are suitable for hypothesis generation in large-scale tracking and correlation applications. New theoretical results include an improved upper bound on the calculation of the JAM and new upper bound inequalities, ε_3 and ε_4, for the permanent of general nonnegative matrices.

The principal conclusion that can be drawn from this chapter is that the ambiguities introduced by a dense environment are extremely difficult and computationally expensive to resolve. Although this chapter examined the most general possible case in which tracking has to be performed in an environment dense with indistinct (other than position) targets, there is little doubt that the constraints imposed by most real-world applications would necessitate some sacrifice of this generality.

Acknowledgments

The author gratefully acknowledges support from the University of Oxford, U.K., and the Naval Research Laboratory, Washington, D.C.

Appendix 9.A   Algorithm for Data Association Experiment

The following algorithm demonstrates how $O(n^3)$ scaling can be obtained for approximating the JAM of an $n \times n$ association matrix using inequality ε_3. A straightforward implementation that scales as $O(n^4)$ can be obtained easily; the key to removing a factor of n comes from the fact that from a precomputed product of row sums, *rprod*, the product of all row sums excluding row i is just *rprod* divided by row sum i. In other words, performing an $O(n)$ step of explicitly computing the product of all row sums except row i is not necessary. Some care must be taken to accommodate row sums that are zero, but the following pseudocode shows that there is little extra overhead incurred by the more efficient implementation. Similar techniques lead to the advertised scaling for the other inequality approaches. (Note, however, that sums of logarithms should be used in place of explicit products to ensure numerical stability in actual implementations.)

E3r(M, P, n)
> r and c are vectors of length n corresponding to the row and column sums, respectively. *rn* is a vector of length n of normalized row sums.
> for $i = 1$ to $n : r_i \leftarrow c_i \leftarrow 0.0$
> Apply iterative renormalization to M
> for $i = 1$ to n:
>> for $j = 1$ to n:
>>> $r_i \leftarrow r_i + M_{ij}$
>>> $c_i \leftarrow c_i + M_{ij}$
>> end
> end
> for $i = 1$ to n:
>> for $l = 1$ to n:
>>> $rn_l \leftarrow 0.0$
>>> for $k = 1$ to n, if $(c_k - M_{ik}) > 0.0$
>>>> then $rn_l \leftarrow rn_l + M_{ik}/(c_k - M_{ik})$
>> end
>> for $j = 1$ to n:
>>> $nprod = 1.0$
>>> for $k = 1$ to n, if $k \neq j$
>>>> then $nprod \leftarrow nprod * (c_k - M_{ik})$
>>> $rprod \leftarrow cprod \leftarrow 1.0$
>>> for $k = 1$ to n, if $k \neq 1$
>>>> if $(c_j - M_{ij}) > 0.0$
>>>>> then $rprod \leftarrow rprod * (rn_k - M_{kj})/(c_j - M_{ij})$
>>>> else $rprod \leftarrow rprod * rn_k$

> end
> $rprod \leftarrow rprod * nprod$
> $P_{ij} \leftarrow M_{ij} * rprod$
>   end
> end
> Apply iterative renormalization to P
> **end.**

References

1. Reid, D.B., An algorithm for tracking multiple targets, *IEEE Trans. Automatic Control*, AC-24(6), 1979.
2. Cox, I.J. and Leonard, J.J., Modeling a dynamic environment using a Bayesian multiple hypothesis approach, *Int'l. J. AI*, 1994.
3. Bar-Shalom, Y. and Li, X.R., *Multitarget-Multisensor Tracking: Principles and Techniques*, YBS Press, Storrs, CT, 1995.
4. Blackman, S., *Multiple-Target Tracing with Radar Applications*, Artech House, Dedham, MA, 1986.
5. Bar-Shalom, Y. and Fortmann, T.E., *Tracking and Data Association*, Academic Press, New York, 1988.
6. Collins, J.B. and Uhlmann, J.K., Efficient gating in data association for multivariate Gaussian distributions, *IEEE Trans. Aerospace and Electronic Systems*, 28, 1990.
7. Ryser, H.J., *Combinatorial Mathematics*, 14, Carus Mathematical Monograph Series, Mathematical Association of America, 1963.
8. Uhlmann, J.K., Algorithms for multiple-target tracking, *American Scientist*, 80(2), 1992.
9. Papadimitrious, C.H. and Steiglitz, K., *Combinatorial Optimization Algorithms and Complexity*, Prentice Hall, Englewood Cliffs, NJ, 1982.
10. Cox, I.J. and Miller, M.L., On finding ranked assignments with application to multitarget tracking and motion correspondence, Submitted to *IEEE Trams. Aerospace and Electronic Systems*, 1993.
11. Nagarajan, V., Chidambara, M.R., and Sharma, R.M., New approach to improved detection and tracking in track-while-scan radars, Part 2: detection, track initiation, and association, *IEEE Proc.*, 134(1), 1987.
12. Brualdi, R.A. and Ryser, H.J., *Combinatorial Matrix Theory*, Cambridge University Press, Cambridge, U.K., 1992.
13. Minc, H., *Permanents, Encyclopedia of Mathematics and its Applications*, Addison-Wesley, 6(4), Pt. F, 1978.
14. Valiant, L.G., The complexity of computing the permanent, *Theoretical Computer Science*, 8, 1979.
15. Nijenhuis, A. and Wilf, H.S., *Combinatorial Algorithms*, Academic Press, New York, 2nd edition, 1978.
16. Uhlmann, J.K., *Dynamic Localization and Map Building: New Theoretical Foundations*, Ph.D. thesis, University of Oxford, 1995.
17. Karmarkar, N. et al., A Monte Carlo algorithm for estimating the permanent, *SIAM J. Comput.*, 22(2), 1993.
18. Jerrum, M. and Vazirani, U., A mildly exponential approximation algorithm for the permanent, Technical Report, Department of Computer Science, University of Edinburgh, Scotland, 1991. ECS-LFCS-91–179.
19. Jerrum, M. and Sinclair, A., Approximating the permanent, *SIAM J. Comput.*, 18(6):1149–1178, 1989.
20. Jurkat, W.B. and Ryser, H.J., Matrix factorizations of determinants and permanents, *J. Algebra*, 3, 1966.

10

A Bayesian Approach to Multiple-Target Tracking*

10.1 Introduction **10-1**
Definition of Bayesian Approach • Relationship to Kalman Filtering

10.2 Bayesian Formulation of the Single-Target Tracking Problem **10-3**
Bayesian Filtering • Problem Definition • Computing the Posterior • Likelihood Functions

10.3 Multiple-Target Tracking without Contacts or Association (Unified Tracking) **10-8**
Multiple-Target Motion Model • Multiple-Target Likelihood Functions • Posterior Distribution • Unified Tracking Recursion

10.4 Multiple-Hypothesis Tracking (MHT) **10-12**
Contacts, Scans, and Association Hypotheses • Scan and Data Association Likelihood Functions • General Multiple-Hypothesis Tracking • Independent Multiple-Hypothesis Tracking

10.5 Relationship of Unified Tracking to MHT and Other Tracking Approaches **10-22**
General MHT Is a Special Case of Unified Tracking • Relationship of Unified Tracking to Other Multiple-Target Tracking Algorithms • Critique of Unified Tracking

10.6 Likelihood Ratio Detection and Tracking **10-23**
Basic Definitions and Relations • Likelihood Ratio Recursion • Log-Likelihood Ratios • Declaring a Target Present • Track-Before-Detect

References **10-30**

Lawrence D. Stone
Metron Inc.

10.1 Introduction

This chapter views the multiple-target tracking problem as a Bayesian inference problem and highlights the benefits this approach. The goal of this chapter is to provide the reader with some insights and perhaps a new view of multiple-target tracking. It is not designed to provide the reader with a set of algorithms for multiple-target tracking.

---

*This chapter is based on *Bayesian Multiple Target Tracking*, by Stone, L. D., Barlow, C. A., and Corwin, T. L., 1999. Artech House, Inc., Norwood, MA. www.artechhouse.com.

The chapter begins with a Bayesian formulation of the single-target tracking problem and then extends this formulation to multiple targets. It then discusses some of the interesting consequences of this formulation, including:

- A mathematical formulation of the multiple-target tracking problem with a minimum of complications and formalisms
- The emergence of likelihood functions as a generalization of the notion of contact and as the basic currency for valuing and combining information from disparate sensors
- A general Bayesian formula for calculating association probabilities
- A method, called unified tracking, for performing multiple-target tracking when the notions of contact and association are not meaningful
- A delineation of the relationship between multiple-hypothesis tracking (MHT) and unified tracking
- A Bayesian track-before-detect methodology called likelihood ratio detection and tracking.

10.1.1   Definition of Bayesian Approach

To appreciate the discussion in this chapter, the reader must first understand the concept of Bayesian tracking. For a tracking system to be considered Bayesian, it must have the following characteristics:

- *Prior Distribution* — There must be a prior distribution on the state of the targets. If the targets are moving, the prior distribution must include a probabilistic description of the motion characteristics of the targets. Usually the prior is given in terms of a stochastic process for the motion of the targets.
- *Likelihood Functions* — The information in sensor measurements, observations, or contacts must be characterized by likelihood functions.
- *Posterior Distribution* — The basic output of a Bayesian tracker is a posterior probability distribution on the (joint) state of the target(s). The posterior at time t is computed by combining the motion updated prior at time t with the likelihood function for the observation(s) received at time t.

These are the basics: prior, likelihood functions, posterior. If these are not present, the tracker is not Bayesian. The recursions given in this chapter for performing Bayesian tracking are all "recipes" for calculating priors, likelihood functions, and posteriors.

10.1.2   Relationship to Kalman Filtering

Kalman filtering resulted from viewing tracking as a least squares problem and finding a recursive method of solving that problem. One can think of many standard tracking solutions as methods for minimizing mean squared errors. Chapters 1 to 3 of Blackman and Popoli[1] give an excellent discussion of tracking from this point of view. One can also view Kalman filtering as Bayesian tracking. To do this, one starts with a prior that is Gaussian in the appropriate state space with a "very large" covariance matrix. Contacts are measurements that are linear functions of the target state with Gaussian measurement errors. These are interpreted as Gaussian likelihood functions and combined with motion updated priors to produce posterior distributions on target state. Because the priors are Gaussian and the likelihood functions are Gaussian, the posteriors are also Gaussian. When doing the algebra, one finds that the mean and covariance of the posterior Gaussian are identical to the mean and covariance of the least squares solution produced by the Kalman filter. The difference is that from the Bayesian point of view, the mean and covariance matrices represent posterior Gaussian distributions on target state. Plots of the mean and characteristic ellipses are simply shorthand representations of these distributions.

Bayesian tracking is not simply an alternate way of viewing Kalman filtering. Its real value is demonstrated when some of the assumptions required for Kalman filtering are not satisfied. Suppose the prior distribution on target motion is not Gaussian, or the measurements are not linear functions of the target state, or the measurement error is not Gaussian. Suppose that multiple sensors are involved and are quite different. Perhaps they produce measurements that are not even in the target state space. This can happen if, for example, one of the measurements is the observed signal-to-noise-ratio at a sensor. Suppose that one has to deal with measurements that are not even contacts (e.g., measurements that are so weak that they fall below the threshold at which one would call a contact). Tracking problems involving these situations do not fit well into the mean squared error paradigm or the Kalman filter assumptions. One can often stretch the limits of Kalman filtering by using linear approximations to nonlinear measurement relations or by other nonlinear extensions. Often these extensions work very well. However, there does come a point where these extensions fail. That is where Bayesian filtering can be used to tackle these more difficult problems. With the advent of high-powered and inexpensive computers, the numerical hurdles to implementing Bayesian approaches are often easily surmounted. At the very least, knowing how to formulate the solution from the Bayesian point of view will allow one to understand and choose wisely the approximations needed to put the problem into a more tractable form.

10.2 Bayesian Formulation of the Single-Target Tracking Problem

This section presents a Bayesian formulation of single-target tracking and a basic recursion for performing single-target tracking.

10.2.1 Bayesian Filtering

Bayesian filtering is based on the mathematical theory of probabilistic filtering described by Jazwinski.[2] Bayesian filtering is the application of Bayesian inference to the problem of tracking a single target. This section considers the situation where the target motion is modeled in continuous time, but the observations are received at discrete, possibly random, times. This is called continuous-discrete filtering by Jazwinski.

10.2.2 Problem Definition

The single-target tracking problem assumes that there is one target present in the state space; as a result, the problem becomes one of estimating the state of that target.

10.2.2.1 Target State Space

Let S be the state space of the target. Typically, the target state will be a vector of components. Usually some of these components are kinematic and include position, velocity, and possibly acceleration. Note that there may be constraints on the components, such as a maximum speed for the velocity component. There can be additional components that may be related to the identity or other features of the target. For example, if one of the components specifies target type, then that may also specify information such as radiated noise levels at various frequencies and motion characteristics (e.g., maximum speeds). In order to use the recursion presented in this section, there are additional requirements on the target state space. The state space must be rich enough that (1) the target's motion is Markovian in the chosen state space and (2) the sensor likelihood functions depend only on the state of the target at the time of the observation. The sensor likelihood functions depend on the characteristics of the sensor, such as its position and measurement error distribution which are assumed to be known. If they are not known, they need to be determined by experimental or theoretical means.

10.2.2.2 Prior Information

Let $X(t)$ be the (unknown) target state at time t. We start the problem at time 0 and are interested in estimating $X(t)$ for $t \geq 0$. The prior information about the target is represented by a stochastic process $\{X(t); t \geq 0\}$. Sample paths of this process correspond to possible target paths through the state space, S. The state space S has a measure associated with it. If S is discrete, this measure is a discrete measure. If S is continuous (e.g., if S is equal to the plane), this measure is represented by a density. The measure on S can be a mixture or product of discrete and continuous measures. Integration with respect to this measure will be indicated by ds. If the measure is discrete, then integration becomes summation.

10.2.2.3 Sensors

There is a set of sensors that report observations at an ordered, discrete sequence of (possibly random) times. These sensors may be of different types and report different information. The set can include radar, sonar, infrared, visual, and other types of sensors. The sensors may report only when they have a contact or on a regular basis. Observations from sensor j take values in the measurement space H_j. Each sensor may have a different measurement space. The probability distribution of each sensor's response conditioned on the value of the target state s is assumed to be known. This relationship is captured in the likelihood function for that sensor. The relationship between the sensor response and the target state s may be linear or nonlinear, and the probability distribution representing measurement error may be Gaussian or non-Gaussian.

10.2.2.4 Likelihood Functions

Suppose that by time t observations have been obtained at the set of times $0 \le t_1 \le \ldots \le t_K \le t$. To allow for the possibility that more than one sensor observation may be received at a given time, let Y_k be the set of sensor observations received at time t_k. Let y_k denote a value of the random variable Y_k. Assume that the likelihood function can be computed as

$$L_k\left(y_k \middle| s\right) = \mathbf{Pr}\left\{Y_k = y_k \middle| X\left(t_k\right) = s\right\} \text{ for } s \in S \qquad (10.1)$$

The computation in Equation 10.1 can account for correlation among sensor responses. If the distribution of the set of sensor observations at time t_k is independent given target state, then $L_k\,(y_k|s)$ is computed by taking the product of the probability (density) functions for each observation. If they are correlated, then one must use the joint density function for the observations conditioned on target state to compute $L_k\,(y_k|s)$.

Let $\mathbf{Y}(t) = (Y_1, Y_2, \ldots, Y_K)$ and $\mathbf{y} = (y_1, \ldots, y_K)$. Define $L(\mathbf{y}|s_1, \ldots, s_K) = \mathbf{Pr}\,\{\mathbf{Y}(t) = \mathbf{y}|X(t_1) = s_1, \ldots, X(t_K) = s_K\}$. Assume

$$\mathbf{Pr}\left\{\mathbf{Y}\left(t\right) = \mathbf{y} \middle| X\left(u\right) = s\left(u\right),\ 0 \le u \le t\right\} = L\left(\mathbf{y} \middle| s\left(t_1\right), \ldots, s\left(t_K\right)\right) \qquad (10.2)$$

Equation 10.2 means that the likelihood of the data $\mathbf{Y}(t)$ received through time t depends only on the target states at the times $\{t_1, \ldots, t_K\}$ and not on the whole target path.

10.2.2.5 Posterior

Define $q(s_1, \ldots, s_K) = \mathbf{Pr}\{X(t_1) = s_1, \ldots, X(t_K) = s_K\}$ to be the prior probability (density) that the process $\{X(t);\ t \ge 0\}$ passes through the states $s_1, \ldots, s_K$ at times $t_1, \ldots, t_K$. Let $p(t_K, s_K) = \mathbf{Pr}\{X(t_K) = s_K|\mathbf{Y}(t_K) = \mathbf{y}\}$. Note that the dependence of p on $\mathbf{y}$ has been suppressed. The function $p(t_K, \cdot)$ is the posterior distribution on $X(t_K)$ given $\mathbf{Y}(t_K) = \mathbf{y}$. In mathematical terms, the problem is to compute this posterior distribution. Recall that from the point of view of Bayesian inference, the posterior distribution on target state represents our knowledge of the target state. All estimates of target state derive from this posterior.

10.2.3 Computing the Posterior

Compute the posterior by the use of Bayes' theorem as follows:

$$p\left(t_k, s_K\right) = \frac{\mathbf{Pr}\left\{\mathbf{Y}\left(t_K\right) = \mathbf{y} \text{ and } X\left(t_K\right) = s_K\right\}}{\mathbf{Pr}\left\{\mathbf{Y}\left(t_K\right) = \mathbf{y}\right\}}$$

$$= \frac{\int L\left(\mathbf{y} \middle| s_1, \ldots, s_K\right) q\left(s_1, s_2, \ldots, s_K\right) ds_1 ds_2 \cdots ds_{K-1}}{\int L\left(\mathbf{y} \middle| s_1, \ldots, s_K\right) q\left(s_1, s_2, \ldots, s_K\right) ds_1 ds_2 \cdots ds_K} \qquad (10.3)$$

Computing $p(t_K, s_K)$ can be quite difficult. The method of computation depends upon the functional forms of q and L. The two most common ways are batch computation and a recursive method.

10.2.3.1 Recursive Method

Two additional assumptions about q and L permit recursive computation of $p(t_K, s_K)$. First, the stochastic process $\{X(t; t \geq 0\}$ must be Markovian on the state space S. Second, for $i \neq j$, the distribution of $Y(t_i)$ must be independent of $Y(t_j)$ given $(X(t_1) = s_1, \ldots, X(t_K) = s_K)$ so that

$$L\left(\mathbf{y}\middle|s_1,\ldots,s_K\right)=\prod_{k=1}^{K}L_k\left(y_k\middle|s_k\right) \tag{10.4}$$

The assumption in Equation 10.4 means that the sensor responses (or observations) at time t_k depend only on the target state at the time t_k. This is not automatically true. For example, if the target state space is position only and the observation is a velocity measurement, this observation will depend on the target state over some time interval near t_k. The remedy in this case is to add velocity to the target state space. There are other observations, such as failure of a sonar sensor to detect an underwater target over a period of time, for which the remedy is not so easy or obvious. This observation may depend on the whole past history of target positions and, perhaps, velocities.

Define the transition function $q_k\left(s_k\middle|s_{k-1}\right) = \mathbf{Pr}\{X(t_k) = s_k | X(t_{k-1}) = s_{k-1}\}$ for $k \geq 1$, and let q_0 be the probability (density) function for $X(0)$. By the Markov assumption

$$q\left(s_1,\ldots,s_K\right)=\int_S\prod_{k=1}^{K}q_k\left(s_k\middle|s_{k-1}\right)q_0\left(s_0\right)ds_0 \tag{10.5}$$

10.2.3.2 Single-Target Recursion

Applying Equations 10.4 and 10.5 to 10.3 results in the basic recursion for single-target tracking given below.

Basic Recursion for Single-Target Tracking

Initialize Distribution: $\qquad p\left(t_0, s_0\right) = q_0\left(s_0\right)$ for $s_0 \in S$ $\qquad\qquad$ (10.6)

For $k \geq 1$ and $s_k \in S$,

Perform Motion Update: $\qquad p^-\left(t_k, s_k\right) = \int q_k\left(s_k\middle|s_{k-1}\right)p\left(t_{k-1}, s_{k-1}\right)ds_{k-1}$ $\qquad$ (10.7)

Compute Likelihood Function L_k from the observation $Y_k = y_k$

Perform Information Update: $\quad p\left(t_k, s_k\right) = \dfrac{1}{C}L_k\left(y_k\middle|s_k\right)p^-\left(t_k, s_k\right)$ $\qquad\qquad$ (10.8)

The motion update in Equation 10.7 accounts for the transition of the target state from time t_{k-1} to t_k. Transitions can represent not only the physical motion of the target, but also changes in other state variables. The information update in Equation 10.8 is accomplished by point-wise multiplication of $p^-(t_k, s_k)$ by the likelihood function $L_k(y_k|s_k)$. Likelihood functions replace and generalize the notion of contacts in this view of tracking as a Bayesian inference process. Likelihood functions can represent sensor information such as detections, no detections, Gaussian contacts, bearing observations, measured signal-to-noise ratios, and observed frequencies of a signal. Likelihood functions can represent and incorporate information in situations where the notion of a contact is not meaningful. Subjective information also can be incorporated by using likelihood functions. Examples of likelihood functions are provided in Section 10.2.4. If there has been no observation at time t_k, then there is no information update, only a motion update.

The above recursion does not require the observations to be linear functions of the target state. It does not require the measurement errors or the probability distributions on target state to be Gaussian. Except in special circumstances, this recursion must be computed numerically. Today's high-powered scientific workstations can compute and display tracking solutions for complex nonlinear trackers. To do this, discretize the state space and use a Markov chain model for target motion so that Equation 10.7 is computed through the use of discrete transition probabilities. The likelihood functions are also computed on the discrete state space. A numerical implementation of a discrete Bayesian tracker is described in Section 3.3 of Stone et al.[3]

10.2.4   Likelihood Functions

The use of likelihood functions to represent information is at the heart of Bayesian tracking. In the classical view of tracking, contacts are obtained from sensors that provide estimates of (some components of) the target state at a given time with a specified measurement error. In the classic Kalman filter formulation, a measurement (contact) Y_k at time t_k satisfies the measurement equation

$$Y_k = \mathbf{M}_k X(t_k) + \varepsilon_k \tag{10.9}$$

where

> Y_k is an r-dimensional real column vector
> $X(t_k)$ is an l-dimensional real column vector
> $\mathbf{M}_k$ is an $r \times l$ matrix
> $\varepsilon_k \sim N(0, \Sigma_k)$

Note that $\sim N(\mu, \Sigma)$ means "has a Normal (Gaussian) distribution with mean μ and covariance Σ." In this case, the measurement is a linear function of the target state and the measurement error is Gaussian. This can be expressed in terms of a likelihood function as follows. Let $L_G(y|x) = \Pr\{Y_k = y | X(t_k) = x\}$. Then

$$L_G\left(y|x\right) = \left(2\pi\right)^{-r/2} \left|\det \Sigma_k\right|^{-1/2} \exp\left(-\frac{1}{2}\left(y - \mathbf{M}_k x\right)^T \Sigma^{-1}\left(y - \mathbf{M}_k x\right)\right) \tag{10.10}$$

Note that the measurement y is data that is known and fixed. The target state x is unknown and varies, so that the likelihood function is a function of the target state variable x. Equation 10.10 looks the same as a standard elliptical contact, or estimate of target state, expressed in the form of multivariate normal distribution, commonly used in Kalman filters. There is a difference, but it is obscured by the symmetrical positions of y and $\mathbf{M}_k x$ in the Gaussian density in Equation 10.10. A likelihood function does not represent an estimate of the target state. It looks at the situation in reverse. For each value of target state x, it calculates the probability (density) of obtaining the measurement y given that the target is in state x. In most cases, likelihood functions are not probability (density) functions on the target state space. They need not integrate to one over the target state space. In fact, the likelihood function in Equation 10.10 is a probability density on the target state space only when Y_k is l-dimensional and $\mathbf{M}_k$ is an $l \times l$ matrix.

Suppose one wants to incorporate into a Kalman filter information such as a bearing measurement, speed measurement, range estimate, or the fact that a sensor did or did not detect the target. Each of these is a nonlinear function of the normal Cartesian target state. Separately, a bearing measurement, speed measurement, and range estimate can be handled by forming linear approximations and assuming Gaussian measurement errors or by switching to special non-Cartesian coordinate systems in which the

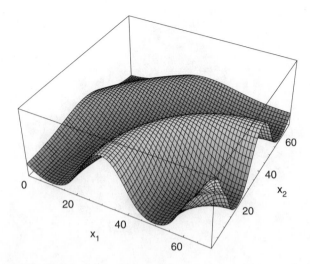

FIGURE 10.1 Detection likelihood function for a sensor at (70,0).

measurements are linear and hopefully the measurement errors are Gaussian. In combining all this information into one tracker, the approximations and the use of disparate coordinate systems become more problematic and dubious. In contrast, the use of likelihood functions to incorporate all this information (and any other information that can be put into the form of a likelihood function) is quite straightforward, no matter how disparate the sensors or their measurement spaces. Section 10.2.4.1 provides a simple example of this process involving a line of bearing measurement and a detection.

10.2.4.1 Line of Bearing Plus Detection Likelihood Functions

Suppose that there is a sensor located in the plane at (70,0) and that it has produced a detection. For this sensor the probability of detection is a function, $P_d(r)$, of the range r from the sensor. Take the case of an underwater sensor such as an array of acoustic hydrophones and a situation where the propagation conditions produce convergence zones of high detection performance that alternate with ranges of poor detection performance. The observation (measurement) in this case is $Y = 1$ for detection and 0 for no detection. The likelihood function for detection is $L_d(1|x) = P_d(r(x))$, where $r(x)$ is the range from the state x to the sensor. Figure 10.1 shows the likelihood function for this observation.

Suppose that, in addition to the detection, there is a bearing measurement of 135 degrees (measured counter-clockwise from the x_1 axis) with a Gaussian measurement error having mean 0 and standard deviation 15 degrees. Figure 10.2 shows the likelihood function for this observation. Notice that, although the measurement error is Gaussian in bearing, it does not produce a Gaussian likelihood function on the target state space. Furthermore, this likelihood function would integrate to infinity over the whole state space. The information from these two likelihood functions is combined by point-wise multiplication. Figure 10.3 shows the likelihood function that results from this combination.

10.2.4.2 Combining Information Using Likelihood Functions

Although the example of combining likelihood functions presented in Section 10.2.4.1 is simple, it illustrates the power of using likelihood functions to represent and combine information. A likelihood function converts the information in a measurement to a function on the target state space. Since all information is represented on the same state space, it can easily and correctly be combined, regardless of how disparate the sources of the information. The only limitation is the ability to compute the likelihood function corresponding to the measurement or the information to be incorporated. As an example, subjective information can often be put into the form of a likelihood function and incorporated into a tracker if desired.

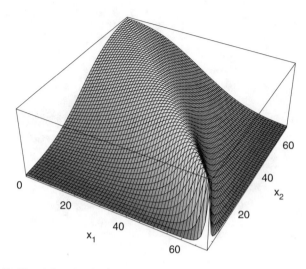

FIGURE 10.2 Bearing likelihood function for a sensor at (70,0).

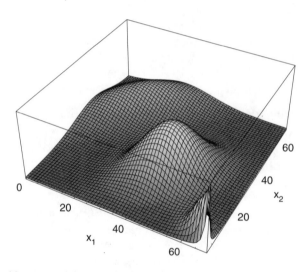

FIGURE 10.3 Combined bearing and detection likelihood function.

10.3 Multiple-Target Tracking without Contacts or Association (Unified Tracking)

In this section, the Bayesian tracking model for a single target is extended to multiple targets in a way that allows multiple-target tracking without calling contacts or performing data association.

10.3.1 Multiple-Target Motion Model

In Section 10.2, the prior knowledge about the single target's state and its motion through the target state space S were represented in terms of a stochastic process $\{X(t); t \geq 0\}$ where $X(t)$ is the target state at time t. This motion model is now generalized to multiple targets.

Begin the multiple-target tracking problem at time $t = 0$. The total number of targets is unknown but bounded by $\overline{N}$, which is known. We assume a known bound on the number of targets because it allows us to simplify the presentation and produces no restriction in practice. Designate a region, $\mathfrak{R}$, which defines the boundary of the tracking problem. Activity outside of $\mathfrak{R}$ has no importance. For example, we might be interested in targets having only a certain range of speeds or contained within a certain geographic region.

Add an additional state ϕ to the target state space S. If a target is not in the region $\mathfrak{R}$, it is considered to be in state ϕ. Let $S^+ = S \cup \{\phi\}$ be the extended state space for a single target and $\mathbf{S}^+ = S^+ \times \cdots \times S^+$ be the joint target state space where the product is taken $\overline{N}$ times.

10.3.1.1 Multiple-Target Motion Process

Prior knowledge about the targets and their "movements" through the state space $\mathbf{S}^+$ is expressed as a stochastic process $\mathbf{X} = \{X(t); t \geq 0\}$. Specifically, let $\mathbf{X}(t) = (X_1(t), \ldots, X_{\overline{N}}(t))$ be the state of the system at time t where $X_n(t) \in S^+$ is the state of target n at time t. The term "state of the system" is used to mean the joint state of all of the the targets. The value of the random variable $X_n(t)$ indicates whether target n is present in $\mathfrak{R}$ and, if so, in what state. The number of components of $\mathbf{X}(t)$ with states not equal to ϕ at time t gives the number of targets present in $\mathfrak{R}$ at time t. Assume that the stochastic process $\mathbf{X}$ is Markovian in the state space $\mathbf{S}^+$ and that the process has an associated transition function. Let $q_k(s_k \mid s_{k-1}) = \mathbf{Pr}\{\mathbf{X}(t_k) = s_k \mid \mathbf{X}(t_{k-1}) = s_{k-1}\}$ for $k \geq 1$, and let q_0 be the probability (density) function for $\mathbf{X}(0)$. By the Markov assumption

$$\mathbf{Pr}\left\{\mathbf{X}(t_1) = s_1, \ldots, \mathbf{X}(t_K) = s_K\right\} = \int \prod_{k=1}^{K} q_k\left(s_k \middle| s_{k-1}\right) q_0\left(s_0\right) ds_0 \tag{10.11}$$

The state space $\mathbf{S}^+$ of the Markov process $\mathbf{X}$ has a measure associated with it. If the process $\mathbf{S}^+$ is a discrete space Markov chain, then the measure is discrete and integration becomes summation. If the space is continuous, then functions such as transition functions become densities on $\mathbf{S}^+$ with respect to that measure. If $\mathbf{S}^+$ has both continuous and discrete components, then the measure will be the product or mixture of discrete and continuous measures. The symbol $d\mathbf{s}$ will be used to indicate integration with respect to the measure on $\mathbf{S}^+$, whether it is discrete or not. When the measure is discrete, the integrals become summations. Similarly, the notation $\mathbf{Pr}$ indicates either probability or probability density as appropriate.

10.3.2 Multiple-Target Likelihood Functions

There is a set of sensors that report observations at a discrete sequence of possibly random times. These sensors may be of different types and may report different information. The sensors may report only when they have a contact or on a regular basis. Let $Z(t, j)$ be an observation from sensor j at time t. Observations from sensor j take values in the measurement space H_j. Each sensor may have a different measurement space.

For each sensor j, assume that one can compute

$$\mathbf{Pr}\left\{Z(t, j) = z \middle| \mathbf{X}(t) = s\right\} \text{ for } z \in H_j \text{ and } s \in \mathbf{S}^+ \tag{10.12}$$

To compute the probabilities in Equation 10.12, one must know the distribution of the sensor response conditioned on the value of the state $\mathbf{s}$. In contrast to Section 10.2, the likelihood functions in this section can depend on the joint state of all the targets. The relationship between the observation and the state $\mathbf{s}$ may be linear or nonlinear, and the probability distribution may be Gaussian or non-Gaussian.

Suppose that by time t, observations have been obtained at the set of discrete times $0 \leq t_1 \leq \ldots \leq t_K \leq t$. To allow for the possibility of receiving more than one sensor observation at a given time, let Y_k be the set of sensor observations received at time t_k. Let y_k denote a value of the random variable Y_k. Extend Equation 10.12 to assume that the following computation can be made

$$L_k\!\left(y_k \mid \mathbf{s}\right) = \mathbf{Pr}\!\left\{Y_k = y_k \mid \mathbf{X}\!\left(t_k\right) = \mathbf{s}\right\} \text{ for } \mathbf{s} \in \mathbf{S}^+ \tag{10.13}$$

$L_k\,(y_k|\cdot)$ is called the *likelihood function* for the observation $Y_k = y_k$. The computation in Equation 10.13 can account for correlation among sensor responses if required.

Let $\mathbf{Y}(t) = (Y_1, Y_2,\ldots, Y_K)$ and $\mathbf{y} = (y_1,\ldots, y_K)$. Define $L(\mathbf{y}|\mathbf{s}_1,\ldots, \mathbf{s}_K) = \mathbf{Pr}\,\{\mathbf{Y}(t) = \mathbf{y}|\mathbf{X}(t_1) = \mathbf{s}_1,\ldots, \mathbf{X}(t_K) = \mathbf{s}_K\}$.

In parallel with Section 10.2, assume that

$$\mathbf{Pr}\!\left\{\mathbf{Y}(t) = \mathbf{y} \mid \mathbf{X}(u) = \mathbf{s}(u),\ 0 \le u \le t\right\} = L\!\left(\mathbf{y} \mid \mathbf{s}\!\left(t_1\right),\ldots,\mathbf{s}\!\left(t_K\right)\right) \tag{10.14}$$

and

$$L\!\left(\mathbf{y} \mid \mathbf{s}_1,\ldots,\mathbf{s}_K\right) = \prod_{k=1}^{K} L_k\!\left(y_k \mid \mathbf{s}_k\right) \tag{10.15}$$

Equation 10.14 assumes that the distribution of the sensor response at the times $\{t_k,\ k = 1,\ldots, K\}$ depends only on the system states at those times. Equation 10.15 assumes independence of the sensor response distributions across the observation times. The effect of both assumptions is to assume that the sensor response at time t_k depends only on the system state at that time.

10.3.3   Posterior Distribution

For unified tracking, the tracking problem is equivalent to computing the posterior distribution on $\mathbf{X}(t)$ given $\mathbf{Y}(t)$. The posterior distribution of $\mathbf{X}(t)$ represents our knowledge of the number of targets present and their state at time t given $\mathbf{Y}(t)$. From this distribution point estimates can be computed, when appropriate, such as maximum *a posteriori* probability estimates or means. Define $q(\mathbf{s}_1,\ldots, \mathbf{s}_K) = \mathbf{Pr}\{\mathbf{X}(t_1) = \mathbf{s}_1,\ldots, \mathbf{X}(t_K) = \mathbf{s}_K\}$ to be the prior probability (density) that the process $\mathbf{X}$ passes through the states $\mathbf{s}_1,\ldots, \mathbf{s}_K$ at times $t_1,\ldots,t_K$. Let q_0 be the probability (density) function for $\mathbf{X}(0)$. By the Markov assumption

$$q\!\left(\mathbf{s}_1,\ldots,\mathbf{s}_K\right) = \int \prod_{k=1}^{K} q_k\!\left(\mathbf{s}_k \mid \mathbf{s}_{k-1}\right) q_0\!\left(\mathbf{s}_0\right) d\mathbf{s}_0 \tag{10.16}$$

Let $p(t, \mathbf{s}) = \mathbf{Pr}\{\mathbf{X}(t) = \mathbf{s}|\mathbf{Y}(t)\}$. The function $p(t,\cdot)$ gives the posterior distribution on $\mathbf{X}(t)$ given $\mathbf{Y}(t)$ By Bayes' theorem,

$$
\begin{aligned}
p(t_K, \mathbf{s}_K) &= \frac{\mathbf{Pr}\!\left\{\mathbf{Y}\!\left(t_K\right) = \mathbf{y} \text{ and } \mathbf{X}\!\left(t_K\right) = \mathbf{s}_K\right\}}{\mathbf{Pr}\!\left\{\mathbf{Y}\!\left(t_K\right) = \mathbf{y}\right\}} \\[2mm]
&= \frac{\int L\!\left(\mathbf{y} \mid \mathbf{s}_1,\ldots,\mathbf{s}_K\right) q\!\left(\mathbf{s}_1,\mathbf{s}_2,\ldots,\mathbf{s}_K\right) d\mathbf{s}_1 \cdots d\mathbf{s}_{K-1}}{\int L\!\left(\mathbf{y} \mid \mathbf{s}_1,\ldots,\mathbf{s}_K\right) q\!\left(\mathbf{s}_1,\mathbf{s}_2,\ldots,\mathbf{s}_K\right) d\mathbf{s}_1 \cdots d\mathbf{s}_K}
\end{aligned}
\tag{10.17}
$$

10.3.4 Unified Tracking Recursion

Substituting Equations 10.15 and 10.16 into Equations 10.17 gives

$$p(t_K, \mathbf{s}_K) = \frac{1}{C'} \int \prod_{k=1}^{K} L_k(y_k | \mathbf{s}_k) \prod_{k=1}^{K} q_k(\mathbf{s}_k | \mathbf{s}_{k-1}) q_0(\mathbf{s}_0) d\mathbf{s}_0 \cdots d\mathbf{s}_{K-1}$$

$$= \frac{1}{C'} L_K(y_K | \mathbf{s}_K) \int q_K(\mathbf{s}_K | \mathbf{s}_{K-1})$$

$$\times \left[\int \prod_{k=1}^{K-1} L_k(y_k | \mathbf{s}_k) q_k(\mathbf{s}_k | \mathbf{s}_{k-1}) q_0(\mathbf{s}_0) d\mathbf{s}_0 \cdots d\mathbf{s}_{K-2} \right] d\mathbf{s}_{K-1}$$

and

$$p(t_K, \mathbf{s}_K) = \frac{1}{C} L_K(y_K | \mathbf{s}_K) \int q_K(\mathbf{s}_K | \mathbf{s}_{K-1}) p(t_{K-1}, \mathbf{s}_{K-1}) d\mathbf{s}_{K-1} \tag{10.18}$$

where C and C' normalize $p(t_K, \cdot)$ to be a probability distribution. Equation 10.18 provides a recursive method of computing $p(t_K, \cdot)$. Specifically,

Unified Tracking Recursion

Initialize Distribution: $\quad p(t_0, \mathbf{s}_0) = q_0(\mathbf{s}_0) \text{ for } \mathbf{s}_0 \in \mathbf{S}^+ \tag{10.19}$

For $k \geq 1$ and $\mathbf{s}_k \in \mathbf{S}^+$,

Perform Motion Update: $\quad p^-(t_k, \mathbf{s}_k) = \int q_k(\mathbf{s}_k | \mathbf{s}_{k-1}) p(t_{k-1}, \mathbf{s}_{k-1}) d\mathbf{s}_{k-1} \tag{10.20}$

Compute Likelihood Function L_k from the observation $Y_k = y_k$

Perform Information Update: $p(t_k, \mathbf{s}_k) = \frac{1}{C} L_k(y_k | \mathbf{s}_k) p^-(t_k, \mathbf{s}_k) \tag{10.21}$

10.3.4.1 Multiple-Target Tracking without Contacts or Association

The unified tracking recursion appears deceptively simple. The difficult part is performing the calculations in the joint state space of the $\bar{N}$ targets. Having done this, the combination of the likelihood functions defined on the joint state space with the joint distribution function of the targets automatically accounts for all possible association hypotheses without requiring explicit identification of these hypotheses. Section 10.4 demonstrates that this recursion produces the same joint posterior distribution as multiple-hypothesis tracking (MHT) does when the conditions for MHT are satisfied. However, the unified tracking recursion goes beyond MHT. One can use this recursion to perform multiple-target tracking when the notions of contact and association (notions required by MHT) are not meaningful. Examples of this are given in Section 5.3 of Stone et al.[3] Another example by Finn[4] applies to tracking two aircraft targets with a monopulse radar when the aircraft become so close together in bearing that their signals become unresolved. They merge inextricably at the radar receiver.

10.3.4.1.1   *Merged Measurements*

The problem tackled by Finn[4] is an example of the difficulties caused by merged measurements. A typical example of merged measurements is when a sensor's received signal is the sum of the signals from all the targets present. This can be the case with a passive acoustic sensor. Fortunately, in many cases the signals are separated in space or frequency so that they can be treated as separate signals. In some cases, two targets are so close in space (and radiated frequency) that it is impossible to distinguish which component of the received signal is due to which target. This is a case when the notion of associating a contact to a target is not well defined. Unified tracking will handle this problem correctly, but the computational load may be too onerous. In this case an MHT algorithm with special approximations could be used to provide an approximate but computationally feasible solution. See, for example, Mori et al.[5]

Section 10.4 presents the assumptions that allow contact association and multiple-target tracking to be performed by using MHT.

10.3.4.2   *Summary of Assumptions for Unified Tracking Recursion*

In summary, the assumptions required for the validity of the unified tracking recursion are

1. The number of targets is unknown but bounded by $\overline{N}$.
2. $S^+ = S \cup \{\phi\}$ is the extended state space for a single target where ϕ indicates the target is not present. $X_n(t) \in S^+$ is the state of the nth target at time t.
3. $\mathbf{X}(t) = (X_1(t),\ldots, X_{\overline{N}}(t))$ is the state of the system at time t, and $\mathbf{X} = \{\mathbf{X}(t); t \geq 0\}$ is the stochastic process describing the evolution of the system over time. The process, $\mathbf{X}$, is Markov in the state space $\mathbf{S}^+ = S^+ \times \cdots \times S^+$ where the product is taken $\overline{N}$ times.
4. Observations occur at discrete (possibly random) times, $0 \leq t_1 \leq t_2 \ldots$ Let $Y_k = y_k$ be the observation at time t_k, and let $\mathbf{Y}(t_K) = \mathbf{y}_K = (y_1,\ldots, y_K)$ be the first K observations. Then the following is true

$$\mathbf{Pr}\left\{\mathbf{Y}\left(t_K\right)=\mathbf{y}_K\middle|\mathbf{X}(u)=\mathbf{s}(u), 0 \leq u \leq t_K\right\}$$

$$= \mathbf{Pr}\left\{\mathbf{Y}\left(t_K\right)=\mathbf{y}_K\middle|\mathbf{X}\left(t_K\right)=\mathbf{s}\left(t_K\right), k=1,\ldots,K\right\}$$

$$= \prod_{k=1}^{K} L_k\left(y_k \mid \mathbf{s}\left(t_K\right)\right)$$

10.4   Multiple-Hypothesis Tracking (MHT)

In classical multiple-target tracking, the problem is divided into two steps: (1) association and (2) estimation. Step 1 associates contacts with targets. Step 2 uses the contacts associated with each target to produce an estimate of that target's state. Complications arise when there is more than one reasonable way to associate contacts with targets. The classical approach to this problem is to form association hypotheses and to use MHT, which is the subject of this section. In this approach, alternative hypotheses are formed to explain the source of the observations. Each hypothesis assigns observations to targets or false alarms. For each hypothesis, MHT computes the probability that it is correct. This is also the probability that the target state estimates that result from this hypothesis are correct. Most MHT algorithms display only the estimates of target state associated with the highest probability hypothesis.

The model used for the MHT problem is a generalization of the one given by Reid[6] and Mori et al.[7] Section 10.4.3.3 presents the recursion for general multiple-hypothesis tracking. This recursion applies to problems that are nonlinear and non-Gaussian as well as to standard linear-Gaussian situations. In this general case, the distributions on target state may fail to be independent of one another (even when conditioned on an association hypothesis) and may require a joint state space representation. This recursion includes a conceptually simple Bayesian method of computing association probabilities. Section 10.4.4 discusses the case where the target distributions (conditioned on an association hypothesis) are independent of one another. Section 10.4.4.2 presents the independent MHT recursion that holds when these independence conditions are satisfied. Note that not all tracking situations satisfy these independence conditions.

Numerous books and articles on multiple-target tracking examine in detail the many variations and approaches to this problem. Many of these discuss the practical aspects of implementing multiple target trackers and compare approaches. See, for example, Antony,[8] Bar-Shalom and Fortman,[9] Bar-Shalom and Li,[10] Blackman,[11] Blackman and Popoli,[1] Hall,[12] Reid,[6] Mori et al.,[7] and Waltz and Llinas.[13] With the exception of Mori et al.,[7] these references focus primarily on the linear-Gaussian case.

In addition to the full or classical MHT as defined by Reid[6] and Mori et al.,[7] a number of approximations are in common use for finding solutions to tracking problems. Examples include joint probabilistic data association (Bar-Shalom and Fortman[9]) and probabilistic MHT (Streit[14]). Rather than solve the full MHT, Poore[15] attempts to find the data association hypothesis (or the n hypotheses) with the highest likelihood. The tracks formed from this hypothesis then become the solution. Poore does this by providing a window of scans in which contacts are free to float among hypotheses. The window has a constant width and always includes the latest scan. Eventually contacts from older scans fall outside the window and become assigned to a single hypothesis. This type of hypothesis management is often combined with a nonlinear extension of Kalman filtering called an interactive multiple model Kalman filter (Yeddanapudi et al.[16]).

Section 10.4.1 presents a description of general MHT. Note that general MHT requires many more definitions and assumptions than unified tracking.

10.4.1 Contacts, Scans, and Association Hypotheses

This discussion of MHT assumes that sensor responses are limited to contacts.

10.4.1.1 Contacts

A *contact* is an observation that consists of a called detection and a measurement. In practice, a detection is called when the signal-to-noise ratio at the sensor crosses a predefined threshold. The measurement associated with a detection is often an estimated position for the object generating the contact. Limiting the sensor responses to contacts restricts responses to those in which the signal level of the target, as seen at the sensor, is high enough to call a contact. Section 10.6 demonstrates how tracking can be performed without this assumption being satisfied.

10.4.1.2 Scans

This discussion further limits the class of allowable observations to scans. The observation Y_k at time t_k is a *scan* if it consists of a set $\mathfrak{C}_k$ of contacts such that each contact is associated with at most one target, and each target generates at most one contact (i.e., there are no merged or split measurements). Some of these contacts may be false alarms, and some targets in $\mathfrak{R}$ might not be detected on a given scan.

More than one sensor group can report a scan at the same time. In this case, the contact reports from each sensor group are treated as separate scans with the same reporting time. As a result, $t_{k+1} = t_k$. A scan can also consist of a single contact report.

10.4.1.3 Data Association Hypotheses

To define a data association hypothesis, h, let

$$\mathfrak{C}_j = \text{set of contacts of the } j\text{th scan}$$

$$\mathfrak{H}(k) = \text{set of all contacts reported in the first } k \text{ scans}$$

Note that

$$\mathfrak{H}(k) = \bigcup_{j=1}^{k} \mathfrak{C}_j$$

A data association hypothesis, h, on $\Re(k)$ is a mapping

$$h : \Re(k) \to \{0,1,\ldots,\overline{N}\}$$

such that

$h(c) = n > 0$ means contact c is associated to target n

$h(c) = 0$ means contact c is associated to a false alarm

and no two contacts from the same scan are associated to the same target.

Let $H(k)$ = set of all data association hypotheses on $\tilde{\mathfrak{H}}(k)$. A hypothesis h on $\Re(k)$ partitions $\Re(k)$ into sets $U(n)$ for $n = 0,1,\ldots,\overline{N}$ where $U(n)$ is the set of contacts associated to target n for $n > 0$ and $U(0)$ is the set of contacts associated to false alarms.

10.4.1.4   Scan Association Hypotheses

Decomposing a data association hypothesis h into scan association hypotheses is convenient. For each scan Y_k, let

M_k = the number of contacts in scan k

Γ_k = the set of all functions $\gamma:\{1,\ldots,M_k\} \to \{0,\ldots,\overline{N}\}$ such that no two contacts are assigned to the same positive number. If $\gamma(m) = 0$, then contact m is associated to a false alarm. If $\gamma(m) = n > 0$, then contact m is associated to target n.

A function $\gamma \in \Gamma_k$ is called a *scan association hypothesis* for the kth scan, and Γ_k is the set of scan association hypotheses for the kth scan. For each contact, a scan association hypothesis specifies which target generated the contact or that the contact was due to a false alarm.

Consider a data association hypothesis $h_K \in H(K)$. Think of h_K as being composed of K scan association hypotheses $\{\gamma_1,\ldots,\gamma_K\}$ where γ_k is the association hypothesis for the kth scan of contacts. The hypothesis $h_K \in H(K)$ is the extension of the hypothesis $h_{K-1} = \{\gamma_1,\ldots,\gamma_{k-1}\} \in H(K-1)$. That is, h_K is composed of h_{K-1} plus γ_K. This can be written as $h_K = h_{K-1} \wedge \gamma_K$.

10.4.2   Scan and Data Association Likelihood Functions

The correctness of the scan association hypothesis γ is equivalent to the occurrence of the event "the targets to which γ associates contacts generate those contacts." Calculating association probabilities requires the ability to calculate the probability of a scan association hypothesis being correct. In particular, we must be able to calculate the probability of the event $\{\gamma \wedge Y_k = y_k\}$, where $\{\gamma \wedge Y_k = y_k\}$ denotes the conjunction or intersection of the events γ and $Y_k = y_k$.

10.4.2.1   Scan Association Likelihood Function

Assume that for each scan association hypothesis γ, one can calculate the scan association likelihood function

$$l_k\left(\gamma \wedge Y_k = y_k \,|\, \mathbf{s}_k\right) = \mathbf{Pr}\left\{\gamma \wedge Y_k = y_k \,\middle|\, \mathbf{X}(t_k) = \mathbf{s}_k\right\}$$

$$= \mathbf{Pr}\left\{Y_k = y_k \,\middle|\, \gamma \wedge \mathbf{X}(t_k) = \mathbf{s}_k\right\}\mathbf{Pr}\left\{\gamma \,|\, \mathbf{X}(t_k) = \mathbf{s}_k\right\} \text{ for } \mathbf{s}_k \in \mathbf{S}^+ \tag{10.22}$$

The factor $\mathbf{Pr}\{\gamma | \mathbf{X}(t_k) = \mathbf{s}_k\}$ is the prior probability that the scan association γ is the correct one. We normally assume that this probability does not depend on the system state $\mathbf{s}_k$, so that one may write

$$l_k\left(\gamma \wedge Y_k = y_k \,|\, \mathbf{s}_k\right) = \mathbf{Pr}\left\{Y_k = y_k \,\middle|\, \gamma \wedge \mathbf{X}\left(t_k\right) = \mathbf{s}_k\right\}\mathbf{Pr}\left\{\gamma\right\} \text{ for } \mathbf{s}_k \in \mathbf{S}^+ \tag{10.23}$$

Note that $l_k(\gamma \wedge Y_k = y_k | \cdot)$ is not, strictly speaking, a likelihood function because γ is not an observation. Nevertheless, it is called a likelihood function because it behaves like one. The likelihood function for the observation $Y_k = y_k$ is

$$L_k\left(y_k | \mathbf{s}_k\right) = \mathbf{Pr}\left\{Y_k = y_k \,\Big|\, \mathbf{X}\left(t_k\right) = \mathbf{s}_k\right\} = \sum_{\gamma \in \Gamma_k} \ell_k\left(\gamma \wedge Y_k = y_k | \mathbf{s}_k\right) \text{ for } \mathbf{s}_k \in \mathbf{S}^+ \tag{10.24}$$

10.4.2.1.1 Scan Association Likelihood Function Example

Consider a tracking problem where detections, measurements, and false alarms are generated according to the following model. The target state, s, is composed of an l-dimensional position component, z, and an l-dimensional velocity component, v, in a Cartesian coordinate space, so that $s = (z, v)$. The region of interest, $\mathfrak{R}$, is finite and has volume V in the l-dimensional position component of the target state space. There are at most $\overline{N}$ targets in $\mathfrak{R}$.

Detections and measurements. If a target is located at z, then the probability of its being detected on a scan is $P_d(z)$. If a target is detected then a measurement Y is obtained where $Y = z + \varepsilon$ and $\varepsilon \sim N(0, \Sigma)$. Let $\eta(y, z, \Sigma)$ be the density function for a $N(z, \Sigma)$ random variable evaluated at y. Detections and measurements occur independently for all targets.

False alarms. For each scan, false alarms occur as a Poisson process in the position space with density ρ. Let Φ be the number of false alarms in a scan, then

$$\mathrm{Pr}\{\Phi = j\} = \frac{(\rho V)^j}{j!} \text{ for } j = 0, 1, \dots$$

Scan. Suppose that a scan of M measurements is received $\mathbf{y} = (y_1, \dots, y_M)$ and γ is a scan association. Then γ specifies which contacts are false and which are true. In particular, if $\gamma(m) = n > 0$, measurement m is associated to target n. If $\gamma(m) = 0$, measurement m is associated to a false target. No target is associated with more than one contact. Let

$\varphi(\gamma)$ = the number of contacts associated to false alarms
$I(\gamma) = \{n : \gamma \text{ associates no contact in the scan to target } n\}$ = the set of targets that have no contacts associated to them by γ

Scan Association Likelihood Function. Assume that the prior probability is the same for all scan associations, so that for some constant G, $\mathbf{Pr}\{\gamma\} = G$ for all γ. The scan association likelihood function is

$$l\left(\gamma \wedge \mathbf{y} \,|\, \mathbf{s} = (z, v)\right) = \frac{G(\rho V)^{\varphi(\gamma)}}{\varphi(\gamma)! V^{\varphi(\gamma)}} e^{-\rho V} \prod_{\{m : \gamma(m) > 0\}} P_d\left(z_{\gamma(m)}\right) \eta\left(y_m, z_{\gamma(m)}, \Sigma\right) \prod_{n \in I(\gamma)} \left(1 - P_d(z_n)\right) \tag{10.25}$$

10.4.2.2 Data Association Likelihood Function

Recall that $Y(t_K) = \mathbf{y}_K$ is the set of observations (contacts) contained in the first K scans and $H(K)$ is the set of data association hypotheses defined on these scans. For $h \in H(K)$, $\mathbf{Pr}\{h \wedge Y(t_K) = \mathbf{y}_K | \mathbf{X}(u) = \mathbf{s}_u, 0 \le u \le t_K\}$ is the likelihood of $\{h \wedge Y(t_K) = \mathbf{y}_K\}$, given $\{\mathbf{X}(u) = \mathbf{s}_u, 0 \le u \le t_K\}$. Technically, this is not a likelihood function either, but it is convenient and suggestive to use this terminology. As with the observation likelihood functions, assume that

$$\mathbf{Pr}\left\{h \wedge Y\left(t_K\right) = \mathbf{y}_K \,\Big|\, \mathbf{X}(u) = \mathbf{s}(u), 0 \le u \le t_K\right\}$$
$$= \mathbf{Pr}\left\{h \wedge Y\left(t_K\right) = \mathbf{y}_K \,\Big|\, \mathbf{X}\left(t_k\right) = \mathbf{s}\left(t_k\right), k = 1, \dots, K\right\} \tag{10.26}$$

In addition, assuming that the scan association likelihoods are independent, the data association likelihood function becomes

$$\mathbf{Pr}\left\{h \wedge \mathbf{Y}(t_K) = \mathbf{y}_K \middle| \mathbf{X}(t_k) = \mathbf{s}_k, k = 1,\ldots,K\right\} = \prod_{k=1}^{K} \ell_k\left(\gamma_k \wedge Y_k = y_k | \mathbf{s}_k\right) \tag{10.27}$$

where $\mathbf{y}_K = (y_1,\ldots y_K)$ and $h = \{\gamma_1,\ldots,\gamma_k\}$.

10.4.3   General Multiple-Hypothesis Tracking

Conceptually, MHT proceeds as follows. It calculates the posterior distribution on the system state at time t_K, given that data association hypothesis h is true, and the probability, $\alpha(h)$, that hypothesis h is true for each $h \in H(K)$. That is, it computes

$$p(t_K, \mathbf{s}_K | h) \equiv \mathbf{Pr}\left\{\mathbf{X}(t_K) = \mathbf{s}_K | h \wedge \mathbf{Y}(t_K) = \mathbf{y}_K\right\} \tag{10.28}$$

and

$$\alpha(h) \equiv \mathbf{Pr}\left\{h \middle| \mathbf{Y}(t_K) = \mathbf{y}_K\right\} = \frac{\mathbf{Pr}\left\{h \wedge \mathbf{Y}(t_K) = \mathbf{y}_K\right\}}{\mathbf{Pr}\left\{\mathbf{Y}(t_K) = \mathbf{y}_K\right\}} \quad \text{for each } h \in H(K) \tag{10.29}$$

Next, MHT can compute the Bayesian posterior on system state by

$$p(t_K, \mathbf{s}_K) = \sum_{h \in H(K)} \alpha(h) p(t_K, \mathbf{s}_K | h) \tag{10.30}$$

Subsequent sections show how to compute $p(t_K, \mathbf{s}_K | h)$ and $\alpha(h)$ in a joint recursion.

A number of difficulties are associated with calculating the posterior distribution in Equation 10.30. First, the number of data association hypotheses grows exponentially as the number of contacts increases. Second, the representation in Equation 10.30 is on the joint $\overline{N}$ fold target state space, a state space that is dauntingly large for most values of $\overline{N}$. Even when the size of the joint state space is not a problem, displaying and understanding the joint distribution is difficult.

Most MHT algorithms overcome these problems by limiting the number of hypotheses carried, displaying the distribution for only a small number of the highest probability hypotheses — perhaps only the highest. Finally, for a given hypothesis, they display the marginal distribution on each target, rather than the joint distribution. (Note, specifying a data association hypothesis specifies the number of targets present in $\mathfrak{R}$.) Most MHT implementations make the linear-Gaussian assumptions that produce Gaussian distributions for the posterior on a target state. The marginal distribution on a two-dimensional target position can then be represented by an ellipse. It is usually these ellipses, one for each target, that are displayed by an MHT to represent the tracks corresponding to an hypothesis.

10.4.3.1   Conditional Target Distributions

Distributions conditioned on the truth of an hypothesis are called *conditional target distributions*. The distribution $p(t_K, \cdot | h)$ in Equation 10.28 is an example of a conditional joint target state distribution. These distributions are always conditioned on the data received (e.g., $\mathbf{Y}(t_K) = \mathbf{y}_K$), but this conditioning does not appear in our notation, $p(t_K, \mathbf{s}_K | h)$.

Let $h_K = \{\gamma_1,\ldots,\gamma_K\}$, then

$$p\left(t_K,\mathbf{s}_K \mid h_K\right) = \Pr\left\{\mathbf{X}\left(t_K\right) = \mathbf{s}_K \mid h_K \wedge \mathbf{Y}\left(t_K\right) = \mathbf{y}_K\right\}$$

$$= \frac{\Pr\left\{\left(\mathbf{X}\left(t_K\right) = \mathbf{s}_K\right) \wedge h_K \wedge \left(\mathbf{Y}\left(t_K\right) = \mathbf{y}_K\right)\right\}}{\Pr\left\{h_K \wedge \mathbf{Y}\left(t_K\right) = \mathbf{y}_K\right\}}$$

and by Equation 10.11 and the data association likelihood function in Equation 10.27,

$$\Pr\left\{\left(\mathbf{X}\left(t_K\right) = \mathbf{s}_K\right) \wedge h_K \wedge \left(\mathbf{Y}\left(t_K\right) = \mathbf{y}_K\right)\right\}$$

$$= \int \left\{\prod_{k=1}^{K} \ell_k\left(\gamma_k \wedge Y_k = y_k \mid \mathbf{s}_k\right) \prod_{k=1}^{K} q_k\left(\mathbf{s}_k \mid \mathbf{s}_{k-1}\right) q_0\left(\mathbf{s}_0\right)\right\} d\mathbf{s}_0 \, d\mathbf{s}_1 \cdots d\mathbf{s}_{K-1} \tag{10.31}$$

Thus,

$$p\left(t_K,\mathbf{s}_K \mid h_K\right) = \frac{1}{C\left(h_K\right)} \int \left\{\prod_{k=1}^{K} \ell_k\left(\gamma_k \wedge Y_k = y_k \mid \mathbf{s}_k\right) q_k\left(\mathbf{s}_k \mid \mathbf{s}_{k-1}\right)\right\} q_0\left(\mathbf{s}_0\right) d\mathbf{s}_0 \, d\mathbf{s}_1 \cdots d\mathbf{s}_{K-1} \tag{10.32}$$

where $C(h_K)$ is the normalizing factor that makes $p(t_K,\cdot \mid h_K)$ a probability distribution. Of course

$$C\left(h_K\right) = \int \left\{\prod_{k=1}^{K} \ell_k\left(\gamma_k \wedge Y_k = y_k \mid \mathbf{s}_k\right) q_k\left(\mathbf{s}_k \mid \mathbf{s}_{k-1}\right)\right\} q_0\left(\mathbf{s}_0\right) d\mathbf{s}_0 \, d\mathbf{s}_1 \cdots d\mathbf{s}_K$$

$$\tag{10.33}$$

$$= \Pr\left\{h_K \wedge \mathbf{Y}\left(t_K\right) = \mathbf{y}_K\right\}$$

10.4.3.2 Association Probabilities

Sections 4.2.1 and 4.2.2 of Stone et al.[3] show that

$$\alpha\left(h\right) = \Pr\left\{h \mid \mathbf{Y}\left(t_K\right) = \mathbf{y}_K\right\} = \frac{\Pr\left\{h \wedge \mathbf{Y}\left(t_K\right) = \mathbf{y}_K\right\}}{\Pr\left\{\mathbf{Y}\left(t_K\right) = \mathbf{y}_K\right\}} = \frac{C\left(h\right)}{\displaystyle\sum_{h' \in H(K)} C\left(h'\right)} \quad \text{for } h \in H\left(K\right) \tag{10.34}$$

10.4.3.3 General MHT Recursion

Section 4.2.3 of Stone et al.[3] provides the following general MHT recursion for calculating conditional target distributions and hypothesis probabilities.

General MHT Recursion

1. *Intialize:* Let $H(0) = \{h_0\}$, where h_0 is the hypothesis with no associations. Set

$$\alpha\left(h_0\right) = 1 \text{ and } p\left(t_0,\mathbf{s}_0 \mid h_0\right) = q_0\left(\mathbf{s}_0\right) \text{ for } \mathbf{s}_0 \in \mathbf{S}^+$$

2. **Compute Conditional Target Distributions:** For $k = 1,2,\ldots$, compute

$$p^-\left(t_k, \mathbf{s}_k \mid h_{k-1}\right) = \int q_k\left(\mathbf{s}_k \mid \mathbf{s}_{k-1}\right) p\left(t_{k-1}, \mathbf{s}_{k-1} \mid h_{k-1}\right) d\mathbf{s}_{k-1} \quad \text{for } h_{k-1} \in H\left(k-1\right) \tag{10.35}$$

For $h_k = h_{k-1} \wedge \gamma_k \in H(k)$, compute

$$\beta\left(h_k\right) = \alpha\left(h_{k-1}\right) \int \ell_k\left(\gamma_k \wedge Y_k = y_k \mid \mathbf{s}_k\right) p^-\left(t_k, \mathbf{s}_k \mid h_{k-1}\right) d\mathbf{s}_k \tag{10.36}$$

$$p\left(t_k, \mathbf{s}_k \mid h_k\right) = \frac{\alpha\left(h_{k-1}\right)}{\beta\left(h_k\right)} \ell_k\left(\gamma_k \wedge Y_k = y_k \mid \mathbf{s}_k\right) p^-\left(t_k, \mathbf{s}_k \mid h_{k-1}\right) \quad \text{for } \mathbf{s}_k \in \mathbf{S}^+ \tag{10.37}$$

3. **Compute Association Probabilities:** For $k = 1,2,\ldots$, compute

$$\alpha\left(h_k\right) = \frac{\beta\left(h_k\right)}{\displaystyle\sum_{h \in H(k)} \beta\left(h\right)} \quad \text{for } h_K \in H\left(K\right) \tag{10.38}$$

10.4.3.4 Summary of Assumptions for General MHT Recursion

In summary, the assumptions required for the validity of the general MHT recursion are

1. The number of targets is unknown but bounded by $\overline{N}$.
2. $\mathbf{S}^+ = S \cup \{\phi\}$ is the extended state space for a single target, where ϕ indicates that the target is not present.
3. $X_n(t) \in S^+$ is the state of the nth target at time t.
4. $\mathbf{X}(t) = (X_1(t),\ldots,X_{\overline{N}}(t))$ is the state of the system at time t, and $\mathbf{X} = \{\mathbf{X}(t); t \geq 0\}$ is the stochastic process describing the evolution of the system over time. The process, $\mathbf{X}$, is Markov in the state space $\mathbf{S}^+ = S^+ \times \cdots \times S^+$ where the product is taken $\overline{N}$ times.
5. Observations occur as contacts in scans. Scans are received at discrete (possibly random) times $0 \leq t_1 \leq t_2\ldots$ Let $Y_k = y_k$ be the scan (observation) at time t_k, and let $\mathbf{Y}(t_K) = \mathbf{y}_K \equiv (y_1,\ldots,y_K)$ be the set of contacts contained in the first K scans. Then, for each data association hypothesis $h \in H(K)$, the following is true:

$$\mathbf{Pr}\left\{h \wedge \mathbf{Y}\left(t_K\right) = \mathbf{y}_K \mid \mathbf{X}(u) = \mathbf{s}(u), 0 \leq u \leq t_K\right\}$$

$$= \mathbf{Pr}\left\{h \wedge \mathbf{Y}\left(t_K\right) = \mathbf{y}_K \mid \mathbf{X}\left(t_K\right) = \mathbf{s}\left(t_K\right), k = 1,\ldots,K\right\}$$

6. For each scan association hypothesis γ at time t_k, there is a scan association likelihood function

$$l_k\left(\gamma \wedge Y_k = y_k \mid \mathbf{s}_k\right) = \mathbf{Pr}\left\{\gamma \wedge Y_k = y_k \mid \mathbf{X}\left(t_k\right) = \mathbf{s}_k\right\}$$

7. Each data association hypothesis, $h \in H(K)$, is composed of scan association hypotheses so that $h = \{\gamma_1,\ldots,\gamma_K\}$ where γ_k is a scan association hypothesis for scan k.
8. The likelihood function for the data association hypothesis $h = \{\gamma_1,\ldots,\gamma_K\}$ satisfies

$$\mathbf{Pr}\left\{h \wedge \mathbf{Y}(t_K) = \mathbf{y}_K \middle| \mathbf{X}(u) = \mathbf{s}(u), 0 \le u \le t_K\right\} = \prod_{k=1}^{K} l_k\left(\gamma_k \wedge Y_k = y_k \middle| \mathbf{s}(t_K)\right)$$

10.4.4 Independent Multiple-Hypothesis Tracking

The decomposition of the system state distribution into a sum of conditional target distributions is most useful when the conditional distributions are the product of independent single-target distributions. This section presents a set of conditions under which this happens and restates the basic MHT recursion for this case.

10.4.4.1 Conditionally Independent Scan Association Likelihood Functions

Prior to this section no special assumptions were made about the scan association likelihood function $l_k(\gamma \wedge Y_k = y_k | \mathbf{s}_k) = \mathbf{Pr}\{\gamma \wedge Y_k = y_k | \mathbf{X}(t_k) = \mathbf{s}_k\}$ for $\mathbf{s}_k \in \mathbf{S}^+$. In many cases, however, the joint likelihood of a scan observation and a data association hypothesis satisfies an independence assumption when conditioned on a system state.

The likelihood of a scan observation $Y_k = y_k$ obtained at time t_k is *conditionally independent*, if and only if, for all scan association hypotheses $\gamma \in \Gamma_k$,

$$l_k\left(\gamma \wedge Y_k = y_k \middle| \mathbf{s}_k\right) = \mathbf{Pr}\left\{\gamma \wedge Y_k = y_k \middle| \mathbf{X}(t_k) = \mathbf{s}_k\right\}$$

$$= g_0^\gamma(y_k) \prod_{n=1}^{\overline{N}} g_n^\gamma\left(y_k, x_n\right) \text{ for } \mathbf{s}_k = \left(x_1, \ldots, x_{\overline{N}}\right) \tag{10.39}$$

for some functions g_n^γ, $n = 0, \ldots, \overline{N}$, where g_0^γ can depend on the scan data but not $\mathbf{s}_k$.

Equation 10.39 shows that conditional independence means that the probability of the joint event $\{\gamma \wedge Y_k = y_k\}$, conditioned on $\mathbf{X}(t_k) = (x_1, \ldots, x_{\overline{N}})$, factors into a product of functions that each depend on the state of only one target. This type of factorization occurs when the component of the response due to each target is independent of all other targets. As an example, the scan association likelihood in Equation 10.25 is conditionally independent. This can be verified by setting

$$g_0^\gamma(\mathbf{y}) = \frac{G(\rho V)^{\varphi(\gamma)}}{\varphi(\gamma)! V^{\varphi(\gamma)}} = \frac{G\rho^{\varphi(\gamma)}}{\varphi(\gamma)!}$$

and for $n = 1, \ldots, \overline{N}$

$$g_n^\gamma(\mathbf{y}, z_n) = \begin{cases} P_d(z_n)\eta(y_m, z_n, \Sigma) & \text{if } \gamma(m) = n \text{ for some } m \\ 1 - P_d(z_n) & \text{if } \gamma(m) \ne n \text{ for all } m \end{cases}$$

Conditional independence implies that the likelihood function for the observation $Y_k = y_k$ is given by

$$L_k\left(Y_k = y_k \middle| \mathbf{X}(t_k) = \left(x_1, \ldots, x_{\overline{N}}\right)\right)$$

$$= \sum_{\gamma \in \Gamma_k} g_0^\gamma(y_k) \prod_{n=1}^{\overline{N}} g_n^\gamma\left(y_k, x_n\right) \text{ for all } \left(x_1, \ldots, x_{\overline{N}}\right) \in \mathbf{S}^+ \tag{10.40}$$

The assumption of conditional independence of the observation likelihood function is implicit in most multiple target trackers. The notion of conditional independence of a likelihood function makes sense only when the notions of contact and association are meaningful. As noted in Section 10.3, there are cases in which these notions do not apply. For these cases, the scan association likelihood function will not satisfy Equation 10.39.

Under the assumption of conditional independence, the Independence Theorem, given below, says that conditioning on a data association hypothesis allows the multiple-target tracking problem to be decomposed into $\overline{N}$ independent single target problems. In this case, conditioning on an hypothesis greatly simplifies the joint tracking problem. In particular, no joint state space representation of the target distributions is required when they are conditional on a data association hypothesis.

10.4.4.1.1 Independence Theorem

Suppose that (1) the assumptions of Section 10.4.3.4 hold, (2) the likelihood functions for all scan observations are conditionally independent, and (3) the prior target motion processes, $\{X_n(t); t \geq 0\}$ for $n = 1,...,\overline{N}$ are mutually independent. Then the posterior system state distribution conditioned on the truth of a data association hypothesis is the product of independent distributions on the targets.

Proof. The proof of this theorem is given in Section 4.3.1 of Stone et al.[3]

Let $\mathbf{Y}(t) = \{Y_1, Y_2,...,y_{K(t)}\}$ be scan observations that are received at times $0 \leq t_1 \leq...,\leq t_k \leq t$, where $K = K(t)$, and let $H(k)$ be the set of all data association hypotheses on the first k scans. Define $p_n(t_k, x_n|h) = \Pr\{X_n(t_k) = x_n|h\}$ for $x_n \in S^+$, $k = 1,..., K$, and $n = 1,...,\overline{N}$. Then by the independence theorem,

$$p\left(t_k, \mathbf{s}_k \,|\, h\right) = \prod_n P_n\left(t_k, x_n \,|\, h\right) \text{ for } h \in H(k) \text{ and } \mathbf{s}_k = \left(x_1,...,x_{\overline{N}}\right) \in S^+ \tag{10.41}$$

Joint and Marginal Posteriors. From Equation 10.30 the full Bayesian posterior on the joint state space can be computed as follows:

$$\begin{aligned} p\left(t_K, \mathbf{s}_K\right) &= \sum_{h \in H(K)} \alpha\left(h\right) p\left(t_K, \mathbf{s}_k \,|\, h\right) \\ &= \sum_{h \in H(K)} \alpha\left(h\right) \prod_n P_n\left(t_K, x_n \,|\, h\right) \text{ for } \mathbf{s}_K = \left(x_1,...,x_{\overline{N}}\right) \in S^+ \end{aligned} \tag{10.42}$$

Marginal posteriors can be computed in a similar fashion. Let $\bar{p}_n(t_K, \cdot)$ be the marginal posterior on $X_n(t_K)$ for $n = 1,..., \overline{N}$. Then

$$\begin{aligned} \bar{p}_n\left(t_K, x_n\right) &= \int \left[\sum_{h \in H(K)} \alpha\left(h\right) \prod_{l=1}^{\overline{N}} P_l\left(t_K, x_l \,|\, h\right) \right] \prod_{l \neq n} dx_l \\ &= \sum_{h \in H(K)} \alpha\left(h\right) p_n\left(t_K, x_n \,|\, h\right) \int \prod_{l \neq n} P_l\left(t_K, x_l \,|\, h\right) dx_l \\ &= \sum_{h \in H(K)} \alpha\left(h\right) p_n\left(t_K, x_n \,|\, h\right) \text{ for } n = 1,..., \overline{N} \end{aligned}$$

Thus, the posterior marginal distribution on target n may be computed as the weighted sum over n of the posterior distribution for target n conditioned on h.

10.4.4.2 Independent MHT Recursion

Let $q_0(n, x) = \mathbf{Pr}\{X_n(0) = x\}$ and $q_k(x|n, x') = \mathbf{Pr}\{X_n(t_k) = x|X_n(t_{k-1}) = x'\}$. Under the assumptions of the independence theorem, the motion models for the targets are independent, and $q_k(\mathbf{s}_k|\mathbf{s}_{k-1}) = \prod_n q_k(x_n|n, x'_n)$, where $\mathbf{s}_k = (x_1,\ldots,x_{\overline{N}})$ and $\mathbf{s}_{k-1} = (x'_1,\ldots,x'_{\overline{N}})$. As a result, the transition density, $q_k(\mathbf{s}_k|\mathbf{s}_{k-1})$, factors just as the likelihood function does. This produces the independent MHT recursion below.

Independent MHT Recursion

1. ***Initialize:*** Let $H(0) = \{h_0\}$ where h_0 is the hypothesis with no associations. Set

$$\alpha(h_0) = 1 \text{ and } p_n(t_0, \mathbf{s}_0 | h_0) = q_0(n, \mathbf{s}_0) \text{ for } \mathbf{s}_0 \in S^+ \text{ and } n = 1,\ldots,\overline{N}$$

2. ***Compute Conditional Target Distributions:*** For $k = 1, 2, \ldots$, do the following: For each $h_k \in H(k)$, find $h_{k-1} \in H(k-1)$ and $\gamma \in \Gamma_k$, such that $h_k = h_{k-1} \wedge \gamma$. Then compute

$$p_n^-(t_k, x | h_{k-1}) = \int_{S^+} q_k(x|n, x') p_n(t_{k-1}, x' | h_{k-1}) dx' \text{ for } n = 1,\ldots,\overline{N}$$

$$p_n(t_k, x | h_k) = \frac{1}{C(n, h_k)} g_n^\gamma(y_k, x) p_n^-(t_k, x | h_{k-1}) \text{ for } x \in S^+ \text{ and } n = 1,\ldots,\overline{N} \tag{10.43}$$

$$p(t_k, \mathbf{s}_k | h_k) = \prod_n p_n(t_k, x_n | h_k) \text{ for } \mathbf{s}_k = (x_1,\ldots,x_{\overline{N}}) \in S^+$$

where $C(n, h_k)$ is the constant that makes $p_n(t_k, \cdot | h_k)$ a probability distribution.

3. ***Compute Association Probabilities:*** For $k = 1, 2, \ldots$, and $h_k = h_{k-1} \wedge \gamma \in H(k)$ compute

$$\beta(h_k) = \alpha(h_{k-1}) \int g_0^\gamma(y_k) \left[\prod_n g_n^\gamma(y_k, x_n) p_n^-(t_k, x_n | h_{k-1}) \right] dx_1 \cdots dx_{\overline{N}}$$

$$= \alpha(h_{k-1}) g_0^\gamma(y_k) \prod_n \int g_n^\gamma(y_k, x_n) p_n^-(t_k, x_n | h_{k-1}) dx_n \tag{10.44}$$

Then

$$\alpha(h_k) = \frac{\beta(h_k)}{\sum_{h'_k \in H(k)} \beta(h'_k)} \text{ for } h_K \in H(K) \tag{10.45}$$

In Equation 10.43, the independent MHT recursion performs a motion update of the probability distribution on target n given h_{k-1} and multiplies the result by $g_n^\gamma(y_k, x)$, which is the likelihood function of the measurement associated to target n by γ. When this product is normalized to a probability distribution, we obtain the posterior on target n given $h_k = h_{k-1} \wedge \gamma$. Note that these computations are all performed independently of the other targets. Only the computation of the association probabilities in Equations 10.44 and 10.45 requires interaction with the other targets and the likelihoods of the measurements associated to them. This is where the independent MHT obtains its power and simplicity. *Conditioned on a data association hypothesis, each target may be treated independently of all other targets.*

10.5 Relationship of Unified Tracking to MHT and Other Tracking Approaches

This section discusses the relationship of unified tracking to other tracking approaches such as general MHT.

10.5.1 General MHT Is a Special Case of Unified Tracking

Section 5.2.1 of Stone et al.[3] shows that the assumptions for general MHT that are given in Section 10.4.3.4 imply the validity of the assumptions for unified tracking given in Section 10.3.4.2. This means that whenever it is valid to perform general MHT, it is valid to perform unified tracking. In addition, Section 5.2.1 of Stone et al.[3] shows that when the assumptions for general MHT hold, MHT produces the same Bayesian posterior on the joint target state space as unified tracking does. Section 5.3.2 of Stone et al.[3] presents an example where the assumptions of unified tracking are satisfied, but those of general MHT are not. This example compares the results of running the general MHT algorithm to that obtained from unified tracking and shows that unified tracking produces superior results. This means that general MHT is a special case of unified tracking.

10.5.2 Relationship of Unified Tracking to Other Multiple-Target Tracking Algorithms

Bethel and Paras,[17] Kamen and Sastry,[18] Kastella,[19-21] Lanterman et al.,[22] Mahler,[23] and Washburn[24] have formulated versions of the multiple-target tracking problem in terms of computing a posterior distribution on the joint target state space. In these formulations the steps of data association and estimation are unified as shown in Section 10.3 of this chapter.

Kamen and Sastry,[18] Kastella,[19] and Washburn[24] assume that the number of targets is known and that the notions of contact and association are meaningful. They have additional restrictive assumptions. Washburn[24] assumes that all measurements take values in the same space. (This assumption appears to preclude sets of sensors that produce disparate types of observations.) Kamen and Sastry[18] and Kastella[19] assume that the measurements are position estimates with Gaussian errors. Kamen and Sastry[18] assume perfect detection capability. Kastella[20] considers a fixed but unknown number of targets. The model in Kastella[20] is limited to identical targets, a single sensor, and discrete time and space. Kastella[21] extends this to targets that are not identical. Bethel and Paras[17] require the notions of contact and association to be meaningful. They also impose a number of special assumptions, such as requiring that contacts be line-of-bearing and assuming that two targets cannot occupy the same cell at the same time.

Mahler's formulation, in Section 3 of Mahler,[23] uses a random set approach in which all measurements take values in the same space with a special topology. Mahler[23] does not provide an explicit method for handling unknown numbers of targets. Lanterman et al.[22] consider only observations that are camera images. They provide formulas for computing posterior distributions only in the case of stationary targets. They discuss the possibility of handling an unknown number of targets but do not provide an explicit procedure for doing so.

In Goodman et al.,[25] Mahler develops an approach to tracking that relies on random sets. The random sets are composed of finite numbers of contacts; therefore, this approach applies only to situations where there are distinguishable sensor responses that can clearly be called out as contacts or detections. In order to use random sets, one must specify a topology and a rather complex measure on the measurement space for the contacts. The approach, presented in Sections 6.1 and 6.2 of Goodman et al.[25] requires that the measurement spaces be identical for all sensors. In contrast, the likelihood function approach presented in Section 10.3 of this chapter, which transforms sensor information into a function on the target state space, is simpler and appears to be more general. For example, likelihood functions and the tracking approach presented Section 10.3 can accommodate situations in which sensor responses are not strong enough to call contacts.

The approach presented in Section 10.3 differs from previous work in the following important aspects:

- The unified tracking model applies when the number of targets is unknown and varies over time.
- Unified tracking applies when the notions of contact and data association are not meaningful.
- Unified tracking applies when the nature (e.g., measurement spaces) of the observations to be fused are disparate. It can correctly combine estimates of position, velocity, range, and bearing as well as frequency observations and signals from sonars, radars, and IR sensors. Unified tracking can fuse any information that can be represented by a likelihood function.
- Unified tracking applies to a richer class of target motion models than are considered in the references cited above. It allows for targets that are not identical. It provides for space-and-time dependent motion models that can represent the movement of troops and vehicles through terrain and submarines and ships though waters near land.

10.5.3 Critique of Unified Tracking

The unified tracking approach to multiple-target tracking has great power and breadth, but it is computationally infeasible for problems involving even moderate numbers of targets. Some shrewd numerical approximation techniques are required to make more general use of this approach.

The approach does appear to be feasible for two targets as explained by Finn.[4] Kock and Van Keuk[26] also consider the problem of two targets and unresolved measurements. Their approach is similar to the unified tracking approach; however, they consider only probability distributions that are mixtures of Gaussian ones. In addition, the target motion model is Gaussian.

A possible approach to dealing with more than two targets is to develop a system that uses a more standard tracking method when targets are well separated and then switches to a unified tracker when targets cross or merge.

10.6 Likelihood Ratio Detection and Tracking

This section describes the problem of detection and tracking when there is, at most, one target present. This problem is most pressing when signal-to-noise ratios are low. This will be the case when performing surveillance of a region of the ocean's surface hoping to detect a periscope in the clutter of ocean waves or when scanning the horizon with an infrared sensor trying to detect a cruise missile at the earliest possible moment. Both of these problems have two important features: (1) a target may or may not be present; and (2) if a target is present, it will not produce a signal strong enough to be detected on a single glimpse by the sensor.

Likelihood ratio detection and tracking is based on an extension of the single-target tracking methodology, presented in Section 10.2, to the case where there is either one or no target present. The methodology presented here unifies detection and tracking into one seamless process. Likelihood ratio detection and tracking allows both functions to be performed simultaneously and optimally.

10.6.1 Basic Definitions and Relations

Using the same basic assumptions as in Section 10.2, we specify a prior on the target's state at time 0 and a Markov process for the target's motion. A set of K observations or measurements $\mathbf{Y}(t) = (Y_1,..,Y_K)$ are obtained in the time interval $[0, t]$. The observations are received at the discrete (possibly random) times $(t_1,...,t_K)$ where $0 < t_1... \leq t_K \leq t$. The measurements obtained at these various times need not be made with the same sensor or even with sensors of the same type; the data from the various observations need not be of the same structure. Some observations may consist of a single number while others may consist of large arrays of numbers, such as the range and azimuth samples of an entire radar scan. However, we do assume that, conditioned on the target's path, the statistics of the observations made at any time by a sensor are *independent* of those made at other times or by other sensors.

The state space in which targets are detected and tracked depends upon the particular problem. Characteristically, the target state is described by a vector, some of whose components refer to the spatial location of the target, some to its velocity, and perhaps some to higher-order properties such as acceleration. These components, as well as others which might be important to the problem at hand, such as target orientation or target strength, can assume continuous values. Other elements that might be part of the state description may assume discrete values. Target class (type) and target configuration (such as periscope extended) are two examples.

As in Section 10.3, the target state space S is augmented with a null state to make $S^+ = S \cup \phi$. There is a probability (density) function, p, defined on S^+, such that $p(\phi) + \int p(s)ds = 1$.

Both the state of the target $X(t) \in S^+$ and the information accumulated for estimating the state probability densities evolve with time t. The process of target detection and tracking consists of computing the posterior version of the function p as new observations are available and propagating it to reflect the temporal evolution implied by target dynamics. Target dynamics include the probability of target motion into and out of S as well as the probabilities of target state changes.

Following the notation used in Section 10.2 for single target Bayesian filtering, let $p(t,s) = \mathbf{Pr}\{X(t) = s | \mathbf{Y}(t) = (Y(t_1),\dots,Y(t_K))\}$ for $s \in S^+$ so that $p(t,\cdot)$ is the posterior distribution on $X(t)$ given all observations received through time t. This section assumes that the conditions that insure the validity of the basic recursion for single-target tracking in Section 10.2 hold, so that $p(t,\cdot)$ can be computed in a recursive manner. Recall that $p^-(t_k, s_k) = \int_{S^+} q(s_k | s_{k-1}) \, p(t_{k-1}, s_{k-1}) \, ds_{k-1}$ for $s_k \in S^+$ is the posterior from time t_{k-1} updated for target motion to time t_k, the time of the kth observation. Recall also the definition of the *likelihood* function L_k. Specifically, for the observation $Y_k = y_k$

$$L_k\left(y_k \,|\, s\right) = \mathbf{Pr}\left\{Y_k = y_k \,|\, X\left(t_k\right) = s\right\} \tag{10.46}$$

where for each $s \in S^+$, $L_k(\cdot|s)$ is a probability (density) function on the measurement space H_k.

According to Bayes' rule,

$$p\left(t_k, s\right) = \frac{p^-\left(t_k, s\right) L_k\left(y_k \,|\, s\right)}{C(k)} \qquad \text{for } s \in S$$

$$\tag{10.47}$$

$$p\left(t_k, \phi\right) = \frac{p^-\left(t_k, \phi\right) L_k\left(y_k \,|\, \phi\right)}{C(k)}$$

In these equations, the denominator is the probability of obtaining the measurement $Y_k = y_k$, that is, $C(k) = p^-(t_k, \phi) \, L_k \, (y_k|\phi) + \int_{s \in S} p^-(t_k,s) L_k(y_k|s) \, ds$.

10.6.1.1   Likelihood Ratio

The ratio of the state probability (density) to the null state probability $p(\phi)$ is defined to be the *likelihood ratio (density)*, $\Lambda(s)$; that is,

$$\Lambda(s) = \frac{p(s)}{p(\phi)} \text{ for } s \in S \tag{10.48}$$

It would be more descriptive to call $\Lambda(s)$ the target likelihood ratio to distinguish it from the measurement likelihood ratio defined below. However, for simplicity, we use the term likelihood ratio for $\Lambda(s)$. The notation for Λ is consistent with that already adopted for the probability densities. Thus, the prior and posterior forms become

$$\Lambda^-(t,s) = \frac{p^-(t,s)}{p^-(t,\phi)} \text{ and } \Lambda(t,s) = \frac{p(t,s)}{p(t,\phi)} \text{ for } s \in S \text{ and } t \geq 0 \tag{10.49}$$

The likelihood ratio density has the same dimensions as the state probability density. Furthermore, from the likelihood ratio density one may easily recover the state probability density as well as the probability of the null state. Since $\int_S L(t,s)\,ds = (1 - p(t,\phi))/p(t,\phi)$, it follows that

$$p(t,s) = \frac{\Lambda(t,s)}{1 + \int_S \Lambda(t,s')ds'} \text{ for } s \in S \quad p(t,\phi) = \frac{1}{1 + \int_S \Lambda(t,s')ds'} \tag{10.50}$$

10.6.1.2 Measurement Likelihood Ratio

The measurement likelihood ratio $\mathcal{L}_k$ for the observation Y_k is defined as

$$\mathcal{L}_k(y|s) = \frac{L_k(y|s)}{L_k(y|\phi)} \text{ for } y \in H_k, s \in S \tag{10.51}$$

$\mathcal{L}_k(y|s)$ is the ratio of the likelihood of receiving the observation $Y_k = y_k$ (given the target is in state s) to the likelihood of receiving $Y_k = y_k$ given no target present. As discussed by Van Trees,[27] the measurement likelihood ratio has long been recognized as part of the prescription for optimal receiver design. This section demonstrates that it plays an even larger role in the overall process of sensor fusion.

Measurement likelihood ratio functions are chosen for each sensor to reflect its salient properties, such as noise characterization and target effects. These functions contain all the sensor information that is required for making optimal Bayesian inferences from sensor measurements.

10.6.2 Likelihood Ratio Recursion

Under the assumptions for which the basic recursion for single-target tracking in Section 10.1 holds, the following recursion for calculating the likelihood ratio holds.

Likelihood Ratio Recursion

Initialize: $\qquad\qquad\qquad\qquad\qquad p(t_0,s) = q_0(s) \text{ for } s \in S^+ \tag{10.52}$

For $k \geq 1$ and $s \in S^+$,

Perform Motion Update: $\qquad\qquad p^-(t_k,s) = \int_{S^+} q_k(s|s_{k-1})p(t_{k-1},s_{k-1})ds_{k-1} \tag{10.53}$

Calculate Likelihood Function: $\quad L_k(y_k|s) = \Pr\{Y_k = y_k | X(t_k) = s\} \tag{10.54}$

Perform Information Update: $p(t_k, s) = \dfrac{1}{C} L_k(y_k \mid s) p^-(t_k, s)$ (10.55)

For $k \geq 1$,

Calculate Likelihood Ratio: $\Lambda(t_k, s) = \dfrac{p(t_k, s)}{p(t_k, \phi)}$ for $s \in S$ (10.56)

The constant, C, in Equation 10.55 is a normalizing factor that makes $p(t_k, \cdot)$ a probability (density) function.

10.6.2.1 Simplified Recursion

The recursion given in Equations 10.52–10.56 requires the computation of the full probability function $p(t_k, \cdot)$ using the basic recursion for single-target tracking discussed in Section 10.2. A simplified version of the likelihood ratio recursion has probability mass flowing from the state ϕ to S and from S to ϕ in such a fashion that

$$p^-(t_k, \phi) = q_k(\phi \mid \phi) p(t_{k-1}, \phi) + \int_S q_k(\phi \mid s) p(t_{k-1}, s) \, ds$$

$$= p(t_{k-1}, \phi)$$

(10.57)

Since

$$p^-(t_k, s_k) = q_k(s_k \mid \phi) p(t_{k-1}, \phi) + \int_S q_k(s_k \mid s) p(t_{k-1}, s) \, ds \ \text{ for } s_k \in S$$

we have

$$\Lambda^-(t_k, s_k) = \frac{q_k(s_k \mid \phi) p(t_{k-1}, \phi) + \int_S q_k(s_k \mid s) p(t_{k-1}, s) \, ds}{p^-(t_k, \phi)}$$

$$= \frac{q_k(s_k \mid \phi) + \int_S q_k(s_k \mid s) \Lambda(t_{k-1}, s) \, ds}{p^-(t_k, \phi) \big/ p(t_{k-1}, \phi)}$$

From Equation 10.57 it follows that

$$\Lambda^-(t_k, s_k) = q_k(s_k \mid \phi) + \int_S q_k(s_k \mid s) \Lambda(t_{k-1}, s) \, ds \ \text{ for } s_k \in S$$ (10.58)

Assuming Equation 10.57 holds, a simplified version of the basic likelihood ratio recursion can be written.

Simplified Likelihood Ratio Recursion

Initialize Likelihood Ratio: $\qquad\qquad\qquad \Lambda(t_0,s) = \dfrac{p(t_0,s)}{p(t_0,\phi)}$ for $s \in S$ $\qquad$ (10.59)

For $k \geq 1$ and $s \in S$,

Perform Motion Update: $\qquad \Lambda^-(t_k,s) = q_k(s|\phi) + \displaystyle\int_S q_k(s|s_{k-1}) \Lambda(t_{k-1}, s_{k-1}) ds_{k-1}$ $\qquad$ (10.60)

Calculate Measurement Likelihood Ratio: $\qquad \mathcal{L}_k(y|s) = \dfrac{L_k(y|s)}{L_k(y|\phi)}$ $\qquad$ (10.61)

Perform Information Update: $\qquad \Lambda(t_k,s) = \mathcal{L}_k(y|s) \Lambda^-(t_k,s)$ $\qquad$ (10.62)

The simplified recursion is a reasonable approximation to problems involving surveillance of a region that may or may not contain a target. Targets may enter and leave this region, but only one target is in the region at a time.

As a special case, consider the situation where no mass moves from state ϕ to S or from S to ϕ under the motion assumptions. In this case $q_k(s|\phi) = 0$ for all $s \in S$, and $p^-(t_k,\phi) = p(t_{k-1},\phi)$ so that Equation 10.60 becomes

$$ \Lambda^-(t_k,s) = \int_S q_k(s|s_{k-1}) \Lambda(t_{k-1}, s_{k-1}) ds_{k-1} \qquad (10.63) $$

10.6.3 Log-Likelihood Ratios

Frequently, it is more convenient to write Equation 10.62 in terms of natural logarithms. Doing so results in quantities that require less numerical range for their representation. Another advantage is that, frequently, the logarithm of the measurement likelihood ratio is a simpler function of the observations than is the actual measurement likelihood ratio itself. For example, when the measurement consists of an array of numbers, the measurement log-likelihood ratio often becomes a linear combination of those data, whereas the measurement likelihood ratio involves a product of powers of the data. In terms of logarithms, Equation 10.62 becomes

$$ \ln \Lambda(t_k,s) = \ln \Lambda^-(t_k,s) + \ln \mathcal{L}_k(y_k|s) \quad \text{for} \quad s \in S \qquad (10.64) $$

The following example is provided to impart an understanding of the practical differences between a formulation in terms of probabilities and a formulation in terms of the logarithm of the likelihood ratios. Suppose there are I discrete target states, corresponding to physical locations so that the target state $X \in \{s_1, s_2, \ldots, s_I\}$ when the target is present. The observation is a vector, $\mathbf{Y}$, that is formed from measurements corresponding to these spatial locations, so that $\mathbf{Y} = (Y(s_1), \ldots, Y(s_I))$, where in the absence of a target in state, s_i, the observation $Y(s_i)$ has a distribution with density function $\eta(\cdot, 0, 1)$, where $\eta(\cdot, \mu, \sigma^2)$ is the density function for a Gaussian distribution with mean μ and variance σ^2. The observations are independent of one another regardless of whether a target is present. When a target is present in the ith state, the mean for $Y(s_i)$ is shifted from 0 to a value r. In order to perform a Bayesian update, the likelihood function for the observation $\mathbf{Y} = \mathbf{y} = (y(s_1),\ldots,y(s_I))$ is computed as follows:

$$L\left(\mathbf{y} \mid s_i\right) = \eta\left(y\left(s_i\right), r, 1\right) \prod_{j \neq i} \eta\left(y\left(s_j\right), 0, 1\right)$$

$$= \exp\left(ry\left(s_i\right) - \frac{1}{2}r^2\right) \prod_{j=1}^{I} \eta\left(y\left(s_j\right), 0, 1\right)$$

Contrast this with the form of the measurement log-likelihood ratio for the same problem. For state i,

$$\ln \mathcal{L}_k\left(\mathbf{y} \mid s_i\right) = ry\left(s_i\right) - \frac{1}{2}r^2$$

Fix s_i and consider $\ln \mathcal{L}_k(\mathbf{Y} \mid s_i)$ as a random variable. That is, consider $\ln \mathcal{L}_k(\mathbf{Y} \mid s_i)$ before making the observation. It has a Gaussian distribution with

$$\mathbf{E}\left[\ln \mathcal{L}\left(\mathbf{Y} \mid s_i\right) \mid X = s_i\right] = +\frac{1}{2}r^2$$

$$\mathbf{E}\left[\ln \mathcal{L}\left(\mathbf{Y} \mid s_i\right) \mid X = \phi\right] = -\frac{1}{2}r^2$$

$$\mathbf{Var}\left[\ln \mathcal{L}\left(\mathbf{Y} \mid s_i\right)\right] = r^2$$

This reveals a characteristic result. *Whereas the likelihood function for any given state requires examination and processing of all the data, the log-likelihood ratio for a given state commonly depends on only a small fraction of the data — frequently only a single datum.* Typically, this will be the case when the observation $\mathbf{Y}$ is a vector of independent observations.

10.6.4   Declaring a Target Present

The likelihood ratio methodology allows the Bayesian posterior probability density to be computed, including the discrete probability that no target resides in S at a given time. It extracts all possible inferential content from the knowledge of the target dynamics, the *a priori* probability structure, and the evidence of the sensors. This probability information may be used in a number of ways to decide whether a target is present. The following offers a number of traditional methods for making this decision, all based on the integrated likelihood ratio. Define

$$p(t,1) = \int_S p(t,s)ds = \mathbf{Pr}\{\text{target present in } S \text{ at time } t\}$$

Then

$$\overline{\Lambda}(t) = p(t,1)/p(t,\phi)$$

is defined to be the *integrated likelihood ratio at time t*. It is the ratio of the probability of the target being present in S to the probability of the target not being present in S at time t.

10.6.4.1 Minimizing Bayes' Risk

To calculate Bayes' risk, costs must be assigned to the possible outcomes related to each decision (e.g., declaring a target present or not). Define the following costs:

$C(1|1)$ if target is declared to be present and it is present
$C(1|\phi)$ if target is declared to be present and it is not present
$C(\phi|1)$ if target is declared to be not present and it is present
$C(\phi|\phi)$ if target is declared to be not present and it is not present

Assume that it is always better to declare the correct state; that is,

$$C(1|1) < C(\phi|1) \text{ and } C(\phi|\phi) < C(1|\phi)$$

The *Bayes' risk* of a decision is defined as the expected cost of making that decision. Specifically the Bayes' risk is

$$p(t,1)C(1|1) + p(t,\phi)C(1|\phi) \text{ for declaring a target present}$$

$$p(t,1)C(\phi|1) + p(t,\phi)C(\phi|\phi) \text{ for declaring a target not present}$$

One procedure for making a decision is to take that action which minimizes the Bayes' risk. Applying this criterion produces the following decision rule. Define the threshold

$$\Lambda_T = \frac{C(1|\phi) - C(\phi|\phi)}{C(\phi|1) - C(1|1)} \tag{10.65}$$

Then declare

$$\text{Target present if } \overline{\Lambda}(t) > \Lambda_T$$

$$\text{Target not present if } \overline{\Lambda}(t) \leq \Lambda_T$$

This demonstrates that the integrated likelihood ratio is a sufficient decision statistic for taking an action to declare a target present or not when the criterion of performance is the minimization of the Bayes' risk.

10.6.4.2 Target Declaration at a Given Confidence Level

Another approach is to declare a target present whenever its probability exceeds a desired confidence level, p_T. The integrated likelihood ratio is a sufficient decision statistic for this criterion as well. The prescription is to declare a target present or not according to whether the integrated likelihood ratio exceeds a threshold, this time given by $\Lambda_T = p_T/(1 - p_T)$.

A special case of this is the *ideal receiver*, which is defined as the decision rule that minimizes the average number of classification errors. Specifically, if $C(1|1) = 0$, $C(\phi|\phi) = 0$, $C(1|\phi) = 1$, and $C(\phi|1) = 1$, then minimizing Bayes' risk is equivalent to minimizing the expected number of miscalls of target present or not present. Using Equation 10.65 this is accomplished by setting $\Lambda_T = 1$, which corresponds to a confidence level of $p_T = \frac{1}{2}$.

10.6.4.3 Neyman-Pearson Criterion for Declaration

Another standard approach in the design of target detectors is to declare targets present according to a rule that produces a specified false alarm rate. Naturally, the target detection probability must still be acceptable at that rate of false alarms. In the ideal case, one computes the distribution of the likelihood ratio with and without the target present and sets the threshold accordingly. Using the Neyman-Pearson approach, a threshold, Λ_T, is identified such that calling a target present when the integrated likelihood ratio is above Λ_T produces the maximum probability of detection subject to the specified constraint on false alarm rate.

10.6.5   Track-Before-Detect

The process of likelihood ratio detection and tracking is often referred to as *track-before-detect*. This terminology recognizes that one is tracking a possible target (through computation of $P(t, \cdot)$) before calling the target present. The advantage of track-before-detect is that it can integrate sensor responses over time on a moving target to yield a detection in cases where the sensor response at any single time period is too low to call a detection. In likelihood ratio detection and tracking, a threshold is set and a detection is called when the likelihood ratio surface exceeds that threshold. The state at which the peak of the threshold crossing occurs is usually taken to be the state estimate, and one can convert the likelihood ratio surface to a probability distribution for the target state.

Section 6.2 of Stone et al.[3] presents an example of performing track-before-detect using the likelihood ratio detection and tracking approach on simulated data. Its performance is compared to a matched filter detector that is applied to the sensor responses at each time period. The example shows that, for a given threshold setting, the likelihood ratio detection methodology produces a 0.93 probability of detection at a specified false alarm rate. In order to obtain that same detection probability with the matched filter detector, one has to suffer a false alarm rate that is higher by a factor of 10^{18}. As another example, Section 1.1.3 of Stone et al.[3] describes the application of likelihood ratio detection and tracking to detecting a periscope with radar.

References

1. Blackman, S.S. and Popoli, R., *Design and Analysis of Modern Tracking Systems*, Artech House Inc., Boston, 1999.
2. Jazwinski, A.H., *Stochastic Processes and Filtering Theory*, Academic Press, New York, 1970.
3. Stone, L.D., Barlow, C.A., and Corwin, T.L., *Bayesian Multiple Target Tracking*, Artech House Inc., Boston, 1999.
4. Finn, M.V., Unified data fusion applied to monopulse tracking, *Proc. IRIS 1999 Nat'l. Symp. Sensor and Data Fusion*, I, 47–61, 1999.
5. Mori, S., Kuo-Chu, C., and Chong, C-Y., Tracking aircraft by acoustic sensors, *Proc. 1987 Am. Control Conf.*, 1099–1105, 1987.
6. Reid, D.B., An algorithm for tracking multiple targets, *IEEE Trans. Automatic Control*, AC-24, 843–854, 1979.
7. Mori, S., Chong, C-Y., Tse, E., and Wishner, R. P., Tracking and classifying multiple targets without *a priori* identification, *IEEE Trans. Automatic Control*, AC-31, 401–409, 1986.
8. Antony, R.T., *Principles of Data Fusion Automation*, Artech House Inc., Boston, 1995.
9. Bar-Shalom, Y. and Fortman, T.E., *Tracking and Data Association*, Academic Press, New York, 1988.
10. Bar-Shalom, Y. and Li, X.L., *Multitarget-Multisensor Tracking: Principles and Techniques*, Published by Yaakov Bar-Shalom, Storrs, CT, 1995.
11. Blackman, S.S., *Multiple Target Tracking with Radar Applications*, Artech House Inc., Boston, 1986.
12. Hall, D.L., *Mathematical Techniques in Multisensor Data Fusion*, Artech House Inc., Boston, 1992.
13. Waltz, E. and Llinas, J., *Multisensor Data Fusion*, Artech House Inc., Boston, 1990.
14. Streit, R.L., *Studies in Probabilistic Multi-Hypothesis Tracking and Related Topics*, Naval Undersea Warfare Center Publication SES-98-101, Newport, RI, 1998.
15. Poore, A.B., Multidimensional assignment formulation of data association problems arising from multitarget and multisensor tracking, *Computational Optimization and Applications*, 3, 27–57, 1994.

16. Yeddanapudi, M., Bar-Shalom, Y., and Pattipati, K.R., IMM estimation for multitarget-multisensor air traffic surveillance, *Proc. IEEE,* 85, 80–94, 1997.

17. Bethel, R.E. and Paras, G.J., A PDF multisensor multitarget tracker, *IEEE Trans. Aerospace and Electronic Systems,* 34, 153–168, 1998.

18. Kamen, E.W. and Sastry, C.R., Multiple target tracking using products of position measurements, *IEEE Trans. Aerospace and Electronics Systems,* 29, 476–493, 1993.

19. Kastella, K., Event-averaged maximum likelihood estimation and mean-field theory in multitarget tracking, *IEEE Trans. Automatic Control,* AC-40, 1070–1074, 1995.

20. Kastella, K, Discrimination gain for sensor management in multitarget detection and tracking, *IEEE-SMC and IMACS Multiconference CESA,* 1–6, 1996.

21. Kastella, K, Joint multitarget probabilities for detection and tracking, *Proc. SPIE, Acquisition, Tracking and Pointing XI,* 3086, 122–128, 1997.

22. Lanterman, A.D., Miller, M.I., Snyder, D.L., and Miceli, W.J., Jump diffusion processes for the automated understanding of FLIR Scenes, *SPIE Proceedings,* 2234, 416–427, 1994.

23. Mahler, R., Global optimal sensor allocation, *Proc. 9th Nat'l. Symp. Sensor Fusion,* 347–366, 1996.

24. Washburn, R.B., A random point process approach to multiobject tracking, *Proc. Am. Control Conf.,* 3, 1846–1852, 1987.

25. Goodman, I.R., Mahler, R.P.S., and Nguyen, H.T., *Mathematics of Data Fusion,* Kluwer Academic Publishers, Boston, 1997.

26. Kock, W. and Van Keuk, G., Multiple hypothesis track maintenance with possibly unresolved measurements, *IEEE Trans. Aerospace and Electronics Systems,* 33, 883–892, 1997.

27. Van Trees, H.L., *Detection, Estimation, and Modulation Theory, Part I: Detection, Estimation, and Linear Modulation Theory,* John Wiley & Sons, New York, 1967.

11

Data Association Using Multiple Frame Assignments

11.1 Introduction 11-1
11.2 Problem Background 11-2
11.3 Assignment Formulation of Some General Data
Association Problems 11-3
11.4 Multiple Frame Track Initiation and Track
Maintenance 11-8
Track Initiation • Track Maintenance Using a Sliding Window
11.5 Algorithms 11-10
Preprocessing • The Lagrangian Relaxation Algorithm for the
Assignment Problem • Algorithm Complexity • Improvement
Methods
11.6 Future Directions 11-14
Other Data Association Problems and Formulations • Frames
of Data • Sliding Windows • Algorithms • Network-
Centric Multiple Frame Assignments
Acknowledgments 11-16
References 11-16

Aubrey B. Poore
Colorado State University

Suihua Lu
Colorado State University

Brian J. Suchomel
Numerica, Inc.

11.1 Introduction

The ever-increasing demand in surveillance is to produce highly accurate target identification and estimation in real time, even for dense target scenarios and in regions of high track contention. Past surveillance sensor systems have relied on individual sensors to solve this problem; however, current and future needs far exceed single sensor capabilities. The use of multiple sensors, through more varied information, has the potential to greatly improve state estimation and track identification. Fusion of information from multiple sensors is part of a much broader subject called data or information fusion, which for surveillance applications is defined as "a multilevel, multifaceted process dealing with the detection, association, correlation, estimation, and combination of data and information from multiple sources to achieve refined state and identity estimation, and complete and timely assessments of situation and threat".[1] (A comprehensive discussion can be found in Waltz and Llinas.)[2] Level 1 deals with single and multisource information involving tracking, correlation, alignment, and association by sampling the external environment with multiple sensors and exploiting other available sources. Numerical processes thus dominate Level 1. Symbolic reasoning involving various techniques from artificial intelligence permeates Levels 2 and 3.

Within Level 1 fusion, architectures for single and multiple platform tracking must also be considered. These are generally delineated into centralized, distributed, and hybrid architectures[3-5] each with its advantages and disadvantages. The architecture most appropriate to the current development is that of a centralized tracking, wherein all measurements are sent to one location and processed with tracks being transmitted back to the different platforms. This architecture is optimal in that it is capable of producing the best track quality (e.g., purity and accuracy) and a consistent air picture.[4,5] Although this architecture is appropriate for single platform tracking, it may be unacceptable for multiple platform tracking for several reasons. For example, communication loading and the single-point-failure problems are important shortcomings. However, this architecture does provide a baseline against which other architectures should be compared. The case of distributed data association is discussed further in Section 11.6.

The methods for centralized tracking follow two different approaches: single and multiple frame processing. The three basic methods in single frame processing are nearest neighbor, joint probabilistic data association (JPDA), and global-nearest neighbor. The nearest neighbor works well when the objects are all far apart and with infrequent clutter. The JPDA works well for a few targets in moderate clutter, but is computationally expensive and may corrupt the target recognition or discrimination information.[6] The global-nearest neighbor approach is posed as a two-dimensional assignment problem (for which there are algorithms that solve the problem optimally in polynomial time) has been successful for cases of moderate target density and light clutter.

Deferred logic techniques consider several data sets or frames of data all at once in making data association decisions. At one extreme is batch processing in which all observations (from all time) are processed together. This method is too computationally intensive for real-time applications. The other extreme is sequential processing. This chapter examines deferred logic methods that fall between these two extremes. The most popular deferred logic method used to track large numbers of targets in low to moderate clutter is called multiple-hypothesis tracking (MHT) in which a tree of possibilities is built, likelihood scores are assigned to each track, an intricate pruning logic is developed, and the data association problem is solved using an explicit enumeration scheme. The use of these enumeration schemes to solve this NP-hard combinatorial optimization problem in real-time is inevitably faulty in dense scenarios, since the time required to solve the problem optimally can grow exponentially with the size of the problem.

Over the last ten years a new formulation and class of algorithms for data association have proven to be superior to all other deferred logic methods.[4,7-14] This formulation is based on multidimensional assignment problems and the algorithms, on Lagrangian relaxation. The use of combinatorial optimization in multitarget tracking is not new; it dates back to the pioneering work of Morefield,[15] who used integer programming to solve set packing and covering problems arising from a data association problem. MHT has been popularized by the fundamental work of Reid.[16] These works are further discussed in Blackman and Popoli,[4] Bar-Shalom and Li,[6] and Waltz and Llinas,[1] all of which also serve as excellent introductions to the field of multitarget tracking and multisensor data fusion. Bar-Shalom, Deb, Kirubarajan, and Pattipati[17-21] have also formulated sensor fusion problems in terms of these multidimensional assignment problems and have developed algorithms as discussed in Section 11.4 of this chapter.

The performance of any tracking system is dependent on a large number of components. Having one component that is superior to all others does not guarantee a superior tracking system. To address some of these other issues, this chapter provides a brief overview of the many issues involved in the design of a tracking system, placing the problem within the context of the more general surveillance and fusion problem in Section 11.2. The formulation of the problem is presented in Section 11.3, and an overview of the Lagrangian relaxation based methods appears in Section 11.4. Section 11.5 contains a summary of some opportunities for future investigation.

11.2 Problem Background

A question that often arises is that of the difference between air traffic control and surveillance. In the former, planes, through their beacon codes, generally identify themselves so that observations can be associated with the correct plane. For the surveillance problem, the objects being tracked do not identify themselves, requiring a figure of merit to be derived for the association of a sequence of observations to a particular target. The term "target" is used rather than "object." This chapter addresses surveillance needs and describes the use of likelihood ratios for track association.

The targets under consideration are classified as point or small targets; measurements or observations of these targets are in the form of kinematic information, such as range, azimuth, elevation, and range rate. Future sensor systems will provide additional feature or attribute information.[4]

A general surveillance problem involves the use of multiple platforms, such as ships, planes, or stationary ground-based radar systems, on which one or more sensors are located for tracking multiple objects. Optimization problems permeate the field of surveillance, particularly in the collection and fusion of information. First, there are the problems of routing and scheduling surveillance platforms and then dynamically retasking the platforms as more information becomes available. For each platform, the scarce sensor resources must be allocated and managed to maximize the information returned. The second area, information fusion, is the subject of this chapter.

Many issues are involved in the design of a fusion system for multiple surveillance platforms, such as fusion architectures, communication links between sensor platforms, misalignment problems, tracking coordinate systems, motion models, likelihood ratios, filtering and estimation, and the data association problem of partitioning reports into tracks and false alarms. The recent book, *Design and Analysis of Modern Tracking Systems*, by Blackman and Popoli[4] presents an excellent overview of these topics and an extensive list of other references.

One aspect of surveillance that is seldom discussed in the literature is the development of data structures required to put all of this information together efficiently. In reality, a tracking system generally is composed of a dynamic search tree that organizes this information and recycles memory for real-time processing. However, the central problem is the data association problem.

To place the current data association problem within the context of the different architectures of multiplatform tracking, a brief review of the architectures is helpful. The first architecture is *centralized fusion*, in which raw observations are sent from the multiple platforms to a central processing unit where they can be combined to give superior state estimation (compared to sensor level fusion).[4] At the other extreme is *track fusion*, wherein each sensor forms tracks along with the corresponding statistical information from its own reports and then sends this preprocessed information to a processing unit that correlates the tracks. Once the correlation is complete, the tracks can be combined and the statistics can be modified appropriately. In reality, many sensor systems are *hybrids* of these two architectures, in which some preprocessed data and some raw data are used and switches between the two are possible. A discussion of the advantages and disadvantages of these architectures is presented in the Blackman and Popoli book.[4] The centralized and hybrid architectures are most applicable to the current data association problem.

11.3 Assignment Formulation of Some General Data Association Problems

The goal of this section is to formulate the data association problem for a large class of multiple-target tracking and sensor fusion applications as a multidimensional assignment problem. This development extracts the salient features and assumptions that occur in a large class of these problems and is a brief update to earlier work.[22] A general class of data association problems was posed as set packing problems by Morefield[15] in 1977. Using an abstracted view of Morefield's work to include set coverings, packings, and partitionings, this section proceeds to formulate the assignment problem.

In tracking, a common surveillance challenge is to estimate the past, current, or future state of a collection of targets (e.g., airplanes in the air, ships on the sea, or automobiles on the ground) from a sequence of scans of the surveillance region by one or more sensors. This work specifically addresses "small" targets[23] for which the sensors generally supply kinematic information such as range, azimuth, elevation, range rate, and some limited attribute or feature information.

Suppose that one or more sensors, either colocated or distributed, survey the surveillance region and produce a stream of observations (or measurements), each with a distinct time tag. These observations are then arranged into sets of observations called *frames of data*. Mathematically, let $Z(k) = \{z_{i_k}^k\}_{i_k=1}^{M_k}$ denote the kth frame of data where each $z_{i_k}^k$ is a vector of noise-contaminated observations with an associated time tag $t_{i_k}^k$. The index k represents the frame number and i_k represents the i_kth observation in frame k. An observation in the frame of data $Z(k)$ may emanate from a true target or may be a false report.

This discussion assumes that each frame of data is a "proper frame," in which each target is seen no more than once. For a rotating radar, one sweep or scan of the field of view generally constitutes a proper frame. For sensors such as electronically scanning phased array radar, wherein the sensor switches from surveillance to tracking mode, the partitioning of the data into proper frames of data is more interesting as there are several choices. More efficient partitioning methods will be addressed in forthcoming work.

The data association problem to be solved is to correctly partition the data into observations emanating from individual targets and false alarms. The combinatorial optimization problem that governs a large number of data association problems in multitarget tracking and multisensor data fusion[1,4,6,7,12-15,18-21,24] is generally posed as

$$\text{Maximize} \left\{ P\left(\Gamma = \gamma \middle| Z^N \right) \middle| \lambda \in \Gamma^* \right\} \tag{11.1}$$

where Z^N represents N data sets (Equation 11.2), γ is a partition of indices of the data (Equations 11.3 and 11.4a-11.4d), Γ^* is the finite collection of all such partitions (Equations 11.4a-11.4d), Γ is a discrete random element defined on Γ^*, γ^0 is a reference partition, and $P(\Gamma = \gamma | Z^N)$ is the posterior probability of a partition γ being true given the data Z^N. Each of these terms must be defined. The objective then becomes formulating a reasonably general class of these data association problems (Equation 11.1) as multidimensional assignment problems (Equation 11.15).

In the surveillance example, the data sets were observations of the objects in the surveillance region, including false reports. Including more general types of data, such as tracks and track-observation combinations, as well as observations, Reid[16] used the term *reports* for the contents of the data sets. Thus, let $Z(k)$ denote a data set of M_k reports $\{z_{i_k}^k\}_{i_k=1}^{M_k}$ and let Z^N denote the cumulative data set of N such sets defined by

$$Z(k) = \left\{ z_{i_k}^k \right\}_{i_k=1}^{M_k} \quad \text{and} \quad Z^N = \left\{ Z(1), ..., Z(N) \right\} \tag{11.2}$$

respectively. In multisensor data fusion and multitarget tracking, the data sets $Z(k)$ may represent different classes of objects. For track initiation in multitarget tracking, the objects are observations that must be partitioned into tracks and false alarms. In this formulation of track maintenance (Section 11.4), one data set is comprised of tracks and the remaining data sets include observations that are assigned to existing tracks, false observations, and observations of new tracks. In sensor level tracking, the objects to be fused are tracks.[1] In centralized fusion,[1,4] the objects can all be observations that represent targets or false reports, and the problem is to determine which observations emanate from a common source.

The next task is to define what is meant by a partition of the cumulative data set Z^N in Equation 11.2. Because this definition is independent of the actual data in the cumulative data set Z^N, a partition of the indices in Z^N must first be defined. Let

$$I^N = \left\{ I(1), I(2), ..., I(N) \right\}, \quad \text{where} \quad I(k) = \left\{ i_k \right\}_{i_k=1}^{M_k} \tag{11.3}$$

denote the indices in the data sets Equation 11.2. A partition γ of I^N and the collection of all such partitions Γ^* is defined by

$$\gamma = \left\{\gamma_1, ..., \gamma_{n(\gamma)} \middle| \gamma_i \neq \emptyset \text{ for each } i\right\} \tag{11.4a}$$

$$\gamma_i \cap \gamma_j = \emptyset \text{ for } i \neq j \tag{11.4b}$$

$$I^N = \cup_{j=1}^{n(\gamma)} \gamma_j \tag{11.4c}$$

$$\Gamma^* = \left\{\gamma \middle| \gamma \text{ satisfies (11.4a)-(11.4b)}\right\} \tag{11.4d}$$

Here, $\gamma_i \subset I^N$ in Equation 11.4a will be called a track, so that $n(\gamma)$ denotes the number of tracks (or elements) in the partition γ. A $\gamma \in \Gamma^*$ is called a *set partitioning* of the indices I^N if the properties in Equations 11.4a–11.4c are valid; a *set covering* of I^N if the property in Equation 11.4b is omitted, but the other two properties Equation 11.4a and Equation 11.4c are retained; and a *set packing* if the property in Equation 11.4c is omitted, but Equations 11.4a and 11.4b are retained.[25] A partition $\gamma \in \Gamma^*$ of the index set I^N induces a partition of the data Z^N via

$$Z_\gamma = \left\{Z_{\gamma_1}, ..., Z_{\gamma_{n(\gamma)}}\right\} \text{ where } Z_{\gamma_i} \subset Z^N \tag{11.5}$$

Clearly, $Z_{\gamma_i} \cap Z_{\gamma_j} = \emptyset$ for $i \neq j$ and $Z^N = \cup_{j=1}^{n(\gamma)} Z_{\gamma_j}$. Each Z_{γ_i} is considered to be a track of data. Note that a Z_{γ_i} need not have observations from each frame of data, $Z(k)$, but it must, by definition, have at least one observation.

Under several independence assumptions between tracks,[22] a probabilistic framework can be established in which

$$P\left(\Gamma = \gamma \middle| Z^N\right) = \frac{C}{p\left(Z^N\right)} \prod_{\gamma_i \in \gamma} p\left(Z_{\gamma_i}\right) G\left(\gamma_i\right) \tag{11.6}$$

where C is a constant and G is a function. This completes the formulation of the general data association problem as presented in the works of Poore[22] and Morefield.[15]

The next objective is to refine this formulation in a way that is amenable to the assignment problem. For notational convenience in representing tracks, add a *zero index* to each of the index sets $I(k)$ $(k = 1, ..., N)$ in Equation 11.3 and a *dummy report* z_0^k to each of the data sets $Z(k)$ in Equation 11.2, and require that each

$$\gamma_i = \left(i_1, ..., i_N\right) \tag{11.7}$$

$$Z_{\gamma_i} = Z_{i_1 ... i_n} \equiv \left(z_{i_1}^1, ..., z_{i_N}^N\right)$$

where i_k and $z_{i_k}^k$ can assume the values of 0 and z_0^k, respectively. The dummy report z_0^k serves several purposes in the representation of missing data, false reports, initiating tracks, and terminating tracks. If Z_{γ_i} is missing an actual report from the data set $Z(k)$, then $\gamma_i = (i_1, ..., i_{k-1}, 0, i_{k+1}, ..., i_N)$ and $Z_{\gamma_i} = \{z_{i_1}^1, ..., z_{i_{k-1}}^{k-1}, z_0^k, z_{i_{k+1}}^{k+1}, ..., z_{i_N}^N\}$. A false report $z_{i_k}^k(i_k > 0)$ is represented by $\gamma_i = (0, ..., 0, i_k, 0, ..., 0)$ and $Z_{\gamma_i} = \{z_0^1, ..., z_0^{k-1}, z_{i_k}^k, z_0^{k+1}, ..., z_0^N\}$ in which there is only one actual report. The partition γ^0 of the data in which all reports are declared to be false reports is defined by

$$Z_{\gamma^0} = \left\{ Z_{0\ldots 0i_k 0\ldots 0} \equiv \left(z_0^1, \ldots, z_0^{k-1}, z_{i_k}^k, z_0^{k+1}, \ldots, z_0^N \right) \Big| i_k = 1, \ldots, M_k; k = 1, \ldots, N \right\} \tag{11.8}$$

If each data set $Z(k)$ represents a "proper frame" of observations, a track that initiates on frame $m > 1$ will contain only the dummy report z_0^k from each of the data sets $Z(k)$ for each $k = 1, \ldots, m - 1$. Likewise, a track that terminates on frame m would have only the dummy report from each of the data sets for $k > m$. These representations are discussed further in Section 11.3 for both track initiation and track maintenance.

The use of the 0–1 variable

$$z_{i_1 \ldots i_N} = \begin{cases} 1 & \text{if } (i_1, \ldots, i_N) \in \gamma \\ 0 & \text{otherwise} \end{cases} \tag{11.9}$$

yields an equivalent characterization of a partition (Equations 11.4a to 11.4d and 11.7) as a solution of the equations

$$\sum_{i_1=0}^{M_1} \cdots \sum_{i_{k-1}=0}^{M_{k-1}} \sum_{i_{k+1}=0}^{M_{k+1}} \cdots \sum_{i_N=0}^{M_N} z_{i_1 \ldots i_N} = 1 \tag{11.10}$$

With this characterization of a partition of the cumulative data set Z^N as a set of equality constraints (Equation 11.10), the multidimensional assignment problem can then be formulated.

Observe that for $\gamma_i = (i_1, \ldots, i_N)$, as in Equation 11.7, and the reference partition (Equation 11.8),

$$\frac{P\left(\Gamma = \gamma | Z^N \right)}{P\left(\Gamma = \gamma^0 | Z^N \right)} \equiv L_\gamma \equiv \prod_{(i_1 \cdots i_N) \in \gamma} L_{i_1 \cdots i_N} \tag{11.11}$$

where

$$L_{i_1 \cdots i_N} = \frac{p\left(Z_{i_1 \cdots i_N} \right) G\left(Z_{i_1 \cdots i_N} \right)}{\prod p\left(Z_{0 \cdots 0 i_k 0 \cdots 0} \right) G\left(Z_{0 \cdots 0 i_k 0 \cdots 0} \right)} \tag{11.12}$$

Here, the index i_k in the denominator corresponds to the kth index of $Z_{i_1 \ldots i_N}$ in the numerator. Next, define

$$c_{i_1 \cdots i_N} = -\ln L_{i_1 \cdots i_N} \tag{11.13}$$

so that

$$-\ln \left[\frac{P\left(\gamma | Z^N \right)}{P\left(\gamma^0 | Z^N \right)} \right] = \sum_{(i_1, \ldots, i_N) \in \gamma} c_{i_1 \cdots i_N} \tag{11.14}$$

Thus, in view of the characterization of a partition (Equations 11.4a to 11.4d and 11.5) specialized by Equation 11.7 as a solution of Equation 11.10, the independence assumptions[22] and the expansion

(Equation 11.6) problem (Equation 11.1) are equivalently characterized as the following N-dimensional assignment problem:

$$\text{Minimize} \sum_{i_1=0}^{M_1} \cdots \sum_{i_N=0}^{M_N} c_{i_1 \cdots i_N} z_{i_1 \cdots i_N} \tag{11.15}$$

$$\text{Subject to} \sum_{i_2=0}^{M_2} \cdots \sum_{i_N=0}^{M_N} z_{i_1 \cdots i_N} = 1, \quad i = 1,\ldots,M_1 \tag{11.16}$$

$$\sum_{i_1=0}^{M_1} \cdots \sum_{i_{k-1}=0}^{M_{k-1}} \sum_{i_{k+1}=0}^{M_{k+1}} \cdots \sum_{i_N=0}^{M_N} z_{i_1 \cdots i_N} = 1 \tag{11.17}$$

for $i_k = 1,\ldots,M_k$ and $k = 2,\ldots,N-1$

$$\sum_{i_1=0}^{M_1} \cdots \sum_{i_{N-1}=0}^{M_{N-1}} z_{i_1 \cdots i_N} = 1, \quad i_N = 1,\ldots,M_N \tag{11.18}$$

$$z_{i_1 \cdots i_N} \in \{0, 1\} \text{ for all } i_1,\ldots,i_N \tag{11.19}$$

where $c_{0 \cdots 0}$ is arbitrarily defined to be zero. Note that the definition of a partition and the 0–1 variable $z_{i_1 \cdots i_N}$ in Equation 11.9 imply $z_{0 \cdots 0} = 0$. (If $z_{0 \cdots 0}$ is not preassigned to zero and $c_{0 \cdots 0}$ is defined arbitrarily, then $z_{0 \cdots 0}$ is determined directly from the value of $c_{0 \cdots 0}$, since it does not enter the constraints other than being a zero-one variable.) Also, each cost coefficient with exactly one nonzero index is zero (i.e., $c_{0 \cdots 0 i_k 0 \cdots 0} = 0$ for all $i_k = 1,\ldots,M_k$ and $k = 1,\ldots,N$) based on the use of the normalizing partition γ^0 in the likelihood ratio in Equations 11.1 and 11.12. Deriving the same problem formulation is possible when not assuming that the cost coefficients with exactly one nonzero index are zero;[22] however, these other formulations can be reduced to the one above using the invariance theorem presented in Section 11.5.1.

Derivation of the assignment problem (Equation 11.15) leads to several pertinent remarks. The definition of a partition in Equations 11.4 and 11.5 implies that each actual report belongs to, at most, one track of reports $Z_{\gamma i}$ in a partition Z_γ of the cumulative data set. This can be modified to allow multiassignments of one, some, or all of the actual reports. The assignment problem changes accordingly. For example, if $z_{i_k}^k$ is to be assigned no more than, exactly, or no less than $n_{i_k}^k$ times, then the "= 1" in the constraint (Equation 11.15) is changed to "$\leq, =, \geq n_{i_k}^k$," respectively. (This allows both set coverings and packings in the formulation.) In making these changes, pay careful attention to the independence assumptions.[22] Inequality constraint problems — in addition to the problem of unresolved closely spaced objects — are common elements of the sensor fusion multiresolution problem.

The likelihood ratio $L_{i_1 \cdots i_N}$ is a complicated expression containing probabilities for detection, termination, model maneuvers, and density functions for an expected number of false alarms and initiating targets. The likelihood that an observation arises from a particular target is also included; this requires target dynamics to be estimated through the corresponding sequence of observations $\{z_{i_1}^1,\ldots,z_{i_k}^k,\ldots,z_{i_N}^N\}$. Filtering such sequences is the most time consuming part of the problem formulation and considerably exceeds the time required to solve the data association problem. Derivations for appropriate likelihood ratios can be found in the work of Poore[22] and Blackman and Popoli.[4]

11.4 Multiple Frame Track Initiation and Track Maintenance

A general expression has been developed for the data association problem arising from tracking. The underlying tracking application is a dynamic one in that information from one or more sensors continually arrives at the processing unit where the data is partitioned into frames of data for the assignment problem. Thus, the dimension of the assignment problem grows with the number of frames of data, N. Processing all of the data at once, called batch processing, eventually becomes computationally unacceptable as the dimension N increases. To circumvent this problem, a sliding window can be used, in which data association decisions are hard prior to the window and soft within the window. The first sliding window (single pane) formulation was presented in 1992[26] and refined in 1997[27] to include a dual pane window. The single pane sliding window and its refinements are described in this section.

The moving window and resulting *search tree* are truly the heart of a tracking system. The importance of the underlying data structures to the efficiency of a tracking system cannot be overemphasized; however, these data structures can be very complex and are not discussed here.

11.4.1 Track Initiation

The pure track initiation problem is to formulate and solve the assignment problem described in the previous section with an appropriate number of frames of data N. The choice of the number of frames N is not trivial. Choices of $N = 4, 5, 6$ have worked well in many problems; however, a good research topic would involve the development of a method that adaptively chooses the number of frames based on the problem complexity.

11.4.2 Track Maintenance Using a Sliding Window

The term *track maintenance* as used in this section includes three functions: (1) extending existing tracks, (2) terminating existing tracks, and (3) initiating new ones. Suppose that the observations on P frames of observations have been partitioned into tracks and false alarms and that K *new* frames of observations are to be added. One approach to solving the resulting data association problem is to formulate it as a track initiation problem with $P + K$ frames. This is the previously mentioned *batch* approach.

The *deferred logic* approach adopted here would treat the track extension problem within the framework of a window sliding over the frames of observations. The P frames are partitioned into two components: the first H frames, in which *hard* data association decisions are made, and the next S frames, in which *soft* decisions are made. The K *new* frames of observations are added, making the number of frames in the sliding window $N = S + K$, while the number of frames in which data association decisions are hard is $H = P - S$. Various sliding windows can be developed including single pane, double pane, and multiple pane windows. The intent of each of these is efficiency in solving the underlying tracking problem.

11.4.2.1 A Single Pane Sliding Window

Assuming $K = 1$, let M_0 denote the number of confirmed tracks (i.e., tracks that arise from the solution of the data association problem) on frame k constructed from a solution of the data association problem utilizing frames up to $k + N - 1$. Data association decisions are fixed on frames up to k. Now a new frame of data is added. Thus, frame k denotes a list of tracks and frames $k + 1$ to $k + N$ denote observations. For $i_0 = 1, \ldots, M_0$, the i_0th such track is denoted by T_{i_0} and the $(N + 1)$-tuple $\{T_{i_0}, z_{i_1}^1, \ldots, z_{i_N}^N\}$ will denote a track T_{i_0} plus a set of observations $\{z_{i_1}^1, \ldots, z_{i_N}^N\}$, actual or dummy, that are feasible with the track T_{i_0}. The $(N + 1)$-tuple $\{T_0, z_{i_1}^1, \ldots, z_{i_N}^N\}$ will denote a track that initiates in the sliding window. A false report in the sliding window is one in which all indices except one are zero in the $(N + 1)$-tuple $\{T_0, z_{i_1}^1, \ldots, z_{i_N}^N\}$.

The hypothesis about a partition $\gamma \in \Gamma^*$ being true is now conditioned on the truth of the M_0 tracks entering the N-scan window. (Thus, the assignments prior to this sliding window are fixed.) The likelihood function is given by $L_\gamma = \prod_{\{T_{i_0}, z_{i_1}, \ldots, z_{i_N}\} \in \gamma} L_{i_0 i_1 \cdots i_N}$, where $L_{i_0 i_1 \cdots i_N} = L_{T_{i_0}} L_{i_1 \cdots i_N}$, $L_{T_{i_0}}$ is the composite likelihood from the discarded frames just prior to the first scan in the window for $i_0 > 0$, $L_{T_0} = 1$, and

$L_{i_1 \cdots i_N}$ is defined as in Equation 11.8 for the N-scan window. ($L_{T_0} = 1$ is used for any tracks that initiate in the sliding window.) Thus, the track extension problem can be formulated as Maximize $\{L_\gamma | \gamma \in \Gamma^*\}$. With the same convention as in Section 11.3, a feasible partition is one that is defined by the properties in Equations 11.4 and 11.7. Analogously, the definition of the zero-one variable

$$
z_{i_0 i_1 \cdots i_N} = \begin{cases} 1 & \text{if } \left\{ T_{i_0}, z_{i_1}^1, \ldots, z_{i_N}^N \right\} \text{ is assigned as a unit} \\ 0 & \text{otherwise} \end{cases}
$$

and the corresponding cost for the assignment of the sequence $\{T_{i_0}, z_{i_1}, \ldots, z_{i_N}\}$ to a track by $c_{i_0 i_1 \cdots i_N} = -\ln L_{i_0 i_1 \cdots i_N}$ yield the following multidimensional assignment formulation of the data association problem for track maintenance:

$$
\text{Minimize} \quad \sum_{i_0=0}^{M_0} \cdots \sum_{i_N=0}^{M_N} c_{i_0 \cdots i_N} z_{i_0 \cdots i_N} \tag{11.20}
$$

$$
\text{Subject to} \quad \sum_{i_1=0}^{M_1} \cdots \sum_{i_N=0}^{M_N} z_{i_0 \cdots i_N} = 1, \quad i_0 = 1, \ldots, M_0 \tag{11.21}
$$

$$
\sum_{i_0=0}^{M_0} \sum_{i_2=0}^{M_2} \cdots \sum_{i_M=0}^{M_N} z_{i_0 i_1 \cdots i_N} = 1, \quad i = 1, \ldots, M_1 \tag{11.22}
$$

$$
\sum_{i_0=0}^{M_0} \cdots \sum_{i_{k-1}=0}^{M_{k-1}} \sum_{i_{k+1}=0}^{M_{k+1}} \cdots \sum_{i_M=0}^{M_N} z_{i_0 \cdots i_N} = 1 \tag{11.23}
$$

$$
\text{for} \quad i_k = 1, \ldots, M_k \text{ and } k = 2, \ldots, N-1
$$

$$
\sum_{i_0=0}^{M_0} \cdots \sum_{i_{N-1}=0}^{M_{N-1}} z_{i_0 \cdots i_N} = 1, \quad i_N = 1, \ldots, M_N \tag{11.24}
$$

$$
z_{i_0 \cdots i_N} \in \left\{ 0, \, 1 \right\} \text{ for all } i_0, \ldots, i_N \tag{11.25}
$$

Note that the association problem involving N frames of observations is an N-dimensional assignment problem for track initiation and an $(N + 1)$-dimensional one for track maintenance.

11.4.2.2 Double and Multiple Pane Window

In the single pane window, the first frame contains a list of tracks and the remaining N frames, observations. The assignment problem is of dimension $N + 1$ where N is the number of frames of observations in front of the existing tracks. The same window is being used to initiate new tracks and continue existing ones. A newer approach is based on the belief that once a track is well established, only a few frames are needed to benefit from the soft decisions on track continuation, while a longer window is needed for track initiation. Thus, if the current frame is numbered k with frames $k + 1, \ldots, k + N$ being for track continuation, one can go back to frames $k - M, \ldots, k$ and allow observations not attached to tracks that exist through frame k to be used to initiate new tracks. Indeed, this concept works extremely well in practice and was initially proposed in the work of Poore and Drummond.[27] The next approach to evolve is that of a multiple pane window, in which the position of hard data association decisions of observations to tracks can vary within the frames $k - 1,\ldots,k,\ldots,k + N$, depending on the difficulty of the problem as measured by track contention. The efficiency of this approach has yet to be determined.

11.5   Algorithms

The multidimensional assignment problem for data association is one of combinatorial optimization and is NP-hard,[28] even for the case $N = 3$. The data association problems arising in tracking are generally sparse, large scale, and noisy with real-time needs. Given the noise in the problem, the objective is generally to solve the problem to within the noise level. Thus, heuristic methods are recommended; however, efficient branch and bound schemes are potentially applicable to this problem, since they provide the baseline against which other heuristic algorithms are judged. Lagrangian relaxation methods have worked extremely well, probably due to the efficiency of nonsmooth optimization methods[29,30] and the fact that the relaxed problem is the two-dimensional assignment problem for which there are some very efficient algorithms, such as the auction[31] and Jonker Volbenant (JV) algorithms.[32] Another advantage is that these relaxation methods provide both a lower and upper bound on the optimal solution and, thus, some measure of closeness to optimality. (This is obviously limited by the duality gap.)

This section surveys some of the algorithms that have been particularly successful in solving the data association problems arising from tracking. There are, however, many potential algorithms that could be used. For example, GRASP (Greedy Randomized Adaptive Local Search Procedure)[33-35] also has been used successfully.

11.5.1   Preprocessing

Two frequently used preprocessing techniques, *fine gating* and *problem decomposition*, are presented in this section. These two methods can substantially reduce the complexity of the multidimensional assignment problem.

11.5.1.1   Fine Gating

The term *fine gating* is used because this method is the last in a sequence of techniques used to reduce the unlikely paths in the layered graph or pairings of combinations of reports. This method is based on the following theorem.[22]

Theorem 1 (Invariance Property)
Let $N > 1$ and $M_k > 0$ for $k = 1,\ldots,N$, and assume $\hat{c}_{0\cdots0} = 0$ and $u_0^k = 0$ for $k = 1,\ldots,N$. Then the minimizing solution and objective function value of the following multidimensional assignment problem are independent of any choice of $u_{i_k}^k$ for $i_k = 1,\ldots,M_k$ and $k = 1,\ldots,N$.

$$\text{Minimize} \quad \sum_{i_1=0}^{M_1} \cdots \sum_{i_N=0}^{M_N} \left(\hat{c}_{i_1\cdots i_N} - \sum_{k=1}^{N} u_{i_k}^k \right) z_{i_0\cdots i_N} + \sum_{k=1}^{N} \sum_{i_k=0}^{M_k} u_{i_k}^k \tag{11.26}$$

$$\text{Subject to} \quad \sum_{i_2=0}^{M_2} \cdots \sum_{i_N=0}^{M_N} z_{i_1\cdots i_N} = 1, \quad i_1 = 1,\ldots,M_1 \tag{11.27}$$

$$\sum_{i_1=0}^{M_1} \cdots \sum_{i_{k-1}=0}^{M_{k-1}} \sum_{i_{k+1}=0}^{M_{k+1}} \cdots \sum_{i_N=0}^{M_N} z_{i_1\cdots i_N} = 1 \tag{11.28}$$

for $i_k = 1,\ldots,M_k$ and $k = 2,\ldots,N-1$

$$\sum_{i_1=0}^{M_1} \cdots \sum_{i_{N-1}=0}^{M_{N-1}} z_{i_1 \cdots i_N} = 1, \quad i_N = 1,\ldots,M_N \tag{11.29}$$

$$z_{i_1 \cdots i_N} \in \{0, 1\} \quad \text{for all} \quad i_1,\ldots,i_N \tag{11.30}$$

If $\hat{c}_{i_1 \cdots i_N} = c_{i_1 \cdots i_N}$ and $u_{i_k}^k c_{0 \cdots 0 i_k 0 \cdots 0}$ is identified in this theorem, then

$$c_{i_1 \cdots i_N} - \sum_{k=1}^{N} c_{0 \cdots 0 i_k 0 \cdots 0} > 0 \tag{11.31}$$

implies that the corresponding zero-one variable $z_{i_1 \cdots i_N}$ and cost $c_{i_1 \cdots i_N}$ can be removed from the problem because a lower cost can be achieved with the use of the variables $z_{0 \cdots 0 i_k 0 \cdots 0} = 1$. (This does not mean that one should set $z_{0 \cdots 0 i_k 0 \cdots 0} = 1$ for $k = 1,\ldots,N$.)

In the special case in which all costs with exactly one nonzero index are zero, this test is equivalent to

$$c_{i_1 \cdots i_N} > 0 \tag{11.32}$$

11.5.1.2 Problem Decomposition

Decomposition of the multidimensional assignment problem into a sequence of disjoint problems can improve the solution quality and the speed of the algorithm, even on a serial machine. The following decomposition method, originally presented in the work of Poore, Rijavec, Barker, and Munger,[36] uses graph theoretic methods.

Decomposition of the multidimensional assignment problem is accomplished by determining the connected components of the associated layered graph. Let

$$\mathcal{Z} = \left\{ z_{i_1 i_2 \cdots i_N} \,\middle|\, z_{i_1 i_2 \cdots i_N} \text{ is not preassigned to zero} \right\} \tag{11.33}$$

denote the set of assignable variables. Define an undirected graph $\mathcal{G}(\mathcal{N}, \mathcal{A})$ where the set of nodes is

$$\mathcal{N} = \left\{ z_{i_k}^k \,\middle|\, i_k = 1,\ldots,M_k; \;\; k = 1,\ldots,N \right\} \tag{11.34}$$

and the set of arcs is

$$\mathcal{A} = \left\{ \left(z_{j_k}^k, z_{j_l}^l \right) \,\middle|\, k \neq 1, \; j_k \neq 0, \; j_l \neq 0 \right.$$
$$\left. \text{and there exists } z_{i_1 i_2 \cdots i_N} \in \mathcal{Z} \text{ such that } j_k = i_k \text{ and } j_l = i_l \right\} \tag{11.35}$$

The nodes corresponding to zero index have not been included in this graph, because two variables that have only the zero index in common can be assigned independently. Connected components of the graph are easily found by constructing a spanning forest via a depth first search. Furthermore, this procedure can be used at each level in the relaxation (i.e., applied to each assignment problem for $k = 3,\ldots,N$). Note that the decomposition algorithm depends only on the problem structure (i.e., the feasibility of the variables) and not on the cost function.

As an aside, this decomposition often yields small problems that are best and more efficiently handled by a branch and bound or an explicit enumeration procedure to avoid the overhead associated with relaxation. The remaining components are solved by relaxation. However, extensive decomposition can be time consuming and limiting the number of components to approximately ten is desirable, unless one is using a parallel machine.

11.5.2   The Lagrangian Relaxation Algorithm for the Assignment Problem

This section presents Lagrangian relaxation algorithms for multidimensional assignment which have proven to be computationally efficient and accurate for tracking purposes. The N dimensional assignment problem has $M_1 + \cdots + M_N$ individual constraints, which can be grouped into N constraint sets. Let $u_k = (u_0^k, u_1^k, \ldots, u_{M_k}^k)$ denote the $M_k + 1$ dimensional Lagrange multiplier vector associated with the kth constraint set, with $u_0^k = 0$ and $k = 1, \ldots, N$. The full set of multipliers is denoted by the vector $u = [u^1, \ldots, u^N]$. The multidimensional assignment problem is relaxed to an n-dimensional assignment problem by incorporating $N - n$ constraint sets into the objective function. There are several choices of n. The case $n = 0$ yields the linear programming dual; $n = 2$ yields a two-dimensional assignment problem and has been highly successful in practice.

Although any constraint sets can be relaxed, sets $n + 1, \ldots, N$ are chosen for convenience. In the tracking problem using a sliding window, these are the correct sets given the data structures that arise from the construction of tracks.

The *relaxed problem* for multiplier vector u is given by

$$L_n(u^{n+1}, \ldots, u^N) = \tag{11.36}$$

$$\text{Minimize} \quad \sum_{i_1=0}^{M_1} \cdots \sum_{i_N=0}^{M_N} \left(c_{i_1 \cdots i_N} + \sum_{k=n+1}^{N} u_{i_k}^k \right) z_{i_1 \cdots i_N} - \sum_{k=n+1}^{N} \sum_{i_k=0}^{M_k} u_{i_k}^k \tag{11.37}$$

$$\text{Subject to:} \quad \sum_{i_2=0}^{M_2} \cdots \sum_{i_N=0}^{M_N} z_{i_1 \cdots i_N} = 1, \quad i_1 = 1, \ldots, M_1 \tag{11.38}$$

$$\sum_{i_1=0}^{M_1} \sum_{i_3=0}^{M_3} \cdots \sum_{i_N=0}^{M_N} z_{i_1 \cdots i_N} = 1, \quad i_2 = 1, \ldots, M_2 \tag{11.39}$$

$$\vdots$$

$$\sum_{i_1=0}^{M_1} \cdots \sum_{i_{n-1}=0}^{M_{n-1}} \sum_{i_{n+1}=0}^{M_{n+1}} \cdots \sum_{i_N=0}^{M_N} z_{i_1 \cdots i_N} = 1, \quad i_n = 1, \ldots, M_n, \tag{11.40}$$

$$z_{i_1 \cdots i_N} \in \{0, 1\} \quad \text{for all} \quad \{i_1, \ldots, i_N\} \tag{11.41}$$

The above problem can be reduced to an n-dimensional assignment problem using the transformation

$$x_{i_1 i_2 \cdots i_n} = \sum_{i_{n+1}=0}^{M_{n+1}} \cdots \sum_{i_N=0}^{M_N} z_{i_1 \cdots i_n i_{n+1} \cdots i_N} \quad \text{for all} \quad i_1, i_2, \ldots, i_n \tag{11.42}$$

$$c_{i_1 i_2 \cdots i_n} = \text{Min}\left\{ c_{i_1 \cdots i_N} + \sum_{k=n+1}^{N} u_{i_k}^u \middle| \text{ for all } i_{n+1}, \ldots, i_N \right\} \tag{11.43}$$

$$c_{0 \cdots 0} = \sum_{i_{n+1}=0}^{M_{n+1}} \cdots \sum_{i_N=0}^{M_N} \text{Min}\left\{ 0, c_{0 \cdots 0 i_{n+1} \cdots i_N} + \sum_{k=n+1}^{N} u_{i_k}^u \right\} \tag{11.44}$$

Thus, the Lagrangian relaxation algorithm can be summarized as follows.

1. Solve the problem

$$\text{Maximize } L_n\left(u^{n+1}, \ldots, u^N \right) \tag{11.45}$$

2. Give an optimal or near optimal solution $(u^{n+1}, \ldots, u^N)$; solve the above assignment problem for this given multiplier vector. This produces an alignment of the first n indices. Let these be enumerated by $\{(i_1^j, \ldots, i_n^j)\}_{j=0}^{J}$ where $(i_1^0, \ldots, i_n^0) = (0, \ldots, 0)$. Then, the variable and cost coefficient

$$x_{j i_{n+1} \cdots i_N} = z_{i_1^j \cdots i_n^j i_{n+1} \cdots i_N} \tag{11.46}$$

$$c_{j i_{n+1} \cdots i_N} = c_{i_1^j \cdots i_n^j i_{n+1} \cdots i_N} \tag{11.47}$$

satisfy the following $N - n + 1$ dimensional assignment problem.

$$\text{Minimize } \sum_{j=0}^{J} \sum_{i_{n+1}=0}^{M_{n+1}} \cdots \sum_{i_N=0}^{M_N} c_{j i_{n+1} \cdots i_N} z_{j i_{n+1} \cdots i_N} \tag{11.48}$$

$$\text{Subject to: } \sum_{i_{n+1}=0}^{M_{n+1}} \cdots \sum_{i_N=0}^{M_N} z_{j i_{n+1} \cdots i_N} = 1, \quad j = 1, \ldots, J \tag{11.49}$$

$$\sum_{j=0}^{J} \sum_{i_{n+2}=0}^{M_3} \cdots \sum_{i_N=0}^{M_N} z_{j i_{n+1} \cdots i_N} = 1, \text{ for } i_{n+1} = 1, \ldots, M_{n+1} \tag{11.50}$$

$$\vdots$$

$$\sum_{j=0}^{J} \cdots \sum_{i_{N-1}=0}^{M_{N-1}} z_{j i_{n+1} \cdots i_N} = 1, \quad i_N = 1, \ldots, M_N \tag{11.51}$$

$$z_{j i_{n+1} \cdots i_N} \in \{0, 1\} \text{ for all } \{j, i_{n+1}, \ldots, i_N\} \tag{11.52}$$

Let the nonzero zero-one variables in a solution be denoted by $z_{j i_{n+1}^j \cdots i_N^j} = 1$, then the solution of the original problem is $z_{i_1^j \cdots i_n^j i_{n+1}^j \cdots i_N^j} = 1$ with all remaining values of z being zero.

In summary, this algorithm is the result of the N-dimensional assignment problem being relaxed to an n-dimensional assignment problem by relaxing $N - n$ constraint sets. The problem of restoring feasibility is defined as an $N - n + 1$ dimensional problem.

Notice that the problem of maximizing $L_n(u^{n+1},\ldots,u^N)$ is one of nonsmooth optimization. Bundle methods[37,38] have proven to be particularly successful for this purpose.

11.5.2.1 A Class of Algorithms

Earlier work[8,12,39,40] involved relaxing an N-dimensional assignment problem to an N-dimensional one, which is NP-hard for $N > 2$, by relaxing one set of constraints. The corresponding dual function $L_n(u^{n+1},\ldots,u^N)$ is piecewise linear and concave, but the evaluation of the function and subgradients, as needed by the nonsmooth maximization, requires an optimal solution of an NP-hard n-dimensional assignment problem when $n > 2$. To address the real-time needs, suboptimal solutions to the relaxed problem must be used; however, suboptimal solutions only provide approximate function and subgradient values. To moderate this difficulty, Poore and Rijavec[40] used a concave, piecewise affine merit function to provide guidance for the function values for the nonsmooth optimization phase. This approach computed approximate subgradients from good quality feasible solutions obtained from multiple relaxation and recovery cycles executed at lower levels in a recursive fashion. (The number of cycles can be any fixed number greater than or equal to one or it can be chosen adaptively by allowing the nonsmooth optimization solver to converge to within user-defined tolerances.) Despite these approximations, the numerical performance of these prior algorithms has been quite good.[40]

A variation on the N-to-$(N-1)$ relaxation algorithm using a one cycle is attributable to Deb, Pattipati, Yeddanapudi, and Bar-Shalom.[17] Similar approximate function and subgradient values are used at each level of the relaxation process. To moderate this difficulty, they modify the accelerated subgradient method of Shor[41] by further weighting the search direction in the direction of violated constraints, and they report improvement over the accelerated subgradient method. They do not, however, use problem decomposition and a merit function as in Poore and Rijavec's previous work.[40]

When relaxing an N-dimensional assignment problem to an n-dimensional one, the one case in which the aforementioned difficulties are resolved is for $n = 2$; this is the algorithm that is currently used in the Poore and Rijavec tracking system.

11.5.3 Algorithm Complexity

In the absence of a complexity analysis, computational experience shows that between 90 and 95 percent of all the computation time is consumed in the search to find the minimizer in the list of feasible arcs that are present in the assignment problem. Thus, the time required to solve an assignment problem appears to be computationally linear in the number of feasible arcs (i.e., tracks) in the assignment problem. Obviously, the gating techniques used in tracking to control the number of feasible tracks are fundamental to managing complexity.

11.5.4 Improvement Methods

The relaxation methods presented above have been enormously successful in providing quality solutions to the assignment problem. Improvement techniques are fundamentally important. Based on the relaxation and a branch and bound framework,[11] significant improvements in the quality of the solution have been achieved for tracking problems. On difficult problems in tracking, the improvement can be significant. Although straight relaxation can produce solutions to within three percent of optimality, the improvement techniques can produce solutions to within one-half (0.5) percent of optimal on at least one class of very difficult tracking problems. Given the ever increasing need for more accurate solutions, further improvements, such as local search methods, should be developed.

11.6 Future Directions

Following the previous sections' overview of the problem formulation and presentation of highly successful algorithms, this section summarizes some of the open issues that should be addressed in the future.

11.6.1 Other Data Association Problems and Formulations

Several alternate formulations — such as more general set packings and coverings of the data — should be pursued. The current formulation is sufficiently general to include other important cases, such as multisensor resolution problems and unresolved closely spaced objects in which an observation is necessarily assigned to more than one target. The assigning of a report to more than one target can be accomplished within the context of the multidimensional assignment problems by using inequality constraints. The formulation, algorithms, and testing have yet to be systematically developed. Although allowing an observation to be assigned to more than one target is easy to model mathematically, allowing a target to be assigned to more than one observation is difficult and may introduce nonlinearity into the objective function due to a loss of the independence assumptions discussed in Section 11.3. Certainly, the original formulation of Morefield[15] is worth revisiting.

11.6.2 Frames of Data

The use of the multidimensional assignment problems to solve the central data association problem rests on the assumption that each target is seen at most once in each frame of data (i.e., the frame is a "proper" frame). For a mechanically rotating radar, this is reasonably easy to approximate as a sweep of the surveillance region. For electronically scanning sensors, which can switch from searching mode to tracking mode, the solution to this partitioning problem is less obvious. Although one such solution has been developed, the formulation and solution are not yet optimal.

11.6.3 Sliding Windows

The batch approach to the data association problem of partitioning the observations into tracks and false alarms is to add a frame of data (observations) to the existing frames and then formulate and solve the corresponding multidimensional assignment problem. The dimension of the assignment problem increases by the number of frames added. To avoid the intractability of this approach, a moving window approach was developed in 1992 wherein the data association problem is resolved over the window of a limited number of frames of data.[26] This was revised in 1996[27] to use different length windows for track continuation (maintenance) and initiation. This approach improves the efficiency of the multiframe processing and maintains the longer window needed for track initiation. Indeed, numerical experiments to date show it to be far more efficient than a single pane window. One can easily imagine using different depths in the window for continuing tracks, depending on the complexity of the problem. The efficiency of this approach in practice has yet to be determined.

A fundamental question relates to determining the dimension of the assignment problem that is most appropriate for a particular tracking problem. The goal of future research will be the development of a method that adapts the dimension to the difficulty of the problem or to the need in the surveillance problem.

11.6.4 Algorithms

Several Lagrangian relaxation methods were outlined in the previous section. The method that has been most successful involves relaxation to a two-dimensional assignment problem, maximization of the resulting relaxed problem with respect to the multipliers, and restoration of feasibility to the original problem by formulating this recovery problem as a multidimensional assignment problem of one dimension lower than the original. The process is then repeated until a two-dimensional assignment problem is reached which can be solved optimally.

Such an algorithm can generally produce solutions that are accurate to within three percent of optimal on very difficult problems and optimal for easy problems. The use of an improvement algorithm based on branch and bound can considerably improve the performance to within one-half (0.5) percent of optimal, at least on some classes of difficult tracking problems.[11] Other techniques, such as local search, should equally improve the solution quality.

Speed enhancements using the decomposition and clustering discussed in the previous section and as presented in 1992[36] can improve the speed of the assignment solvers by an order of magnitude on large-scale and difficult problems. Further work on distributed and parallel computing should enhance both the solution quality and speed.

Another direction of work is the computation of K-near optimal solutions similar to K-best solutions for two dimensional assignment problems.[11] The K-near optimal solutions provide important information about the reliability of the computed tracks which is not available with K-best solutions.

Finally, other approaches to the assignment problem, such as GRASP,[33-35] have also been successful, but not to the extent that relaxation has been.

11.6.5 Network-Centric Multiple Frame Assignments

Multiple platform tracking, like single platform multiple-sensor tracking, also has the potential to significantly improve track estimation by providing geometric diversity and sensor variety. The architecture for data association methods discussed in the previous sections can be applied to multiple platform tracking in a centralized manner, wherein all measurements are sent to one location for processing and then tracks are transmitted back to the different platforms. The centralized architecture is probably optimal in that it is capable of producing the best track quality (e.g., purity and accuracy) and a consistent air picture; however, it is unacceptable in many applications as a result of issues such as communication loading and single-point-failure. Thus, a distributed architecture is needed for both estimation/fusion[3] and data association. While much has been achieved in the area of distributed fusion, few efforts have been extended to distributed, multiple frame, data association.

The objectives for a distributed multiple frame data association approach to multiple platform tracking are to achieve a performance approaching that of the centralized architecture and to achieve a consistent or single integrated air picture (SIAP) across multiple platforms while maintaining communications loads to within a practical limit. Achieving these objectives will require researchers to address a host of problems or topics, including (1) distributed data association and estimation; (2) single integrated air picture; (3) management of communication loading using techniques such as data pruning, data compression (e.g., tracklets,[5] push/request schemes, and target prioritization; (4) network topology of the communication architecture, including the design of new communication architectures and the incorporation of legacy systems; (5) types of information (e.g., measurements, tracks, tracklets) sent across the network; (6) sensor location and registration errors (i.e. "gridlock"); (7) pedigree problems; and, (8) out-of-order, latent, and missing data caused by both sensor and communication problems.

The network-centric algorithm architecture of the Navy's Cooperative Engagement Capability and Joint Composite Tracking Network provides a consistent or single integrated air picture across multiple platforms. This approach limits the communications loads to within a practical limit.[5] This was designed with single frame data association in mind and has not been extended to multiple frame approaches. Thus, the development of a "network multiple frame assignment" approach to data association remains an open and fundamentally important problem.

Acknowledgments

This work was partially supported by the Air Force Office of Scientific Research through AFOSR Grant Numbers F49620-97-1-0273 and F49620-00-1-0108 and by the Office of Naval Research through ONR Grant Number N00014-99-1-0118.

References

1. Waltz, E. and Llinas, J., *Multisensor Data Fusion*, Artech House, Boston, MA, 1990.
2. Hall, D.L., *Mathematical Techniques in Multisensor Data Fusion*, Artech House, Boston, MA, 1992.
3. Drummond, O.E., A hybrid fusion algorithm architecture and tracklets, *Proc. SPIE Conf. Signal and Data Proc. of Small Targets*, Vol. 3136, San Diego, CA, 1997, 485.

4. Blackman, S. and Popoli, R., *Design and Analysis of Modern Tracking Systems*, Artech House Inc., Norwood, MA, 1999.

5. Moore, J.R. and Blair, W.D., Practical aspects of multisensor tracking, *Multitarget-Multisensor Tracking: Applications and Advances III*, Bar-Shalom, Y., and Blair, W.D., Eds., Artech House Inc., Norwood, MA, 2000.

6. Bar-Shalom, Y. and Li X.R., *Multitarget-Multisensor Tracking: Principles and Techniques*, OPAMP Tech. Books, Los Angeles, 1995.

7. Poore, A.B. and Rijavec, N., Multitarget tracking and multidimensional assignment problems, *Proc. 1991 SPIE Conf. Signal and Data Processing of Small Targets*, Vol. 1481, 1991, 345.

8. Poore, A.B. and Rijavec, N., A Lagrangian relaxation algorithm for multi-dimensional assignment problems arising from multitarget tracking, *SIAM J. Optimization*, 3(3), 545, 1993.

9. Poore, A.B. and Robertson, A.J. III, A new class of Lagrangian relaxation based algorithms for a class of multidimensional assignment problems, *Computational Optimization and Applications*, 8(2), 129, 1997.

10. Shea, P.J. and Poore, A.B., Computational experiences with hot starts for a moving window implementation of track maintenance, *Proc. 1998 SPIE Conf.: Signal and Data Processing of Small Targets*, Vol. 3373, 1998.

11. Poore, A.B. and Yan, X., Some algorithmic improvements in multi-frame most probable hypothesis tracking, *Signal and Data Processing of Small Targets*, SPIE, Drummond, O.E., Ed, 1999.

12. Barker, T.N., Persichetti, J.A., Poore, A.B. Jr., and Rijavec, N., Method and system for tracking multiple regional objects, U.S. Patent No. 5,406,289, issued April 11, 1995.

13. Poore, A.B. Jr., Method and system for tracking multiple regional objects by multi-dimensional relaxation, U.S. Patent No. 5,537,119, issued July 16, 1996.

14. Poore, A.B. Jr., Method and system for tracking multiple regional objects by multi-dimensional relaxation, CIP, U.S. Patent No. 5,959,574, issued on September 28, 1999.

15. Morefield, C.L., Application of 0-1 integer programming to multitarget tracking problems, *IEEE Trans. Automatic Control*, 22(3), 302, 1977.

16. Reid, D.B., An algorithm for tracking multiple targets, *IEEE Trans. Automatic Control*, 24(6), 843, 1996.

17. Deb, S., Pattipati, K.R., Bar-Shalom, Y., and Yeddanapudi, M., A generalized s-dimensional assignment algorithm for multisensor multitarget state estimation, *IEEE Trans. Aerospace and Electronic Systems*, 33, 523, 1997.

18. Kirubarajan, T., Bar-Shalom, Y., and Pattipati, K.R., Multiassignment for tracking a large number of overlapping objects, *Proc. SPIE Conf. Signal & Data Proc. of Small Targets*, Vol. 3136, 1997.

19. Kirubarajan, T., Wang, H., Bar-Shalom, Y., and Pattipati, K.R., Efficient multisensor fusion using multidimensional assignment for multitarget tracking, *Proc. SPIE Conf. Signal Processing, Sensor Fusion and Target Recognition*, 1998.

20. Popp, R., Pattipati, K., Bar-Shalom, Y., and Gassner, R., An adaptive m-best assignment algorithm and parallelization for multitarget tracking, *Proc. 1998 IEEE Aerospace Conf.*, Snowmass, CO, 1998.

21. Chummun, M., Kirubarajan, T., Pattipati, K.R., and Bar-Shalom, Y., Efficient multidimensional data association for multisensor-multitarget tracking using clustering and assignment algorithms, *Proc. 2nd Internat'l. Conf. Information Fusion*, 1999.

22. Poore, A.B., Multidimensional assignment formulation of data association problems arising from multitarget tracking and multisensor data fusion, *Computational Optimization and Applications*, 3, 27, 1994.

23. Drummond, O.E., Target tracking, *Wiley Encyclopedia of Electrical and Electronics Engineering*, Vol. 21, John Wiley & Sons, NY, 1999, 377.

24. Sittler, R.W., An optimal data association problem in surveillance theory, *IEEE Trans. Military Electronics*, 8(2), 125, 1964.

25. Nemhauser, G.L. and Wolsey, L.A., *Integer and Combinatorial Optimization*, John Wiley & Sons, NY, 1988.

26. Poore, A.B., Rijavec, N., and Barker, T., Data association for track initiation and extension using multiscan windows, *Proc. SPIE Signal and Data Processing of Small Targets*, Vol. 1698, 1992, 432.

27. Poore, A.B. and Drummond, O.E., Track initiation and maintenance using multidimensional assignment problems, *Lecture Notes in Economics and Mathematical Systems*, Vol. 450, Pardalos, P.M., Hearn, D., and Hager, W., Eds, Springer-Verlag, New York, 1997, 407.

28. Garvey, M. and Johnson, D. *Computers and Intractability: A Guide to the Theory of NP-Completeness*, W.H. Freeman & Co., San Francisco, CA, 1979.

29. Hiriart-Urruty, J.B. and Lemaréchal, C., *Convex Analysis and Minimization Algorithms I*, Springer-Verlag, Berlin, 1993.

30. Hiriart-Urruty, J.-B. and Lemaréchal, C., *Convex Analysis and Minimization Algorithms II*, Springer-Verlag, Berlin, 1993.

31. Bertsekas, D.P., *Network Optimization*, Athena Scientific, Belmont, MA, 1998.

32. Jonker, R. and Volgenant, A., A shortest augmenting path algorithm for dense and sparse linear assignment problems, *Computing*, 38, 325, 1987.

33. Murphey, R., Pardalos, P., and Pitsoulis, L., A GRASP for the multi-target multi-sensor tracking problem, *Networks, Discrete Mathematics and Theoretical Computer Science Series*, Vol. 40, American Mathematical Society, 1998, 277.

34. Murphey, R., Pardalos, P., and Pitsoulis, L., A parallel GRASP for the data association multidimensional assignment problem, *Parallel Processing of Discrete Problems*, IMA Volumes in Mathematics and its Applications, Vol. 106, Springer-Verlag, New York, 1998, 159.

35. Robertson, A., A set of greedy randomized adaptive local search procedure (GRASP) implementations for the multidimensional assignment problem, *Computational Optimization and Applications*, 2001.

36. Poore, A.B., Rijavec, N., Barker, T., and Munger, M., Data association problems posed as multidimensional assignment problems: numerical simulations, *Proc. SPIE: Signal and Data Processing of Small Targets*, Vol. 1954, Drummond, O.E., Ed., 1993, 564.

37. Schramm, H. and Zowe, J., A version of the bundle idea for minimizing a nonsmooth function: Conceptual idea, convergence analysis, numerical results, *SIAM J. Optimization*, 2, 121, 1992.

38. Kiwiel, K.C., Methods of descent for nondifferentiable optimization, *Lecture Notes in Mathematics*, Vol. 1133, Springer-Verlag, Berlin, 1985.

39. Poore, A.B. and Rijavec, N., Multidimensional assignment problems, and Lagrangian relaxation, *Proc. SDI Panels on Tracking*, 2, Institute for Defense Analyses, 1991, 3-51 to 3-74.

40. Poore, A.B. and Rijavec, N., A numerical study of some data association problems arising in multitarget tracking, *Large Scale Optimization: State of the Art*, Hager, W.W., Hearn, D.W., and Pardalos, P.M., Eds., Kluwer Academic Publishers B.V., Boston, MA, 1994, 339.

41. Shor, N.Z., *Minimization Methods for Non-Differentiable Functions*, Springer-Verlag, New York, 1985.

12

General Decentralized Data Fusion with Covariance Intersection (CI)

12.1 Introduction **12-1**
12.2 Decentralized Data Fusion **12-2**
12.3 Covariance Intersection **12-5**
Problem Statement • The Covariance Intersection Algorithm
12.4 Using Covariance Intersection for Distributed Data
Fusion... **12-8**
12.5 Extended Example... **12-10**
12.6 Incorporating Known Independent Information **12-13**
Example Revisited
12.7 Conclusions **12-19**
Acknowledgments... **12-21**
Appendix 12.A   The Consistency of CI **12-21**
Appendix 12.B   MATLAB Source Code............................... **12-23**
Conventional CI • Split CI
References... **12-24**

Simon Julier
IDAK Industries

Jeffrey K. Uhlmann
University of Missouri

12.1 Introduction

One of the most important areas of research in the field of control and estimation is decentralized (or distributed) data fusion. The motivation for decentralization is that it can provide a degree of scalability and robustness that cannot be achieved with traditional centralized architectures. In industrial applications, decentralization offers the possibility of producing plug-and-play systems in which sensors can be slotted in and out to optimize a tradeoff between price and performance. This has significant implications for military systems as well because it can dramatically reduce the time required to incorporate new computational and sensing components into fighter aircraft, ships, and other types of platforms.

The benefits of decentralization are not limited to sensor fusion onboard a single platform; decentralization also can allow a network of platforms to exchange information and coordinate activities in a flexible and scalable fashion that would be impractical or impossible to achieve with a single, monolithic platform. Interplatform information propagation and fusion form the crux of the network centric warfare (NCW) vision for the U.S. military. The goal of NCW is to equip all battlespace entities — aircraft, ships, and even individual human combatants — with communication and computing capabilities to allow each to represent a node in a vast decentralized command and control network. The idea is that each

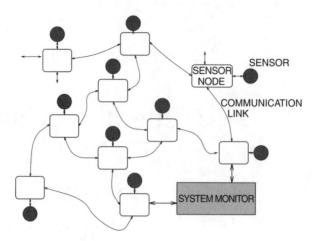

FIGURE 12.1 A distributed data fusion network. Each box represents a fusion node. Each node possesses 0 or more sensors and is connected to its neighboring nodes through a set of communication links.

entity can dynamically establish a communications link with any other entity to obtain the information it needs to perform its warfighting role.

Although the notion of decentralization has had strong intuitive appeal for several decades, achieving its anticipated benefits has proven extremely difficult. Specifically, implementers quickly discovered that if communications paths are not strictly controlled, pieces of information begin to propagate redundantly. When these pieces of information are reused (i.e., double-counted), the fused estimates produced at different nodes in the network become corrupted. Various approaches for avoiding this problem were examined, but none seemed completely satisfactory. In the mid-1990s, the redundant information problem was revealed to be far more than just a practical challenge; it is a manifestation of a fundamental theoretical limitation that could not be surmounted using traditional Bayesian control and estimation methods such as the Kalman filter.[1] In response to this situation, a new data fusion framework, based on covariance intersection (CI), was developed. The CI framework effectively supports all aspects of general decentralized data fusion.

The structure of this chapter is as follows: Section 12.2 describes the decentralized data fusion (DDF) problem. The CI algorithm is described in Section 12.3. Section 12.4 demonstrates how CI supports distributed data fusion and describes one such distribution architecture. A simple example of a network with redundant links is presented in Section 12.5. Section 12.6 shows how to exploit known information about network connectivity and/or information proliferation within the CI framework. This chapter concludes with a brief discussion of other applications of CI.

12.2   Decentralized Data Fusion

A decentralized data fusion system is a collection of processing nodes, connected by communication links (Figure 12.1), in which none of the nodes has knowledge about the overall network topology. Each node performs a specific computing task using information from nodes with which it is linked, but no "central" node exists that controls the network. There are many attractive properties of such decentralized systems,[2] including

- Decentralized systems are reliable in the sense that the loss of a subset of nodes and/or links does not necessarily prevent the rest of the system from functioning. In a centralized system, however, the failure of a common communication manager or a centralized controller can result in immediate catastrophic failure of the system.

- Decentralized systems are flexible in the sense that nodes can be added or deleted by making only local changes to the network. For example, the addition of a node simply involves the establishment of links to one or more nodes in the network. In a centralized system, however, the addition of a new node can change the topology in such a way as to require massive changes to the overall control and communications structure.

The most important class of decentralized networks involves nodes associated with sensors or other information sources. Information from distributed sources propagates through the network so that each node obtains the data relevant to its own processing task. In a battle management application, for example, one node might be associated with the acquisition of information from reconnaissance photographs, another with ground-based reports of troop movements, and another with the monitoring of communications transmissions. Information from these nodes could then be transmitted to a node that estimates the position and movement of enemy troops. The information from this node could then be transmitted back to the reconnaissance photo node, which would use the estimated positions of troops to aid in the interpretation of ambiguous features in satellite photos.

In most applications, the information propagated through a network is converted to a form that provides the estimated state of some quantity of interest. In many cases, especially in industrial applications, the information is converted into means and covariances that can be combined within the framework of Kalman-type filters. A decentralized network for estimating the position of a vehicle, for example, could combine acceleration estimates from nodes measuring wheel speed, from laser gyros, and from pressure sensors on the accelerator pedal. If each independent node provides the mean and variance of its estimate of acceleration, fusing the estimates to obtain a better filtered estimate is relatively easy.

The most serious problem arising in decentralized data fusion networks is the effect of redundant information.[3] Specifically, pieces of information from multiple source cannot be combined within most filtering frameworks unless they are independent or have a known degree of correlation (i.e., known cross covariances). In the battle management example described above, the effect of redundant information can be seen in the following scenario, sometimes referred to as the "whispering in the hall" problem:

1. The photoreconnaissance node transmits information about potentially important features. This information then propagates through the network, changing form as it is combined with information at other nodes in the process.
2. The troop position estimation node eventually receives the information in some form and notes that one of the indicated features could possibly represent a mobilizing tank battalion at position *x*. There are many other possible interpretations of the feature, but the possibility of a mobilizing tank battalion is deemed to be of such tactical importance that it warrants the transmission of a low confidence hypothesis (a "heads up" message). Again, the information can be synopsized, augmented, or otherwise transformed as it is relayed through a sequence of nodes.
3. The photoreconnaissance photo node receives the low confidence hypothesis that a tank battalion may have mobilized at position *x*. A check of available reconnaissance photos covering position *x* reveals a feature that is consistent with the hypothesis. Because the node is unaware that the hypothesis was based on that same photographic evidence, it assumes that the feature that it observes is an independent confirmation of the hypothesis. The node then transmits high confidence information that a feature at position *x* represents a mobilizing tank battalion.
4. The troop position node receives information from the photoreconnaissance node that a mobilizing tank battalion has been identified with high confidence. The troop position node regards this as confirmation of its early hypothesis and calls for an aggressive response to the mobilization. The obvious problem is that the two nodes are exchanging redundant pieces of information but are treating them as independent pieces of evidence mounting in support of the hypothesis that a tank battalion has mobilized. The end result is that critical resources may be diverted in reaction to what is, in fact, a low probability hypothesis.

A similar situation can arise in a decentralized monitoring system for a chemical process:

1. A reaction vessel is fitted with a variety of sensors, including a pressure gauge.
2. Because the bulk temperature of the reaction cannot be measured directly, a node is added that uses pressure information, combined with a model for the reaction, to estimate temperature.
3. A new node is added to the system that uses information from the pressure and temperature nodes.

Clearly, the added node will always be using redundant information from the pressure gauge. If the estimates of pressure and temperature are treated as independent, then the fact that their relationship is always exactly what is predicted by the model might lead to over confidence in the stability of the system. This type of inadvertent use of redundant information arises commonly when attempts are made to decompose systems into functional modules. The following example is typical:

1. A vehicle navigation and control system maintains one Kalman filter for estimating position and a separate Kalman filter for maintaining the orientation of the vehicle.
2. Each filter uses the same sensor information.
3. The full vehicle state is determined (for prediction purposes) by combining the position and orientation estimates.
4. The predicted position covariance is computed essentially as a sum of the position and orientation covariances (after the estimates are transformed to a common vehicle coordinate frame).

The problem in this example is that the position and orientation errors are not independent. This means that the predicted position covariance will underestimate the actual position error. Obviously, such overly confident position estimates can lead to unsafe maneuvers.

To avoid the potentially disastrous consequences of redundant data on Kalman-type estimators, covariance information must be maintained. Unfortunately, maintaining consistent cross covariances in arbitrary decentralized networks is not possible.[1] In only a few special cases, such as tree and fully connected networks, can the proliferation of redundant information be avoided. These special topologies, however, fail to provide the reliability advantage because the failure of a single node or link results in either a disconnected network or one that is no longer able to avoid the effects of redundant information. Intuitively, the redundancy of information in a network is what provides reliability; therefore, if the difficulties with redundant information are avoided by eliminating redundancy, then reliability will be also be eliminated.

The proof that cross covariance information cannot be consistently maintained in general decentralized networks seems to imply that the purported benefits of decentralization are unattainable. However, the proof relies critically on the assumption that some knowledge of the degree of correlation is necessary in order to fuse pieces of information. This is certainly the case for all classical data fusion mechanisms (e.g., the Kalman filter and Bayesian nets), which are based on applications of Bayes' rule. Furthermore, independence assumptions are also implicit in many ad hoc schemes that compute averages over quantities with intrinsically correlated error components.*

The problems associated with assumed independence are often side stepped by artificially increasing the covariance of the combined estimate. This heuristic (or filter "tuning") can prevent the filtering process from producing nonconservative estimates, but substantial empirical analysis and "tweaking" is required to determine how much to increase the covariances. Even with this empirical analysis, the integrity of the Kalman filter framework is compromised, and reliable results cannot be guaranteed. In many applications, such as in large decentralized signal/data fusion networks, the problem is much more acute and no amount of heuristic tweaking can avoid the limitations of the Kalman filter framework.[2] This is of enormous consequence, considering the general trend toward decentralization in complex military and industrial systems.

In summary, the only plausible way to simultaneously achieve robustness, exibility, and consistency in a general decentralized network is to exploit a data fusion mechanism that does not require independence assumptions. Such a mechanism, called Covariance Intersection (CI), satisfies this requirement.

---

*Dubious independence assumptions have permeated the literature over the decades and are now almost taken for granted. The fact is that statistical independence is an extremely rare property. Moreover, concluding that an approach will yield good approximations when "almost independent" is replaced with "assumed independent" in its analysis is usually erroneous.

12.3 Covariance Intersection

12.3.1 Problem Statement

Consider the following problem. Two pieces of information, labeled A and B, are to be fused together to yield an output, C. This is a very general type of data fusion problem. A and B could be two different sensor measurements (e.g., a batch estimation or track initialization problem), or A could be a prediction from a system model, and B could be sensor information (e.g., a recursive estimator similar to a Kalman filter). Both terms are corrupted by measurement noises and modeling errors, therefore, their values are known imprecisely and A and B are the random variables $\mathbf{a}$ and $\mathbf{b}$, respectively. Assume that the true statistics of these variables are unknown. The only available information are estimates of the means and covariances of $\mathbf{a}$ and $\mathbf{b}$ and the cross-correlations between them. These are $\{\mathbf{a}, \mathbf{P}_{aa}\}$, $\{\mathbf{b}, \mathbf{P}_{bb}\}$, and 0, respectively.*

$$\overline{\mathbf{P}}_{aa} = E\left[\tilde{\mathbf{a}}\tilde{\mathbf{a}}^T\right] \quad \overline{\mathbf{P}}_{ab} = E\left[\tilde{\mathbf{a}}\tilde{\mathbf{b}}^T\right] \quad \overline{\mathbf{P}}_{bb} = E\left[\tilde{\mathbf{b}}\tilde{\mathbf{b}}^T\right] \tag{12.1}$$

where $\tilde{\mathbf{a}} \triangleq \mathbf{a} - \overline{\mathbf{a}}$ and $\tilde{\mathbf{b}} \triangleq \mathbf{b} - \overline{\mathbf{b}}$ are the true errors imposed by assuming that the means are $\mathbf{a}$ and $\mathbf{b}$. Note that the cross-correlation matrix between the random variables, $\overline{\mathbf{P}}_{ab}$, is unknown and will not, in general, be 0.

The only constraint that we impose on the assumed estimate is consistency. In other words,

$$\mathbf{P}_{aa} - \overline{\mathbf{P}}_{aa} \geq 0,$$
$$\mathbf{P}_{bb} - \overline{\mathbf{P}}_{bb} \geq 0. \tag{12.2}$$

This definition conforms to the standard definition of consistency.[4] The problem is to fuse the consistent estimates of A and B together to yield a new estimate C, $\{\mathbf{c}, \mathbf{P}_{cc}\}$, which is guaranteed to be consistent:

$$\mathbf{P}_{cc} - \overline{\mathbf{P}}_{cc} \geq 0 \tag{12.3}$$

where $\tilde{\mathbf{c}} \triangleq \mathbf{c} - \overline{\mathbf{c}}$ and $\overline{\mathbf{P}}_{cc} = E\left[\tilde{\mathbf{c}}\tilde{\mathbf{c}}^T\right]$.

12.3.2 The Covariance Intersection Algorithm

In its generic form, the CI algorithm takes a convex combination of mean and covariance estimates that are represented information (inverse covariance) space. The intuition behind this approach arises from a *geometric* interpretation of the Kalman filter equations. The general form of the Kalman filter equation can be written as

$$\overline{\mathbf{c}} = \mathbf{W}_a \overline{\mathbf{a}} + \mathbf{W}_b \overline{\mathbf{b}} \tag{12.4}$$

$$\mathbf{P}_{cc} = \mathbf{W}_a \mathbf{P}_{aa} \mathbf{W}_a^T + \mathbf{W}_a \mathbf{P}_{ab} \mathbf{W}_b^T$$
$$+ \mathbf{W}_b \mathbf{P}_{ba} \mathbf{W}_a^T + \mathbf{W}_b \mathbf{P}_{bb} \mathbf{W}_b^T \tag{12.5}$$

where the weights $\mathbf{W}_a$ and $\mathbf{W}_b$ are chosen to minimize the trace of $\mathbf{P}_{cc}$. This form reduces to the conventional Kalman filter if the estimates are independent ($\mathbf{P}_{ab} = \mathbf{0}$) and generalizes to the Kalman filter with colored noise when the correlations are known.

---

*Cross correlation can also be treated as a nonzero value. For brevity, we do not discuss this case here.

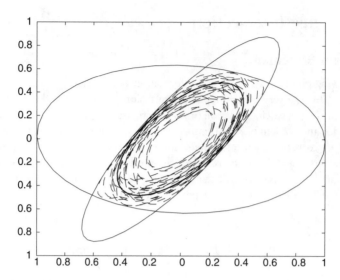

FIGURE 12.2   The shape of the updated covariance ellipse. The variances of $\mathbf{P}_{aa}$ and $\mathbf{P}_{bb}$ are the outer solid ellipses. Different values of $\mathbf{P}_{cc}$ that arise from different choices of $\mathbf{P}_{ab}$ are shown as dashed ellipses. The update with truly independent estimates is the inner solid ellipse.

These equations have a powerful geometric interpretation: If one plots the covariance ellipses (for a covariance matrix $\mathbf{P}$ this is the locus of points $\{\mathbf{p} : \mathbf{p}^T \mathbf{P}^{-1} \mathbf{p} = c\}$ where c is a constant), $\mathbf{P}_{aa}$, $\mathbf{P}_{bb}$, and $\mathbf{P}_{cc}$ for all choices of $\mathbf{P}_{ab}$, $\mathbf{P}_{cc}$ always lies within the *intersection* of $\mathbf{P}_{aa}$ and $\mathbf{P}_{bb}$. Figure 12.2 illustrates this for a number of different choices of $\mathbf{P}_{ab}$.

This interpretation suggests the following approach: if $\mathbf{P}_{cc}$ lies within the intersection of $\mathbf{P}_{aa}$ and $\mathbf{P}_{bb}$ for any possible choice of $\mathbf{P}_{ab}$, then an update strategy that finds a $\mathbf{P}_{cc}$ which encloses the intersection region must be consistent even if there is no knowledge about $\mathbf{P}_{ab}$. The tighter the updated covariance encloses the intersection region, the more effectively the update uses the available information.*

The intersection is characterized by the convex combination of the covariances, and the Covariance Intersection algorithm is:[5]

$$\mathbf{P}_{cc}^{-1} = \omega\mathbf{P}_{aa}^{-1} + \left(1 - \omega\right)\mathbf{P}_{bb}^{-1} \tag{12.6}$$

$$\mathbf{P}_{cc}^{-1}\mathbf{c} = \omega\mathbf{P}_{aa}^{-1}\mathbf{a} + \left(1 - \omega\right)\mathbf{P}_{bb}^{-1}\mathbf{b} \tag{12.7}$$

where $\omega \in [0, 1]$. Appendix 12.A proves that this update equation is consistent in the sense given by Equation 12.3 for all choices of $\mathbf{P}_{ab}$ and ω.

As illustrated in Figure 12.3, the free parameter ω manipulates the weights assigned to **a** and **b**. Different choices of ω can be used to optimize the update with respect to different performance criteria, such as minimizing the trace or the determinant of $\mathbf{P}_{cc}$. Cost functions, which are convex with respect to ω, have only one distinct optimum in the range $0 \leq \omega \leq 1$. Virtually any optimization strategy can be used, ranging from Newton-Raphson to sophisticated semidefinite and convex programming[6] techniques, which can minimize almost any norm. Appendix 12.B includes source code for optimizing ω for the fusion of two estimates.

Note that some measure of covariance size must be minimized at each update in order to guarantee nondivergence; otherwise an updated estimate could be larger than the prior estimate. For example, if

---

*Note that the discussion of "intersection regions" and the plotting of particular covariance contours should not be interpreted in a way that confuses CI with ellipsoidal bounded region filters. CI does not exploit error bounds, only covariance information.

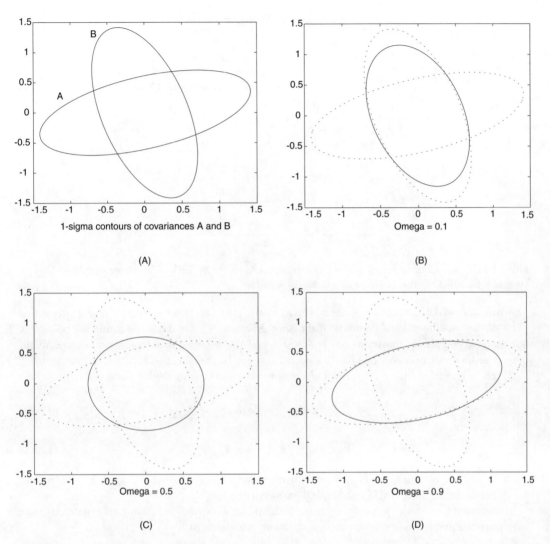

FIGURE 12.3 The value of ω determines the relative weights applied to each information term. (A) Shows the 1-sigma contours for 2-D covariance matrices A and B. (B)–(D) show the updated covariance C (drawn in a solid line) for several different values of ω. For each value of ω, C passes through the intersection points of A and B.

one were always to use $\omega = 0.5$, then the updated estimate would simply be the Kalman updated estimate with the covariance inflated by a factor of two. Thus, an update with an observation that has a very large covariance could result in an updated covariance close to twice the size of the prior estimate. In summary, the use of a fixed measure of covariance size with the CI equations leads to the nondivergent CI filter.

An example of the tightness of the CI update can be seen in Figure 12.4 for the case when the two prior covariances approach singularity:

$$\left\{ \mathbf{a}, \mathbf{A} \right\} = \left\{ \begin{bmatrix} 1 \\ 0 \end{bmatrix}, \begin{bmatrix} 1.5 & 0.0 \\ 0.0 & \varepsilon \end{bmatrix} \right\} \tag{12.8}$$

$$\left\{ \mathbf{b}, \mathbf{B} \right\} = \left\{ \begin{bmatrix} 0 \\ 1 \end{bmatrix}, \begin{bmatrix} \varepsilon & 0.0 \\ 0.0 & 1.0 \end{bmatrix} \right\} \tag{12.9}$$

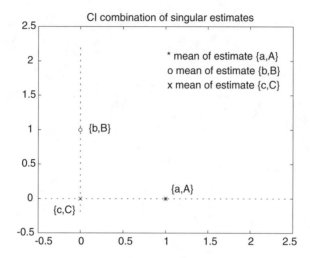

FIGURE 12.4   The CI update {c,C} of two 2-D estimates {a,A} and {b,B}, where A and B are singular, defines the point of intersection of the colinear sigma contours of A and B.

The covariance of the combined estimate is proportional to ε, and the mean is centered on the intersection point of the one-dimensional contours of the prior estimates. This makes sense intuitively because, if one estimate completely constrains one coordinate, and the other estimate completely constrains the other coordinate, there is only one possible update that can be consistent with both constraints.

CI can be generalized to an arbitrary number of $n > 2$ updates using the following equations:

$$\mathbf{P}_{cc}^{-1} = \omega_1 \mathbf{P}_{a_1 a_1}^{-1} + \cdots + \omega_n \mathbf{P}_{a_n a_n}^{-1} \tag{12.10}$$

$$\mathbf{P}_{cc}^{-1} \mathbf{c} = \omega_1 \mathbf{P}_{a_1 a_1}^{-1} \mathbf{a}_1 + \cdots + \omega_n \mathbf{P}_{a_n a_n}^{-1} \mathbf{a}_n \tag{12.11}$$

where $\Sigma_{i=1}^{n}\ \omega_i = 1$. For this type of batch combination of large numbers of estimates, efficient codes, such as the public domain MAXDET[7] and SPDSOL[8] are available.

In summary, CI provides a general update algorithm that is capable of yielding an updated estimate even when the prediction and observation correlations are unknown.

12.4   Using Covariance Intersection for Distributed Data Fusion

Consider again the data fusion network that is illustrated in Figure 12.1. The network consists of N nodes whose connection topology is completely arbitrary (i.e., it might include loops and cycles) and can change dynamically. Each node has information only about its local connection topology (e.g., the number of nodes with which it directly communicates and the type of data sent across each communication link). Assuming that the process and observation noises are independent, the only source of unmodeled correlations is the distributed data fusion system itself. CI can be used to develop a distributed data fusion algorithm which directly exploits this structure. The basic idea is illustrated in Figure 12.5. Estimates that are propagated from other nodes are correlated to an unknown degree and must be fused with the state estimate using CI. Measurements taken locally are known to be independent and can be fused using the Kalman filter equations.

Using conventional notation,[9] the estimate at the ith node is $\hat{\mathbf{x}}_i(k|k)$ with covariance $\mathbf{P}_i(k|k)$. CI can be used to fuse the information that is propagated between the different nodes. Suppose that, at time step $k + 1$, node i locally measures the observation vector $\mathbf{z}_i(k|k)$. A distributed fusion algorithm for propagating the estimate from timestep k to timestep $k + 1$ for node i is:

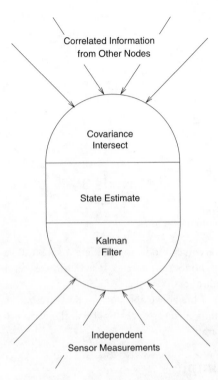

FIGURE 12.5   A canonical node in a general data fusion network that constructs its local state estimate using CI to combine information received from other nodes and a Kalman filter to incorporate independent sensor measurements.

1. Predict the state of node i at time $k + 1$ using the standard Kalman filter prediction equations.
2. Use the Kalman filter update equations to update the prediction with $z_i(k + 1)$. This update is the distributed estimate with mean $\hat{\mathbf{x}}_i^*(k + 1|k + 1)$ and covariance $\mathbf{P}_i^*(k + 1|k + 1)$. It is not the final estimate, because it does not include observations and estimates propagated from the other nodes in the network.
3. Node i propagates its distributed estimate to all of its neighbors.
4. Node i fuses its prediction $\hat{\mathbf{x}}_i(k + 1|k)$ and $\mathbf{P}_i(k + 1|k)$ with the distributed estimates that it has received from all of its neighbors to yield the partial update with mean $\hat{\mathbf{x}}_i^+(k + 1|k + 1)$ and covariance $\mathbf{P}_i^+(k + 1|k + 1)$. Because these estimates are propagated from other nodes whose correlations are unknown, the CI algorithm is used. As explained above, if the node receives multiple estimates for the same time step, the batch form of CI is most efficient. Finally, node i uses the Kalman filter update equations to fuse $z_i(k + 1)$ with its partial update to yield the new estimate $\hat{\mathbf{x}}_i(k + 1|k + 1)$ with covariance $\mathbf{P}_i(k + 1|k + 1)$. The node incorporates its observation last using the Kalman filter equations because it is known to be independent of the prediction or data which has been distributed to the node from its neighbors. Therefore, CI is unnecessary. This concept is illustrated in Figure 12.5.

An implementation of this algorithm is given in the next section. This algorithm has a number of important advantages. First, all nodes propagate their most accurate partial estimates to all other nodes without imposing any unrealistic requirements for perfectly robust communication. Communication paths may be uni- or bidirectional, there may be cycles in the network, and some estimates may be lost while others are propagated redundantly. Second, the update rates of the different filters do not need to be synchronized. Third, communications do not have to be guaranteed — a node can broadcast an estimate without relying on other nodes' receiving it. Finally, each node can use a different observation model: one node may have a high accuracy model for one subset of variables of relevance to it, and

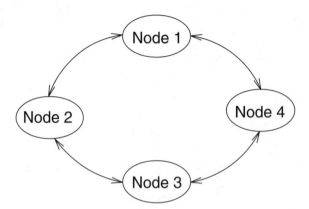

FIGURE 12.6   The network layout for the example.

another node may have a high accuracy model for a different subset of variables, but the propagation of their respective estimates allows nodes to construct fused estimates representing the union of the high accuracy information from both nodes.

The most important feature of the above approach to decentralized data fusion is that it is provably guaranteed to produce and maintain consistent estimates at the various nodes.* Section 5 demonstrates this consistency in a simple example.

12.5   Extended Example

Suppose the processing network, shown in Figure 12.6, is used to track the position, velocity and acceleration of a one-dimensional particle. The network is composed of four nodes. Node 1 measures the position of the particle only. Nodes 2 and 4 measure velocity and node 3 measures acceleration. The four nodes are arranged in a ring. From a practical standpoint, this configuration leads to a robust system with built-in redundancy: data can flow from one node to another through two different pathways. However, from a theoretical point of view, this configuration is extremely challenging. Because this configuration is neither fully connected nor tree-connected, optimal data fusion algorithms exist only in the special case where full knowledge of the network topology and the states at each node is known.

The particle moves using a nominal constant acceleration model with process noise injected into the jerk (derivative of acceleration). Assuming that the noise is sampled at the start of the timestep and is held constant throughout the prediction step, the process model is

$$X_{(k+1)} = \mathbf{F}x_{(k)} + \mathbf{G}v_{(K+1)} \tag{12.12}$$

where

$$\mathbf{F} = \begin{bmatrix} 1 & \Delta T & \Delta T^2/2 \\ 0 & 1 & \Delta T \\ 0 & 0 & 1 \end{bmatrix} \text{ and } \mathbf{G} = \begin{bmatrix} \Delta T^3/6 \\ \Delta T^2/2 \\ \Delta T \end{bmatrix}$$

---

*The fundamental feature of CI can be described as consistent estimates in, consistent estimates out. The Kalman filter, in contrast, can produce an inconsistent fused estimate from two consistent estimates if the assumption of independence is violated. The only way CI can yield an inconsistent estimate is if a sensor or model introduces an inconsistent estimate into the fusion process. In practice this means that some sort of fault-detection mechanism needs to be associated with potentially faulty sensors.

$v(k)$ is an uncorrelated, zero-mean Gaussian noise with variance $\sigma_v^2 = 10$ and the length of the time step $\Delta T = 0.1s$.

The sensor information and the accuracy of each sensor is given in Table 12.1.

Assume, for the sake of simplicity, that the structure of the state space and the process models are the same for each node and the same as the true system. However, this condition is not particularly restrictive and many of the techniques of model and system distribution that are used in optimal data distribution networks can be applied with CI.[10]

The state at each node is predicted using the process model:

TABLE 12.1 Sensor Information and Accuracy for Each Node from Figure 12.6

Node	Measures	Variance
1	x	1
2	$\dot{x}$	2
3	$\ddot{x}$	0.25
4	$\dot{x}$	3

$$\hat{\mathbf{x}}_i\left(k+1\big|k\right) = \mathbf{F}\hat{\mathbf{x}}_i\left(k\big|k\right)$$

$$\mathbf{P}_i\left(k+1\big|k\right) = \mathbf{F}\mathbf{P}_i\left(k+1\big|k\right)\mathbf{F}^T + \mathbf{Q}\left(k\right)$$

The partial estimates $\hat{\mathbf{x}}_i^*\,(k + 1|k + 1)$ and $\mathbf{P}_i^*\,(k + 1|k + 1)$ are calculated using the Kalman filter update equations. If R_i is the observation noise covariance on the ith sensor, and H_i is the observation matrix, then the partial estimates are

$$v_i\left(k+1\right) = \mathbf{z}_i\left(k+1\right) - \mathbf{H}_i\hat{\mathbf{x}}_i\left(k+1\big|k\right) \tag{12.13}$$

$$\mathbf{S}_i\left(k+1\right) = \mathbf{H}_i\mathbf{P}_i\left(k+1\big|k\right)\mathbf{H}_i^T + \mathbf{R}_i\left(k+1\right) \tag{12.14}$$

$$\mathbf{W}_i\left(k+1\right) = \mathbf{P}_i\left(k+1\big|k\right)\mathbf{H}_i^T\mathbf{S}_i^{-1}\left(k+1\right) \tag{12.15}$$

$$\hat{\mathbf{x}}_i^*\left(k+1\big|k+1\right) = \hat{\mathbf{x}}_i\left(k+1\big|k\right) + \mathbf{W}_i\left(k+1\right)v_i\left(k+1\right) \tag{12.16}$$

$$\mathbf{P}_i^*\left(k+1\big|k+1\right) = \mathbf{P}_i\left(k+1\big|k\right) - \mathbf{W}_i\left(k+1\right)\mathbf{S}_i\left(k+1\right)\mathbf{W}_i^T\left(k+1\right) \tag{12.17}$$

Examine three strategies for combining the information from the other nodes:

1. The nodes are disconnected. No information flows between the nodes and the final updates are given by

$$\hat{\mathbf{x}}_i\left(k+1\big|k+1\right) = \hat{\mathbf{x}}_i^*\left(k+1\big|k+1\right) \tag{12.18}$$

$$\mathbf{P}_i\left(k+1\big|k+1\right) = \mathbf{P}_i^*\left(k+1\big|k+1\right) \tag{12.19}$$

2. Assumed independence update. All nodes are assumed to operate independently of one another. Under this assumption, the Kalman filter update equations can be used in Step 4 of the fusion strategy described in the last section.
3. CI-based update. The update scheme described in Section 12.4 is used.

The performance of each of these strategies was assessed using a Monte Carlo of 100 runs.

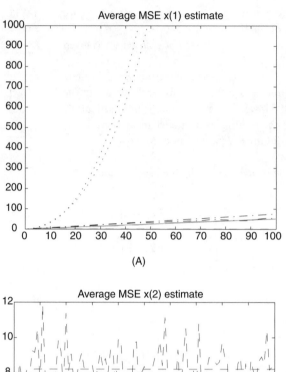

FIGURE 12.7 Disconnected nodes. (A) Mean squared error in x. (B) Mean squared error in ẋ. (C) Mean squared error in ẍ. Mean squared errors and estimated covariances for all states in each of the four nodes. The curves for Node 1 are solid, Node 2 are dashed, Node 3 are dotted, and Node 4 are dash-dotted. The mean squared error is the rougher of the two lines for each node.

The results from the first strategy (no data distribution) are shown in Figure 12.7. As expected, the system behaves poorly. Because each node operates in isolation, only Node 1 (which measures x) is fully observable. The position variance increases without bound for the three remaining nodes. Similarly, the velocity is observable for Nodes 1, 2, and 4, but it is not observable for Node 3.

The results of the second strategy (all nodes are assumed independent) are shown in Figure 12.8. The effect of assumed independence observations is obvious: all of the estimates for all of the states in all of the nodes (apart from x for Node 3) are inconsistent. This clearly illustrates the problem of double counting.

Finally, the results from the CI distribution scheme are shown in Figure 12.9. Unlike the other two approaches, all the nodes are consistent and observable. Furthermore, as the results in Table 12.2 indicate, the steady-state covariances of all of the states in all of the nodes are smaller than those for case 1. In other words, this example shows that this data distribution scheme successfully and usefully propagates data through an apparently degenerate data network.

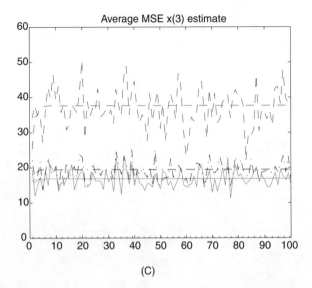

(C)

FIGURE 12.7   (continued).

This simple example is intended only to demonstrate the effects of redundancy in a general data distribution network. CI is not limited in its applicability to linear, time invariant systems. Furthermore, the statistics of the noise sources do not have to be unbiased and Gaussian. Rather, they only need to obey the consistency assumptions. Extensive experiments have shown that CI can be used with large numbers of platforms with nonlinear dynamics, nonlinear sensor models, and continuously changing network topologies (i.e., dynamic communications links).[11]

12.6   Incorporating Known Independent Information

CI and the Kalman filter are diametrically opposite in their treatment of covariance information: CI conservatively assumes that no estimate provides statistically independent information, and the Kalman filter assumes that every estimate provides statistically independent information. However, neither of these two extremes is representative of typical data fusion applications. This section demonstrates how the CI framework can be extended to subsume the generic CI filter and the Kalman filter and provide a completely general and optimal solution to the problem of maintaining and fusing consistent mean and covariance estimates.[22]

The following equation provides a useful interpretation of the original CI result. Specifically, the estimates {**a, A**} and {**b, B**} are represented in terms of their joint covariance:

$$\left\{ \begin{bmatrix} \mathbf{a} \\ \mathbf{b} \end{bmatrix}, \begin{bmatrix} \mathbf{A} & \mathbf{P}_{ab} \\ \mathbf{P}_{ab}^{T} & \mathbf{B} \end{bmatrix} \right\} \tag{12.20}$$

where in most situations the cross covariance, $\mathbf{P}_{ab}$, is unknown. The CI equations, however, support the conclusion that

$$\begin{bmatrix} \mathbf{A} & \mathbf{P}_{ab} \\ \mathbf{P}_{ab}^{T} & \mathbf{B} \end{bmatrix} \leq \begin{bmatrix} \omega \mathbf{A}^{-1} & 0 \\ 0 & (1-\omega)\mathbf{B}^{-1} \end{bmatrix}^{-1} \tag{12.21}$$

because CI must assume a joint covariance that is conservative with respect to the true joint covariance. Evaluating the inverse of the right-hand-side (RHS) of the equation leads to the following consistent/conservative estimate for the joint system:

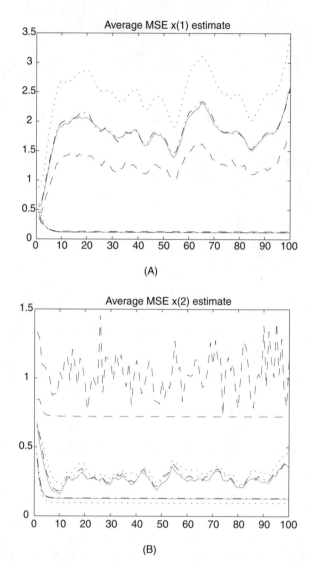

FIGURE 12.8   All nodes assumed independent. (A) Mean squared error in x. (B) Mean squared error in $\dot{x}$. (C) Mean squared error in $\ddot{x}$. Mean squared errors and estimated covariances for all states in each of the four nodes. The curves for Node 1 are solid, Node 2 are dashed, Node 3 are dotted, and Node 4 are dash-dotted. The mean squared error is the rougher of the two lines for each node.

$$\left\{ \begin{bmatrix} \mathbf{a} \\ \mathbf{b} \end{bmatrix}, \begin{bmatrix} \frac{1}{\omega}\mathbf{A} & 0 \\ 0 & \frac{1}{1-\omega}\mathbf{B} \end{bmatrix} \right\} \tag{12.22}$$

From this result, the following generalization of CI can be derived:*

CI with Independent Error: Let $a = a_1 + a_2$ and $b = b_1 + b_2$, where a_1 and b_1 are correlated to an unknown degree, while the errors associated with a_2 and b_2 are completely independent of all others.

---

*In the process, a consistent estimate of the covariance of $\mathbf{a} + \mathbf{b}$ is also obtained, where a and b have an unknown degree of correlation, as $\frac{1}{\omega}\mathbf{A} + \frac{1}{1-\omega}\mathbf{B}$. We refer to this operation as *covariance addition* (CA).

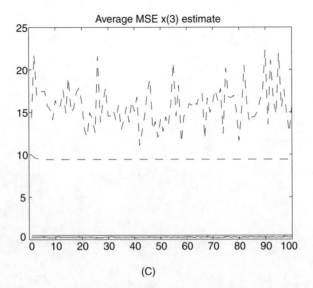

FIGURE 12.8 (continued).

Also, let the respective covariances of the components be A_1, A_2, B_1, and B_2. From the above results, a consistent joint system can be formed as:

$$\left\{ \begin{bmatrix} a_1 + a_2 \\ b_1 + b_2 \end{bmatrix}, \begin{bmatrix} \frac{1}{\omega} A_1 + A_2 & 0 \\ 0 & \frac{1}{1-\omega} B_1 + B_2 \end{bmatrix} \right\} \tag{12.23}$$

Letting $A = \frac{1}{\omega} A_1 + A_2$ and $B = \frac{1}{1-\omega} B_1 + B_2$, gives the following generalized CI equations:

$$C = \left[A^{-1} + B^{-1} \right]^{-1} = \left[\left(\tfrac{1}{\omega} A_1 + A_2 \right)^{-1} + \left(\tfrac{1}{1-\omega} B_1 + B_2 \right)^{-1} \right]^{-1} \tag{12.24}$$

$$c = \left[A^{-1} a + B^{-1} b \right]^{-1} = C \left[\left(\tfrac{1}{\omega} A_1 + A_2 \right)^{-1} a + \left(\tfrac{1}{1-\omega} B_1 + B_2 \right)^{-1} b \right] \tag{12.25}$$

where the known independence of the errors associated with a_2 and b_2 is exploited.

Although the above generalization of CI exploits available knowledge about independent error components, further exploitation is impossible because the combined covariance C is formed from *both* independent and correlated error components. However, CI can be generalized even further to produce and maintain separate covariance components, C_1 and C_2, reflecting the correlated and known-independent error components, respectively. This generalization is referred to as Split CI.

If we let $\tilde{a}_1$ and $\tilde{a}_2$ be the correlated and known-independent error components of a, with $\tilde{b}_1$ and $\tilde{b}_2$ similarly defined for b, then we can express the errors $\tilde{c}_1$ and $\tilde{c}_2$ in information (inverse covariance) form as

$$C^{-1} \left(\tilde{c}_1 + \tilde{c}_2 \right) = A^{-1} \left(\tilde{a}_1 + \tilde{a}_2 \right) + B^{-1} \left(\tilde{b}_1 + \tilde{b}_2 \right) \tag{12.26}$$

from which the following can be obtained after premultiplying by C:

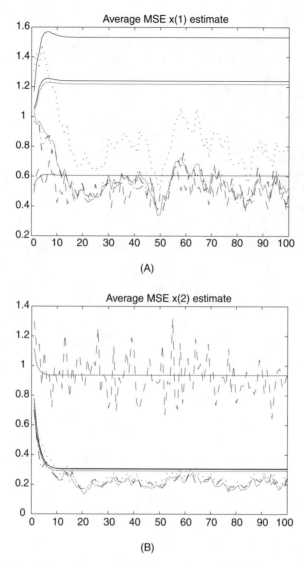

FIGURE 12.9 CI distribution scheme. (A) Mean squared error in x. (B) Mean squared error in $\dot{x}$. (C) Mean squared error in $\ddot{x}$. Mean squared errors and estimated covariances for all states in each of the four nodes. The curves for Node 1 are solid, Node 2 are dashed, Node 3 are dotted, and Node 4 are dash-dotted. The mean squared error is the rougher of the two lines for each node.

$$\left(\tilde{\mathbf{c}}_1 + \tilde{\mathbf{c}}_2\right) = \mathbf{C}\left[\mathbf{A}^{-1}\left(\tilde{\mathbf{a}}_1 + \tilde{\mathbf{a}}_2\right) + \mathbf{B}^{-1}\left(\tilde{\mathbf{b}}_1 + \tilde{\mathbf{b}}_2\right)\right] \tag{12.27}$$

Squaring both sides, taking expectations, and collecting independent terms* yields:

$$\mathbf{C}_2 = \left(\mathbf{A}^{-1} + \mathbf{B}^{-1}\right)^{-1}\left(\mathbf{A}^{-1}\mathbf{A}_2\mathbf{A}^{-1} + \mathbf{B}^{-1}\mathbf{B}_2\mathbf{B}^{-1}\right)\left(\mathbf{A}^{-1} + \mathbf{B}^{-1}\right)^{-1} \tag{12.28}$$

---

*Recall that $\mathbf{A} = \frac{1}{\omega}\mathbf{A}_1$ and $\mathbf{B} = \frac{1}{1-\omega}\mathbf{B}_1 + \mathbf{B}_2$.

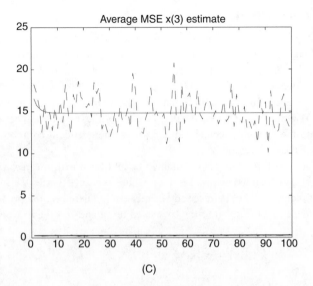

FIGURE 12.9 (continued).

TABLE 12.2 The Diagonal Elements of the
Covariance Matrices for Each Node at the End
of 100 Timesteps for Each of the Consistent
Distribution Schemes

Node	Scheme	σ_x^2	σ_x^2	q_x^2
1	NONE	0.8823	8.2081	37.6911
	CI	0.6055	0.9359	14.823
2	NONE	50.5716*	1.6750	16.8829
	CI	1.2186	0.2914	0.2945
3	NONE	77852.3*	7.2649*	0.2476
	CI	1.5325	0.3033	0.2457
4	NONE	75.207	2.4248	19.473
	CI	1.2395	0.3063	0.2952

Note: NONE – no distribution, and CI – the CI
algorithm). The asterisk denotes that a state is *unob-
servable* and its variance is increasing without bound.

where the nonindependent part can be obtained simply by subtracting the above result from the overall
fused covariance $\mathbf{C} = (\mathbf{A}^{-1} + \mathbf{B}^{-1})^{-1}$. In other words,

$$\mathbf{C}_1 = \left(\mathbf{A}^{-1} + \mathbf{B}^{-1}\right)^{-1} - \mathbf{C}_2 \qquad (12.29)$$

Split CI can also be expressed in batch form analogously to the batch form of original CI. Note that the
covariance addition equation can be generalized analogously to provide Split CA capabilities.

The generalized and split variants of CI optimally exploit knowledge of statistical independence. This
provides an extremely general filtering, control, and data fusion framework that completely subsumes
the Kalman filter.

12.6.1 Example Revisited

The contribution of generalized CI can be demonstrated by revisiting the example described in Section 12.5. The scheme described earlier attempted to exploit information that is independent in the observations. However, it failed to exploit one potentially very valuable source of information — the fact that the distributed estimates ($\hat{\mathbf{x}}_i^*$ $(k+1|k+1)$ with covariance $\mathbf{P}_i^*$ $(k+1|k+1)$) contain the observations taken at time step $k+1$. Under the assumption that the measurement errors are uncorrelated, generalized CI can be exploited to significantly improve the performance of the information network. The distributed estimates are split into the (possibly) correlated and known independent components, and generalized CI can be used to fuse the data remotely.

The estimate of node i at time step k is maintained in split form with mean $\hat{\mathbf{x}}_i$ $(k|k)$ and covariances $\mathbf{P}_{i,1}$ $(k|k)$ and $\mathbf{P}_{i,2}$ $(k|k)$. As explained below, it is not possible to ensure that $\mathbf{P}_{i,2}$ $(k|k)$ will be independent of the distributed estimates that will be received at time step k. Therefore, the prediction step combines the correlated and independent terms into the correlated term, and sets the independent term to 0:

$$\hat{\mathbf{x}}_i\left(k+1\big|k\right)=\mathbf{F}\hat{\mathbf{x}}_i\left(k\big|k\right)$$

$$\mathbf{P}_{i,1}\left(k+1\big|k\right)=\mathbf{F}\left(\mathbf{P}_{i,1}\left(k+1\big|k\right)+\mathbf{P}_{i,2}\left(k+1\big|k\right)\right)\mathbf{F}^T+\mathbf{Q}\left(k\right) \qquad (12.30)$$

$$\mathbf{P}_{i,2}\left(k+1\big|k\right)=0$$

The process noise is treated as a correlated noise component because each sensing node is tracking the same object. Therefore, the process noise that acts on each node is perfectly correlated with the process noise acting on all other nodes.

The split form of the distributed estimate is found by applying split CI to fuse the prediction with $\mathbf{z}_i$ $(k + 1)$. Because the prediction contains only correlated terms, and the observation contains only independent terms ($\mathbf{A}_2 = 0$ and $\mathbf{B}_1 = 0$ in Equation 12.24) the optimized solution for this update occurs when $\omega = 1$. This is the same as calculating the normal Kalman filter update and explicitly partitioining the contributions of the predictions from the observations. Let $\mathbf{W}_i^*(k + 1)$ be the weight used to calculate the distributed estimate. From Equation 12.30 its value is given by,

$$\mathbf{S}_i^*\left(k+1\right)=\mathbf{H}_i\mathbf{P}_{i,1}\left(k+1\big|k\right)\mathbf{H}_i^T+\mathbf{R}_i\left(k+1\right) \qquad (12.31)$$

$$\mathbf{W}_i^*\left(k+1\right)=\mathbf{P}_{i,1}\left(k+1\big|k\right)\mathbf{H}_i^T\,\mathbf{S}_i^*\left(k+1\right)^{-1} \qquad (12.32)$$

Note that the Covariance Addition equation can be generalized analogously to provide Split CA capabilities.

Taking outer products of the prediction and observation contribution terms, the correlated and independent terms of the distributed estimate are

$$\mathbf{P}_{i,1}^*\left(k+1\big|k+1\right)=\mathbf{X}\left(k\right)+\mathbf{1}\mathbf{P}_i\left(k+1\big|k\right)\mathbf{X}^T\left(k+1\right)$$

$$\mathbf{P}_{i,2}^*\left(k+1\big|k+1\right)=\mathbf{W}\left(k+1\right)+\mathbf{1}\mathbf{R}\left(k\right)+\mathbf{1}\mathbf{W}^T\left(k+1\right) \qquad (12.33)$$

where $\mathbf{X}(k + 1) = \mathbf{I} - \mathbf{W}_i^*(k + 1)\mathbf{H}(k + 1)$.

The split distributed updates are propagated to all other nodes where they are fused with split CI to yield a split partial estimate with mean $\hat{\mathbf{x}}_i^+$ $(k + 1|k + 1)$ and covariances $\mathbf{P}_{i,1}^+$ $(k + 1|k + 1)$ and $\mathbf{P}_{i,2}^+(k + 1|k + 1)$.

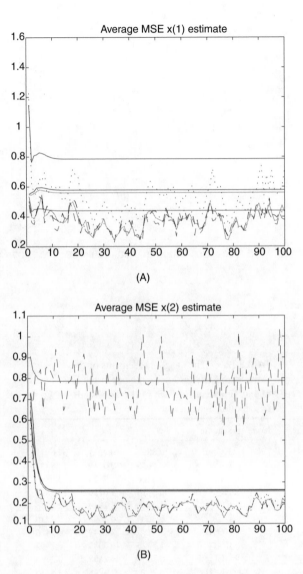

FIGURE 12.10 Mean squared errors and estimated covariances for all states in each of the four nodes. (A) Mean squared error in x. (B) Mean squared error in ẋ. (C) Mean squared error in ẍ. The curves for Node 1 are solid, Node 2 are dashed, Node 3 are dotted, and Node 4 are dash-dotted. The mean squared error is the rougher of the two lines for each node.

Split CI can now be used to incorporate $z(k)$. However, because the observation contains no correlated terms ($B_1 = 0$ in Equation 12.24), the optimal solution is always $\omega = 1$.

The effect of this algorithm can be seen in Figure 12.10 and in Table 12.3. As can be seen, the results of generalized CI are dramatic. The most strongly affected node is Node 2, whose position variance is reduced almost by a factor of 3. The least affected node is Node 1. This is not surprising, given that Node 1 is fully observable. Even so, the variance on its position estimate is reduced by more than 25%.

12.7 Conclusions

This chapter has considered the extremely important problem of data fusion in arbitrary data fusion networks. It described a general data fusion/update technique that makes no assumptions about the

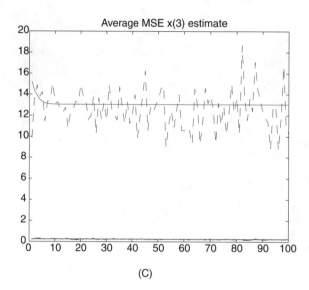

(C)

FIGURE 12.10   (continued).

TABLE 12.3   The Diagonal Elements of the
Covariance Matrices for Each Node at the End
of 100 Timesteps for Each of the Consistent
Distribution Schemes

Node	Scheme	σ_x^2	$\sigma_{\dot{x}}^2$	$\sigma_{\ddot{x}}^2$
1	NONE	0.8823	8.2081	37.6911
	CI	0.6055	0.9359	14.823
	GCI	0.4406	0.7874	13.050
2	NONE	50.5716*	1.6750	16.8829
	CI	1.2186	0.2914	0.2945
	GCI	0.3603	0.2559	0.2470
3	NONE	77852.3*	7.2649*	0.2476
	CI	1.5325	0.3033	0.2457
	GCI	0.7861	0.2608	0.2453
4	NONE	75.207	2.4248	19.473
	CI	1.2395	0.3063	0.2952
	GCI	0.5785	0.2636	0.2466

Note: NONE — no distribution; CI — the CI algorithm; GCI — generalized CI algorithm, which is described
in Section 12.6. An asterisk denotes that a state is *unobserv-
able* and its variance is increasing without bound. The
covariance used for the GCI values is $P_i\,(k|k) = P_{i,1}\,(k|k) + P_{i,2}\,(k|k)$.

independence of the estimates to be combined. The use of the covariance intersection framework to combine mean and covariance estimates without information about their degree of correlation provides a direct solution to the distributed data fusion problem.

However, the problem of unmodeled correlations reaches far beyond distributed data fusion and touches the heart of most types of tracking and estimation. Other application domains for which CI is highly relevant include:

- *Multiple model filtering* — Many systems switch behaviors in a complicated manner, so that a comprehensive model is difficult to derive. If multiple approximate models are available that capture different behavioral aspects with different degrees of fidelity, their estimates can be combined to achieve a better estimate. Because they are all modeling the same system, however, the different estimates are likely to be highly correlated.[12,13]
- *Simultaneous map building and localization for autonomous vehicles* — When a vehicle estimates the positions of landmarks in its environment while using those same landmarks to update its own position estimate, the vehicle and landmark position estimates become highly correlated.[5,14]
- *Track-to-track data fusion in multiple-target tracking systems* — When sensor observations are made in a dense target environment, there is ambiguity concerning which tracked target produced each observation. If two tracks are determined to correspond to the same target, assuming independence may not be possible when combining them, if they are derived from common observation information.[11,12]
- *Nonlinear filtering* — When nonlinear transformations are applied to observation estimates, correlated errors arise in the observation sequence. The same is true for time propagations of the system estimate. Covariance intersection will ensure nondivergent nonlinear filtering if every covariance estimate is conservative. Nonlinear extensions of the Kalman filter are inherently flawed because they require independence regardless of whether the covariance estimates are conservative.[5,15-20]

Current approaches to these and many other problems attempt to circumvent troublesome correlations by heuristically adding "stabilizing noise" to updated estimates to ensure that they are conservative. The amount of noise is likely to be excessive in order to guarantee that no covariance components are underestimated. Covariance intersection ensures the best possible estimate, given the amount of information available. The most important fact that must be emphasized is that the procedure makes no assumptions about independence, nor the underlying distributions of the combined estimates. Consequently, covariance intersection likely will replace the Kalman filter in a wide variety of applications where independence assumptions are unrealistic.

Acknowledgments

The authors gratefully acknowledge the support of IDAK Industries for supporting the development of the full CI framework and the Office of Naval Research (Contract N000149WX20103) for supporting current experiments and applications of this framework. The authors also acknowledge support from RealityLab.com and the University of Oxford.

Appendix 12.A   The Consistency of CI

This appendix proves that covariance intersection yields a consistent estimate for any value of ω and $\bar{\mathbf{P}}_{ab}$ providing that a and b are consistent.[21]

The CI algorithm calculates its mean using Equation 12.7. The actual error in this estimate is

$$\tilde{\mathbf{c}} = \mathbf{P}_{cc}\left\{\omega\mathbf{P}_{aa}^{-1}\tilde{\mathbf{a}} + \left(1-\omega\right)\mathbf{P}_{bb}^{-1}\tilde{\mathbf{b}}\right\} \tag{12.34}$$

By taking outer products and expectations, the actual mean squared error which is committed by using Equation 12.7 to calculate the mean is

$$E\left[\tilde{\mathbf{c}}\tilde{\mathbf{c}}^T\right] = \mathbf{P}_{cc}\left\{\omega^2\mathbf{P}_{aa}^{-1}\bar{\mathbf{P}}_{aa}\mathbf{P}_{aa}^{-1} + \omega\left(1-\omega\right)\mathbf{P}_{aa}^{-1}\bar{\mathbf{P}}_{ab}\mathbf{P}_{bb}^{-1}\right.$$
$$\left. + \omega\left(1-\omega\right)\mathbf{P}_{bb}^{-1}\bar{\mathbf{P}}_{ba}\mathbf{P}_{aa}^{-1} + \left(1-\omega\right)^2\mathbf{P}_{bb}^{-1}\bar{\mathbf{P}}_{bb}\mathbf{P}_{bb}^{-1}\right\}\mathbf{P}_{cc} \tag{12.35}$$

Because $\bar{\mathbf{P}}_{ab}$ is not known, the true value of the mean squared error cannot be calculated. However, CI implicitly calculates an upper bound of this quantity. If Equation 12.35 is substituted into Equation 12.3, the consistency condition can be written as

$$\mathbf{P}_{cc} - \mathbf{P}_{cc}\left\{\omega^2 \mathbf{P}_{aa}^{-1}\bar{\mathbf{P}}_{aa}\mathbf{P}_{aa}^{-1} + \omega\left(1-\omega\right)\mathbf{P}_{aa}^{-1}\bar{\mathbf{P}}_{ab}\mathbf{P}_{bb}^{-1}\right.$$
$$\left. + \omega\left(1-\omega\right)\mathbf{P}_{bb}^{-1}\bar{\mathbf{P}}_{ba}\mathbf{P}_{aa}^{-1} + \left(1-\omega\right)^2 \mathbf{P}_{bb}^{-1}\bar{\mathbf{P}}_{bb}\mathbf{P}_{bb}^{-1}\right\}\mathbf{P}_{cc} \geq 0 \qquad (12.36)$$

Pre- and postmultiplying both sides by $\mathbf{P}_{cc}^{-1}$ and collecting terms, gives

$$\mathbf{P}_{cc}^{-1} - \omega^2 \mathbf{P}_{aa}^{-1}\bar{\mathbf{P}}_{aa}\mathbf{P}_{aa}^{-1} - \omega\left(1-\omega\right)\mathbf{P}_{aa}^{-1}\bar{\mathbf{P}}_{ab}\mathbf{P}_{bb}^{-1}$$
$$- \omega\left(1-\omega\right)\mathbf{P}_{bb}^{-1}\bar{\mathbf{P}}_{ba}\mathbf{P}_{aa}^{-1} - \left(1-\omega\right)^2 \mathbf{P}_{bb}^{-1}\bar{\mathbf{P}}_{bb}\mathbf{P}_{bb}^{-1}\right\}\mathbf{P}_{cc} \geq 0 \qquad (12.37)$$

An upper bound on $\mathbf{P}_{cc}^{-1}$, which can be found and expressed using $\mathbf{P}_{aa}$, $\mathbf{P}_{bb}$, $\bar{\mathbf{P}}_{aa}$, and $\bar{\mathbf{P}}_{bb}$. From the consistency condition for **a**,

$$\mathbf{P}_{aa} - \bar{\mathbf{P}}_{aa} \geq 0 \qquad (12.38)$$

or, by pre- and postmultiplying by $\mathbf{P}_{aa}^{-1}$,

$$\mathbf{P}_{aa}^{-1} \geq \mathbf{P}_{aa}^{-1}\bar{\mathbf{P}}_{aa}\mathbf{P}_{aa}^{-1} \qquad (12.39)$$

A similar condition exists for b and, substituting these results in Equation 12.6,

$$\mathbf{P}_{cc}^{-1} = \omega \mathbf{P}_{aa}^{-1} + \left(1-\omega\right)\mathbf{P}_{bb}^{-1} \qquad (12.40)$$

$$\geq \omega \mathbf{P}_{aa}^{-1}\bar{\mathbf{P}}_{aa}\mathbf{P}_{aa}^{-1} + \left(1-\omega\right)\mathbf{P}_{bb}^{-1}\bar{\mathbf{P}}_{bb}\mathbf{P}_{bb}^{-1} \qquad (12.41)$$

Substituting this lower bound on $\mathbf{P}_{cc}^{-1}$ into Equation 12.37 leads to

$$\omega\left(1-\omega\right)\left(\mathbf{P}_{aa}^{-1}\bar{\mathbf{P}}_{aa}\mathbf{P}_{aa}^{-1} - \mathbf{P}_{aa}^{-1}\bar{\mathbf{P}}_{ab}\mathbf{P}_{bb}^{-1} - \mathbf{P}_{bb}^{-1}\bar{\mathbf{P}}_{ba}\mathbf{P}_{aa}^{-1}\bar{\mathbf{P}}_{bb}\mathbf{P}_{bb}^{-1}\right) \geq 0 \qquad (12.42)$$

or

$$\omega\left(1-\omega\right)\mathbf{E}\left[\left\{\mathbf{P}_{aa}^{-1}\tilde{\mathbf{a}} - \mathbf{P}_{bb}^{-1}\tilde{\mathbf{b}}\right\}\left\{\mathbf{P}_{aa}^{-1}\tilde{\mathbf{a}} - \mathbf{P}_{bb}^{-1}\tilde{\mathbf{b}}\right\}^T\right] \geq 0 \qquad (12.43)$$

Clearly, the inequality must hold for all choices of $\bar{\mathbf{P}}_{ab}$ and $\omega \in [0, 1]$.

Appendix 12.B   MATLAB Source Code

This appendix provides source code for performing the CI update in MATLAB.

12.B.1   Conventional CI

```
function [c,C,omega]=CI(a,A,b,B,H)
%
% function [c,C,omega]=CI(a,A,b,B,H)
%
% This function implements the CI algorithm and fuses two estimates
% (a,A) and (b,B) together to give a new estimate (c,C) and the value
% of omega which minimizes the determinant of C.  The observation
% matrix is H.

Ai=inv(A);
Bi=inv(B);

% Work out omega using the matlab constrained minimiser function
% fminbnd().
f=inline('1/det(Ai*omega+H''*Bi*H*(1-omega))', ...
         'omega', 'Ai', 'Bi', 'H');
omega=fminbnd(f,0,1,optimset('Display','off'),Ai,Bi,H);

% The unconstrained version of this optimisation is:
% omega = fminsearch(f,0.5,optimset('Display','off'),Ai,Bi,H);
% omega = min(max(omega,0),1);

% New covariance
C=inv(Ai*omega+H'*Bi*H*(1-omega));

% New mean
nu=b-H*a;
W=(1-omega)*C*H'*Bi;
c=a+W*nu;
```

12.B.2   Split CI

```
function [c,C1,C2,omega] = SCI(a,A1,A2,b,B1,B2,H)
%
% function [c,C1,C2,omega] = SCI(a,A1,A2,b,B1,B2,H)
%
% This function implements the split CI algorithm and fuses two
% estimates (a,A1,A2) and (b,B1,B2) together to give a new estimate
% (c,C1,C2) and the value of omega which minimizes the determinant of
% (C1+C2). The observation matrix is H.
%

% Work out omega using the matlab constrained minimiser function
% fminbnd().

f=inline('1/det(omega*inv(A1+omega*A2)+(1-omega)*H''*inv(B1+(1-
omega)*B2)*H)', ...
         'omega', 'A1', 'A2', 'B1', 'B2', 'H');
omega = fminbnd(f,0,1,optimset('Display','off'),A1,A2,B1,B2,H);
```

```
% The unconstrained version of this optimisation is:
% omega = fminsearch(f,0.5,optimset('Display','off'),A1,A2,B1,B2,H);
% omega = min(max(omega,0),1);

Ai=omega*inv(A1+omega*A2);
HBi=(1-omega)*H'*inv(B1+(1-omega)*B2);

% New covariance
C=inv(Ai+HBi*H);
C2=C*(Ai*A2*Ai'+HBi*B2*HBi')*C;
C1=C-C2;

% New mean
nu=b-H*a;
W=C*HBi;
c=a+W*nu;
```

References

1. Utete, S.W., Network management in decentralised sensing systems, Ph.D. thesis, Robotics Research Group, Department of Engineering Science, University of Oxford, 1995.
2. Grime, S. and Durrant-Whyte H., Data fusion in decentralized sensor fusion networks, *Control Engineering Practice*, 2(5), 849, 1994.
3. Chong, C., Mori, S., and Chan, K., Distributed multitarget multisensor tracking, *Multitarget Multisensor Tracking*, Artech House Inc., Boston, 1990.
4. Jazwinski, A.H., *Stochastic Processes and Filtering Theory*, Academic Press, New York, 1970.
5. Uhlmann, J.K., Dynamic map building and localization for autonomous vehicles, Ph.D. thesis, University of Oxford, 1995/96.
6. Vandenberghe, L. and Boyd, S., Semidefinite programming, *SIAM Review*, March 1996.
7. Wu, S.P., Vandenberghe, L., and Boyd, S., Maxdet: Software for determinant maximization problems, alpha version, Stanford University, April 1996.
8. Boyd, S. and Wu, S.P., *SDPSOL: User's Guide*, November 1995.
9. Bar-Shalom, Y. and Fortmann, T.E., *Tracking and Data Association*, Academic Press, New York, 1988.
10. Mutambara, A.G.O., *Decentralized Estimation and Control for Nonlinear Systems*, CRC Press, 1998.
11. Nicholson, D. and Deaves, R., Decentralized track fusion in dynamic networks, in *Proc. 2000 SPIE Aerosense Conf.*, 2000.
12. Bar-Shalom, Y. and Li, X.R., *Multitarget-Multisensor Tracking: Principles and Techniques*, YBS Press, Storrs, CT, 1995.
13. Julier, S.J. and Durrant-Whyte, H., A horizontal model fusion paradigm, *Proc. SPIE Aerosense Conf.*, 1996.
14. Uhlmann, J., Julier, S., and Csorba, M., Nondivergent simultaneous map building and localization using covariance intersection, in *Proc. 1997 SPIE Aerosense Conf.*, 1997.
15. Julier, S.J., Uhlmann, J.K., and Durrant-Whyte, H.F., A new approach for the nonlinear transformation of means and covariances in linear filters, *IEEE Trans. Automatic Control*, 477, March 2000.
16. Julier, S.J., Uhlmann, J.K., and Durrant-Whyte, H.F., A new approach for filtering nonlinear systems, in *Proc. American Control Conf.*, Seattle, WA, 1995, 1628.
17. Julier, S.J. and Uhlmann, J.K., A new extension of the Kalman filter to nonlinear systems, in *Proc. AeroSense: 11th Internat'l. Symp. Aerospace/Defense Sensing, Simulation and Controls*, SPIE, 1997.
18. Julier, S.J. and Uhlmann, J.K., A consistent, debiased method for converting between polar and Cartesian coordinate systems, in *Proc. of AeroSense: 11th Internat'l. Symp. Aerospace/Defense Sensing, Simulation and Controls*, SPIE, 1997.

19. Juliers, S.J., A skewed approach to filtering, *Proc. AeroSense: 12th Internat'l. Symp. Aerospace/Defense Sensing, Simulation and Controls*, SPIE, 1998.

20. Julier, S.J., and Uhlmann, J.K., A General Method for Approximating Nonlinear Transformations of Probability Distributions, published on the Web at http://www.robots.ox.ac.uk/~siju, August 1994.

21. Julier, S.J. and Uhlmann, J.K., A non-divergent estimation algorithm in the presence of unknown correlations, *American Control Conf.*, Albuquerque, NM, 1997.

22. Julier, S.J. and Uhlmann, J.K., Generalized and split covariance intersection and addition, Technical Disclosure Report, Naval Research Laboratory, 1998.

13

Data Fusion in Nonlinear Systems

13.1 Introduction 13-1
13.2 Estimation in Nonlinear Systems................................. 13-2
   Problem Statement • The Transformation of Uncertainty
13.3 The Unscented Transformation (UT)........................... 13-5
   The Basic Idea • An Example Set of Sigma Points •
   Properties of the Unscented Transformation
13.4 Uses of the Transformation............................. 13-8
   Polar to Cartesian Coordinates • A Discontinuous
   Transformation
13.5 The Unscented Filter (UF) 13-12
13.6 Case Study: Using the UF with Linearization Errors 13-13
13.7 Case Study: Using the UF with a High-Order
   Nonlinear System 13-15
13.8 Multilevel Sensor Fusion 13-18
13.9 Conclusions 13-20
Acknowledgments.. 13-21
References.. 13-21

Simon Julier
IDAK Industries

Jeffrey K. Uhlmann
University of Missouri

13.1 Introduction

The extended Kalman filter (EKF) has been one of the most widely used methods for tracking and estimation based on its apparent simplicity, optimality, tractability, and robustness. However, after more than 30 years of experience with it, the tracking and control community has concluded that the EKF is difficult to implement, difficult to time, and only reliable for systems that are almost linear on the time scale of the update intervals. This chapter reviews the unscented transformation (UT), a mechanism for propagating mean and covariance information through nonlinear transformations, and describes its implications for data fusion. This method is more accurate, is easier to implement, and uses the same order of calculations as the EKF. Furthermore, the UT permits the use of Kalman-type filters in applications where, traditionally, their use was not possible. For example, the UT can be used to rigorously integrate artificial intelligence-based systems with Kalman-based systems.

Performing data fusion requires estimates of the state of a system to be converted to a common representation. The mean and covariance representation is the *lingua franca* of modern systems engineering. In particular, the covariance intersection (CI)[1] and Kalman filter (KF)[2] algorithms provide mechanisms for fusing state estimates defined in terms of means and covariances, where each mean vector defines the nominal state of the system and its associated error covariance matrix defines a lower bound on the squared error. However, most data fusion applications require the fusion of mean and covariance estimates defining the state of a system in different coordinate frames. For example, a tracking system might maintain estimates in a global Cartesian coordinate frame, while observations of the tracked objects are generated in the local coordinate frames of various sensors. Therefore, a transformation must be applied to convert between the global coordinate frame and each local coordinate frame.

If the transformation between coordinate frames is linear, the linearity properties of the mean and covariance makes the application of the transformation trivial. Unfortunately, most tracking sensors take measurements in a local polar or spherical coordinate frame (i.e., they measure range and bearings) that is not linearly transformable to a Cartesian coordinate frame. Rarely are the natural coordinate frames of two sensors linearly related. This fact constitutes a fundamental problem that arises in virtually all practical data fusion systems.

The UT, a mechanism that addresses the difficulties associated with converting mean and covariance estimates from one coordinate frame to another, can be applied to obtain mean and covariance estimates from systems that do not inherently produce estimates in that form. For example, this chapter describes how the UT can allow high-level artificial intelligence (AI) and fuzzy control systems to be integrated seamlessly with low-level KF and CI systems.

The structure of this chapter is as follows: Section 13.2 describes the nonlinear transformation problem within the Kalman filter framework and analyzes the KF prediction problem in detail. The UT is introduced and its performance is analyzed in Section 13.3. Section 13.4 demonstrates the effectiveness of the UT with respect to a simple nonlinear transformation (polar to Cartesian coordinates with large bearing uncertainty) and a simple discontinuous system. Section 13.5 examines how the transformation can be embedded into a fully recursive estimator that incorporates process and observation noise. Section 13.6 discusses the use of the UT in a tracking example, and Section 13.7 describes its use with a complex process and observation model. Finally, Section 13.8 shows how the UT ties multiple levels of data fusion together into a single, consistent framework.

13.2   Estimation in Nonlinear Systems

13.2.1   Problem Statement

Minimum mean squared error (MMSE) estimators can be broadly classified into linear and nonlinear estimators. Of the linear estimators, by far the most widely used is the Kalman filter.[2]* Many researchers have attempted to develop suitable nonlinear MMSE estimators. However, the optimal solution requires that a complete description of the conditional probability density be maintained,[3] and this exact description requires a potentially unbounded number of parameters. As a consequence, many suboptimal approximations have been proposed in the literature. Traditional methods are reviewed by A. H. Jazwinski[4] and P. S. Maybeck.[5] Recent algorithms have been proposed by F. E. Daum,[6] N. J. Gordon et al.,[7] and M. A. Kouritzin.[8] Despite the sophistication of these and other approaches, the extended Kalman filter (EKF) remains the most widely used estimator for nonlinear systems.[9,10] The EKF applies the Kalman filter to nonlinear systems by simply linearizing all of the nonlinear models so that the traditional linear Kalman filter equations can be applied. However, in practice, the EKF has three well-known drawbacks:

1. Linearization can produce highly unstable filters if the assumption of local linearity is violated. Examples include estimating ballistic parameters of missiles[11-14] and some applications of computer vision.[15] As demonstrated later in this chapter, some extremely common transformations that are used in target tracking systems are susceptible to these problems.

---

*Researchers often (and incorrectly) claim that the Kalman filter can be applied only if the following two conditions hold: (i) all probability distributions are Gaussian and (ii) the system equations are linear. The Kalman filter is, in fact, the minimum mean squared *linear* estimator that can be applied to *any* system with *any* distribution, provided the first two moments are known. However, it is only the globally optimal estimator under the special case that the distributions are all Gaussian.

2. Linearization can be applied only if the Jacobean matrix exists, and the Jacobian matrix exists only if the system is differentiable at the estimate. Although this constraint is satisfied by the dynamics of continuous physical systems, some systems do not satisfy this property. Examples include jump-linear systems, systems whose sensors are quantized, and expert systems that yield a finite set of discrete solutions.

3. Finally, the derivation of the Jacobian matrices is nontrivial in most applications and can often lead to significant implementation difficulties. In P. A. Dulimov,[16] for example, the derivation of a Jacobian requires six pages of dense algebra. Arguably, this has become less of a problem, given the widespread use of symbolic packages such as Mathematica[17] and Maple.[18] Nonetheless, the computational expense of calculating a Jacobian can be extremely high if the expressions for the terms are nontrivial.

Appreciating how the UT addresses these three problems requires an understanding of some of the mechanics of the KF and EKF.

Let the state of the system at a time step k be the state vector $\mathbf{x}(k)$. The Kalman filter propagates the first two moments of the distribution of $\mathbf{x}(k)$ recursively and has a distinctive "predictor-corrector" structure. Let $\hat{\mathbf{x}}\,(i|j)$ be the estimate of $\mathbf{x}(i)$ using the observation information up to and including time j, $\mathbf{Z}^j = [\mathbf{z}(1),...,\mathbf{z}(j)]$. The covariance of this estimate is $\mathbf{P}(i|j)$. Given an estimate $\hat{\mathbf{x}}(k|k)$, the filter first predicts what the future state of the system will be using the process model. Ideally; the predicted quantities are given by the expectations

$$\hat{\mathbf{x}}\left(k+1|k\right)=E\left[\mathbf{f}\left[\mathbf{x}\left(k\right),\mathbf{u}\left(k\right),\mathbf{v}\left(k\right),k\right]\Big|\mathbf{Z}^k\right] \tag{13.1}$$

$$\mathbf{P}\left(k+1|k\right)=E\left[\left\{\mathbf{x}\left(k+1\right)-\hat{\mathbf{x}}\left(k+1|k\right)\right\}\left\{\mathbf{x}\left(k+1\right)-\hat{\mathbf{x}}\left(k+1|k\right)\right\}^T\Big|\mathbf{Z}^k\right] \tag{13.2}$$

When $\mathbf{f}[\cdot]$ and $\mathbf{h}[\cdot]$ are nonlinear, the precise values of these statistics can be calculated only if the distribution of $\mathbf{x}(k)$ is perfectly known. However, this distribution has no general form, and a potentially unbounded number of parameters are required. Therefore, in most practical algorithms these expected values must be approximated.

The estimate $\hat{\mathbf{x}}(k+1|k+1)$ is found by updating the prediction with the current sensor measurement. In the Kalman filter, a linear update rule is specified and the weights are chosen to minimize the mean squared error of the estimate.

$$\hat{\mathbf{x}}\left(k+1|k+1\right)=\hat{\mathbf{x}}\left(k+1|k\right)+\mathbf{W}\left(k+1\right)v\left(k+1\right)$$

$$\mathbf{P}\left(k+1|k+1\right)=\mathbf{P}\left(k+1|k\right)-\mathbf{W}\left(k+1\right)\mathbf{P}_{vv}\left(k+1|k\right)\mathbf{W}^T\left(k+1\right)$$

$$v\left(k+1\right)=\mathbf{z}\left(k+1\right)-\hat{\mathbf{z}}\left(k+1|k\right)$$

$$\mathbf{W}\left(k+1\right)=\mathbf{P}_{xv}\left(k+1|k\right)\mathbf{P}_{vv}^{-1}\left(k+1|k\right) \tag{13.3}$$

Note that these equations are only a function of the predicted values of the first two moments of $\mathbf{x}(k)$ and $\mathbf{z}(k)$. Therefore, the problem of applying the Kalman filter to a nonlinear system is the ability to predict the first two moments of $\mathbf{x}(k)$ and $\mathbf{z}(k)$.

13.2.2   The Transformation of Uncertainty

The problem of predicting the future state or observation of the system can be expressed in the following form. Suppose that $\mathbf{x}$ is a random variable with mean $\bar{\mathbf{x}}$ and covariance $\mathbf{P}_{xx}$. A second random variable, $\mathbf{y}$, is related to $\mathbf{x}$ through the nonlinear function

$$\mathbf{y} = \mathbf{f}\left[\mathbf{x}\right] \tag{13.4}$$

The mean $\bar{\mathbf{y}}$ and covariance $\mathbf{P}_{yy}$ of $\mathbf{y}$ must be calculated.

   The statistics of $\mathbf{y}$ are calculated by (1) determining the density function of the transformed distribution and (2) evaluating the statistics from that distribution. In some special cases, exact, closed form solutions exist (e.g., when $\mathbf{f}[\cdot]$ is linear or is one of the forms identified in F. E. Daum[6]). However; as explained above, most data fusion problems do not possess closed-form solutions and some kind of an approximation must be used. A common approach is to develop a transformation procedure from the Taylor series expansion of Equation 13.4 about $\bar{\mathbf{x}}$. This series can be expressed as

$$\mathbf{f}\left[\mathbf{x}\right] = \mathbf{f}\left[\bar{\mathbf{x}} + \delta\mathbf{x}\right]$$
$$= \mathbf{f}\left[\bar{\mathbf{x}}\right] + \nabla\mathbf{f}\delta\mathbf{x} + \frac{1}{2}\nabla^2\mathbf{f}\delta\mathbf{x}^2 + \frac{1}{3!}\nabla^3\mathbf{f}\delta\mathbf{x}^3 + \frac{1}{4!}\nabla^4\mathbf{f}\delta\mathbf{x}^4 + \cdots \tag{13.5}$$

where $\delta\mathbf{x}$ is a zero mean Gaussian variable with covariance $\mathbf{P}_{xx}$ and $\nabla^n\mathbf{f}\delta\mathbf{x}^n$ is the appropriate nth order term in the multidimensional Taylor Series. The transformed mean and covariance are

$$\bar{\mathbf{y}} = \mathbf{f}\left[\bar{\mathbf{x}}\right] + \frac{1}{2}\nabla^2\mathbf{f}\mathbf{P}_{xx} + \frac{1}{2}\nabla^4\mathbf{f}\,E\left[\delta\mathbf{x}^4\right] + \cdots \tag{13.6}$$

$$\mathbf{P}_{yy} = \nabla\mathbf{f}\mathbf{P}_{xx}\left(\nabla\mathbf{f}\right)^T + \frac{1}{2\times 4!}\nabla^2\mathbf{f}\left(E\left[\delta\mathbf{x}^4\right] - E\left[\delta\mathbf{x}^2\mathbf{P}_{yy}\right] - E\left[\mathbf{P}_{yy}\delta\mathbf{x}^2\right] + \mathbf{P}_{yy}^2\right)\left(\nabla^2\mathbf{f}\right)^T +$$
$$\frac{1}{3!}\nabla^3\mathbf{f}\,E\left[\delta\mathbf{x}^4\right]\left(\nabla\mathbf{f}\right)^T + \cdots \tag{13.7}$$

In other words, the nth order term in the series for $\bar{\mathbf{x}}$ is a function of the nth order moments of $\mathbf{x}$ multiplied by the nth order derivatives of $\mathbf{f}[\cdot]$ evaluated at $\mathbf{x} = \bar{\mathbf{x}}$. If the moments and derivatives can be evaluated correctly up to the nth order, the mean is correct up to the nth order as well. Similar comments hold for the covariance equation, although the structure of each term is more complicated. Since each term in the series is scaled by a progressively smaller and smaller term the lowest-order terms in the series are likely to have the greatest impact. Therefore, the prediction procedure should be concentrated on evaluating the lower order terms.

   The EKF exploits linearization. Linearization assumes that the second- and higher-order terms of $\delta\mathbf{x}$ in Equation 13.5 can be neglected. Under this assumption,

$$\bar{\mathbf{y}} = \mathbf{f}\left[\bar{\mathbf{x}}\right] \tag{13.8}$$

$$\mathbf{P}_{yy} = \nabla\mathbf{f}\mathbf{P}_{xx}\left(\nabla\mathbf{f}\right)^T \tag{13.9}$$

   However, in many practical situations, linearization introduces significant biases or errors. These cases require more accurate prediction techniques.

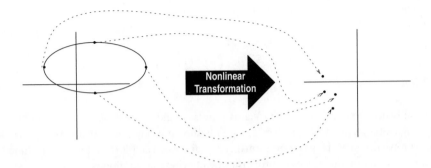

FIGURE 13.1   The principle of the unscented transformation.

13.3   The Unscented Transformation (UT)

13.3.1   The Basic Idea

The UT is a method for calculating the statistics of a random variable that undergoes a nonlinear transformation. This method is founded on the intuition that *it is easier to approximate a probability distribution than it is to approximate an arbitrary nonlinear function or transformation.*[19] The approach is illustrated in Figure 13.1. A set of points (*sigma points*) is chosen with sample mean and sample covariance of the nonlinear function is $\bar{\mathbf{x}}$ and $\mathbf{P}_{xx}$. The nonlinear function is applied to each point, in turn, to yield a cloud of transformed points; $\bar{\mathbf{y}}$ and $\mathbf{P}_{yy}$ are the statistics of the transformed points.

Although this method bears a superficial resemblance to Monte Carlo-type methods, there is an extremely important and fundamental difference. The samples are not drawn at random; they are drawn according to a specific, deterministic algorithm. Since the problems of statistical convergence are not relevant, high-order information about the distribution can be captured using only a very small number of points. For an n-dimensional space, only $n + 1$ points are needed to capture any given mean and covariance. If the distribution is known to be symmetric, $2n$ points are sufficient to capture the fact that the third- and all higher-order odd moments are zero for any symmetric distribution.[19]

The set of sigma points, S, consists of l vectors and their appropriate weights, $S = \{i = 0, 0,..., l - 1 :$ $X_i, W_i\}$. The weights W_i can be positive or negative but must obey the normalization condition

$$\sum_{i=0}^{l-1} W_i = 1 \tag{13.10}$$

Given these points, $\bar{\mathbf{y}}$ and $\mathbf{P}_{yy}$ are calculated using the following procedure:

1. Instantiate each point through the function to yield the set of transformed sigma points,

$$\boldsymbol{y}_i = \mathbf{f}\left[\boldsymbol{X}_i\right]$$

2. The mean is given by the weighted average of the transformed points,

$$\bar{y} = \sum_{i=0}^{l-1} W_i \boldsymbol{y}_i \tag{13.11}$$

3. The covariance is the weighted outer product of the transformed points,

$$\mathbf{P}_{yy} = \sum_{i=0}^{l-1} W_i \{ \mathbf{y}_i - \bar{\mathbf{y}} \} \{ \mathbf{y}_i - \bar{\mathbf{y}} \}^T \qquad (13.12)$$

The crucial issue is to decide how many sigma points should be used, where they should be located, and what weights they should be assigned. The points should be chosen so that they capture the "most important" properties of $\mathbf{x}$. This can be formalized as follows. Let $P_x(\mathbf{x})$ be the density function of $\mathbf{x}$. The sigma points capture the necessary properties by obeying the condition

$$\mathbf{g}\left[S, p_x(\mathbf{x})\right] = 0$$

The decision as to which properties of $\mathbf{x}$ are to be captured precisely and which are to be approximated is determined by the demands of the particular application in question. Here, the *moments* of the distribution of the sigma points are matched with those of $\mathbf{x}$. This is motivated by the Taylor series expansion, given in Section 13.2.2, which shows that matching the moments of $\mathbf{x}$ up to the nth order means that Equations 13.11 and 13.12 capture $\bar{\mathbf{y}}$ and $\mathbf{P}_{yy}$ up to the nth order as well.[20]

Note that the UT is distinct from other efforts published in the literature. First, some authors have considered the related problem of assuming that the distribution takes on a particular parameterized form, rather than an entire, arbitrary distribution. Kushner, for example, describes an approach whereby a distribution is approximated at each time step by a Gaussian.[21] However, the problem with this approach is that it does not address the fundamental problem of calculating the mean and covariance of the nonlinearly transformed distribution. Second, the UT bears some relationship to quadrature, which has been used to approximate the integrations implicit in statistical expectations. However, the UT avoids some of the difficulties associated with quadrature methods by approximating the unknown distribution. In fact, the UT is most closely related to perturbation analysis. In a 1989 article, Holztmann introduced a noninfinitesimal perturbation for a scalar system.[22] Holtzmann's solution corresponds to that of the symmetric UT in the scalar case, but their respective generalizations (e.g., to higher dimensions) are not equivalent.

13.3.2   An Example Set of Sigma Points

A set of sigma points can be constructed using the constraints that they capture the first three moments of a symmetric distribution: $\mathbf{g}\,[S, p_x(\mathbf{x})] = [\mathbf{g}_1\,[S, p_x(\mathbf{x})]\;\mathbf{g}_2\,[S, p_x(\mathbf{x})]\;\mathbf{g}_3\,[S, p_x(\mathbf{x})]]^T$ where

$$\mathbf{g}_1\left[S, p_x(\mathbf{x})\right] = \sum_{i=0}^{p} W_i \mathbf{X}_i - \hat{\mathbf{x}} \qquad (13.13)$$

$$\mathbf{g}_2\left[S, p_x(\mathbf{x})\right] = \sum_{i=0}^{p} W_i (\mathbf{X}_i - \bar{\mathbf{x}})^2 - \mathbf{P}_{xx} \qquad (13.14)$$

$$\mathbf{g}_3\left[S, p_x(\mathbf{x})\right] = \sum_{i=0}^{p} W_i (\mathbf{X}_i - \bar{\mathbf{x}})^3 \qquad (13.15)$$

The set is[23]

$$X_0\left(k\big|k\right) = \hat{\mathbf{x}}\left(k\big|k\right)$$

$$W_0 = \kappa\big/\left(n+\kappa\right)$$

$$X_i\left(k\big|k\right) = \hat{\mathbf{x}}\left(k\big|k\right) + \left(\sqrt{\left(n+\kappa\right)\mathbf{P}\left(k\big|k\right)}\right)_i$$

$$W_i = 1\big/\left\{2\left(n+\kappa\right)\right\}$$ \hfill (13.16)

$$X_{i+n}\left(k\big|k\right) = \hat{\mathbf{x}}\left(k\big|k\right) - \left(\sqrt{\left(n+\kappa\right)\mathbf{P}\left(k\big|k\right)}\right)_i$$

$$W_{i+n} = 1\big/\left\{2\left(n+\kappa\right)\right\}$$

where κ is a real number, $\left(\sqrt{(n+\kappa)\mathbf{P}(k|k)}\right)_i$ is the ith row or column* of the matrix square root of $(n + \kappa)$ $\mathbf{P}\,(k|k)$, and W_i is the weight associated with the ith point.

13.3.3 Properties of the Unscented Transform

Despite its apparent similarity to other efforts described in the data fusion literature, the UT has a number of features that make it well suited for the problem of data fusion in practical problems:

- The UT can predict with the same accuracy as the second-order Gauss filter, but without the need to calculate Jacobians or Hessians. The reason is that the mean and covariance of **x** are captured precisely up to the second order, and the calculated values of the mean and covariance of **y** also are correct to the second order. This indicates that the mean is calculated to a higher order of accuracy than the EKF, whereas the covariance is calculated to the same order of accuracy.
- The computational cost of the algorithm is the same order of magnitude as the EKF. The most expensive operations are calculating the matrix square root and determining the outer product of the sigma points to calculate the predicted covariance. However, both operations are $O(n^3)$, which is the same cost as evaluating the $n \times n$ matrix multiplies needed to calculate the predicted covariance.**
- The algorithm naturally lends itself to a "black box" filtering library. The UT calculates the mean and covariance using standard vector and matrix operations and does not exploit details about the specific structure of the model.
- The algorithm can be used with distributions that are not continuous. Sigma points can straddle a discontinuity. Although this does not precisely capture the effect of the discontinuity, its effect is to spread the sigma points out such that the mean and covariance reflect the presence of the discontinuity.
- The UT can be readily extended to capture more information about the distribution. Because the UT captures the properties of the distribution, a number of refinements can be applied to improve greatly the performance of the algorithm. If only the first two moments are required, then $n + 1$ sigma points are sufficient. If the distribution is assumed or is known to be symmetric, then $n + 2$

---

*If the matrix square root **A** of **P** is of the form $\mathbf{P} = \mathbf{A}^T\mathbf{A}$, then the sigma points are formed from the *rows* of **A**. However, for a root of the form $\mathbf{P} = \mathbf{AA}^T$, the *columns* of **A** are used.

**The matrix square root should be calculated using numerically efficient and stable methods such as the Cholesky decomposition.[24]

sigma points are sufficient. Therefore, the total number of calculations required for calculating the new covariance is $O(n^3)$, which is the same order as that required by the EKF. The transform has also been demonstrated to propagate successfully the fourth-order moment (or kurtosis) of a Gaussian distribution[25] and that it can be used to propagate the third-order moments (or skew) of an arbitrary distribution.[26]

13.4   Uses of the Transformation

This section demonstrates the effectiveness of the UT with respect to two nonlinear systems that represent important classes of problems encountered in the data fusion literature — coordinate conversions and discontinuous systems.

13.4.1   Polar to Cartesian Coordinates

One of the most important transformations in target tracking is the conversion from polar to Cartesian coordinates. This transformation is known to be highly susceptible to linearization errors. D. Lerro and Y. Bar-Shalom, for example, show that the linearized conversion can become inconsistent when the standard deviation in the bearing estimate is less than a degree.[27] This subsection illustrates the use of the UT on a coordinate conversion problem with extremely high angular uncertainty.

Suppose a mobile autonomous vehicle detects targets in its environment using a range-optimized sonar sensor. The sensor returns polar information (range, r, and bearing, θ), which is converted to estimate Cartesian coordinates. The transformation is

$$\begin{pmatrix} x \\ y \end{pmatrix} = \begin{pmatrix} r\cos\theta \\ r\sin\theta \end{pmatrix} \text{ with } \nabla\mathbf{f} = \begin{bmatrix} \cos\theta & -r\sin\theta \\ \sin\theta & r\cos\theta \end{bmatrix}$$

The real location of the target is (0, 1). The difficulty with this transformation arises from the physical properties of the sonar. Fairly good range accuracy (with 2cm standard deviation) is traded off to give a very poor bearing measurement (standard deviation of 15°).[28] The large bearing uncertainty causes the assumption of local linearity to be violated.

To appreciate the errors that can be caused by linearization, compare its values for the statistics of (x, y) with those of the true statistics calculated by Monte Carlo simulation. Due to the slow convergence of random sampling methods, an extremely large number of samples (3.5×10^6) were used to ensure that accurate estimates of the true statistics were obtained. The results are shown in Figure 13.2(a). This figure shows the mean and 1σ contours calculated by each method. The 1σ contour is the locus of points $\{\mathbf{y} : (\mathbf{y} - \bar{\mathbf{y}}) \, \mathbf{P}_y^{-1} (\mathbf{y} - \bar{\mathbf{y}}) = 1\}$ and is a graphical representation of the size and orientation of $\mathbf{P}_{yy}$. The figure demonstrates that the linearized transformation is biased and inconsistent. This is most pronounced along the y-axis, where linearization estimates that the position is 1m, whereas in reality it is 96.7cm. In this example, linearization errors effectively introduce an error which is over 1.5 times the standard deviation of the range measurement. Since it is a bias that arises from the transformation process itself, the same error with the same sign will be committed each time a coordinate transformation takes place. Even if there were no bias, the transformation would still be inconsistent because its ellipse is not sufficiently extended along the y-axis.

In practice, this inconsistency can be resolved by introducing additional stabilizing noise that increases the size of the transformed covariance. This is one possible explanation of why EKFs are difficult to tune — sufficient noise must be introduced to offset the defects of linearization. However, introducing stabilizing noise is an undesirable solution because the estimate remains biased and there is no general guarantee that the transformed estimate remains consistent or efficient.

The performance benefits of using the UT can be seen in Figure 13.2(b), which shows the means and 1σ contours determined by the different methods. The mismatch between the UT mean and the true mean is extremely small (approximately 6×10^{-4}). The transformation is consistent, ensuring that the

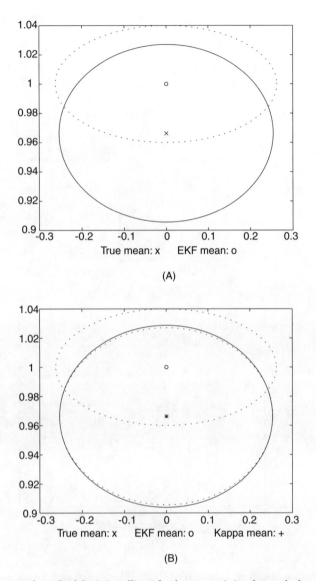

FIGURE 13.2 The mean and standard deviation ellipses for the true statistics, those calculated through linearization and those calculated by the unscented transformation. (A) Results from linearization. The true mean is at × and the uncertainty ellipse is solid. Linearization calculates the mean at ○ and the uncertainty ellipse is dashed. (B) Results from the UT. The true mean is at × and the uncertainty ellipse is dotted. The UT mean is at + (overlapping the position of the true mean) and is the solid ellipse. The linearized mean is at ○ and its ellipse is also dotted.

filter does not diverge. As a result, there is no need to introduce artificial noise terms that would degrade performance even when the angular uncertainty is extremely high.

13.4.2 A Discontinuous Transformation

Consider the behavior of a two-dimensional particle whose state consists of its position $\mathbf{x}\,(k) = [x(k),$ $y(k)]^T$. The projectile is initially released at time 1 and travels at a constant and known speed, v_x, in the x direction. The objective is to estimate the mean position and covariance of the position at time 2, $[x(2),$ $y(2)]^T$, where $\Delta T \stackrel{\Delta}{=} t_2 - t_1$. The problem is made difficult by the fact that the path of the projectile is obstructed by a wall that lies in the "bottom right quarter-plane" ($x \geq 0, y \leq 0$). If the projectile hits the wall, a perfectly elastic collision will occur, and the projectile will be reflected back at the same velocity

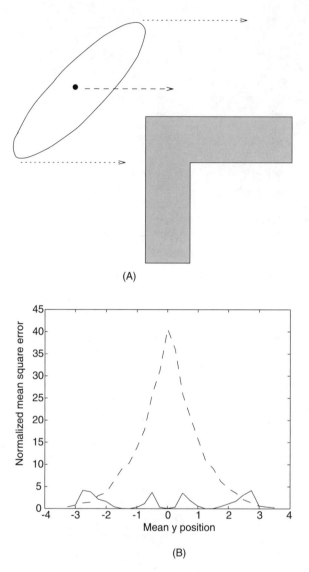

FIGURE 13.3   A discontinuous system example: a particle can either strike a wall and rebound, or continue to move in a straight line. The experimental results show the effect of using different start values for *y*.

as it traveled forward. This situation is illustrated in Figure 13.3(A), which also shows the covariance ellipse of the initial distribution.

The process model for this system is

$$x(2) = \begin{cases} x(1) + \Delta T v_x & y(1) \geq 0 \\ x(1) - \Delta T v_x & y(1) < 0 \end{cases} \tag{13.17}$$

$$y(2) = y(1) \tag{13.18}$$

At time 1, the particle starts in the left half-plane ($x \leq 0$) with position $[x(1), y(1)]^T$. The error in this estimate is Gaussian, has zero mean, and has covariance $\mathbf{P}(1|1)$. Linearized about this start condition, the system appears to be a simple constant velocity linear model.

1. The set of sigma points are created by applying a sigma point selection algorithm (e.g., Equation 13.16) to the augmented system given by Equation 13.21.

2. The transformed set is given by instantiating each point through the process model,

$$\mathbf{X}_i\left(k+1|k\right)=\mathbf{f}\left[\mathbf{X}_i\left(k|k\right),\mathbf{u}\left(k\right),k\right]$$

3. The predicted mean is computed as

$$\hat{\mathbf{x}}\left(k+1|k\right)=\sum_{i=0}^{2n^a}W_i\mathbf{X}_i\left(k+1|k\right)$$

4. And the predicted covariance is computed as

$$\mathbf{P}\left(k+1|k\right)\sum_{i=0}^{2n^a}\left\{W_i\left(\mathbf{X}_i\left(k+1|k\right)-\hat{\mathbf{x}}\left(k+1|k\right)\right)\right\}\left\{\mathbf{X}_i\left(k+1|k\right)-\hat{\mathbf{x}}\left(k+1|k\right)\right\}^T$$

5. Instantiate each of the prediction points through the observation model,

$$\mathbf{Z}_i\left(k+1|k\right)=\mathbf{h}\left[\mathbf{X}_i\left(k+1|k\right),\mathbf{u}\left(k\right),k\right]$$

6. The predicted observation is calculated by

$$\hat{\mathbf{z}}\left(k+1|k\right)=\sum_{i=1}^{2n^a}W_i\mathbf{Z}_i\left(k+1|k\right)$$

7. The innovation covariance is

$$\mathbf{P}_{vv}\left(k+1|k\right)=\sum_{i=0}^{2n^a}W_i\left\{\mathbf{Z}_i\left(k|k-1\right)-\hat{\mathbf{z}}\left(k+1|k\right)\right\}\left\{\mathbf{Z}_i\left(k|k-1\right)-\hat{\mathbf{z}}\left(k+1|k\right)\right\}^T$$

8. The cross-correlation matrix is determined by

$$\mathbf{P}_{xz}\left(k+1|k\right)=\sum_{i=0}^{2n^a}W_i\left\{\mathbf{X}_i\left(k|k-1\right)-\hat{\mathbf{x}}\left(k+1|k\right)\right\}\left\{\mathbf{Z}_i\left(k|k-1\right)-\hat{\mathbf{z}}\left(k+1|k\right)\right\}^T$$

9. Finally, the update can be performed using the normal Kalman filter equations:

$$\hat{\mathbf{x}}\left(k+1|k+1\right)=\hat{\mathbf{x}}\left(k+1|k\right)+\mathbf{W}\left(k+1\right)v\left(k+1\right)$$

$$\mathbf{P}\left(k+1|k+1\right)=\mathbf{P}\left(k+1|k\right)-\mathbf{W}\left(k+1\right)\mathbf{P}_{vv}\left(k+1|k\right)\mathbf{W}^T\left(k+1\right)$$

$$v\left(k+1\right)=\mathbf{z}\left(k+1\right)-\hat{\mathbf{z}}\left(k+1|k\right)$$

$$\mathbf{W}\left(k+1\right)=\mathbf{P}_{xv}\left(k+1|k\right)\mathbf{P}_{vv}^{-1}\left(k+1|k\right)$$

FIGURE 13.4   A general formulation of the Kalman filter using the unscented transformation. As explained in the text, there is a significant scope for optimizing this algorithm.

The true conditional mean and covariance was determined using Monte Carlo simulation for different choices of the initial mean of y. The mean squared error calculated by the EKF and by the UT for different values is shown in Figure 13.3(B). The UT estimates the mean very closely, suffering only small spikes as the translated sigma points successively pass the wall. Further analysis shows that the covariance for the filter is only slightly larger than the true covariance, but conservative enough to account for the deviation of its estimated mean from the true mean. The EKF, however, bases its entire estimate of the conditional mean on the projection of the prior mean; therefore, its estimates bear no resemblance to the true mean, except when most of the distribution either hits or misses the wall and the effect of the discontinuity is minimized.

13.5 The Unscented Filter

The UT can be used as the cornerstone of a recursive Kalman-type of estimator. The transformation processes that occur in a Kalman filter (Equation 13.3) consist of the following steps:

1. Predict the new state of the system, $\hat{\mathbf{x}}\,(k+1|k)$, and its associated covariance, $\mathbf{P}\,(k+1|k)$. This prediction must take into account the effects of process noise.
2. Predict the expected observation, $\hat{\mathbf{z}}\,(k+1|k)$, and the innovation covariance, $\mathbf{P}_{vv}\,(k+1|k)$. This prediction should include the effects of observation noise.
3. Finally, predict the cross-correlation matrix, $\mathbf{P}_{xz}\,(k+1|k)$.

These steps can be easily accommodated by slightly restructuring the state vector and process and observation models. The most general formulation augments the state vector with the process and noise terms to give an $n^a = n + q + r$ dimensional vector,

$$\mathbf{x}^a\left(k\right) = \begin{bmatrix} \mathbf{x}\left(k\right) \\ \mathbf{v}\left(k\right) \\ \mathbf{w}\left(k\right) \end{bmatrix} \tag{13.19}$$

The process and observation models are rewritten as a function of $\mathbf{x}^a\,(k)$,

$$\mathbf{x}\left(k+1\right) = \mathbf{f}^a\!\left[\mathbf{x}^a\left(k\right), \mathbf{u}\left(k\right), k\right]$$
$$\mathbf{z}\left(k+1\right) = \mathbf{h}^a\!\left[\mathbf{x}^a\left(k+1\right), \mathbf{u}\left(k\right), k\right] \tag{13.20}$$

and the UT uses sigma points that are drawn from

$$\hat{\mathbf{x}}^a\left(k|k\right) = \begin{pmatrix} \hat{\mathbf{x}}\left(k|k\right) \\ 0_{q \times 1} \\ 0_{m \times 1} \end{pmatrix} \text{ and } \mathbf{P}^a\left(k|k\right) = \begin{bmatrix} \mathbf{P}\left(k|k\right) & 0 & 0 \\ 0 & \mathbf{Q}\left(k\right) & 0 \\ 0 & 0 & \mathbf{R}\left(k\right) \end{bmatrix} \tag{13.21}$$

The matrices on the leading diagonal are the covariances and the off-diagonal sub-blocks are the correlations between the state errors and the process noises.* Although this method requires the use of additional sigma points, it incorporates the noises into the predicted state with the same level of accuracy as the propagated estimation errors. In other words, the estimate is correct to the second order and no Jacobians, Hessians, or other numerical approximations must be calculated.

---

*If correlations exist between the noise terms, Equation 13.21 can be generalized to draw the sigma points from the covariance matrix

$$P^a(k|k) = \begin{bmatrix} \mathbf{P}(k|k) & \mathbf{P}_{xv}\,(k|k) & \mathbf{P}_{xw}\,(k|k) \\ \mathbf{P}_{vx}\,(k|k) & \mathbf{Q}(k) & \mathbf{P}_{vw}\,(k|k) \\ \mathbf{P}_{wx}\,(k|k) & \mathbf{P}_{wv}\,(k|k) & \mathbf{R}(k) \end{bmatrix}$$

Such correlation structures commonly arise in algorithms such as the Schmidt-Kalman filter.[29]

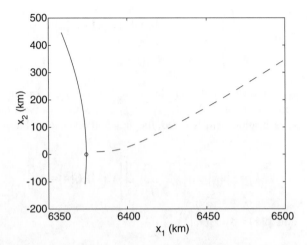

FIGURE 13.5   The reentry problem. The dashed line is the sample vehicle trajectory and the solid line is a portion of the Earth's surface. The position of the radar is marked by a ∘.

The full unscented filter is summarized in Figure 13.4. However, recall that this is the most general form of the UF and many optimizations can be made. For example, if the process model is linear, but the observation model is not, the normal linear Kalman filter prediction equations can be used to calculate $\hat{\mathbf{x}}\,(k+1|k)$ and $\mathbf{P}\,(k+1|k)$. The sigma points would be drawn from the prediction distribution and would only be used to calculate $\hat{\mathbf{z}}\,(k+1|k)$, $\mathbf{P}_{xv}\,(k+1|k)$, and $\mathbf{P}_{vv}\,(k+1|k)$.

The following two sections describe the application of the unscented filter to two case studies. The first demonstrates the accuracy of the recursive filter, and the second considers the problem of an extremely involved process model.

13.6   Case Study: Using the UF with Linearization Errors

This section considers the problem that is illustrated in Figure 13.5: a vehicle entering the atmosphere at high altitude and at very high speed. The position of the body is to be tracked by a radar which accurately measures range and bearing. This type of problem has been identified by a number of authors[11-14] as being particularly stressful for filters and trackers, based on the strongly nonlinear nature of three types of forces that act on the vehicle. The most dominant is aerodynamic drag, which is a function of vehicle speed and has a substantial nonlinear variation in altitude. The second type of force is gravity, which accelerates the vehicle toward the center of the earth. The final type of force is random buffeting. The effect of these forces gives a trajectory of the form shown in Figure 13.5. Initially the trajectory is almost ballistic; however, as the density of the atmosphere increases, drag effects become important and the vehicle rapidly decelerates until its motion is almost vertical. The tracking problem is made more difficult by the fact that the drag properties of the vehicle could be only very crudely known.

In summary, the tracking system should be able to track an object which experiences a set of complicated, highly nonlinear forces. These depend on the current position and velocity of the vehicle, as well as on certain characteristics that are not precisely known. The filter's state space consists of the position of the body (x_1 and x_2), its velocity (x_3 and x_4), and a parameter of its aerodynamic properties (x_5). The vehicle state dynamics are

$$\dot{x}_1\big(k\big) = x_3\big(k\big)$$

$$\dot{x}_2\big(k\big) = x_4\big(k\big)$$

$$\dot{x}_3\big(k\big) = D\big(k\big)x_3\big(k\big) + G\big(k\big)x_1\big(k\big) + v_1\big(k\big)$$

(13.22)

$$\dot{x}_4(k) = D(k)x_4(k) + G(k)x_2(k) + v_2(k)$$

$$\dot{x}_5(k) = v_3(k)$$

where $D(k)$ is the drag-related force term, $G(k)$ is the gravity-related force term, and $v_1(k)$, $v_2(k)$, and $v_3(k)$ are the process noise terms. Defining $R(k) = \sqrt{x_1^2(k) + x_2^2(k)}$ as the distance from the center of the Earth and $V(k) = \sqrt{x_3^2(k) + x_4^2(k)}$ as absolute vehicle speed, the drag and gravitational terms are

$$D(k) = -\beta(k)\exp\left\{\frac{[R_0 - R(k)]}{H_0}\right\}V(k), \quad G(k) = -\frac{Gm_0}{r^3(k)}$$

and $\beta(k) = \beta_0 \exp x_5(k)$

For this example, the parameter values are $\beta_0 = -0.59783$, $H_0 = 13.406$, $Gm_0 = 3.9860 \times 10^5$, and $R_0 = 6374$, and they reflect typical environmental and vehicle characteristics.[13] The parameterization of the ballistic coefficient, $\beta(k)$, reflects the uncertainty in vehicle characteristics.[12] β_0 is the ballistic coefficient of a "typical vehicle," and it is scaled by $\exp x_5(k)$ to ensure that its value is always positive. This is vital for filter stability.

The motion of the vehicle is measured by a radar located at (x_r, y_r). It can measure range r and bearing θ at a frequency of 10Hz, where

$$r_r(k) = \sqrt{\left(x_1(k) - x_r\right)^2 + \left(x_2(k) - y_r\right)^2} + w_1(k)$$

$$\theta(k) = \tan^{-1}\left(\frac{x_2(k) - y_r}{x_1(k) - x_r}\right) + w_2(k)$$

$w_1(k)$ and $w_2(k)$ are zero mean uncorrelated noise processes with variances of lm and 17mrad, respectively.[30] The high update rate and extreme accuracy of the sensor results in a large quantity of extremely high quality data for the filter.

The true initial conditions for the vehicle are

$$x(0) = \begin{pmatrix} 6500.4 \\ 349.14 \\ -1.8093 \\ -6.7967 \\ 0.6932 \end{pmatrix} \text{ and } P(0) = \begin{bmatrix} 10^{-6} & 0 & 0 & 0 & 0 \\ 0 & 10^{-6} & 0 & 0 & 0 \\ 0 & 0 & 10^{-6} & 0 & 0 \\ 0 & 0 & 0 & 10^{-6} & 0 \\ 0 & 0 & 0 & 0 & 0 \end{bmatrix}$$

In other words, the vehicle's coefficient is twice the nominal coefficient.

The vehicle is buffeted by random accelerations,

$$Q(k) = \begin{bmatrix} 2.4064 \times 10^{-5} & 0 & 0 \\ 0 & 2.4064 \times 10^{-5} & 0 \\ 0 & 0 & 0 \end{bmatrix}$$

The initial conditions assumed by the filter are

$$\hat{\mathbf{x}}(0|0) = \begin{pmatrix} 6500.4 \\ 349.14 \\ -1.8093 \\ -6.7967 \\ 0 \end{pmatrix} \text{ and } \mathbf{P}(0|0) = \begin{bmatrix} 10^{-6} & 0 & 0 & 0 & 0 \\ 0 & 10^{-6} & 0 & 0 & 0 \\ 0 & 0 & 10^{-6} & 0 & 0 \\ 0 & 0 & 0 & 10^{-6} & 0 \\ 0 & 0 & 0 & 0 & 1 \end{bmatrix}$$

The filter uses the nominal initial condition and, to offset for the uncertainty, the variance on the initial estimate is 1.

Both filters were implemented in discrete time and observations were taken at a frequency of 10Hz. However, as a result of the intense nonlinearities of the vehicle dynamics equations, the Euler approximation of Equation 13.22 was valid only for small time steps. The integration step was set to be 50ms, which meant that two predictions were made per update. For the unscented filter, each sigma point was applied through the dynamics equations twice. For the EKF, an initial prediction step and relinearization had to be performed before the second step.

The performance of each filter is shown in Figure 13.6. This figure plots the estimated mean squared estimation error (the diagonal elements of $\mathbf{P}(k|k)$) against actual mean squared estimation error (which is evaluated using 100 Monte Carlo simulations). Only x_1, x_3, and x_5 are shown. The results for x_2 are similar to x_1, and the x_4 and x_3 results are the same. In all cases, the unscented filter estimates its mean squared error very accurately, maximizing the confidence of the filter estimates. The EKF, however, is highly inconsistent; the peak mean squared error in x_1 is 0.4km², whereas its estimated covariance is over 100 times smaller. Similarly, the peak mean squared velocity error is 3.4×10^{-4}km²s⁻², which is more than five times the true mean squared error. Finally, x_5 is highly biased, and this bias decreases only slowly over time. This poor performance shows that, even with regular and high quality updates, linearization errors inherent in the EKF can be sufficient to cause inconsistent estimates.

13.7 Case Study: Using the UF with a High-Order Nonlinear System

In many tracking applications, obtaining large quantities of accurate sensor data is difficult. For example, an air traffic control system might measure the location of an aircraft only once every few seconds. When information is scarce, the accuracy of the process model becomes extremely important for two reasons. First, if the control system must obtain an estimate of the target state more frequently than the tracker updates, a predicted tracker position must be used. Different models can have a significant impact on the quality of that prediction.[31] Second, to optimize the performance of the tracker, the limited data from the sensors must be exploited to the greatest degree possible. Within the Kalman filter framework, this can only be achieved by developing the most accurate process model that is practical. However, such models can be high order and nonlinear. The UF greatly simplifies the development and refinement of such models.

This section demonstrates the ease with which the UF can be applied to a prototype Kalman filter-based localization system for a conventional road vehicle. The road vehicle, shown in Figure 13.7, undertakes maneuvers at speeds in excess of 45mph (15ms⁻¹). The position of the vehicle is to be estimated with submeter accuracy. This problem is made difficult by the paucity of sensor information. Only the following sources are available: an inertial navigation system (which is polled only at 10Hz), a set of encoders (also polled at 10Hz), and a bearing-only sensor (rotation rate of 4Hz) which measures bearing to a set of beacons. Because of the low quality of sensor information, the vehicle localization system can meet the performance requirements only through the use of an accurate process model. The model that was developed is nonlinear and incorporates kinematics, dynamics and slip due to tire deformation. It also contains a large number of process noise terms. This model is extremely cumbersome to work with, but the UF obviates the need to calculate Jacobians, greatly simplifying its use.

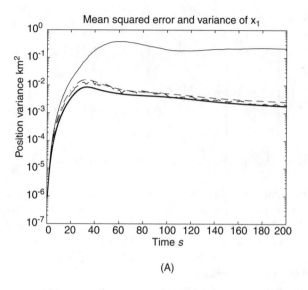

(A)

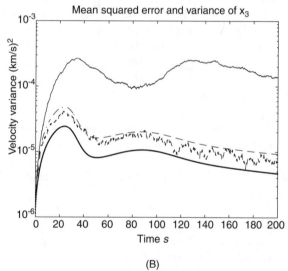

(B)

FIGURE 13.6 The mean squared errors and estimated covariances calculated by an EKF and an unscented filter. (A) Results for x_1. (B) Results for x_3. (C) Results for x_5. In all of the graphs, the solid line is the mean squared error calculated by the EKF, and the dotted line is its estimated covariance. The dashed line is the unscented mean squared error and the dot-dashed line is its estimated covariance. In all diagrams, the EKF estimate is inconsistent but the UT estimate is not.

The model of vehicle motion is developed from the two-dimensional "fundamental bicycle" which is shown in Figure 13.7.[32-34] This approximation, which is conventional for vehicle ride and handling analysis, assumes that the vehicle consists of front and rear virtual wheels.* The vehicle body is the line *FGR* with the front axle affixed at *F* and the rear axle fixed at *R*. The center of mass of the vehicle is located at *G*, a distance *a* behind the front axle and *b* in front of the rear axle. The length of the wheel base is $B = a + b$. The wheels can slip laterally with slip angles α_f and α_r, respectively. The control inputs are the steer angle, δ, and angular speed, ω, of the front virtual wheel.

---

*Each virtual wheel lumps together the kinematic and dynamic properties of the pairs of wheels at the front and rear axles.

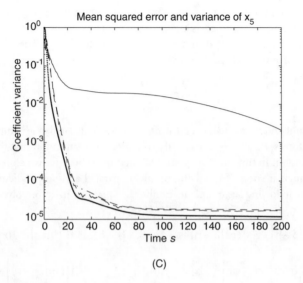

(C)

FIGURE 13.6 (continued).

(A)   (B)

FIGURE 13.7 The actual experimental vehicle and the "fundamental bicycle model" representation used in the design of the vehicle process model. (A) The host vehicle at the test site with sensor suite. (B) Vehicle kinematics.

The filter estimates the position of F, (X_F, Y_F) the orientation of FGR, ψ, and the effective radius of the front wheel, R, (defined as the ratio of vehicle velocity to the rotation rate of the front virtual wheel). The speed of the front wheel is V_F and the path curvature is ρ_F. From the kinematics. the velocity of F is

$$\dot{X}_F = V_F \cos\left[\psi + \delta - \alpha_f\right] \qquad \rho_F = \left(\sin\left[\delta - \alpha_f\right] + \cos\left[\delta - \alpha_f\right]\tan\alpha_r\right)\Big/B$$

$$\dot{Y}_F = V_F \cos\left[\psi + \delta - \alpha_f\right] \quad \text{where} \quad V_F = R\omega\cos\alpha_f$$

$$\dot{\psi} = \rho_F V_F - \dot{\delta} - \dot{\alpha}_f$$

$$\dot{R} = 0$$

The slip angle (α_f) plays an extremely important role in determining the path taken by the vehicle and a model for determining the slip angle is highly desirable. The slip angle is derived from the properties of the tires. Specifically, tires behave as if they are linear torsional springs. The slip ankle on each wheel is proportional to the force that acts on the wheel[33]

$$\alpha_f = F_{yf}/C_{\alpha_f} \qquad \alpha_r = F_{yr}/C_{\alpha_r} \qquad (13.23)$$

C_{α_f} and C_{α_r} are the front and rear wheel lateral stiffness coefficients (which are imprecisely known). The front and rear lateral forces, F_{yf} and F_{yr}, are calculated under the assumption that the vehicle has reached a steady state; at any instant in time the forces are such that the vehicle moves along an arc with constant radius and constant angular speed.[35] Resolving moments parallel and perpendicular to OG, and taking moments about G, the following simultaneous nonlinear equations must be solved:

$$\rho_G m V_G^2 = F_{xf} \sin[\delta-\beta] + F_{yf}\cos[\delta-\beta] + F_{yr}\cos\beta, \quad \beta = \tan^{-1}\left(\left\{b\tan[\delta-\alpha_f] - \alpha\tan\alpha_r\right\}/B\right)$$

$$0 = F_{xf}\cos[\delta-\beta] - F_{yf}\sin[\delta-\beta] + F_{yr}\sin\beta, \quad \rho_G = \cos\beta\left\{\tan[\delta-\alpha_f] + \tan\alpha_r\right\}/B$$

$$0 = F_{xf}\sin\delta + aF_{yf}\cos\delta - F_{yr}, \qquad V_G = V_F\cos[\delta-\alpha_f]\sec\beta$$

m is the mass of the vehicle, V_G and ρ_G are the speed and path curvature of G, and β is the attitude angle (illustrated in Figure 13.7). These equations are solved using a conventional numerical solver[24] to give the tire forces. Through Equation 13.23, these determine the slip angles and, hence, the path of the vehicle. Since C_{α_f} and C_{α_r} must account for modeling errors (such as the inaccuracies of a linear force-slip angle relationship), these were treated as states and their values were estimated.

As this section has shown, a comprehensive vehicle model is extremely complicated. The state space consists of six highly interconnected states. The model is made even more complicated by the fact that that it possesses twelve process noise terms. Therefore, 18 terms must be propagated through the nonlinear process model. The observation models are also very complex. (The derivation and debugging of such Jacobians proved to be extremely difficult.) However, the UF greatly simplified the implementation, tuning, and testing of the filter. An example of the performance of the final navigation system is shown in Figure 13.8. Figure 13.8(a) shows a "figure of eight" route that was planned for the vehicle. This path is highly dynamic (with continuous and rapid changes in both vehicle speed and steer angle) and contains a number of well-defined landmarks (which were used to validate the algorithm). There is extremely good agreement between the estimated and the actual paths, and the covariance estimate ($0.25m^2$ in position) exceeds the performance requirements.

13.8 Multilevel Sensor Fusion

This section discusses how the UT can be used in systems that do not inherently use a mean and covariance description to describe their state. Because the UT can be applied to such systems, it can be used as a consistent framework for multilevel data fusion. The problem of data fusion has been decomposed into a set of hierarchical domains.[36] The lowest levels, Level 0 and Level 1 (object refinement), are concerned with quantitative data fusion problems such as the calculation of a target track. Level 2 (situation refinement) and Level 3 (threat refinement) apply various high-level data fusion and pattern recognition algorithms to attempt to glean strategic and tactical information from these tracks.

The difficulty lies in the fundamental differences in the representation and use of information. On the one hand, the low-level tracking filter provides only mean and covariance information. It does not specify an exact kinematic state from which an expert system could attempt to infer a tactical state. On the other hand, an expert system may be able to predict accurately the behavior of a pilot under a range of situations.

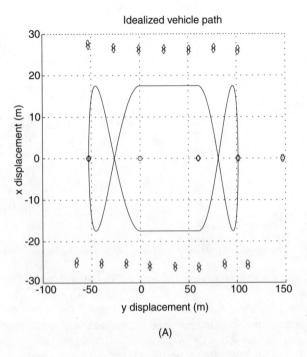

(A)

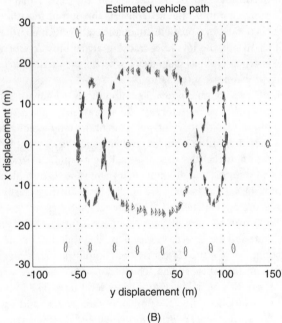

(B)

FIGURE 13.8 The positions of the beacons can be seen in (A) and (B) as the row of ellipses at the top and bottom of the figures.

However, the system does not define a rigorous low-level framework for fusing its predictions with raw sensor information to obtain high-precision estimates suitable for reliable tracking. The practical solution to this problem has been to take the output of standard control and estimation routines, discretize them into a more symbolic form (e.g., "slow" or "fast"), and process them with an expert/fuzzy rule base. The

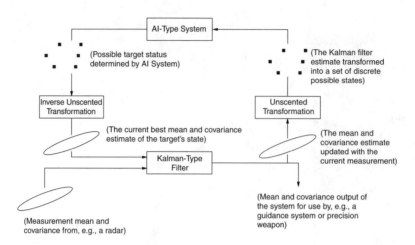

FIGURE 13.9 A possible framework for multilevel information fusion using the unscented transformation.

results of such processing are then converted into forms that can be processed by conventional process technology.

One approach for resolving this problem, illustrated in Figure 13.9, is to combine the different data fusion algorithms together into a single, composite data fusion algorithm that takes noise-corrupted raw sensor data and provides the inferred high-level state. From the perspective of the track estimator, the higher level fusion rules are considered to be arbitrary, nonlinear transformations. From the perspective of the higher level data fusion algorithms, the UT converts the output from the low-level tracker into a set of vectors. Each vector is treated as a possible kinematic state, which is processed by the higher-level fusion algorithms. In other words, the low-level tracking algorithms do not need to understand the concept of higher-level constructs, such as maneuvers, whereas the higher level algorithms do not need to understand or produce probabilistic information.

Consider the problem of tracking an aircraft. The aircraft model consists of two components — a kinematic model, which describes the trajectory of the aircraft for a given set of pilot inputs, and an expert system, which attempts to infer current pilot intentions and predict future pilot inputs. The location of the aircraft is measured using a tracking system, such as a radar.

Some sigma points might imply that the aircraft is making a rapid acceleration, some might indicate a moderate acceleration, and yet others might imply that there is no discernible acceleration. Each of the state vectors produced from the UT can be processed individually by the expert system to predict various possible future states of the aircraft. For some of the state vectors, the expert system will signal air evasive maneuvers and predict the future position of the aircraft accordingly. Other vectors, however, will not signal a change of tactical state and the expert system will predict that the aircraft will maintain its current speed and bearing. The second step of the UT consists of computing the mean and covariance of the set of predicted state vectors from the expert system. This mean and covariance gives the predicted state of the aircraft in a form that can then be fed back to the low-level filter. The important observation to be made is that this mean and covariance reflect the probability that the aircraft will maneuver *even though the expert system did not produce any probabilistic information and the low-level filter knows nothing about maneuvers.*

13.9 Conclusions

This chapter has described some of the important issues arising from the occurrence of nonlinear transformations in practical data fusion applications. Linearization is the most widely used approach for dealing with nonlinearities, but linearized approximations have been shown to yield relatively poor results. In response to this and other deficiencies of linearization, a new technique based on the UT has been developed for directly applying nonlinear transformations to discrete point distributions having specified statistics (such as mean and covariance). An analysis of the new approach reveals that

- The UT is demonstrably superior to linearization in terms of expected error for all absolutely continuous nonlinear transformations. The UT can be applied with non-differentiable functions in which linearization is not possible.
- The UT avoids the derivation of Jacobian (and Hessian) matrices for linearizing nonlinear kinematic and observation models. This makes it conducive to the creation of efficient, general purpose "black box" code libraries.
- Empirical results for several nonlinear transformations that typify those arising in practical applications clearly demonstrate that linearization yields very poor approximations compared to those of the UT.

Beyond analytic claims of unproved accuracy, the UT offers a black box solution to a wide variety of problems arising in both low- and high-level data fusion applications. In particular, it offers a mechanism for seamlessly integrating the benefits of high-level methodologies — such as artificial intelligence, fuzzy logic and neural networks — with the low-level workhorses of modern engineering practice — such as covariance intersection and the Kalman filter.

Acknowledgments

The authors gratefully acknowledge support from IDAK Industries and the University of Oxford.

References

1. Julier, S.J. and Uhlmann, J.K., A non-divergent estimation algorithm in the presence of unknown correlations, *American Control Conf.*, 4, 2369–2373, 1997.
2. Kalman, R.E., A new approach to linear filtering and prediction problems, *Trans. ASME, J. Basic Engineering*, 82, 34–45, 1960.
3. Kushner, H.J., Dynamical equations for optimum non-linear filtering, *J. Differential Equations*, 3, 179–190, 1967.
4. Jazwinski, A.H., *Stochastic Processes and Filtering Theory*, Academic Press, New York, 1970.
5. Maybeck, P.S., *Stochastic Models, Estimation, and Control*, Vol. 2, Academic Press, New York, 1982.
6. Daum, F.E., New exact nonlinear filters, *Bayesian Analysis of Time Series and Dynamic Models*, J.C. Spall, ed., Marcel Drekker, Inc., New York, 199–226, 1988.
7. Gordon, N.J., Salmond, D.J., and Smith, A.F.M., Novel approach to nonlinear/non-Gaussian Bayesian state estimation, *IEEE Proc.-F.*, 140(2), 107–113, 1993.
8. Kouritzin, M.A., On exact filters for continuous signals with discrete observations. Technical report, University of Minnesota, Institute for Mathematics and its Applications, May 1996.
9. Uhlmann, J.K., Algorithms for multiple target tracking, *American Scientist*, 80(2), 128–141, 1992.
10. Sorenson, H.W., Ed. *Kalman Filtering: Theory and Application*, IEEE Press, 1985.
11. Athans, M., Wishner, R.P., and Bertolini, A., Suboptimal state estimation for continuous time nonlinear systems from discrete noisy measurements, *IEEE Trans. Automatic Control*, TAC-13(6), 504–518, 1968.
12. Costa, P.J., Adaptive model architecture and extended Kalman-Bucy filters, *IEEE Trans. Aerospace and Electronic Systems*, AES-30(2), 525–533, 1994.
13. Austin. J.W. and Leondes, C.T., Statistically Linearized Estimation of Reentry Trajectories, *IEEE Trans. Aerospace and Electronic Systems*, AES-17(1), 54–61, 1981.
14. Mehra, R. K. A Comparison of Several Nonlinear Filters for Reentry Vehicle Tracking, *IEEE Trans. Automatic Control*, AC-16(4), 307–319, 1971.
15. Viéville, T. and Sander, P., Using pseudo Kalman filters in the presence of constraints application to sensing behaviours, Technical report, INRIA, April 1992.

16. Dulimov, P.A., Estimation of Ground Parameters for the Control of a Wheeled Vehicle, Master's thesis, The University of Sydney, 1997.

17. Wolfram, S., *The Mathematica Book,* Wolfram Research, 4th edition, 1999.

18. Redfern, D., *Maple V Handbook — Release 4,* 1996.

19. Uhlmann, J.K., Simultaneous map building and localization for real-time applications, technical report, Transfer thesis. University of Oxford, 1994.

20. Julier, S.J. and Uhlmann, J.K. A general method for approximating nonlinear transformations of probability distributions, http://www.robots.ox.ac.uk/~siju, August 1994.

21. Kushner, H.J., Approximations to optimal nonlinear filters, *IEEE Trans. Automatic Control,* AC-12(5), 546–556, 1967.

22. Holtzmann, J., On using perturbation analysis to do sensitivity analysis: derivatives vs. differences, *IEEE Conf. Decision and Control,* 2018–2023, 1989.

23. Julier, S.J., Uhlmann, J.K., and Durrant-Whyte, H.F., A new approach for the nonlinear transformation of means and covariances in linear filters, *IEEE Trans. Automatic Control,* 477–482, March 2000.

24. Press, W.H., Teukolsky, S.A., Vetterling, W.T., and Flannery B.P., *Numerical Recipes in C: The Art of Scientific Computing,* Cambridge University Press, 2nd edition, 1992.

25. Julier, S.J. and Uhlmann, J.K., A consistent, debiased method for converting between polar and Cartesian coordinate systems, *Proc. AeroSense: Acquisition, Tracking and Pointing XI,* SPIE, 3086, 110–121, 1997.

26. Julier, S.J., A skewed approach to filtering, *Proc. of AeroSense: The 12th International Symp. Aerospace/Defense Sensing, Simulation, and Controls,* 3373, 54–65. SPIE, 1998. Signal and Data Processing of Small Targets.

27. Lerro, D. and Bar-Shalom, Y.K., Tracking with Debiased Consistent Converted Measurements vs. EKF. *IEEE Trans. Aerospace and Electonics Systems,* AES-29(3), 1015–1022, 1993.

28. Leonard, J., *Directed Sonar Sensing for Mobile Robot Navigation,* Kluwer Academic Press, Boston, 1991.

29. Schmidt, S.F., Applications of state space methods to navigation problems, *Advanced Control Systems,* C.T. Leondes, Ed., Academic Press, New York, Vol. 3, 293–340, 1966.

30. Chang, C.B., Whiting, R.H., and Athans, M. On the state and parameter estimation for maneuvering reentry vehicles, *IEEE Trans. Automatic Control,* AC-22, 99–105, 1977.

31. Julier, S.J., *Comprehensive Process Models for High-Speed Navigation,* Ph.D. thesis, University of Oxford, 1997.

32. Ellis, J.R., *Vehicle Handling Dynamics,* Mechanical Engineering Publications, London, 1994.

33. Wong, J.Y., *Theory of Ground Vehicles,* John Wiley & Sons, New York, 2nd edition, 1993.

34. Dixon, J.C., *Tyres, Suspension and Handling,* Cambridge University Press, Cambridge, U.K., 1991.

35. Julier, S.J. and Durrant-Whyte, H.F., Navigation and parameter estimation of high speed road vehicles, *Robotics and Automation Conf.,* 101–105, 1995.

36. Klein, L., *Sensor and Data Fusion Concepts and Applications,* SPIE, 2nd edition, 1965.

14

Random Set Theory for Target Tracking and Identification

14.1 Introduction **14**-3
  The "Bayesian Iceberg": Models, Optimality, Computability •
  Why Multisource, Multitarget, Multi-Evidence Problems Are
  Tricky • Finite-Set Statistics (FISST) • Why Random Sets?

14.2 Basic Statistics for Tracking and Identification............ **14**-10
  Bayes Recursive Filtering • Constructing Likelihood
  Functions from Sensor Models • Constructing Markov
  Densities from Motion Models • Optimal State Estimators

14.3 Multitarget Sensor Models **14**-12
  Case I: No Missed Detections, No False Alarms • Case II:
  Missed Detections • Case III: Missed Detection and False
  Alarms • Case IV: Multiple Sensors

14.4 Multitarget Motion Models... **14**-14
  Case I: Target Number Does Not Change • Case II: Target
  Number Can Decrease • Case III: Target Number Can Increase
  and Decrease

14.5 The FISST Multisource-Multitarget Calculus **14**-15
  The Belief-Mass Function of a Sensor Model • The Belief-
  Mass Function of a Motion Model • The Set Integral and Set
  Derivative • "Turn-the-Crank" Rules for the FISST Calculus

14.6 FISST Multisource-Multitarget Statistics...................... **14**-19
  Constructing True Multitarget Likelihood Functions •
  Constructing True Multitarget Markov Densities • Multitarget
  Prior Distributions • Multitarget Posterior Distributions •
  Expected Values and Covariances • The Failure of the Classical
  State Estimators • Optimal Multitarget State Estimators •
  Cramér-Rao Bounds for Multitarget State Estimators •
  Multitarget Miss Distance

14.7 Optimal-Bayes Fusion, Tracking, ID............................. **14**-23
  Multisensor-Multitarget Filtering Equations • A Short History
  of Multitarget Filtering • Computational Issues in Multitarget
  Filtering • Optimal Sensor Management

14.8 Robust-Bayes Fusion, Tracking, ID.............................. **14**-26
  Random Set Models of "Ambiguous" Data • Forms of
  Ambiguous Data • True Likelihood Functions for Ambiguous
  Data • Generalized Likelihood Functions • Posteriors
  Conditioned on Ambiguous Data • Practical Generalized
  Likelihood Functions • Unified Multisource-Multitarget Data
  Fusion

Ronald Mahler
Lockheed Martin

14.9   Summary and Conclusions **14-30**
          Acknowledgments... **14-30**
          References.. **14-30**

The subject of this chapter is finite-set statistics (FISST), which is described extensively in *Mathematics of Data Fusion*,[1] and summarized in the recent article, "An Introduction to Multisource-Multitarget Statistics and its Applications."[2] FISST provides a fully unified, scientifically defensible, probabilistic foundation for the following aspects of multisource, multitarget, multiplatform data fusion: (1) multisource integration (i.e., detection, identification, and tracking) based on Bayesian filtering and estimation;[3-7] (2) sensor management using control theory;[5,8,9] (3) performance evaluation using information theory;[10-13] (4) expert systems theory (e.g., fuzzy logic, the Dempster-Shafer theory of evidence, and rule-based inference);[14-16] (5) distributed fusion;[17] and (5) aspects of situation/threat assessment.[18]

The core of FISST is a multisource-multitarget differential and integral calculus based on the fact that belief-mass functions are the multisensor-multitarget counterparts of probability-mass functions. One purpose of this calculus is to enable signal processing engineers to directly generalize conventional, engineering-friendly statistical reasoning to multisensor, multitarget, multi-evidence applications. A second purpose is to extend Bayesian (and other probabilistic) methodologies so that they are capable of dealing with (1) imperfectly characterized data and sensor models and (2) true sensor models and true target models for multisource-multitarget problems. One consequence is that FISST encompasses certain expert-system approaches that are often described as "heuristic" — fuzzy logic, the Dempster-Shafer theory of evidence, and rule-based inference — as special cases of a single probabilistic paradigm.

FISST has attracted a great deal of positive attention — and two distinct forms of criticism — since it was first described in 1994. "An Introduction to Multisource-Multitarget Statistics and its Applications"[2] was written as a response to both reactions, and this chapter is essentially a distillation of that monograph.

On the one hand, many engineering researchers have complained that *Mathematics of Data Fusion* is difficult to understand. "An Introduction to Multisource-Multitarget Statistics and its Applications" explains that if the foundations outlined in that book are taken for granted, FISST can — by design — be reduced to a relatively simple "Statistics 101" formalism suitable for real applications using real data. Some applications currently being investigated at the applied-research level are[2]

- Multisource datalink fusion and target identification,[19]
- Naval passive-acoustic anti-submarine fusion and target identification,[20]
- Air-to-ground target identification using SAR,[21]
- Scientific performance evaluation of multisource-multitarget data fusion algorithms,[13,22,23]
- Unified detection and tracking using true multitarget likelihood functions and approximate nonlinear filters,[24]
- Joint tracking, pose estimation, and target identification using HRRR.[25]

In recent years, FISST has been the target of a few attacks that can be summarized by the following statement: "The multisource-multitarget engineering problems addressed by FISST actually require nothing more complicated than Bayes' rule, which means that FISST is of mere theoretical interest at best and, at worst, is nothing more than pointless mathematical obfuscation." "Multisource-Multitarget Statistics" demonstrates that this assertion is extraordinary — less in the ignorance that it displays regarding FISST than in the ignorance that it displays regarding Bayes' rule. Venturing away from standard applications addressed by standard textbooks and applying Bayesian approaches in a "cookbook" fashion (rather than using a FISST approach) can create severe difficulties. This chapter highlights these difficulties and shows how and why FISST resolves them.

The chapter is organized as follows. Section 14.1 introduces the basic practical issues underlying Bayesian tracking and identification. Section 14.2 summarizes the basic statistical foundations of single-sensor, single-target tracking and identification. Section 14.3 introduces the concept of a multisource-multitarget measurement model; multitarget motion models are likewise introduced in Section 14.4. The mathematical core of the approach — the FISST multisource-multitarget integral and differential calculus — is summarized in Section 14.5. The basic concepts of multisource-multitarget statistics — especially those of true multitarget measurement models and true multitarget motion models — are described in Section 14.6. Optimal-Bayes and Robust-Bayes multisource-multitarget fusion, detection, tracking, and identification (including sensor management) are described in Sections 14.7 and 14.8, respectively. Conclusions are presented in Section 14.9.

14.1 Introduction

This section describes the problems that FISST is meant to address and summarizes the FISST approach for addressing them. The section is organized as follows. Sections 14.1.1 and 14.1.2 describe some basic engineering issues in single-target and multitarget Bayesian inference — the "Bayesian Iceberg." The FISST approach is summarized in in Section 14.1.3, and Section 14.1.4 shows why this approach is necessary if the "Bayesian Iceberg" is to be confronted successfully.

14.1.1 The "Bayesian Iceberg": Models, Optimality, Computability

Recursive Bayesian nonlinear filtering and estimation has become the most accepted standard for developing algorithms that are optimal, theoretically defensible, and practical. However, the success of any optimal-Bayes technique hinges on a number of subtle issues. In recent years, shortcomings in conventional Bayesian thinking — and corresponding desires to expand the conceptual scope of Bayesian methods to deal with them — have become evident. The purpose of this section is to explain why this is the case. Our jumping-off point is Bayes' rule:

$$f_{posterior}\left(\mathbf{x}\,\middle|\,\mathbf{z}\right) = \frac{f\left(\mathbf{x}\,\middle|\,\mathbf{z}\right)f_{prior}\left(\mathbf{x}\right)}{f\left(\mathbf{z}\right)} \tag{14.1}$$

where:

> $\mathbf{x}$ denotes the unknown quantities of interest (e.g., position, velocity, and target class)
> $f_{prior}\left(\mathbf{x}\right)$, the prior distribution, encapsulates our previous knowledge about $\mathbf{x}$
> $\mathbf{z}$ represents new data
> $f\left(\mathbf{x}\,\middle|\,\mathbf{z}\right)$, the likelihood function, describes the generation of data
> $f_{posterior}\left(\mathbf{x}\,\middle|\,\mathbf{z}\right)$, the posterior distribution, encapsulates our current knowledge about $\mathbf{x}$
> $f\left(\mathbf{z}\right)$ is the normalization constant

If time has expired before collection of the new information, $\mathbf{z}$, then $f_{prior}\left(\mathbf{x}\right)$ cannot be immediately used in Bayes' rule. It must first be extrapolated to a new prior $f_{prior}^{+}\left(\mathbf{x}\right)$ that accounts for the uncertainties caused by possible interim target motion. This extrapolation is accomplished using either the Markov time-prediction integral[26]

$$f_{prior}^{+}\left(\mathbf{x}\right) = \int f^{+}\left(\mathbf{x}\,\middle|\,\mathbf{y}\right)f_{prior}\left(\mathbf{y}\right)d\mathbf{y} \tag{14.2}$$

or solution of the Fokker-Planck equation (FPE).[27] In Equation 14.2, the density $f^{+}\left(\mathbf{x}\,\middle|\,\mathbf{y}\right)$ is the Markov transition density that describes the likelihood that the target will have state $\mathbf{x}$ if it previously had state $\mathbf{y}$.

The density, $f_{posterior}\left(\mathbf{x}\,\middle|\,\mathbf{z}\right)$, contains all relevant information about the unknown state variables (i.e., position, velocity, and target identity) contained in $\mathbf{x}$. One could, in principle, plot graphs or contour maps of $f_{posterior}(\mathbf{x}|\mathbf{z})$ in real time. However, this presupposes the existence of a human operator trained to interpret them. Since current operational reality in most data fusion and tracking applications consists of fewer and fewer operators overwhelmed by more and more data, full automation is a necessity. To render the information in $f_{posterior}(\mathbf{x}|\mathbf{z})$ available for fully automated real-time applications, we need a Bayes-optimal state estimator that extracts an estimate $\hat{\mathbf{x}}$ of the actual target state from the posterior. The maximum *a posteriori* (MAP) and expected *a posteriori* (EAP) estimators

$$\hat{\mathbf{x}}^{MAP} = \arg\max_x f_{posterior}\left(\mathbf{x}|\mathbf{z}\right) \qquad \hat{\mathbf{x}}^{EAP} = \int x f_{posterior}\left(\mathbf{x}|\mathbf{z}\right) d\left(\mathbf{x}\right) \qquad (14.3)$$

are the most familiar Bayes-optimal state estimators. (Here, $\hat{\mathbf{x}} = \arg\max_x f(\mathbf{x})$ means that $f(\hat{\mathbf{x}})$ is the largest possible value of $f(\mathbf{x})$.)

The current near-sacrosanct status of the Bayesian approach is due in large part to the fact that it leads to provably optimal algorithms within a simple conceptual framework — one that imposes no great mathematical demands on its practitioners. Since engineering mathematics is a tool and not an end in itself, this simplicity is a great strength — but also a great weakness. Recent years have seen the emergence of a "cookbook Bayesian" viewpoint, which seems to consist of the belief that it is possible to appear deeply authoritative about any engineering research and development problem — and at the same time avoid careful thinking — by (1) writing down Bayes' rule, (2) uttering the sacred incantation "Bayes-optimal" (usually without a clue as to what this phrase actually means), (3) declaring victory, and then (4) portraying complacency toward possible unexpected difficulties as a sign of intellectual superiority. What such a belief reveals is a failure to grasp the fact that in and of itself, Bayes' rule is nearly content-free — its proof requires barely a single line. Its power derives from the fact that it is merely the visible tip of a conceptual iceberg, the existence of which tends to be forgotten precisely because the rest of the iceberg is taken for granted. In particular, both the optimality and the simplicity of the Bayesian frame-work can be taken for granted only within the confines of standard applications addressed by standard textbooks. When one ventures out of these confines one must exercise proper engineering prudence — which includes verifying that standard textbook assumptions still apply.

14.1.1.1 The Bayesian Iceberg: Sensor Models

Bayes' rule exploits, to the best possible advantage, the high-fidelity knowledge about the sensor contained in the likelihood function $f(\mathbf{z}|\mathbf{x})$. If $f(\mathbf{z}|\mathbf{x})$ is too imperfectly understood, then an algorithm will "waste" a certain amount N_{sens} of data trying (and perhaps failing) to overcome the mismatch between model and reality. Many forms of data (e.g., generated by tracking radars) are well enough characterized that $f(\mathbf{z}|\mathbf{x})$ can be constructed with sufficient fidelity. Other kinds of data (e.g., synthetic aperture radar (SAR) images or high range resolution radar (HRRR) range profiles) are proving to be so difficult to simulate that there is no assurance that sufficiently high fidelity will ever be achieved, particularly in real-time operation. This is the point at which "cookbook Bayesian" complacency enters the scene: we jot down Bayes' rule and declare victory. In so doing, we have either failed to understand that there is a potential problem — that our algorithm is Bayes-optimal with respect to an imaginary sensor unless we have the provably true likelihood $f(\mathbf{z}|\mathbf{x})$ — or we are playing a shell game. That is, we are avoiding the real algorithmic issue (what to do when likelihoods cannot be sufficiently well characterized) and instead implicitly "passing the buck" to the data simulation community.

Finally, there are types of data — features extracted from signatures, English-language statements received over link, and rules drawn from knowledge bases — that are so ambiguous (i.e., poorly understood from a statistical point of view) that probabilistic approaches are not obviously applicable. Rather than seeing this as a gap in Bayesian inference that needs to be filled, the "cookbook Bayesian" viewpoint sidesteps the problem and frowns upon those who attempt to fill the gap with heuristic approaches such as fuzzy logic.

Even if $f(\mathbf{z}|\mathbf{x})$ can be determined with sufficient fidelity, multitarget problems present a new challenge. We will "waste" data — or worse — unless we find the corresponding provably true multitarget likelihood — the specific function $f(\mathbf{z}_1,\ldots,\mathbf{z}_m|\mathbf{x}_1,\ldots,\mathbf{x}_n)$ that describes, with the same high fidelity as $f(\mathbf{z}|\mathbf{x})$, the likelihood that the sensor will collect observations $\mathbf{z}_1,\ldots,\mathbf{z}_m$ (m random) given the presence of targets with states $\mathbf{x}_1,\ldots\mathbf{x}_n$ (n random). Again, "cookbook Bayesian" complacency overlooks the problem — that our boast of "Bayes-optimality" is hollow unless we can construct the provably true $f(\mathbf{z}_1,\ldots,\mathbf{z}_m|\mathbf{x}_1,\ldots\mathbf{x}_n)$ — or encourages us to play another shell game, constructing a heuristic multitarget likelihood and implicitly assuming that it is true.

14.1.1.2 The Bayesian Iceberg: Motion Models

Much of what has been said about likelihoods $f(\mathbf{z}|\mathbf{x})$ applies with equal force to Markov densities $f^+(\mathbf{x}|\mathbf{y})$. The more accurately that $f^+(\mathbf{x}|\mathbf{y})$ models target motion, the more effectively Bayes' rule will do its job. Otherwise, a certain amount N_{targ} of data must be expended in overcoming poor motion-model selection. Once again, however, what does one do in the multitarget situation if $f^+(\mathbf{x}|\mathbf{y})$ is truly accurate? We must find the provably true multitarget Markov transition density — i.e., the specific function $f(\mathbf{x}_1,\ldots,\mathbf{x}_n|\mathbf{y}_1,\ldots\mathbf{y}_{n'})$ that describes, with the same high fidelity as $f^+(\mathbf{x}|\mathbf{y})$ how likely it is that a group of targets that previously were in states $\mathbf{y}_1,\ldots\mathbf{y}_{n'}$ (n' random) will now be found in states $\mathbf{x}_1,\ldots\mathbf{x}_n$ (n random). The "cookbook Bayesian" viewpoint encourages us to simply assume that

$$f(\mathbf{x}_1,\ldots,\mathbf{x}_n|\mathbf{y}_1,\ldots,\mathbf{y}_{n'}) = f^+(\mathbf{x}_1|\mathbf{y}_1)\ldots f^+(\mathbf{x}_n|\mathbf{y}_{n'})$$

meaning, in particular, that the number of targets is constant and target motions are uncorrelated. However, in real-world scenarios, targets can appear (e.g., MIRVs and decoys emerging from a ballistic missile re-entry vehicle) or disappear (e.g., aircraft that drop beneath radar coverage) in correlated ways. Consequently, multitarget filters that assume uncorrelated motion and/or constant target number may perform poorly against dynamic multitarget environments, for the same reason that single-target trackers that assume straight-line motion may perform poorly against maneuvering targets. In either case, data is "wasted" in trying to overcome — successfully or otherwise — the effects of motion-model mismatch.

14.1.1.3 The Bayesian Iceberg: Output Generation

When we are faced with the problem of extracting an "answer" from the posterior distribution, complacency may encourage us to blindly copy state estimators from textbooks — e.g. the MAP and EAP estimators of Equation 14.3 — or define ad hoc ones. Great care must be exercised in the selection of a state estimator, however. If the state estimator has unrecognized inefficiencies, then a certain amount, N_{est}, of data will be unnecessarily "wasted" in trying to overcome them, though not necessarily with success. For example, the EAP estimator plays an important role in theory, but often produces erratic and inaccurate solutions when the posterior is multimodal.

Another example involves joint-estimation applications in which the state $\mathbf{x}$ has the form $\mathbf{x} = (\mathbf{u}, \mathbf{v})$, where $\mathbf{u}$, $\mathbf{v}$ are two different kinds of state variables — e.g. kinematic state variables $\mathbf{u}$ versus target-identity state variables $\mathbf{v}$. In this case, we may be tempted to treat $\mathbf{u}$, $\mathbf{v}$ as nuisance parameters and compute separate marginal-MAP estimates:

$$\hat{\mathbf{u}}^{MAP} = \arg\max_{\mathbf{u}} \int f_{posterior}\left(\mathbf{u},\mathbf{v}|\mathbf{z}\right)d\mathbf{v} \qquad \hat{\mathbf{v}}^{MAP} = \arg\max_{\mathbf{v}} \int f_{posterior}\left(\mathbf{u},\mathbf{v}|\mathbf{z}\right)d\mathbf{u} \qquad (14.4)$$

Because integration loses information about the state variable being regarded as a nuisance parameter, estimators of this type can converge more slowly than a joint estimator. They will also produce noisy, unstable solutions when $\mathbf{u}$, $\mathbf{v}$ are correlated and the signal-to-noise ratio is not large.[28]

In the multitarget case, the dangers of taking state estimation for granted become even more acute. For example, we may fail to notice that the multitarget versions of the standard MAP and EAP estimators are not even defined, let alone provably optimal. The source of this failure is, moreover, not some abstruse point in theoretical statistics. Rather, it is a "cookbook Bayesian" carelessness that encourages us to overlook unexpected difficulties caused by a familiar part of everyday engineering practice: *units of measurement.*

14.1.1.4 The Bayesian Iceberg: Formal Optimality

The failure of the standard Bayes-optimal state estimators in the multitarget case has far-reaching consequences for *optimality.* One of the reasons that the "cookbook Bayesian" viewpoint has become so prevalent is that — because well-known textbooks have already done the work for us — we can, without expended effort or cost, proclaim the "optimality" of whatever algorithms we propose. A key point that is often overlooked, however, is that many of the classical Bayesian optimality results depend on certain seemingly esoteric mathematical concerns. In perhaps ninety-five percent of all engineering applications — which is to say, standard applications covered by the standard textbooks — these concerns can be safely ignored. Danger awaits the "cookbook Bayesian" viewpoint in the other five percent. Such is the case with multitarget applications. Because the standard Bayes-optimal state estimators fail in the multitarget case, we must construct new multitarget state estimators and prove that they are well behaved. This means that words like "topology" and "measurable" can no longer be swept under the rug.

14.1.1.5 The Bayesian Iceberg: Computability

If the simplicity of Equations 14.1 and 14.3 cannot be taken for granted in so far as modeling and optimality are concerned, then such is even more the case when it comes to computational tractability.[29] The prediction integral and Bayes normalization constant must be computed using numerical integration, and — since an infinite number of parameters are required to characterize $f_{posterior}(\mathbf{x}|\mathbf{y})$ in general — approximation is unavoidable. A "cookbook Bayesian" viewpoint may tempt us to apply blindly (deceptively) easy-to-understand textbook techniques that seem to promise high (i.e., $O(N)$ or $O(N \log N)$) computational efficiencies per data-collection cycle. Naive approximations, however, create the same difficulties as model-mismatch problems. An algorithm must "waste" a certain amount N_{appx} of data overcoming — or failing to overcome — accumulation of approximation error, numerical instability, etc.

One example is the use of central finite-difference schemes to solve the Fokker-Planck equation (FPE) for f_{prior}^{+}. In filtering problems, the convection term of the FPE dominates the diffusion term. Under such circumstances, central differencing results in loss of probability mass (as well as creation of negative probability), not only at the boundaries but throughout the region of interest, often resulting in poor solutions and numerical instability. This fact has long been well known in the computational fluid dynamics community.[30] Not only has this problem been cited as one of "Seven Deadly Sins of Numerical Computation,"[31] it is such a well-known error that it is cited as such in *Numerical Recipes in C.*[32]

14.1.1.6 The Bayesian Iceberg: Robustness

The engineering issues addressed in the previous sections (measurement-model mismatch, motion-model mismatch, inaccurate or slowly-convergent state estimators, accumulation of approximation error, numerical instability) can be collectively described as problems of robustness — i.e., the "brittleness" that Bayesian approaches can exhibit when reality deviates too greatly from assumption. One might be tempted to argue that, in practical application, these difficulties can be overcome by simple *brute force* — i.e., by assuming that the data-rate is high enough to permit a large number of computational cycles per unit time. In this case — or so the argument goes — the algorithm will override its internal inefficiencies because the total amount N_{data} of data that is collected is much larger than the amount $N_{ineffic} \stackrel{\Delta}{=} N_{sens}+N_{targ}+N_{est}+N_{apps}$ of data required to overcome those inefficiencies.

If this were the case, there would be few tracking and target identification problems left to solve. Most current challenging problems are challenging *either because data rates are not sufficiently high or because brute force computation cannot be accomplished in real time.* This unpleasant reality is particularly evident in target-classification problems. An optimal-Bayes target identification algorithm that is forced to use less-than-high-fidelity sensor models has a tendency to be *highly confident about incorrect identifications.*

This is because it treats its imperfect target models as though they are perfect, thus potentially interpreting a weak or spurious match (between data and target model) as a high-likelihood match. Because data is either too sparse or too low-quality, not enough information is available to compensate for avoidable algorithm inefficiencies. The brute force approach is likely to fail even when data rates are large. In tracking applications, for example, Bayes filtering algorithms will usually be *barely tractable for real-time operation even if they are ideal* (i.e., $N_{ineffic} = 0$ and even if they have $O(N)$ computational efficiency.)[2] This is because N (which typically has the form $N = N_0^d$ where d is the dimensionality of the problem) is *already very large* — which means that the algorithm designer must use all possible means to reduce computations, not increase them. In such situations, brute force means the same thing as *non-real-time operation*. Such difficulties will be even more pronounced in multitarget situations where d will be quite large.

14.1.2 Why Multisource, Multitarget, Multi-Evidence Problems Are Tricky

Given the technical community's increased understanding of the actual complexity of real signature and other kinds of data, it is no longer credible to invoke Bayesian filtering and estimation as a cookbook panacea. Likewise, given the technical community's increased understanding of the actual complexity of multitarget problems, it is not credible to propose yet another ad hoc multitarget tracking approach. One needs systematic and fully probabilistic methodologies for

- Modeling uncertainty in poorly characterized likelihoods
- Modeling ambiguous data and likelihoods for such data
- Constructing multisource likelihoods for ambiguous data
- Efficiently fusing data from all sources (ambiguous or otherwise)
- Constructing provably true (as opposed to heuristic) multisource-multitarget likelihood functions from the underlying sensor models of the sensors
- Constructing provably true (as opposed to heuristic) multitarget Markov densities, from the underlying motion models of the targets, that account for target motion correlations and changes in target number
- Constructing stable, efficient, and provably optimal multitarget state estimators that address the failure of the classical Bayes-optimal estimators — in particular, that simultaneously determine target number, target kinematics, and target identity without resort to operator intervention or optimal report-to-track association.

14.1.3 Finite-Set Statistics (FISST)

One of the major goals of finite-set statistics (FISST) is to address the "Bayesian Iceberg" issues described in Sections 14.1.1 and 14.1.2. FISST deals with imperfectly characterized data and/or measurement models by extending Bayesian approaches in such a way that they are robust with respect to these ambiguities. This *robust-Bayes* methodology is described more fully in Section 14.1.7. FISST deals with the difficulties associated with multisource and/or multitarget problems by *directly* extending engineering-friendly single-sensor, single-target statistical calculus to the multisensor-multitarget realm. This *optimal-Bayes* methodology is described in Sections 14.3 through 14.7. Finally, FISST provides mathematical tools that may help address the formidable computational difficulties associated with multisource-multitarget filtering (whether optimal or robust). Some of these ideas are discussed in Section 14.7.3.

The basic approach is as follows. A suite of known sensors transmits, to a central data fusion site, the observations they collect regarding targets whose number, positions, velocities, identities, threat states, etc. are unknown. Then

1. Reconceptualize the sensor suite as a single sensor (a "global sensor").
2. Reconceptualize the target set as a single target (a "global target") with multitarget state $X = \{\mathbf{x}_1,\ldots,\mathbf{x}_n\}$ (a "global state").

3. Reconceptualize the set $Z = \{\mathbf{z}_1,\ldots,\mathbf{z}_m\}$ of observations, collected by the sensor suite at approximately the same time, as a single measurement (a "global measurement") of the "global target" observed by the "global sensor."

4. Represent statistically ill-characterized ("ambiguous") data as random closed subsets Θ of (multisource) observation space (see Section 14.8). Thus, in general, $Z = \{\mathbf{z}_1,\ldots,\mathbf{z}_m, \Theta_1,\ldots,\Theta_{m'}\}$.

5. Just as single-sensor, single-target data can be modeled using a measurement model $Z = h(\mathbf{x}, \mathbf{W})$, model multitarget multisensor data using a multisensor-multitarget measurement model — a randomly varying finite set $\Sigma = T(X) \cup C(X)$ (See Section 14.3.).

6. Just as single-target motion can be modeled using a motion model $X_{k+1} = \Phi_k(\mathbf{x}_k,\mathbf{V}_k)$, model the motion of multitarget systems using a multitarget motion model — a randomly varying finite set $\Gamma_{k+1} = \Phi_k(X_k,V_k) \cup B_k(X_k)$. (See Section 14.4.)

Given this, we can *mathematically reformulate multisensor, multitarget estimation problems as single-sensor, single-target problems*. The basis of this reformulation is the concept of belief-mass. Belief-mass functions are nonadditive generalizations of probability-mass functions. (Nevertheless, they are not heuristic: they are equivalent to probability-mass functions on certain abstract topological spaces.[1]) That is,

- Just as the probability-mass function $p\,(S|\mathbf{x}) = \Pr(\mathbf{Z} \in S)$ is used to describe the generation of conventional data, $\mathbf{z}$, use belief-mass functions of the general form $\rho(\Theta|\mathbf{x}) = \Pr(\Theta \subseteq \Sigma|\mathbf{x})$ to describe the generation of ambiguous data Θ. (See Section 14.8.)

- Just as the probability-mass function $p(S|\mathbf{x}) = \Pr(\mathbf{Z} \in S)$ of a single-sensor, single-target measurement model, $\mathbf{Z}$, is used to describe the statistics of ordinary data, use the belief-mass function $\beta(S|X) = \Pr(\Sigma \subseteq S)$ of a multisource-multitarget measurement model Σ to describe the statistics of multisource-multitarget data.

- Just as the probability-mass function $p_{k+1|k}(S|\mathbf{x}_k) = \Pr(\mathbf{X}_k \in S)$ of a single-target motion model $\mathbf{X}_k$ is used to describe the statistics of single-target motion, use the belief-mass function $\beta_{k+1|k}(S|X_k) = \Pr(\Gamma_{k+1} \subseteq S)$ of a multitarget motion model Γ_{k+1} to describe multitarget motion.

The FISST multisensor-multitarget differential and integral calculus is what transforms these mathematical abstractions into a form that can be used in practice.

- Just as the likelihood function, $f(\mathbf{z}|\mathbf{x})$, can be derived from $p(S|\mathbf{x})$ via differentiation, so the true multitarget likelihood function $f(Z|X)$ can be derived from $\beta(S|X)$ using a generalized differentiation operator called the *set derivative*. (See Section 14.6.1.)

- Just as the Markov transition density $f_{k+1|k}(\mathbf{x}_{k+1}|\mathbf{x}_k)$ can be derived from $p_{k+1|k}(S|\mathbf{x}_k)$ via differentiation, so the true multitarget Markov transition density index can be derived from $\beta_{k+1|k}(S|X_k)$ via set differentiation.

- Just as $f(\mathbf{z}|\mathbf{x})$ and $p(S|\mathbf{x})$ are related by $p(S|\mathbf{x}) = \int_S f(\mathbf{z}|\mathbf{x})dx$, so $f(Z|X)$ and $\beta(Z|X)$ are related by $\beta(Z|X) = \int_S f(Z|X)\delta Z$ where the integral is now a multisource-multitarget *set integral*.

Accordingly, let $Z^{(k)} = \{Z_1,\ldots,Z_k\}$ be a time sequence of multisource-multitarget observations. This enables true *multitarget posterior distributions* $f_{k|k}(X|Z^{(k)})$ to be created from the true multisource, multitarget likelihood (see Section 14.6.4):

$$f_{k|k}\left(\varnothing\,\big|Z^{(k)}\right) = \text{posterior likelihood that no targets are present}$$

$$f_{k|k}\left(\{\mathbf{x}_1,\ldots,\mathbf{x}_n\}\,\big|Z^{(k)}\right) = \text{posterior likelihood of } n \text{ targets with states } \mathbf{x}_1,\ldots,\mathbf{x}_n$$

From these distributions, simultaneous, provably optimal estimates of target number, kinematics, and identity can be computed without resorting to the optimal report-to-track assignment characteristic of

TABLE 14.1   Mathematical Parallels between Single-Sensor, Single-Target Statistics and Multisensor, Multitarget Statistics

Random Vector, Z	Finite Random Set, Σ
Sensor, $\odot$	Global sensor, $\odot^{*}$
Target, $\otimes$	Global target, $\otimes^{*}$
Observation, $\mathbf{z}$	Observation-set, Z
Parameter, $\mathbf{x}$	Parameter-set, X
Sensor model, $\mathbf{z} = h(\mathbf{x}, \mathbf{W})$	Multitarget sensor model, $Z = T(X) \cup C(X)$
Motion model, $\mathbf{x}_{k+1} = \Phi_k\left(\mathbf{x}_k, \mathbf{V}_k\right)$	Multitarget, multi-motion, $X_{k+1} = \Phi_k\left(X_k, \mathbf{V}_k\right) \cup B_k(X_k)$
Differentiation, $\dfrac{dp}{dz}$	Set differentiation, $\dfrac{\delta\beta}{\delta Z}$
Integration, $\int_s f\left(\mathbf{z}\|\mathbf{x}\right) d\mathbf{z}$	Set integration, $\int_s f\left(Z\|X\right)\delta Z$
Probability-mass function, $p\left(S\|\mathbf{x}\right)$	Belief-mass function, $\beta\left(S\|X\right)$
Likelihood function, $f\left(\mathbf{z}\|\mathbf{x}\right)$	Multitarget likelihood function, $f\left(Z\|X\right)$
Posterior density, $f_{k\|k}\left(\mathbf{x}\|Z^k\right)$	Multitarget posterior, $f_{k\|k}\left(X\|Z^{(k)}\right)$
Markov densities, $f_{k+1\|k}\left(\mathbf{x}_{k+1}\|\mathbf{x}_k\right)$	Multitarget Markov densities, $f_{k+1\|k}\left(X_{k+1}\|X_k\right)$

multihypothesis approaches (see Section 14.6.7). Finally, these fundamentals enable both optimal-Bayes and robust-Bayes multisensor-multitarget data fusion, detection, tracking, and identification (see Sections 14.7.1 and 14.8).

Table 14.1 summarizes the direct mathematical parallels between the world of single-sensor, single-target statistics and the world of multisensor, multitarget statistics. This parallelism is so close that general statistical methodologies can, with a bit of prudence, be directly translated from the single-sensor, single-target case to the multisensor-multitarget case. That is, the table can be thought of as a dictionary that establishes a direct correspondence between the words and grammar in the random-vector language and cognate words and grammar of the random-set language. Consequently, any "sentence" (any concept or algorithm) phrased in the random-vector language can, in principle, be directly "translated" into a corresponding sentence (corresponding concept or algorithm) in the random-set language. This process can be encapsulated as a general methodology for attacking multisource-multitarget data fusion problems:

Almost-Parallel Worlds Principle (APWOP) — Nearly any concept or algorithm phrased in random-vector language can, in principle, be directly translated into a corresponding concept or algorithm in the random-set language.[4,5,10]

The phrase "almost-parallel" refers to the fact that the correspondence between dictionaries is not precisely one-to-one. For example, vectors can be added and subtracted, whereas finite sets cannot. Nevertheless, the parallelism is complete enough that, provided some care is taken, 100 years of accumulated knowledge about single-sensor, single-target statistics can be directly brought to bear on multisensor-multitarget problems.

The following simple example has been used to illustrate the APWOP since 1994. The performance of a multitarget data fusion algorithm can be measured by constructing information-based measures of effectiveness, as shown in the following multitarget generalization of the Kullback-Leibler cross-entropy.[1,4,10]

Single-sensor, single-target   Multisensor, multitarget

$$K(f,g) \overset{\Delta}{=} \int f(\mathbf{x}) \log\left(\frac{f(\mathbf{x})}{g(\mathbf{x})}\right) d\mathbf{x} \quad \rightarrow \quad K(f,g) \overset{\Delta}{=} \int f(X) \log\left(\frac{f(X)}{g(X)}\right) \delta X \tag{14.5}$$

Here the APWOP replaces conventional statistical concepts with their FISST multisensor, mulitarget counterparts. The ordinary densities f,g and the ordinary integral on the left are replaced by the multi-target densities f,g and the set integral (Section 14.5.3) on the right.

14.1.4 Why Random Sets?

Random set theory was systematically formulated by Mathéron in the mid-1970s.[33] Its centrality as a unifying foundation for expert systems theory and ill-characterized evidence has become increasingly apparent since the mid-1980s.[34-38] Its centrality as a unifying foundation for data fusion applications has been promoted since the late-1970s by I.R. Goodman. The basic relationships between random set theory and the Dempster-Shafer theory of evidence were established by Shafer,[39] Nguyen,[40] and others.[41] The basic relationships between random set theory and fuzzy logic can be attributed to Goodman,[34,42] Orlov,[43] and Hohle.[44] Mahler developed relationships between random set theory and rule-based evidence.[45,46] Mori, Chong, et al. first proposed random set theory as a potential foundation for multitarget detection, tracking, and identification (although within a multihypothesis framework).[1,47] (Since 1995, Mori has published a number of very interesting papers based on random set ideas.[48-51])

FISST, whose fundamental ideas were codified in 1993 and 1994, builds upon this existing body of research by showing that random set theory provides a unified framework for both expert systems theory and multisensor-multitarget fusion detection, tracking, identification, sensor management, and performance estimation. FISST is unique in that it provides, under a single probabilistic paradigm, a unified and relatively simple and familiar statistical calculus for addressing all of the "Bayesian Iceberg" problems described earlier.

14.2 Basic Statistics for Tracking and Identification

This section summarizes those aspects of conventional statistics that are most pertinent to tracking and target identification. The foundation of applied tracking and identification — the recursive Bayesian nonlinear filtering equations — are described in Section 14.2.1. The procedure for constructing provably true sensor likelihood functions from sensor models, and provably true Markov transition densities from target motion models, is described in Sections 14.2.2 and 14.2.3, respectively. Bayes-optimal state estimation is reviewed in Section 14.2.4. In all four sections, data "vectors" have the form $y = (y_1,\ldots, y_n, w_1,\ldots, w_n)$ where $y_1,\ldots, y_n$ are continuous variables and $w_1,\ldots,w_n$ are discrete variables.[1] Integrals of functions of such variables involve both summations and continuous integrals.

14.2.1 Bayes Recursive Filtering

Most signal processing engineers are familiar with the Kalman filtering equations. Less well known is the fact that the Kalman filter is a special case of the Bayesian discrete-time recursive nonlinear filter.[26,27,52] This more general filter is nothing more than Equations 14.6, 14.7, and 14.8 applied recursively:

$$f_{k+1|k}\left(\mathbf{x}_{k+1}\middle|Z^k\right) = \int f_{k+1|k}\left(\mathbf{x}_{k+1}\middle|\mathbf{x}_k\right) f_{k|k}\left(\mathbf{x}_k\middle|Z^k\right) d\mathbf{x}_k \tag{14.6}$$

$$f_{k+1|k+1}\left(\mathbf{x}_{k+1}\middle|Z^{k+1}\right) = \frac{f\left(\mathbf{z}_{k+1}\middle|\mathbf{x}_{k+1}\right) f_{k+1|k}\left(\mathbf{x}_{k+1}\middle|Z^k\right)}{f\left(\mathbf{z}_{k+1}\middle|Z^k\right)} \tag{14.7}$$

$$\hat{\mathbf{x}}_{k|k}^{MAP} = \arg\max_{\mathbf{x}} f_{k|k}\left(\mathbf{x}|Z^k\right) \qquad \hat{\mathbf{x}}_{k|k}^{EAP} = \int \mathbf{x}\, f_{k|k}\left(\mathbf{x}|Z^k\right)d\mathbf{x} \tag{14.8}$$

where

$f(\mathbf{z}_{k+1}|Z^k) = \int f(\mathbf{z}_{k+1}|\mathbf{y}_{k+1})f_{k+1|k}\,(\mathbf{y}_{k+1}|Z^k)d\mathbf{y}_{k+1}$ is the Bayes normalization constant

$f(\mathbf{z}|\mathbf{x})$ is the sensor likelihood function

$f_{k+1|k}\,(\mathbf{x}_{k+1}|Z^k)$ is the target Markov transition density

$f_{k|k}\,(\mathbf{x}_k|Z^k)$ is the posterior distribution conditioned on the data-stream $Z^k = \{\mathbf{z}_1,...,\mathbf{z}_k\}$

$f_{k+1|k}\,(\mathbf{x}_{k+1}|Z^k)$ is the time-prediction of the posterior $f_{k|k}\,(\mathbf{x}_k|Z^k)$ to time-step $k+1$

The practical success of Equations 14.6 and 14.7 relies on the ability to construct effectively the likelihood function, $f(\mathbf{z}|\mathbf{x})$, and the Markov transition density, $f_{k+1|k}\,(\mathbf{x}_{k+1}|\mathbf{x}_k)$. Although likelihood functions sometimes are constructed via direct statistical analysis of data, more typically they are constructed from sensor measurement models. Markov densities typically are constructed from target motion models. In either case, differential and integral calculus must be used, as shown in the following two sections.

14.2.2 Constructing Likelihood Functions from Sensor Models

Suppose that a target with randomly varying state $\mathbf{X}$ is interrogated by a sensor that generates observations of the form $\mathbf{Z} \triangleq \mathbf{Z}|_{\mathbf{X}=\mathbf{x}} \triangleq \mathbf{h}(\mathbf{x}) + \mathbf{W}$ (where $\mathbf{W}$ is a zero-mean random noise vector with density $f_w\,(\mathbf{w})$) but does not generate missed detections or false alarms. The statistical behavior of $\mathbf{Z}$ is characterized by its *likelihood function*, $f(\mathbf{z}|\mathbf{x}) \triangleq f_{\mathbf{Z}|\mathbf{X}}\,(\mathbf{z}|\mathbf{x})$, which describes the likelihood that the sensor will collect measurement $\mathbf{z}$ given that the target has state $\mathbf{x}$. How is this likelihood function computed? Begin with the *probability mass function* of the sensor model: $p\,(S|\mathbf{x}) \triangleq p(\mathbf{z}|\mathbf{x}) = pr(\mathbf{Z} \in S)$. This is the total probability that the random observation $\mathbf{Z}$ will be found in any given region S if the target has state $\mathbf{x}$. The total probability mass, $p\,(S|\mathbf{x})$, in a region S is the sum of all of the likelihoods in that region: $p\,(S|\mathbf{x}) = \int_S f(\mathbf{y}|\mathbf{x})d\mathbf{y}$. So,

$$p\left(E_{\mathbf{z}}|\mathbf{x}\right) = \int_{E_{\mathbf{z}}} f\left(\mathbf{y}|\mathbf{x}\right)d\mathbf{y} \cong f\left(\mathbf{z}|\mathbf{x}\right)\cdot\lambda\left(E_{\mathbf{z}}\right)$$

when $E_{\mathbf{z}}$ is some very small region surrounding the point $\mathbf{z}$ with (hyper) volume $V = \lambda\,(E_{\mathbf{z}})$. (For example, $E_{\mathbf{z}} = B_{\varepsilon,\mathbf{z}}$ is a hyperball of radius ε centered at $\mathbf{z}$.) So,

$$\frac{p\left(E_{\mathbf{z}}|\mathbf{x}\right)}{\lambda\left(E_{\mathbf{z}}\right)} \cong f\left(\mathbf{z}|\mathbf{x}\right)$$

where the smaller the value of $\lambda(E_{\mathbf{z}})$, the more accurate the approximation. Stated differently, the likelihood function, $f\,(\mathbf{z}|\mathbf{x})$, can be constructed from the probability measure, $p\,(S|\mathbf{x})$, via the limiting process

$$f\left(\mathbf{z}|\mathbf{x}\right) = \frac{\delta p}{\delta \mathbf{z}} \triangleq \lim_{\lambda(E_{\mathbf{z}}) \searrow 0} \frac{p\left(E_{\mathbf{z}}|\mathbf{x}\right)}{\lambda\left(E_{\mathbf{z}}\right)} \tag{14.9}$$

The resulting equations

$$p\left(S|\mathbf{x}\right) = \int_S \frac{\delta p}{\delta \mathbf{z}}d\mathbf{z} \qquad \frac{\delta}{\delta \mathbf{z}}\int_S f\left(\mathbf{y}|\mathbf{x}\right)d\mathbf{y} = f\left(\mathbf{z}|\mathbf{x}\right) \tag{14.10}$$

are the relationships that show that $f(\mathbf{z}|\mathbf{x}) \triangleq \delta p/\delta z$ is the provably true likelihood function (i.e., the density function that faithfully describes the measurement model $\mathbf{Z} = \mathbf{h}(\mathbf{x}, \mathbf{W})$). For this particular problem, the "true" likelihood is, therefore:

$$f\left(\mathbf{z}\Big|\mathbf{x}\right)=\lim_{\varepsilon\searrow 0}\frac{\Pr\left(\mathbf{Z}\in B_{\varepsilon,\mathbf{z}}\right)}{\lambda\left(B_{\varepsilon,\mathbf{z}}\right)}=\lim_{\varepsilon\searrow 0}\frac{p_{\mathbf{w}}\left(B_{\varepsilon,\mathbf{z}-h(\mathbf{x})}\right)}{\lambda\left(B_{\varepsilon,\mathbf{z}-h(\mathbf{x})}\right)}=f_{\mathbf{w}}\left(\mathbf{z}-\mathbf{h}\left(\mathbf{x}\right)\right)$$

14.2.3   Constructing Markov Densities from Motion Models

Suppose that, between the kth and $(k + 1)^{\text{st}}$ measurement collection times, the motion of the target is best modeled by an equation of the form $\mathbf{X}_{k+1} = \Phi_k(\mathbf{x}_k) + \mathbf{V}_k$ where $\mathbf{V}_k$ is a zero-mean random vector with density $f_{\mathbf{V}_k}(\mathbf{v})$. That is, if the target had state $\mathbf{x}_k$ at time-step k, then it will have state $\Phi_k(\mathbf{x}_k)$ at time-step $k + 1$ — except possible error in this belief is accounted for by appending the random variation $\mathbf{V}_k$. How would $f_{k+1|k}(\mathbf{X}_{k+1}|\mathbf{X}_k)$ be constructed? This situation parallels that of Section 14.2.2. The probability mass function $p_{k+1}(S|\mathbf{x}_k) \triangleq \Pr(\mathbf{X}_{k+1} \in S)$ is the total probability that the target will be found in region S at time-step $k + 1$, given that it had state $\mathbf{x}_k$ at time-step k. So,

$$f_{k+1|k}\left(\mathbf{x}_{k+1}\Big|\mathbf{x}_k\right)=\lim_{\varepsilon\searrow 0}\frac{p_{k+1}\left(B_{\varepsilon,\mathbf{x}_{k+1}}\Big|\mathbf{x}_k\right)}{\lambda\left(B_{\varepsilon,\mathbf{x}_{k+1}}\right)}=f_{\mathbf{V}_k}\left(x_{k+1}-\Phi_k\left(\mathbf{x}_k\right)\right)$$

is the true Markov density associated with the motion model $\mathbf{X}_{k+1} = \Phi_k(\mathbf{X}_k) + \mathbf{V}_k$ more generally, the equations

$$p_{k+1|k}\left(S\Big|\mathbf{x}_k\right)=\int_S\frac{\delta p_{k+1|k}}{\delta \mathbf{x}_{k+1}}d\mathbf{x}_{k+1},\quad \frac{\delta}{\delta \mathbf{x}_{k+1}}\int_S f_{k+1|k}\left(\mathbf{x}_{k+1}\Big|\mathbf{x}_k\right)d\mathbf{x}_{k+1}=f_{k+1|k}\left(\mathbf{x}_{k+1}\Big|\mathbf{x}_k\right)$$

are the relationships that show that $f_{k+1|k}(\mathbf{x}_{k+1}|\mathbf{x}_k) \triangleq \delta p_{k+1|k}/\delta \mathbf{x}_{k+1}$ is the provably true Markov density — i.e., the density function that faithfully describes the motion model $\mathbf{X}_{k+1} = \Phi_k(\mathbf{x}_k, \mathbf{V}_k)$.

14.2.4   Optimal State Estimators

An estimator of the state $\mathbf{x}$ is any family $\hat{\mathbf{x}}(\mathbf{z}_1,\ldots,\mathbf{z}_m)$ of state-valued functions of the (static) measurements $\mathbf{z}_1,\ldots,\mathbf{z}_m$. "Good" state estimators $\hat{\mathbf{x}}$ should be Bayes-optimal in the sense that, in comparison to all other possible estimators, they minimize the Bayes risk

$$R_C\left(\hat{\mathbf{x}},m\right)=\int C\left(\mathbf{x},\hat{\mathbf{x}}\left(\mathbf{z}_1,\ldots,\mathbf{z}_m\right)\right)f\left(\mathbf{z}_1,\ldots,\mathbf{z}_m\Big|\mathbf{x}\right)f\left(\mathbf{x}\right)d\mathbf{x}\,d\mathbf{z}_1\cdots d\mathbf{z}_m$$

for some specified cost (i.e., objective) function $C(\mathbf{x}, \mathbf{y})$ defined on states $\mathbf{x}, \mathbf{y}$.[53] Secondly, they should be statistically consistent in the sense that $\hat{\mathbf{x}}(\mathbf{z}_1,\ldots,\mathbf{z}_m)$ converges to the actual target state as $m \to \infty$. Other properties (e.g., asymptotically unbiased, rapidly convergent, stably convergent, etc.) are desirable as well. The most common "good" Bayes state estimators are the maximum *a posteriori* (MAP) and expected *a posteriori* (EAP) estimators described earlier.

14.3   Multitarget Sensor Models

In the single-target case, probabilistic approaches to tracking and identification (and Bayesian approaches in particular) depend on the ability to construct sensor models and likelihood functions that model the data-generation process with enough fidelity to ensure optimal performance. As argued in Section 14.1.1, multitarget tracking and identification have suffered from the lack of a statistical calculus of the kind that is readily available in the single-target case. The purpose of this section is to show how to construct FISST multitarget measurement models. These models, and the FISST statistical calculus to be introduced in Section 14.5, will be integrated in Section 14.6.1 to construct true multitarget likelihood functions.

The following sections illustrate the process of constructing multitarget measurement models for the following successively more realistic situations: (1) multitarget measurement models with no missed detections and no false alarms/clutter; (2) multitarget measurement models with missed detections; (3) multitarget measurement models with missed detections and false alarms or clutter; and (4) multi-target measurement models for the multiple-sensor case. The problem of constructing single-target and multitarget measurement models for data that is ambiguous is discussed in Section 14.8.

14.3.1 Case I: No Missed Detections, No False Alarms

Suppose that two targets with states $\mathbf{x}_1$ and $\mathbf{x}_2$ are interrogated by a single sensor that generates observations of the form $\mathbf{Z} = h(\mathbf{x}) + \mathbf{W}_1$, where $\mathbf{W}_1$ is a random noise vector with density $f_w(\mathbf{w})$. Assume also that there are no missed detections or false alarms, and that observations within a scan are independent. Then the *multitarget measurement* is the *randomly varying two-element observation set*.

$$\Sigma = \left\{ \mathbf{Z}_1, \mathbf{Z}_2 \right\} = \left\{ h\left(\mathbf{x}_1\right) + \mathbf{W}_1, h\left(\mathbf{x}_2\right) + \mathbf{W}_2 \right\} \tag{14.11}$$

where $\mathbf{W}_1, \mathbf{W}_2$ are independent random vectors with density $f_w(\mathbf{w})$. We assume that individual targets produce unique observations only for the sake of clarity. Clearly, we could just as easily produce models for other kinds of sensors — for example, sensors that detect only superpositions of the signals produced by multiple targets. One such measurement model is

$$\Sigma = \left\{ \mathbf{Z} \right\} = \left\{ h_1\left(\mathbf{x}_1\right) + h_2\left(x_2\right) + \mathbf{W} \right\}$$

14.3.2 Case II: Missed Detections

Suppose that the sensor of Section 14.3.1 has a probability of detection $p_D < 1$. In this case, observations can have not only the form $Z = \{\mathbf{z}_1, \mathbf{z}_2\}$, but also $Z = \{\mathbf{z}\}$ or $Z = \emptyset$ (missed detection). The more complex observation model $\Sigma = T_1 \cup T_2$ is needed, which has T_1, T_2 observation sets with the following properties: (a) $T_i = \emptyset$ (i.e., missed detection) with probability $1 - p_D$ and (b) T_i is nonempty with probability p_D, in which case, $T_i = \{Z_i\}$. If the sensor has a specific field of view, then $p_D = p_D(\mathbf{x}, \mathbf{y})$ will be a function of both the target state $\mathbf{x}$ and the state $\mathbf{y}$ of the sensor.

14.3.3 Case III: Missed Detection and False Alarms

Suppose the sensor of Section 14.3.2 has probability of false alarm $p_{FA} \neq 0$. Then we need an observation model of the form

$$\Sigma = \overbrace{T}^{\text{target}} \cup \overbrace{C}^{\text{clutter}} = T_1 \cup T_2 \cup C$$

where C models false alarms and/or (possibly state-dependent) point clutter. As a simple example, C could have the form $C = C_1 \cup \ldots \cup C_m$, where each C_j is a clutter generator — meaning that there is a probability, p_{FA}, that C_j will be nonempty (i.e., generator of a clutter-observation) — in which case $C = \{C_i\}$ where C_i is some random noise vector with density $f_{Ci}(\mathbf{z})$. (Note: $\Sigma = T_1 \cup C$ models the single-target case.)

14.3.4   Case IV: Multiple Sensors

In this case, observations will have the form $\mathbf{z}^{[s]} \triangleq (\mathbf{z}, s)$ where the integer tag s identifies which sensor originated the measurement. A two-sensor multitarget measurement will have the form $\Sigma = \Sigma^{[1]} \cup \Sigma^{[2]}$ where $\Sigma^{[s]}$ for $s = 1, 2$ is the random multitarget measurement-set collected by the sensor with tag s and can have any of the forms previously described.

14.4   Multitarget Motion Models

This section shows how to construct multitarget motion models in relation to the construction of single-target motion models. These models, combined with the FISST calculus (Section 14.5), enables the construction of true multitarget Markov densities (Section 14.6.2). In the single-target case, the construction of Markov densities from motion models strongly parallels the construction of likelihood functions from sensor measurement models. In like fashion, the construction of multitarget Markov densities strongly resembles the construction of multisensor-multitarget likelihood functions.

This section illustrates the process of constructing multitarget motion models by considering the following increasingly more realistic situations: (1) multitarget motion models assuming that target number does not change; (2) multitarget motion models assuming that target number can decrease; and (3) multitarget motion models assuming that target number can decrease or increase.

14.4.1   Case I: Target Number Does Not Change

Assume that the states of individual targets have the form $\mathbf{x} = (\mathbf{y}, c)$ where $\mathbf{y}$ is the kinematic state and c is the target type. Assume that each target type has an associated motion model $\mathbf{Y}_{c,k+1} = \Phi_{c,k}(\mathbf{y}_k) + \mathbf{W}_{c,k}$. Define

$$\mathbf{X}_{k+1,\mathbf{x}} = \Phi_{k+1}\left(\mathbf{x}, \mathbf{W}_k\right) \stackrel{\Delta}{=} \left(\Phi_{c,k}\left(\mathbf{y}\right) + \mathbf{W}_{c,k}, c\right)$$

where $\mathbf{W}_k$ denotes the family of random vectors $\mathbf{W}_{c,k}$.

To model a multitarget system in which two targets never enter or leave the scenario, the obvious multitarget extension of the single-target motion model would be $\Gamma_{k+1} = \Phi_k (X_k, \mathbf{W}_k)$, where Γ_{k+1} is the randomly varying parameter set at time step $k + 1$. That is, for the cases $X = \emptyset, X = \{\mathbf{x}\}, or\ X = \{\mathbf{x}_1, \mathbf{x}_2\}$, respectively, the multitarget state transitions are:

$$\Gamma_{k+1} = \emptyset$$

$$\Gamma_{k+1} \stackrel{\Delta}{=} \left\{X_{k+1,\mathbf{x}}\right\} = \left\{\Phi_k\left(\mathbf{x}, \mathbf{W}_k\right)\right\}$$

$$\Gamma_{k+1} \stackrel{\Delta}{=} \left\{X_{k+1,\mathbf{x}_1}, X_{k+1,\mathbf{x}_2}\right\} = \left\{\Phi_k\left(\mathbf{x}_1, \mathbf{W}_k\right), \Phi_k\left(\mathbf{x}_2, \mathbf{W}_k\right)\right\}$$

14.4.2   Case II: Target Number Can Decrease

Modeling scenarios in which target number can decrease (but not increase) is analogous to modeling multitarget observations with missed detections. Suppose that no more than two targets are possible, but that one or more of them can vanish from the scene. One possible motion model would be $\Gamma_{k+1|k} = \Phi (X_{k|k}, \mathbf{W}_k)$ where, for the cases $X = \emptyset, X = \{\mathbf{x}\}, or\ X = \{\mathbf{x}_1, \mathbf{x}_2\}$, respectively,

$$\Gamma_{k+1} = \emptyset, \ \Gamma_{k+1} = T_{k,x}, \ \Gamma_{k+1} = T_{k,x_1} \cup T_{k,x_2}$$

where $T_{k,x}$ is a track-set with the following properties: (a) $T_{k,x} \neq \emptyset$ with probability p_v, in which case $T_{k,x} = \{X_{k+1,x}\}$, and (b) $T_{k,x} = \emptyset$ (i.e., target disappearance), with probability $1 - p_v$. In other words, if no targets are present in the scene, this will continue to be the case. If, however, there is one target in the scene, then either this target will persist (with probability p_v) or it will vanish (with probability $1 - p_v$). If there are two targets in the scene, then each will either persist or vanish in the same manner. In general, one would model $\Phi_k(\{x_1,\ldots,x_n\}) = T_{k,x_1} \cup \ldots \cup Tk,x_n$.

14.4.3 Case III: Target Number Can Increase and Decrease

Modeling scenarios in which target number can decrease and/or increase is analogous to modeling multitarget observations with missed detections and clutter. In this case, one possible model is

$$\Phi_k\left(\left\{x_1,\ldots,x_n\right\}\right) = T_{k,x_1} \cup \ldots \cup T_{k,x_n} \cup B_k$$

where B_k is the set of birth targets (i.e, targets that have entered the scene).

14.5 The FISST Multisource-Multitarget Calculus

This section introduces the mathematical core of FISST — the FISST multitarget integral and differential calculus. That is, it shows that the belief-mass function $\beta(S)$ of a multitarget sensor or motion model plays the same role in multisensor-multitarget statistics that the probability-mass function $p(S)$ plays in single-target statistics. The integral $\int_s f(z)dz$ and derivative dp/dz — which can be computed using elementary calculus — are the mathematical basis of conventional single-sensor, single-target statistics. We will show that the basis of multisensor-multitarget statistics is a multitarget integral $\int_s f(z)\delta z$ and a multitarget derivative $\delta\beta/\delta Z$ that can also be computed using "turn-the-crank" calculus rules. In particular we will show that, using the FISST calculus,

- True multisensor-multitarget likelihood functions can be constructed from the measurement models of the individual sensors, and
- True multitarget Markov transition densities can be constructed from the motion models of the individual targets.

Section 14.5.1 defines the belief-mass function of a multitarget sensor measurement model and Section 14.5.2 defines the belief-mass function of a multitarget motion model. The FISST multitarget integral and differential calculus is introduced in Section 14.5.3. Section 14.5.4 lists some of the more useful rules for using this calculus.

14.5.1 The Belief-Mass Function of a Sensor Model

Just as the statistical behavior of a random observation vector Z is characterized by its probability mass function $p(S|x) = \Pr(Z \in S)$, the statistical behavior of the random observation-set Σ is characterized by its *belief-mass function*:[1]

$$\beta(S|X) \overset{\Delta}{=} \beta_{\Sigma|\Gamma}(S|X) \overset{\Delta}{=} \Pr(\Sigma \subseteq S)$$

where Γ is the random multitarget state. The belief mass is the total probability that all observations in a sensor (or multisensor) scan will be found in any given region S, if targets have multitarget state X.

For example, if $X = \{x\}$ and $\Sigma = \{Z\}$ where Z is a random vector, then

$$\beta(S|X) = \Pr(\Sigma \subseteq S) = \Pr(Z \in S) = p(S|x)$$

In other words, the belief mass of a *random vector* is equal to its probability mass. On the other hand, for a single-target, missed-detection model $\Sigma = T_1$,

$$\beta(S|X) = \Pr(T_1 \subseteq S) = \Pr(T_1 = \emptyset) + \Pr(T_1 \neq \emptyset, Z \in S)$$

$$= \Pr(T_1 = \emptyset) + \Pr(T_1 \neq \emptyset)\Pr(Z \in S) = 1 - p_D + p_D\, p_Z(S|\mathbf{x}) \qquad (14.12)$$

and for the two-target missed-detection model $\Sigma = \{T_1 \cup T_2\}$ (Section 14.3.2),

$$\beta(S|X) = \Pr(T_1 \subseteq S)\Pr(T_2 \subseteq S) = \left[1 - p_D + p_D p(S|\mathbf{x}_1)\right]\left[1 - p_D + p_D\, p(S|\mathbf{x}_2)\right]$$

where $p(S|\mathbf{x}) \overset{\Delta}{=} \Pr(T_i \subseteq S | T_i \neq \emptyset)$ and $X = \{\mathbf{x}_1, \mathbf{x}_2\}$. Setting $p_D = 1$ yields

$$\beta(S|X) = p(S|\mathbf{x}_1)\, p(S|\mathbf{x}_2) \qquad (14.13)$$

which is the belief-mass function for the model $\Sigma = \{\mathbf{Z}_1, \mathbf{Z}_2\}$ of Equation 14.11. Suppose next that two sensors with identifying tags $s = 1,2$ collect observation-sets $\Sigma = \Sigma^{[1]} \cup \Sigma^{[2]}$. The corresponding belief-mass function has the form $\beta_\Sigma (S^{[1]} \cup S^{[2]}|X) = \Pr(\Sigma^{[1]} \subseteq S^{[1]}, \Sigma^{[2]} \subseteq S^{[2]})$ where $S^{[1]}, S^{[2]}$ are (measurable) subsets of the measurement spaces of the respective sensors. If the two sensors are independent then the belief-mass function has the form

$$\beta_\Sigma\left(S^{[1]} \cup S^{[2]}\middle|X\right) = \beta_{\Sigma^{[1]}}\left(S^{[1]}\middle|X\right)\beta_{\Sigma^{[2]}}\left(S^{[2]}\middle|X\right) \qquad (14.14)$$

14.5.2 The Belief-Mass Function of a Motion Model

In single-target problems, the statistics of a motion model $\mathbf{X}_{k+1} = \Phi_k(X_k, \mathbf{W}_k)$ are described by the probability-mass function $p_{\mathbf{X}k+1}(S|\mathbf{x}_k) = \Pr(\mathbf{X}_{k+1} \in S)$, which is the probability that the target-state will be found in the region S if it previously had state $\mathbf{x}_k$. Similarly, suppose that $\Gamma_{k+1} = \Phi_k(X_k, \mathbf{W}_k)$ is a multitarget motion model (Section 14.4). The statistics of the finitely varying random state-set Γ_{k+1} can be described by its belief-mass function:

$$\beta_{\Gamma_{k+1}}\left(S\middle|X_k\right) \overset{\Delta}{=} \Pr\left(\Gamma_{k+1} \subseteq S\right)$$

This is the total probability of finding all targets in region S at time-step $k+1$ if, in time-step k, they had multitarget state $X_k = \{\mathbf{x}_{k,1}, \ldots, \mathbf{x}_{k,n(k)}\}$.

For example, the belief-mass function for the multitarget motion model of Section 14.4.1 is $\beta_{k+1|k}(S|\mathbf{x}_k, \mathbf{y}_k) = \Pr(\Phi_k(\mathbf{x}_k) + V_k \in S)\Pr(\Phi_k(\mathbf{y}_k) + W_k \in S) = pW_k(S|\mathbf{x}_k)\, pW_k(S|\mathbf{y}_k)$. This is entirely analogous to Equation 14.13, the belief-mass function for the multitarget measurement model of Section 14.3.1.

14.5.3 The Set Integral and Set Derivative

Equation 14.9 showed that single-target likelihood can be computed from probability-mass functions using an operator $\delta/\delta\mathbf{z}$ inverse to the integral. Multisensor-multitarget likelihoods can be constructed from belief-mass functions in a similar manner.

14.5.3.1 Basic Ideas of the FISST Calculus

For example, convert the missed-detection model $\Sigma = T_1 \cup T_2$ of Section 14.3.1 into a multitarget likelihood function

$$f\left(Z|X\right)=\begin{cases} f\left(\emptyset|\mathbf{x}_1,\mathbf{x}_2\right) \text{ if } Z=\emptyset \\ f\left(z|\mathbf{x}_1,\mathbf{x}_2\right) \text{ if } Z=\{\mathbf{z}\} \\ 2f\left(\mathbf{z}_1,\mathbf{z}_2|\mathbf{x}_1,\mathbf{x}_2\right) \text{ if } Z=\{\mathbf{z}_1,\mathbf{z}_2\} \\ \quad 0 \quad \text{ if otherwise} \end{cases}$$

(Notice that two alternative notations for a multitarget likelihood function were used — a *set* notation and a *vector* notation. In general, the two notations are related by the relationship $f(\{\mathbf{z}_1,\dots,\mathbf{z}_m\}|X) = m! \, f(\mathbf{z}_1,\dots,\mathbf{z}_m|X)$.)

The procedure required is suggested by analogy to ordinary probability theory. Consider the measurement model $\Sigma = \{\mathbf{Z}_1, \mathbf{Z}_2\}$ of Section 14.3.1 and assume that we have constructed a multitarget likelihood. Then the total probability that Σ will be in the region S should be the sum of all the likelihoods that individual observation-vectors $(\mathbf{z}_1, \mathbf{z}_2)$ will be contained in $S \times S$:

$$\beta(S|X)= \int_{S\times S} f\left(\mathbf{z}_1,\mathbf{z}_2|\mathbf{x}_1,\mathbf{x}_2\right)d\mathbf{z}_1 d\mathbf{z}_2 = \frac{1}{2}\int_{S\times S} f\left(\{\mathbf{z}_1,\mathbf{z}_2\}|\{\mathbf{x}_1,\mathbf{x}_2\}\right)d\mathbf{z}_1 d\mathbf{z}_2$$

Likewise, for the missed-detection model $\Sigma = T_1 \cup T_2$ of Section 14.3.2, the possible observation-acts are, respectively, $\Sigma = \emptyset$ (missed detections on both targets), $\{\mathbf{z}\} \subseteq S$ (missed detection on one target), and $\{\mathbf{z}_1, \mathbf{z}_2\} \subseteq S$ (no missed detections). Consequently, the total probability that Σ will be in the region S should be the sum of the likelihoods of all of these possible observations:

$$\beta(S|X)=\Pr\left(\Sigma=\emptyset\right)+\Pr\left(\Sigma\subseteq S, |\Sigma|=1\right)+\Pr\left(\Sigma\subseteq S, |\Sigma|=2\right)$$

$$=f\left(\emptyset|X\right)+\int_S f\left(\{\mathbf{z}\}|X\right)d\mathbf{z}+\frac{1}{2}\int_{S\times S} f\left(\{\mathbf{z}_1,\mathbf{z}_2\}|X\right)d\mathbf{z}_1 d\mathbf{z}_2 \qquad (14.15)$$

$$\overset{\Delta}{=}\int_S f\left(Z|X\right)\delta Z$$

where $|Y|=k$ means that the set Y has k elements and where the quantity $\int_S f_\Sigma(Z|X)\delta Z$ is a set integral. The equation $\beta(S|X) = \int_S f(Z|X)\delta Z$ is the multitarget analog of the usual probability-summation equation $p(S|\mathbf{x}) = \int_S f(\mathbf{z}|\mathbf{x})d\mathbf{z}.$

14.5.3.2 The Set Integral

Suppose a function $F(Y)$ exists for a finite-set variable Y. That is, $F(Y)$ has the form

$$F(\emptyset) \;=\; \text{probability that } Y = \emptyset$$

$$F\left(\{\mathbf{y}_1,\dots,\mathbf{y}_j\}\right) \;=\; j!F\left(\mathbf{y}_1,\dots,\mathbf{y}_j\right) \;=\; \text{likelihood that } Y=\{\mathbf{y}_1,\dots\mathbf{y}_j\}$$

In particular, F could be a multisource-multitarget likelihood $F(Z)=f(Z|X)$ or a multitarget Markov density $F(X)=f_{k+1|k}(X|X_k)$ or a multitarget prior or posterior $F(X)=F_{k|k}(X|Z^{(k)})$. Then the set integral of F is[12]

$$\int_S F(Y)\delta Y \overset{\Delta}{=} F(\emptyset) + \sum_{j=1}^{\infty} \frac{1}{j!} \underset{\underbrace{S \times \cdots \times S}_{j\ times}}{\int} F(\{\mathbf{y}_1, \ldots, \mathbf{y}_j\})d\mathbf{y}_1 \ldots d\mathbf{y}_i \qquad (14.16)$$

for any region S.

14.5.3.3 The Set Derivative

Constructing the multitarget likelihood function $f(Z\,|\,X)$ of a multisensor-multitarget sensor model (or the multitarget Markov transition density $f_{k+1|k}(X'\,|\,X)$ of a multitarget motion model) requires an operation that is the inverse of the set integral — the set derivative. Let $\beta(S)$ be any function whose arguments S are arbitrary closed subsets (typically this will be the belief-mass function of a multisensor-multitarget measurement model or of a multitarget motion model). If $Z=\{\mathbf{z}_1,\ldots,\mathbf{z}_m\}$ with $\mathbf{z}_1,\ldots,\mathbf{z}_m$ distinct, the set derivative[1] is the following generalization of Equation 14.9:

$$\frac{\delta\beta}{\delta\mathbf{z}}(S) \overset{\Delta}{=} \frac{\delta}{\delta\mathbf{z}}\beta(S) \overset{\Delta}{=} \lim_{\lambda(E_\mathbf{z})\searrow 0} \frac{\beta(S\cup E_\mathbf{z}) - \beta(S)}{\lambda(E_\mathbf{z})}$$

$$\frac{\delta\beta}{\delta Z}(S) \overset{\Delta}{=} \frac{\delta^m\beta}{\delta\mathbf{z}_1\cdots\delta\mathbf{z}_m}(S) \overset{\Delta}{=} \frac{\delta}{\delta\mathbf{z}_1}\cdots\frac{\delta}{\delta\mathbf{z}_m}\beta(S) \qquad (14.17)$$

$$\frac{\delta\beta}{\delta\emptyset}(S) \overset{\Delta}{=} \beta(S)$$

14.5.3.4 Key Points on Multitarget Likelihoods and Markov Densities

The set integral and the set derivative are inverse to each other:

$$\beta(S) = \int_S \frac{\delta\beta}{\delta X}(\emptyset)\delta X \qquad\qquad F(X) = \left[\frac{\delta}{\delta X}\int_S F(Y)\delta Y\right]_{S=0}$$

These are the multisensor-multitarget analogs of Equation 14.10. They yield two fundamental points of the FISST multitarget calculus[1]:

- The provably true likelihood function $f(Z\,|\,X)$ of a multisensor-multitarget problem is a set derivative of the belief-mass function $\beta(S\,|\,X)$ of the corresponding sensor (or multisensor) model:

$$f(Z|X) = \frac{\delta\beta}{\delta Z}(\emptyset|X) \qquad (14.18)$$

- The provably true Markov transition density $f_{k+1|k}(X_{k+1}\,|\,X_k)$ of a multitarget problem is a set derivative of the belief-mass function $\beta_{k+1|k}(S\,|\,X_k)$ of the corresponding multitarget motion model:

$$f_{k+1|k}(X_{k+1}|X_k) = \frac{\delta\beta_{k+1|k}}{\delta X_{k+1}}(\emptyset|X_k) \qquad (14.19)$$

14.5.4 "Turn-the-Crank" Rules for the FISST Calculus

Engineers usually find it possible to apply ordinary Newtonian differential and integral calculus by applying the "turn-the-crank" rules they learned as college freshman. Similar "turn-the-crank" rules exist for the FISST calculus, for example:

$$\frac{\delta}{\delta Z}\left[a_1\beta_1(S)+a_2\beta_2(S)\right]=a_1\frac{\delta\beta_1}{\delta Z}(S)+a_2\frac{\delta\beta_2}{\delta Z}(S) \qquad \text{(sum rule)}$$

$$\frac{\delta}{\delta z}\left[\beta_1(S)\beta_2(S)\right]=\frac{\delta\beta_1}{\delta z}(S)\beta_2(S)+\beta_1(S)\frac{\delta\beta_2}{\delta z}(S)$$

$$\frac{\delta}{\delta Z}\left[\beta_1(S)\beta_2(S)\right]=\sum_{W\subseteq z}\frac{\delta\beta_1}{\delta W}(S)\frac{\delta\beta_2}{\delta(Z-W)}(S) \qquad \text{(product rules)}$$

$$\frac{\delta}{\delta z}f\left(\beta_1(S),\dots,\beta_n(S)\right)=\sum_{i=1}^{n}\frac{\partial f}{\partial x_i}\left(\beta_1(S),\dots,\beta_n(S)\right)\frac{\delta\beta_i}{\delta z}(S) \qquad \text{(chain rule)}$$

14.6 FISST Multisource-Multitarget Statistics

Thus far this chapter has described the multisensor-multitarget analogs of measurement and motion models, probability mass functions, and the integral and differential calculus. This section shows how these concepts join together to produce a direct generalization of ordinary statistics to multitarget statistics. Section 14.6.1 illustrates how true multitarget likelihood functions can be constructed from multitarget measurement models using the "turn-the-crank" rules of the FISST calculus. Section 14.6.2 shows how to similarly construct true multitarget Markov densities from multitarget motion models. The concepts of multitarget prior distribution and multitarget posterior distribution are introduced in Sections 14.6.3 and 14.6.4. The failure of the classical Bayes-optimal state estimators in multitarget situations is described in Section 14.6.6. The solution of this problem — the proper definition and verification of Bayes-optimal multitarget state estimators — is described in Section 14.6.7. The remaining two subsections summarize a Cramér-Rao performance bound for vector-valued multitarget state estimators and a "multitarget miss distance."

14.6.1 Constructing True Multitarget Likelihood Functions

Let us apply the turn-the-crank formulas of Subsection 5.4 to the belief-mass function $\beta(S|X)=p(S|x_1)$ $p(S|x_2)$ corresponding to the measurement model $\Sigma=\{Z_1,Z_2\}$ of Equation 3.2, where $X=\{x_1,x_2\}$. We get

$$\frac{\delta\beta}{\delta z_1}(S|X)=\frac{\delta}{\delta z_1}\beta(S|X)=\frac{\delta}{\delta z_1}\left[p(S|x_1)p(S|x_2)\right]$$

$$=\frac{\delta p}{\delta z_1}(S|x_1)p(S|x_2)+p(S|x_1)\frac{\delta p}{\delta z_1}(S|x_2)$$

$$=f(z_1|x_1)p(S|x_2)+p(S|x_1)f(z_1|x_2)$$

$$\frac{\delta^2\beta}{\delta z_2\delta z_1}(S|X)=\frac{\delta}{\delta z_2}\frac{\delta\beta}{\delta z_1}(S|X)=f(z_1|x_1)\frac{\delta}{\delta z_2}p(S|x_2)+\frac{\delta}{\delta z_2}p(S|x_1)f(z_1|x_2)$$

$$=f(z_1|x_1)f(z_2|x_2)+f(z_2|x_1)f(z_1|x_2)$$

$$\frac{\delta^3\beta}{\delta z_3\delta z_2\delta z_1}(S|X)=\frac{\delta}{\delta z_3}\frac{\delta^2\beta}{\delta z_2\delta z_1}(S|X)=0$$

and the higher-order derivatives vanish identically. The multitarget likelihood is

$$f\left(\emptyset|X\right)=\frac{\delta\beta}{\delta\emptyset}\left(\emptyset\right)=\beta\left(\emptyset|X\right)=0$$

$$f\left(\mathbf{z}|X\right)=\frac{\delta\beta}{\delta\mathbf{z}}\left(\emptyset|X\right)=0 \qquad\qquad (14.20)$$

$$f\left(\{\mathbf{z}_1,\mathbf{z}_2\}|X\right)=\frac{\delta\beta}{\delta\mathbf{z}_2\delta\mathbf{z}_1}\left(\emptyset|X\right)=f\left(\mathbf{z}_1|\mathbf{x}_1\right)f\left(\mathbf{z}_2|\mathbf{x}_2\right)+f\left(\mathbf{z}_2|\mathbf{x}_1\right)f\left(\mathbf{z}_1|\mathbf{x}_2\right)$$

where $f(Z|X) = 0$ identically if Z contains more than two elements. More general multitarget likelihoods can be computed similarly.[2]

14.6.2   Constructing True Multitarget Markov Densities

Multitarget Markov densities[1,7,53] are constructed from multitarget motion models in much the same way that multisensor-multitarget likelihood functions were constructed from multisensor-multitarget measurement models in Section 14.6.1. First, construct a multitarget motion model $\Gamma_{k+1} = \Phi_k(X_k,\mathbf{W}_k)$ from the underlying motion models of the individual targets. Second, build the corresponding belief-mass function $\beta_{k+1|k}(S|X_k)=\Pr(\Gamma_{k+1}\subseteq S)$. Finally, construct the multitarget Markov density $f_{k+1|k}(X_{k+1}|X_k)$ from the belief-mass function using the turn-the-crank formulas of the FISST calculus.

For example, the belief measure $\beta_{k+1|k}(S|\mathbf{x}_k,\mathbf{y}_k)=p_{wk}(S|\mathbf{x}_k)\,p_{wk}(S|\mathbf{y}_k)$ for the multitarget motion model of Section 14.4.1 has the same form as the multitarget measurement model in Equation 14.20. Consequently, its multitarget Markov density is[33]

$$f_{k+1|k}\left(\{\mathbf{x}_{k+1},\mathbf{y}_{k+1}\}|\{\mathbf{x}_k,\mathbf{y}_k\}\right)=f_{k+1|k}\left(\mathbf{x}_{k+1}|\mathbf{x}_k\right)f_{k+1|k}\left(\mathbf{y}_{k+1}|\mathbf{y}_k\right)$$

$$+f_{k+1|k}\left(\mathbf{y}_{k+1}|\mathbf{x}_k\right)f_{k+1|k}\left(\mathbf{x}_{k+1}|\mathbf{y}_k\right)$$

14.6.3   Multitarget Prior Distributions

The initial states of the targets in a multitarget system are specified by a *multitarget prior* of the form $f_0(X) = f_{0|0}(X)$,[1,4] where $\int f_0(X)\delta X = 1$ and where the integral is a set integral. Suppose that states have the form $\mathbf{x} = (\mathbf{y},c)$ where $\mathbf{y}$ is the kinematic state variable restricted to some bounded region D of (hyper) volume $\lambda(D)$ and c the discrete state variable(s), drawn from a universe C with N possible members. In conventional statistics, the uniform distribution $u(\mathbf{x}) = \lambda(D)^{-1}N^{-1}$ is the most common way of initializing a Bayesian algorithm when nothing is known about the initial state of the target. The concepts of prior and uniform distributions carry over to multitarget problems, but in this case there is an additional dimension that must be taken into account — target number.

For example, suppose that there can be no more than M possible targets in a scene.[1,4] If $X = \{\mathbf{x}_1,\ldots,\mathbf{x}_n\}$, the multitarget uniform distribution is

$$u(X)=\begin{cases} n!\,N^{-n}\lambda(D)^{-n}\left(M+1\right)^{-1} & \text{if } X\subseteq D\times C \\ 0 & \text{if otherwise} \end{cases}$$

14.6.4   Multitarget Posterior Distributions

Given a multitarget likelihood $f(Z|X)$ and a multitarget prior $f_0(X|Z_1,\ldots,Z_k)$, the *multitarget posterior* is

$$f\left(X\middle|Z_1,...,Z_{k+1}\right)=\frac{f\left(Z_{k+1}\middle|X\right)f_0\left(X\middle|Z_1,...,Z_k\right)}{f\left(Z_{k+1}\middle|Z_1,...,Z_k\right)} \tag{14.21}$$

where $f(X|Z_1,...,Z_{k+1}) = \int f(Z_{k+1}|X) \, f_0 (X|Z_1,...,Z_k)\delta X$ is a set integral.[1,4] Like multitarget priors, multitarget posteriors are normalized multitarget densities: $\int f(X|Z_1,...,Z_k)\delta X = 1$, where the integral is a set integral.

Multitarget posteriors and priors, like multitarget density functions in general, have one peculiarity that sets them apart from conventional densities: their behavior with respect to units of measurement.[1,6,28] In particular, when continuous state variables are present, *the units of a multitarget prior or posterior $f(X)$ vary with the cardinality of X.*

As a simple example of multitarget posteriors and priors, suppose that a scene is being observed by a single sensor with probability of detection p_D and no false detections.[1] This sensor collects a single observation $Z = \emptyset$ (missed detection) or $Z = \{z_0\}$. Let the multitarget prior be

$$F_0(X)= \begin{cases} 1-\pi_0 \text{ if } X=\emptyset \\ \pi_0\,\pi(x) \text{ if } X=\{x\} \\ 0 \text{ if } |X|\geq 2 \end{cases}$$

where $\pi(x)$ denotes the conventional prior. That is, there is at most one target in the scene. There is prior probability $1 - \pi_0$ that there are no targets at all. The prior density of there being exactly one target with state x is $\pi_0\,\pi\,(x)$. The nonvanishing values of the corresponding *multitarget* posterior can be shown to be:

$$F\left(\emptyset\middle|\emptyset\right) = \frac{1-\pi_0}{1-\pi_0 p_D} \qquad F\left(\{x\}\middle|\emptyset\right)=\frac{\pi_0-w_0 p_D}{1-\pi_0 p_D}\,\pi(x)$$

$$F\left(\{x\}\middle|\{z_0\}\right)=f\left(x\middle|z_0\right)$$

where $f(x|z)$ is the conventional posterior. That is, the fact that nothing is observed (i.e., $\emptyset$) may be attributable to the fact that no target is actually present (with probability $F(\emptyset|\emptyset)$) or that a target is present, but was not observed because of a missed detection (with probability $1 - F(\emptyset|\emptyset)$).

14.6.5 Expected Values and Covariances

Suppose that $\Sigma_1,...,\Sigma_m$ are finite random sets and the $F(Z_1,...,Z_m)$ is a function that transforms finite sets into vectors. The expected value and covariance of the random vector $\mathbf{X} = F(\Sigma_1,...,\Sigma_m)$ are[1]

$$E_X \mathbf{X} \overset{\Delta}{=} \int \mathbf{F}\left(Z_1,...,Z_m\right)f\left(Z_1,...,Z_m\middle|X\right)\delta Z_1\cdots\delta Z_m$$

$$C_{\mathbf{x},\mathbf{x}} \overset{\Delta}{=} \int \left(F\left(Z_1,...,Z_m\right)-E_X\left[\mathbf{X}\right]\right)\left(\mathbf{F}\left(Z_1,...,Z_m\right)-E_X\left[\mathbf{X}\right]\right)^T -f\left(Z_1,...,Z_m\middle|X\right)\delta Z_1\cdots\delta Z_m$$

14.6.6 The Failure of the Classical State Estimators

The material in this section has been described in much greater detail in a recent series of papers.[6,7,28] In general, in multitarget situations (i.e., the number of targets is unknown and at least one state variable is continuous) the classical Bayes-optimal estimators cannot be defined. This can be explained using a simple example.[2] Let

$$f(X) = \begin{cases} 1/2 & \text{if } X=0 \\ 1/2\, N_{\sigma^2}(x-1) & \text{if } X=\{x\} \\ 0 & \text{if } |X| \geq 2 \end{cases}$$

where the variance σ^2 has units km^2. To compute that classical MAP estimate, find the state $X = 0$ or $X = \{x\}$ that maximizes $f(X)$. Because $f(0) = 1/2$ is a unitless probability and $f(\{1\}) = 1/2\sqrt{2\pi}\sigma$ has units of $1/km$, the classical MAP would compare the values of two quantities that are incommensurable because of mismatch of units. As a result, the numerical value of $f(\{1\})$ can be arbitrarily increased or decreased — thereby getting $X^{MAP} = 0$ (no target in the scene) or $X^{MAP} \neq 0$ (target in the scene) — simply by changing units of measurement. The posterior expectation also fails. If it existed, it would be

$$\int Xf(X)\delta X = 0 \cdot f(0) + \int xf(x)dx = \frac{1}{2}(0 + 1km)$$

Notice that, once again, there is the problem of mismatched units — the unitless quantity 0 must be added to the quantity $1\ km$. Even assuming that the continuous variable x is discrete (to alleviate this problem disappears) still requires the quantity 0 be added to the quantity 1. If $0 + 1 = 0$, then $1 = 0$, which is impossible. If $0 + 1 = 1$ then $0 = 0$, resulting in the same mathematical symbol representing two different states (the no-target state 0 and the single-target state $x = 0$). The same problem occurs if $0 + a = b_a$ is defined for *any* real numbers a, b_a since then $0 = b_a - a$.

Thus, it is false to assert that if "the target space is discretized into a collection of cells [then] in the continuous case, the cell probabilities can be replaced by densities in the usual way."[57] General continuous/discrete-state multitarget statistics are not blind generalizations of discrete-state special cases. Equally false is the assertion that "The [multitarget] posterior distribution … constitutes the Bayes estimate of the number and state of the targets … From this distribution we can compute other estimates when appropriate, such as maximum *a posteriori* probability estimates or means."[55,56] Posteriors are not "estimators" of state variables like target number or target position/velocity; the multitarget MAP can be defined only when state space is discretized and a multitarget posterior expectation cannot be defined at all.

14.6.7 Optimal Multitarget State Estimators

Section 14.6.6 asserted that the classical Bayes-optimal state estimators do not exist in general multitarget situations; therefore, new estimators must be defined and demonstrated to be statistically well behaved.

In conventional statistics, the maximum likelihood estimator (MLE) is a special case of the MAP estimator (assuming that the prior is uniform) and, as such, is optimal and convergent. In the multitarget case, this does not hold true. If $f(Z|X)$ is the multitarget likelihood function, the units of measurement for $f(Z|X)$ are determined by the observation-set Z (which is fixed) and not the multitarget state X. Consequently, in multitarget situations, the classical MLE is defined,[4] although the classical MAP is not

$$\{\hat{\mathbf{x}}_1, \ldots, \hat{\mathbf{x}}_n\}^{MLE} \overset{\Delta}{=} \arg \max_{n, \mathbf{x}_1, \ldots, \mathbf{x}_n} f\left(Z \middle| \{\mathbf{x}_1, \ldots, \mathbf{x}_n\}\right)$$

or in condensed notation, $\hat{X}^{MLE} = \arg \max_X f(Z|X)$. The multitarget MLE will converge to the correct answer if given enough data.[1]

Because the multitarget MLE is not a Bayes estimator, new multitarget Bayes state estimators must be defined and their optimality must be demonstrated. In 1995 two such estimators were introduced, the "Marginal Multitarget Estimator (MaME)" and the "Joint Multitarget Estimator (JoME)."[1,28] The JoME is defined as

$$\{\hat{x}_1, \ldots, \hat{x}_n\}^{JoME} \overset{\Delta}{=} \arg \max_{n, \mathbf{x}_1, \ldots, \mathbf{x}_n} f_{k|k}\left(\{\mathbf{x}_1, \ldots, \mathbf{x}_n\} \middle| Z^{(k)}\right) \cdot \frac{c^n}{n!}$$

where c is a fixed constant whose units have been chosen so the $f(X) = c^{-|X|}$ is a multitarget density. Or, in condensed notation, $\hat{X}^{JoME} = \arg\max_X f_{k|k}(X|Z^{(k)}) \cdot c^{|X|}/|X|!$. One of the consequences of this is that both the JoME and the multitarget MLE estimate the number $\hat{n}$ and the identities/kinematics $\hat{x}_1, \ldots, \hat{x}_{\hat{n}}$ of targets optimally and simultaneously without resort to optimal report-to-track association. In other words, these multitarget estimators optimally resolve the conflicting objectives of detection, tracking, and identification.

14.6.8   Cramér-Rao Bounds for Multitarget State Estimators

The purpose of a performance bound is to quantify the theoretically best-possible performance of an algorithm. The most well-known of these is the Cramér-Rao bound, which states that no unbiased state estimator can achieve better than a certain minimal accuracy (covariance) defined in terms of the likelihood function $f(\mathbf{z}|\mathbf{x})$. This bound can be generalized to estimators J_m of vector-valued outputs of multisource-multitarget algorithms:[1,2,4]

$$\left(\mathbf{v}, C_{J_{m,x}}\right) \cdot \left(\mathbf{w}, L_{X,\mathbf{x},m}(\mathbf{w})\right) \geq \left(\mathbf{v}, \frac{\partial}{\partial_x \mathbf{w}} E_X[J_m]\right)$$

14.6.9   Multitarget Miss Distance

FISST provides a natural generalization of the concept of "miss distance" to multitarget situations, defined by

$$d_{Haus}(G,X) = \max\left\{d_0(G,X), d_0(X,G)\right\} \quad d_0(G,X) = \max_{g \in G} \max_{x \in X} \|g - x\|$$

14.7   Optimal-Bayes Fusion, Tracking, ID

Section 14.6 demonstrated that conventional single-sensor, single-target statistics can be directly generalized to multisensor-multitarget problems. This section shows how this leads to simultaneous multisensor-multitarget fusion, detection, tracking, and identification based on a suitable generalization of nonlinear filtering Equations 14.6 and 14.7. This approach is optimal because it is based on true multitarget sensor models and true multitarget Markov densities, which lead to true multitarget posterior distributions and, hence, optimal multitarget filters.

Section 14.7.1 summarizes the FISST approach to optimal multisource-multitarget detection, tracking, and target identification. Section 14.7.2 is a brief history of multitarget recursive Bayesian nonlinear filtering. Section 14.7.3 summarizes a "para-Gaussian" approximation that may offer a partial solution to computational issues. Section 14.7.4 suggests how optimal control theory can be directly generalized to multisensor-multitarget sensor management.

14.7.1   Multisensor-Multitarget Filtering Equations

Bayesian multitarget filtering is inherently nonlinear because multitarget likelihoods $f(Z|X)$ are, in general, highly non-Gaussian even for a Gaussian sensor.[2] Therefore, *multitarget nonlinear filtering is unavoidable if the goal is optimal-Bayes tracking of multiple, closely spaced targets.*

Using FISST, nonlinear filtering Equations 14.6 and 14.7 of Section 14.2 can be generalized to multisensor, multitarget problems. Assume that a time-sequence $Z^{(k)} = \{Z_1, \ldots, Z_k\}$ of *precise* multisensor-multitarget observations, $Z_k = \{\mathbf{z}_{j;1}, \ldots, \mathbf{z}_{j;m(j)}\}$, has been collected. Then the state of the multitarget system is described by the true multitarget posterior density $f_{k|k}(X_k|Z^{(k)})$. Suppose that, at any given time instant $k + 1$, we wish to update $f_{k|k}(X_k|Z^{(k)})$ to a new multitarget posterior, $f_{k+1|k+1}(X_{k+1}|Z^{(k+1)})$, on the basis of a new observation-set Z_{k+1}. Then nonlinear filtering Equations 14.6 and 14.7 become[1,5]

TABLE 14.2   History of Multitarget, Bayesian Nonlinear Filtering

Date	Author(s)	Theoretical Basis
1991	Miller et al.[59-62] "Jump Diffusion"	Stochastic PDEs
1994	Bethel and Paras[63]	Discrete filtering
1994	Mahler[1,3-5] "Finite-Set Statistics"	FISST
1996	Stone et al.[56,57,64] "Unified Data Fusion"	Heuristic
1996	Mahler-Kastella[55,65] "Joint Multitarget Probabilites"	FISST[9]
1997	Portenko et. al.[66]	Point processes

$$f_{k+1|k}\left(X_{k+1}\middle|Z^{(k)}\right)=\int\left[\begin{array}{c}f_{k+1|k}\left(X_{k+1}\middle|X_k\right)\\ \cdot f_{k|k}\left(X_k\middle|Z^{(k)}\right)\end{array}\right]\delta X_k$$

$$f_{k+1|k+1}\left(X_{k+1}\middle|Z^{(k+1)}\right)\propto f\left(Z_{k+1}\middle|X_{k+1}\right)f_{k+1|k}\left(X_{k+1}\middle|Z^{(k)}\right)$$

(14.22)

with normalization constant $f(Z_{k+1}|Z^{(k)}) = \int f(Z_{k+1}|Y)\, f_{k+1|k}(Y|Z^{(k)})\delta Y$ and where the two integrals now are set integrals.

14.7.2   A Short History of Multitarget Filtering

The concept of multitarget Bayesian nonlinear filtering is a relatively new one. For situations where the number of targets is known, the earliest exposition appears to be attributable to Washburn[58] in 1987. (For more information, see "Why Multisource, Multitarget Data Fusion is Tricky"[28]) Table 14.2 summarizes the history of the approach when the number of targets n is not known and must be determined in addition to the individual target states. The earliest work in this case appears to have originated with Miller, O'Sullivan, Srivastava, and others. Their very sophisticated approach is also the only approach that addresses the continuous evolution of the multitarget state. (All other approaches listed in the table assume discrete state-evolution.)

Mahler[3,4] was the first to systematically deal with the general discrete state-evolution case (Bethel and Paras[63] assumed discrete observation and state variables). Portenko et al.[66] used branching-process concepts to model changes in target number. Kastella's[8,9,55] "joint multitarget probabilities (JMP)," introduced at LM-E in 1996, was a renaming of a number of early core FISST concepts (i.e., set integrals, multitarget Kullback Leibler metrics, multitarget posteriors, joint multitarget state estimators, and the APWOP) that were devised two years earlier.

Stone et al. provided a valuable contribution by clarifying the relationship between multitarget Bayes filtering and multihypothesis correlation.[1,2] Nevertheless, their approach, which cites multitarget filtering Equation 14.22,[57] is described as "heuristic" in the table. This is because (1) its theoretical basis is so imprecisely formulated that the authors have found it possible to both disparage and implicitly assume a random set framework; (2) its Bayes-optimality and "explicit procedures" are both frequently asserted but never actually justified or spelled out with precision; (3) its treatment of certain basic technical issues in Bayes multitarget filtering — specifically the claim to have an "explicit procedure" for dealing with an unknown number of targets — is erroneous (see Section 14.6.6); and (4) the only justifications offered in support of its claim to be "simpler and … more general"[57] are false assertions about the supposed theoretical deficiencies of earlier research — particularly other researchers' alleged lack of an "explicit procedure" for dealing with an unknown number of targets.[56,57] (See *An Introduction to Multisource-Multitarget Statistics and Its Applications* for more details.[2])

14.7.3   Computational Issues in Multitarget Filtering

The single-sensor, single-target Bayesian nonlinear filtering Equations 14.6 and 14.7 are already computationally demanding. Computational difficulties can get only worse when attempting to implement the multitarget nonlinear filtering Equation 14.22. This section summarizes some ideas (first proposed in *Mathematics of Data Fusion*[1,7]) for approximate multitarget nonlinear filtering.

14.7.3.1 The Gaussian Approximation in Single-Target Filtering

A possible strategy is suggested by drawing an analog with the single-target nonlinear filtering Equations 14.6 and 14.7. The Gaussian approximation uses the identity[27]

$$N_A(\mathbf{x}-\mathbf{a})N_B(\mathbf{x}-\mathbf{b})=N_{A+B}(\mathbf{a}-\mathbf{b})N_C(\mathbf{x}-\mathbf{c}) \tag{14.23}$$

(where $C^{-1} \overset{\Delta}{=} A^{-1} + B^{-1}$ and $C^{-1}\mathbf{c} \overset{\Delta}{=} A^{-1}\mathbf{a} + B^{-1}\mathbf{b}$). This shows that Bayes' rule is closed with respect to Gaussians and also that the prediction and Bayes-normalization integrals satisfy the *closed-form integrability property*:

$$\int N_A(\mathbf{x}-\mathbf{a})N_B(\mathbf{x}-\mathbf{b})d\mathbf{x} = N_{A+B}(\mathbf{a}-\mathbf{b}) \tag{14.24}$$

Suppose that $f(Z|X)$ is the multitarget likelihood for a Gaussian sensor (taking into account both missed detections and false alarms). There is a family of multitarget distributions that has a closed-form integrability property analogous to Equation 14.24. Using this family, computational tractability may be feasible even if the motion models used do not assume that the number of targets is fixed.

14.7.3.2 Para-Gaussian Multitarget Distributions

Suppose that a single Gaussian sensor has missed detections. From the FISST multitarget calculus, the multitarget likelihood is[1]

$$f\left(\{\mathbf{z}_1,\ldots,\mathbf{z}_m\}\big|\{\mathbf{x}_1,\ldots,\mathbf{x}_n\}\right)=p_D^m(1-p_D)^{n-m} \cdot \sum_{1\leq i_1 \neq \ldots \neq i_m \leq n} N_Q(\mathbf{z}_1-B\mathbf{x}_{i_1})\cdots N_Q(\mathbf{z}_m-B\mathbf{x}_{i_m})$$

for $Z=(\mathbf{z}_1,\ldots,\mathbf{z}_m)$ and $X=(\mathbf{x}_1,\ldots,\mathbf{x}_n)$. If the Gaussian sensor also is corrupted by a statistically independent, state-independent clutter process with density $\kappa(Z)$, the multitarget likelihood is

$$f_{clutter}(Z|X)=\sum_{W\subseteq Z} f(W|X)\kappa(Z-W)$$

For $X=(\mathbf{x}_1,\ldots,\mathbf{x}_n)$ and $X'=(\mathbf{x}'_1,\ldots,\mathbf{x}'_{n'})$ and $C_{n',n}=n'!/(n'-n)!n!$, define

$$N_{Q,q}(X|X')\overset{\Delta}{=}q(n|n')C_{n',n}^{-1}\sum_{1\leq i_1 \neq \ldots \neq i_n \leq n'} N_Q(\mathbf{x}_1-\mathbf{x}'_{i_1})\cdots N_Q(\mathbf{x}_n-\mathbf{x}'_{i_n})$$

$$N_{Q,q,\kappa}(X|X')\overset{\Delta}{=}\sum_{W\subseteq X} N_{Q,q}(W|X')\kappa(X-W)$$

where $q(n|n')\geq 0$ for all j and where $q(n|n')\geq 0$ (if $n>n'$ or $n<0$) and $\sum_{n=0}^{n'}q(n|n')=1$. The distribution $N_{Q,q,\kappa}(X|X')$ is called a *para-Gaussian density*.[1] The multitarget likelihood of a Gaussian sensor is a para-Gaussian.

The set integral of certain products of para-Gaussians *can be evaluated in closed form*:[1,7]

$$\int N_{P,p,\kappa}(Z|X)N_{Q,q}(Z|X)\delta X=N_{P+Q,p\otimes q,\kappa}(Z|X)$$

where $(p\otimes q)(k|i)\overset{\Delta}{=}\sum_{j=k}^i p(k|j)q(j|i)$. (In fact, this result can be generalized to much more general multitarget Markov densities.[1]) Consequently, both the multitarget prediction integral and the multitarget Bayes normalization constant can be evaluated in closed form if the densities in the integrands are suitable para-Gaussians (see below). The resulting computational advantage suggests a multitarget analog of the Gaussian approximation. Two publications[2,54] provide a more detailed discussion of this approach.

14.7.4 Optimal Sensor Management

Sensor management has been usefully described as the process of directing the right platforms and the right sensors on those platforms, to the right targets at the right times. FISST allows multiplatform-multisensor sensor management to be reformulated as a direct generalization of optimal (nonlinear) control theory, based on the multitarget miss distance of Section 14.6.9. See *An Introduction to Multi-source-Multitarget Statistics and Its Application* for more details.[2]

14.8 Robust-Bayes Fusion, Tracking, ID

This section addresses the question of how to extend Bayesian (or other forms of probabilistic) inference to situations in which likelihood functions and/or data are imperfectly understood. The optimal-Bayes techniques described in previous sections can be extended to robust-Bayes techniques designed to address such issues. The basic approach, which was summarized in Section 14.1.3, is as follows:

1. Represent statistically ill-characterized ("ambiguous") data as random closed subset 1 of (multi-source) observation space.
2. Thus, in general, multisensor-multitarget observations will be randomly varying finite sets of the form $Z = \{\mathbf{z}_1,...,\mathbf{z}_m,\Theta_1,...,\Theta_{m'}\}$, where $\mathbf{z}_1,...,\mathbf{z}_m$ are conventional data and $\Theta_1,...,\Theta_{m'}$ are "ambiguous" data.
3. Just as the probability-mass function $p(S|\mathbf{x}) = \Pr(\mathbf{Z} \in S)$ is used to describe the generation of conventional data $\mathbf{z}$, use "generalized likelihood functions" such as $\rho(\Theta|\mathbf{x}) = \Pr(\Theta \subseteq \Sigma|\mathbf{x})$ to describe the generation of ambiguous data.
4. Construct single-target posteriors $f_{k|k}(\mathbf{x}|Z^k)$ and multitarget posteriors $f_{k|k}(X|Z^{(k)})$ conditioned on all data, whether "ambiguous" or otherwise.
5. Proceed essentially as in Section 14.7.

Section 14.8.1 discusses the concept of "ambiguous" data and why random set theory provides a useful means of mathematically representing such data. Section 14.8.2 discusses the various forms of ambiguous data — imprecise, vague (fuzzy), and contingent — and their corresponding random set representations. Section 14.8.3 defines the concept of a true Bayesian likelihood function for ambiguous data and argues that true likelihoods of this kind may be impossible to construct in practice. Section 14.8.4 proposes an engineering compromise — the concept of a generalized likelihood function. The concept of a posterior distribution conditioned on ambiguous data is introduced in Section 14.8.5. Section 14.8.6 shows how to construct practical generalized likelihood functions, based on the concept of geometric model-matching. Finally, the recursive Bayesian nonlinear filtering equations — Equations 14.6 and 14.7 — are generalized to the multisource-multitarget case in Section 14.8.7.

14.8.1 Random Set Models of "Ambiguous" Data

The FISST approach to data that is difficult to statistically characterize is based on the key notion that *ambiguous data can be probabilistically represented as random closed subsets of (multisource) measurement space.*[1]

Consider the following simple example. (For a more extensive discussion see Mahler.[1,2]) Suppose that $\mathbf{z} = C\mathbf{x} + \mathbf{W}$ where $\mathbf{x}$ is target state, $\mathbf{W}$ is random noise, and C is an invertible matrix. Let B be an "ambiguous observation" in the sense that it is a subset of measurement space that constrains the possible values of $\mathbf{z}$: $B \ni \mathbf{z}$. Then the random variable Γ defined by $\Gamma = \{C^{-1}(\mathbf{z} - \mathbf{W})|\mathbf{z} \in B\}$ is the randomly varying subset of all target states that are consistent with this ambiguous observation. That is, the ambiguous observation B also *indirectly* constrains the possible target *states*.

Suppose, on the other hand, that the validity of the constraint $\mathbf{z} \in B$ is uncertain; there may be many possible constraints — of varying plausibility — on $\mathbf{z}$. This ambiguity could be modeled as a randomly varying subset Θ of measurements, where the probability $\Pr(\Theta = B)$ represents the degree of belief in the plausibility of the specific constraint B. The random subset of all states that are consistent with Θ would then be $\Gamma = \{C^{-1}(\mathbf{z} - \mathbf{W}) | \mathbf{z} \in \Theta\}$. (Caution: The random closed subset Θ is a model of *a single observation collected by a single source*. Do not confuse this subset with a multisensor, multitarget observation set, Σ, whose instantiations $\Sigma = Z$ are finite sets of the form $Z = \{\mathbf{z}_1,...,\mathbf{z}_m, \Theta_1,...,\Theta_{m'}\}$ where $\mathbf{z}_1,...,\mathbf{z}_m$ are individual conventional observations and $\Theta_1,...,\Theta_{m'}$ are random-set models of individual ambiguous observations.)

14.8.2   Forms of Ambiguous Data

Recognizing that random sets provide a common probabilistic foundation for various kinds of statistically ill-characterized data is not enough to tell us how to construct practical random set representations of such data. This section shows how three kinds of ambiguous data — imprecise, vague, and contingent — can be represented probabilistically by random sets.

14.8.2.1   Vague Data: Fuzzy Logic

A fuzzy membership function on some (finite or infinite) universe U is a function that assigns a number $f(u)$ between zero and one to each member u of U. The random subset $\Sigma_A(f)$, called the *canonical random set representation* of the fuzzy subset f, is defined by

$$\Sigma_A(f) \overset{\Delta}{=} \{u \in U \,|\, A \le f(u)\} \tag{14.25}$$

14.8.2.2   Imprecise Data: Dempster-Shafer Bodies of Evidence

A Dempster-Shafer body of evidence B on some space U consists of nonempty subsets $B : B_1,..., B_b$ of U and nonnegative weights $b_1,...,b_b$ that sum to one. Define the random subset Σ of U by $p(\Sigma = B_i) = b_i$ for $i = 1,...,b$. Then Σ is the random set representation of B and $B = B^\Sigma$.[34-36,40,41] The Dempster-Shafer theory can be generalized to the case when the B_i are *fuzzy membership functions*.[67] Such "fuzzy bodies of evidence" can also be represented in random set form.

14.8.2.3   Contingent Data: Conditional Event Algebra

Knowledge-based rules have the form $X \Rightarrow S =$ "*if X then S*" where S, X are subsets of a (finite) universe U. There is at least one way to represent knowledge-based rules in random set form.[45,46] Specifically, let Φ be a uniformly distributed random subset of U — that is, one whose probability distribution is $p(\Phi = S) = 2^{-|U|}$ for all $S \subseteq U$. A random set representation $\Sigma_\Phi(X \Rightarrow S)$ of the rule $X \Rightarrow S$ is

$$\Sigma_\Phi(X \Rightarrow S) \overset{\Delta}{=} (S \cap X) \cup (X^c \cap \Phi)$$

14.8.3   True Likelihood Functions for Ambiguous Data

The next step in a strict Bayesian formulation of the ambiguous-data problem is to specify a *likelihood function for ambiguous evidence* that models the understood likelihood that a specific ambiguous datum Θ will be observed, given that a target of state $\mathbf{x}$ is present. This is where practical problems are encountered. The required likelihood function must have the form

$$f(\Theta|\mathbf{x}) = \frac{\Pr(\mathfrak{R}=\Theta, \mathbf{X}=\mathbf{x})}{\Pr(\mathbf{X}=\mathbf{x})}$$

where $\mathfrak{R}$ is a random variable that ranges over all random closed subsets Θ of measurement space. However, $f(\Theta|\mathbf{x})$ cannot be a likelihood function unless it satisfies a normality equation of the form $\int f(\Theta|\mathbf{x})d\Theta = 1$ where $\int \mathbf{f}(\Theta|\mathbf{x})d\Theta$ is an integral that sums over all closed random subsets of measurement space. No clear means exists for constructing a likelihood function $f(\Theta|\mathbf{x})$ that not only models a particular real-world situation but also provably integrates to unity. If $f(\Theta|\mathbf{x})$ could be specified with sufficient exactitude, a high-fidelity *conventional* likelihood $f(\mathbf{z}|\mathbf{x})$ could be constructed.

14.8.4 Generalized Likelihood Functions

To address this problem, FISST employs an engineering compromise based on the fact that Bayes' rule is very general — it applies to all events, *not just those having the specific Bayesian form $X = x$ or $R = 1$.* That is, Bayes' rule states that $\Pr(E_1|E_2)\Pr(E_2) = \Pr(E_2|E_1)\Pr(E_1)$ for any events E_1, E_2. Consequently, let E_Θ be any event with some specified functional dependence on the ambiguous measurement 1 — for example, $E_\Theta : \Theta \supseteq \Sigma \, or \, E_\Theta : \Theta \cap \Sigma \neq \emptyset$, where Θ, Σ are random closed subsets of observation space. Then

$$f(\mathbf{x}|E_\Theta) = \frac{\Pr(X=\mathbf{x}, E_\Theta)}{\Pr(E_\Theta)} = \frac{\rho(\Theta|\mathbf{x}) f_o(\mathbf{x})}{\Pr(E_\Theta)}$$

where $f_o(\mathbf{x}) = \Pr(X = \mathbf{x})$ is the prior distribution on $\mathbf{x}$ and where $\rho(\Theta|\mathbf{x}) \triangleq \Pr(E_\Theta|X = x)$ is considered to be a *generalized likelihood function*. Notice that $\rho(\Theta|\mathbf{x})$ will usually be unnormalized because events E_Θ are not mutually exclusive. Joint generalized likelihood functions can be defined in the same way.

For example, suppose that evidence consists of a fuzzy Dempster-Shafer body of evidence $B : B_1, \ldots, B_b$; $b_1, \ldots, b_b$ on state space V.[68,69] Let $q(v)$ be a prior probability distribution on V and $q(B_i) \triangleq \Sigma_v B_i(v) q(v)$. The FISST likelihood for B can be shown to be

$$q(B|v) = q(B) \sum_{i=1}^{b} \frac{b_i}{q(B_i)} B_i(v) \qquad q(B) = \left[\sum_{i=1}^{b} \frac{b_i}{q(B_i)} \right]^{-1}$$

14.8.5 Posteriors Conditioned on Ambiguous Data

Bayes' rule can be used to compute the following posterior distribution, conditioned on the ambiguous data modeled by the closed random subsets $\Theta_1, \ldots, \Theta_m$:

$$f(\mathbf{x}|\Theta_1, \ldots, \Theta_m) \propto \rho(\Theta_1, \ldots, \Theta_m|\mathbf{x}) f_0(\mathbf{x}) \qquad (14.26)$$

with proportionality constant $p(\Theta_1, \ldots, \Theta_m) \triangleq \int \rho(\Theta_1, \ldots, \Theta_m|\mathbf{x}) f_0(\mathbf{x}) d\mathbf{x}$. For example,[7,25] suppose that evidence consists of a fuzzy Dempster-Shafer body of evidence $B : B_1, \ldots, B_b$; $b_1, \ldots, b_b$ on state space V. Let $q(v)$ be a prior probability distribution on V and $q(B_i) \triangleq \Sigma_v B_i(v) q(v)$. Then the FISST posterior distribution conditioned on B can be shown to be:

$$q(v|B) = q(v) \sum_{i=1}^{b} \frac{b_i}{q(B_i)} B_i(v)$$

14.8.6 Practical Generalized Likelihood Functions

How can generalized likelihood functions be produced that are usable in application? To address this problem, FISST recognizes that *generalized likelihood functions can be constructed using the concept of "model-matching" between observations and model signatures.*

This point can be illustrated by showing that in the conventional Bayesian case, geometric model-matching yields the conventional Bayesian measurement model.[2] Lack of space prevents a detailed illustration here; however, this section will address the general case. Let Θ be the random closed subset of measurement space U that models a particular piece of evidence about the unknown target. Let Σ be another random closed subset of U that models the generation of observations. A *conditional random subset of U* exists, denoted $\Sigma|_{X=x}$, such that $\Pr(\Sigma|_{X=x} = T) = \Pr(\Sigma = T|X = x)$. What is the probability that the observed (ambiguous) evidence Θ matches a particular (ambiguous) model signature $\Sigma|_{X=x}$? Different definitions of geometric matching — for example, $\Theta \supseteq \Sigma|_{X=x}$ (complete consistency between observation and model) or $\Theta \cap \Sigma|_{X=z} \neq \emptyset$ (noncontradiction between observation and model) — will yield different generalized likelihood functions:

$$\rho_1\left(\Theta|x\right) \overset{\Delta}{=} \Pr\left(\left.\Sigma\right|_{X=x} \cap \Theta \neq \emptyset\right) \qquad \rho_2\left(\Theta|x\right) \overset{\Delta}{=} \Pr\left(\left.\Sigma\right|_{X=x} \subseteq \Theta\right)$$

Practical generalized likelihood functions can be constructed by choosing suitable random-set model signatures $\Sigma_x \overset{\Delta}{=} \Sigma|_{X=x}$.

For example, let g be a fuzzy observation and f_x a fuzzy signature model (where both g and f_x are fuzzy membership functions on measurement space).[1] Let $\Theta_g \overset{\Delta}{=} \Sigma_A(g)$ and $\Sigma_x \overset{\Delta}{=} \Sigma_A(f_x)$ and $\rho(g|x) \overset{\Delta}{=} \Pr(\Sigma|_{X=x} \cap \Theta_g \neq \emptyset)$ where $\Sigma_A(h)$ is defined as in Equation 14.25. Then

$$\rho\left(g|x\right) = \max_z \min\left\{g(z), f_x(z)\right\} \tag{14.27}$$

14.8.7 Unified Multisource-Multitarget Data Fusion

Suppose that there are a number of independent sources, some of which supply conventional data and others that supply ambiguous data. As in Section 14.6.1, a multisource-multitarget joint generalized likelihood can be constructed of the form:

$$\rho\left(Z|X\right) = \rho\left(Z^{[1]}, \ldots, Z^{[m]}, \Theta^{[m]}, \ldots, \Theta^{[m']}|X\right)$$

$$Z = Z^{[1]} \cup \cdots \cup Z^{[e]} \cup \Theta^{[e+1]} \cup \cdots \cup \Theta^{[e+e']}$$

where $Z^{[s]} = \{z_1^{[s]}, \ldots, z_{m(s)}^s\}$ denotes a multitarget observation collected by a conventional sensor with identifier $s = 1, \ldots, e$, and where $\Theta^{[s]} = \{\Theta_1^{[s]}, \ldots, \Theta_{m'(s)}^s\}$ denotes a multitarget observation supplied by a source with identifier $s = e + 1, \ldots, e + e'$ that collects ambiguous data. Given this, the data can be fused using Bayes' rule: $\rho(X|Z) \propto \rho(Z|X) f(X)$. Robust multisource-multitarget detection, tracking, and identification can be accomplished by using the joint generalized likelihood function with the multitarget recursive Bayesian nonlinear filtering Equation 14.22 of Section 14.7. In the event that data sources are independent, these become

$$f_{k+1|k}\left(X_{k+1}|Z^{(k)}\right) = \int f_{k+1|k}\left(X_{k+1}|X_k\right) f_{k|k}\left(X_k|Z^{(k)}\right) \delta X_k$$

$$f_{k+1|k+1}\left(X_{k+1}|Z^{(k+1)}\right) \propto f_{(s)}\left(Z_{k+1}^{(s)}|X_{k+1}\right) f_{k+1|k}\left(X_{k+1}|Z^{(k)}\right)$$

$$f_{k+1|k+1}\left(X_{k+1}|Z^{(k+1)}\right) \propto \rho_{(s)}\left(\Theta_{k+1}^{[s]}|X_{k+1}\right) f_{k+1|k}\left(X_{k+1}|Z^{(k)}\right)$$

where $Z^{(k)}$ denotes the time series of data, ambiguous or otherwise, collected from all sources. Mahler provides an example of nonlinear filtering using fuzzy data.[2,16]

14.9　Summary and Conclusions

Finite-set statistics (FISST) were created, in part, to address the issues in probabilistic inference that the "cookbook Bayesian" viewpoint overlooks. These issues include

- Dealing with poorly characterized sensor likelihoods
- Dealing with ambiguous data
- Constructing likelihoods for ambiguous data
- Constructing true likelihoods and true Markov transition densities for multitarget problems
- Dealing with the dimensionality in multitarget problems
- Providing a single, fully probabilistic, systematic, and genuinely unified foundation for multi-source-multitarget detection, tracking, identification, data fusion, sensor management, performance estimation, and threat estimation and prediction
- Accomplishing all of these objectives within the framework of a direct, relatively simple, and engineering-friendly generalization of "Statistics 101."

During the last two years, FISST has begun to emerge from the realm of basic research and is being applied, with some preliminary indications of success, to a range of practical engineering research applications. This chapter has described the difficulties associated with the "cookbook Bayesian" viewpoint and summarized how and why FISST resolves them.

Acknowledgments

The core concepts underlying the work reported in this chapter were developed under internal research and development funding in 1993 and 1994 at the Eagan, MN, division of Lockheed Martin Corporation (LM-E). This work has been supported at the basic research level since 1994 by the U.S. Army Research Office, which also funded preparation of this chapter. Various aspects have been supported at the applied research level by the U.S. Air Force Research Laboratory, SPAWARSYSCOM, and the Office of Naval Research via SPAWAR Systems Center.[33] The content does not necessarily reflect the position or the policy of the Government. No official endorsement should be inferred.

The author of this chapter extends his appreciation to the following individuals for their roles in helping to transform the ideas presented in this chapter into practical application: Dr. Marty O'Hely of the University of Oregon; Dr. Alexsandar Zatezalo of the University of Minesota; Dr. Adel Al-Fallah, Dr. Mel Huff, Dr. Raman Mehra, Dr. Constantino Rago, Dr. Ravi Ravichandran, and Dr. Ssu-Hsin Yu of Scientific Systems Co., Inc.; Ron Allen, John Honeycutt, Robert Myre, and John Werner of Summit Research Corp.; and Trent Brundage, Dr. Keith Burgess, John Hatlestad, Dr. John Hoffman, Paul Leavitt, Dr. Paul Ohmann, Craig Poling, Eric Sorensen, Eric Taipale, and Dr. Tim Zajic of LM-E.

References

1. Goodman, I.R. and Nguyen, H.T., *Uncertainty Models for Knowledge Based Systems*, North-Holland, Amsterdam, 1985.
2. Mahler, R., An Introduction to multisource-multitarget statistics and its applications, Lockheed Martin Technical Monograph, March 15, 2000.
3. Mahler, R., Global integrated data fusion, in *Proc. 7th Nat'l Symp. on Sensor Fusion*, I (Unclass), ERIM, Ann Arbor, MI, 1994, 187.
4. Mahler, R., A unified approach to data fusion, in *Proc. 7th Joint Data Fusion Symp.*, 1994, 154, and *Selected Papers on Sensor and Data Fusion*, Sadjadi, P.A., Ed., SPIE, MS-124, 1996, 325.
5. Mahler, R., Global optimal sensor allocation, in *Proc. 1996 Nat'l. Symp. on Sensor Fusion*, I (Unclass), 1996, 347.

6. Mahler, R., Multisource-multitarget filtering: a unified approach, in *SPIE Proc.*, 3373, 1998, 296.

7. Mahler, R., Multitarget Markov motion models, in *SPIE Proc.*, 3720, 1999, 47.

8. Mahler, R., Global posterior densities for sensor management, in *SPIE Proc.*, 3365, 1998, 252.

9. Musick, S., Kastella, K., and Mahler, R., A practical implementation of joint multitarget probabilities, in *SPIE Proc.*, 3374, 1998, 26.

10. Mahler, R., Information theory and data fusion, in *Proc. 8th Nat'l. Symp. on Sensor Fusion*, I (Unclass), ERIM, Ann Arbor, MI, 1995, 279.

11. Mahler, R., Unified nonparametric data fusion, in *SPIE Proc.*, 2484, 1995, 66.

12. Mahler, R., Information for fusion management and performance estimation, in *SPIE Proc.*, 3374, 1998, 64.

13. Zajic, T. and Mahler, R., Practical multisource-multitarget data fusion performance estimation, in *SPIE Proc.*, 3720, 1999, 92.

14. Mahler, R., Measurement models for ambiguous evidence using conditional random sets, in *SPIE Proc.*, 3068, 1997, 40.

15. Mahler, R., Unified data fusion: fuzzy logic, evidence, and rules, in *SPIE Proc.*, 2755, 1996, 226.

16. Mahler, R. et al., Nonlinear filtering with really bad data, in *SPIE Proc.*, 3720, 1999, 59.

17. Mahler, R., Optimal/robust distributed data fusion: a unified approach, in *SPIE Proc.*, 4052, 2000.

18. Mahler, R., Decisions and data fusion, in *Proc. 1997 IRIS Nat'l. Symp. on Sensor and Data Fusion*, I (unclass), M.I.T. Lincoln Laboratories, 1997, 71.

19. El-Fallah, A. et al., Adaptive data fusion using finite-set statistics, in *SPIE Proc.*, 3720, 1999, 80.

20. Allen, R. et al., Passive-acoustic classification system (PACS) for ASW, in *Proc 1998 IRIS Nat'l. Symp. on Sensor and Data Fusion*, 1998, 179.

21. Mahler, R. et al., Application of unified evidence accrual methods to robust SAR ATR, in *SPIE Proc.*, 3720, 1999, 71.

22. Zajic, T., Hoffman, J., and Mahler, R., Scientific performance metrics for data fusion: new results, in *SPIE Proc.*, 4052, 2000, 173.

23. El-Fallah, A. et al., Scientific performance evaluation for sensor management, in *SPIE Proc.*, 4052, 2000, 183.

24. Mahler, R., Multisource-multitarget detection and acquisition: a unified approach, in *SPIE Proc.*, 3809, 1999, 218.

25. Mahler, R. et al., Joint tracking, pose estimation, and identification using HRRR data, in *SPIE Proc.*, 4052, 2000, 195.

26. Bar-Shalom, Y. and Li, X.-R., *Estimation and Tracking: Principles, Techniques, and Software*, Artech House, Ann Arbor, MI, 1993.

27. Jazwinski, A.H., *Stochastic Processes and Filtering Theory*, Academic Press, New York, 1970.

28. Mahler, R., Why multi-source, multi-target data fusion is tricky, in *Proc. 1999 IRIS Nat'l. Symp. On Sensor and Data Fusion*, 1 (Unclass), Johns Hopkins APL, Laurel, MD, 1995, 135.

29. Sorenson, H.W., Recursive estimation for nonlinear dynamic systems, *Bayesian Analysis of Statistical Time Series and Dynamic Models*, Spall, J.C., Ed., Marcel Dekker, New York, 1988.

30. Fletcher, C.A.J., *Computational Techniques for Fluid Dynamics: Fundamental and General Techniques*, Vol. 1, Springer-Verlag, New York, 1988.

31. McCartin, B. J., Seven deadly sins of numerical computation, *American Mathematical Monthly*, December 1998, 929.

32. Press, W. H. et al., *Numerical Recipes in C: The Art of Scientific Computing*, 2nd ed., Cambridge University Press, Cambridge, U.K., 1992.

33. Matheron, G., *Random Sets and Integral Geometry*, John Wiley & Sons, New York, 1975.

34. Goodman, I.R., Mahler, R.P.S., and Nguyen, H.T., *Mathematics of Data Fusion*, Kluwer Academic Publishers, Dordrecht (Holland), 1997.

35. Kruse, R., Schwencke, E., and Heinsohn, J., *Uncertainty and Vagueness in Knowledge-Based Systems*, Springer-Verlag, New York, 1991.

36. Quinio, P. and Matsuyama, T., Random closed sets: a unified approach to the representation of imprecision and uncertainty, *Symbolic and Quantitative Approaches to Uncertainty*, Kruse, R. and Siegel, P., Eds., Springer-Verlag, New York, 1991, 282.

37. Goutsias, J., Mahler, R., and H. T. Nguyen, *Random Sets: Theory and Application*, Springer-Verlag, New York, 1997.

38. Grabisch, M., Nguyen, H.T., and Waler E.A., *Fundamentals of Uncertainty Calculus with Applications to Fuzzy Inference*, Kluwer Academic Publishers, Dordrecht (Holland), 1995.

39. Shafer, G., and Logan, R., Implementing Dempster's rule for hierarchical evidence, *Artificial Intelligence*, 33, 271, 1987.

40. Nguyen, H.T., On random sets and belief functions, *J. Math. Anal. and Appt.*, 65, 531, 1978.

41. Hestir,K., Nguyen, H.T., and Rogers, G.S. A random set formalism for evidential reasoning, *Conditional Logic in Expert Systems*, Goodman, I. R., Gupta, M.M., Nguyen, H.T., and Rogers, G.S., eds., North-Holland, 1991, 309.

42. Goodman, I.R., Fuzzy sets as equivalence classes of random sets, *Fuzzy Sets and Possibility Theory*, Yager, R., Ed., Permagon, Oxford, U.K., 1982, 327.

43. Orlov, A.L., Relationships between fuzzy and random sets: fuzzy tolerances, *Issledovania po Veroyatnostnostatishesk*, Medelironvaniu Realnikh System, Moscow, 1977.

44. Hohle, U., A mathematical theory of uncertainty: fuzzy experiments and their realizations, *Recent Developments in Fuzzy Set and Possibility Theory*, Yager, R.R., Ed., Permagon Press, Oxford, U.K., 1981, 344.

45. Mahler, R., Representing rules as random sets, I: Statistical correlations between rules, *Information Sciences*, 88, 47, 1996.

46. Mahler, R., Representing rules as random sets, II: Iterated rules, *Int'l. J. Intelligent Sys.*, 11, 583, 1996.

47. Mori, S. et al., Tracking and classifying multiple targets without *a priori* identification, *IEEE Trans. Auto Contr.*, Vol. AC-31, 401. 1986.

48. Mori, S., Multi-target tracking theory in random set formalism, in *1st Int'l. Conf. on Multisource-Multisensor Information Fusion*, 1998.

49. Mori, S., Random sets in data fusion problems, in *Proc. 1997 SPIE*, 1997.

50. Mori, S., A theory of informational exchanges-random set formalism, in *Proc. 1998 IRIS Nat'l. Symp. on Sensor and Data Fusion*, I (Unclass), ERIM, 1998, 147.

51. Mori, S., Random sets in data fusion: multi-object state-extination as a foundation of data fusion theory, *Random Sets: Theory and Application*, Goutsias, J. Mahler, R.P.S., and Nguyen, H.T., Eds., Springer-Verlag, New York, 1997.

52. Ho, Y.C. and Lee, R.C.K., A Bayesian approach to problems in stockastic estimation and control, *IEEE Trans. AC*, AC-9, 333, 1964.

53. Van Trees, H.L., Detection, *Estimation, and Modulation Theory, Part I: Detection, Estimation, and Linear Modulation Theory*, John Wiley & Sons, New York, 1968.

54. Mahler, R., The search for tractable Bayes multitarget filters, in *SPIE Proc.*, 4048, 2000, 310.

55. Kastella, K., Joint multitarget probabilities for detection and tracking, in *SPIE Proc.*, 3086, 1997, 122.

56. Stone, L.D., Finn, M.V., and Barlow, C.A., Unified data fusion, submitted for journal publication in 1997 (manuscript copy, dated May 22, 1997, provided by L. D. Stone).

57. Stone, L.D., Barlow, C.A., and Corwin, T. L., *Bayesian Multiple Target Tracking*, Artech House Inc., 1999.

58. Washburn, R.B., A random point process approach to multi-object tracking, in *Proc. Amer. Contr. Conf.*, 3, 1846, 1987.

59. Lanterman, A. D. et al., Jump-diffusion processes for the automated understanding of FLIR scenes, in *SPIE Proc.*, 2234, 416, 1994.

60. Srivastava, A., Inferences on transformation groups generating patterns on rigid motions, Ph.D. dissertation, Washington University (St. Louis), Dept. of Electrical Engineering, 1996.

61. Srivastava, A., Miller, M.I., and Grenander, U., Jump-diffusion processes for object. tracking and direction finding, in *Proc. 29th Allerton Conf. on Communication, Control, and Computing*, Univ. of Illinois-Urbana, 563, 1991.

62. Srivastava, A. et al., Multitarget narrowband direction finding and tracking using motion dynamics, in *Proc. 30th Allerton Conf. on Communication, Control, and Computation*, Monticello, IL, 279, 1992.

63. Bethel, R.E. and Paras, G.J., A PDF multitarget-tracker, *IEEE Trans AES*, 30, 386, 1994.

64. Barlow, C.A., Stone, L.D., and Finn, M.V., Unified data fusion, in *Proc. 9th Nat'l. Symp. on Sensor Fusion*, Vol. I (Unclassified), Naval Pre-graduate School, Monterey, CA, March 11–13, 1996, 321.

65. Kastella, K., Discrimination gain for sensor management in multitarget detection and tracking, in *Proc. 1996 IMAC5 Multiconf. on Comp. and Eng. Appl. (CE5A '96)*, Symp. on Contr., Opt., and Supervision, Lille, France, 1996, 167.

66. Portenko, N., Salehi, H., and Skorokhod, A., On optimal filtering of multitarget tracking systems based on point processes observations, *Random Operators and Stochastic Equations*, 1, 1, 1997.

67. Mahler, R., Combining ambiguous evidence with respect to ambiguous *a priori* knowledge, part II: fuzzy logic, *Fuzzy Sets and Systems*, 75, 319, 1995.

68. Fixsen, D. and Mahler, R., The modified Dempster-Shafer approach to classification, *IEEE Trans. on Systems, Man and Cybernetics — Part A*, 27, 96, 1997.

69. Mahler, R.P.S., Combining ambiguous evidence with respect to ambiguous *a priori* knowledge, part I: Boolean logic, in *IEEE Trans. SMC, Part A*, 26, 27, 1996.

Systems Engineering and Implementation

15 **Requirements Derivation for Data Fusion Systems**   *Ed Waltz and David L. Hall* 15-1
Introduction • Requirements Analysis Process • Engineering Flow-Down Approach • Enterprise Architecture Approach • Comparison of Approaches

16 **A Systems Engineering Approach for Implementing Data Fusion Systems**
Christopher L. Bowman and Alan N. Steinberg 16-1
Scope • Architecture for Data Fusion • Data Fusion System Engineering Process • Fusion System Role Optimization

17 **Studies and Analyses with Project Correlation: An In-Depth Assessment of Correlation Problems and Solution Techniques**   *James Llinas, Lori McConnel, Christopher L. Bowman, David L. Hall, and Paul Applegate* 17-1
Introduction • A Description of the Data Correlation (DC) Problem • Hypothesis Generation • Hypothesis Evaluation • Hypothesis Selection • Summary

18 **Data Management Support to Tactical Data Fusion**   *Richard Antony* 18-1
Introduction • Database Management Systems • Spatial, Temporal, and Hierarchical Reasoning • Database Design Criteria • Object Representation of Space • Integrated Spatial/Nonspatial Data Representation • Sample Application • Summary and Conclusions

19 **Removing the HCI Bottleneck: How the Human-Computer Interface (HCI) Affects the Performance of Data Fusion Systems**   *Mary Jane M. Hall, Sonya A. Hall, and Timothy Tate* 19-1
Introduction • A Multimedia Experiment • Summary of Results • Implications for Data Fusion Systems

20 **Assessing the Performance of Multisensor Fusion Processes**   *James Llinas* 20-1
Introduction • Test and Evaluation of the Data Fusion Process • Tools for Evaluation: Testbeds, Simulations, and Standard Data Sets • Relating Fusion Performance to Military Effectiveness — Measures of Merit • Summary

21 **Dirty Secrets in Multisensor Data Fusion**   *David L. Hall and Alan N. Steinberg* 21-1
Introduction • The JDL Data Fusion Process Model • Current Practices and Limitations in Data Fusion • Research Needs • Pitfalls in Data Fusion • Summary

15

Requirements Derivation for Data Fusion Systems

Ed Waltz
Veridian Systems

David L. Hall
The Pennsylvania State University

15.1 Introduction 15-1
15.2 Requirements Analysis Process.. 15-2
15.3 Engineering Flow-Down Approach 15-3
15.4 Enterprise Architecture Approach 15-5
    The Three Views of the Enterprise Architecture
15.5 Comparison of Approaches.. 15-6
References... 15-8

15.1 Introduction

The design of practical systems requires the translation of data fusion theoretic principles, practical constraints, and operational requirements into a physical, functional, and operational architecture that can be implemented, operated, and maintained. This translation of principles to practice demands a discipline that enables the system engineer or architect to perform the following basic functions:

- Define user requirements in terms of functionality (qualitative description) and performance (quantitative description),
- Synthesize alternative design models and analyze/compare the alternatives in terms of requirements and risk,
- Select optimum design against some optimization criteria,
- Allocate requirements to functional system subelements for selected design candidates,
- Monitor the as-designed system to measure projected technical performance, risk, and other factors (e.g., projected life cycle cost) throughout the design and test cycle,
- Verify performance of the implemented system against top- and intermediate-level requirements to ensure that requirements are met and to validate the system performance model.

The discipline of system engineering, pioneered by the aerospace community to implement complex systems over the last four decades, has been successfully used to implement both research and development and large-scale data fusion systems. This approach is characterized by formal methods of requirement definition at a high level of abstraction, followed by decomposition to custom components, that can then be implemented. More recently, as information technology has matured, the discipline of enterprise architecture design has also developed formal methods for designing large-scale enterprises using commercially available and custom software and hardware components. Both of these disciplines contribute sound methodologies for implementing data fusion systems.

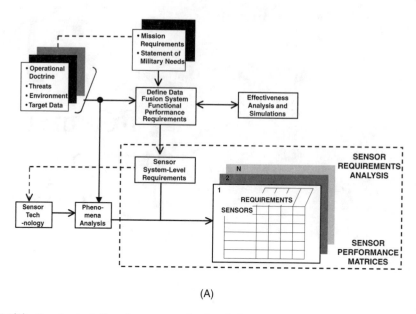

(A)

FIGURE 15.1(A)   Requirements flow-down process for data fusion.

This chapter introduces each approach before comparing the two to illustrate their complementary nature and utility of each. The approaches are not mutually exclusive, and methods from both may be applied to translate data fusion principles to practice.

15.2   Requirements Analysis Process

Derivation of requirements for a multisensor data fusion system must begin with the recognition of a fundamental principal: *there is no such thing as a data fusion system*. Instead, there are applications to which data fusion techniques can be applied. This implies that generating requirements for a generic data fusion system is not particularly useful (although one can identify some basic component functions). Instead, the particular application or mission to which the data fusion is addressed drives the requirements. This concept is illustrated in Figures 15.1(A) and 15.1(B).[1]

Figure 15.1(A) indicates that the requirements analysis process begins with an understanding of the overall mission requirements. What decisions or inferences are sought by the overall system? What decisions or inferences do the human users want to make? The analysis and documentation of this is illustrated at the top of the figure. An understanding of the anticipated targets supports this analysis, the types of threats anticipated, the environment in which the observations and decisions are to be made, and the operational doctrine. For Department of Defense (DoD) applications — such as automated target recognition — this would entail specifying the types of targets to be identified (e.g., army tanks and launch vehicles) and other types of entities that could be confused for targets (e.g., automobiles and school buses). The analysis must specify the environment in which the observations are made, the conditions of the observation process, and sample missions or engagement scenarios. This initial analysis should clearly specify the military or mission needs and how these would benefit from a data fusion system.

From this initial analysis, system functions can be identified and performance requirements associated with each function. The Joint Directors of Laboratories (JDL) data fusion process model can assist with this step. For example, the functions related to communications/message processing could be specified. What are the external interfaces to the system? What are the data rates from each communications link or sensor? What are the system transactions to be performed?[2] These types of questions assist in the formulation of the functional performance requirements. For each requirement, one must also specify

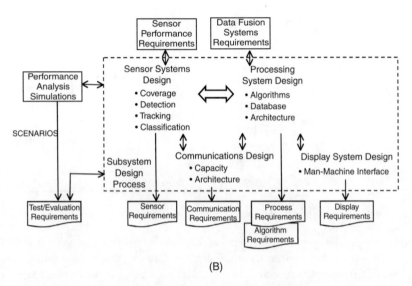

(B)

FIGURE 15.1(B) Requirements flow-down process for data fusion (continued).

how the requirement can be verified or tested (e.g., via simulations, inspection, and effectiveness analysis). A requirement is vague (and not really a requirement) unless it can be verified via a test or inspection.

Ideally, the system designer has the luxury of analyzing and selecting a sensor suite. This is shown in the middle of the diagram that appears in Figure 15.1(A). The designer performs a survey of current sensor technology, analyzes the observational phenomenology (i.e., how the inferences to be made by the fusion system can be mapped to observable phenomena such as infrared spectra and radio frequency measurements). The result of this process is a set of sensor performance measures that link sensors to functional requirements, and an understanding of how the sensors could perform under anticipated conditions. In many cases, of course, the sensors have already been selected (e.g., when designing a fusion system for an existing platform such as a tactical aircraft). Even in such cases, the designer should perform the sensor analysis in order to understand the operation and contributions of each sensor in the sensor suite.

The flow-down process continues as shown in Figure 15.1(B). The subsystem design/analysis process is shown within the dashed frame. At this step, the designer explicitly begins to allocate requirements and functions to subsystems such as the sensor subsystem, the processing subsystem, and the communications subsystem. These must be considered together because the design of each subsystem affects the design of the others. The processing subsystem design entails the further selection of algorithms, the specific elements of the database required, and the overall fusion architecture (i.e., the specification of where in the process flow the fusion actually occurs).

The requirement analysis process results in well-defined and documented requirements for the sensors, communications, processing, algorithms, displays, and test and evaluation requirements. If performed in a systematic and careful manner, this analysis provides a basis for an implemented fusion system that supports the application and mission.

15.3   Engineering Flow-Down Approach

Formal systems engineering methods are articulated by the U.S. DoD in numerous classical military standards[3] and defense systems engineering guides. A standard approach for development of complex hardware and software systems is the *waterfall approach* shown in Figure 15.2.

This approach uses a sequence of design, implementation, and test and evaluation phases or steps controlled by formal reviews and delivery of documentation. The waterfall approach begins at the left side of Figure 15.2 with system definition, subsystem design, preliminary design, and detailed design. In

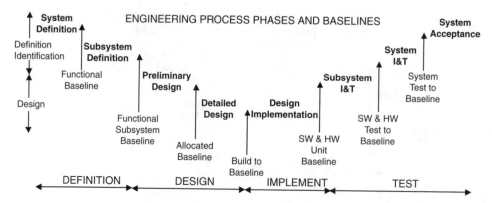

FIGURE 15.2 System engineering methodology.

this approach, the high-level system requirements are defined and partitioned into a hierarchy of increasingly smaller subsystems and components. For software development, the goal is to partition the requirements to a level of detail so that they map to individual software modules comprising no more than about 100 executable lines of code. Formal reviews, such as a requirement review, preliminary design review (PDR), and a critical design review (CDR), are held with the designers, users, and sponsors to obtain agreement at each step in the process. A baseline control process is used, so that requirements and design details developed at one phase cannot be changed in a subsequent phase without a formal change/modification process.

After the low-level software and hardware components are defined, the implementation begins. (This is shown in the middle of Figure 15.2.) Small hardware and software units are built and aggregated into larger components and subsystems. The system development proceeds to build small units, integrate these into larger entities, and test and evaluate evolving subsystems. The test and integration continues until a complete system is built and tested (as shown on the right side of Figure 15.2). Often a series of *builds* and tests are planned and executed.

Over the past 40 years, numerous successful systems have been built in this manner. Advantages of this approach include:

- The ability to systematically build large systems by decomposing them into small, manageable, testable units.
- The ability to work with multiple designers, builders, vendors, users, and sponsoring organizations.
- The capability to perform the development over an extended period of time with resilience to changes in development personnel.
- The ability to define and manage risks by identifying the source of potential problems.
- Formal control and monitoring of the system development process with well-documented standards and procedures.

This systems engineering approach is certainly not suitable for all system developments. The approach is most applicable for large-scale hardware and software system. Basic assumptions include the following:

- The system to be developed is of sufficient size and complexity that it is not feasible to develop it using less formal methods.
- The requirements are relatively stable.
- The requirements can be articulated via formal documentation.
- The underlying technology for the system development changes relatively slowly compared to the length of the system development effort.
- Large teams of people are required for the development effort.
- Much of the system must be built from scratch rather than purchased commercially.

Over the past 40 years, the formalism of systems engineering has been very useful for developing large-scale DoD systems. However, recent advances in information technology have motivated the use of another general approach.

15.4   Enterprise Architecture Approach

The rapid growth in information technology has enabled the construction of complex computing networks that integrate large teams of humans and computers to accept, process, and analyze volumes of data in an environment referred as the "enterprise." The development of enterprise architectures requires the consideration of functional operations and the allocation of these functions in a network of human (cognitive), hardware (physical), or software components.

The enterprise includes the collection of people, knowledge (tacit and explicit), and information processes that deliver critical knowledge (often called "intelligence") to analysts and decision-makers to enable them to make accurate, timely, and wise decisions. This definition describes the enterprise as a process that is devoted to achieving an objective for its stakeholders and users. The enterprise process includes the production, buying, selling, exchange, and/or promotion, of an item, substance, service, and/or system. The definition is similar to that adopted by DaimlerChrysler's extended virtual enterprise, which encompasses its suppliers:

> A DaimlerChrysler coordinated, goal-driven process that unifies and extends the business relationships of suppliers and supplier tiers in order to reduce cycle time, minimize systems cost and achieve perfect quality.[4]

This all-encompassing definition brings the challenge of describing the full enterprise, its operations, and its component parts. Zachman has articulated many perspective views of an enterprise information architecture and has developed a comprehensive framework of descriptions to thoroughly describe an entire enterprise.[5,6] The following section describes a subset of architecture views that can represent the functions in most data fusion enterprises.

15.4.1   The Three Views of the Enterprise Architecture

The enterprise architecture is described in three views (as shown in Figure 15.3), each with different describing products. These three, interrelated perspectives or architecture views are outlined by the DoD in their description of the Command, Control, Communication, Computation, Intelligence, Surveillance, and Reconnaissance (C4ISR) framework.[7] They include:

1. **Operational architecture (OA)** is a description (often graphical) of the operational elements, business processes, assigned tasks, workflows, and information flows required to accomplish or support the C4ISR function. It defines the type of information, the frequency of exchange, and tasks supported by these information exchanges. This view uniquely describes the human role in the enterprise and the interface of human activities to automated (machine) processes.
2. **Systems architecture (SA)** is a description, including graphics, of the systems and interconnections providing for or supporting functions. The SA defines the physical connection, location, and identification of the key nodes, circuits, networks, and war-fighting platforms, and it specifies system and component performance parameters. It is constructed to satisfy operational architecture requirements per standards defined in the technical architecture. The SA shows how multiple systems within a subject area link and interoperate and may describe the internal construction or operations of particular systems within the architecture.
3. **Technical architecture (TA)** is a minimal set of rules governing the arrangement, interaction, and interdependence of the parts or elements whose purpose is to ensure that a conformant system satisfies a specified set of requirements. The technical architecture identifies the services, interfaces, standards, and their relationships. It provides the technical guidelines for implementation of

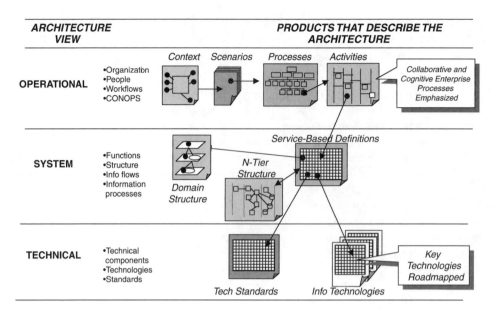

FIGURE 15.3   Three architecture views are described in a variety of products.

systems upon which engineering specifications are based, common building blocks are built, and product lines are developed.

The primary products that describe the three architecture views (Figure 15.3) include the following:

1. **Context diagram:** the intelligence community context that relates shareholders (owners, users, and producers).
2. **Scenarios:** selected descriptions of problems the enterprise must solve that represent the wide range of situations expected to be confronted by the enterprise.
3. **Process hierarchy:** tree diagrams that relate the intelligence community business processes and describe the functional processes that are implemented as basic services.
4. **Activity diagrams:** sequential relationships between business processes that are described in activity sequence diagrams.
5. **Domain operation:** the structure of collaborative domains of human virtual teams (user community).
6. **Service descriptions:** define core and special software services required for the enterprise — most commercial and some custom.
7. **N-tier structure diagram:** the *n*-tier structure of the information system architecture is provided at a top level (e.g., two-tier systems are well-known client-server tiers; three-tier systems are partitioned into data warehouse, business logic, and presentation layers).
8. **Technical standards:** critical technical standards that have particular importance to data fusion business processes, such as data format standards and data service standards (e.g., SQL and XML).
9. **Information technology roadmaps:** projected technology needs and drivers that influence system growth and adoption of emerging technologies are critical components of an enterprise; they recognize the highly dynamic nature of both the enterprise and information technology as a whole.

15.5   Comparison of Approaches

The two design methods are complementary in nature and both provide helpful approaches for decomposing problems into component parts and developing data fusion solutions. Comparing the major distinguishing characteristics of the approaches (Table 15.1) illustrates the strengths of each approach for data fusion system implementation:

TABLE 15.1 Comparison of System-Level Design Approaches

	System Engineering	System Architecting
Perspective	Design from first principles Optimize design solution to meet all functional requirements; quantify risk to life cycle cost and to implementation cost and schedule Provide traceability from requirements to implementation	Design from best standard components Optimize design solution to implement use cases; quantify risk to operational performance and enterprise future growth (scalability, component upgrade) Provide traceability between multiple architecture views
Starting assumptions	• Top-level problem requirements exist and are quantified • Requirements are specific, technical, and quantified • Emphasis on functional models	• Functional components (e.g., software components) exist • Requirements tend to be general, user-oriented applications, and subjective • Emphasis on use case models
Methodology	• Structured problem decomposition • Requirements flow-down (deriving and allocating requirements) • Design, integrate custom components	• Use case modeling • Data and functional modeling • Multiple architecture views construction (multiple functional perspectives) • Design structure, integrate standard components
Basis of risk analysis	Implementation (cost, schedule) and performance risk as a function of alternative design approaches	Operational utility (over life cycle) risk as a function of implementation cost
Design variable perspective	Requirements are fixed; cost is variable	Cost is fixed; requirements are variable
Applicable data fusion systems	• One-of-a-kind intelligence and military surveillance systems • Mission-critical, high reliability systems	• General business data analysis systems • Highly networked component-based, service-oriented systems employing commercial software components

- **Perspective:** System engineering seeks specific (often custom) solutions to meet all specific functional requirements; system architecting begins with components and seeks to perform the set of use cases (accepting requirements' flexibility) with the optimum use of components (minimizing custom designed components).

- **Starting assumptions:** System engineering assumes that top-level system requirements exist and are quantifiable. The requirements are specific and can be documented along with performance specifications. System engineering emphasizes functional models. By contrast, system architecting assumes that functional components exist and that the requirements are general and user oriented. The emphasis is on use-case models.

- **Methodology:** The methodology of system engineering involves structured problem decomposition and requirements flow-down (as described in Section 15.2). The design and implementation proceeds in accordance with a waterfall approach. System architecting involves use-case modeling and data and functional modeling. Multiple functional perspectives may be adopted, and the focus is on integration of standard components.

- **Risk analysis:** Systems engineering addresses risks in system implementation by partitioning the risks to subsystems or components. Risks are identified and addressed by breaking the risks into manageable smaller units. Alternative approaches are identified to address the risks. In the system architecting approach, risk is viewed in terms of operational utility over a life cycle. Alternative components and architectures address risks.

- **Design variables:** System engineering assumes that the system requirements are fixed but that development cost and schedule may be varied to meet the requirements. By contrast, system architecting assumes that the cost is fixed and the requirements may be varied or traded off to meet cost constraints.
- **Application:** The system engineering approach provides a formal means of deriving, tracking, and allocating requirements to permit detailed performance analysis and legal contract administration. This is often applied on one-of-a-kind systems, critical systems, or unique applications. The architectural approach is appropriate for the broader class of systems, where many general approaches can meet the requirements (e.g., many software products may provide candidate solutions).

The two basic approaches described here are complimentary. Both hardware and software developments will tend to evolve toward a hybrid utilization of systems engineering and architecting engineering. The rapid evolution of information technology and the appearance of numerous commercial-off-the-shelf tools provide the basis for the use of methods such as architecting engineering. New data fusion systems will likely involve combinations of traditional systems engineering and architecting engineering approaches, which will provide benefits to both the implementation and the user communities.

References

1. Waltz, E. and Llinas, J., *Multisensor Data Fusion*, Artech House, Inc., Norwood, MA 1990.
2. Finkel, D., Hall, D.L., and Beneke, J., Computer performance evaluation: the use of a time-line queuing method throughout a project life cycle, *Modeling and Simulation*, 13, 729–734, 1982.
3. Classical DoD military standards for systems engineering include: MIL-STD-499B, Draft Military Standard: Systems Engineering, HQ/AFSC/EN, Department of Defense, "For Coordination Review" draft, May 6, 1992, and NSA/CSS Software Product Standards Manual, NSAM 81-3, National Security Agency. More recent documents that organize the principles of systems engineering include: EIA/IS 632, Interim Standard: Systems Engineering, Electronic Industries Alliance, December 1994; Systems Engineering Capability Assessment Model SECAM (version 1.50), INCOSE< June 1996; and the ISO standard for system life cycle processes, ISO 15288.
4. DaimlerChrysler Extended Enterprise, see http://supplier.chrysler.com/purchasing/extent/index.html.
5. Zachman, J.A., A framework for information systems architecture, *IBM Syst. J.*, 26(3), 1987.
6. Sowa, J.F. and Zachman, J.A., Extending and formalizing the framework for information systems architecture, *IBM Syst. J.*, 31(3), 1992.
7. Joint Technical Architecture, Version 2.0, Department of Defense, October 31, 1997 (see paragraph 1.1.5 for definitions of architecture and the three architecture views).

16

A Systems Engineering Approach for Implementing Data Fusion Systems

16.1 Scope 16-1
16.2 Architecture for Data Fusion... 16-2
   Role of Data Fusion in Information Processing Systems • Open
   System Environment • Layered Design • Paradigm-Based
   Architecture
16.3 Data Fusion Systems Engineering Process 16-7
   Data Fusion Engineering Methodology • The Process of
   Systems Engineering
16.4 Fusion System Role Optimization 16-17
   Fusion System Requirements Analysis • Fusion System Tree
   Optimization • Fusion Tree Node Optimization • Detailed
   Design and Development
References 16-38

Christopher L. Bowman
Consultant

Alan N. Steinberg
Utah State University

16.1 Scope

This chapter defines a systematic process for developing data fusion systems and intends to provide a common, effective foundation for the design and development of such systems. It also provides guidelines for selecting among design alternatives for specific applications.

This systems engineering approach has been developed to provide

- A standard model for representing the requirements, design, and performance of data fusion systems, and
- A methodology for developing multisource data fusion systems and for selecting system architecture and technique alternatives for cost-effective satisfaction of system requirements.

This systems engineering approach builds on a set of data fusion engineering guidelines that were developed in 1995–96 as part of the U.S. Air Force Space Command's Project Correlation.[1,2]* The present work extends these guidelines by proposing a formal model for systems engineering, thereby establishing the basis for rigorous problem decomposition, system design, and technique application.

---

*A closely related set of guidelines[3] for selecting among data correlation and association techniques, developed as part of the same project, is discussed in Chapter 17.

Integral to the guidelines is the use of a functional model for characterizing diverse system architectures and processing and control functions within a data fusion process. This architecture paradigm has been found to capture successfully the salient operating characteristics of the diverse automatic and manual approaches that have been employed across a great diversity of data fusion applications.

This fusion system paradigm can be implemented in human, as well as automated, processes and in hybrids involving both. Also, the node paradigm can be employed in both fusion system development and reverse engineering for the purpose of applying countermeasures.

The recommended architecture concept represents data fusion systems as networks — termed *data fusion trees* — of processing nodes that are amenable to a standard representation. The tight coupling of a data fusion tree with a resource management tree is characteristic of many successful system designs. The close relationship between fusion and management — both in interactive operation and in their underlying design principles — will play an important part in effective system design. It will also be important in defining effective engineering methods for achieving those designs.

The guidelines recommend a four-phase process for developing data fusion functionality within an information processing system. Design and development decisions flow from overall system requirements and constraints to a specification of the role for data fusion within the system. Further partitioning results in a specification of a data fusion tree structure and corresponding nodes. Pattern analysis of the requirements for each node allows selection of appropriate techniques, based on analysis and experience.

The largely heuristic methods presented in the Project Correlation Data Fusion Engineering Guidelines are amenable to more rigorous treatment. Systems engineering — and, specifically, data fusion systems engineering — can be viewed as a *planning* (i.e., resource management) process. Available techniques and design resources are applied to meet an objective criterion: cost-effective system performance. Therefore, the techniques of optimal planning can be applied to building optimal systems.

Furthermore, the formal duality between data fusion and resource management — first propounded by Bowman[4] — enables data fusion design principles to be applied to the corresponding problem of resource management. As systems engineering is itself a resource management problem, the principles of data fusion can be used for building data fusion systems. The formal relationship between Data Fusion and Resource Management is discussed in Section 16.3.2.

16.2 Architecture for Data Fusion

16.2.1 Role of Data Fusion in Information Processing Systems

As shown in Figure 16.1, data fusion characteristically is used to provide assessments of an observed environment to users who further assess and manage responses to the assessed environment. A system's *resource management* function uses the fused assessments and user directives to plan and control the

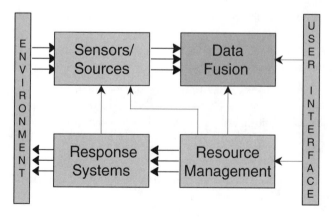

FIGURE 16.1   Characteristic role of data fusion in an information management system.

available resources. This can include platform navigation, weapon deployment, countermeasures, and other active responses, and allocation of sensor and processing resources (including data fusion and response management processing resources) in support of mission objectives.

16.2.2 Open System Environment

The open system environment (OSE) has become the preferred system environment for major command, control, communications, computers, and intelligence (C4I) architectures. This environment comprises a comprehensive set of interfaces, services, supporting formats, and user aspects. The OSE is intended to reduce development, integration, and software maintenance costs. OSE architecture features that were designed to provide the desired system interoperability and portability include

- Distributed processing
- Open-layered architecture
- Near-real-time data distribution
- Data source independence
- Software modularity
- Application independence
- Flexible reconfiguration
- Scalability

16.2.3 Layered Design

A *layered* design approach facilitates system design by hierarchically partitioning a complex design problem. It applies the principle of "divide and conquer" to the systems engineering problem, much like a fan-in fusion tree applies this principle to batching large, diverse input data. The relationship between systems engineering and data fusion processes is described in Section 16.3.

The data fusion systems engineering process applies an open, layered, paradigm-based architecture to permit a cost-effective method for achieving a required level of performance. The process uses the data fusion tree paradigm as its basis, to provide a common methodology for addressing the design and development of fusion systems.[5]

Both the fusion and management tree architectures (i.e., components, interfaces, and utilization guidelines) are implemented in a system's applications layer, as depicted in Figure 16.2. As a result, the user of the recommended data fusion architecture can apply any standard lower-layer architecture (e.g., GCCS, JMCIS, TBMCS).

16.2.4 Paradigm-Based Architecture

The recommended system development process applies a paradigm-based architecture to the specific issues involved in data fusion systems. Systems are specified in terms of a broadly applicable model for the fusion process. The employed paradigm defines a data fusion system in terms of a *fusion tree*, which is a network of *fusion nodes*, each of which is specified according to a standard functional paradigm that describes system components and interfaces.

A fusion tree typically takes the form of a fan-in (or "bottom-up") tree for fusion and a fan-out (or "top-down") tree for management, as illustrated in Figure 16.3. (The characteristic inseparability of the data fusion and resource management aspects of system design prompts the new name "data fusion and resource management dual node architecture.") All such data-flow networks are called *fusion and management trees* (regardless of whether they are actually configured as trees) since tree configurations are the norm. Note that any feedback (e.g., of more accurate central tracks to lower level fusion processes) is performed through process management as part of the resource management feedback tree.

The engineering process discussed in this chapter involves a hierarchical decomposition of system-level requirements and constraints. Goals and constraints are refined by feedback of performance assessments.

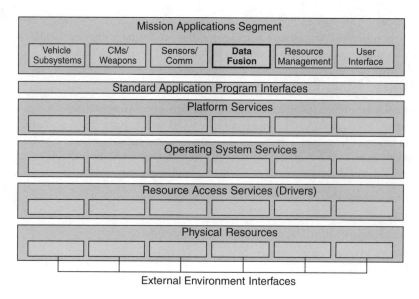

FIGURE 16.2 Data fusion processing in the applications layer of an open-layered architecture.

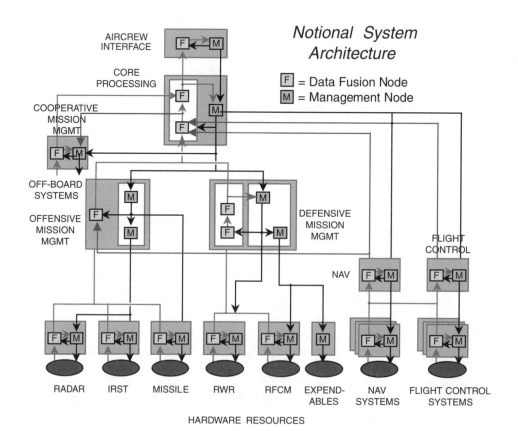

FIGURE 16.3 Representative interlaced data fusion and resource management trees.

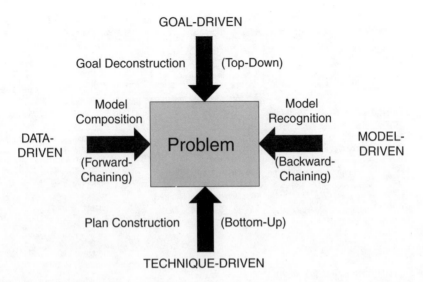

FIGURE 16.4 Types of problem-solving approaches.

TABLE 16.1 Categorization of Problem-Solving Approaches

Type	Procedure	Applicable Techniques	Example
Goal-driven	Deconstruction: Analyze goal into subgoals	• Back-chaining • Parsing	• Trip planning • Collection management
Technique-driven	Construction: Assemble plan	• Genetic algorithms • Synthetic annealing • Trial-and-error	• Trapped fly • Jigsaw puzzle
Model-driven	Deduction: Recognize problem type	• Back-chaining • Case-based reasoning • Operant conditioning • Supervised learning nn	• Medical diagnosis • Fault diagnostics • Target recognition
Data-driven	Induction/Abduction: Discover problem characteristics	• Forward-chaining • Principal-component analysis • Clustering • Unsupervised learning NN	• Fight-flight reaction • Technical intelligence

This is an example of a *goal-driven* problem-solving approach, which is the typical approach in systems engineering. A more general approach to problem-solving is presented in Reference 6. As depicted in Figure 16.4 with amplification in Table 16.1, goal-driven approaches are but one of four approaches applicable to various sorts of problems. In cases where it is difficult to state system goals — as in very experimental systems or in those expected to operate in poorly understood environments ∇ the goal-driven systems engineering approach will incorporate elements of data-, technique-, or model-driven methods.

Fusion system design involves

- selecting the data flow among the fusion nodes (i.e., how data is to be batched for association and fusion processing), and
- selecting the methods to be used within each fusion node for processing input batches of data to refine the estimate of the observed environment.

The fusion node paradigm involves the three basic functions of data alignment, data association, and entity state estimation functions, as shown in Figure 16.5. The means for implementing these functions and the data and control flow among them will vary from node to node and from system to system.

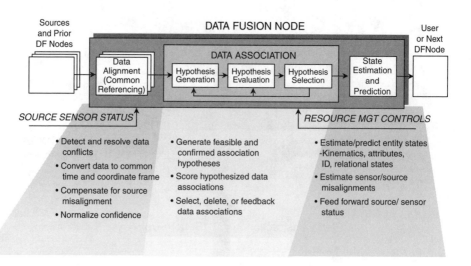

FIGURE 16.5   Data fusion node paradigm.

Nonetheless, the given paradigm has proven to be a useful model for characterizing, developing, and evaluating automatic, human, and other biological fusion systems.

Data alignment (sometimes termed *data preparation* or *common referencing*) preprocesses data received by a node to permit comparisons and associations among the data. Alignment involves functions for

- Common formatting
- Spatio-temporal alignment
- Confidence normalization

Data association assigns data observations — received in the form of sensor measurements or reports — to hypothesized entities (we commonly refer to the resulting entity estimates based on this association as *tracks*). Before a sensor observation and its associated information are fused with an existing track, the hypothesis that the observation is of the same entity as that represented by the track information must be postulated, tested, and then accepted or rejected.*

The association process is accomplished via the following three functions:

- *Hypothesis generation*: identifying sets of sensor reports and existing tracks (report associations) that are feasible candidates for association;
- *Hypothesis evaluation*: determining the confidence (i.e., quality) of the report to track association or non-association as defined by a particular metric;
- *Hypothesis selection*: determining which of the report-to-track associations, track propagations or deletions, and report track initiations or false alarm declarations the fusion system should retain and use for state estimation.

The *state estimation and prediction function* uses a selected batch of data to refine the estimates reported in the data and to infer entity attributes and relations. For example, kinematic measurements or tracks can be filtered to refine kinematic states. Kinematic and other measurements can be compared with entity and situation models to infer features, identity, and relationships of entities in the observed environment. This can include inferring states of entities other than those directly observed.

---

*Significant progress has been made in developing multisensor/multitarget data fusion systems that do not depend on explicit association of observations to tracks (see Chapter 14). Once again, the data fusion node paradigm is meant to be comprehensive; every system should be describable in terms of the functions and structures of the DF and RM Dual Node Architecture. This does not imply that every node or every system need include every function.

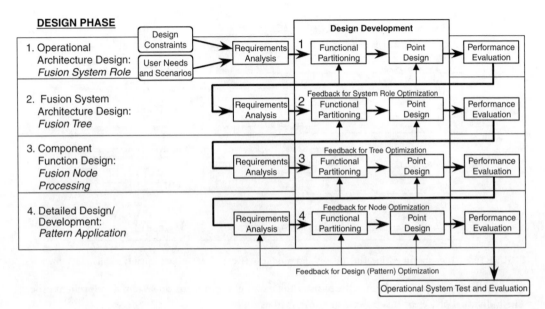

FIGURE 16.6   Data fusion system development process.

16.3   Data Fusion Systems Engineering Process

16.3.1   Data Fusion Engineering Methodology

The Data Fusion Engineering Guidelines provide direction for selecting among fusion trees and fusion node design alternatives in developing a fusion system for specific applications, as discussed in Sections 16.4 through 16.7 below.

The guidelines for the fusion systems engineering process are intended to help improve the cost-effectiveness of fusion system development, integration, test, and evaluation. These guidelines define a methodology for selecting among alternative approaches in the development of a multisource data fusion system. In most applications, fusion processes systems are tightly coupled with management systems that determine how to apply the available resources to achieve the overall system objectives.

The data fusion tree paradigm is used as the basis of the data fusion systems engineering process. This paradigm-based process provides a common methodology for addressing the design and development of fusion systems. The appropriate set of common process objects and their interfaces can be defined in terms of the paradigm, which provides a set of broadly applicable prior objects upon which solution "patterns" can be built in future fusion software developments.

This data fusion system development process, as originally presented in the *Project Correlation Data Fusion Engineering Guidelines*,[1] is depicted in Figure 16.6. This process provides a common methodology for addressing the design and development of data fusion and resource management systems. It applies an open, layered, paradigm-based architecture to permit a cost-effective method for achieving a required level of performance. The process involves four distinct phases, which are amenable to implementation via waterfall, spiral, or other software development processes:

1. *Fusion system role design*: analyze system requirements to determine the relationship between a proposed data fusion system and the other systems with which it interfaces.
2. *Fusion tree design*: define how the data is batched to partition the fusion problem.
3. *Fusion node design*: define the data and control flow within the nodes of a selected fusion tree.
4. *Fusion algorithm design*: define processing methods for the functions to be performed within each fusion node (e.g., data association, data alignment, and state estimation/prediction).

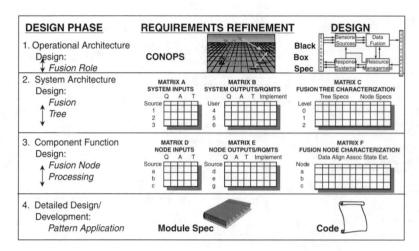

FIGURE 16.7   Data fusion engineering phases and products. (Note: Q A T = Data quality, availability, timeliness.)

The data fusion tree paradigm is the basis of the paradigm for the fusion system development process. The individual phases of this process are described as follows:

1. Fusion System Role Optimization (Section 16.4),
2. Fusion System Tree Optimization (Section 16.5),
3. Fusion Tree Node Optimization (Section 16.6),
4. Detailed Design and Development (Section 16.7).

The requirements and design decisions developed in each phase are documented in a set of canonical products, depicted in Figure 16.7. These requirements and design products are defined in subsequent sections. In general, they provide standardized communications media for

- Coordinating the design of data fusion processes,
- Comparing and contrasting alternative designs,
- Determining the availability of suitable prior designs, and
- Integrating fusion processes into C4ISR and other distributed information systems.

Figure 16.8 indicates the mapping of these products to the C4ISR Architecture Framework developed by the U.S. DoD C4ISR Integrated Architecture Panel (CISA).

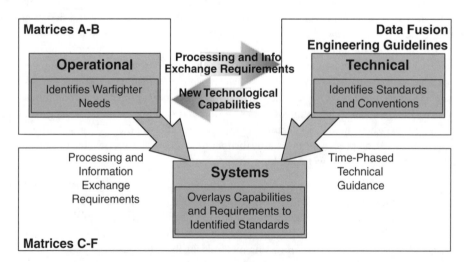

FIGURE 16.8   The data fusion engineering processes.

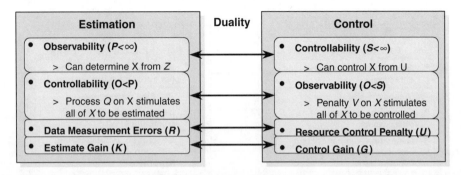

FIGURE 16.9 Formal duality between estimation and control. (Note: Separation principle for linear stochastic systems enables optimal estimator to be cascaded with optimal controller.)

16.3.2 The Process of Systems Engineering

This section discusses formal principles for systems engineering in general and for data fusion engineering specifically. A key insight, first presented in Reference 8, is in formulating the systems engineering process as a resource management problem, allowing the application of Bowman's[4] model of the duality between data fusion and resource management.

16.3.2.1 Data Fusion and Resource Management

A resource management process combines multiple available actions* (e.g., allocation of multiple available resources) over time to maximize some objective function. Such a process must contend with uncertainty in the current situational state and in the predictive consequences of any candidate action. A resource management process will

- Develop candidate response plans to respond to estimated world states,
- Estimate the effects of candidate actions on mission objectives,
- Identify conflicts for resource usage or detrimental side effects of candidate actions, and
- Resolve conflicts to assemble composite resource assignment plans, based on the estimated net impact on mission attainment.

16.3.2.2 Functional Model for Resource Management

The duality between data fusion and resource management, evident in the previous definition, expands on the well-known duality between estimation and control processes. The formal correspondence between estimation and control, shown in Figure 16.9, has been useful in permitting techniques to be applied directly from one to the other. This duality can be extended to include the architectures and remaining functionality of data fusion and resource management, shown in Figure 16.10.

Figure 16.3 depicts the characteristic intertwining of a fan-in data fusion tree with a corresponding fan-out resource management tree.

Given this formal duality, the maturity of data fusion can be used to "bootstrap" the less mature resource management field, much like the duality of estimation did for control over 30 years ago. In so doing, the fusion system development process, shown in Figure 16.5, becomes intertwined with the resource management system development within each design phase shown.

As multinodal data fusion trees are useful in partitioning the data association and state estimation problems, so are resource management trees useful in partitioning planning and control problems. A data fusion tree performs an association/estimation process; a resource management tree performs a planning/execution process. Both of these trees — one synthetic (i.e., constructive) and the other analytic (i.e., decompositional) — are characteristically recursive and hierarchical.

---

*As noted in Section 16.2.4, not all problems have clear, immutable goals. Thus, resource management is often an opportunistic blend of data-, technique-, model-, and goal-driven processes.

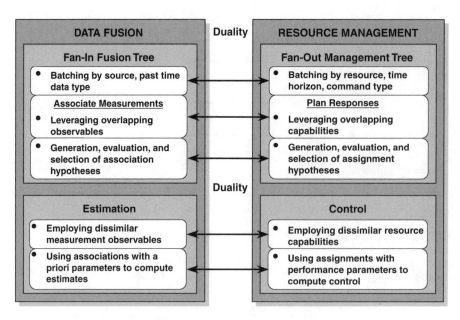

FIGURE 16.10   Extending the duality to data fusion and resource management.

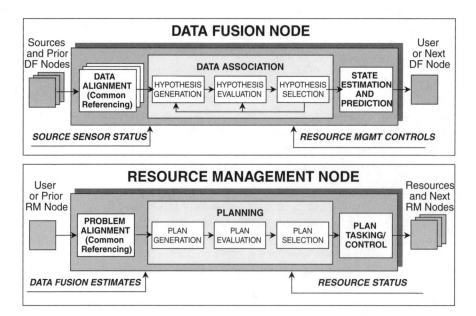

FIGURE 16.11   Dual paradigm for data fusion and resource management nodes.

As depicted in Figure 16.11, a resource management node involves functions that directly correspond to those of a data fusion node:

1. *Problem alignment* (common referencing) — normalizing performance metrics of the given problem and normalizing performance models of available resources, as well as any control format and spatio-temporal alignment.
2. *Planning:*
   • *Plan generation* – candidate partitioning of the problem into subordinate problems and candidate assignment of resources.

- *Plan evaluation* – evaluating the conditional net cost (e.g., Bayesian cost).
- *Plan selection* – determining a decision strategy.

3. *Control* (i.e., plan execution) — generating the control commands to implement the selected resource allocation plan.

Planning is analogous to *data association* in data fusion. Planning functions that correspond to association hypothesis generation, evaluation, and selection involve: (a) plan generation (i.e., searching over a number of possible actions for assembly into candidate plan segments), (b) plan evaluation, and (c) plan selection.

Plan generation, plan evaluation, and plan selection offer challenges similar to those encountered in designing the corresponding data fusion functions. As with hypothesis generation in data fusion, *plan generation* involves potentially massive searches, which must be constrained in practical systems. The objective is to reduce the number of feasible plans for which a detailed evaluation is required. Challenges for *plan evaluation* lie in the derivation of efficient scoring schemes capable of reflecting the expected utility of alternative response plans.

Candidate plans — i.e., schedules of tasking for system resources — are assembled recursively. The level of planning is adapted on the basis of (1) the assessed utility relative to the mission goals of the given plan segment as currently developed and (2) the time available for further planning. By (1), near-term plan segments tend to be constructed in greater detail than far-term ones, for which the added expenditure in planning resources may outweigh the confidence that the plan will still be appropriate at execution time.

Deeper planning is accomplished by recursively partitioning a goal into candidate sets of subgoals (*plan generation*) and combining them into a composite higher-level plan (*plan selection*). At each level, candidate plans are evaluated with regard to their effectiveness in achieving assigned goals, the global value of each respective goal, and the cost of implementing each candidate plan (*plan evaluation*). By evaluating higher-level cost/payoff impacts, the need for deeper planning or for selecting alternate plans is determined. In many applications, plan selection can be an NP-hard problem — a search of multiple resource allocations over n future time intervals.

Contentions for assigning available resources are resolved by prioritized rescheduling on the basis of time sensitivity and predicted utility of contending allocations. Each resource management node presents a candidate plan segment for higher-level evaluation.

Interlaced data fusion nodes estimate potential side effects of the plan. For example, in countermeasures management in a combat aircraft, evaluation of self-protection actions must consider detectable signature changes, flight path changes, or emissions that could interfere with another sensor. Higher-level nodes respond by estimating the impact of such effects on their respective higher-level goals. In this way, plans responsive to global mission goals are assembled in a hierarchical fashion.

16.3.2.3 Systems Engineering as a Resource Management Problem

Resource management is a process for determining a mapping from a problem space to a solution space. Systems engineering is such a process, in which the problem is to build a system to meet a set of requirements. Fundamental to the systems engineering process (as in all resource management processes) is a method for representing the structure of a problem in a way that is amenable to a patterned solution.

The model for resource management permits a powerful general method for systems engineering (e.g., for data fusion systems engineering). It does so by providing a standardized formal representation that allows formal resource allocation theory and methods to be applied.

Issues pertaining to data fusion systems engineering include

- Selecting feature sets for exploitation,
- Discovering exploitable context,
- Modeling problem variability,

- Discovering patterns that allow solution generalization,
- Predicting technique and system performance,
- Predicting development cost and schedule.

As a resource management process, systems engineering can be implemented as a hierarchical, recursive planning and execution process.

16.3.2.4 Data Fusion Systems Engineering

The data fusion (DF) engineering guidelines can be thought of as the design specification for a resource management (RM) process, the "phases" depicted in Figure 16.1 being levels in a hierarchical RM tree. A tree-structured RM process is used to build, validate, and refine a system concept which, in this application, may be a tree-structured RM or DF process — something of a Universal Turing machine. The relationship between system design and system operation is indicated in Table 16.2.

The relationship between the operation of sensor management, data fusion and systems engineering can be dramatized by recasting Figures 16.3 and 16.6 as Figures 16.12 and 16.13, respectively. The systems engineering process (Figure 16.13) is a fan-out (top-down) resource management tree interlaced with a fan-in (bottom-up) data fusion tree, similar to that shown in Figure 16.12. Design goals and constraints flow down from the system level to allocations over successively finer problem partitionings. At each level, a design phase constitutes a grouping of situations (e.g., batches of data) for which responses (design approaches) are coordinated.

Each RM node in a systems engineering process involves functions that are characteristic of all RM nodes:

- *Problem alignment (common referencing)*: normalizing requirements for the given design (sub-)problem and normalizing performance models of available resources (e.g., DF tree types or DF techniques) using problem-space/solution-space matrices.
- *Planning*: generating, evaluating, and selecting design alternatives (partitioned according to the four design phases shown in Figure 16.13).
- *Control (plan execution)*: building or evaluating a DF tree, node, or component technique.

The systems engineering process builds, evaluates, and selects candidate designs for the system and its components via a hierarchical, recursive process that permits simultaneous reasoning at multiple levels of depth. The recursive planning process provides the ability to optimize a design against a given set of requirements and redesign as requirements change.

The systems engineering process distributes the problem solution into multiple design (i.e., management) nodes. Nodes communicate to accumulate incremental evidence for or against each plausible solution. At any given stage, therefore, the systems engineering process will provide the best design plan for achieving current goals, consistent with the available situational and procedural knowledge and the available resources (e.g., design patterns, development and test environments, engineers, time, and money).

TABLE 16.2   Developmental and Operational RM and DF Characteristics

	Resource Management		Data Fusion	
Problem Characteristic	Sensor Management	System Engineering	Sensor Data Fusion	System Evaluation
Process Sequence	Fan-Out Tree	Fan-Out Tree	Fan-In Tree	Fan-In Tree
Solution determination • Generation • Evaluation • Solution	Plan development (problem decomposition)	Design development (problem decomposition)	Data association	Performance diagnosis fusion
Solution filter	Control (plan implementation)	Build (design implementation)	Observed state estimation	Performance estimation

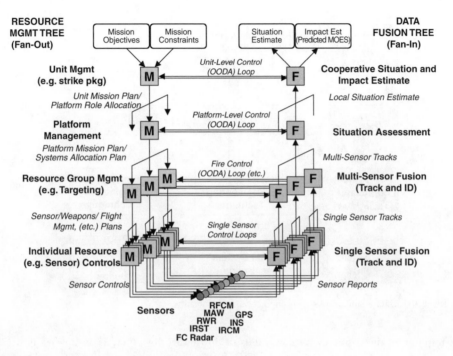

FIGURE 16.12 Dual DF/RM tree architecture for fighter aircraft.

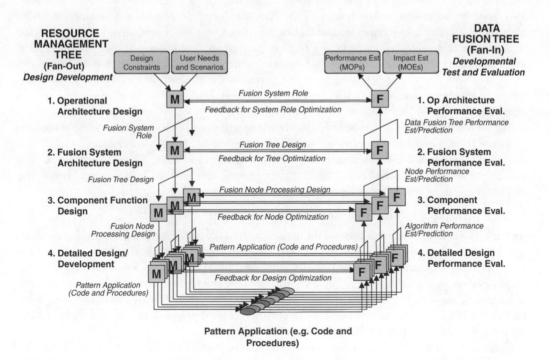

FIGURE 16.13 Dual DF/RM tree architecture for DF systems engineering.

This procedure is hierarchical and recursive. Subordinate resource management nodes are activated to develop more detailed, candidate design segments when a higher-level node determines that more detailed design is both feasible (in terms of estimated resource cost and development time) and beneficial (in terms of the likelihood of attaining the assigned design goal).

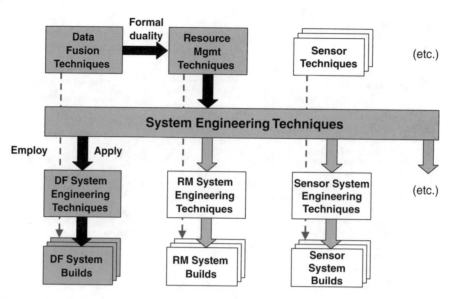

FIGURE 16.14   Data fusion and systems engineering relationships.

In response to the requirements/constraints of downward flow, there is an upward flow consisting of evaluated candidate design segments (i.e., proposed technique assignments and controls plus estimates of the performance of proposed design segments relative to their assigned goals). A higher-level node evaluates cost versus benefit of the received design proposals (in terms of the node's higher-level goal) and may repartition its higher goal into a new set of subgoals if the initial allocations are not achievable. For example, if a required performance against a particular system requirement cannot be met, a new criterion for batching data must be developed. A data fusion tree interleaved with the resource management tree performs the upward flow of evaluated plan segments. In the systems engineering tree depicted, a unique evaluation process can correspond to each design node.

The process of developing data fusion, resource management, and other types of systems is depicted in Figure 16.14. Resource management methods are applied to develop systems engineering and, specifically, data fusion systems engineering techniques. These, in turn, allow the employment of data fusion technology in building systems. Except for the special "dual" relationship between DF and RM, the same is true of other types of systems engineering.

16.3.2.5   Adaptive Information Acquisition and Fusion

An important avenue of research involves the development of adaptive data fusion techniques, which are used by a system's resource management process during run time to select the data to be processed and the processing techniques to be applied. In effect, the data fusion tree and nodes are constructed adaptively, based on the system's assessed current information state and the predicted effectiveness of available techniques to move to a desired information state. Significant work in this area was conducted under the U.S. DARPA DMIF (Dynamic Multi-User Information Fusion) project and continued under the U.S. Air Force Research Laboratory's Adaptive Sensor Fusion project.

Figure 16.15 shows the concept for adaptive response to a dynamic mission environment. This extends the adaptive sensor fusion concept to incorporate all manner of adaptive response, involving the coordinated use of sensors, communications, processing (including data fusion and resource management), and response systems (e.g., weapons, countermeasures, and trajectory management). For lack of a generally accepted term, we refer to such processes as *information acquisition management* processes, to include both sensor management and data fusion process management (i.e., the JDL Model's Level 4).

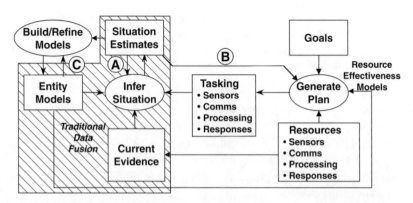

FIGURE 16.15 Adaptive information exploitation.

Traditional data fusion (indicated by the shaded area to the left of Figure 16.15) involves the feedback loop labeled **A** in the figure; estimates of the observed situation, along with prior models are used to interpret new data.

The adaptive sensor fusion concept adds two more feedback loops, labeled **B** and **C**. Resources are allocated based on the current estimated knowledge state and a desired knowledge state. The process develops and refines action plans (**B**) with a goal of mapping from partitionings of possible world state space into decision space. Additionally, the system refines its library of models — target and background models, as well as models of resource performance — as their performance is assessed in mission (**C**).

Following the method developed by Moore and Whinston,[9-11] information acquisition management can be modeled as a process of choosing a sequence of information acquisition actions (i.e., a strategy, α, and a decision function, $\delta:B \to D$, with a goal to maximize the expected net payoff, Ω^*:

$$\Omega^*\left(\alpha, B, \delta\right) = \sum_{B \in B} \sum_{x \in B} \Phi\left(x\right) \omega^*\left[x, \delta\left(B\right), C\left(B\right)\right] \tag{16.1}$$

where

$B = \{B_1, B_2, \ldots, B_q\}$	An exhaustive partitioning of possible world states $x \in X$; $B_i \cap B_j = \phi$ for $i \neq j$
$\Phi:X \to [0,1]$	Probability density function on possible world states
D	The system's available response decisions
$C:B \to R$	Cost function for possible utility results R
$\omega^*:X \times D \times R \to R$	Payoff function

A cost function, C, in an information acquisition system includes such costs as those involved with the allocation of resources, the physical and security risks from such an allocation, processor and communications loading and latency, and resource contentions and mutual interference.

An information acquisition action, $a \in A$, yields an information set, Y_a. A mapping function, $\eta_a:X \to Y_a$, induces an information structure, M_a, on X:

$$M_a = \left\{M \big| \exists y \left[M = \eta_a^{-1}\left(\{y\}\right)\right]\right\}. \tag{16.2}$$

A sequence of such actions creates a sequence, B, of partitionings on X, each a refinement on its predecessor:

$$R(B, \alpha) = \left\{r \big| \exists M M \in M_a \,\& \, \exists j \left(B_j \in B \,\& \, r = B_j \cap M\right)\right\} \tag{16.3}$$

The goal of the information acquisition system is to determine such a sequence for which the final element, B_q, maps into the system's set of feasible responses, D.

Model revision (**C** in Figure 16.15) can be an integral component of an information acquisition system. Such a system will characterize off-normal measurements in terms of four components:

- Random process noise affecting the observations of an individual target entity.
- Random process noise affecting entire classes of entities (e.g., random behavioral or design variability).
- Nonstationary processes in the individual target entity (e.g., kinematic maneuver or signature changes caused by cumulative heating, damage, or intentional action).
- Nonstationary processes affecting related entities (e.g., coordinated maneuvers, collateral damage, or doctrinal or design changes).

Similar to a target hypothesis, a target *model* is ultimately a state estimate representing the association of a multiplicity of data. Target models are *abstract* hypotheses; regular target hypotheses are *concrete* association hypotheses. As accumulated data refines the support and state estimate of concrete hypotheses, so do accumulated data refine the support to abstract hypotheses. Thus, abstract hypotheses — often treated as static databases — are, in fact, adaptive to the sensed environment.

The data fusion paradigm permits this adaptivity to be defined in a rigorous way as an organic part of an integrated data fusion. In other words, the model is not an external part of the system. Indeed, the system's set of models combine with its set of concrete hypotheses at all levels of aggregation to form the fusion system's global hypothesis — its estimation and prediction of the state history of the sensed world.

16.3.2.6 Coordinated Multilevel Data Fusion and Resource Management

The recasting of the systems engineering process as a class of resource management processes — coupled with the formal duality between resource management and data fusion — permits a systematic integration of a wide range of engineering and analysis efforts. In effect, this insight allows data fusion techniques to be used in the systematic design and validation of data fusion systems. It also paves the way for a rigorous approach to the entire discipline of systems engineering. Finally, the search for general methods that span systems engineering and in-mission resource management, should lead to a coordinated multi-level approach to real-world problems. This concept is depicted in Figure 16.16, in the form of the well-known Boyd OODA control loop.

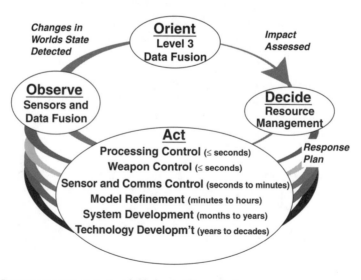

FIGURE 16.16   Resource management at multiple levels of granularity.

As shown, perceived changes in the world state can elicit a diversity of responses, depending on

- The assessed impact (i.e., the cost if no change in the current action plan is taken),
- Timeliness and other physical constraints,
- Resource capability available over time.

An integrated resources management process will employ such factors in considering, evaluating, and selecting one or more of the action types shown in Figure 16.16. For example, an unanticipated capability in a threat weapon system may be inferred through the sensing and fusion processes. Reaction can include a direct response with weapons, other countermeasures, or passive defenses. Other reactions could include refining the object and situation assessment by controlling the data fusion process, or modifying the sensor management plan, or reallocating communications resources to provide the desired new observation data.

These are all "near real-time" responses, with action plans spanning milliseconds to a few minutes. Modifications in the threat system model may also be needed, even within the course of a tactical engagement. The same new assessment may precipitate longer-term responses and the development of system and technology capabilities to address the newly perceived situation.*

16.4   Fusion System Role Optimization

16.4.1   Fusion System Requirements Analysis

16.4.1.1   System Requirements Definition

The fusion systems engineering development process begins with an understanding of needs expressed in the concept of operations (CONOPS) for the system. The CONOPS and its resulting design criteria and constraints define the problem for the fusion system design development.

The CONOPS provides the following:

- Mission needs statement:
  - Current deficiencies.
  - Objectives for the planned system design.
- Operational requirements for the planned system:
  - Operational scenarios: environment and mission objectives.
  - Measures of effectiveness.
- System design constraints:
  - Budget, schedule, processing, and communications environment and standards.

16.4.1.2   Fusion System Functional Role Development

The second step in fusion system role optimization is that of developing the functional role of the fusion system as a "black box" within the system environment. This is an iterative process that culminates in a fusion system specification. The fusion system specification includes

- *System functional capabilities*: requirements inferred from the CONOPS.
- *Data sources*: types of available sensors and other information sources (e.g., Internet and other electronic data sources, human information sources, documents, and databases).

---

*In much the same way that resource management can build coordinated action plans operating at diverse levels of "action granularity", data fusion can be coordinated at diverse levels of "estimation granularity." For example, if a mission objective is to characterize and track a tank column, characterizing and tracking each individual vehicle may not be necessary. By associating observations and estimating entities consistently at the appropriate level of granularity, the combinatorial complexity can be reduced and, potentially, features may be recognized that only emerge at higher levels of aggregation.

- *External interfaces*: descriptions of physical interfaces, communication media, nature of data, security issues, support environments, databases, and users.
- *Hardware/software environment*: physical environmental characteristics, computer equipment, and support software specified for use by the system.
- *Fusion system quality factors*: reliability, redundancy, maintainability, availability, accuracy, portability, flexibility, integrity, reserve capacity, trustworthiness, and robustness.
- *Personnel/training*: numbers, skills, and training responsibilities of available personnel.
- *Documentation*: requirements for manuals, test plans, procedures, training, and other descriptive materials in various hard copy, electronic, and audio/video forms.
- *Logistics*: hardware, software, and database maintenance; supplies and facility impacts.

This black-box design determines the relationship of the proposed system with respect to the other supporting systems, as shown at a high level in Figure 16.1. The design development may use functional partitioning to define design trades, or it may use an object-oriented approach that specifies system charter, function statements, and key use cases. In either approach, the process of iterative feedback development of the fusion system role is employed to optimize system effectiveness.

16.4.1.3   System Effectiveness/Performance Evaluation

The third and final step in fusion system role optimization involves evaluating alternate roles for the fusion system to refine the system requirements and the design. This system requirements analysis provides the effectiveness criteria for this top-level feedback process, which accomplishes the fusion system role optimization. As illustrated in Figure 16.17, the overall performance of a fusion process will be affected by

- The performance of sensors and communications
- The application of prior models
- The nature of process control
- The fusion process per se
- The nature of the operational scenario being examined

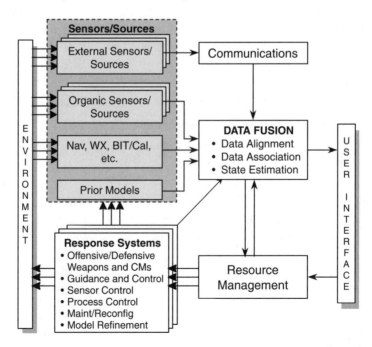

FIGURE 16.17   Data fusion role optimization.

The performance evaluation to optimize the role for the proposed fusion system balances affordability with system effectiveness per the CONOPS.

16.4.2 Fusion System Tree Optimization

After the role for the fusion system is defined (allowing for subsequent iterative refinement), the design of the fusion process itself is optimized using a feedback process similar to that described above. Specifically, requirements are decomposed in a way that enables a fusion tree design solution to be developed and refined.

16.4.2.1 Fusion Tree Requirements Analysis

The requirements refinement process generates detailed descriptions of system elements and operational environments. A sufficiently detailed refinement enables fusion tree design trades regarding performance versus cost/complexity. When using object-oriented design methods, this requirement analysis will include the following elements:

- *Class specifications* with relationships, operations, attributes, and inheritances,
- *Object-scenario diagrams* with interactions between system functions,
- A *data dictionary* with all entity abstractions.

The resulting object and dynamic models, which support both requirements and design, include

- *Observation models* that describe the availability and coverage for each of the data sources and their reported data accuracy, rates, and latency.
- *Contextual models* that describe applicable physical processes, physical and cultural geography, and history.
- *Dynamic environment models* that describe the range of scenarios of interest in terms of objects, their behaviors, and their relationships.
- *Process functional models* that describe available engineering techniques, including libraries of fusion tree and fusion node models with fusion tree and node class diagrams.
- *System environment models* that describe common operating environments, applications interfaces to migration architecture layers, and reporting outputs.
- *User presentation models* that describe graphical user interfaces (GUIs) and application program interfaces (APIs).

The following subsections present a categorization of the input and output requirements refinement for data fusion systems.

16.4.2.1.1 *Fusion System Input Categorization*

The inputs that the fusion system is required to process are categorized in a matrix, which we conventionally refer to as Matrix A and which is shown in Table 16.3. In applying these guidelines in practice, Matrix A lists the assumed salient characteristics of each data source to the data fusion system. For each such source, Matrix A describes

- Reported continuous kinematic data types (e.g., location, altitude, course, speed, angle-only, slant range),
- Reported parametric data types (e.g., RF, PRF, PW, scan rate, RCS, and intensity),
- Reported discrete entity identification inputs (e.g., entity class and type, individual ID and IFF), and
- Reported entity attributes (e.g., track number and scan type).

The elements of Matrix A are general (often qualitative) specifications of input data types and their respective accuracy, timeliness, and availability. For each sensor/source the elements of this matrix are categorized as follows:

TABLE 16.3 Matrix A: Data Fusion System Input Requirements

Sensors/ Sources	Quality				Availability				Timeliness			
	Accuracy		Resolution		Coverage		Reporting		Responsiveness		Adequacy	
	Kinematic/ Param % separation	ID/Attributes % correct	Entity separability	% of reports true entities	% of area of interest	% of entities of interst	% of total no. of reports	% sole entity coverage	Report latency	Update rate	For Situation Warning	For Entity CM (lethal/non)
Sensors												
E-O												
Radar												
(etc.)												
Databases												
External												
Internal												
(etc.)												

TABLE 16.4 Matrix B: Data Fusion System Output Requirements

Mission Area	Functions Supported	Entity Types	Quality		Availability		Timeliness	Other Factors			
			Accuracy	Resolution	Coverage	Responsiveness	Adequacy	System Constraints	User Interface		
			(As in Matrix A, Table 16.3)					H/W & S/W Constraints	On-line Adaptivity	Data Presentation	Analyst Inputs
TAD											
	Entity Acquisition	Bomber Fighter Transport (etc.)									
	Fire Control	(etc.)									
	BDA	(etc.)									
	(etc.)										
TMD											
(etc.)											

1. Classes of reported entities (e.g., aircraft, surface targets, weather patterns, and terrain features).
2. Data aggregation level (e.g., measurement, feature, contact, object report, track, cluster, complex).
3. Source diversity (number of sources and spectral or spatio-temporal diversity).
4. Reporting quality.
5. Confidence reporting (e.g., in terms of statistical, possibilistic, or heuristic representations).
6. Reporting rate.
7. Data availability (e.g., periodicity, aperiodic schedule, and contingencies).
8. Report latency.

Note that data sources include available databases and other data stores (e.g., documents and Web sites), as well as sensors and other "live" sources.

16.4.2.1.2 Fusion System Output Requirements

A fusion system's output requirements are categorized in a Matrix B, with format following the pattern of Table 16.4. The requirements matrix columns include specifications for the various applications mission areas that the system is required to support. Specifications include descriptions of

- Operational mission entities to be reported,
- Fused output data capabilities: data types and associated quality, availability and timeliness, as in Matrix A,
- System functional capabilities: cost, adaptability, robustness, human/computer interfaces, inference explanation, and misalignment issues,
- Software and hardware system environment, and
- User interface requirements.

16.4.2.2   Fusion Tree Design

This section describes a broadly applicable paradigm for representing the solution space of fusion tree types. The paradigm is useful for describing, comparing, and designing alternative fusion tree approaches. A corresponding set of guidelines for designing fusion trees is also described in this section. Based on the data fusion problem space description outlined in Tables 16.3 and 16.4, data fusion trees are designed to partition the problem to achieve a required performance versus cost/complexity.

The smaller the batches of data to be fused, the less complex the overall processing; however, less data generally results in poorer performance. The least complex approach is to process each datum singularly, which occurs often in current fusion systems. However, this serial approach tends to reduce association performance due to the lack of perspective: the process either must allow only high confidence associations or must force data association decisions that may be wrong. For example, shifts in entity geolocation data as a result of sensor misalignments or drift cannot be detected using the one-at-a-time processing approach because the comparison of association hypotheses do not occur over a large enough sample to observe a registration shift pattern.

The maximal performance data fusion tree would have only a single, central node where all available data (i.e., data of *all* types, received from *all* sources over *all* past time) are considered for association and assessment/estimation. However, such a fully batched approach would be unrealistically complex and costly in most real-world applications of interest. Thus, a partitioned approach to fusion system tree design is employed in most practical systems.

The object-oriented system design steps to be accomplished in this fusion tree development phase are as follows:

1. Define a fusion tree type (subject to subsequent refinement or revision).
2. Determine the order and concurrency of fusion nodes (which may include fixed, selectable, or cooperating fusion nodes).
3. Define the data stores strategy (data structures, internal and external files, and databases).
4. Allocate processor and other resources to fusion nodes.
5. Define the software control approach (local, central, distributed, client-server, etc.).
6. Define the implementation architecture (processing hardware, software, and communications architectures; tools; services; technology trades; and standards).

16.4.2.2.1   *Fusion Tree Solution Space*

There are many alternatives available in partitioning the fusion process. Diverse partitioning schemes are appropriate for different data mixes, ambiguities, and densities, and system requirements for information types, quality, availability, and timeliness. Table 16.5 provides examples of the diverse types of partitioning methods for DF trees, including batching by

- Aggregation level – first processing signal or pixel data, then features, object reports, tracks, aggregates (object clusters and complexes).
- Sensor/source – by source type, sensor platform type, or communications medium.
- Data type – by platform class, priority class, or air/surface/subsurface/space objects.
- Time – by observation or reporting time.

TABLE 16.5   Examples of Diverse Partitioning Methods for DF Trees

Partitioning Principle	Fusion Sequence Example			Comments
• Aggregation Level	Signals →	Objects →	Aggregates →	Reduced downstream processing and comms load. Allows localized expertise in sensor data
• Source	On-Board Radar →	On-Board IRST →	Off-Board Tracks → (etc.)	Tailored to expected source data availability (QAT) and processing and comms loading
• Data Type	E-O Imagery →	Radar →	ELINT/ESM → (etc.)	Sequence by expected utility. Allows localized expertise, reduced comms loading.
• Time	Time Step 1 →	Time Step 2 →	Time Step 3 → (etc.)	Reduces data storage and search. Sensitive to process noise and to unmodeled change
• Association Type	Track/Track →	Report/Track →	Track/Track Relationship → (etc.)	Exploits expected high confidence local trackers to reduce downstream processing and comms load.
• Model	Target 1 →	Target 2 →	Target 3 → (etc.)	Quick, high-probability detection of event states of interest. Possibly high false alarm rate.

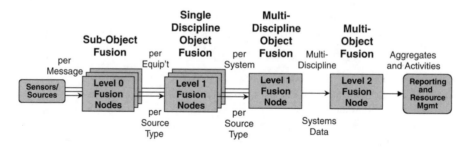

FIGURE 16.18   Fusion tree example.

- Association type – associating tracks across sources (or databases); reports to tracks; and nonidentity relations between tracks (clustering, subordination, interaction, etc.).
- Model – associating reports directly with object models (e.g., within prior ambiguity sets that include entity types of interest).

In cases where the process is data driven (i.e., where there is a need to account for all incoming reports, regardless of source or significance), partitioning by aggregation level is often appropriate. This situation is illustrated in Figure 16.18, where fusion is performed by source and aggregation level, with different types of processing used at each level.

At the other extreme, in a system whose mission is the rapid recognition of a small set of high-value events (e.g., nuclear detonations or ballistic missile launches), model-driven batching may be preferable. In many cases, hybrid schemes for batching are used.

In general, the goal is to reduce the curse of dimensionality with minimum sacrifice in estimation performance. This is accomplished by

- Reducing communications and downstream processing loads by beginning with sources or tracks that are expected to provide the greatest reduction from input to output data rates; these will tend to be sensors with the highest track update rates and high probability of correct observation/track association.
- Localizing the use of specialized data — fusing all ELINT data together before combining it with IMINT data, so that sensor-specific parameters (RF, PRI characteristics, etc.) can be exploited using sensor-specific models, and results being expressed as features or state variables that are commensurate with other classes of sensors (e.g., target type and location).
- Balancing the processing and communications burden within the system by adjusting the operating characteristics among fusion nodes in terms of track impurity and track fragmentation.*

16.4.2.2.2   Fusion Tree Topology Optimization

In fusion tree design, an iterative process is used to define the tree topology (i.e., the data and control flow among nodes) and the processing requirements of each node; this process involves a tradeoff between data availability and processing complexity. These two interrelated design processes are discussed in the following subsections.

In all cases, the fusion tree design process should begin with as fine a partitioning as possible, to reduce the complexity and processing load on individual fusion nodes. Subsequently, additional data are incorporated into nodes, as necessary, to achieve the needed performance. In many cases, fusion is performed sequentially under some batching scheme (i.e., fusion tree).

In a simple fan-in tree, data are taken, either one report or prior hypothesis at a time or in batches, then fused sequentially with a *central* track file. In a variant of this architecture, received data and the central track file are partitioned (e.g., by sensor/source or data type) before fusion.

Another type of fusion tree handles *distributed* track files. These track files are generated when data are batched into several fusion nodes (e.g., by sensor/source, data type, and/or time). This occurs, for example, when different platforms or sensors maintain their own locally generated track files. Key distributed fusion issues are

- The maintenance of a consistent situational estimates among platforms,
- Updating the confidences in the track output,
- Compensating for misalignments, bandwidth limitations, and data latencies.

Simpler and generally preferred distributed fusion trees partition the incoming data and the fusion node products using a fan-in fusion tree. In such systems, the input data is not used more than once, and the fusion products are not fed back into prior fusion nodes to reappear as inputs to later fusion nodes for reprocessing. This reduces the complexity of needing to discount for reuse of data or for self-induced "rumors" (i.e., self-reinforcing inferences via positive feedback).

However, in some fusion systems (e.g., those that process broadcast communication data) there is no straightforward way to prevent such reuse. In some cases, reuse can be detected and rumors can be ignored. In other cases a "pedigree" (i.e., a track's collection and processing history) must be passed to eliminate repeated use of contributing data. Reduced forms of these pedigrees are often required in fan-in trees as well (e.g., to account for the correlated ID from two sources using the same type of ID observables).

In other fusion trees, a very limited track state feedback (e.g., via process management) is used specifically to reduce the computational burden of a prior fusion node. This type of feedback is

---

*These are the type 1 and type 2 tracking errors analogous to the familiar types of detection errors (false alarms and missed detections, respectively). Fusion node optimization is discussed in Section 16.4.3.

differentiated from cues passed through sensor management processes; fusion feedback can directly corrupt (i.e., initiate correlated state errors in) the fusion state estimation, whereas cues enable more effective collection of new data or eliminate the processing of tracks that would be pruned by downstream processing nodes.

Finally, there are fusion trees that permit feedback of tracks into nodes that have already processed data incorporated in the feedback tracks. These tend to be more difficult to design and to control, given the complexities of compensating for the error correlations in the shared data. Nonetheless, in many applications, fusion nodes are distributed like nodes in the World Wide Web. For these applications, the pedigree of the data required to remove the error correlations becomes significant.

16.4.2.2.3   *Fusion Tree Design Issues*

As discussed above, the data fusion tree design determines how data are batched for association and fusion. In some systems, all input batches are fused to a single centralized track file. In other systems, input batches are fused to multiple distributed track files. In the latter case, a fusion tree specifies the order for the various fusion processes, as well as how the inputs are to be batched (e.g., by sensor, time, and data type as described above) to support federated track generation.

The design of an effective fusion tree requires an understanding of the fusion data inputs, assessment of the outputs, and the fusion system requirements. The tree structure provides the designer with a formal mechanism for understanding the difficulty of a given fusion problem, and methods for achieving required performance (e.g., in terms of accuracy, timeliness, throughput, etc.) under required processing and cost constraints.

To reduce complexity and/or cost, the system should accomplish as much as possible as early and as easily as possible in the processing sequence. That is, relatively unambiguous, high confidence decisions should be made as soon as possible in the fusion tree. For example, a sensor that can generate high confidence tracks based on its own data should be allocated its own node of the tree early in the process, using only that sensor's data. Other data that can be fused sequentially to existing tracks with high confidence should also be done early in the fusion tree (e.g., fusion of data from off-board sources reporting unique entity identifiers).

The fusion trees with the simpler fusion node processing are those that use the smaller batches of data (e.g., one report at a time is fused with a centralized set of tracks). However, in many applications, such sequential processing does not yield sufficient association performance, including those involving dense scenarios or highly maneuverable targets and/or low observation updates rates, or with sensors subject to unmodeled misalignment.

In other cases, sequential processing does not yield a sufficiently broad perspective for estimation — for example, for situation assessment or threat prediction where the relationship to contextual data must be considered. Thus, the size of the batches to be selected involves a tradeoff between performance and computational complexity and cost.

The size of the batch of data input to a single fusion node should be only as large as necessary to enable the consideration of the other data needed to achieve sufficiently accurate association and estimation results. Otherwise, smaller batches could be used with no loss in performance and reduced overall complexity.

Figure 16.19 illustrates this process of developing the fusion tree in order to achieve the knee-of-the-curve in performance versus cost/complexity in available data batching dimensions. The resulting fusion tree can be distributed across multiple processors operating in parallel at different points in the quality/quantity surveillance space. Typically, quality and timeliness are reduced at the higher levels of the hierarchy of fusion nodes, which offer broader perspectives (e.g., for situation awareness). This type of fusion tree enables increased speed and accuracy for local tracking/ID (e.g., for weapon handover), as well as high-level situation assessment, all using the same data fusion tree paradigm.

In typical integrated information management systems, the data fusion and resource management trees are highly intertwined, with nodes distributed and coupled at each level (as illustrated in

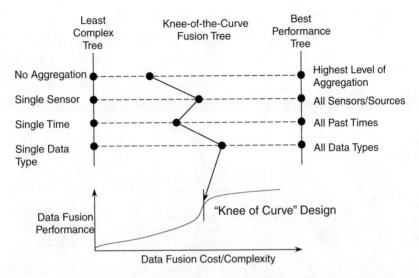

FIGURE 16.19 Selecting data aggregation and batching criteria to achieve desired performance vs. cost.

Figure 16.3). This interaction between fusion and management nodes enables more centralized decisions (and, therefore, generally better decisions), as well as local feedback for higher rate responses. Such tight coupling between estimation and control processes allows rapid autonomous feedback, with centralized coordination commands requiring less communications bandwidth. These cascaded data fusion and resource management trees also enable the data fusion tree to be selected online via resource management (Level 4 data fusion). This can occur when, for example, a high priority set of data arrives or when observed operating conditions differ significantly from those for which the given fusion tree was designed.

In summary, the data fusion tree specifies

- How the data can be batched (e.g., by level, sensor/source, time, and report/data type), and
- The order that the batches (i.e., fusion tree nodes) are to be processed.

The success of a particular fusion tree will be determined by the fidelity of the model of the data environment and of the required decision space used in developing the tree. A tree based on high-fidelity models will be more likely to associate received data effectively (i.e., sufficient to resolve state estimates that are critical to making response decisions). This is the case when the data that is batched in early fusion nodes are of sufficient accuracy and precision in common dimensions to create association hypotheses that closely reflect the actual report-to-entity causal relationships.

For poorly understood or highly dynamic data environments, a reduced degree of partitioning may be warranted. Alternately, the performance of any fusion process can generally be improved by making the process adaptive to the estimated sensed environment. This can be accomplished by integrating the data fusion tree with a dual fan-out resource management tree to provide more accurate local feedback at higher rates, as well as slower situational awareness and mission management with broader areas of concern.[12] A more dynamic approach is to permit the fusion tree itself to be adaptively reconstructed in response to the estimated environment and to the changing data needs.[10,11]

16.4.2.2.4 *Fusion Tree Design Categorization*

The fusion tree architecture used by the data fusion system indicates the tree topology and data batching in the tree nodes. These design decisions are documented in the left half of a Matrix C, illustrated in Table 16.6, and include the following partial categorization:

TABLE 16.6　Matrix C: Fusion Tree Categorization (example)

Fusion System [Per Level]	Fusion Tree Types				Fusion Node Types			
	Centralized Tracks		Distributed Tracks		Batching Types			
	One/Batch	SSQ/PSQ	One/Batch	Fan/FB	Source	Time	Data Types	(etc.)
SYSTEM 1								
Level 1			BATCH	FAN	SR	AT	S/M Act	
Level 2			BATCH	FB	SR	AT	V/H O	
SYSTEM 2								
Level 0	(1HC)	(PSQ)				IT	ID	
Level 1	1HC	(PSQ)				IT	ID	
SYSTEM 3								
Level 0	1HC	PSQ			SR		MF	
Level 1	1HC	PSQ/SSQ			SR		MF	
Level 2	1HC	SSQ					MF	

- *Centralized Track File:* all inputs are fused to a view of the world based on all prior associated data.
- *Batching*
  - *1HC:* one input at a time, high-confidence-only fusion to a centralized track file.
  - *1HC/BB:* one input at a time, high-confidence-only fusion nodes that are followed by batching of ambiguously associated inputs for fusion to a centralized track file.
  - *Batch:* batching of input data by source, time, and/or data type before fusion to a centralized track file.
- *Sequencing*
  - *SSQ:* input batches are fused in a single sequential set of fusion nodes, each with the previous resultant central track file.
  - *PSQ:* input data and the central track file are partitioned before sequential fusion into non-overlapping views of the world, with the input data independently updating each centralized track file.
- *Distributed Track Files:* different views of the world from subsets of the inputs are maintained (e.g., radar-only tracking) and then fused (e.g., onboard to offboard fusion).
- *Batching*
  - *1HC:* one input at a time, high-confidence-only fusion to distributed track files that are later fused.
  - *1HC/BB:* one input at a time, high-confidence-only fusion nodes followed by batching of ambiguously associated inputs for fusion to corresponding distributed track files.
  - *Batch:* batching of input data by source, time, and/or data type for fusion to a corresponding distributed track file.
- *Sequencing*
  - *FAN:* a fan-in fusion of distributed fusion nodes.
  - *FB:* a fusion tree with feedback of tracks into fusion nodes that have already processed a portion of the data upon which the feedback tracks are based.

Fusion tree nodes are characterized by the type of input batching and can be categorized according to combinations in a variety of dimensions, as illustrated above and documented per the right half of Matrix C (Table 16.6). A partial categorization of fusion nodes follows.

- *Sensor/source*
  - *BC:* batching by communications type (e.g., RF, WA, Internet, press)
  - *SR:* batching by sensor type (e.g., imagery, video, signals, text)
  - *PL:* batching by collector platform
  - *SB:* batching by spectral band
- *Data type*
  - *VC:* batching by vehicle class/motion model (e.g., air, ground, missile, sea)
  - *Loc:* batching by location (e.g., around a named area of interest)
  - *S/M Act:* batching into single and/or multiple activities modeled per object (e.g., transit, setup, launch, hide activities for a mobile missile launcher)
  - *V/H O:* batching for vertical and/or horizontal organizational aggregation (e.g., by unit subordination relationships or by sibling relations)
  - *ID:* batching into identification classes (e.g., fixed, relocatable, tracked vehicle or wheeled vehicle)
  - *IFF:* batching into priority class (e.g., friend, foe, and neutral classes)
  - *PARM:* batching by parametrics type
- *Time*
  - *IT:* batching by observation time of data
  - *AT:* batching by arrival time of data
  - *WIN:* time window of data
- *Other.*

In addition to their use in the system design process, these categorizations can be used in object-oriented software design, allowing instantiations as hierarchical fusion tree-type and fusion node-type objects.

16.4.2.3 Fusion Tree Evaluation

This step evaluates the alternative fusion trees to enable the fusion tree requirements and design to be refined. The fusion tree requirement analysis provides the effectiveness criteria for this feedback process, which results in fusion tree optimization. Effectiveness is a measure of the achievement of goals and of their relative value. Measures of effectiveness (MOEs) specific to particular types of missions areas will relate system performance to measures of effectiveness and permit traceability from measurable performance attributes of intelligence association/fusion systems.

The following list of MOEs provides a sample categorization of useful alternatives, after Llinas, Johnson, and Lome.[13] There are, of course, many other metrics appropriate to diverse system applications.

- *Entity nomination rate:* rate at which an information system recognizes and characterizes entities relative to mission response need.
- *Timeliness of information:* effectiveness in supporting response decisions.
- *Entity leakage:* fraction of entities against which no adequate response is taken.
- *Data quality:* measurement accuracy sufficiency for response decision (e.g., targeting or navigation).
- *Location/tracking errors:* mean positional error achieved by the estimation process.
- *Entity feature resolution:* signal parameters, orientation/dynamics as required to support a given application.
- *Robustness:* resistance to degradation caused by process noise or model error.

Unlike MOEs, measures of performance (MOPs) are used to evaluate system operation independent of operational utility and are typically applied later in fusion node evaluation.

16.4.3   Fusion Tree Node Optimization

The third phase of fusion systems engineering optimizes the design of each node in the fusion tree.

16.4.3.1   Fusion Node Requirements Analysis

Versions of Matrices A and B are expanded to a level of detail sufficient to perform fusion node design trades regarding performance versus cost and complexity. Corresponding to system-level input/output Matrices A and B, the requirements for each node in a data fusion tree are expressed by means of quantitative input/output Matrices D and E (illustrated in Tables 16.7 and 16.8, respectively). In other words, the generally qualitative requirements obtained via the fusion tree optimization are refined quantitatively for each node in the fusion tree.

Matrix D expands Matrix A to indicate the quality, availability, and timeliness (QAT in the example matrices) for each essential element of information provided by each source to the given fusion node. The scenario characteristics of interest include densities, noise environment, platform dynamics, viewing geometry, and coverage.

Categories of expansion pertinent to Matrix E include

- Software life-cycle cost and complexity (i.e., affordability)
- Processing efficiency (e.g., computations/sec/watt)
- Data association performance (i.e., accuracy and consistency sufficient for mission)
- Ease of user adaptability (i.e., operational refinements)
- Ease of system tuning to data/mission environments (e.g., training set requirements)
- Ease of self-coding/self-learning (e.g., system's ability to learn how to evaluate hypotheses on its own)
- Robustness to measurement errors and modeling errors (i.e., graceful degradation)
- Result explanation (e.g., ability to respond to queries to justify hypothesis evaluation)
- Hardware timing/sizing constraints (i.e., need for processing and memory sufficient to meet timeliness and throughput requirements).

From an object-oriented analysis, the node-specific object and dynamic models are developed to include models of physical forces, sensor observation, knowledge bases, process functions, system environments, and user presentation formats. A common model representation of these environmental and system factors across fusion nodes is important in enabling inference from one set of data to be applied to another. Only by employing a common means of representing data expectations and uncertainty in hypothesis scoring and state estimates can fusion nodes interact to develop globally consistent inferences.

Fusion node requirements are derived requirements since fusion performance is strongly affected by the availability, quality, and alignment of source data, including sensors, other live sources, and prior databases. Performance metrics quantitatively describe the capabilities of system functionality in non-mission-specific terms. Figure 16.20 shows some MOPs that are relevant to information fusion. The figure also indicates dependency relations among MOPs for fusion and related system functions: sensor, alignment, communications and response management performance, and prior model fidelity. These dependencies form the basis of data fusion performance models.

16.4.3.2   Fusion Node Design

Each fusion node performs some or all of the following three functions:

- *Data Alignment*: time and coordinate conversion of source data.
- *Data Association*: typically, associating reports with tracks.
- *State Estimation*: estimation and prediction of entity kinematics, ID/attributes, internal and external relationships, and track confidence.

The specific design and complexity of each of these functions will vary with the fusion node level and type.

TABLE 16.7 Matrix D Components: Sample Fusion Node Input Requirements

Input	Continuous Parameters									Discrete Attributes									
	Kinematics				Signal Parameters					ID					Attributes			Relational	
Source	Pos	Vel	Angle	Range	RF	AM	FM	PM	Scan	IFF	Class	Type	ELNOT	SEI	Track #	Channel	Mode	Subordination	Role
Source 1	QAT	QAT	QAT	QAT		...													
Source 2	...	...	...	...															
Source 3	...																		

TABLE 16.8 Matrix E Components: Sample Fusion Node Output Requirements

Continuous Parameters									Discrete Attributes									
Kinematics				Signal Parameters					ID					Attributes			Relational	
Pos	Vel	Angle	Range	RF	AM	FM	PM	Scan	IFF	Class	Type	ELNOT	SEI	Track #	Channel	Mode	Subordination	Role
QAT	QAT	QAT	QAT		...													
...	...	...	...															

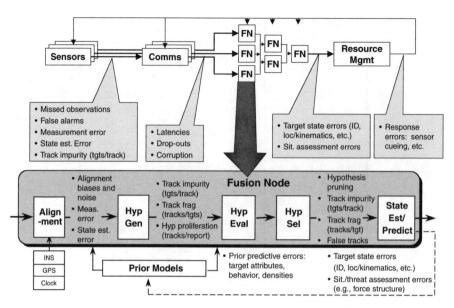

FIGURE 16.20   System-level performance analysis of data fusion.

16.4.3.2.1   Data Alignment

Data alignment (also termed *common referencing* and *data preparation*) includes all processes required to test and modify data received by a fusion node so that multiple items of data can be compared and associated. Data alignment transforms all of the data that has been input to a fusion node to consistent formats, spatio-temporal coordinate frame, and confidence representations, with compensation for estimated misalignments in any of these dimensions.

Data from two or more sensors/sources can be effectively combined only if the data are compatible in format and consistent in frame of reference and in confidence assignments. Alignment procedures are designed to permit association of multisensor data either at the decision, feature, or pixel level, as appropriate to the given fusion node.

Five processes are involved in data alignment:

- *Common Formatting:* testing and transforming the data to system-standard data types and units.
- *Time Propagation:* extrapolating old track location and kinematic data to a current update time.
- *Coordinate Conversion:* translating data received in various coordinate systems (e.g., platform referenced systems) to a common spatial reference system.
- *Misalignment Compensation:* correcting for known misalignments or parallax between sensors.
- *Evidential Conditioning:* assigning or normalizing likelihood, probability, or other confidence values associated with data reports and individual data attributes.

Common formatting. This function performs the data preparation needed for data association, including parsing, fault detection, format and unit conversions, consistency checking, filtering (e.g., geographic, strings, polygon inclusion, date/time, parametric, or attribute/ID), and base-banding.

Time propagation. Before new sensor reports can be associated with the fused track file, the latter must be updated to predict the expected location/kinematic states of moving entities. Filter-based techniques for track state prediction are used in many applications.[15,16]

Coordinate conversion. The choice of a standard reference system for multisensor data referencing depends on

- The standards imposed by the system into which reporting is to be made
- The degree of alignment attainable and required in the multiple sensors to be used

- The sensors' adaptability to various reference standards
- The dynamic range of measurements to be obtained in the system (with the attendant concern for unacceptable quantization errors in the reported data).

A standard coordinate system does not imply that each internetted platform will perform all of its tracking or navigational calculations in this reference frame. The frame selected for internal processing is dependent on what is being solved. For example, when an object's trajectory needs to be mapped on the earth, the WGS84 is a natural frame for processing. On the other hand, ballistic objects (e.g., spacecraft, ballistic missiles, and astronomical bodies) are most naturally tracked in an inertial system, such as the FK5 system of epoch J2000.0. Each sensor platform will require a set of well-defined transformation matrices relating the local frame to the network standard one (e.g., for multi-platform sensor data fusion).[17]

Misalignment compensation. Multisensor data fusion processing enables powerful alignment techniques that involve no special hardware and minimal special software. Systematic alignment errors can be detected by associating reports on entities of opportunity from multiple sensors. Such techniques have been applied to problems of mapping images to one another (or for rectifying one image to a given reference system). *Polynomial warping* techniques can be implemented without any assumptions concerning the image formation geometry. A linear least squares mapping is performed based on known correspondences between a set of points in the two images.

Alignment based on entities of opportunity presupposes correct association and should be performed only with high confidence associations. A high confidence in track association of point source tracks is supported by

- A high degree of correlation in track state, given a constant offset
- Reported attributes (features) that are known *a priori* to be highly correlated and to have reasonable likelihood of being detected in the current mission context
- A lack of comparable high kinematic and feature correlation in conflicting associations among sensor tracks.

Confidence normalization (evidence conditioning). In many cases, sensors/sources provide some indication of the confidence to be assigned to their reports or to individual data fields. Confidence values can be stated in terms of likelihoods, probabilities, or ad hoc methods (e.g., figures of merit). In some cases, there is no reporting of confidence values; therefore, the fusion system must often normalize confidence values associated with a data report and its individual data attributes. Such evidential conditioning uses models of the data acquisition and measurement process, ideally including factors relating to the entity, background, medium, sensor, reference system, and collection management performance.

16.4.3.2.2 Data Association

Data association uses the commensurate information in the data to determine which data should be associated for improved state estimation (i.e., which data belongs together and represents the same physical object or collaborative unit, such as for situation awareness). This section summarizes the top-level data association functions.

The following overview summarizes the top-level data association functions. Mathematically, deterministic data association is a labeled set-covering decision problem: given a set of prepared input data, the problem is to find the best way to sort the data into subsets where each subset contains the data to be used for estimating the state of a hypothesized entity. This collection of subsets must cover all the input data and each must be labeled as an actual target, false alarm, or false track.

The hypothesized groupings of the reports into subsets describe the objects in the surveillance area. Figure 16.21(a) depicts the results of a single scan by each of three sensors, **A**, **B**, and **C**. Reports from each sensor — e.g., reports A_1 and A_2 — are presumed to be related to different targets (or one or both may be false alarms).

Figure 16.21(a) indicates two hypothesized coverings of a set, each containing two subsets of reports — one subset for each target hypothesized to exist. Sensor resolution problems are treated by allowing the report subsets to overlap wherever one report may originate from two objects; e.g., the sensor C_1 report

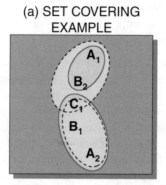 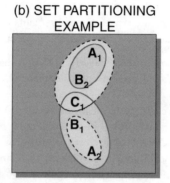

A_n, B_n, C_n = Report n from Sensors A, B, C

⬭ = Association Hypotheses under 1st Global Hypothesis

⬭⬭ = Association Hypotheses under 2nd Global Hypothesis

FIGURE 16.21   Set covering and set partitioning representations of data association.

in Figure 16.21(a). When there is no overlap allowed, data association becomes a labeled *set partitioning* problem, illustrated in Figure 16.21(b).

Data association is segmented into three subfunctions:

1. **Hypothesis generation:** data are used to generate association hypotheses (tracks) via feasibility gating of prior hypotheses (tracks) or via data clustering.
2. **Hypothesis evaluation:** these hypotheses are evaluated for self consistency using kinematic, parametric, attribute, ID, and *a priori* data.
3. **Hypothesis selection:** a search is performed to find the preferred hypotheses based either on the individual track association scores or on a global score (e.g., MAP likelihood of set coverings or partitionings).

In cases where the initial generation or evaluation of all hypotheses is not efficient, hypothesis selection schemes can provide guidance regarding which new hypotheses to generate and score.

In hypothesis selection, a stopping criterion is eventually applied, and the best (or most unique) hypotheses are selected as a basis for entity state estimation, using either probabilistic or deterministic association.

The functions necessary to accomplish data association are as presented in the following sections.

Hypothesis generation. Hypothesis generation reduces the search space for the subsequent functions by determining the feasible data associations. Hypothesis generation typically applies spatio-temporal relational models to gate, prune, combine, cluster, and aggregate the data (i.e., kinematics, parameters, attributes, ID, and weapon system states).

Because track-level hypothesis generation is intrinsically a suboptimizing process (eliminating from consideration low value, thought possible, data associations), it should be conservative, admitting more false alarms rather than eliminating possible true ones. The hypothesis generation process should be designed so that the mean computational complexity of the techniques is significantly less than in the hypothesis evaluation or selection functions.

Hypothesis evaluation. Hypothesis evaluation assigns scores to optimize the selection of the hypotheses resulting from hypothesis generation. The scoring is used in hypothesis selection to compute the overall objective function which guides efficient selection searching. Success in designing hypothesis evaluation techniques resides largely in the means for representing uncertainty. The representational problem involves assigning confidence in models of the deterministic and random processes that generate data.

Concepts for representing uncertainty include

- Fisher likelihood
- Bayesian probabilities
- Evidential mass
- Fuzzy membership
- Information theoretic and other nonparametric similarity metrics
- Neural network
- Ad hoc (e.g., figures of merit or other templating schemes)
- Finite set statistics

These models of uncertainty — described and discussed in References 3, 10, and 18–21 and elsewhere — differ in

- Degree to which they are empirically supportable or supportable by analytic or physical models
- Ability to draw valid inferences with little or no direct training
- Ease of capturing human sense of uncertainty
- Ability to generate inferences that agree either with human perceptions or with truth
- Processing complexity

The average complexity usually grows linearly with the number of feasible associations; however, fewer computations are required per feasible hypothesis than in hypothesis selection.

Hypothesis evaluation is sometimes combined with state estimation, with the uncertainty in the state estimate used as evaluation scores. For example, one may use the likelihood ratio $\lambda(Z)$ as an association score for a set of reports Z with $\lambda(Z)$ being determined as a function of the probability distribution in continuous and discrete state space, x_c and x_d, respectively:

$$\lambda(Z) = \prod_{z \in Z} \left[G_z(x_c) \cdot \sum_{x_d} p(x_d|z) p_B(x_c|x_d) \lambda_B(x_d) \right]$$

where

$\lambda_B(x_d)$ is the prior for a discrete state,

x_d and $p_B(x_c|x_d)$ is a conditioned prior on the expected continuous state x_c,

$G_z(x_c)$ is a density function — possibly Gaussian — on x_c conditioned on Z.[22]

This evaluation would generally be followed by hypothesis selection and updating of track files with the selected state estimates.

Hypothesis selection. Hypothesis selection involves searching the scored hypotheses to select one or more to be used for state estimation. Hypothesis selection eliminates, splits, combines, retains, and confirms association hypotheses to maintain or delete tracks, reports, and/or aggregated objects.

Hypothesis selection can operate at the track level, e.g., using greedy techniques. Preferably, hypothesis selection operates globally across all feasible set partitionings (or coverings). Optimally, this involves searching for a partitioning (or covering) R of reports that maximizes the global score, e.g., the global likelihood $\prod_{Z \in R} \lambda(Z)$.

The full assignment problem, either in set-partitioning or, worse, in set-covering schemes, is of exponential complexity. Therefore, it is common to reduce the search space to that of associating only a current data scan, or just a few, to previously accepted tracks.

This problem is avoided altogether in techniques that estimate multitarget states directly, without the medium of observation-to-track associations.[21,23,24]

16.4.3.2.3 State Estimation

State estimation involves estimating and predicting states, both discrete and continuous, of entities hypothesized to be the referents of sets of sensor/source data. Discrete states include (but are not limited to) values for entity type and specific ID, and activity and discrete parametric attributes (e.g., modulation types).

Depending on the types of entities of interest and the mission information needs, state estimation can include kinematic tracking with misalignment estimation, parametric estimation (e.g., signal modulation characteristics, intensity, size, and cross sections), and resolution of discrete attributes and classification (e.g., by nationality, IFF, class, type or unique identity). State estimation often applies more accurate models to make these updates than are used in data association, especially for the kinematics, parametrics, and their misalignments. Techniques for discrete state estimation are categorized as logical or symbolic, statistical, possibilistic, or ad hoc methods.

State estimation does not necessarily update a track with a unique (i.e., deterministic) data association; however, it can smooth over numerous associations according to their confidences of association (e.g., probabilistic data association filter,[16,25] or global tracking.[21,26]). Also, object and aggregate classification confidences can be updated using probabilistic or possibilistic[27] knowledge combination schemes.

In level 2 data fusion nodes, the state estimation function may perform estimation of relations among entities to include the following classes of relations:

- Spatio-temporal
- Causal
- Organizational
- Informational
- Intentional

In level 3 data fusion, state estimation involves estimation or prediction of costs associated with estimated situations. In a threat environment, these can include assessment of adversaries' intent and impact on one's assets (these topics are treated in Reference 28 and in Chapter 2).

16.4.3.2.4 Fusion Node Component Design Categorization

For each node, the pertinent functions for data alignment, data association, and state estimation are designed. The algorithmic characterizations for each of these three functions can then be determined. The detailed techniques or algorithms are not needed or desired at this point; however, the type of filtering, parsing, gating, scoring, searching, tracking, and identification in the fusion functions can be characterized.

The emphasis in this stage is on achieving balance within the nodes for these functions in their relative computational complexity and accuracy. It is at this point, for example, when the decision to perform deterministic or probabilistic data association is made, as well as what portions of the data are to be used for feasibility gating and for association scoring. The detailed design and development (e.g., the actual algorithms) are not specified until this node processing optimization balance is achieved. For object-oriented design, common fusion node objects for the above functions can be utilized to initiate these designs.

Design decisions are documented in Matrix F, as illustrated in Table 16.9 for a notional data fusion system. The primary fusion node component types used to compare alternatives in Matrix F are listed in the following subsections.

Data Alignment (Common Referencing)

CC: Coordinate conversion (e.g., UTM or ECI to lat/long)
TP: Time propagation (e.g., propagation of last track location to current report time)
SC: Scale and/or mode conversion (e.g., emitter RF base-banding)
FC: Format conversion and error detection and correction

TABLE 16.9 Matrix F: Data Fusion Node Design Characterization (system examples)

| Fusion Nodes | Data Alignment | | | Fusion Tree Node Components | | | | State Estimation | | |
| | | | | Data Association | | | | | | |
	Kinematics	Parametrics	ID/Attributes	Hypothesis Generation	Hypothesis Evaluation	Hypothesis Selection	Kinematics	Parametrics	ID/Attributes
1 (Single Disc/Level 1)	CC, TP	SC	FC	K/P Gate	Bay/LKL/CHI	P,ND,SP,S, MHT	KF	Bay	Bay
2 (Single Disc/Level 1)	CC, TP	SC	FC	K/P Gate	Bay/LKL/CHI	P,ND,SP,S, MHT	KF	Bay	Bay
3 (Single Disc/Level 1)	CC, TP		FC	K Gate	AH	D,1D,SP,S, SHT	KF		AH
4 (Multi-Disc/Level 1)	CC, TP			KP Gate	Bay/LKL	P,2D,SP,S, SHT	AH	Bay	Bay
5 (Multi-Disc/Level 2)				K/P Gate	Bay/BN	P,2D,SC,S, SHT	AH	Bay	Bay
6 (Multi-Disc/Level 2)				K/P Gate, ST	AH	D,2D,SC,S, SHT	KF		L/S

Data Association (Association)

Hypothesis generation — generating feasible association hypotheses
   K gate: Kinematic gating
   K/P gate: Kinematic and parametric gating
   KB: Knowledge-based methods (e.g., logical templating and scripting)
   SM: Syntactic methods
   ST: Script templating based on doctrine
   OT: Organizational templating based on doctrine
   KC: Kinematic clustering
   K/PC: Kinematic and parametric clustering
   PR: Pattern recognition
   GL: Global association
Hypothesis evaluation — assigning scores to feasible association hypotheses
   Bay: Conditional Bayesian scoring, including *a posteriori*, likelihoods, chi-square, Neyman-
      Pearson, and Bayesian nets
   L/S: Logic and symbolic scoring including case-based reasoning, semantic distance, scripts/frames,
      expert system rules, ad hoc, and hybrids
   Poss: Possibilistic scoring (e.g., evidential mass or fuzzy set membership), as well as non-paramet-
      ric, conditional event algebras, and information theoretic; often used in combination with L/S
      techniques, particularly with highly uncertain rules
   NN: Neural networks, including unsupervised and supervised feed-forward and recurrent
Hypothesis selection — selecting one or more association hypotheses for state estimation, based on
   the overall confidenceof the hypotheses.
   D/P: Deterministic or probabilistic data association (i.e., select highest scoring hypothesis or
      smooth over many associations)
   S/M HT: Single- or multiple-hypothesis testing (i.e., maintain best full-scene hypothesis or many
      alternative scenes/world situation views)
   SC/SP: Solve as a set covering or as a set partitioning problem (i.e., allow or disallow one-to-many
      report/track associations)
   2D/ND: Solve as a two-dimensional or as an *n*-dimensional association problem (i.e., associating
      a single batch of data or more scans of data with a single track file)
   S/O: Use suboptimal or optimal search schemes

State Estimation and Prediction

KF: Kalman filter
EKF: Extended Kalman filter to include linearization of fixed and adaptive Kalman gains
MM: Multiple model linear filters using either model averaging (IMM) or selection based on the
   residuals
NL: Nonlinear filtering to include nonlinear templates and Daumís methods
AH: Ad hoc estimation methods without a rigorous basis (not including approximations to rigorous
   techniques)
LS: Least squares estimation and regression
L/S: Logic and symbolic updates, especially for ID/attributes such as case-based reasoning, semantic
   distance, scripts/frames, expert system rules, ad hoc, and hybrids
Prob: Probabilistic ID/attribute updates including Bayesian nets and entity class trees
Poss: Possibilistic (e.g., evidential reasoning and fuzzy logic), as well as nonparametric, conditional
   event algebras, and information theoretic methods

16.4.3.3   *Fusion Node Performance Evaluation*

This step evaluates the alternative fusion node functions to enable the refinement of the fusion node
requirements analysis, and the fusion node design development. The performance evaluation is generally

fed back to optimize the design and interfaces for each function in each node in the trees. This feedback enables balancing of processing loads among nodes and may entail changes in the fusion tree node structure in this optimization process. The fusion node requirements analysis provides the MOPs for this feedback process.

A systems performance relative to these MOPs will be a function of several factors, many external to the association/fusion process, as depicted in Figure 16.20. These factors include

- Alignment errors between sensors
- Sensor estimation errors
- Lack of fidelity in the *a priori* models of the sensed environment and of the collection process
- Restrictions in data availability caused by communications latencies, data compression, or data removal
- Performance of the data association and fusion process.

Measurement and alignment errors can affect the statistics of hypothesis generation (e.g., the probability that an observation will fall within the validation gate of particular tracks) and of hypothesis evaluation (e.g., the likelihood estimated for a set of observations to relate to a single entity). These impacts will, in turn, affect

- Hypothesis selection performance (e.g., the probability that a given assignment hypothesis will be selected for higher-level situation assessment or response, or the probability that a given hypothesis will be pruned from the system).
- State estimation performance (e.g., the probabilities assigned to various entity classification, activity states or kinematic states).[14]

As described above, dependencies exist among MOPs for information fusion node functions, as well as those relating to sensor, alignment, communications and response management performance, and prior model fidelity. Covariance analytic techniques can be employed to predict performance relating to hypothesis generation and evaluation. Markov chain techniques can be used to predict hypothesis selection, state estimation and cueing performance.[29]

16.4.4   Detailed Design and Development

The final phase determines the detailed design of the solution *patterns* for each subfunction of each node in the fusion tree. There is a further flowdown of the requirements and evaluation criteria for each of the subfunctions down to the pattern level. Each pattern contains the following:

1. Name and definition of the problem class it addresses (e.g., hypothesis evaluation of fused MTI/ESM tracks).
2. Context of its application within the data fusion tree paradigm (node and function indices).
3. Requirements and any constraint violations in combining them.
4. The design rationale (prioritization of requirements and constraint relaxation rationale).
5. The design specification.
6. Performance prediction and assessment of requirements satisfaction.

The resulting pattern language can provide an online aid for rapid data fusion solution development. Indeed, given the above higher-level design process, a nonexpert designer should be able to perform the pattern application, if provided a sufficiently rich library of legacy patterns organized into directed graphs.

In an object-oriented system design process, this is the step in which the rapid prototyping executable releases are planned and developed. In addition, the following are completed:

- Class-category diagrams
- Object-scenario diagrams
- Class specifications

This paradigm-based software design and development process is compatible with an iterative enhancement and rapid prototyping process at this point. The performance evaluation feedback drives the pattern optimization process through multiple evolutionary stages of detailed development and evaluation. The paradigm provides the structure above the applications interface (API) in the migration architecture over which it is embedded.

The system development process can further ease this rapid prototyping development and test evolution. The first round of this process is usually performed using workstations driven by an open-loop, non-real-time simulation of the operational environment. The resulting software can then progress through the traditional testing stages: closed-loop, then real-time, followed by man-in-the-loop, and, finally, operational environment test and evaluation.

Software developed at any cycle in the design process can be used in the corresponding hardware-in-the-loop testing stage; ultimately leading to operational system testing. The results of the operational environment and hardware-in-the-loop testing provide the verification and validation for each cycle. Results are fed back to improve the simulations and for future iterations of limited data testing (e.g., against recorded scenarios).

References

1. *Engineering Guidelines*, SWC Talon-Command Operations Support Technical Report 96-11/4, 1997.
2. Alan N. Steinberg and Christopher L. Bowman, Development and application of data fusion engineering guidelines, *Proc. Tenth Nat'l. Symp. Sensor Fusion*, 1997.
3. *Engineering Guidelines for Data Correlation Algorithm Characterization*, TENCAP SEDI Contractor Report, SEDI-96-00233, 1997.
4. Christopher L. Bowman, The data fusion tree paradigm and its dual, *Proc. 7th Nat'l. Symp. Sensor Fusion*, 1994.
5. Christopher L. Bowman, Affordable information fusion via an open, layered, paradigm-based architecture, *Proc. 9th Nat'l. Symp. Sensor Fusion*, 1996.
6. Alan N. Steinberg, Approaches to problem-solving in multisensor fusion, *forthcoming*, 2001.
7. *C4ISR Architecture Framework, Version 1.0*, C4ISR ITF Integrated Architecture Panel, CISA-0000-104-96, June 7, 1996.
8. Alan N. Steinberg, Data fusion system engineering, *Proc. Third Internat'l. Symp. Information Fusion*, 2000.
9. James C. Moore and Andrew B. Whinston, A model of decision-making with sequential information acquisition, *Decision Support Systems*, 2, 1986: 285–307; 3, 1987: 47–72.
10. Alan N. Steinberg, Adaptive data acquisition and fusion, *Proc. Sixth Joint Service Data Fusion Symp.*, 1, 1993, 699–710.
11. Alan N. Steinberg, Sensor and data fusion, *The Infrared and Electro-Optical Systems Handbook*, Vol. 8, 1993, 239–341.
12. Christopher L. Bowman, The Role of process management in a defensive avionics hierarchical management tree, *Tri-Service Data Fusion Symp. Proc.*, John Hopkins University, June 1993.
13. James Llinas, David Johnson and Louis Lome, Developing robust and formal automated approaches to situation assessment, presented at *Situation Awareness Workshop*, Naval Research Laboratory, September 1996.
14. Alan N. Steinberg, Sensitivities to reference system performance in multiple-aircraft sensor fusion, *Proc. 9th Nat'l. Symp. Sensor Fusion*, 1996.
15. S. S. Blackman, *Multiple Entity Tracking with Radar Applications*, Artech House, Inc., Norwood, MA, 1986.
16. Y. Bar-Shalom and X.-R. Li, *Estimation and Tracking: Principles, Techniques, and Software*, Artech House Inc., Boston, 1993.

17. Carl W. Clawson, On the choice of coordinate systems for fusion of passive tracking data, *Proc. Data Fusion Symp.*, 1990.
18. Edward Waltz and James Llinas, *Multisensor Data Fusion*, Artech House Inc., Boston, 1990.
19. Jay B. Jordan and How Hoe, A comparative analysis of statistical, fuzzy and artificial neural pattern recognition techniques, *Proc. SPIE Signal Processing, Sensor Fusion, and Entity Recognition*, 1699, 1992.
20. David L. Hall, *Mathematical Techniques in Multisensor Data Fusion*, Artech House, Boston, 1992.
21. I.R. Goodman, H.T. Nguyen, H.T., and R. Mahler, *Mathematics of Data Fusion (Theory and Decision Library*. Series B, Mathematical and Statistical Methods, Vol. 37), Kluwer Academic Press, Boston, 1998.
22. Alan N. Steinberg and Robert B. Washburn, Multi-level fusion for War Breaker intelligence correlation, *Proc. 8th NSSF*, 1995, 137–156.
23. K. Kastella, Joint multitarget probabilities for detection and tracking, *SPIE* 3086, 1997, 122–128.
24. L. D. Stone, C. A. Barlow, and T. L. Corwin, *Bayesian Multiple Target Tracking*, Artech House Inc., Boston, 1999.
25. Y. Bar-Shalom and T.E. Fortman, *Tracking and Data Association*, Academic Press, San Diego, 1988.
26. Ronald Mahler, The random set approach to data fusion, *Proc. SPIE*, 2234, 1994.
27. Bowman, C. L., Possibilistic verses probabilistic trade-off for data association, *Proc. SPIE*, 1954, April 1993.
28. Alan N. Steinberg, Christopher L. Bowman, and Franklin E., White, Revisions to the JDL Data Fusion Model, *Proc. Third NATO/IRIS Conf.*, 1998.
29. Judea Pearl, *Probabilistic Reasoning in Intelligent Systems*, Morgan Kaufman, 1988.

17

Studies and Analyses within Project Correlation: An In-Depth Assessment of Correlation Problems and Solution Techniques[*]

James Llinas
State University of New York

Capt. Lori McConnell
USAF/Space Warfare Center

Christopher L. Bowman
Consultant

David L. Hall
The Pennsylvania State University

Paul Applegate
Consultant

17.1 Introduction **17-1**
    Background and Perspectives on This Study Effort
17.2 A Description of the Data Correlation (DC) Problem **17-3**
17.3 Hypothesis Generation **17-4**
    Characteristics of Hypothesis Generation Problem Space •
    Solution Techniques for Hypothesis Generation • HG Problem
    Space to Solution Space Map
17.4 Hypothesis Evaluation **17-8**
    Characterization of the HE Problem Space • Mapping of the
    HE Problem Space to HE Solution Techniques
17.5 Hypothesis Selection **17-9**
    The Assignment Problem • Comparisons of Hypothesis
    Selection Techniques • Engineering an HS Solution
17.6 Summary **17-17**
References **17-18**

17.1 Introduction

The "correlation" problem is one in which both measurements from multiple sensors and additional inputs from multiple nonsensor sources must be optimally allocated to estimation processes that produce (through data/information fusion techniques) fused parameter estimates associated with hypothetical

---

[*]This chapter is based on a paper by James Llinas et al., Studies and analyses within project correlation: an in-depth assessment of correlation problems and solution techniques, in *Proceedings of the 9th National Symposium on Sensor Fusion*, March 12–14, 1996, pp. 171–188.

targets and events of interest. In the most general sense, this problem is one of combinatorial optimization, and the solution strategies involve application and extension of existent methods of this type.

This chapter describes a study effort, "Project CORRELATION," which involved stepping back from the many application-specific and system-specific solutions and the extensively described theoretical approaches to conduct an assessment and develop guidelines for moving from "problem space" to "solution space." In other words, the project's purpose was to gain some understanding of the engineering design approaches for solution development and assess the scaleability and reusability of solution methods according to the nature of the problem.

Project CORRELATION was a project within the U.S. Air Force Tactical Exploitation of National Capabilities (AFTENCAP) program. The charter of AFTENCAP was to "exploit all space and national system capabilities for warfighter support." It was not surprising therefore that the issue of how to cost-effectively correlate such multiple sources of data/information is of considerable importance. Another AFTENCAP charter tenet was to "influence new national system design and operations"; it was in the context of this tenet that Project CORRELATION sought to obtain the generic/reusable engineering guidelines for effective correlation problem solution.

17.1.1 Background and Perspectives on This Study Effort

The functions and processes of correlation are part of the functions and processes of data fusion. (See Waltz and Llinas, 1990, and Hall, 1992, for reviews of data fusion concepts and mathematics.[1,2]) As a component of data fusion processing, correlation suffers from some of the same problems as other parts of the overall data fusion process (which has been maturing for approximately 20 years): a lack of an adequate, scientifically based foundation of knowledge to serve as the basis for engineering guidelines with which to approach and effectively solve problems. In part, this lack of this knowledge is the result of relatively few comparative studies that assess and contrast multiple solution methodologies on an equitable basis. A search for modern literature on correlation and related subjects, for example, revealed a small number of such comparative studies and many singular efforts for specialized algorithms. In part, the goal of the effort described in this chapter was to attempt to overlay or map onto these prior works an equitable basis for comparing and assessing the problem spaces in which these (individually described) algorithms work reasonably well. The lack of an adequate literature base of quantitative comparative studies forced such judgments to become subjective, at least to some degree. As a result, an experienced team was assembled to cooperatively form these judgments in the most objective way possible; none of the evaluators has a stake in, or has been in the business of, correlation algorithm development. Moreover, as an augmentation of this overall study, peer reviews of the findings were conducted via a conference and open session in January 1996 and a workshop and presentation at the National Symposium on Sensor Fusion in April 1996.

Others have attempted such characterizations, at least to some degree. For example, Blackman describes the Tracker-Correlator problem space with two parameters: sampling interval and intertarget spacing.[3] This example is, as Blackman remarks, "simplified but instructive." Figure 17.1, from Blackman, shows three regions in this space:

- The upper region of "unambiguous correlation" — characterized by widely spaced targets and sufficiently short sampling intervals.
- An "unstable region" — in which targets are relatively close (in relation to sensor resolution) and miscorrelation occurs regardless of sampling rate.
- A region where miscorrelation occurs without track loss — consisting of very closely spaced targets and where miscorrelation occurs, however, measurements are assigned to some track, resulting in no track loss but degradations in accuracy.

As noted in Figure 17.1, the boundaries separating these regions are a function of the two parameters and are also affected by other aspects of the processing. For example, detection probability (Pd) is known to strongly affect correlation performance, so that alterations in Pd can alter the shape of these regions. For the unstable region, Blackman cites some related studies that show that this region may occur for

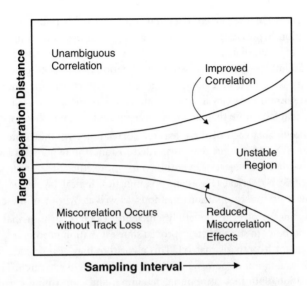

FIGURE 17.1 Interpretation of MTT correlation in a closely spaced target environment.

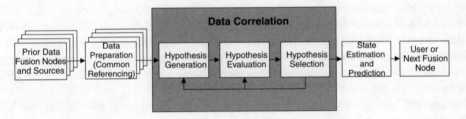

FIGURE 17.2 Data fusion tree node.

target angular separations of about two to five times the angular measurement standard deviation. Other studies quoted by Blackman show that unambiguous tracking occurs for target separations of about five times the measurement error standard deviation. These boundaries are also affected by the specifics of correlation algorithms, all of which have several components.

17.2 A Description of the Data Correlation (DC) Problem

One way to effectively architect a data fusion process is to visualize the process as a tree-type structure with each node of the fusion tree process having a configuration such as that shown in Figure 17.2. The partitioning strategy for such a data fusion tree is beyond the scope of this chapter and is discussed by Bowman.[4]

The processing in each data fusion node is partitioned into three distinguishable tasks:

- Data preparation (common referencing) — time and coordinate conversion of source data, and the correction for known misalignments and data conflicts;
- Data correlation (association) — associates data to "objects";
- State estimation and prediction — updates and predicts the "object" state (e.g., kinematics, attributes, and ID).

This study focused on the processes in the shaded region in Figure 17.2 labeled "Data Correlation." Note data correlation is segmented into three parts, each involved with the association hypotheses that cluster reports together to relate them to the "objects":

- In *hypothesis generation*, the current data and prior selected hypotheses are used to generate the current correlation hypotheses via feasibility gating, pruning, combining, clustering, and object aggregation. That is, alternate hypotheses are defined which represent feasible associations of the input data, including, for example, existing information (e.g., tracks, previous reports, or *a priori* data). Feasibility is defined in a manner that effectively reduces the number of candidates for evaluation and selection (e.g., by a region centered on a time normalized track hypothesis where measurements that fall within that region are accepted as being possibly associated with that track).

- In *hypothesis evaluation*, each of these feasible correlation hypotheses are evaluated using kinematic, attribute, ID, and *a priori* data as needed to rank the hypotheses (with a score reflecting "closeness" to a candidate object or hypothesis) for more efficient selection searching.[5] Evaluation techniques include numerical (Bayesian, possibilistic), logical (symbolic, nonmonotonic), and neural (nonlinear pattern recognition) methods, as well as hybrids of these methods.[6]

- *Hypothesis selection* involves a (hopefully efficient) search algorithm that selects one or more association hypotheses to support an improved state estimation process (e.g., to resolve "overlaps" in the measurement/hypothesis matrix). This algorithm may also provide feedback to aid in the generation and evaluation of new hypotheses to initiate the next search. The selection functions include elimination, splitting, combining, retaining, and confirming correlation hypotheses in order to maintain tracks and/or aggregated objects.

Most simply put, hypothesis generation nominates a set of hypotheses to which observations (based on domain/problem insight) can be associated. The hypothesis evaluation step develops and computes a metric, which reflects the degree to which any observation is associable (accounting for various errors and other domain effects) to that hypothesis. In spite of the use of such metrics, ambiguities can remain in deciding on how best to allocate the observations. As a result, a hypothesis selection function (typically an "assignment" problem solution algorithm) is used to achieve an optimal or near-optimal allocation of the observations (e.g., maximum likelihood based).

Note that the "objects" discussed here are, first of all, hypothetical objects — on the basis of some signal threshold being exceeded, an object, or perhaps more correctly, some "causal factor" is believed to have produced the signal. Typically, the notion of an object's existence is instantiated by starting, in software, an estimation algorithm that attempts to compute (and predict) parameters of interest regarding the hypothetical "object." In the end, the goal is to correlate the "best" (according to some optimization criteria) ensemble of measurements from multiple input sources to the estimation processes, so that, by using this larger quantity of information, improved estimates result.

17.3 Hypothesis Generation

17.3.1 Characteristics of Hypothesis Generation Problem Space

The characterization of the hypothesis generation (HG) problem involves many of the same issues related to hypothesis evaluation (HE) discussed in the next section. These issues include the nature of the input data available, knowledge of target characteristics and behavior, target density, characteristics of the sensors and our knowledge about their performance, the processing time frame, and characteristics of the mission. These are summarized in Table 17.1.

17.3.2 Solution Techniques for Hypothesis Generation

The approach to developing a solution for hypothesis generation can be separated into two aspects: (1) hypothesis enumeration, and (2) identification of feasible hypotheses. A summary of HG solution techniques is provided in Table 17.2. Hypothesis enumeration involves developing a global set of possible hypotheses based on physical, statistical, or explicit knowledge about the observed environment. Hypothesis

TABLE 17.1 Aspects of the Hypothesis Generation Problem Space

Problem Characteristics	Impact on HG	Comments
Input data available	Characteristics of the input data (e.g., availability of locational information, observed entity attributes, entity identity information, etc.) affect factors that can be used to distinguish among entities and, hence, define alternate hypotheses.	Reliability and availability of the input data impacts the hypothesis generation function.
Knowledge of target characteristics/ behavior	The extents to which the target characteristics are known to affect HG. In particular, if the target's kinematic behavior is predictable, then the predicted positions of the target may be used to establish kinematic gates for eliminating unlikely observation-entity pairings. Similarly, identity and attribute data can be used, if known, to reduce the combinatorics.	The definition of what constitutes a target (or entity), clearly affects the HG problem. Definition of complex targets, such as a military unit (e.g., a SAM entity), may entail observation of target components (e.g., an emitter) that must be linked hierarchically to the defined complex entity. Hence, hierarchical syntactic reasoning may be needed to generate a potential hypothesis.
Target density	The target density (i.e., intertarget spacings) relative to sensor accuracy affects the level of ambiguity about potential observation-entity assignments.	If targets are widely spaced relative to sensor accuracy, identifying multiple hypotheses may not be necessary.
Sensor knowledge/ characteristics	The characteristics and knowledge of the sensor characteristics affect HG. Knowledge of sensor uncertainty may improve the predictability of the observation residuals (i.e., the difference between a predicted observation and actual observation, based on the hypothesis that a particular known object is the "cause" of an observation). The number and type of sensors affect viability of HG approaches.	The more accurately the sensor characteristics are known, the more accurately feasible hypotheses can be identified.
Processing time frame	The time available for hypothesis generation and evaluation affects HG. If the data can be processed in a batch mode (i.e., after all data are available), then an exhaustive technique can be used for HG/HE. Alternatively, hypotheses can be generated after sets of data are available. In extremely time-constrained situations, HG may be based on sequential evaluation of individual hypotheses.	The processing timeframe also affects the allowable sophistication of HG processing (e.g., multiple vs. single hypotheses) and the complexity of the HE metric (i.e., probability models, etc.).
Mission characteristics	Mission requirements and constraints affect HG. Factors such as the effect (or penalties) for miscorrelations and required tracking accuracy affect which techniques may be used for HG.	Algorithm requirements for HG (and for any other data fusion technique) are driven and motivated by mission constraints and characteristics.

feasibility assessment provides an initial screening of the total possible hypotheses to define the feasible hypotheses of subsequent processing (i.e., for HE and hypothesis selection or HS processing). These functions are described in the following two sections.

17.3.2.1 Hypothesis Enumeration

1. *Physical models* — Physical models can be used to assist in the definition of potential hypotheses. Examples include intervisibility models to determine the possibility of a given sensor observing a specified target (with specified target-sensor interrelationships, environmental conditions, etc.), models for target motion (i.e., to "move" a target from one location in time to the time of a received observation via dynamic equations of motion, terrain models, etc.), and models of predicted target signature for specific types of sensors.

2. *Syntactic models* — Models can be developed to describe how targets or complex entities are constructed. This is analogous to the manner in which syntactic rules are used to describe how sentences can be correctly constructed in English. Syntactical models may be developed to identify

TABLE 17.2   Solution Techniques for Hypothesis Generation

Processing Function	Solution Techniques	Description	References
Hypothesis enumeration	Physical models	Models of sensor performance, signal propagation, target motion, intervisibility, etc., to identify possible hypotheses	2
	Syntax-based models	Use of syntactical representations to describe the make-up (component entities, interrelationships, etc.) of complex entities such as military units	2
	Doctrine-based models	Definition of tactical scenarios, enemy doctrine, anticipated targets, sensors, engagements, etc. to identify possible hypotheses	1
	Probabilistic models	Probabilistic models of track initiation, track length, birth/death probabilities, etc. to describe possible hypotheses	3, 7
	Ad hoc models	Ad hoc descriptions of possible hypotheses to explain available data; may be based on exhaustive enumeration of hypotheses (e.g., in a batch processing approach)	
Identification of feasible hypotheses	Pattern recognition	Use of pattern recognition techniques, such as cluster analysis, neural networks, or gestalt methods to identify "natural" groupings in input data	8, 9
	Gating techniques	Use of *a priori* parametric boundaries to identify feasible observation pairings and eliminate unlikely pairs; techniques include kinematic gating, probabilistic gating, and parametric range gates	10, 11
	Logical templating	Use of prespecified logical conditions, temporal conditions, causal relations, entity aggregations, etc. for feasible hypotheses	12, 13
	Knowledge-based methods	Establishment of explicit knowledge via rules, scripts, frames, fuzzy relationships, Bayesian belief networks, and neural networks	14, 15

the necessary components (e.g., emitters, platforms, sensors, etc.) that comprise more complex entities, such as a surface-to-air (SAM) missile battalion.

3. *Doctrine-based scenarios* — Models or scenarios can also be developed to describe anticipated conditions and actions for a tactical battlefield or battle space. Thus, anticipated targets, emitters, relationships among entities, entity behavior (e.g., target motion, emitter operating anodes, sequences of actions, etc.), engagement scenarios, and other information can be represented.

4. *Probabilistic models* — Probabilistic models can be developed to describe possible hypotheses. These models can be developed based on a number of factors, such as *a priori* probability of the existence of a target or entity, expected number of false correlations or false alarms, and the probability of a track having a specified length.

5. *Ad hoc models* — In the absence of other knowledge, ad hoc methods may be used to enumerate potential hypotheses, including (if all else fails) an exhaustive enumeration of all possible report-to-report and report-to-track pairs.

The result of hypothesis enumeration is the definition or identification of possible hypotheses for subsequent processing. This step is key to all subsequent processing. Failure to identify realistic possible causes (or "interpretations") for received data (e.g., such as countermeasures and signal propagation phenomena) cannot be recovered in subsequent processing (i.e., the subsequent processing is aimed at reducing the number of hypotheses and ultimately selecting the most likely or feasible hypotheses from the superset produced at this step), at least in a deductive, or model-based approach. It may be possible, in association processes involving learning-based methods, to adaptively create association hypotheses in real time.

TABLE 17.3 Mapping Between HG Problem Space and Solution Space

	Solution Space								
	Hypothesis Enumeration					I.D. of Feasible Hypothesis			
Problem Space	PM	SM	DM	PrM	Ad Hoc	PR	GT	LT	KB
Input Data Categories									
I.D. attributes	N	Y		Y	Y	Y	Y		Y
Kinematic data	Y			Y	Y	Y	Y		Y
Parameter attributes	Y	Y		Y	Y	Y	Y		Y
A priori sensor/scenario			Y	Y	Y				Y
Linguistic data		Y	Y	N	Y				Y
Space-time patterns		Y	Y	Y	Y	Y			Y
High uncertainty			Y	Y	Y		N		Y
Unknown structures			N	N	Y	Y	N		Y
Error PDF	Y			Y		Y	Y		Y
Target characteristics	Y	Y	Y	Y			Y		Y
Signal propagation models	Y			Y					Y
Output Data									
Report-to-report	Y	N	Y	Y	Y	Y	Y	N	N
Report-to-track	Y	N	Y	Y	Y	Y	Y	Y	Y
Track-to-track	Y		Y	Y	Y	Y	Y	Y	Y
Spatio-temporal	Y		Y	Y	Y	Y	Y	N	N
Multi-spectral	Y		Y	Y	Y	Y	Y	N	N
Cross-level	Y	Y	Y	Y	Y	N	Y	Y	Y
Multisite sources	Y		Y	Y	Y		Y	Y	Y
Multiscenes	Y		Y	Y	Y		Y	Y	Y
2-D set partitioning	Y		Y	Y	Y		Y	Y	Y
N-D set partitioning	Y		Y	Y	Y		Y	Y	Y
Requirements/Constraints									
Single scan (N=1)	Y	N	N	Y	Y	Y	Y	N	N
Multiple scans (N=n)	Y	Y	Y	Y	Y	Y	Y	Y	Y
Batch (N=~)	Y	Y	Y	Y	Y	Y	Y	Y	Y
Limited processing	Y	Y	N	Y	Y	Y	Y	N	N
Short decision time	Y	Y	N	Y	Y	Y	Y	N	N

17.3.2.2 Identification of Feasible Hypotheses

The second function required for hypothesis generation involves reducing the set of possible hypotheses to a set of feasible hypotheses. This involves eliminating unlikely report-to-report or report-to-track pairs (hypotheses) based on physical, statistical, explicit knowledge, or ad hoc factors. The challenge is to reduce the number of possible hypotheses to a limited set of feasible hypotheses, without eliminating any viable alternatives that may be useful in subsequent HE and HS processing. A number of automated techniques are used for performing this initial "pruning." These are listed in Table 17.2; space limitations prevent further elaboration.

17.3.3 HG Problem Space to Solution Space Map

A mapping between the hypothesis generation problem space and solution space is summarized in Table 17.3. The matrix shows a relationship between characteristics of the hypothesis generation problem (e.g., input data and output data) and the classes of solutions. Note that this matrix is not especially prescriptive in the sense of allowing a clear selection of solution techniques based on the character of the HG problem. Instead, an engineering design process,[16] must be used to select the specific HG approach applicable to a given problem.

17.4 Hypothesis Evaluation

17.4.1 Characterization of the HE Problem Space

The HE problem space is described for each batch of data (i.e., fusion node) by the characteristics of the data inputs, the type of score outputs, and the measures of desired performance. The selection of HE techniques is based on these characteristics. This section gives a further description of each element of the HE problem space.

17.4.1.1 Input Data Characteristics

The inputs to HE are the feasible associations with pointers to the corresponding input data parameters. The input data are categorized according to the available data type, level of its certainty, and commonality with the other data being associated, as shown in Table 17.4. Input data includes both recently sensed data and *a priori* source data. All data types have a measure of certainty, albeit possibly highly ambiguous, corresponding to each data type.

17.4.1.2 Output Data Characteristics

The characteristics of the HE outputs needed by hypothesis selection, HS, also drive the selection of the HE techniques. Table 17.5 describes these HE output categories, which are partitioned according to the output variable type: logical, integer, real, *N*-dimensional, functional, or none. Most HE outputs are real valued scores reflecting the confidence in the hypothesized association. However, some output a discrete confidence level (e.g., low, medium, or high), while others output multiple scores (e.g., one per data category) or scores with higher order statistics (e.g., fuzzy, evidential, or random set). For some batches of data, no HE scoring is performed, and only a yes/no decision on the feasibles is output for HS. "No explicit association" refers to those rare cases where the data association function is not performed (i.e.,

TABLE 17.4 Input Data Characteristics

Input Data Categories	Description	Examples of Inputs
Identity (ID)	Discrete/integer valued	IFF, class, type of platform/emitter
Kinematics	Continuous-geographical	Position, velocity, angle, range
Attributes/features	Continuous-non-geographical	RF, PRI, PW, size, intensity, signature
A priori sensor/scenario data	Association hypothesis stats	P_D, P_{FA}, birth/death statistics, coverage
Linguistic	Syntactic/semantic	Language, HUMINT, message content
Object aggregation in space-time	Scripts/frames/rules	Observable sequences, object aggregations
High uncertainty-in-uncertainty	Possibilistic (fuzzy, evidential)	Free text, *a priori* and measured object ID
Unknown structure/patterns	Analog/discrete signals	Pixel intensities, RF signatures
Partially known error statistics	Nonparametric data	P(R|H) only partially known
Partial and conflicting data	Missing, incompatible, incorrect	Wildpoints, closed vs. open world, stale
Differing dimensions	Multi-dim discrete/continuous	3-D and 2-D evaluated against N-D track
Differing resolutions	Coarseness of discernability	Sensor resolution differences: radar + IR
Differing data types	Hybrid types and uncertainties	Probabilistic, possibilistic, symbolic

TABLE 17.5 Output Data Characteristics

Score Output Categories	Description	Examples of Outputs
Yes/no association	0/1 logic (no scoring)	High confidence only association
Discrete association levels	Integer score	Low/medium/high confidence levels
Numerical association score	Continuous real-valued score	Association probability/confidence
Multi-scores per association	N-D (integer or real per dim)	Separate score for each data group
Confidence function with score	Score uncertainty functional	Fuzzy membership or density function
No explicit association	State estimates directly on data	No association representation

the data is only implicitly associated in performing state estimation directly on all of the data). An example in image processing is the estimation of object centroid or other features, based on intensity patterns without first clustering the pixel intensity data.

17.4.2   Mapping of the HE Problem Space to HE Solution Techniques

This section describes the types of problems for which the solution techniques are most applicable (i.e., mapping problem space to solution space). A preliminary mapping of this type is shown in Table 17.6; final guidelines were developed by Llinas.[16] The ad hoc techniques are used when the problem is easy (i.e., performance requirements are easy to meet) or the input data errors are ill-defined. Probabilistic techniques are selected according to the error statistics of the input data. Namely, maximum likelihood (ML) techniques are applied when there is no useful *a priori* data; otherwise max *a posteriori* (MAP) are considered. Chi-square (CHI) techniques are applied for data with Gaussian statistics (e.g., without useful ID data), especially when there is data of differing dimensions where ML and MAP would have to use expected values to maintain constant dimensionality. Neyman-Pearson techniques are statistically powerful and are used as the basis for nonparametric techniques (e.g., sign test and Wilcoxon test). Conditional event algebra (CAE) techniques are useful when the input data is given, conditioned on different events (e.g., linguistic data). Rough sets (Rgh) are used to combine/score data of differing resolutions. Information/entropy (Inf) techniques are used to select the density functions and score data whose error statistics are not known. Further discussion of the various implications of problem-to-solution mappings is provided by Llinas.[16]

17.5   Hypothesis Selection

When the initial clustering, gating, distance/closeness metric selection, and fundamental approach to hypothesis evaluation have been completed, the overall correlation process has reached a point where the "most feasible" set of both multisensor measurements and multisource inputs exist. The inputs have been "filtered," in essence, by the preprocessing operations and the remaining inputs will be allocated or "assigned" to the appropriate estimation processes that can exploit them for improved computation and prediction of the states of interest. This process is hypothesis selection, in which the hypothesis set comprises all of the possible/feasible assignment "patterns" (set permutations) of the inputs to the estimation processes; thus, any single hypothesis is one of the set of feasible assignment patterns. This chapter focuses on position and identity estimation from such assigned inputs as the states of interest. However, the hypothesis generation-evaluation-selection process is also relevant to the estimation processes at higher levels of abstraction (e.g., wherein the states are "situational states" or "threat states"), and the state estimation processes, unlike the highly numeric methods used for Level 1 estimates, are reasoning processes embodied in symbolic computer-based operations.

So, what exists as *input to the hypothesis selection* process in effect is, at this point, a matrix (or matrices) where the typical dimensions are the indexed input data/information/measurement set on one hand, and the indexed state estimation systems or processes, along with the allowed ambiguity states, on the other hand (i.e., the "other" states or conditions, beyond those state estimates being maintained, to which the inputs may be assigned). Simply put, for the problems of interest described here, the two dimensions are the indexed measurements and the indexed position or identity state estimation processes. (Note, however, as discussed later in this chapter, that assignment problems can involve more than two dimensions.)

In any case, the matrix/matrices are populated with the closeness measures, which could be considered "costs" of assigning any single measurement to any single estimator (resulting from the HE solution). Despite all of the effort devoted to optimizing the HG and HE solutions, considerable ambiguity (many feasible hypotheses) can still result. The costs in these matrices may directly be the values of the "distance" or scoring metrics selected for a particular approach to correlation, or a newly-developed cost function specifically defined for the hypothesis selection step. The usual strategy for defining the optimal assignment

TABLE 17.6 Mapping HE Problem Space to HE Solution Techniques

Solution Space Problem Space	Ad Hoc	Probabilistic							Possibilistic			Logic Symbolic					Neural		Hybrid
	Hoc	ML	MAP	NP	CHI	CAE	RGH	INF	DS	Fuzzy	S/F	NM	ES	C-B	PR	PD	HC	Super	RS
Input Data Categories																			
Identity (ID)	Y	Y	Y		N														Y
Kinematics	Y	Y	Y		Y														Y
Attributes/features	Y	Y	Y		Y														Y
A priori sensor data	N	N	Y		N														Y
Linguistic						Y				Y	Y	Y							
Object aggregation									Y		Y								
High uncertainty									Y	Y									
Unknown structure								Y									N		
Nonparametric data	Y		Y					Y										Y	
Partial data					Y										Y				
Differing dimension																Y			
Differing resolution							Y												
Differing data types																			Y
Score Output Categories																			
Yes/no association	Y																		
Discrete scores	Y																		
Numerical scores		Y	Y	Y	Y	Y	Y												
Multi-scores per								Y											
Confidence functional									Y										
No explicit scores										Y							Y	Y	
Performance Measures																			
Low cost software	Y	Y	Y	Y	Y	N	N	N	N	N				N	N	N	N	Y	N
Compute efficiency	N	Y	Y	N	Y	Y											Y	Y	
Score accuracy		Y	Y	N	Y	Y											N	N	
User adaptability											Y		Y	Y	Y		N	Y	
Self-trainable														Y			N	Y	
Self-coding																	N	Y	
Robustness to error											Y	Y	Y	Y	Y	Y	N	Y	
Result explanation											Y	Y	Y	Y	Y	Y	N	N	
Solutions		ML	MAP	NP	CHI	CAE	RGH	INF	DS	Fuzzy	S/F	NM	ES	C-B	PR	PD	HC	Super	RS

ML — Max. likelihood   MAP — Max. *a priori*   NP — Neyman-Pearson   CHI — Chi-square   CAE — Conditional algebra event   RGH — Rough sets

INF — Information/entropy   DS — Dempster-Shafer   Fuzzy — Fuzzy set   S/F — Scripts/frames/rules   NM — Nonmonotonic   ES — Expert systems

C-B — Case-based   PR — Partitioned representations   PD — Power domains   HC — Hard-coded   Super — Supervised   RS — Random set

(i.e., selecting the optimal hypothesis) is to find that hypothesis with the lowest total cost of assignment. Recall, however, that there are generally two conditions wherein such matrices develop: (1) when the input systems (e.g., sensors) are initiated (turned on) and (2) when the dynamic state estimation processes of interest is being maintained in a recursive or iterative mode.

As noted above, these assignment or association matrices, despite the careful preprocessing of the HG and HE steps, may still involve ambiguities in how to best assign the inputs to the state estimators. That is, the cost of assigning any given input to any of a few or several estimators may be reasonable or allowable within the definition of the cost function and its associated thresholds of acceptable costs. If this condition exists across many of the inputs, identifying the total-lowest-cost assignments of the inputs becomes a complex problem. The *central problem* to be solved in hypothesis selection is that of defining a way to select the hypothesis with minimum total cost from all feasible/permissible hypotheses for any given case; often, this involves large combinations of possibilities and leads to a problem in *combinatorial optimization*. In particular, this problem — called the *assignment problem* in the domain of combinatorial optimization — is applicable to many cases other than the measurement assignment problem presented in this chapter and has been well studied by the mathematical and operations research community, as well as by the data fusion community.

17.5.1 The Assignment Problem

The goal of the assignment problem is to obtain an optimal way of assigning various available N resources (in this case, typically measurements) to various N or M ($M<> = N$) "processes" that require them (in our case estimation processes, typically). Each such feasible assignment of the $N \times N$ problem (a permutation of the set N) has a cost associated with it, and the usual notion of optimality equates to minimum cost as mentioned above. Although special cases allow for multiassignment (in which resources are shared), for many problems, a typical constraint allows only one-to-one assignments; these problems are sometimes called "bipartite matching" problems.

Solutions for the typical and special variations of these problems are provided by mathematical programming and optimization techniques. (Historically, some of the earliest applications were to multiworker/multitask problems and many operations research and mathematical programming texts motivate assignment problem discussions in this context.) This problem is also characterized in the related literature according to the nature of the mathematical programming or optimization techniques used as *solutions* applied for each special case. Not surprisingly, because the underlying problem model has broad applicability to many specific and real problems, the literature describes certain variations of the assignment problem and its solution in different (and sometimes confusing) ways. For example, assignment-type problems also arise in analyzing *flows in networks*. Ahuja et al., in describing network flows, divide their discussion into six topic areas:[17]

1. Applications
2. Basic properties of network flows
3. Shortest path problems
4. Maximum flow problems
5. Minimum cost flow problems
6. Assignment problems

In their presentation, they characterize the assignment problem as a "minimum cost flow problem on a network."[17] This characterization, however, is exactly the same as asserted in other applications. In network parlance, however, the assignment problem is now called a "variant of the shortest path problem" (which involves determining directed paths of smallest cost from any node X to all other nodes). Thus, "successive shortest path" algorithms solve the assignment problem as a sequence of N shortest path problems (where N = number of resources = number of processes in the ("square") assignment problems).[17] In essence, this is the bipartite matching problem re-stated in a different way.

17.5.2   Comparisons of Hypothesis Selection Techniques

Many technical, mathematical aspects comprise the assignment problem that, given space limitations, are not described in this chapter. For example, there is the crucial issue of solution complexity in the formal sense (i.e., in the sense of "big O" analyses), and there is the dilemma of choosing between multidimensional solutions and two dimensional solutions and all that is involved in such choices; in addition, many other topics remain for the process designer. The solution space of assignment problems at Level 1 can be thought of as comprising: (a) *linear and nonlinear mathematical programming* (with some emphasis on integer programming), (b) *dynamic programming and branch-and-bound methods* as part of the family of methods employing implicit enumeration strategies, and (c) *approximations and heuristics*. Although this generalization is reasonable, note that the assignment problems of the type experienced for Level 1 data fusion problems arise in many different application areas; as a result, many specific solution types have been developed over the years, making broad generalizations difficult.

Table 17.7 summarizes the conventional methods used for the most frequently structured versions of assignment problems for data fusion Level 1 (i.e., deterministic, 2-D, set-partitioned problem formulations). Furthermore, these are solutions for the linear case (i.e., linear objective or cost functions) and linear constraints. Without doubt, this is the most frequently discussed case in the literature and the solutions and characteristics cited in Table 17.7 represent a reasonable benchmark in the sense of applied solutions (but not in the sense of improved optimality; see the ND solution descriptions below).

TABLE 17.7   Frequently Cited Level 1 Assignment Algorithms
(Generally: Deterministic, 2-D, Set Partitioned)

Algorithms	Applicability (Problem Space)	Processing Characteristics	Runtime Performance*
Hungarian[22]	Square matrices; optimal pair-wise algorithm	Primal-dual; steepest descent	O(nS(n,m,C)); or O([# trks + # msmts]**3)
Munkres[23]	Square matrices; optimal pair-wise algorithm		
Bourgeois-Lassalle;[24] see also References 25 and 26	Rectangular matrices; optimal pair-wise algorithm	B-L faster than squaring-off method of Kaufmann	
Stephans-Krupa[26,27]	Sparse matrices; optimal pair-wise algorithm		
JVC[28,29]	Sparse matrices; optimal pair-wise algorithm	Appears to be the fastest of the traditional methods; S-K second fastest to NC; sparse Munkres third; JV is augmenting cycle approach	
Auction types[32]	Optimal pair-wise algorithm	Primal-dual, coordinate descent; among the fastest algorithms; parallelizable versions developed; appears much faster than N algorithm (as does JVC)	O(n**2mC); scaled version O(nmlognC); others show O(n** 1/2mlogC)
Primal simplex/ alternating basis[30]	Applied to relatively large matrices (1000–4500 nodes); optimal pair-wise algorithm	Moderate speed for large problems	
Signature methods[31]	Optional pair-wise algorithm	Dual simplex approach	O(n**3)

* O = worst case, n = # of nodes, m = # of arcs, C = upper bound on costs, S = successive shortest path solution time for given parameters.

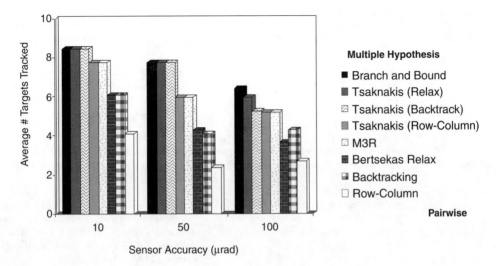

FIGURE 17.3 Functional comparison of correlation algorithms.

Note that certain formulations of the assignment problem can lead to nonlinear formulations, such as the quadratic assignment. Additionally, the problem can be formulated as a "multiassignment" problem as for a set-covering approach, although these structures usually result in linear multiassignment formulations if the cost/objective function can be treated as separable. Otherwise, it, too, will typically be a nonlinear, quadratic-type formulation.[18] No explicitly related nonlinear formulations for Level 1 type applications were observed in the literature; however, Finke et al. is a useful source for solutions to the quadratic problem.[19] The key work in multiassignment structures for relevant fusion type problems is Tsaknakis, et al., who form a solution analogous to the auction-type approach.[18]

This chapter introduces additional material focused on comparative assessments of assignment algorithms. These materials were drawn, in part, from works completed during the Strategic Defense Initiative, or SDI ("Star Wars"), era and reported at the "SDI Tracking Panels" proceedings assembled by the Institute for Defense Analysis, and, in part, from a variety of other sources. Generally, the material dates to about 1990 and is, therefore, reasonably recent but could benefit from updating.

One of the better sources, in the sense of its focus on the same issues/problems of interest to this study, is Washburn.[20] In one segment of a larger report, Washburn surveys and summarizes various aspects of what he calls "Data Partitioning," "Gating," and "Pairwise"; "Multiple Hypothesis"; and "Specialized" Object Correlation algorithms. *Data partitioning and gating* equate to the terms *hypothesis generation and evaluation* used here, and the term *correlation* equates to the term *hypothesis selection* (input data set assignment) used here. Washburn's assessments of multiple hypothesis class of correlation algorithms he reviewed are presented in Table 17.8. Washburn repeats a critical remark related to comparisons of these algorithms: for those cases where a multiscan or multiple data set approach is required to achieve necessary performance levels (this was perhaps typical on SDI, especially for midcourse and terminal engagements), the (exact) optimal solution is unable to be computed, and so comparisons to the true solution are infeasible, and a "relaxed" approach to comparison and evaluation must be taken. This is, of course, the issue of developing an optimal solution to the ND case, which is an NP-hard problem. This raises again the key question of what to do if an exact solution is not feasible.

17.5.2.1 2-D vs. ND Performance

One of the only comparisons of 2-D vs. ND for a common problem that was captured in this study was a reference in the Washburn, 1992 report to Allen, et al. (1988) that examined comparative performance for a 10-target/3-passive sensor case.[20,21] These results are shown in Figure 17.3 from Washburn.[20] Essentially, "pairwise" — 2-D — solutions performed considerably worse than the 3-D approaches; branch-and-bound, RELAX, and backtrack all achieved average tracking of 8 of 10 targets — far below perfection.

TABLE 17.8   Multiple Hypothesis Object Correlation Algorithms Summary

Algorithm	Functional Performance	Processing Requirements
Multidimensional Row-Column (MRC)[33,34]	Worst functional performance of multiple hypothesis algorithms.	$O([n_s - 1] \cdot m_G)$ operations. Selection process is sequential, but nonsequential approaches could be applied for parallel processing.
Multidimensional Maximum Marginal Return[35,36]	Performs better than pairwise correlation and better than MRC. Worse than MST or backtracking M³R approach, but may have acceptable performance in sparse scenarios.	$O(n_s - 1] \cdot m_G \log m_G)$ operations. Parallel algorithm has been developed for binary tree processor.
Backtracking and Lookahead M³R [36]	Performed significantly better than M³R by using backtracking heuristic to improve solutions.	$O([n_s - 1] \cdot m_G \log m_G)$ operations. More complicated logic was found difficult to parallelize.
Monte Carlo M³R [36]	Uses randomization to change effect of branch ordering on M³R and improve solution. Annealing approach converges to optimal solution (as computed by Branch & Bound).	Has $O(n_{MC}[n_s - 1] \cdot m_G \log m_G)$ processing requirements where n_{MC} Monte Carlo iterations are required. This number may be very large and evaluations need to determine how small for adequate performance. Parallelizes completely over n_{MC} Monte Carlo iterations.
Minimal Spanning Tree (MST)[35] Tsaknakis	Performance depends on bipartite assignment used. With optimal, MST obtained best performance in Reference 35 evaluations against alternative correlation algorithms (including Branch & Bound) for small scenarios.	Processing depends on assignment algorithm used. Optimal Auction algorithm (which can be parallelized), processing is $O(n_s^2[n_T + n_M] \cdot m_G \log[n_T + n_M] \cdot C)$. Parallelizes over n_s^2 factor, as well as over bipartite assignment algorithm.
Vlterbl Correlation[34,37]	Special cost structure gives limited applicability or poor performance if approximations are used. No evaluations for SDI tracking problems.	Processing requirement is unacceptably large for large scenarios.
Branch & Bound[35,36,38,39,40]	Optimal multiple hypothesis algorithm. Obtained best performance in Reference 35 evaluations against alternative correlation algorithms with same scan depth n_s.	Uses various backtracking heuristic to speed up search, but potentially searches all possible solutions. Has very large requirements for even small dense scenarios. May be feasible if data partitioning produces small groups of data. Parallel algorithm developed in Reference 39.
Relaxation[41]	Iterative algorithm that generates feasible solution at each iteration and approaches optimal solution. Little evaluation in realistic scenarios to determine convergence rate.	Processing depends on bipartite assignment algorithm used in each iteration and on the number of iterations required. $O(n_R[n_s - 1] \cdot [n_T + n_M] \cdot m_G \log[n_T + n_M] \cdot C)$ operations using Auction with n_R relaxation iterations.

This effort involved examining several other comparison and survey studies. To summarize, the research found that, among other factors, performance is affected by angular accuracy-to-angular target separation, number of targets, and average target density in the HG process gates, coding language, degree of parallelization, and available reaction time.

17.5.3   Engineering an HS Solution

As mentioned at the outset, this project sought to develop a set of engineering guidelines to guide data fusion process engineers in correlation process design. Such guidelines have, in fact, been developed,[16] and an example for the HS process is provided in this section. In the general case, a fairly complex feasible hypothesis matrix exists. The decision tree guideline for engineering these cases is shown in Figure 17.4. Although the order of the questions or criteria shown in the figure is believed to be correct in the sense of maximal ordered partitioning of the problem/solution spaces, the decisions could possibly be determined by some other sequence. Because the approaches are so basically different and involve such different philosophies, the first question is whether the selected methodology is *deterministic* or *probabilistic*. If it is deterministic, then the engineering choices follow the *top* path, and vice versa. Some of the tradeoff

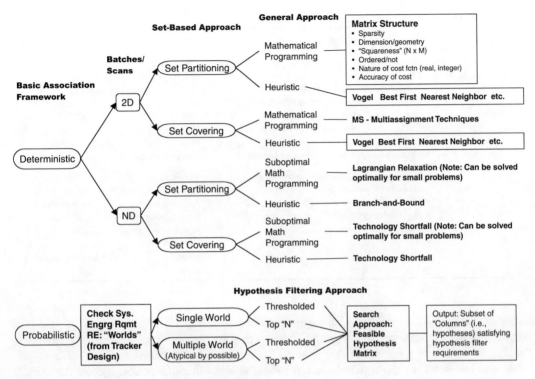

FIGURE 17.4 Representative HS decision flow diagram.

TABLE 17.9 Tradeoff 1

Design Decision	Positives	Negatives
Deterministic	Wide array of formal mathematical methods	Potentially incorrect specific assignments
Probabilistic	• Proven useful for few tracks plus high clutter • "Finesses" specific assignments by probabilistically using all measurements in a gate (an "all-neighbors" approach)	• Not widely applied to other than special cases as noted • Assignment processing not based on proven formalized methods; removes assignment from conventional correlation processing, embeds it in tracker • For many problems, not conceptually correct; measurements truly only come from one source in most (but not all) cases

factors affecting this choice are shown in Table 17.9, which concentrates on the deterministic family of methods.

17.5.3.1 Deterministic Approaches

If the approach is deterministic, the next crucial decision relates to the quantity of input data that will be dealt with at any one time. In the tracking world, this usually reflects the *number of scans* or *batches of data* that will be considered within a processing interval. As mentioned above, these issues reflect the more complex case of correlating and assigning *ensembles* of data at a time rather than single contacts at a time. In the typical cases where data ensembles are processed, they are frequently delineated by source type — often implying the type of sensor — or they are grouped by scan time. In a multisensor or multisource (or multitime) architecture, the, (near-globally optimal) "natural" formulation for assignment processing would be "*n*-dimensional," with *n* designating either the number of batch segments

TABLE 17.10 Tradeoff 2

Design Decision	Positives	Negatives
2D- (~Single Scan)	• Each 2-D solution is optimal for the data processed (i.e., for the given assignment matrix) • Most, if not all, traditional solutions exhibit cubic (in no. of inputs) polynomial-time, worst-case runtime behavior • Existing public-domain codes available; easier to code • Requires less processing power and memory	• Primarily that optimality is local in the sense of a single-scans worth of data, and that more globally optimal result comes from processing WEI data (ideal is batch solution with all data) • Not easy to retrospectively adjust for errors in processing; can require carrying alternative scenes
ND (~Multi-Scan)	• More globally optimal in the sense of multiple data sets considered • Methods exist to permit self-evaluation of closeness to optimality, and many show good (near-optimal) performance	• ND problem is NP-hard so solutions an exponential in worst-case runtime behavior • Apart from branch-and-bound methods or other explicit enumeration techniques, solutions are strictly suboptimal • More difficult to code; some codes nonpublic; more demanding of computing resources

from a single source, for example, or the specific-scan segments from each of n sources. In the strictest sense of attempting to achieve near-global optimality, these data should be processed in an ND approach. The ND solutions usually applied in practice, because they are computationally demanding, attempt to process the most data possible but fall well short of an all-data (i.e., true batch), truly optimal approach. The "window width" that such techniques can handle is yet another specific design choice to be made in the ND case. Because of the computational aspects, in part, there are applications where the data are segmented into source-specific sets, and a sequence of 2-D assignment solutions are applied to obtain a satisficing ("good enough," knee-of-the-curve performance), but strictly suboptimal result. The second choice is depicted in Table 17.10. Obviously this design decision separates the assignment solutions into quite different categories; the basic tradeoff is in the nature of the optimality sought versus the solution complexity and runtime behavior.

For either the 2-D or ND cases, the next decision has to do with whether the problem formulation is of the set-partitioning or set-covering type. This decision, as will be seen, also separates the potential solution types into considerably different categories, because these formulations are entirely different views of the problem. Recall that the set-partitioning scheme is one that divides the notional allocations of inputs into mutually exclusive sets with single inputs eventually being assigned, exclusively, to a particular state. The set-covering approach also divides the input data ensemble, but into nonexclusive sets such that single inputs can be assigned to more than one state. Only in the sense of permitting multiple assignments, does the set-covering construct have some similarity to the notion of probabilistic assignment. However, the general set-covering approach makes assignments deterministically, as opposed to probabilistically, and also takes a different approach to defining allocable sets. The probabilistic approach permits multiassignment as determined largely by the gating process defined by HG; all measurements in a gate are probabilistically assigned to the track, and to the degree that gates overlap, multiassignment of specific measurements exists. Generalized set-covering would ideally consider a number of factors on which to define its overlapping sets and is conceptually more robust than the probabilistic approach. So what is at issue here is one's "view of the world," in the sense of how the data should be grouped, which, in turn, affects the eventual structure of the assignment matrix or other construct that provides the input to the assignment problem and its solution.

As noted in Table 17.11 and shown in Figure 17.4, the set-partitioning view of the problem leads to a situation where there are many more known and researched solutions, so the fusion node designer will have wider solution choices if the problem is set up this way.

The next design choice is whether to employ mathematical programming-type solutions or some type of heuristic approach. The mathematical programming solution set includes linear programming (LP)

TABLE 17.11 Tradeoff 3

Design Decision	Positives	Negatives
Set-Partitioning	• For many, if not most, cases, this is the true view of the world in that singular data truly associate to only one causal factor or object (see Tradeoff 1, Deterministic). This assertion is conditioned on several factors, including sensor resolution and target spacing, but represents a frequent case. • Much larger set of researched and evaluated solution types	• Lacks flexibility in defining feasible data sets • Does not solve multispectral or differing-resolution or crossing-object-type problems
Set-Covering	More flexible in formulating feasible data groupings; often the true representation of real-world data sets	More complex formulation and solution approach

methods, integer and dynamic programming, and a wide variety of variations of these types. While mathematically elegant and formal, these methods can become computationally demanding for problems of a large scale, whether they are 2-D or ND. So, in many practical situations, an approximation-based or heuristic approach — which will develop a reasonably accurate solution and require considerably less computing resources — is preferable. Often heuristics can be embedded as subroutines within otherwise formal methods.

17.6 Summary

Data correlation can be viewed as a three-step process of hypothesis generation, hypothesis evaluation, and hypothesis selection. The first step limits the number of possible associations by permitting or "gating through" only those that are feasible. These potential associations are evaluated and quantitatively scored in the second step and assigned in the third.

Each of these stages can be performed in a variety of ways, depending on the specific "nature" of the data correlation problem. This "nature" has less to do with the specific military application than its required characteristics, such as processing efficiency and types of input data and output data, knowledge of how the characteristics relate to the statistics of the input data and the contents of the supporting database. The data fusion systems engineer needs to consider the system's user-supplied performance-level requirements, implementation restrictions (if any), and characteristics of the system data. In this way, the application can be considered as "encoded" into an engineering-level "problem space." This problem space is then mapped into a solution space via the guidelines of the type discussed in this chapter.

Beginning with hypothesis generation, the designer needs to consider how the input data gets batched (e.g., single-input, single-sensor scan, multiple-sensor scan, or seconds between update "frames") and what metric(s) to use. The goal should be a process that is less computationally intensive than the hypothesis evaluation step and coincides with available input data (position, attribute, ID), and the list of possible hypotheses data (e.g., "old track," "new track," "false alarm," "ECM/deception," or "anomalous propagation ghosting,"). Section 17.3 discusses these design factors and provides a preliminary mapping of gating techniques to problem space.

The next function, hypothesis evaluation, considers much of the same material as hypothesis generation; however, where the previous function evaluated candidate hypotheses on a pass/fail basis, this function must grade the goodness of unrejected hypotheses. Again, the type of input data must be considered together with its uncertainty characteristics. Probabilistic, possibilistic, and script-based techniques can be applied to generate scores that are, for example, real, integer, or discrete. These outputs are then used by the selection stage to perform the final assignment.

The selection stage accepts the scores to determine an "optimum" allocation of the gated-through subset of a batch of data to targets, other measurements, or other possible explanations for the data (e.g., false alarms). Optimum, in this case, means that the cost (value) of the set of associations (measurement-track, measurement-measurement), when taken as a whole, is minimized (maximized). The tools that handle ambiguous measurements are solutions to what is called the "assignment problem." Solutions to this problem can be computationally intensive; therefore, efficient solutions that cover both the 2-D and ND matrices of varying geometries are required. Section 17.5 provides a brief tutorial to the assignment problem and lists the numerical characteristics of popular techniques.

As mentioned previously in this chapter, the evolving guidelines and mappings presented in Sections 17.3, 17.4, and 17.5 were presented to an open audience at the government-sponsored Combat Information/Intelligence Symposium in January 1996 and at the (U.S.) National Symposium on Sensor Fusion in April 1996; however, further peer review is encouraged. The basis of such reviews should be the version of the Engineering Guidelines as presented in Llinas (1997).[16] Further, this chapter has treated these three individual processes as isolated parts, rather than part of a larger, integrated notion of correlation processing. Thus, additional work remains to be performed in order to assess the interprocess sensitivities in design choices, so that an effective, integrated solution result from the application of these guidelines.

References

1. Waltz, E. and Llinas, J., *Multi-sensor Data Fusion*, Artech House, Inc., Norwood, MA, 1990.
2. Hall, D.L., *Mathematical Techniques in Multisensor Data Fusion*, Artech House, Inc., Norwood, MA, 1992.
3. Blackman, S., *Multiple Target Tracking with Radar Applications*, Artech House, Inc., Norwood, MA, 1986.
4. Bowman, C.L., The data fusion tree paradigm and its dual, in *Proc. 7th Nat'l Fusion Symp.*, invited paper, Sandia Labs, NM, March 1994.
5. Bowman, C., Max likelihood track correlation for multisensor integration, in *18th IEEE Conference on Decision and Control*, December 1979.
6. Bowman, C.L., Possibilistic versus probabilistic trade-off for data association, in *Proc. Signal and Data Processing of Small Targets* (Session 1954), SPIE, April 1993.
7. Bar-Shalom, Y. and Tse, E., Tracking in a cluttered environment with probabilistic data association, *Automatica*, Vol. 2, 451–460, September 1975.
8. Aldenderfer, M.S. and Blashfield, R.K., Cluster analysis, in *Quantitative Applications in the Social Sciences*, Paper No. 07-044, London, UK, 1984.
9. Fukanaga, K., *Introduction to Statistical Pattern Recognition*, 2nd ed., Academic Press, New York, 1990.
10. Blackman, S., *Multitarget Tracking with Radar Applications*, Artech House, Norwood, CT, 1985.
11. Uhlmann, J.K. and Zuniga, M.R., Results of an efficient gating algorithm for large-scale tracking scenarios, *Naval Research Reviews*, 1:24–29, 1991.
12. Hall, D.L. and Linn, R.J., Comments on the use of templating for multisensor data fusion, in *Proc. 1989 Tri-Service Data Fusion Symp.*, Vol, 1, 345, May 1989.
13. Noble, D.F., Template-based data fusion for situation assessment, in *Proc. 1987 Tri-Service Data Fusion Symp.*, Vol, 1, 152, June 1987.
14. Jackson, P., *Introduction to Expert Systems*, Addison-Wesley, Reading, MA, 1986.
15. Pearl, J., *Probabilistic Reasoning in Intelligent Systems: Networks of Plausible Inference*, Morgan Kaufmann, San Mateo, CA, 1988.
16. Llinas, J. et al., Engineering Guidelines for Data Correlation Algorithm Characterization, Vol. 3 of Final Reports on Project Correlation, June 1996.
17. Ahuja, R.K. et al., Network flows, in *Handbook in OR and MS*, Vol. I, GL Nemhauser et al., Eds., Elsevier Science Publishers, North-Holland, 1989, Chapter IV.
18. Tsaknakis, H., Doyle, Buckley, M., and Washburn, R.B., Tracking closely spaced objects using multiassignment algorithms, in *Proc. DFS 91*, 1991.
19. Finke, G. et al., Quadratic assignment problems, in *Annals of Discrete Mathematics*, 31, Elsevier Science Pub., 1987.
20. Washburn, R.B., Tracking algorithms, Section 2, in a report on the SDI Algorithm Architecture program, provided to TRW as prime, September 1992.

21. Allen, T.G. et al., Multiple Infomation Set Tracking Correlator (MISTC), TR-406, Final Report, Alphatech Inc., Burlington, MA, September 1988.
22. Kuhn, H. W., The Hungarian Method for the Assignment Problem, *Naval Res. Log Quar.* 2, 1955.
23. Munkres, J. Algorithms for the Assignment and Transportation Problems, *J. SIAM* 5, 1957.
24. Bourgeois, F. and J. Lassale, An Extension of the Munkres Algorithm for the Assignment Problem to Rectangular Matrices, *Comm ACM*, 14, 12, December 1971.
25. Silver, R., An Algorithm for the Assignment Problem, *Comm ACM*, 3, November 1960.
26. Kaufmann, A., *Introduction a la Combinatorique en Veu de ses Applications*, Dwrod Pub, Paris, 1968.
27. Salazar, D. L., Application of Optimization Techniques to the Multi-Target Tracking Problem, Master's Thesis, University of Alabama at Huntsville, AL, 1980.
28. Drummond, D. E., D. A. Castanon, and M. S. Bellovin, Comparison of 2-D Assignment Algorithms for Sparse, Rectangular, Floating Point Cost Matrices, self-published, 1990.
29. Junker, R. and A. Volgenant, A Shortest Augmenting Path Algorithm for Dense and Sparse Linear Assignment Problems, *Computing*, 38, 1987.
30. Barr, R. S. et al., The Alternating Basis Algorithms for Assignment Problems, *Math Prog*, 13, 1977.
31. Balinski, M. L., Signature Methods for the Assignment Problem, *Oper. Res.*, 33, 1985.
32. Bersakas, D. P. and J. Eckstein, Dual Coordinate Step Methods for Linear Network Flow Problems, *Math Prog.*, Series B, 1988.
33. Liggins, M. and T. Kurien, Report-to-target assignment in multitarget tracking, *Proceedings of the 1988 IEEE Conference on Decision and Control*, Austin, TX, December 1988.
34. Wolf, J.K., A.M. Viterbi, and G.S. Dixon, Finding the best set of K paths through a trellis with application to multitarget tracking, *IEEE Transactions on Aerospace Electronics and Systems*, Vol. AES-25, 287–296.
35. Allen, T.G., L.B. Feinberg, R.O. LaMaire, K.R. Pattipati, H. Tsaknakis, R.B. Washburn, W. Wren, T. Dobbins, and P. Patterson, Multiple information set tracking correlator (MISTC), TR-406, Final Report, ALPHATECH, Inc., Burlington, MA, and Honeywell, Inc., Space and Strategic Avionics Division, September 1988, USASDC Contract No. DASG60-87-C-0015.
36. Parallel processing of battle management algorithms, Final Report, CDRL Item A003, AT&T Bell Labs, Whippany, NJ, December 1989.
37. Castanon, D.A., Efficient algorithms for finding the K best paths through a trellis, *IEEE Transactions on Aerospace Electronics and Systems*, May 1989.
38. Erbacher, J.A., J. Todd, and C.H. Hopkins, SDI tracker/correlator algorithm implementation document and algorithm design document, R-041-87, Contract No. N00014-86-C-2129, VERAC, San Diego, CA, April 1987.
39. Allen, T.G., G. Cybenko, C. Angelli, and J. Polito, Parallel processing for multitarget surveillance and tracking, ALPHATECH Technical Report TR-360, Burlington, MA, January 1988.
40. Nagarajan, V., M.R. Chidambara, and R.N. Sharma, Combinatorial problems in multitarget tracking — a comprehensive solution, *IEEE Proceedings*, Vol. 134, 1987, 113–118.
41. Pattipati, K.R., S. Deb, Y. Bar-Shalom, and R.B. Washburn, Passive multisensor data association using a new relaxation algorithm, *IEEE Trans. Automatic Control*, February 24, 1989.

18

Data Management Support to Tactical Data Fusion

18.1 Introduction **18**-1
18.2 Database Management Systems **18**-2
18.3 Spatial, Temporal, and Hierarchical Reasoning **18**-3
18.4 Database Design Criteria **18**-6
   Intuitive Algorithm Development • Efficient Algorithm
   Performance • Data Representation Accuracy • Database
   Performance Efficiency • Spatial Data Representation
   Characteristics • Database Design Tradeoffs
18.5 Object Representation of Space **18**-14
   Low-Resolution Spatial Representation • High-Resolution
   Spatial Representation • Hybrid Spatial Feature Representation
18.6 Integrated Spatial/Nonspatial Data Representation **18**-16
18.7 Sample Application **18**-17
   Problem-Solving Approach • Detailed Example
18.8 Summary and Conclusions **18**-25
Acknowledgment **18**-25
Reference **18**-25

Richard Antony
VGS Inc.

18.1 Introduction

Historically, data fusion automation directed at tactical situation awareness applications employed database management systems (DBMS) primarily for storage and retrieval of sensor-derived parametric and text-based data, fusion products, and algorithm components such as templates and exemplar sets. With the increased emphasis over the last decade on multimedia data sources, such as imagery, video, and graphic overlays, the role of database management systems has expanded dramatically. As a consequence, DBMS are now widely recognized as a critical, and perhaps limiting component of the overall system design.

To enhance situation awareness capability, fusion algorithms will increasingly seek to emulate the problem-solving proficiency of human analysts by employing deep problem domain knowledge that is sensitive to problem context. In tactical applications, such contextual knowledge includes existing weather conditions, local natural domain features (e.g., terrain/elevation, surface materials, vegetation, rivers, and drainage regions), and manmade features (e.g., roads, airfields, and mobility barriers). These data sets represent largely *a priori* information. Thus, the vast majority of sensor-derived and *a priori* knowledge bases consist of spatially organized information. For large-scale applications, these data sets must be stored, searched, and manipulated to support a spectrum of real-time fusion algorithms. At present, such capability is beyond the state of the art.

The objective of this chapter is to provide a brief description of the problem and indicate a feasible development path. Section 18.2 introduces DBMS requirements. Section 18.3 discusses spatial, temporal, and hierarchical reasoning that represent key underlying requirements of advanced data fusion automation. Section 18.4 discusses critical database design criteria. Section 18.5 presents the concept of an object-oriented representation of space, showing that it is fully analogous to the traditional object representation of entities that exist within a domain. Section 18.6 briefly describes a composite database system consisting of an integrated representation of both spatial and nonspatial objects. Section 18.7 discusses reasoning approaches and presents a comprehensive example to demonstrate the application of the proposed architecture, and Section 18.8 offers a brief summary.

18.2   Database Management Systems

In general, data fusion applications can require access to data sets that are maintained in a variety of forms, including

- Text
- Tables (e.g., track files, equipment characteristics, and logistical databases)
- Entity-relationship graphs (e.g., organizational charts, functional flow diagrams)
- Maps (e.g., natural and cultural features)
- Images (e.g., optical, forward-looking infrared radar, and synthetic aperture radar)
- Three-dimensional physical models (e.g., terrain, buildings, and vehicles).

Perhaps the simplest data representation form is the *flat file*, so named because it lacks an organizational structure. With no access-specific organizational structure, data access normally requires some form of exhaustive search. Database *indexing* seeks to overcome the inefficiency of the exhaustive search. A database index is analogous to a book index in the sense that it affords direct access to information. Just as a book might use multiple index dimensions, such as a subject index organized alphabetically and a figure index organized numerically, a DBMS can provide multiple, distinct indexes for data sets. Numerous data representation schemes exist, including hierarchical, network, and relational data models. Each of these models support some form of indexing.

Following the development of the *relational data model* in 1970, relational database management system (RDBMS) development experienced explosive growth for more than two decades. The relational model maintains data sets in tables. Each row of a table stores one occurrence of an entity, and columns maintain the values of an entity's attributes. To facilitate rapid search, tables can be indexed with respect to either a particular attribute or a linear combination of attributes. Multiple tables that share a primary *key* (a unique attribute) can be viewed together as a composite table (e.g., linking personnel data and corporate records through an employee's social security number). Because the relational model fundamentally supports only linear search dimensions, it affords inefficient access to data that exhibit significant dependencies across multiple dimensions. As a consequence, a RDBMS tends to be suboptimal for managing two- (2-D) or three-dimensional (3-D) spatially organized information.

To overcome this limitation, commercial geographic information systems (GIS) were developed that offered direct support to the management and manipulation of spatially-organized data. A GIS typically employs vector and/or grid-based 2-D representations of points, lines, and regions, as well as 3-D representations of surfaces stored in triangulated irregular networks (TIN).* A limited number of GIS support 3-D spatial data structures such as octrees. As the utility of GIS systems became more evident,

---

*Although the discussions throughout this chapter focus on 2-D spatial representations, the concepts also apply to 3-D.

hybrid data management systems were built by combining a GIS and a RDBMS. Although well intentioned, such approaches to data management tended to be both inefficient and difficult to maintain.

During the past decade, object-oriented reasoning became the dominant reasoning paradigm for large-scale software development programs. In this paradigm, objects

- Contain data, knowledge bases, and procedures,
- Inherit properties from parent objects based on an explicitly represented object hierarchy,
- Communicate with and control other objects.

As a natural response to the widespread nature of the object-oriented development environment, numerous commercial object-oriented database management systems (OODBMS) were developed to offer enhanced support to object-oriented reasoning approaches. In general, OODBMS permit users to define new data types as needed (extensibility), while hiding implementation details from the user (encapsulation). In addition, such databases allow explicit relationships to be defined between objects. As a result, the OODBMS offers more flexible data structures and more "semantic expressiveness" than the strictly table-based relational data model.

To accommodate some of the desirable attributes of the object-oriented data model, numerous extensions to the relational data model have been proposed and developed in recent years. Today, most commercial RDBMS have been extended to support the *object-relational* model and are referred to as object-relational database management systems (ORDBMS).

18.3 Spatial, Temporal, and Hierarchical Reasoning

To initiate the database requirements discussion, consider the following spatially-oriented query:

Find all roads west of River 1, south of Road 10, and not in Forest 11.

Given the data representation depicted in Figure 18.1, humans can readily identify all roads that lie within the specified query window. However, if the data set were instead presented in a vector (i.e., tuple-based) form, considerable analysis would be required to answer the query. Thus, although the two representations might be technically "equivalent," the representation in Figure 18.1 permits a human to readily perceive all the relevant spatial relationships about the various features, while a vector representation does not. In addition, because humans can perceive a boundary-only representation of a region as representing the boundary *plus* the associated enclosed area, humans can perform a two-dimensional set operation virtually "by inspection." With a vector-based representation form, on the other hand, all spatial relationships among features must be discovered by computational means. Data representation also can dramatically impact the efficiency of machine-based reasoning.

To capitalize on a human's facility for spatial reasoning, military analysts have historically plotted sensor reports and analysis products on clear acetates overlaid on appropriately scaled maps and map

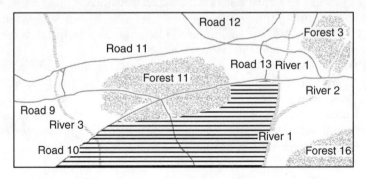

FIGURE 18.1   Two-dimensional map-based representation supporting search for all roads that meet a set of spatial constraints.

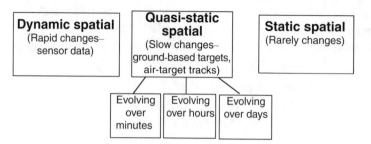

FIGURE 18.2   Three general classes of spatio-temporal data.

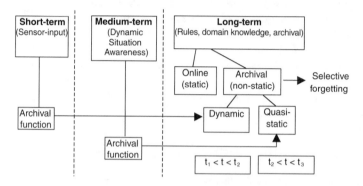

FIGURE 18.3   Possible implementation strategy for temporally sensitive spatial data.

products. The highly intuitive nature of such data representations supports spatial "focus of attention," effectively limiting the size of the search space. Because only those acetates containing features relevant to a particular stage of analysis need be considered, the paradigm also supports search space reduction along nonspatial dimensions. All analysis products generated are in a standard, fully registered form and, thus, are directly usable in subsequent analysis. With its many virtues, the acetate overlay-reasoning paradigm provides a natural metaphor for studying machine-based approaches to spatial reasoning.

In the tactical situation awareness problem domain, spatially organized information often possesses a dynamic, time-varying character. As a result of the intimate relationship between the three spatial dimensions and time, a DBMS that supports data fusion applications must provide a combined spatial and temporal (spatio-temporal) representation.

The retrieval of data that is close in both space *and* time to new sensor-derived information, for example, represents a key database search operation. If domain entity locations are maintained as discrete 3-tuples (x_i, y_i, t_i), indexing or sorting along any *individual* dimension is straightforward.* Although a traditional RDBMS proves to be inefficient when searching across multiple dependent search dimensions, modern GIS and DBMS provide at least limited support to spatio-temporal data. However, these systems tend to support rather myopic approaches to reasoning.

Considerable research into spatio-temporal representation and reasoning has been conducted during the last two decades primarily under university and government sponsorship and also by GIS and database system vendors. Rather than reviewing the literature, this chapter focuses on a conceptual level view of the problem and related issues. Figure 18.2 offers temporal sensitivity taxonomy for spatial data, and Figure 18.3 maps the three classes of temporal sensitivity to a human memory metaphor. Later discussions will focus on a potential implementation strategy based on Figure 18.3.

---

*In order to simplify the discussion, only 2-D spatial data structures are addressed throughout this chapter. Extensions to 3-D are straightforward.

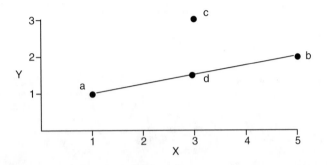

FIGURE 18.4 Two-dimensional depiction of data set showing the spatial relationships between the various features.

On the surface, treating time as an independent variable or dimension seems reasonable; after all, time advances independently. However, treating time as an independent variable requires that the spatial dimensions of a spatio-temporal representation be treated as dependent dimensions. Because we seek a representation that supports object-oriented reasoning, and objects are typically associated with physical or aggregate entities that exist within a 2-D or 3-D "world," such an arrangement is unacceptable. Certainly in tactical data fusion applications, the spatial attributes of physical objects (e.g., location, shape, velocity, and path) represent key problem dimensions, with time playing a more subordinate role.

The benefits of treating time as a dependent variable and employing a true *n*-dimensional spatial data structure* can be illustrated by the following simple example. Suppose an existing fusion database contains the *temporally-sorted* entity locations:

$$(x_a, y_a, t_a) = (1, 1, 1)$$

$$(x_c, y_c, t_c) = (3, 3, 5)$$

$$(x_b, y_b, t_b) = (5, 2, 10)$$

Assume a new sensor report, indicating a target detection at $(x_d, y_d, t_d) = (3, 1.5, 5)$, must be fused with the existing data set. Direct comparison of this new report with the existing database suggests that query point **d** is "near" point **c** (i.e., $t_c = t_d$) and "not near" either points **a** or **b**. If the database was sorted by all three dimensions, with the y-coordinate selected as the primary index dimension, point **d** would appear to be "near" point **a** and "not near" either points **b** or **c**. Suppose that the detections occurring at points **a** and **b** represent successive detections of the same entity (i.e., the endpoints of a track segment). As illustrated in Figure 18.4, which depicts the existing data set and all new detections in a *true* 2-D spatial representation (rather than a list-oriented, tuple-based representation), query point **d** is readily discovered to be "spatially distant" from all three points, but "spatially close" to the line segment **a-b**.

To accommodate the temporal dimension, the distance between (x_d, y_d) and the *interpolated* target position along the line segment **a-b** (at time t_d) can be readily computed. If the target under track is known to be a wheeled vehicle moving along a road network, data associations can be based on a *road-constrained* trajectory, rather than a *linear* trajectory.

Although target reports are typically collected at discrete times (e.g., at radar revisit rates), the actual trajectory of any physical object is a continuous function in both space and time. Despite the fact that the closest line segment to an arbitrary point in space can be derived from tuple-based data, much more efficient data retrieval is possible if all spatial features are explicitly represented within the database. Thus, by employing a *time-coded, explicit* spatial representation (and not just preserving line segment endpoints), a database can effectively support a search along continuous dimensions in both space and time.

---

*A structure that preserves both the natural spatial search dimensions and the true character of the data (e.g., area for image-like data, volume for 3-D data).

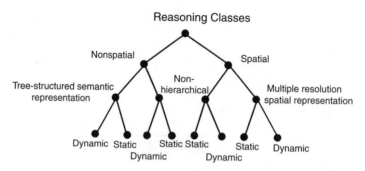

FIGURE 18.5 Taxonomy of the three principal reasoning classes.

With such a representation, candidate track segments can be found through highly localized (true 2-D) spatial searches about point **d**.

Representing moving targets by continuous, time-referenced spatially organized trajectories not only match the characteristics of the physical phenomenon, it also supports highly effective database search dimensions. Equally important, such representations preserve spatial relationships between dynamic objects and other static and dynamic spatially organized database entities. Thus, multiple justifications exist for subordinating temporal indexing to spatial indexing.

Both semantic and spatial reasoning are intrinsically hierarchical. Semantic object reasoning relies heavily on explicitly represented object hierarchies; similarly, spatial reasoning can benefit from multiple resolution spatial representations. Time is a monotonically increasing function; therefore, temporal phenomena possess a natural sequential ordering which can be treated as a specialized hierarchical representation. These considerations attest to the appropriateness of subordinating hierarchical reasoning to both semantic and spatial reasoning and, in turn, subordinating temporal reasoning to hierarchical reasoning.

Figure 18.5 depicts the hierarchical relationship that exists between the three reasoning classes. At the highest level of abstraction, reasoning can be treated as either *spatial* or *semantic* (nonspatial). Each of these reasoning classes can be handled using either *hierarchical* or *nonhierarchical* approaches. Hierarchical spatial reasoning employs *multiple resolution spatial* representations and hierarchical nonspatial reasoning uses *tree-structured semantic* representations. Each of these classes, in turn, may or may not be temporally sensitive.

If we refer to *temporal reasoning* as *dynamic reasoning* and *nontemporal reasoning* as *static reasoning*, there exist four classes of *spatial reasoning* and four classes of *nonspatial reasoning*: (1) dynamic, hierarchical; (2) static, hierarchical; (3) dynamic, nonhierarchical; and (4) static, nonhierarchical. To support effectively data fusion automation, a database must accommodate each of these reasoning classes. Because nonhierarchical reasoning is effectively a special case of hierarchical reasoning, and static reasoning is just a special case of dynamic reasoning, data fusion applications are adequately served by a DBMS that provides *dynamic hierarchical spatial representations* and *dynamic hierarchical semantic representations*. Thus, supporting these two key reasoning classes is the primary database design criterion.

18.4 Database Design Criteria

This section addresses key database design criteria to support advanced algorithm development. Data representation can have a profound impact on algorithm development. The preceding discussion regarding spatial, temporal, and hierarchical reasoning highlighted the benefit of representations that provide natural search dimensions and that preserve significant relationships among domain objects. In addition to supporting key database operations, such as search efficiency, highly intuitive data representations can facilitate the development of sophisticated algorithms that emulate the top-down reasoning process of human analysts.

18.4.1 Intuitive Algorithm Development

In analyzing personnel data, for example, table-based representations are much more natural (and useful) than tree-based data structures. Hierarchical data structures are generally more appropriate for representing the organization of a large company than a purely spatially organized representation. Raster representations of imagery tend to support more intuitive image processing algorithms than do vector-based representations of point, line, and region boundary-based features.

The development of sophisticated machine-based reasoning benefits from full integration of both the semantic and spatial data representations, as well as integration among multiple representations of a given data type. For example, a river possesses both semantic attributes (class, nominal width, *and* flow-rate) *and* spatial attributes (beginning location, ending location, shape description, and depth profile). At a reasonably low resolution, a river's shape might be best represented as a *lineal* feature; at a higher resolution, a *region-based* representation could be more appropriate.

18.4.2 Efficient Algorithm Performance

For algorithms that require large supporting databases, data representation form can significantly affect algorithm performance efficiency. For business applications that demand access to massive table-based data sets, the relational data model supports more efficient algorithm performance than a semantic network model. Two-dimensional template matching algorithm efficiency tends to be higher if true 2-D (map-like) spatial representations are used rather than vector-based data structures. Multiple level-of-abstraction semantic representations and multiple resolution spatial representations tend to support more efficient problem solving than do nonhierarchical representations.

18.4.3 Data Representation Accuracy

In general, data representation accuracy must be adequate to support the widest possible range of data fusion applications. For finite resolution spatially-organized representations, accuracy depends on the data sampling method. In general, data sampling can be either *uniform* or *nonuniform*. Uniformly sampled spatial data are typically maintained as integers, while nonuniformly sampled spatial data are typically represented using floating-point numbers. Although the pixel-based representation of a region boundary is an integer-based representation, a vector-based representation of the same boundary maintains the vertices of the piecewise linear approximation of the boundary as a floating-point list. For a given memory size, nonuniformly sampled representations tend to provide higher accuracy than uniformly sampled representations.

18.4.4 Database Performance Efficiency

Algorithm performance efficiency relies on the six key database efficiency classes described in the following paragraphs.

18.4.4.1 Storage Efficiency

Storage efficiency refers to the relative storage requirements among alternative data representations. Figure 18.6 depicts two similar polygons and their associated vector, raster, pixel boundary, and quadtree representations. With a raster representation (Figures 18.6(c) and (d)), A/Δ^2 nodes are required to store the region, where A is the area of the region, and Δ is the spatial width of a (square) resolution cell (or pixel). Accurate replication of the region shown in Figure 18.6(a) requires a resolution cell size four times smaller than that required to replicate the region in Figure 18.6(b). Because the pixel boundary representation (Figures 18.6(e) and (f)) maintains only the boundary nodes of the region, the required node count is proportional to P/Δ, where P is the perimeter of the region. For fixed Δ, the ratio of the node count for the raster representation relative to the pixel boundary representation is $A/(\Delta P)$. Although a quadtree representation (Figures 18.6(g) and (h)) stores both the region boundary and its interior, the

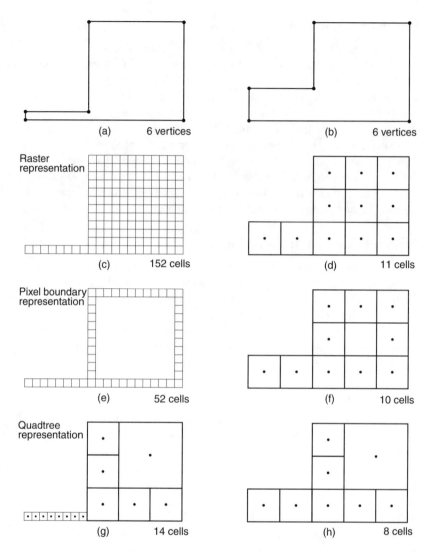

FIGURE 18.6 Storage requirements for four spatial data representations of two similarly-shaped regions.

required node count, for large regions, is proportional to the region perimeter rather than its area.[1] Thus, for grid-based spatial decompositions, the region node count depends on the maximum resolution cell size and the overall size of the region (either its area or perimeter).

Whereas storage requirements for grid-based representations tend to be dependent on the size of the region, storage requirements for nonuniform decompositions are sensitive to the small-scale feature details. Consequently, for nonuniform representations, no simple relationship exists between a region's size and its storage requirements. For example, two regions of the same physical size possessing nearly identical uniform decomposition storage requirements might have dramatically different storage requirements under a nonuniform decomposition representation. Vector-based polygon representations are perhaps the most common nonuniform decomposition. In general, when the level of detail in the region boundary is high, the polygon tuple count will be high; when the level of detail is low, few vertices are required to represent accurately the boundary. Because the level of detail of the boundaries in both Figures 18.6(a) and (b) are identical, both require the same number of vertices. In general, for a given accuracy, nonuniform decompositions have significantly lower storage requirements than uniform decompositions.

18.4.4.2 Search Efficiency

Achieving *search efficiency* requires effective control of the search space size for the range of typical database queries. In general, search efficiency can be improved by storing or indexing data sets along effective search dimensions. For example, for vector-represented spatial data sets, associating a bounding box (i.e., vertices of the rectangle that forms the smallest outer boundary of a line or region) with the representation provides a simple indexing scheme that affords significant search reduction potential. Many kinds of data possess natural representation forms that, if preserved, facilitate the search. 2-D representations that preserve the essential character of maps, images, and topographic information permit direct access to data along very natural 2-D search dimensions. Multiple resolution spatial representations potentially support highly efficient top-down spatial search and reasoning.

18.4.4.3 Overhead Efficiency

Database *overhead efficiency* includes both *indexing* efficiency and *data maintenance* efficiency. Indexing efficiency refers to the cost of creating data set indices, while data maintenance efficiency refers to the efficiency of re-indexing and reorganization operations, including tree-balancing required following data insertions or deletions. Because natural data representations do not require separate indexing structures, such representations tend to support overhead efficiency. Although relatively insignificant for static databases, database maintenance efficiency can become a significant factor in highly dynamic data sets.

18.4.4.4 Association Efficiency

Association efficiency refers to the efficient determination of relationships among data sets (e.g., inclusion, proximity). "Natural" data representations tend to enhance significantly association efficiency over vector-based spatial representations because they tend to preserve the inherent organizational characteristics of data. Although relational database tables can be joined (via a static, single-dimension explicit key), the relational model does not support efficient data set association for data that possess correlated attributes (e.g., combined spatial and temporal proximity).

18.4.4.5 Complex Query Efficiency

Complex query efficiency includes both set operation efficiency and complex clause evaluation efficiency. *Set operation efficiency* demands efficient Boolean and fuzzy set operations among point, line, and region features. *Complex clause evaluation efficiency* requires query optimization for compound query clauses, including those with mixed spatial and semantic constraints.

18.4.4.6 Implementation Efficiency

Implementation efficiency is enhanced by a database architecture and associated data structures that support the effective distribution of data, processing, *and* control. Figure 18.7 summarizes these 12 key design considerations.

18.4.5 Spatial Data Representation Characteristics

Many spatial data structures and numerous variants exist. The taxonomy depicted in Figure 18.8 provides an organizational structure that is useful for comparing and contrasting spatial data structures. At the highest level-of-abstraction, sampled representations of 2-D space can employ either uniform (regular) or nonuniform (nonregular) decompositions. Uniform decompositions generate *data-independent* representations, while nonuniform decompositions produce *data-dependent* representations.

With the exception of fractal-like data, low-resolution representations of spatial features tend to be uniformly distributed and high-resolution representations of spatial data tend to be nonuniformly distributed. Consequently, uniform decompositions tend to be most appropriate for storing relatively low-resolution spatial representations, as well as supporting both data registration and efficient spatial indexing. Conversely, nonuniform decompositions support memory-efficient high-resolution spatial data representations.

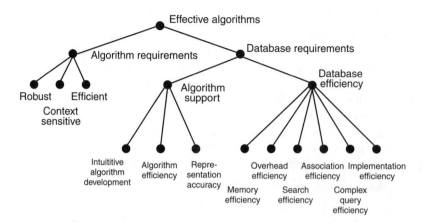

FIGURE 18.7 High-level summary of the database design criteria.

Hierarchical decompositions of 2-D spatial data support the continuum from highly global to highly local spatial reasoning. Global analysis typically begins with relatively low-resolution data representations and relatively global constraints. The resulting analysis products are then progressively refined through higher-resolution analysis and the application of successively more local constraints. For example, a U.S. Interstate road map provides a low-resolution representation that supports the first stage of a route-planning strategy. Once a coarse resolution plan is established, various state, county, and city maps can be used to refine appropriate portions of the plan to support excursions from the Interstate highway to visit tourist attractions or to find lodging and restaurants. Humans routinely employ such top-down reasoning, permitting relatively global (often near-optimal) solutions to a wide range of complex tasks.

Data decompositions that preserve the inherent 2-D character of spatial data are defined to be *2-D data structures*. 2-D data structures provide natural spatial search dimensions, eliminating the need to maintain separate search indices. In addition, such representations preserve Euclidean distance metrics and other spatial relationships. Thus, a raster representation is considered a *2-D data structure*, while tuple-based representations of lines and region boundaries are *non-2-D data structures*.

Explicit data representations literally depict a set of data, while *implicit* data representations maintain information that permits reconstruction of the data set. Thus, a raster representation is an explicit data representation because it explicitly depicts line and region features. A vector representation, which maintains only the end-point pairs of piecewise continuous lines and region boundaries, is considered an implicit representation. In general, implicit representations tend to be more memory-efficient than explicit representations.

Specific feature spatial representations maintain individual point, line, and region features, and *composite feature* spatial representations store the presence or absence of multiple features. A raster representation that maintains a code describing all features in a given cell is a composite spatial representation; the pixels associated with the boundary list of a particular region is a specific spatial representation. Specific spatial representations are most effective for performing spatial operations with respect to individual features; composite representations are most effective for representing classes of spatial features.

Areal-based representations maintain a region's boundary and interior. *Nonareal-based* representations explicitly store only a region's boundary. Boolean set operations among 2-D spatially organized data sets inherently involve both boundaries and region interiors. As a result, areal-based representations tend to support more efficient set operation generation than nonareal-based representations among all classes of spatial features.

Data representations that are both regular and hierarchical support efficient tree-oriented spatial search. Regular decompositions utilize a fixed grid size at each decomposition level and provide a fixed resolution relationship between levels, enabling registered data sets to be readily associated at all resolution

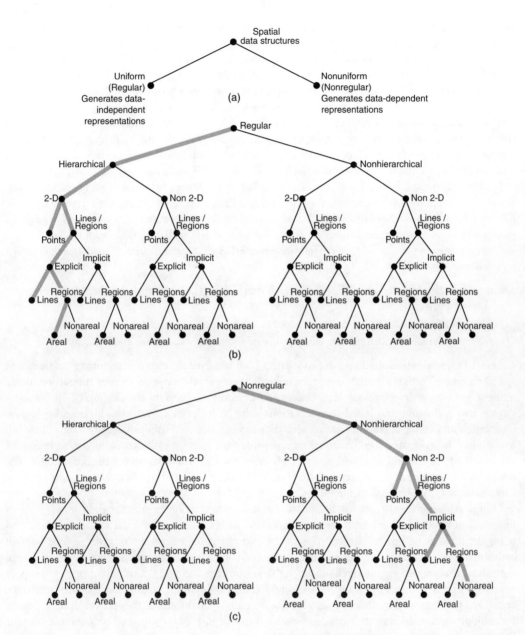

FIGURE 18.8 Spatial data structure taxonomies with the recommended representation branches outlined in bold gray lines.

levels. Registered data representations that are both regular and areal-based support efficient set operations among all classes of 2-D spatial features. Because raster representations are both regular and areal-based, set intersection generation requires only simple Boolean AND operations between respective raster cells. Similarly, set union can be generated by computing the Boolean OR between the respective cells.

Spatial representations that are both regular and 2-D preserve spatial adjacency among all classes of spatial features. They employ a data-independent, regular decomposition and, therefore, do not require extensive database rebuilding operations following insertions and deletions. In addition, because no separate spatial index is required, re-indexing operations are unnecessary. Spatially local changes tend to have highly localized effects on the representation; as a result, spatial decompositions that are both regular and 2-D are relatively dynamic data structures. Consequently, database maintenance efficiency is generally enhanced by regular 2-D spatial representations.

Data fusion involves the composition of both dynamic (sensor data, existing fusion products) and static (tables of equipment, natural and cultural features) data sets; consequently, data fusion algorithm efficiency is enhanced by employing compatible data representation forms for both dynamic and static data. For example, suppose the (static) road network was maintained in a nonregular vector-based representation, while the (dynamic) target track file database was maintained in a regular 2-D data structure. The disparate nature of the representations would make the fusion of these data sets both cumbersome and inefficient. Thus, maintaining static and dynamic information with identical spatial data structures potentially enhances both database *association efficiency* and *maintenance efficiency.*

Spatial representations that are *explicit, regular, areal,* and *2-D* support efficient set operation generation, offer natural search dimensions for fully registered spatial data, and represent relatively dynamic data structures. Because they possess key characteristics of analog 2-D spatial representations (e.g., paper maps and images), the representations are defined to be *true 2-D* spatial representations. Spatial representations possessing one or more of the following properties violate the requirements of true 2-D spatial representations:

- Spatial data is not stored along 2-D search dimensions (i.e., spatial data stored as list or table-based representations).
- Region representations are nonareal-based (i.e., bounding polygons).
- Nonregular decompositions are employed (e.g., vector representations and R-trees).

True 2-D spatial representations tend to support both intuitive algorithm development and efficient spatial reasoning. Such representations preserve key 2-D spatial relationships, thereby supporting highly intuitive spatial search operations (e.g., "Northwest of point A," "beyond the city limits," or "inside a staging area"). Because grid-based representations store both the boundary and interior of a region, intersection and union operations among such data can be generated using straightforward cell-by-cell operations. The associated computational requirements scale linearly with data set size independent of the complexity of the region data sets (e.g., embedded holes or multiple disjoint regions). Conversely, set operation generation among regions based on non-true-2-D spatial representations tend to require combinatorial computational requirements.

Computational geometry approaches to set intersection generation require the determination of all line segment intersection points; therefore, computational complexity is of order $\mathbf{m} \times \mathbf{n}$, where $\mathbf{m}$ and $\mathbf{n}$ are the number of vertices associated with the two polygons. When one or both regions contain embedded holes, dramatic increases in computational complexity can result. For analysis and fusion algorithms that rely on set intersections among large complex regions, algorithm performance can be adversely impacted. Although the use of bounding rectangles (bounding boxes) can significantly enhance search and problem solving efficiency for vector-represented spatial features, they provide limited benefit for multiple-connected regions, extended lineal features (e.g., roads, rivers, and topographic contour lines), and directional and proximity-based searches.

Specific spatial representations that are *explicit, regular, areal, 2-D,* and *hierarchical* (or hierarchical true 2-D) support highly efficient top-down spatial search and top-down areal-based set operations among specific spatial features. With a quadtree-based region representation, for example, set intersection can be performed using an efficient top-down, multiple resolution process that capitalizes on two key characteristics of Boolean set operations. First, the intersection product is a proper subset of the smallest region being intersected. Second, the intersection product is both *commutative* and *associative* and, therefore, independent of the order in which the regions are composed. Thus, rather than exhaustively ANDing all nodes in all regions, intersection generation can be reformulated as a search problem where the smallest region is selected as a variable resolution spatial search "window" for interrogating the representation of the next larger region.[1] Nodes in the second smallest region that are determined to be within this search window are known to be in the intersection product. Consequently, the balance of the nodes in the larger region need not be tested.

If three or more regions are intersected, the product that results from the previous intersection product becomes the search window for interrogating the next larger region. Consequently, while the computational complexity of polygon intersection is a function of the product of the number of vertices in each polygon, the computational complexity for region intersection using a regular grid-based representation is related to the number of tuples that occur within cells of the physically smallest region. Composite spatial representations that are *explicit, regular, areal, 2-D,* and *hierarchical* support efficient top-down spatial search among classes of spatial features.

Based on the relationships between spatial data decomposition classes and key database efficiency classes, a number of generalizations can be formulated:

- Hierarchical, true 2-D spatial representations support the development of highly intuitive algorithms.
- For tasks that require relatively low-resolution spatial reasoning, algorithm efficiency is enhanced by hierarchical, true 2-D spatial representations.
- Implicit, nonregular spatial representations support representation accuracy.
- Specific spatial representations are most appropriate for reasoning about specific spatial features; conversely, composite spatial representations are preferable for reasoning about classes of spatial features. For example, a true 2-D composite road network representation would support more efficient determination of the closest road to a given point in space than a representation maintaining individual named roads. With the latter representation, a sizable portion of the road database would have to be interrogated.
- Finite resolution data decompositions that are implicit, nonregular, nonhierarchical, nonareal, and non-2-D tend to be highly storage-efficient.
- Spatial indexing efficiency is enhanced by hierarchical, true 2-D representations that support natural, top-down search dimensions.
- Spatially local changes require only relatively local changes to the underlying representation; therefore, database maintenance efficiency is enhanced by regular, dynamic, 2-D spatial representations.
- For individual spatial features, database search efficiency is enhanced by spatial representations that are specific, hierarchical, and true 2-D, and that support distributed search.
- Search efficiency for classes of spatial features is enhanced by composite hierarchical spatial feature representations.
- True 2-D spatial representations preserve spatial relationships among data sets; therefore, for individual spatial features, database association efficiency is enhanced by hierarchical true 2-D spatial representations of specific spatial features. For classes of spatial features, association efficiency is enhanced by hierarchical, true 2-D composite feature representations.
- For specific spatial features, complex query efficiency is enhanced by hierarchical, true 2-D representations of individual features; for classes of spatial features, complex query efficiency is enhanced by hierarchical, composite, true 2-D spatial representations.
- Finally, database implementation efficiency is enhanced by data structures that support the distribution of data, processing, and control.

These general observations are summarized in Table 18.1.

18.4.6 Database Design Tradeoffs

Object-oriented reasoning potentially supports the construction of robust, context-sensitive fusion algorithms, enabling data fusion automation to benefit from the development of an effective OODBMS. This section explores design tradeoffs for an object-oriented database that seek to achieve an effective compromise among the algorithm support and database efficiency issues listed in Table 18.1. As previously discussed, the principal database design requirement is support for *dynamic, hierarchical spatial reasoning* and *dynamic, hierarchical semantic reasoning*. Consequently, an optimal database must provide data structures that facilitate storage and access to both temporally and hierarchically organized spatial and nonspatial information.

TABLE 18.1 Spatial Representation Attributes Supporting Nine Spatial Database Requirements

Spatial database requirements \ Spatial representation attributes	Regular	Nonregular	Hierarchical	Nonhierarchical	Areal	Nonareal	2-D	Non-2-D	Static structures	Dynamic data structures	Distributed	Nondistributed	Explicit feature representation	Implicit feature representation	Specific feature based	Composite feature based
Intuititive algorithm development	●		●		●		●						●		●	●
Spatial reasoning algorithm efficiency																
- Individual features	●		●		●		●						●		●	
- Feature classes	●		●		●		●						●			●
Representation accuracy		●						●							●	
DB storage efficiency		●		●		●		●							●	
DB overhead efficiency																
- Indexing	●		●		●		●						●		●	
- Maintenance	●						●			●			●		●	
DB search efficiency																
- Individual features	●		●		●		●				●		●		●	
- Feature classes	●		●		●		●				●		●			●
DB association efficiency																
- Individual features	●		●		●		●						●	●	●	
- Feature classes	●		●		●		●						●	●		●
DB query efficiency																
- Individual features	●		●		●		●						●	●	●	
- Feature classes	●		●		●		●						●	●		●
DB implementation efficiency													●			

Nonspatial (or semantic) declarative knowledge can be represented as n-tuples, arrays, tables, transfer functions, frames, trees, and graphs. Modern semantic object databases provide effective support to complex, multiple level-of-abstraction problems that (1) possess extensive parent/child relationships, (2) benefit from problem decomposition, and/or (3) demand global solutions. Because object-oriented representations permit the use of very general internal data structures, and the associated reasoning paradigm fully embraces hierarchical representations at the semantic object level, conventional object databases intrinsically support hierarchical semantic reasoning. Semantic objects can be considered relatively dynamic data structures because temporal changes associated with a specific object tend to affect only that object or closely related objects.

The character and capabilities of a *spatial object* database are analogous to those of a semantic object database. A spatial object database must support top-down, multiple level-of-abstraction (i.e., multiple resolution) reasoning with respect to classes of spatial objects, as well as permit efficient reasoning with specific spatial objects. Just as the semantic object paradigm requires an explicitly represented semantic object hierarchy, a spatial object database requires an equivalent spatial object hierarchy. Finally, just as specific entities in conventional semantic object databases possess individual semantic object representations, specific spatial objects require individual spatial object representations.

18.5 Object Representation of Space

Consider the query:

Determine the class-1 road closest to query point (x_1, y_1).

In a database that maintains only implicit representations of individual spatial features, the above *feature-class query* could potentially require the interrogation of all class-1 road representations. Just as a hierarchical

TABLE 18.2   Key Representation Characteristics Required by the Three Distinct Spatial Data Representation Classes

Representation characteristics	Key spatial data types	Object representation of 2-D space	Low-resolution spatial representation	High-resolution spatial resolution
1	Finite resolution	●	●	●
2	Regular/Nonregular	Regular	Regular	Nonregular
3	Areal/Nonareal	Areal	Areal	Nonareal
4	2-D/Non-2-D	2-D	2-D	Non-2-D
5	Explicit/Implicit	Explicit	Explicit	Implicit
6	Hierarchical/Nonhierarchical	Hierarchical	Hierarchical	Nonhierarchical
7	Specific features/Composite features	Composite features	Specific features	Specific features
8	Relatively dynamic data structure	●	●	●
9	Distributed reprsentation potential	●	●	●
10	Low/high precision	Low	Low	High

representation of semantic objects permits efficient class-oriented queries, a hierarchical representation of space supports efficient queries with respect to classes of spatial objects. At the highest level-of-abstraction, an *object representation of space* consists of a single object that characterizes the entire area of interest (Asia, a single map sheet, a division's area of interest). At each successively lower level of the spatial object hierarchy, space is decomposed into progressively smaller regions that identify spatial objects (specific entities or entity classes) associated with that region. Just as higher-order semantic objects possess more general properties than their offspring, higher-order object representations of space characterize the general properties of 2-D space.

Based on the principles outlined in the last section, an object representation of 2-D space must satisfy the properties summarized in Table 18.2. As previously mentioned, a true 2-D spatial representation possesses the first five characteristics listed in Table 18.2. With the addition of the sixth property, the spatial object hierarchy provides a uniform hierarchical spatial representation for classes of point, line, and region features. The *pyramid* data structure fully satisfies these first six properties. Because multiple classes of spatial objects (e.g., roads, waterways, and soil-type) can be maintained within each cell, a pyramid representation is well suited to maintain a composite feature representation. In a pyramid representation, all the following are true:

1. Spatially local changes require only relatively local changes to the data structure.
2. Limited re-indexing is required following updates.
3. Extensive tree-balancing operations are not required following insertions or deletions.

As a result, the pyramid is a relatively *dynamic data structure*. In addition, the hierarchical and grid-based character of the pyramid data structure enables it to readily accommodate data distribution. Therefore, a pyramid data structure fully satisfies all nine requirements for an object representation of 2-D space.

18.5.1   Low-Resolution Spatial Representation

As summarized in Table 18.2, an effective low-resolution spatial representation must possess ten key properties. With the exception of the composite feature representation property, the low-resolution spatial representation requirements are identical to those of the object representation of 2-D space. Whereas a composite-feature-based representation supports efficient spatial search with respect to classes of spatial features, a specific feature-based representation supports effective search and manipulation of specific point, line, and region features. A regular region quadtree possesses all of the properties presented in Table 18.2, column 3.

18.5.2   High-Resolution Spatial Representation

An effective high-resolution spatial representation must possess the ten properties indicated in the last column of Table 18.2. Vector-based spatial representations clearly meet the first four criteria. Because

Feature	Hybrid representation	Example
Point	Node location and within node offset	
Line	Node location of each associated node and list of vertices (node offset form) of piecewise line segments within each node	
Region	List of minimal region quadtree nodes; line feature representation within all boundary leaf nodes	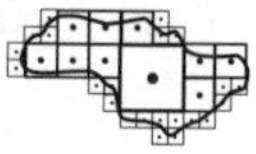

FIGURE 18.9 Quadtree-indexed vector spatial representation for points, lines, and regions.

they use nonhierarchical representations of specific features and employ implicit piecewise linear representations of lines and region boundaries, vector representations also satisfy properties 5 through 7. Changes to a feature require a modification of the property list of a single feature; therefore, vector representations tend to be relatively dynamic data structures. A representation of individual features is self contained, so vector representations can be processed in a highly distributed manner. Finally, vector-based representations inherently provide high precision. Thus, vector-based representations satisfy all of the requirements of the high-resolution spatial representation.

18.5.3 Hybrid Spatial Feature Representation

Traditionally, spatial database design has involved the selection of either a single spatial data structure or two or more alternative, but substantially independent, data structures (e.g., vector and quadtree). Because of the additional degrees of freedom it adds to the design process, the use of a hybrid spatial representation offers the potential for achieving a near-optimal compromise across the spectrum of design requirements. Perhaps the most straightforward design approach for such a hybrid data structure is to directly integrate a multiple-resolution, low-resolution spatial representation and a memory-efficient, high-resolution spatial representation.

As indicated in Figure 18.9, the quadtree data structure can form the basis of an effective hybrid data structure, serving the role of both a low-resolution spatial data representation and an efficient spatial index into high accuracy vector representations of point, line, and region boundaries. Table 18.3 offers a coarse-grain evaluation of the effectiveness of the vector, raster, pixel boundary, region quadtree, and the recommended hybrid spatial representation based on the database design criteria. Table 18.4 summarizes the key characteristics of the recommended spatial object representation and demonstrates that it addresses the full spectrum of spatial data design issues.

18.6   Integrated Spatial/Nonspatial Data Representation

To effectively support data fusion applications, spatial and nonspatial data classes must be fully integrated. Figure 18.10 depicts a high level view of the resultant semantic/spatial database kernel depicting both explicit and implicit links between the various data structures. Because a pyramid data structure can be viewed as a *complete* quadtree, the pyramid and the low-resolution spatial representation offer a unified structure, with the latter effectively an extension to the former. Therefore, the quadtree data structure

TABLE 18.3 Comparison of Spatial Data Representations Based on Their Ability to Support Database Design Criteria

Design criteria \ Representations	Vector based*	Raster based	Pixel boundary	Region quadtree	Hybrid spatial representation
• *Intuitive algorithm development*	Poor	Good	Moderate	Good	Good
• *Computationally efficient algorithms*	Poor	Moderate	Poor	Good	Good
• *Representation accuracy*	Good	Poor	Poor	Poor	Good
• *Data storage efficiency*	Good	Poor	Moderate	Moderate	Moderate
• *DB overhead efficiency*	Good	Good	Good	Moderate	Moderate
• *Spatial search efficiency*					
Specific features	Moderate	Good	Moderate	Good	Good
Feature classes	Poor	Good	Poor	Moderate	Good
• *Complex Boolean query efficiency*					
Specific features	Moderate	Good	Moderate	Good	Good
Feature classes	Poor	Moderate	Poor	Moderate	Good
• *DB implementation efficiency*	Good	Good	Good	Good	Good

* Assumes the use of bounding boxes.

TABLE 18.4 Summary Characteristics of the Three Spatial Representation Classes

Database element \ Spatial representation attributes	Regular	Nonregular	Hierarchical	Nonhierarchical	Areal	Nonareal	2-D	Non-2-D	Explicit	Implicit	Specific	Composite
Object representation of 2-D space (pyramid)	●		●		●		●					●
Hybrid representation of specific object												
Low-resolution index	●		●		●		●				●	
High-resolution representation		●		●		●		●		●	●	
Composite representation: pyramid & hybrid spatial representation	●	●	●	●	●	●	●	●	●	●	●	●

serves as the link between the pyramid and the individual spatial feature representations and provides a hierarchical spatial index to high-resolution vector-represented point, line, and region features.

The integrated spatial and semantic object representation permits efficient top-down search for domain objects based on a combination of semantic and spatial constraints. The fully integrated data structures support efficient and effective manipulation of spatial and nonspatial relations, and the computation of spatial and semantic distance metrics among domain objects. The quadtree-indexed data structure preserves the precision of the highest fidelity spatial data without compromising the memory efficiency of the overall system.

18.7 Sample Application

Using a reasonably simple, easily visualized path-planning application, this section illustrates key benefits of the database system design just presented. Humans routinely plan simple tasks in a nearly subconscious manner (e.g., walking from one room to another and picking up a glass of water); however, other tasks require considerable deliberation. Consider, for instance, planning an automobile trip from Washington,

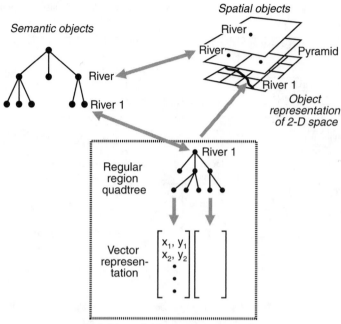

FIGURE 18.10   High-level view of the integrated semantic/spatial object database.

D.C. to Portland, OR. Adding constraints to the problem, such as stops at nearby national parks and other attractions along the way, can greatly complicate the task of generating an optimal plan. One approach to planning such a trip would be to accumulate all the county road maps through which such a route might pass; however, the amount of data would be overwhelming. The task would be much less daunting if the planner first acquired a map of the U.S. Interstate highway system and used this map to select a coarse resolution plan. Then, a small collection of state maps would probably suffice for planning routes to specific attractions located some distance from an interstate highway. Finally, for locating a friend's residence, a particular historical landmark, or some other point of interest, detailed county or even city maps could be used to refine the details of the overall plan.

Clearly, the top-down path development approach is both more natural and much more efficient than a single level of abstraction approach, accommodating the continuum from global to highly local objectives and evaluation metrics. Given time and resource constraints, there will likely be a need to apply weighted values to indicate the relative importance of competing subtasks: travelers with different interests and background would probably rate attractions such as the Grand Tetons and the Eisenhower Library quite differently.

In path planning, as with virtually all decision making, effective evaluation criteria are needed to select among large numbers of candidate solutions. Path selection metrics could be as simple as seeking to minimize the overall path length, minimize travel time, or maximize travel speed. On the other hand, the planner may want to minimize or maximize a weighted combination of several individual metrics or to apply fuzzier (more subjective) measures of performance. Whereas crisp metrics can be treated as *absolute* metrics (e.g., the shortest path length or the fastest speed), fuzzy metrics are most appropriate for representing *relative* metrics or constraints (e.g., selecting a *moderately* scenic route that is *not much longer* than the shortest route). In addition to selection criteria, various problem constraints could be appropriate at the different levels of the planning process. These might include regular stops for fuel, meals, and nightly motel stays.

In addition to developing effective *a priori* plans, real-world tasks often require the ability to perform real-time replanning. Replanning can span the spectrum from global modifications to highly local changes. As an illustration of the latter, suppose a detailed plan called for travel in the right-hand lane of Route 32, a four-lane divided highway. During plan execution, an accident is encountered blocking further movement in the right-most lane requiring the initial plan to be locally adjusted to avoid this impasse. More global modifications would be required if, prior to reaching Route 32, the driver learned of a recent water main break that was expected to keep the road closed to traffic the entire day.

In general, top-down planning begins by establishing a coarse-grain skeletal solution that is then recursively refined to the appropriate level of detail. Multiple level-of-abstraction planning tends to minimize the size of both the search and decision spaces, making it highly effective. Based on both its efficacy and efficiency, top-down planning readily accommodates replanning, a critical component in complex, dynamic problem domains.

18.7.1 Problem-Solving Approach

This section outlines a hierarchical path-planning algorithm that seeks to develop one or more routes between two selected locations for a wheeled military vehicle capable of both "on-road" and "off-road" movement. Rather than seeking a mathematically optimal solution, we strive for a near-optimal global solution by emulating a human-oriented approach to path selection and evaluation.

In general, path selection is potentially sensitive to a wide range of geographical features (i.e., terrain, elevation changes, soil type, surface roughness, and vegetation), environmental features (i.e., temperature, time of day, precipitation rate, visibility, snow depth, and wind speed), and cultural features (i.e., roads, bridges, the location of supply caches, and cities). Ordering the applicable domain constraints from the *most* to the *least* significant effectively treats path development as a top-down, constraint-satisfaction task.

In general, achieving "high quality" global solutions requires that global constraints be applied before more local constraints. Thus, in Stage 1 (the highest level of abstraction), path development focuses on route development, considering only *extended barriers* to ground travel. Extended barriers are those cultural and manmade features that cannot be readily circumnavigated. As a natural consequence, they tend to have a more profound impact on route selection than smaller-scale features. Examples of extended features include rivers, canyons, neutral zones, ridges, and the Great Wall in China. Thus, in terms of the supporting database, first stage analysis involves searching for extended barriers that lie between the path's *start* and *goal* state.

If an extended barrier is discovered, candidate barrier crossing locations (e.g., bridges or fording locations for a river or passes for a mountain range) must be sought. One or more of these barrier-crossing sites must be selected as candidate high-level path subgoals. As with all stages of the analysis, an evaluation metric is required to either "prune" or rank candidate subgoals. For example, the selected metric might retain the **n** closest bridges possessing adequate weight-carrying capacity. For each subgoal that satisfies the selection metric, the high-level barrier-crossing strategy is reapplied, searching for extended barriers and locating candidate barrier crossing options until the final goal state is reached. The product of Stage 1 analysis is a set of coarse resolution candidate paths represented as a set of subgoals that satisfy the high-level path evaluation metrics and global domain constraints.

Because the route is planned for a wheeled vehicle, the existing road network has the next most significant impact on route selection. Thus, in Stage 2, road connectivity is established between all subgoal pairs identified in Stage 1. For example, the shortest road path or the shortest **m** road paths are generated for those subgoals discovered to be *on* or *near* a road. When the subgoal is not near a road or when path evaluation indicates that all candidate paths provide only low-confidence solutions, overland travel between the subgoals would be indicated. Upon completion of Stage 2 analysis, the coarse resolution path sets developed during Stage 1 will have been refined with the appropriate road-following segments.

In Stage 3, overland paths are developed between all subgoal pairs not already linked by high-confidence, road-following paths. Whereas *extended barriers* were associated with the most global constraints on mobility, *nonextended barriers* (e.g., hills, small lakes, drainage ditches, fenced fields, or forests) represent the most significant mobility constraints for overland travel. Ordering relevant nonextended barriers from stronger constraints (i.e., larger barriers and no-go regions) to weaker constraints (i.e., smaller barriers and slow-go regions) will extend the top-down analysis process. At each successive level of refinement, a selection metric, sensitive to progressively more local path evaluation constraints, is applied to the candidate path sets. The path refinement process terminates when one or more candidate paths have been generated that satisfy all path evaluation constraints. Individual paths are then rank-ordered against selected evaluation metrics.

While traditional path development algorithms generate plans based on brute force optimization by minimizing "path resistance" or other similar metric, the hierarchical constraint-satisfaction-based approach just outlined emulates a more human-like approach to path development. Rather than using simple, single level-of-abstraction evaluation metrics (path resistance minimization), the proposed approach supports more powerful reasoning, including concatenated metrics (e.g., "maximal conceal-ment from one or more vantage points" plus "minimal travel time to a goal state"). A path that meets both of these requirements might consist of a set of road segments not visible from specified vantage points, as well as high mobility off-road path segments for those sections of the roadway that *are* visible from those vantage points. Hierarchical constraint-based reasoning captures the character of human problem-solving approaches, achieving the spectrum from global to more local subgoals, producing intuitively satisfying solutions. In addition, top-down, recursive refinement tends to be more efficient than approaches that attempt to directly generate high-resolution solutions.

18.7.2   Detailed Example

This section uses a detailed example of the top-down path-planning process to illustrate the potential benefits of the integrated semantic and spatial database discussed in Section 18.6. Because the database provides both natural and efficient access to both hierarchical semantic information and multiple-resolution spatial data, it is well suited to problems that are best treated at multiple levels of abstraction. The tight integration between semantic and spatial representation allows effective control of both the search space and the solution set size.

The posed problem is to determine one or more "good" routes for a wheeled vehicle from the start to the goal state depicted in Figure 18.11. Stage 1 begins by performing a spatially anchored search (i.e., anchored by both the start and goal states) for extended mobility barriers associated with both the cultural and geographic feature database. As shown in Figure 18.12, the highest level-of-abstraction representation

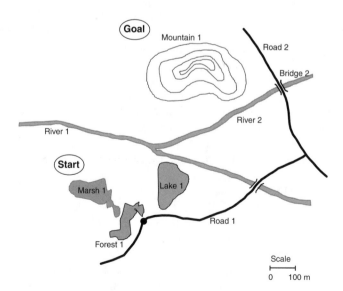

FIGURE 18.11   Domain mobility map for path development algorithm.

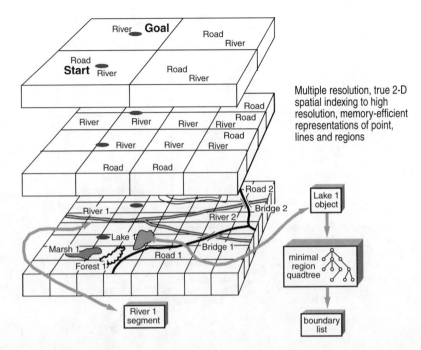

FIGURE 18.12 Top-down multiple resolution spatial search, from the start toward the goal node, reveals the existence of a river barrier.

of the object representation of space (i.e., the top-level of the pyramid) indicates that a river, which represents an extended ground-mobility barrier, exists in the vicinity of both the start and the goal states. At this level of abstraction, it cannot be determined whether the extended barrier lies between the two points.

The pyramid data structure supports highly focused, top-down searching to determine whether ground travel between the start and goal states is blocked by a river. At the next higher resolution level, however, ambiguity remains. Finally, at the third level of the pyramid, it can be confirmed that a river lies between the start and goal states. Therefore, an efficient, global path strategy can be pursued that requires breaching the identified barrier. Consequently, bridges, suitable fording locations, or bridging operations become candidate subgoals.

If, on the other hand, no extended barrier had been discovered in the cells shared by the start and goal states (or in any intervening cells) at the outset of Stage 1 analysis, processing would terminate without generating any intervening subgoals. In this case, Stage 1 analysis would indicate that a direct path to the goal is feasible.

While conventional path planning algorithms operate strictly in the spatial domain, a flexible top-down path-planning algorithm supported by an effectively integrated semantic and spatial database can operate across both the semantic and spatial domains. For example, suppose nearby bridges are selected as the primary subgoals. Rather than perform spatial search, direct search of the semantic object (River 1) could determine nearby bridges. Figure 18.13 depicts attributes associated with that semantic object, including the location of a number of bridges that cross the river. To simplify the example, only the closest bridge (Bridge 1) will be selected as a candidate subgoal (denoted $SG_{1,1}$). Although this bridge could have been located via spatial search in both directions along the river (from the point at which a line from the start to the goal state intersects River 1), a semantic-based search is more efficient.

To determine if one or more extended barriers lie between $SG_{1,1}$ and the goal state, a spatial search is reinitiated from the subgoal in the direction of the goal state. High-level spatial search within the pyramid data structure reveals another *potential* river barrier. Top-down spatial search once again verifies the

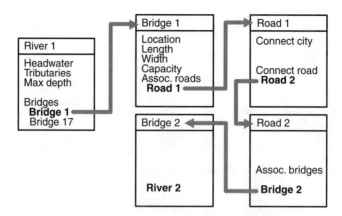

FIGURE 18.13   Semantic object database for the path development algorithm.

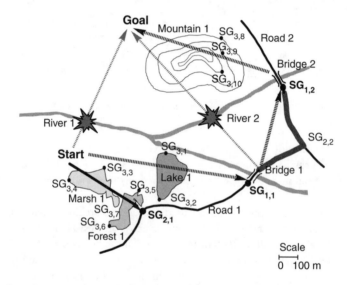

FIGURE 18.14   Sub-goals associated with all three stages of the path development algorithm.

existence of a second extended barrier (River 2). Just as before, the closest bridging location, denoted as $SG_{1,2}$, is identified by evaluating the bridge locations maintained by the semantic object (River 2). Spatial search from Bridge 2 toward the goal state reveals no extended barriers that would interfere with ground travel between $SG_{1,2}$ and the goal state.

As depicted in Figure 18.14, the first stage of the path development algorithm generates a single path consisting of the three subgoal pairs (start, $SG_{1,1}$), ($SG_{1,1}$, $SG_{1,2}$), ($SG_{1,2}$, goal) satisfying the global objective of reaching the goal state by breaching all extended barriers. Thus, at the conclusion of Stage 1, the primary alternatives to path flow have been identified.

In Stage 2, road segments connecting adjacent subgoals that are on or near the road network must be identified. The semantic object representation of the bridges identified as subgoals during the Stage 1 analysis also identify their road association; therefore, a road network solution potentially exists for the subgoal pair ($SG_{1,1}$, $SG_{1,2}$). Algorithms are widely available for efficiently generating minimum distance paths within a road network. As a result of this analysis, the appropriate segments of Road 1 and Road 2 are identified as members of the candidate solution set (shown in bold lines in Figure 18.14).

Next, the paths between the start state and $SG_{1,1}$ are investigated. $SG_{1,1}$ is known to be on a road and the start state is not; therefore, determining whether the start state is near a road is the next objective.

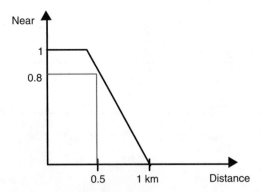

FIGURE 18.15   Membership function for fuzzy metric "near."

Suppose the assessment is based on the fuzzy qualifier *near* shown in Figure 18.15. Because the detailed spatial relations between features cannot be economically maintained with a semantic representation, spatial search must be used. Based on the top-down, multiple-resolution object representation of space, a road is determined to exist within the vicinity of the start node. A top-down spatially localized search within the pyramid efficiently reveals the closest road segment to the start node. Computing the Euclidean distance from that segment to the start node, the state node is determined to be near a road with degree of membership 0.8.

Because the start node has been determined to be near a road, in addition to direct overland travel toward Bridge 1 (*start*, $SG_{1,1}$), an alternative route exists based on overland travel to the nearest road (subgoal $SG_{2,1}$) followed by road travel to Bridge 1 ($SG_{2,1}$, $SG_{1,1}$). Although a spectrum of variants exists between direct travel to the bridge and direct travel to the closest road segment, at this level of abstraction only the primary alternatives must be identified. Repeating the analysis for the path segment ($SG_{1,2}$, goal), the goal node is determined to be not near any road. Consequently, overland route travel is required for the final leg of the route.

In Stage 3, all existing nonroad path segments are refined based on more local evaluation criteria and mobility constraints. First, large barriers, such as lakes, marshes, and forests are considered. Straight-line search from the start node to $SG_{1,1}$ reveals the existence of a large lake. Because circumnavigation of the lake is required, two subgoals are generated ($SG_{3,1}$ and $SG_{3,2}$) as shown in Figure 18.14, one representing clockwise travel and the other counter-clockwise travel around the barrier. In a similar manner, spatial search from the start state toward $SG_{2,1}$ reveals a large marsh, generating, in turn, two additional subgoals ($SG_{3,3}$ and $SG_{3,4}$).

Spatial search from both $SG_{3,3}$ toward $SG_{2,1}$ reveals a forest obstacle (Forest 1). Assuming that the forest density precludes wheeled vehicle travel, two more subgoals are generated representing a northern route ($SG_{3,5}$) and a southern route ($SG_{3,6}$) around the forest. Because a road might pass through the forest, a third strategy must be explored (road travel through the forest). The possibility of a road through the forest can be investigated by testing containment or generating the intersection between Forest 1 and the road database.

The integrated spatial/semantic database discussed in Section 18.6 provides direct support to containment testing and intersection operations. With a strictly vector-based representation of roads and regions, intersection generation might require interrogation of a significant portion of the road database; however, the quadtree-indexed vector spatial representation presented permits direct spatial search of that portion of the road database that is within Forest 1.[1] Suppose a dirt road is discovered to intersect the forest. Since no objective criterion exists for evaluating the "best" subpath(s) at this level of analysis, an additional subgoal ($SG_{3,7}$) is established. To illustrate the benefits of deferring decision making, consider the fact that although the length of the road through the forest could be shorter than the travel distance around the forest, the road may not enter and exit the forest at locations that satisfy the overall path selection criteria.

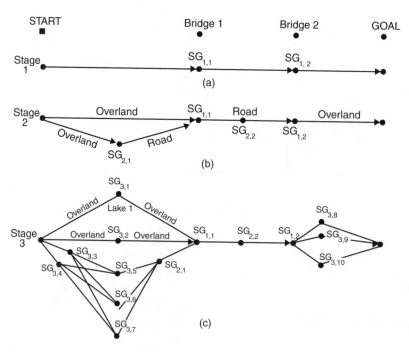

FIGURE 18.16   Path development graph following (a) stage 1, (b) stage 2, and (c) stage 3.

Continuing with the last leg of the path, spatial search from $SG_{1,2}$ to the goal state identifies a mountain obstacle. Because of the inherent flexibility of a constraint-satisfaction-based problem-solving paradigm, a wide range of local path development strategies can be considered. For example, the path could be constrained to employ one or more of the following strategies:

1. Circumnavigate the obstacle ($SG_{3,8}$).
2. Remain below a specified elevation ($SG_{3,9}$).
3. Follow a minimum terrain gradient $SG_{3,10}$.

Figure 18.16 shows the path-plan subgoal graph following Stage 1, Stage 2, and Stage 3. Proceeding in a top-down fashion, detailed paths between all sets of subgoals can be recursively refined based on the evaluation of progressively more local evaluation criteria and domain constraints. Path evaluation criteria at this level of abstraction might include (1) the minimum mobility resistance, (2) minimum terrain gradient, or (3) maximal speed paths.

Traditional path planning algorithms generate global solutions by using highly local nearest-neighbor path extension strategies (e.g., gradient descent), requiring the generation of a combinatorial number of paths. Global optimization is typically achieved by rank ordering all generated paths against an evaluation metric (e.g., shortest distance or maximum speed). Supported by the semantic/spatial database kernel, the top-down path-planning algorithm just outlined requires significantly smaller search spaces when compared to traditional, single-resolution algorithms. Applying a single high-level constraint that eliminates the interrogation of a single 1 km × 1 km resolution cell, for example, could potentially eliminate search-and-test of as many as 10,000 10 m × 10 m resolution cells. In addition to efficiency gains, due to its reliance on a hierarchy of constraints, a top-down approach potentially supports the generation of more robust solutions. Finally, because it emulates the problem-solving character of humans, the approach lends itself to the development of sophisticated algorithms capable of generating intuitively appealing solutions.

In summary, the hierarchical path development algorithm

1. Employs a reasoning approach that effectively emulates manual approaches,
2. Can be highly robust because constraint sets are tailored to a specific vehicle class,
3. Is dynamically sensitive to the current domain context, and
4. Generates efficient global solutions.

The example outlined in this section demonstrates the utility of the database kernel presented in Section 18.6. By facilitating the efficient, top-down, spatially anchored search and fully integrated semantic and the spatial object search, the spatial/semantic database provides direct support to a wide range of demanding, real-world problems.

18.8 Summary and Conclusions

Situation awareness development for remote sensing applications relies on the effective combination of a wide range of data and knowledge sources, including the maximal use of relevant sensor-derived (e.g., imagery, overlays, and video) and nonsensor-derived information (e.g., topographic features; cultural features; and past, present, and future weather conditions). Sensor-supplied information provides dynamic information that feeds the analysis process; however, relatively static domain-context knowledge provides equally valuable information that constrains the interpretation of sensor-derived information. Due to the potentially large volume of both sensor and nonsensor-derived databases, the character and capability of the supporting database management system can significantly impact both the effectiveness and the efficiency of machine-based reasoning.

This chapter outlined a number of top-down design considerations for an object database kernel that supports the development of both effective and efficient data fusion algorithms. At the highest level of abstraction, the near-optimal database kernel consists of two classes of objects: semantic and spatial. Because conventional OODBMS provide adequate support to semantic object representations, the chapter focused on the design for the spatial object representation.

A spatial object realization consisting of an object representation of 2-D space integrated with a hybrid spatial representation of individual point, line, and region features was shown to achieve an effective compromise across all design criteria. An object representation of 2-D space provides a spatial object hierarchy metaphorically similar to a conventional semantic object hierarchy. Just as a semantic object hierarchy supports top-down semantic reasoning, a spatial object hierarchy supports top-down spatial reasoning. A hybrid spatial representation, the quadtree-indexed vector representation, supports an efficient top-down search and analysis and high-precision refined analysis of individual spatial features. Both the object representation of 2-D space and the multiple-resolution representation of individual spatial features employ the identical quadtree decomposition. Therefore, the quadtree-indexed vector representation is a natural extension of the object representation of 2-D space.

Acknowledgment

Preparation of this chapter was funded by the U.S. Army CECOM I2WD, Fort Monmouth, NJ.

Reference

1. R. Antony, *Principles of Data Fusion Automation*, Artech House Inc., Boston, 1995.

19

Removing the HCI Bottleneck: How the Human-Computer Interface (HCI) Affects the Performance of Data Fusion Systems*

Mary Jane M. Hall
TECH REACH Inc.

Capt. Sonya A. Hall
Minot AFB

Timothy Tate
Naval Training Command

19.1 Introduction 19-1
19.2 A Multimedia Experiment.. 19-3
    SBIR Objective • Experimental Design and Test Approach • CBT Implementation
19.3 Summary of Results.. 19-5
19.4 Implications for Data Fusion Systems........................... 19-9
Acknowledgment 19-10
References.. 19-11

19.1 Introduction

During the past two decades, an enormous amount of effort has focused on the development of automated multisensor data systems.[1-3] These systems seek to combine data from multiple sensors to improve the ability to detect, locate, characterize, and identify targets. Since the early 1970s, numerous data fusion systems have been developed for a wide variety of applications, such as automatic target recognition, identification-friend-foe-neutral (IFFN), situation assessment, and threat assessment.[4] At this time, an extensive legacy exists for department of defense (DoD) applications. That legacy includes a hierarchical process model produced by the Joint Directors of Laboratories (shown in Figure 19.1), a taxonomy of algorithms,[5] training material,[6] and engineering guidelines for algorithm selection.[7]

The traditional approach for fusion of data progresses from the sensor data (shown on the left side of Figure 19.1) toward the human user (on the right side of Figure 19.1). Conceptually, sensor data are preprocessed using signal processing or image processing algorithms. The sensor data are input to a Level 1 fusion process that involves data association and correlation, state vector estimation, and identity

---

*This chapter is based on a paper by Mary Jane Hall et al., Removing the HCI bottleneck: How the human computer interface (HCI) affects the performance of data fusion systems, *Proceedings of the 2000 MSS National Symposium on Sensor and Data Fusion, Vol. II,* June 2000, pp. 89–104.

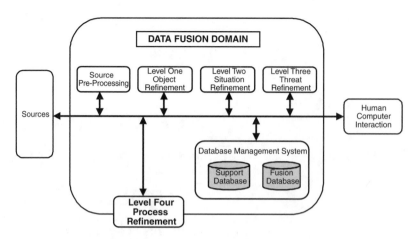

FIGURE 19.1   Joint directors of laboratories (JDL) data fusion process model.

estimation. The Level 1 process results in an evolving database that contains estimates of the position, velocity, attributes, and identities of physically constrained entities (e.g., targets and emitters). Subsequently, automated reasoning methods are applied in an attempt to perform automated situation assessment and threat assessment. These automated reasoning methods are drawn from the discipline of artificial intelligence.

Ultimately, the results of this dynamic process are displayed for a human user or analyst (via a human-computer interface (HCI) function). Note that this description of the data fusion process has been greatly simplified for conceptual purposes. Actual data fusion processing is much more complicated and involves an interleaving of the Level 1 through Level 3 (and Level 4) processes. Nevertheless, this basic orientation is often used in developing data fusion systems: the sensors are viewed as the information source and the human is viewed as the information user or sink. In one sense, the rich information from the sensors (e.g., the radio frequency time series and imagery) is compressed for display on a small, two-dimensional computer screen.

Bram Ferran, the vice president of research and development at Disney Imagineering Company, recently pointed out to a government agency that this approach is a problem for the intelligence community. Ferran[8] argues that the broadband sensor data are funneled through a very narrow channel (i.e., the computer screen on a typical workstation) to be processed by a broadband human analyst. In his view, the HCI becomes a bottleneck or very narrow filter that prohibits the analyst from using his extensive pattern recognition and analytical capability. Ferran suggests that the computer bottleneck effectively defeats one million years of evolution that have made humans excellent data gatherers and processors. Interestingly, Clifford Stoll[9,10] makes a similar argument about personal computers and the multimedia misnomer.

Researchers in the data fusion community have not ignored this problem. Waltz and Llinas[3] noted that the overall effectiveness of a data fusion system (from sensing to decisions) is affected by the efficacy of the HCI. Llinas and his colleagues[11] investigated the effects of human trust in aided adversarial decision support systems, and Hall and Llinas[12] identified the HCI area as a key research need for data fusion. Indeed, in the past decade, numerous efforts have been made to design visual environments, special displays, HCI toolkits, and multimedia concepts to improve the information display and analysis process. Examples can be found in the papers by Neal and Shapiro,[13] Morgan and Nauda,[14] Nelson,[15] Marchak and Whitney,[16] Pagel,[17] Clifton,[18] Hall and Wise,[19] Kerr et al.,[20] Brendle,[21] and Steele, Marzen, and Corona.[22]

A particularly interesting antisubmarine warfare (ASW) experiment was reported by Wohl et al.[23] Wohl and his colleagues developed some simple tools to assist ASW analysts in interpreting sensor data. The tools were designed to overcome known limitations in human decision making and perception. Although very basic, the support tools provided a significant increase in the effectiveness of the ASW analysis. The experiment suggested that cognitive-based tools might provide the basis for significant improvements in the effectiveness of a data fusion system.

FIGURE 19.2   Example of a full immersion 3-D (HCI).

In recent years, there have been enormous advances in the technology of human computer interfaces. Advanced HCI devices include environments such as:

- A three-dimensional full immersion NCSA CAVE™, illustrated in Figure 19.2, which was developed at the University of Illinois, Champaign-Urbana campus (http://www.ncsa.uiuc.edu/VEG/ncsa-CAVE.html).
- Haptic interfaces to allow a person to touch and feel a computer display.[24]
- Wearable computers for augmented reality.[25]

The technology exists to provide very realistic displays and interaction with a computer. Such realism can even be achieved in field conditions using wearable computers, heads-up displays, and eye-safe laser devices that paint images directly on the retina.

Unfortunately, advances in understanding of human information needs and how information is processed have not progressed as rapidly. There is still much to learn about cognitive models and how humans access, reason with, and are affected by information.[26-29] That lack of understanding of cognitive-based information access and the potential for improving the effectiveness of data fusion systems motivated the research described in this chapter.

19.2   A Multimedia Experiment

19.2.1   SBIR Objective

Under a Phase II SBIR effort (Contract No. N00024-97-C-4172), Tech Reach, Inc. (a small company located in State College, PA) designed and conducted an experiment to determine if a multimode information access approach improves learning efficacy. The basic concept involved the research hypothesis that computer-assisted training, which adapts to the information access needs of individual students, significantly improves training effectiveness while reducing training time and costs.

The Phase II effort included

- Designing, implementing, testing, and evaluating a prototype computer-based training (CBT) system that presents material in three formats (emphasizing aural, visual, and kinesthetic presentations of subject material);
- Selecting and testing an instrument to assess a student's most effective learning mode;
- Developing an experimental design to test the hypothesis;
- Conducting a statistical analysis to affirm or refute the research hypothesis;
- Documenting the results.

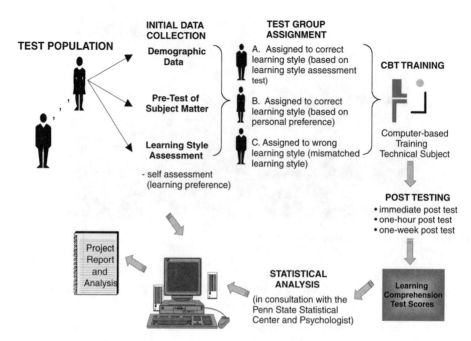

FIGURE 19.3   Overview of a test concept.

19.2.2   Experimental Design and Test Approach

The basic testing concept for this project is shown in Figure 19.3 and described in detail by M. J. Hall.[30] The selected sample consisted of approximately 100 Penn State ROTC students, 22 selected adult learners (i.e., post-secondary adult education students), and 120 U.S. Navy (USN) enlisted personnel at the USN Atlantic Fleet Training Center at DAM NECK (Virginia Beach, VA). This sample was selected to be representative of the population of interest to the U.S. Navy sponsor.

As shown in Figure 19.3, the testing was conducted using the following steps:

1. **Initial Data Collection:** Data were collected to characterize the students in the sample (including demographic information, a pretest of the students' knowledge of the subject matter, and a learning style assessment using standard test instruments).
2. **Test Group Assignment:** The students were randomly assigned to one of three test groups. The first group used the CBT that provided training in a mode that matched their learning preference mode as determined by the CAPSOL learning styles inventory instrument.[31] The second group trained using the CBT that emphasized their learning preference mode as determined by the student's self-selection. Finally, the third group was trained using the CBT that emphasized a learning preference mode that was deliberately mismatched with the student's preferred mode (e.g., utilization of aural emphasis for a student whose learning preference is known to be visual).
3. **CBT Training:** Each student was trained on the subject matter using the interactive computer-based training module (utilizing one of the three information presentation modes: visual, aural, or kinesthetic).
4. **Post-testing:** Post-testing was conducted to determine how well the students mastered the training material. Three post-tests were conducted: (a) an immediate post-test after completion of the training material, (b) an identical comprehension test administered one hour after the training session, and (c) an identical comprehensive test administered one week after the initial training session.

The test subjects were provided with a written explanation of the object of the experiment and its value to the DoD. Testing was conducted in four locations, as summarized in Table 19.1. Test conditions

TABLE 19.1 Summary of Conducted Tests

Test	Objective	Test Subjects	Date/Location
Benchmark	Correlate learning style inventories to determine whether the CAP-SOL is sufficiently reliable and valid in comparison to the Canfield LSI	50 Penn State University U.S. Navy (USN) ROTC Cadets	22 Jan. 1998 / Wagner Building The Pennsylvania State Univ. (PSU) University Park, PA
Concept Testing	Determine if the use of alternative modes of information access (i.e., aural, visual, and kinethestic emphasized presentation styles) provides enhanced learning using a computer-based training (CBT) delivery system	54 PSU USN ROTC Cadets 47 PSU U.S. Air Force ROTC Cadets 12 Altoona Career and Technology Center students 5 South Hills Business School students	5 and 12 Feb. 1998 / Wagner Building, PSU 17 and 24 Feb. 1998 / Wagner Building, PSU 21 and 28 April 1998 / Altoona Career and Technology Center, Altoona, PA 11 and 18 May 1998 / FCTLANT, DAM NECK Virginia Beach, VA
Operational Testing	Determine if the use of alternative modes of information access (i.e., aural, visual, and kinethestic emphasized presentation styles) provides enhanced learning using a computer based training (CBT) delivery system	87 U.S. Navy enlistees	3 and 10 August 1998 / FCTLANT, DAM NECK Virginia Beach, VA

were controlled to minimize extraneous variations (e.g., the use of different rooms for pre- and post-tests, different time of day for learning versus the one-week post-test, or different instructions provided to test subjects).

19.2.3   CBT Implementation

The computer-based training (CBT) module for this experiment was a training module that described the functions and use of an oscilloscope. The training module was implemented using interactive multimedia software for operation on personal computers. The commercial authoring shell, *Toolbook*, developed by Asymetrix Corporation, was used for the implementation. DoD standards were followed for the design and implementation of the CBT module. An example of the CBT display screens is shown in Figures 19.4, 19.5, and 19.6.

The subject matter selected — operation and functions of an oscilloscope — was chosen for several reasons. First, the subject matter is typical of the training requirements for military personnel involved in equipment operation, maintenance, and repair. Second, the subject could be trained in a coherent, yet small, CBT module. Third, the likelihood that a significant number of test participants would have *a priori* knowledge of the subject matter was small. Finally, the subject matter was amenable to implementation with varied emphasis on aural, visual, and kinesthetic presentation styles. All of the CBT screens, aural scripts, and logic for the implemented CBT modules are described by M. J. Hall.[32]

19.3   Summary of Results

The details of the analysis and the results of the multimedia experiment are provided by M. J. Hall[33] and S. A. Hall.[34] For brevity, this chapter contains only a brief summary focusing on the results for the Penn State ROTC students.

During February 1998, 101 Pennsylvania State University USAF and USN enlisted personnel were tested at the Wagner Building, University Park, PA. In particular, 54 USN ROTC students were tested on February 5 and 12. Similarly, 47 USAF ROTC students were tested on February 17 and 24. During the first session, demographic data were collected, along with information on learning preference (via the CAPSOL and

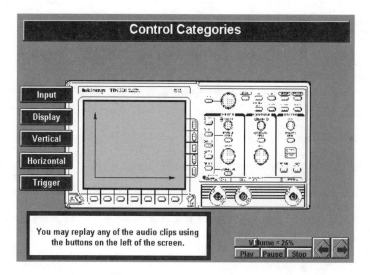

FIGURE 19.4   Example of an aural CBT display screen.

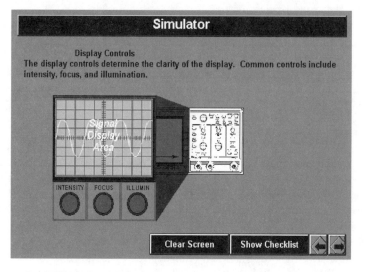

FIGURE 19.5   Example of an kinesthetic CBT display screen.

the self-assessment preference statements). During the initial session, a subject matter pretest was administered and followed by the CBT. Immediately after the training, the subject matter test was given, followed by the one-hour test. One week later, another post-test was administered.

The Penn State ROTC students represented a relatively homogeneous and highly motivated group (as judged by direct observation and by the anonymous questionnaire completed by each student). This group of undergraduate students was closely grouped in age, consisted primarily of Caucasian males, and was heavily oriented toward scientific and technical disciplines. The students were also highly computer literate. These students seemed to enjoy participating in an educational experiment and were pleased to be diverted from their usual leadership laboratory assignments.

A sample of the test results is shown in Figure 19.7. The figure shows the average number of correct answers obtained from students based on the subject matter pretest (presumably demonstrating the student's *a priori* knowledge of the subject), followed by the immediate post-test (demonstrating the amount learned based on the CBT), followed by the one-hour post-test and, finally, the one-week post-test. The latter two tests sought to measure the retention of the learned subject matter. Figure 19.7 shows

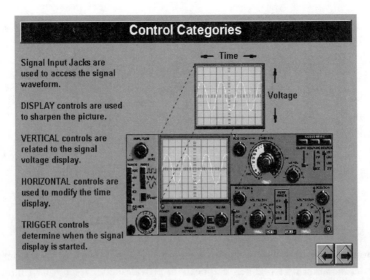

FIGURE 19.6   Example of a visual CBT display screen.

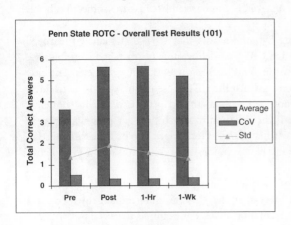

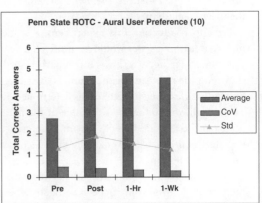

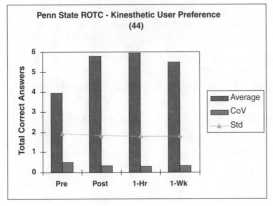

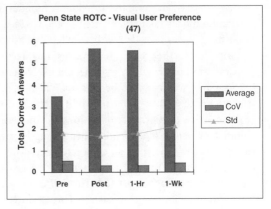

FIGURE 19.7   Sample test results for ROTC students — presentation mode.

four histograms: (1) the overall test results in the upper left side of the figure, (2) results for users who preferred the aural mode in the upper right corner, (3) results of students who preferred the kinesthetic mode, in the lower right corner, and finally, (4) students who preferred the visual mode of presentation.

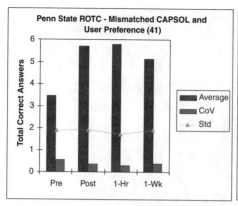
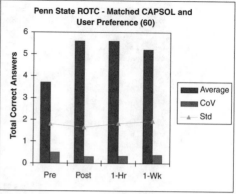

FIGURE 19.8 Matched vs. mismatched learning (ROTC).

Although these results show only a small difference in learning and learning retention based on presentation style, the aural user preference appears to provide better retention of information over a one-week period. An issue yet to be investigated is whether this effect is caused by an unintended secondary mode re-enforcement (i.e., whether the material is emphasized because the subject both *sees* the material and *hears* it presented).

Prior to conducting this experiment, a working hypothesis was that CBT that matched learners' preferred learning styles would be more effective than CBT that deliberately mismatched learning style preferences. Figure 19.8 shows a comparison of learning and learning retention in two cases. The case shown on the right side of the figure is the situation in which the CBT presentation style is matched to the student's learning preference (as determined by both self-assessment and by the CAPSOL instrument). The case shown on the left-hand side of the figure is the situation in which the CBT presentation style is mismatched to the student's learning preference. At first glance, there appears to be little difference in the effect of the matched versus the mismatched cases. However, long-term retention seems to be better for matched training.

Before concluding that match versus mismatch of learning style has little effect on training efficacy, a number of factors need to be investigated. First, note that the CBT module implemented for this experiment exhibited only a limited separation of presentation styles. For example, the aural presentation style was not solely aural but provided a number of graphic displays for the students. Hence, the visual and aural modes (as implemented in the CBT) were not mutually exclusive. Thus, a visually oriented student who was provided with aurally emphasized training could still receive a significant amount of information via the graphic displays. A similar factor was involved in the kinesthetic style. A more extreme separation of presentation styles would likely show a greater effect on the learning efficacy.

Second, a number of other factors had a significant effect on learning efficacy. Surprisingly, these unanticipated factors overshadowed the effect of learning style match versus mismatch. One factor in particular has significant implications for the design of data fusion systems: whether a user considers himself a *group* or *individual* learner. Group learners prefer to learn in a group setting, while individual learners prefer to learn in an exploratory mode as an individual. Figure 19.9 shows a comparison of the learning retention results by individual versus group learning styles. The figure shows the change in score from the pretest to the post-test. The figure also shows the change from the pretest to the one-hour post-test, and to the one-week post-test. These values are shown for group learners, individual learners, and learners who have no strong preference between group or individual learning. The figure shows that individual learners (and students who have no strong preference) exhibited a significant increase in both learning and learning retention over students who consider themselves group learners. In effect, students who consider themselves to be group learners gain very little from the CBT training. This is simply one of the personal factors that affect the efficacy of computer-based training. M. J. Hall[33] and S. A. Hall[34] provide a more complete discussion of these factors.

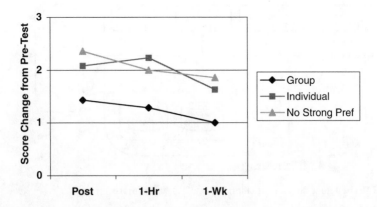

FIGURE 19.9 Learning retention by individual vs. group learning styles.

19.4 Implications for Data Fusion Systems

The experiment described in this chapter was a very basic experiment using a homogeneous, highly motivated group of ROTC students. All of these students were highly computer literate. Although preliminary, the results indicate that factors such as group versus individual learning style can significantly affect the ability of an individual to comprehend and obtain information from a computer. This suggests that efforts to create increasingly sophisticated computer displays may have little or no effect on the ability of some users to understand and use the presented data. Many other factors, such as user stress, the user's trust in the decision-support system, and preferences for information access style, also affect the efficacy of the human-computer interface.

Extensive research is required in this area. Instead of allowing the HCI for a data fusion system to be driven by the latest and greatest display technology, researchers should examine this subject more fully to develop adaptive interfaces that encourage human-centered data fusion. This theme is echoed by Hall and Garga.[35] This approach could break the HCI bottleneck (especially for nonvisual, group-oriented individuals) and leverage the human cognitive abilities for wide-band data access and processing. This area should be explicitly recognized by creating a Level 5 process in the JDL data fusion process model. This concept is illustrated in Figure 19.10.

In this concept, HCI processing functions are explicitly augmented by functions to provide a cognitive-based interface. What functions should be included in the new Level 5 process? The following are examples of new types of algorithms and functions for Level 5 processing (based on discussions with D. L. Hall):[36]

- **Deliberate synesthesia:** Synesthesia is a neurological disorder in humans in which the senses are cross-wired.[37] For example, one might associate a particular taste with the color red. Typically, this disorder is associated with schizophrenia or drug abuse. However, such a concept could be deliberately exploited for normal humans to translate visual information into other types of representations, such as sounds (including direction of the sound) or haptic cues. For example, sound might offer a better means of distinguishing between closely spaced emitters than overlapping volumes in feature space. Algorithms could be implemented to perform sensory cross-translation to improve understanding.

- **Time compression/expansion:** Human senses are especially oriented to detecting change. Development of time compression and time expansion replay techniques could assist the understanding of an evolving tactical situation.

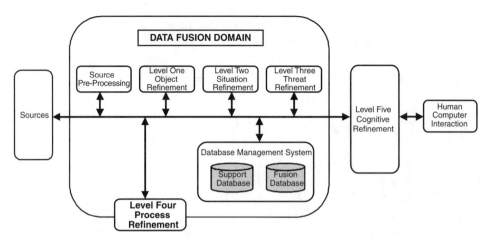

FIGURE 19.10   Revised joint directors of laboratories (JDL) fusion process model.

- **Negative reasoning enhancement:** For many types of diagnosis, such as mechanical fault diagnosis or medical pathology, experts explicitly rely on negative reasoning.[38] This approach explicitly considers what information is not present which would confirm or refute a hypothesis. Unfortunately, however, humans have a tendency to ignore negative information and only seek information that confirms a hypothesis (see Piattelli-Palmarini's description of the three-card problem[39]). Negative reasoning techniques could be developed to overcome the tendency to seek confirmatory evidence.
- **Focus/defocus of attention:** Methods could be developed to systematically assist in directing the attention of an analyst to consider different aspects of data. In addition, methods might be developed to allow a user to de-focus his attention in order to comprehend a broader picture. This is analogous to how experienced Aikido masters deliberately blur their vision in order to avoid distraction by an opponent's feints.[40]
- **Pattern morphing methods:** Methods could be developed to translate patterns of data into forms that are more amenable for human interpretation (e.g., the use of Chernoff faces to represent varying conditions or the use of Gabor-type transformations to leverage our natural vision process.[41]
- **Cognitive aids:** Numerous cognitive aids could be developed to assist human understanding and exploitation of data. Experiments should be conducted along the lines initiated by Wohl et al.[23] Tools could also be developed along the lines suggested by Rheingold.[29]
- **Uncertainty representation:** Finally, visual and oral techniques could be developed to improve the representation of uncertainty. An example would be the use of three-dimensional icons to represent the identity of a target. The uncertainty in the identification could be represented by blurring or transparency of the icon.

These areas only touch the surface of the human-computer interface improvements. By rethinking the HCI for data fusion, we may be able to re-engage the human in the data fusion process and leverage our evolutionary heritage.

Acknowledgment

Funding for this research was provided under a Phase II SBIR by NAVSEA, Contract No. N00024-97-C-4172, September 1997.

References

1. Hall, D.L. and Llinas, J., An introduction to multisensor data fusion, in *Proc. IEEE*, Vol. 85, No. 1, January 1997.
2. Hall, D.L., *Mathematical Techniques in Multisensor Data Fusion*, Artech House Inc., Norwood, MA, 1992.
3. Waltz, E. and Llinas, J., *Multisensor Data Fusion*, Artech House Inc., Norwood, MA, 1990.
4. Hall, D.L., Linn, R.J., and Llinas, J., A survey of data fusion systems, in *Proc. SPIE Conf. on Data Structures and Target Classification*, 1490, SPIE, 1991, 13.
5. Hall, D.L. and Linn, R.J., A taxonomy of algorithms for multisensor data fusion, in *Proc. 1990 Joint Service Data Fusion Symp.*, Johns Hopkins Applied Research Laboratory, Laurel, MD, 1990, 593.
6. Hall, D.L., *Lectures in Multisensor Data Fusion*, Artech House, Inc., Norwood, MA, 2000.
7. Hall, D.L. and R.J. Linn, Algorithm selection for data fusion systems, in *Proc. of the 1987 Tri-Service Data Fusion Symp.*, Johns Hopkins Applied Physics Laboratory, Laurel, MD, 1987, 100.
8. Ferran, B., presentation to the U.S. Government, Spring 1999 (available on videotape).
9. Stoll, C., *Silicon Snake Oil: Second Thoughts on the Information Highway*, Doubleday, New York, 1995.
10. Stoll, C., *High Tech Heretic: Why Computers Don't Belong in the Classroom and Other Reflections by a Computer Contrarian*, Doubleday, New York, 1999.
11. Llinas, J. et al., Studies and analyses of vulnerabilities in aided adversarial decision-making, technical report, State University of New York at Buffalo, Dept. of Industrial Engineering, February 1997.
12. Hall, D.L and Llinas, J., A challenge for the data fusion community I: research imperatives for improved processing, in *Proc. 7th Nat. Symp. on Sensor Fusion*, Albuquerque, NM, 1994.
13. Neal, J.G. and Shapiro, S.C., Intelligent integrated interface technology, in *Proc. 1987 Tri-Service Data Fusion Symp.*, Johns Hopkins University, Applied Physics Laboratory, Laurel, MD, 1987, 428.
14. Morgan, S.L. and Nauda, A., A user-system interface design tool, in *Proc. 1988 Tri-Service Data Fusion Symp.*, Johns Hopkins University, Applied Physics Laboratory, Laurel, MD, 1988, 377.
15. Nelson, J.B., Rapid prototyping for intelligence analyst interfaces, in *Proc. 1989 Tri-Service Data Fusion Symp.*, Johns Hopkins University, Applied Physics Laboratory, Laurel, MD, 1989, 329.
16. Marchak, F.M. and Whitney, D.A., Rapid prototyping in the design of an integrated sonar processing workstation, in *Proc. 1991 Joint Service Data Fusion Symp.*, 1, Johns Hopkins University, Applied Physics Laboratory, Laurel, MD, 1991, 606.
17. Pagel, K., Lessons learned from HYPRION, JNIDS Hypermedia authoring project, in *Proc. Sixth Joint Service Data Fusion Symp.*, 1, Johns Hopkins University, Applied Physics Laboratory, Laurel, MD, 1993, 555.
18. Clifton III, T.E., ENVOY: An analyst's tool for multiple heterogeneous data source access, in *Proc. Sixth Joint Service Data Fusion Symposium*, 1, Johns Hopkins University, Applied Physics Laboratory, Laurel, MD, 1993, 565.
19. Hall, D.L. and Wise, J.H., The use of multimedia technology for multisensor data fusion training, *Proc. Sixth Joint Service Data Fusion Symp.*, 1, Johns Hopkins University, Applied Physics Laboratory, Laurel, MD, 1993, 243.
20. Kerr, R. K. et al., TEIA: Tactical environmental information agent, in *Proc. Intelligent Ships Symp. II*, The American Society of Naval Engineers Delaware Valley Section, Philadelphia, PA, 1996, 173.
21. Brendle Jr., B. E., Crewman's associate: interfacing to the digitized battlefield, in *Proc. SPIE: Digitization of the Battlefield II*, 3080, Orlando, FL, 1997, 195.
22. Steele, A., V. Marzen, and B. Corona, Army Research Laboratory advanced displays and interactive displays Fedlab technology transitions, in *Proc. SPIE: Digitization of the Battlespace IV*, 3709, Orlando, Florida, 1999, 205.
23. Wohl, J.G. et al., Human cognitive performance in ASW data fusion, in *Proc. 1987 Tri-Service Data Fusion Symp.*, Johns Hopkins University, Applied Physics Laboratory, Laurel, MD, 1987, 465.

24. Ellis, R.E., Ismaeil, O.M., and Lipsett, M., Design and evaluation of a high-performance haptic interface, *Robotica*, 14, 321, 1996.

25. Gemperle, F. et al., Design for wearability, in *Proc. Second International Symp. on Wearable Computers*, Pittsburgh, PA, 1998, 116.

26. Pinker, S., *How the Mind Works*, Penguin Books Ltd., London, 1997.

27. Claxton, G., *Hare Brain Tortoise Mind: Why Intelligence Increases When You Think Less*, The Ecco Press, Hopewell, NJ, 1997.

28. J. St. B. T. Evans, S. E. Newstead, and R. M. J. Byrne, *Human Reasoning: The Psychology of Deduction*, Lawrence Erlbaum Associates, 1993.

29. Rheingold, H., *Tools for Thought: The History and Future of Mind-Expanding Technology*, 2nd ed., MIT Press, Cambridge, MA, 2000.

30. Hall, M.J., R&D test and acceptance plan, SBIR Project N95-171, Report Number A009, Contract Number N00024-97-C-4172, prepared for the Naval Sea Systems Command, December 1998.

31. CAP-WARE: Computerized Assessment Program, Process Associates, Mansfield, OH, 1987.

32. Hall, M.J., Product drawings and associated lists, SBIR Project N95-171, Report Number A012, Contract Number N00024-97-C-4172, prepared for the Naval Sea Systems Command, September 1998.

33. Hall, M.J., Adaptive human computer interface (HCI) for improved learning in the electronic classroom, final report, Phase II SBIR Project N95-171, Contract No. N00024-97-C-4172, NAVSEA, Arlington, VA, September 1998c.

34. Hall, S.A., An investigation of factors that affect the efficacy of human-computer interaction for military applications, MS thesis, Aeronautical Science Department, Embry-Riddle University, December 2000.

35. Hall, D.L. and Garga, A.K., Pitfalls in data fusion (and how to avoid them), in *Proc. 2nd Int. Conf. on Information Fusion (Fusion 99)*, Sunnyvale, CA, 1999.

36. Hall, D.L., Private communication to M.J. Hall, April 23, 2000.

37. Bailey, D., Hideaway: ongoing experiments in synthesia, Research Initiative Grant (SRIS), University of Maryland, Baltimore, Graduate School, 1992, http://www.research.umbc.edu/~bailey/filmography.htm.

38. Hall, D.L., Hansen, R.J., and Lang, D.C., The negative information problem in mechanical diagnostics, *Transactions of the ASME*, 119, 1997, 370.

39. Piattelli-Palmarini, M., *Inevitable Illusions: How Mistakes of Reason Rule over Minds*, John Wiley & Sons, New York, 1994.

40. Dobson, T. and Miller, V. (Contributor), *Aikido in Everyday Life: Giving in to Get Your Way*, North Atlantic Books, reprint edition, March 1993.

41. Olshausen, B.A. and Field, D.J., Vision and coding of natural images, *American Scientist*, 88, 2000, 238.

20

Assessing the Performance of Multisensor Fusion Processes

20.1 Introduction 20-1
20.2 Test and Evaluation of the Data Fusion Process........... 20-3
Establishing the Context for Evaluation • T&E
Philosophies • T&E Criteria • Approach to T&E • The
T&E Process — A Summary
20.3 Tools for Evaluation: Testbeds, Simulations, and
Standard Data Sets 20-7
20.4 Relating Fusion Performance to Military
Effectiveness — Measures of Merit............................... 20-11
20.5 Summary... 20-17
References 20-17

James Llinas
State University of New York

20.1 Introduction

In recent years, numerous prototypical systems have been developed for multisensor data fusion. A paper by Hall, Linn, and Llinas[1] describes over 50 such systems developed for DoD applications even some 10 years ago. Such systems have become ever more sophisticated. Indeed, many of the prototypical systems summarized by Hall, Linn, and Llinas[1] utilize advanced identification techniques such as knowledge-based or expert systems, Dempster-Shafer interface techniques, adaptive neural networks, and sophisticated tracking algorithms.

While much research is being performed to develop and apply new algorithms and techniques, much less work has been performed to formalize the techniques for determining how well such methods work or to compare alternative methods against a common problem. The issues of system performance and system effectiveness are keys to establishing, first, how well an algorithm, technique, or collection of techniques performs in a technical sense and, second, the extent to which these techniques, as part of a system, contribute to the probability of success when that system is employed on an operational mission. An important point to remember in considering the evaluation of data fusion processes is that those processes are either a component of a system (if they were designed-in at the beginning) or they are enhancements to a system (if they have been incorporated with the intention of performance enhancement). Said otherwise, it is not usual that the data fusion processes are "the" system under test; data fusion processes are said to be designed *into* systems rather than being systems in their own right. What is important to understand in this sense is that the data fusion processes contribute a marginal or piecewise improvement to the overall system, and if the contribution of the DF process per se wants to be calculated, it must be done while holding other factors fixed. If the DF processes under examination are enhancements, another important point is that such performance must be evaluated in comparison to an agreed-to baseline (e.g., without DF capability, or presumably a "lesser" DF capability). More will be said on these points later.

Another early point to be made is that our discussion here is largely about *automated* DF processing (although we will make some comments about human-in-the-loop aspects later), and by and large such processes are enabled through software. Thus, it should be no surprise that remarks made herein draw on or are similar to concerns for test and evaluation of complex software processes.

System performance at Level 1, for example, focuses on establishing how well a system of sensors and data fusion algorithms may be utilized to achieve estimates of or inferences about location, attributes, and identity of platforms or emitters. Particular measures of performance (MOPs) may characterize a fusion system by computing one or more of the following:

- Detection probability — probability of detecting entities as a function of range, signal-to-noise ratio, etc.
- False alarm rate — rate at which noisy or spurious signals are incorrectly identified as valid targets.
- Location estimate accuracy — the accuracy with which the position of an entity is determined.
- Identification probability — probability of correctly identifying an entity as a target.
- Identification range — the range between a sensing system and target at which the probability of correct identification exceeds an established threshold.
- Time from transmission to detect — time delay between a signal emitted by a target (or by an active sensor) and the detection by a fusion system.
- Target classification accuracy — ability of a sensor suite and fusion system to correctly identify a target as a member of a general (or particular) class or category.

These MOPs measure the ability of the fusion process as an information process to transform signal energy either emitted by or reflected from a target, to infer the location, attributes, or identity of the target. MOPs are often functions of several dimensional parameters used to quantify, in a single variable, a measure of operational performance.

Conversely, measures of effectiveness (MOEs) seek to provide a measure of the ability of a fusion system to assist in completion of an operational mission. MOEs may include

- Target nomination rate — the rate at which the system identifies and nominates targets for consideration by weapon systems.
- Timeliness of information — timeline of availability of information to support command decisions.
- Warning time — time provided to warn a user of impending danger or enemy activity.
- Target leakage — percent of enemy units or targets that evade detection.
- Countermeasure immunity — ability of a fusion system to avoid degradation by enemy countermeasures.

At an even higher level, measures of force effectiveness (MOFE) quantify the ability of the total military force (including the systems having data fusion capabilities) to complete its mission. Typical MOFEs include rates and ratios of attrition, outcomes of engagement, and functions of these variables. In the overall mission definition other factors such as cost, size of force, force composition, etc. may also be included in the MOFE.

This chapter presents both top-down, conceptual and methodological ideas on the test and evaluation of data fusion processes and systems, describes some of the tools available and needed to support such evaluations, and discusses the spectrum of measures of merit useful for quantification of evaluation results.

20.2   Test and Evaluation of the Data Fusion Process

Although, as has been mentioned above, the DF process is frequently part of a larger system process (i.e., DF is often a "subsystem" or "infrastructure" process to a larger whole) and thereby would be subjected to an organized set of system-level test procedures, this section develops a stand-alone, top-level model of the test and evaluation (T&E) activity for a general DF process. This characterization is considered the proper starting point for the subsequent detailed discussions on metrics and evaluation because it establishes a viewpoint or framework (a context) for those discussions, and also because it challenges the DF process architect to formulate a global and defendable approach to T&E.

In this discussion, it is important to understand the difference between the terms "test" and "evaluation." One distinction (according to *Webster's* dictionary) is that testing forms a basis for evaluation. Alternately, testing is a process of conducting trials in order to prove or disprove a hypothesis — here, a hypothesis regarding the characteristics of a procedure within the DF process. Testing is essentially laboratory experimentation regarding the active functionality of DF procedures and, ultimately, the overall process (*active* meaning during their execution — not statically analyzed).

On the other hand, evaluation takes its definition from its root word: value. Evaluation is thus a process by which the value of DF procedures is determined. Value is something measured in context; it is because of this that a context must be established.

The view taken here is that the T&E activities will both be characterized as having the following components:

- A *philosophy* that establishes or emphasizes a particular point of view for the tests and/or evaluations that follow. The simplest example of this notion is reflected in the so-called "black box" or "white box" viewpoints for T&E, from which either external (I/O behaviors) or internal (procedure execution behaviors) are examined (a similar concern for software processes in general, as noted above). Another point of view revolves about the research or development goals established for the program. The philosophy establishes the high-level statement the context mentioned above and is closely intertwined with the program goals and objectives, as discussed below.
- A set of *criteria* according to which the quality and correctness of the T&E results or inferences will be judged.
- A set of *measures* through which judgments on criteria can be made, and a set of *metrics* upon which the measures depend and, importantly, *which can be measured during T&E experiments.*
- An *approach* through which tests and/or analyses can be defined and conducted that
  - Are consistent with the philosophy, and
  - Produce results (measures and metrics) that can be effectively judged against the criteria.

20.2.1   Establishing the Context for Evaluation

Assessments of delivered value for defense systems must be judged in light of system or program goals and objectives. In the design and development of such systems, many translations of the stated goals and objectives occur as a result of the systems engineering process, which both analyses (decomposes) the goals into functional and performance requirements and synthesizes (reassembles) system components intended to perform in accordance with these requirements. Throughout this process, however, the program goals and objectives must be kept in view because they establish the context in which value will be judged.

Context therefore reflects what the program and the DF process or system within it are trying to achieve — i.e., what the research or developmental goals (the purposes of building the system at hand) are. Such goals are typically reflected in the program name, such as a "Proof of Concept" program or "Production Prototype" program. Many recent programs involve "demonstrations" or "experiments" of some type or other, with these words reflecting in part the nature of such program goals or objectives.

Several translations must occur for the T&E activities themselves. The first of these is the translation of goals and objectives into T&E philosophies; i.e., philosophies follow from statements about goals and objectives. Philosophies primarily establish points of view or perspectives for T&E that are consistent with, and can be traced to, the goals and objectives: they establish the *purpose* of investing in the T&E process. Philosophies also provide guidelines for the development of T&E criteria, for the definition of meaningful T&E cases and conditions, and, importantly, a sense of a "satisfaction scale" for test results and value judgments that guides the overall investment of precious resources in the T&E process. That is, T&E philosophies, while generally stated in nonfinancial terms, do in fact establish economic philosophies for the commitment of funds and resources to the T&E process. In today's environment (it makes sense categorically in any case), notions of affordability must be considered for any part of the overall systems engineering approach and for system development, to include certainly the degree of investment to be made in T&E functions.

20.2.2   T&E Philosophies

Establishing a philosophy for T&E of a DF process is also tightly coupled to the establishment of what the DF process boundaries are. In general, it can argued that the T&E of any process within a system should attempt the longest extrapolation possible in relating process behavior to program goals; i.e., the evaluation should endeavor to relate process test results to program goals *to the extent possible*. This entails first understanding the DF process boundary, and then assessing the degree to which DF process results can be related to superordinate processes; for defense systems, this means assessing the degree to which DF results can be related to mission goals. Philosophies aside, certain "acid tests" should always be conducted:

- Results with and without fusion (e.g., multisensor vs. single sensor or some "best" sensor).
- Results as a function of the number of sensors or sources involved (e.g., single sensor, 2, 3,...,N sensor results for a common problem).

These last two points are associated with defining some type of *baseline* against which the candidate fusion process is being evaluated. Said otherwise, these points address the question, "Fusion as compared to what?" If it is agreed that data fusion processing provides a marginal benefit, then that gain must be evaluated in comparison to the "unenhanced" or baseline system. That comparison also provides the basis for the cost-effectiveness tradeoff in that the relative costs of the baseline and fusion-enhanced systems can be compared to the relative performance of each.

Other philosophies could be established, however, such as

- *Organizational*: A philosophy that examines the benefits of DF products accruing to the system-owning organization and, in turn, subsequent superordinate organizations in the context of organizational purposes, goals, and objectives (no "platform" or "mission" may be involved; the benefits may accrue to an organization).
- *Economic*: A philosophy that is explicitly focused on some sense of economic value of the DF results (weight, power, volume, etc.) or cost in a larger sense, such as the cost of weapons expended, etc.
- *Informal*: The class of philosophies in which DF results are measured against some human results or expectations.
- *Formal*: The class of philosophies in which the evaluation is carried out according to appropriate formal techniques that prove or otherwise rigorously validate the program results or internal behaviors (e.g., proofs of correctness, formal logic tests. formal evaluations of complexity).

The list is not presented as complete but as representative; further consideration would no doubt uncover still other perspectives.

20.2.3   T&E Criteria

Once having espoused one or another of the philosophies, there exists a perspective from which to select various criteria, which will collectively provide a basis for evaluation. It is important at this step to realize the full meaning and subsequent relationships impacted by the selection of such criteria.

There should be a functionally complete hierarchy that emanates from each criterion as follows:

- **Criterion** — a standard, rule, or test upon which a judgment or decision can be made (this is a formal dictionary definition),

which leads to the definition of

- **Measures** — the "dimensions" of a criterion, i.e., the factors into which a criterion can be divided

and, finally,

- **Metrics** — those attributes of the DF process or its parameters or processing results which are considered easily and straightforwardly quantifiable or able to be defined categorically, which are relatable to the measures, and which are *observable*.

Thus, there is, in the most general case, a functional relationship as:

$$\text{Criterion} = \text{fct} \left[(\text{Measure}_i = \text{fct} (\text{Metric}_i, \text{Metric}_j \ldots), \text{Measure}_{j\,=}\, \text{fct} (\text{Metric}_k, \text{Metric}_i \ldots), \text{etc.} \right]$$

Each metric, measure, and criterion also has a scale that must be considered. Moreover, the scales are often incongruent so that some type of normalized *figure of merit* approach may be necessary in order to integrate metrics on disparate scales and construct a unified, quantitative parameter for making judgments.

One reason to establish these relationships is to provide for *traceability* of the logic applied in the T&E process. Another rationale, which argues for the establishment of these relationships, is in part derived from the requirement or desire to estimate, even roughly, predicted system behaviors against which to compare actual results. Such prediction must occur at the metric level; predicted and actual metrics subsequently form the basis for comparison and evaluation. The prediction process must be functionally consistent with this hierarchy. For Level 1 numeric processes, prediction of performance expectations can often be done, to a degree, on an analytical basis. (It is assumed here that in many T&E frameworks the "truth" state is known; this is certainly true for simulation-based experimentation but may not be true during operational tests, in which case comparisons are often done against consensus opinions of experts.) For Level 2 and 3 processes, which generally employ heuristics and relatively complex lines of reasoning, the ability to predict the metrics with acceptable accuracy must usually be developed from a sequence of exploratory experiments. Failure to do so may in fact invalidate the overall approach to the T&E process because the fundamental traceability requirement being described here would be confounded.

Representative criteria focused on the DF process per se are listed below for the numerically dominated Level 1 processes, and the symbolic-oriented Level 2 and 3 processes.

Level 1 Criteria	Level 2, 3 Criteria
• Accuracy	• Correctness in reasoning
• Repeatability/consistency	• Quality or relevance of decisions/advice/recommendations
• Robustness	• Intelligent behavior
• Computational complexity	• Adaptability in reasoning (robustness)

Criteria such as computational efficiency, time-critical performance, and adaptability are applicable to all levels whereas certain criteria reflect either the largely numeric or largely symbolic processes which distinguish these fusion-processing levels.

Additional conceptual and philosophical issues regarding what constitutes "goodness" for software of any type can, more or less, alter the complexity of the T&E issue. For example, there is the issue of reliability versus trustworthiness. Testing oriented toward measuring reliability is often "classless"; i.e., it occurs without distinction of the type of failure encountered. Thus, reliability testing often derives an unweighted likelihood of failure, without defining the class or, perhaps more importantly, a measure of the severity of the failure. This perspective derives from a philosophy oriented to the unweighted conformance of the software with the software specifications, a common practice within the DoD and its contractors.

It can be asserted, based on the argument that exhaustive path testing is infeasible for complex software, that trustworthiness of software is a more desirable goal to achieve via the T&E process. Trustworthiness can be defined as a measure of the software's likelihood of failing catastrophically. Thus, the trustworthiness characteristic can be described by a function that yields the probability of occurrence for all significant levels of severe failures. This probabilistic function provides the basis for the estimation of a confidence interval for trustworthiness. The system designer/developer (or customer) can thus have a basis for assuring that the level of failures will not, within specified probabilistic limits, exceed certain levels of severity.

20.2.4   Approach to T&E

The final element of this framework is called the *approach* element of the T&E process. In this sense, approach means a set of activities, which are both procedural and analytical, that generates the "measure" results of interest (via analytical operations on the observed metrics) as well as provides the mechanics by which decisions are made based on those measures and in relation to the criteria. The approach consists of two components as described below:

- A *procedure*, which is a metric-gathering paradigm; it is an experimental procedure.
- An *experimental design*, which defines (1) the test cases, (2) the standards for evaluation, and (3) the analytical framework for assessing the results.

Aspects of experimental design include the formal methods of classical, statistical experimental, design.[2] Few if any DF research efforts in the literature have applied this type of formal strategy, presumably as a result of cost limitations. Nevertheless, there are the serious questions of sample size and confidence intervals for estimates, among others, to deal with in the formulation of any T&E program, since simple comparisons of mean values, etc. under unstructured test conditions may not have very much statistical significance in comparison to the formal requirements of a rigorous experimental design. Such DF efforts should at least recognize the risks associated with such analyses.

This latter point relates to a fundamental viewpoint taken here about the T&E of DF processes: the DF process can be considered a function that operates on random variables (the noise-corrupted measurements or other uncertain inputs, i.e., those which have a statistical uncertainty) to produce estimates which are themselves random variables and therefore which have a distribution. Most would agree that the inputs to the DF process are stochastic in nature (sensor observation models are nearly always based on statistical models); if this is agreed, then any operation on those random variables produces random variables. It could be argued that the data fusion processes, separated from the sensor systems (and their noise effects), are deterministic "probability calculators"; in other words, processes which, given the same input — the same random variable — produce the same output — the same output random variable.[3] In this constrained context, we would certainly want and expect a data fusion algorithm, if no other internal stochastic aspects are involved, to generate the same output when given a fixed input. It could therefore be argued that some portion of the T&E process should examine such repeatability. But DeWitt[3] also agrees that the proper approach for a "probabilistic predictor" involves stochastic methods such as those that examine the closeness of distributions. (DeWitt raises some interesting epistemological views about evaluating such processes, but, as in his report, we also do not wish to "plow new ground in that area," although recognizing its importance.) Thus, we argue here for T&E techniques that somehow account for and consider the stochastic nature of the DF results when exposed to "appropriately representative" input, such as by employment of Monte Carlo-based experiments, analysis of variance methods, distributional closeness, and statistically designed experiments.

20.2.5   The T&E Process — A Summary

This section has suggested a framework for the definition and discussion of the T&E process for the DF process and DF-enhanced systems; this framework is summarized in Figure 20.1. Much of the rationale and many of the issues raised are derived from good systems engineering concepts but are intended to sensitize DF researchers to the need for formalized T&E methods to quantify or otherwise evaluate the marginal contributions of the DF process to program/system goals. This formal framework is consistent with the formal and structured methods for the T&E of C3 systems in general — see, for example References 4 and 5. Additionally, since fusion processes at Levels 2 and 3 typically involve the application of knowledge-based systems, further difficulties involving the T&E of such systems or processes can also complicate the approach to evaluation since, in effect, human reasoning strategies (implemented in software), not mathematical algorithms, are the subject of the tests. Improved formality of the T&E process for knowledge-based systems, using a framework similar to that proposed here, is described in Reference 6. Little, if any, formal T&E work of this type, with statistically qualified results, appears in the DF literature. As DF procedures, algorithms, and technology mature, the issues raised here will have to be dealt with, and the development of guidelines and standards for DF process T&E undertaken. The starting point for such efforts is an integrated view of the T&E domain — the proposed process is one such view, providing a framework for discussion among DF researchers.

20.3   Tools for Evaluation: Testbeds, Simulations, and Standard Data Sets

Part of the overall T&E process just described involves the decision regarding the means for conducting the evaluation of the DF process at hand. Generally, there is a cost vs. quality/fidelity tradeoff in making this choice, as is depicted in Figure 20.2.[7] Another characterization of the overall spectrum of possible tools is shown in Table 20.1.

Over the last several years, the defense community has built up a degree of testbed capability for studying various components of the DF process. In general, these testbeds have been associated with a particular program and its range of problems, and — except in one or two instances — the testbeds have permitted *parametric-level* experimentation but not *algorithm-level* experimentation. That is, these testbeds, as software systems, were built from "point" designs for a given application wherein normal control parameters could be altered to study attendant effects, but these testbeds could not (at least easily) permit replacement of such components as a tracking algorithm. Recently, some new testbed designs are moving in this direction. One important consequence of building testbeds that permit algorithm-level test and replacement is of course that such testbeds provide a consistent basis for system evolution over time, and in principle such testbeds, in certain cases, could be shared by a community of researcher-developers. In an era of tight defense research budgets, algorithm-level shareable testbeds, it is suspected and hoped, will become the norm for the DF community. A snapshot of some representative testbeds and experimental capabilities is shown in Table 20.2.

An inherent difficulty (or at least an issue) in testing data fusion algorithms warrants discussion because it fundamentally results from the inherent complexity of the DF process: the complexity of the DF process may make it infeasible or unaffordable to evolve, through experimentation, DF processing strategies that are optimal for other than Level 1 applications. This issue depends on the philosophy with which one approaches testbed design. Consider that even in algorithmic-replaceable testbeds, the "test article" (a term for the algorithm under test) will be tested *in the framework of the surrounding algorithms available from the testbed "library."* Hence, a tracking algorithm will be tested while using a separate detection algorithm, a particular strategy for track initiation, etc. Table 20.3 shows some of the testable (replaceable)

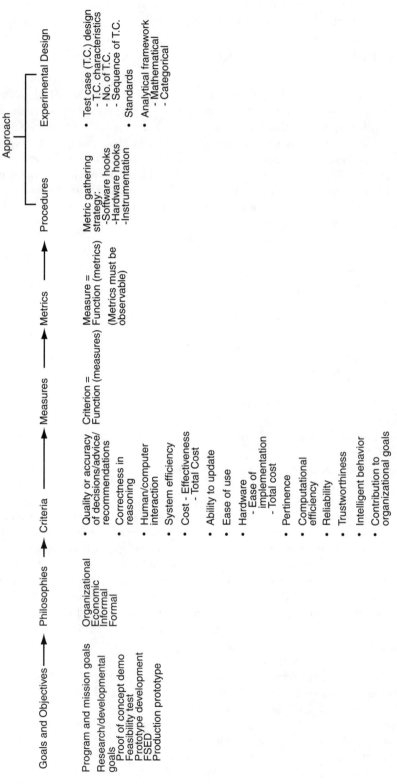

FIGURE 20.1   Test and evaluation activities for the data fusion process.

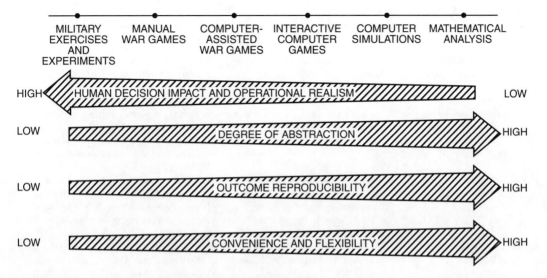

FIGURE 20.2 Applicability of modeling technique (see Reference 7).

TABLE 20.1 Generic Spectrum of Evaluation Tools

Toolset	Characteristics
I Digital Simulations	
• Level 1: Engineering models	• Relatively high fidelity; explore physics and 1 vs. 1 problem
• Level 2: 1 vs N	⎫ ⎧ Explore engagement effects
• Level 3: M vs N	⎬ Engagement models ⎨ Fidelity decreases with Level
• Level 4: Organizaitonal Level	⎭ ⎩ Engagement complexity increases with Level
• Level 5: Theatre Level	• Individualized, ad hoc simulations for tracking, ID, detection
• Numberous DF Process Models	Statistical qualification usually feasible
II Hybrid simulations	
Man-in-the-loop and/or equipment-in-the-loop	Important effects of real humans and equipment; more costly; statistical qualification often unaffordable
III Specialized field data collection/calibration	Real-world physics, phenomenology; relatively costly; often used to verify/validate digital simulations; good for phenomenological modeling but not for behavior modeling; statistically controlled in most cases
IV Test range data collection	Real-world physics, humans, equipment; relatively costly; can do limited engagement effects studies; some behavioral effected modeled; statistically uncontrolled
V Military exercises	Real-world physics, humans, equipment, and tactics/doctrine; costly; data difficult to collect/analyze; extended engagement effects studies at least feasible; extended behavioral effects modeled; statistically uncontrolled
VI Combat operations	For example, Desert Storm; actual combat effects with real adversaries; data difficult to collect/analyze; high fidelity enemy behavioral data; statistically uncontrolled

DF functions for the SDI Surveillance Testbed developed during the SDI program. Deciding on the granularity of the test articles (i.e., the plug-replaceable level of fidelity of algorithms to be tested) is a significant design decision for a DF testbed designer. Even if the testbed has, for example, multiple detection algorithms in its inventory, cost constraints will probably not permit combinatorial testing for optimality. The importance therefore of clearly thinking about and expressing the T&E philosophy, goals, and objectives, as described in Section 20.2, becomes evident relative to this issue. In many real-world situations, it is likely therefore that T&E of DF processes will espouse a "satisficing" philosophy, i.e., be based on developing solutions that are good enough because cost and other practical constraints will likely prohibit extensive testing of wide varieties of algorithmic combinations.

TABLE 20.2   Representative Multisensor Data Fusion Testbeds

Testbed	General Purpose	Characteristics
(1) Multisensor, Multitarget Data Fusion Testbed (Rome Lab, Air Force, Ref. 8)	Compare the performance of various "Level 1" fusion algorithms (association, tracking, identification).	Three major components: Scenario generator, platform simulator, data analysis and display. Monte Carlo capability.
(2) Advanced Sensor Exploitation Testbed ("ASET") (Rome Lab, Air Force, Ref. 9)	Large testbed to study "Level 2" fusion algorithms primarily for order of battle estimation.	Six major components: Scenario generator, Sensor Simulators, C^{3}I Simulator, Fusion Element, Timing and Control, Evaluation, Nominal man-in-the-loop capability, Parametric flexibility.
(3) National Testbed (SDIO, Ref. 10)	Evaluate candidate fusion concepts for SDI/"Level 1" tracking and discrimination functions. Study effects of inter-element communication links.	Large-scale, multi-Cray (plus other large computers) type environment to simulate broad range of functions in SDI problem. Purposeful man-in-the-loop capability.
(4) Surveillance Testbed (SDIO, Ref. 11)	High fidelity background and sensor (radar, optical) simulator for broad range of SDI "Level 1" fusion algorithm testing.	Algorithm-level test and replacement design. "Framework" and "Driver" concept which separates simulation and analysis activities.
(5) NATO Data Fusion Demonstrator (NATO, Ref. 12)	Initial configuration is to study "Level 1-2-3" fusion processes for airland battle (Initially Army) applications.	"Client-Server" design concept to permit algorithm-level test and replacement.
(6) All-Source Systems evaluation Testbed ("ASSET") (Univ. of Va., Ref. 13)	To evaluate data association and correlation algorithms for ASAS-type airland battle applications.	Algorithm-level test and replacement deisgn; standardized metrics; connected to Army C^{2}SW.
(7) AWACS Fusion Evaluation Testbed ("FET") (Mitre, Ref. 14)	Provide analysis and evaluation support of multisensor integration (MSI) algorithms for AWACS Level 1 applications.	Algorithm-level test and replacement design; permits live and simulate data as driver; standardized MOE's; part of Air Defense C^2 (ADC^2) Lab.

TABLE 20.3   SDI Surveillance Testbed: Testable Level 1 Fusion Functions

Test Article	Function
Bulk Filter	Reject measurement data from nonlethal objects
Track Initialization	Data association and filters for starting tracks from measurement data only
Track Continuation	Data association and filters to improve existing track information using new measurement data
Differential Initialization	Data association and filters for starting new tracks using measurement data and existing track information
Cluster Track	Data association and filters that maintain group information for multiple objects
Track Fusion	Data association and filters for combining track information from multiple sources
Track File Editing	Reject improbable measurements from a sequence of measurements
Feature Calculation	Calculate discrimination features from measurement data
Classification	Provide an estimate of an object's lethality
Discrimination Data Fusion	Combine discrimination information from multiple sources
Radar Search and Acquisition	Search of a handover volume
Sensor Tasking/Radar Scheduling	Scheduling of sensor resources

Standardized challenge problems and associated data sets are another goal that the DF community should be seeking. To this date, very little calibrated and simultaneously collected data on targets of interest with known truth conditions exist. Some contractors have invested in the collection of such data, but those data often become proprietary. Alternatives to this situation include artificially synthesizing multisensor data from individual sensor data collected under nonstandard conditions (not easy to do in a convincing manner), or employing high-fidelity sensor and phenomenological simulators.

Some attempts have been made to collect such data for community-wide application. One of the earliest of such activities was the 1987 DARPA HI-CAMP experiments that collected pulse Doppler radar and long wavelength infrared data on fairly large U.S. aircraft platforms under a limited set of observation conditions. The Army (Night Vision Laboratory) has also collected data (ground-based sensors, ground targets) for community use under a 1989 program called Multisensor Fusion Demonstration, which collected carefully ground-truthed and calibrated multisensor data for DF community use. More recently, DARPA, in combination with the Air Force Research Laboratory Sensors Directorate, made available a broad set of synthetic aperture radar (SAR) data for ground targets under a program known as MSTAR.[8] Nevertheless, the availability of such data to support algorithm development and DF system prototypes is extremely limited and represents a serious detriment and cost driver to the DF community.

However, similar to the DF process and its algorithms, the tool sets and data sets for supporting DF research and development are just beginning to mature. Modern designs of true testbeds permitting flexible algorithm-level test-and-replace capability for scientific experimentation are beginning to appear and are at least usable within certain subsets of the DF community; it would be encouraging to at least see *plans* to share such facilities on a broader basis as short-term, prioritized program needs are satisfied — i.e., in the long term, these facilities should enter a national inventory. The need for data sets from real sensors and targets, even though such sensor-target pairs may be representative for only a variety of applications, is a more urgent need of the community. Programs whose focus is on algorithm development are having to incur redundant costs of data collection for algorithm demonstrations with real data when, in many cases, representative real data would suffice. Importantly, the availability of such data sets provides a natural framework for comparative analyses when various techniques are applied against a common or baseline problem as represented by the data. Comparative analyses set the foundation for developing deeper understanding of what methods work where, for what reasons, and for what cost.

20.4 Relating Fusion Performance to Military Effectiveness — Measures of Merit

Because sensors and fusion processes are contributors to improved information accuracy, timeliness, and content, a major objective of many fusion analyses is to determine the effect of these contributions to military effectiveness. This effectiveness must be quantified, and numerous quantifiable measures of merit can be envisioned; for conventional warfare such measures might be engagement outcomes, exchange ratios (the ratio of blue-red targets killed), total targets serviced, and so on as previously mentioned. The ability to relate data fusion performance to military effectiveness is difficult because of the many factors that relate improved information to improved combat effectiveness and the uncertainty in modeling them. These factors include

- Cumulative effects of measurement errors that result in targeting errors
- Relations between marginal improvements in data and improvements in human decision making
- Effects of improved threat assessment on survivability of own forces.

These factors and the hierarchy of relationships between data fusion performance and military effectiveness must be properly understood in order for researchers to develop measures and models that relate them. Said otherwise, there is a large conceptual distance between the value of improved information quality, as provided by data fusion techniques, and its effects on military effectiveness; this large distance is what makes such evaluations difficult. The Military Operations Research Society[9] has recommended a hierarchy of measures that relate performance characteristics of C3 systems (including fusion) to military effectiveness (see Table 20.4).

Dimensional parameters are the typical properties or characteristics that directly define the elements of the data fusion system elements, such as sensors, processors, communication channels, etc. (These are equivalent to the "metrics" defined in Section 20.2.) They directly describe the behavior or structure of the system and should be considered to be typical measurable specification values (bandwidth, bit-error rates, physical dimensions, etc.).

TABLE 20.4   Four Categories of Measures of Merit[9]

Measure	Definition	Typical Examples
Measure of Force Effectiveness (MOFE)	Measure of how a C³ system and the force (sensors, weapons, C³ system) of which it is a part perform military missions	Outcome of battle Cost of system Survivability Attrition rate Exchange ratio Weapons on targets
Measures of Effectiveness (MOE)	Measure of how a C³ system performs its functions within an operational environment	Target nomination rate Timeliness of information Accuracy of information Warning time Target leakage Countermeasure immunity Communications survivability
Measures of Performance (MOP)	Measures closely related to dimensional parameters (both physical and structural) but measure attributes of behavior	Detection probability False alarm rate Location estimate accuracy Identification probability Identification range Time from detect to transmission Communication time delay Sensor spatial coverage Target classification accuracy
Dimensional Parameters	The properties or characteristics inherent in the physical entities whose values determine system behavior and the structure under question, even when not operating	Signal-to-noise ratio Operations per second Operations per second Aperture dimensions Bit error rates Resolution Sample rates Antijamming margins Cost

Measures of performance (MOPS) are measures that describe the important behavioral attributes of the system. MOPs are often functions of several dimensional parameters to quantify, in a single variable, a significant measure of operational performance. Intercept and detection probabilities, for example, are important MOPS that are functions of several dimensional parameters of both the sensors and detailed signal processing operations, data fusion processes, and the characteristics of the targets being detected.

Measures of effectiveness (MOEs) gauge the degree to which a system or militarily significant function was successfully performed. Typical examples, as shown in Table 20.4, are target leakage and target nomination rate.

Measures of force effectiveness (MOFEs) are the highest-level measures that quantify the ability of the total military force (including the data fusion system) to complete its mission. Typical MOFEs include rates and ratios of attrition, outcome of engagements, and functions of these variables. To evaluate the overall mission, factors other than outcome of the conflict (e.g., cost, size of force, composition of force) may also be included in the MOFE.

Figure 20.3 depicts the relationship between a set of *surveillance* measures for a two-sensor system, showing the typical functions that relate lower-level dimensional parameters upward to higher level measures. In this example, sensor coverages (spatial and frequency), received signal-to-noise ratios, and detection thresholds define sensor-specific detection and false alarm rate MOPs labeled "Measures of Detection Processing Performance (MODPP)" on the figure.

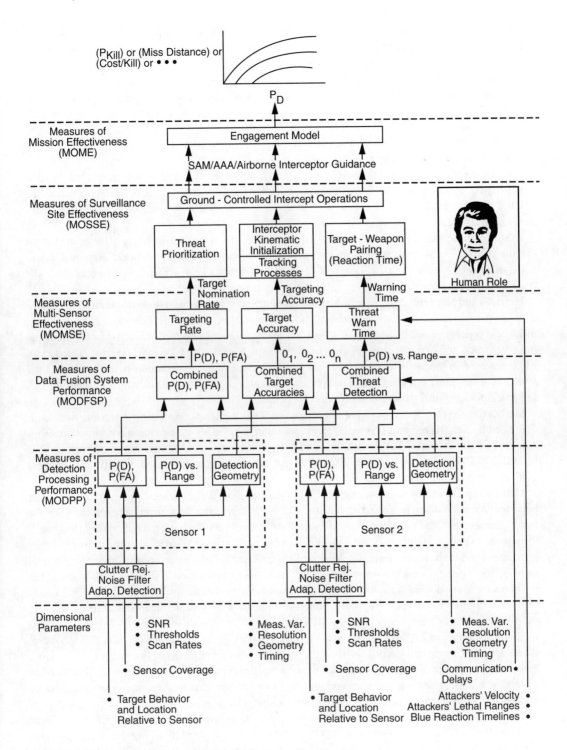

FIGURE 20.3 Hierarchical relationship among fusion measures.

Alternately, the highest level measures are those that relate P_D (or other detection-specific parameters) to mission effectiveness. Some representative metrics are shown at the top of Figure 20.3, such as P_{Kill}, Cost/Kill, Miss Distance, etc. These metrics could be developed using various computer models to simulate end-game activities, while driving the detection processes with actual sensor data. As mentioned earlier, there is a large "conceptual distance" between the lowest-level and highest-level measures. Forming the computations linking one to the other requires extensive analyses, data and parameters, simulation tools, etc., collectively requiring possibly significant investments.

The next level down the hierarchy represents the viewpoint of studying surveillance site effectiveness; these measures are labeled "MOSSE" in Figure 20.3. Note, too, that at this level there is a human role that can enter the evaluation; of frequent concern is the workload level to which an operator is subjected. That is, human factors-related measures enter most analyses that range over this evaluation space, adding yet other metrics and measures into the evaluation process; these are not elaborated on here, but they are recognized as important and possibly critical.

Lower levels measure the effectiveness of employing multiple sensors in generating target information (labeled "MOMSE-measures" in Figure 20.3), the effectiveness of data combining or fusing per se (labeled "MODSFP-measures"), and the effectiveness of sensor-specific detection processes (labeled "MODPP-measures"), as mentioned above.

In studying the literature on surveillance and detection processes, it is interesting to see analogous varying perspectives for evaluation. Table 20.5 shows a summary of some of the works examined, where a hierarchical progression can be seen, and compares favorably with the hierarchy of Figure 20.3; the correspondence is shown in the "Level in Hierarchy" column in Table 20.5.

Metrics a and b in Table 20.5 reflect a surveillance system-level viewpoint; these two metrics are clearly dependent on detection process performance, and such interactive effects could be studied.[10] Track purity, metric c, a concept coined by Mori, et al.,[11] assesses the percentage of correctly associated measurements in a given track, and so it evaluates the association/tracking boundary (MOSSE/MOMSE of Figure 20.3). As commented in the table, this metric is not explicitly dependent on detection performance but the setting of association gates (and thus the average innovations standard deviation, which depends on P_D), so a track purity-to-detection process connection is clear.

Metrics d and e, the system operating characteristic (SOC) and tracker operating characteristic (TOC), developed by Bar-Shalom and others,[12,13] form quantitative bases for connecting track initiation, SOC, and tracker performance, TOC, with P_D, and thus detection threshold strategy performance (the MOSSE/MOMSE boundary in Figure 20.3). SOC evaluates a composite track initiation logic, whereas TOC evaluates the state error covariance, each as connected to, or a function of, single-look P_D.

Metric f is presented by Kurniawan et al.[14] as a means to formulate optimum energy management or pulse management strategies for radar sensors. The work develops a semiempirical expression for the mean square error (MSE) as a function of controllable parameters (including the detection threshold), thereby establishing a framework for optimum control in the sense of MSE. Nagarajan and others[15] formulate a similar but more formally developed basis for sensor parameter control, employing several metrics. Relationships between the metrics and P_D/threshold levels are established in both cases, so the performance of the detection strategy can be related to, among other things, MSE for the tracker process in a fashion not unlike the TOC approach. Note that these metrics evaluate the interrelationships across two hierarchical levels, relating MOSSE to MODFSP.

Metric h, developed by Hashlamoun and Varshney[16] for a distributed binary decision fusion system, is based on developing expressions for the Min (probability of error, POE) at the global (fusion) level. Employing the Blackwell theorem, this work then formulates expressions for optimum decision making (detection threshold setting) by relating Min (POE) to various statistical distance metrics (distance between H_0, H_1 conditional densities), which directly affect the detection process.

The lowest levels of metrics, as mentioned above, are those intimately related to the detection process. These are the standard probabilistic measures P_D and P_{fa} and, for problems involving clutter backgrounds, the metrics that comprise the set known as "clutter filter performance measures." This latter group has a standard set of IEEE definitions and has been the subject of study of the Surface Radar Group of the AES Radar Panel.[17] The set comprises

- MTI improvement factor
- Signal-to-clutter ratio improvement

TABLE 20.5 Alternative Concepts and Metrics for Evaluation

Evaluation Point of View	Metric Name	Level in Hierarchy	Calculation	Remarks	Ref.
(a) Tracking Accuracy/ Reliability/Survivability	Avg. Track (Target) "Exposure"	MOSSE	Average time during which N Surv. system signals are on target	Simulation-based, scenario-dependent analysis	10
(b) Site and System Computational Workload	Signal Loading	MOSSE	Compute sampling distribution for received signal density at (site, system)	Simulation-based scenario-dependent analysis	10
(c) Track Continuity	Track Purity	MOSSE/MOMSE	Percent correctly associated msmts. in a given track; fct (avg. innov. std. dev., target density)	Deterministic target dynamics	11
(d) Composite (M/N) Track Init. Logic	System Operating Characteristic (SOC)	MOSSE/MOMSE	Compute $\bar{P}_{D_t}$, target track detection probability over several scans	Markov chain model, includes effects of FA	12
(e) Steady-State RMS Position Error	Tracker Operating Characteristic (TOC)	MOSSE/MOMSE	Compute numerically steady-state $\bar{P}$ (P_D, P_F)	PDA tracker (Poisson clutter)	13
(f) "Energy management" for optimum target tracking (optimal pulse sharing, sampling/ integration periods)	Steady-State MSE	MOSSE/MODFSP	Compute MSE as fct (parameters)	Empirical formulation, square-law detector, Swerling II (Single Sensor)	14
(g) Optimum Threshold Management (track init, delete, continuation)	"Nagarajan"	MOSSE/MODFSP	Various Metrics	CFAR assumed (Single Sensor)	15
(h) Detection Error Control (threshold control)	Min Prob. (Error) (global)	MODPP	Relate MinPOE to interhypothesis distance matric	Distributed (binary) decision fusion; Blackwell Theorem	16

- Subclutter visibility
- Interclutter visibility
- Filter mismatch loss
- Clutter visibility factor

In the context of the DF process, it has been those researchers working at Level 1 who have been most active in the definition and nomination of measures and metrics for evaluation. In particular, the tracking research community has offered numerous measures for evaluation of both tracking, association, and assignment functions.[18-22] In the U.S., the Automatic Target Recognizer Working Group (ATRWG), involved with Level 1 classification processing, has also been active in recommending standards for various measures.[23]

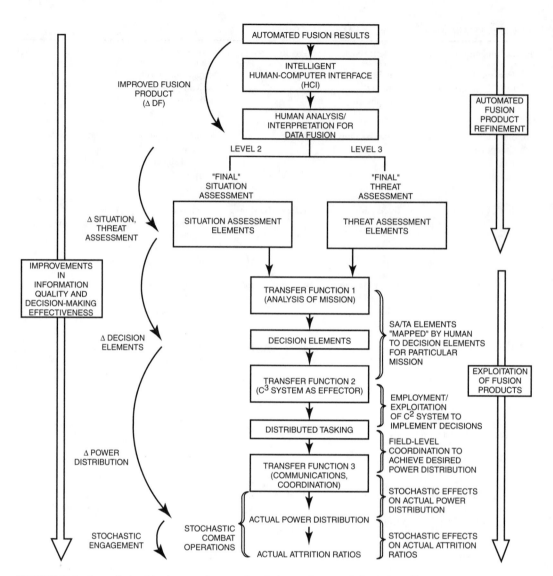

FIGURE 20.4 Interrelationships between human data fusion and system and mission effectiveness.

The challenges in defining measures to have a reasonably complete set across the DF process clearly lie in the Level 2–Level 3 areas. Exactly how to assess the "goodness" of a situation or threat assessment is admittedly difficult but certainly not intractable; for example, analogous concepts employed for assessing the quality of images come to mind as possible candidates. A complicating factor for evaluation at these levels is that the "final" determinations of situations or threats typically involve *human* data fusion — i.e., final interpretation of the automated data fusion products by a human analyst. The human is therefore the final interpreter and effecter/decision maker in many data fusion systems, and understanding the interrelationship of MOEs will require understanding a group of "transfer functions" which characterize the translation of information about situation and threat elements into eventual engagement outcomes; one depiction of these interrelationships is shown in Figure 20.4. This figure begins, at the top, with the final product of the automated data fusion process; all algorithmic and symbolic processing associated with fusion has occurred by this point. That product is communicated to a human through an HCI for both Levels 2 and 3 as shown. Given this cognitive interpretation (of the displayed automated results), the human must "transfer" this interpretation, via considerations of

- Decision elements associated with decisions for the given mission (transfer function 1), and
- Considerations of the effectiveness of the C3 system, and communications coordination aspects, to influence the power distribution of his forces on the battlefield (transfer functions 2 and 3).

The implemented decisions (via the C3 and communications systems) result in revisions to force deployment, which then (stochastically) yield an engagement outcome, and the consequent force-effectiveness measures and results. Thus, if all the aspects of human decision making are to be formally accounted for, the conceptual distance mentioned earlier is extended even further, as described in Figure 20.4.

There is yet one more, and important, aspect to think about in all this: ideally, the data fusion process is implemented as a dynamic (i.e., runtime dynamic) adaptive feedback process — that is, we have Level 4, process refinement, at work during runtime, in some way. Such adaptation could involve adaptive sensor management, enabling some intelligent logic to dynamically improve the quality of data input, or it could involve adaptive algorithm management, enabling a logic that switches algorithms in some optimal or near-optimal way. Hence, the overall fusion process is not fixed during runtime execution, so that temporal effects need to be considered. Control theorists talk about evaluating alternative control trajectories when discussing evaluation of specific control laws that enable some type of adaptive logic or equations. Data fusion process analysts will therefore have to think in a similar way and evaluate performance and effectiveness as a function of time. This hints at an evaluation approach that focuses on different phases of a mission and the prioritized objectives at each time-phase point or region. Combined with the stochastic aspects of fusion process evaluation, this temporal dependency only further complicates the formulation of a reasonable approach.

20.5 Summary

Developing an understanding of the relationships among these various measures is a difficult problem, labeled here the "interconnectivity of MOEs" problem. The toolkit/testbed needed to generate, compare, and evaluate such measures can be quite broad in scope and itself represent a development challenge as described in Section 20.3.

Motivated in part by the need for continuing maturation of the DF process and the amalgam of techniques employed, and in part by expected reductions in defense research budgets, the data fusion community must consider strategies for the sharing of resources for research and development. Part of the spectrum of such resources includes standardized (i.e., with approved models) testbed environments, which will offer an economical basis not only for testing of DF techniques and algorithms, but, importantly, a means to achieve optimality or at least properly satisfying performance of candidate methods under test. However, an important adjunct to shareable testbeds is the standardization of both the overall approach to evaluation and the family of measures involved. This chapter has attempted to offer some ideas for discussion on several of these very important matters.

References

1. Linn, R.J., Hall, D.L., and Llinas, J., A Survey of Multisensor Data Fusion Systems, presented at the SPIE, Sensor Fusion Conference, Orlando, FL, April 1991.
2. Villars, D.S., *Statistical Design and Analysis of Experiments for Development Research,* Wm. C. Brown Co., Dubuque, Iowa, 1951.
3. DeWitt, R.N., Principles for Testing a Data Fusion System, PSR (Pacific-Sierra Research, Inc.) Internal Report, March 1998.
4. **Anon,** A., Methodology for the test and evaluation of C^{3}I systems, Vol. 1 — Methodology Description, Report 1ST-89-R-003A, prepared by Illgen Simulation Technologies, Santa Barbara, CA, May 1990.

5. *International Test and Evaluation Association,* Special Journal Issue on Test and Evaluation of C³IV, Vol. VIII, 2, 1987.

6. Llinas, J. et al., The test and evaluation process for knowledge-based systems, Rome Air Development Center Report RADC-TR-87-181, November 1987.

7. Przemieniecki, J.S., Introduction to mathematical methods in defense analysis, *Amer. Inst. Of Aero. and Astro.,* Washington, D.C., 1985.

8. Bryant, M. and Garber, F., SVM classifier applied to the MSTAR public data set, *Proc. SPIE,* Vol. 3721, 1999.

9. Sweet, R., Command and control evaluation workshop, MORS C² MOE Workshop, Military Operations Research Society, January 1985.

10. Summers, M.S. et al., Requirements on a fusion/tracker process of low observable vehicles, Final Report, Rome Laboratory/OCTM Contract No. F33657-82-C-0118, Order 0008, January 1986.

11. Mori, S. et al., Tracking performance evaluation: Prediction of track purity, SPIE, Vol. 1096, Signal and Data Processing of Small Targets, 1989.

12. Bar-Shalom, Y. et al., From receiver operating characteristic to system operating characteristic: Evaluation of a large-scale surveillance system, *Proc. Of EASCON87,* Washington, D.C., October 1987.

13. Fortmann, T.E. and Bar-Shalom, Y., Detection thresholds for tracking in clutter-A connection between estimation and signal processing, *IEEE Transactions on Automatic Control,* Vol. 30, March 1985.

14. Kurniawan, Z. et al., Parametric optimization for an integrated radar detection and tracking system, *IEEE Proc. (F),* 132, 1, February 1985.

15. Nagarajan, V. et al., New approach to improved detection and tracking performance in track-while-scan-radars; Part 3: Performance prediction, optimization, and testing, *IEEE Proc. (F),* 134, 1, February 1987.

16. Hashlamoon, W.A. and Varshney, P.K., Performance aspects of decentralized detection structures, *Proc. of 1989 Int'l. Conf. Of SMC,* Cambridge, MA, November 1989.

17. Ward, H.R., Clutter filter performance measures, *Proc. of 1980 International Radar Conference,* Arlington, VA, April 1980.

18. Wiener, Howard L., Willman, Joseph H., Kullback, Joseph H., and Goodman, Irwin R., *Naval Ocean-Surveillance Correlation Handbook 1978,* Naval Research Laboratory, 1978.

19. Schweiter, G.A., MOPs for the SDI tracker-correlator, Daniel H. Wagner Associates Interim Memorandum to Naval Research Laboratory, 1978.

20. Mifflin, T., Ocean-surveillance tracker-correlator measures of performance, Naval Research Laboratory Internal Memorandum, 1986.

21. Askin, K., SDI-BM tracker/correlator measures of performance, Integrated Warfare Branch Information Technology Division, April 4, 1989.

22. Belkin, B., Proposed operational measures of performance, Daniel H. Wagner Associates Internal Memorandum, August 8, 1988.

23. **Anon,** Target recognizer definitions and performance measures, prepared under the auspices of the Automatic Target Recognizer Working Group Data Collection Guidelines Subcommittee of the Data Base Committee, ATRWG No. 86-001, approved February 1986.

21

Dirty Secrets in Multisensor Data Fusion*

21.1 Introduction 21-1
21.2 The JDL Data Fusion Process Model............................. 21-2
21.3 Current Practices and Limitations in Data Fusion........ 21-2
   Level 1: Object Refinement • Level 2: Situation Refinement •
   Level 3: Threat Refinement • Level 4: Process Refinement •
   Human-Computer Interface (HCI) • Database Management
21.4 Research Needs 21-7
   Data Sources • Source Preprocessing • Level 1: Object
   Refinement • Level 2: Situation Refinement and Level 3:
   Threat Refinement • Human-Computer Interface (HCI) •
   Database Management • Level 4: Processing • Infrastructure
   Needs
21.5 Pitfalls in Data Fusion 21-9
21.6 Summary.. 21-10
References 21-10

David L. Hall
The Pennsylvania State University

Alan N. Steinberg
Utah State University

21.1 Introduction

Over the past two decades, an enormous amount of Department of Defense (DoD) funding has been applied to the problem of data fusion systems, and a large number of prototype systems have been implemented.[1] The data fusion community has developed a data fusion process model,[2] a data fusion lexicon,[3] and engineering guidelines for system development.[4] Although a significant amount of progress has been made,[5,6] much work remains to be done. Hall and Garga,[7] for example, identified a number of pitfalls or problem areas in implementing data fusion systems. Hall and Llinas[8] described some shortcomings in the use of data fusion systems to support individual soldiers, and M. J. Hall, S. A. Hall, and Tate[9] addressed issues related to the effectiveness of human-computer interfaces for data fusion systems.

This chapter summarizes recent progress in multisensor data fusion research and identifies areas in which additional research is needed. In addition, it describes some issues — or *dirty secrets* — in the current practice of data fusion systems.

---

*This chapter is based on a paper by David L. Hall and Alan N. Steinberg, Dirty secrets of multisensor data fusion, *Proceedings of the 2000 MSS National Symposium on Sensor Data Fusion*, Vol. 1, pp. 1–16, June 2000, San Antonio, TX.

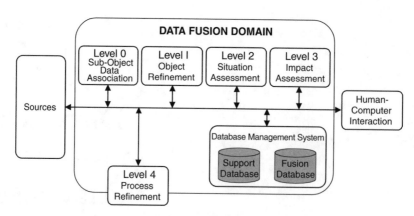

FIGURE 21.1   Top-level view of the JDL data fusion process model.

21.2   The JDL Data Fusion Process Model

The Joint Directors of Laboratories (JDL) Data Fusion Working Group was established in 1986 to assist in coordinating DoD activities in data fusion and to improve communications among different DoD research and development groups. Led by Frank White NOSC, the JDL working group performed a number of activities, including (1) development of a data fusion process model,[2] (2) creation of a lexicon for data fusion,[3] (3) development of engineering guidelines for building data fusion systems,[4] and (4) organization and sponsorship of the Tri-Service Data Fusion Conference from 1987 to 1992. The JDL Data Fusion Working Group has continued to support community efforts in data fusion, leading to the annual National Symposium on Sensor Data Fusion and the initiation of a Fusion Information Analysis Center (FUSIAC).[10]

The JDL Data Fusion Process Model[2,6,11] was introduced in Chapter 1 and was described in detail in Chapter 2. A top-level view of the model is illustrated in Figure 21.1, and a summary of the processes is shown in Table 21.1.

The JDL model is a two-layer hierarchical model that identifies fusion processes, processing functions, and processing techniques to accomplish the primary data fusion functions. The model was intended to facilitate communications among data fusion researchers and implementation engineers, rather than to serve as a prescription for implementing a fusion system or an exhaustive enumeration of fusion functions and techniques. The model has evolved since its original exposition to the data fusion community. Steinberg and Bowman,[12] for example, have recommended the inclusion of a new Level 0 processing to account for processing such as predetection fusion and coherent signal processing of multisensor data. In addition, they suggest renaming and re-interpreting the Level 2 and Level 3 processes to focus on understanding the external world environment (rather than a military-oriented situation and threat focus). Morefield[13] has suggested that the distinction between Level 2 and Level 3 is artificial, and that these processes should be considered a single process. Bowman has suggested that the JDL model can be detrimental to communications if systems engineers focus on the model rather than take a systematic architecture analysis and decomposition approach. Many of these comments have merit. However, this chapter utilizes the JDL model as a middle ground for describing the current state of practice and limitations.

21.3   Current Practices and Limitations in Data Fusion

A summary of the current state and limitations of data fusion is provided in Table 21.2. It is an update of a similar figure originally introduced by Hall and Llinas in 1993[5] and subsequently updated by them in 1997.[6] For each of the key components of the JDL process, the figure summarizes the current practices and limitations.

TABLE 21.1 Summary of JDL Processes and Functions

Process Components	Process Description	Functions
Sources of information	Local and remote sensors accessible to the data fusion system Information from reference systems and human inputs	Local and distributed sensors External data sources Human inputs
HCI	Provides an interface to allow a human to interact with the fusion system	Graphical displays Natural language processing
Source preprocessing	Processing of individual sensor data to extract information, improve signal to noise, and prepare the data for subsequent fusion processing	Signal and image processing Canonical transformations Feature extraction and data modeling
Level 1 processing: Object refinement	Association, correlation, and combination of information to detect, characterize, locate, track, and identify objects (e.g., tanks, aircraft, emitters)	Data alignment Correlation Position, kinematic, attribute estimation Object identity estimation
Level 2 processing: Situation refinement	Development of a description of the current relationships among objects and events in the context of their environment	Object aggregation Event and activity interpretation Context-based reasoning
Level 3 processing: Threat refinement	Projection of the current situation into the future to draw inferences about enemy threats, friendly and enemy vulnerabilities, and opportunities for operations	Aggregate force estimation Intent prediction Multiperspective analysis Temporal projections
Level 4 processing: Process refinement	A metaprocess that seeks to optimize the ongoing data fusion process (e.g., to improve accuracy of inferences, utilization of communication and computer resources)	Performance evaluation Process control Source requirement determination Mission management
Data management	Provide access to and management of dynamic data including sensor data, target state vectors, environmental information, doctrine, and physical models	Data storage and retrieval Data mining Archiving Compression Relational queries and updates

21.3.1 Level 1: Object Refinement

Level 1 processing seeks to combine information about the location and attributes of entities (such as tanks or aircraft) to detect, locate, characterize, track, and identify the entities. Level 1 processing involves data assignment/correlation, estimation of the state of an entity, and an estimate of the entity's identity. The typical data fusion system partitions the object refinement problem into three basic components: (1) data assignment/correlation, (2) estimate of a state vector (e.g., for target tracking), and (3) estimation of a target's identity.

Object refinement is relatively easy when there are relatively few, widely separated targets moving in predictable paths. Target identity classification can generally be performed when there are observable target attributes (e.g., size, shape, and spectral signature) that can be uniquely mapped to target class or identity. This requires either an accurate model to link attributes with target identity, or a very large set of training data to train a pattern classification algorithm.

However, when these observing conditions are violated, the problem becomes much more challenging. Closely spaced, rapidly maneuvering targets, for example, are difficult to track because sensor measurements cannot be easily associated to the appropriate targets. Furthermore, because acceleration cannot be observed directly, maneuvering targets cause a potential loss of track, because the future position of the targets cannot accurately be predicted. Complex observing environments, involving multipath signal propagation clutter, dispersion, or other effects on signal-to-noise, can cause difficulties in data associ-

TABLE 21.2 Summary of Current State of Multisensor Data Fusion

JDL Process	Current Practices	Limitations and Challenges
Level 1: Object refinement	Sensor preprocessing using standard signal and image processing methods	Dense target environments
	Explicit separation of correlation and estimation problem	Rapidly maneuvering targets
		Complex signal propagation
	Multiple target tracking using MHT,[22] JPDA,[23] etc.	Codependent sensor observations
	Use of ad hoc maneuver models	Background clutter
	Object ID dominated by feature-based methods[24]	Context-based reasoning
	Pattern recognition using ANN[25]	Integration of identity and kinematic data
	Emerging guidelines for selection of correlation algorithms[4,26]	Lack of available ANN training data (for target identification)[25]
	Promising work by Poore,[14] Mahler,[27] Barlow et al.[28]	No true fusion of image and nonimage data (at the data level)
Level 2: Situation refinement	Numerous prototype systems[1]	Very limited operational systems
	Dominance of rule-based KBS	No experience in scaling up prototypes to operational systems
	Variations include blackboard systems,[29] logical templating,[30] and case-based reasoning[31]	Very limited cognitive models[9]
	Emerging use of fuzzy logic[32] and agent-based systems[33]	Perfunctory test and evaluation against *toy* problems[1]
		No proven technique for knowledge engineering[11]
Level 3: Threat refinement	Same as Level 2 processing	Same as Level 2
	Limited advisory status	Difficult to quantify intent[34]
	Limited deployment experience	Models require established enemy doctrine
	Dominated by ad hoc methods	Difficult to model rapidly evolving situations
	Doctrine-specific, fragile implementations	
Level 4: Process refinement	Robust methods for single-sensor systems	Difficult to incorporate mission constraints
	Formulations based on operations research[11]	Scaling problem when many sensors (10^N) and adaptive systems[37]
	Limited context-based reasoning	
	Focus on measures of performance (MOP) versus measures of effectiveness (MOE)[35]	Difficult to optimally use noncommensurate sensors
		Very difficult to link human information needs to sensor control[37]
HCI	HCI dominated by the technology of the week	Very little research has been performed to understand how human analysts' process data and make accurate inferences
	Focus on ergonomic versus cognitive-based design	
	Numerous graphics-based displays and systems[38,39]	
	Advanced, 3-D full immersion HCI available[40] and haptic interfaces[41,42]	Creative HCI is needed to adapt to individual users and to provide mitigation of known cognitive biases and illusions[9,43]
Database management	Extensive use of 4th and 5th generation COTS DBMS	Need a generalized DBMS capability for text, signal data, images, and symbolic information
	DBMS individually optimized for text, signal data, imagery, or symbolic information (but not the intersection of any two)	Need a software solution to multilevel security
	DBMS requires extensive tailoring for individual data fusion systems	

ation and state estimation (because often no accurate model is available that can link the value of a target state vector to predicted observations). Combining data from sensors that are codependent (i.e., for which the sensor data are not statistically independent) is also difficult. Finally, complex targets without distinguishing attributes are difficult to classify or identify.

Current Level 1 processing is dominated by estimation techniques such as Kalman filters,[11] multiple-hypothesis tracking (MHT),[22] joint probabilistic data association (JPDA) filters,[23] or related techniques. The problem of identity declaration is generally performed using a feature-based, pattern recognition approach.[11,24] This involves representing the sensor data using extracted features (e.g., spectral peaks in a radar cross-section observation) and mapping the feature vector to a location in feature-space that can be uniquely identified with a target class or identity. Typical techniques include artificial neural networks (ANN) or cluster algorithms.[24] This identification process works well when there is a unique map between the observed features and the target class but requires a significant amount of training data. However, the methods fail when training data is lacking[25] or there is ambiguity in the feature-to-target class mapping. Emerging methods include both model-based techniques and syntactic methods that develop descriptions of a target in terms of its elementary components.

Poore[14] and Mahler[27] developed two promising methods for Level 1 fusion. Poore revisited the approach of separating the problems of object correlation, target tracking, and identity estimation. Poore relinks these problems into a single optimization problem with multiple constraints (e.g., find the set of state vectors — including the association between observations and tracks — that best fits the observational data). While this larger problem is even more difficult than the original subproblems, Poore has developed approximation methods to improve the computational feasibility. By contrast, Mahler developed applications of random set theory to address the joint problem of data association and state estimation. In addition, Barlow, Stone, and Finn[28] used a unified method based on Bayesian inference to simultaneously estimate target state, identity, and association of the data. Finally, an extensive survey of methods for data correlation has been performed by Llinas et al.[26]

21.3.2 Level 2: Situation Refinement

Level 2 processing seeks to understand the entities' relationships with other entities and with their environment. This process involves recognition of patterns, context-based reasoning, and understanding of spatial, temporal, causal, and functional relationships. Although this is a difficult problem, numerous prototype systems have been developed for DoD applications.[1] The predominant methods involve knowledge-based systems utilizing production rules,[11] fuzzy logic,[32] logical templates,[30] or case-based reasoning.[23] Emerging systems are beginning to utilize agent-based approaches[31] and blackboard architectures.[29]

While this is a very active area of research, the results to date are relatively disappointing. Very few operational systems have been deployed. Many of the prototype systems have addressed limited or *toy* problems with little or no test and evaluation. There is little experience on how to scale these small prototype systems to larger scale operational systems. A key problem for Level 2 and Level 3 processing is the lack of cognitive models for performing situation assessment. Current cognitive models are pathetic. Researchers simply do not know how to model the reasoning process to perform a Gestalt-type of situation assessment. Numerous ad hoc methods (e.g., rules, frames, fuzzy logic, decision trees, scripts, and templates) have been applied. One difficulty pertains to how knowledge engineering can be performed to identify key information, interrelationships, and associated uncertainty information. Again, Mahler's random set theory[27] provides a basis for a unified calculus of uncertainty. However, the application to realistic problems is far from routine. A general implementation approach has not yet been developed.

Improvements to Level 2 processing likely will emerge from an improved understanding of how to select and use existing methods for knowledge representation (e.g., rules, frames, scripts, and fuzzy logic), coupled with a better understanding of the strengths and weaknesses of human cognition for these types of tasks. One example would be the incorporation of negative information in reasoning. Negative information involves reasoning about information that has not been observed but would be expected for a hypothesized situation. The use of negative reasoning appears to be a key element of successful diagnosis and inference in many areas, such as medical diagnosis or diagnosis of mechanical faults.[15]

Another promising area for research involves the development of aids that are capable of compensating for known human cognitive biases and shortcomings (e.g., confirmation bias in which humans seek information that confirms a proposed hypothesis, rather than evidence that refutes the hypothesis).[9,35] Research by Wohl[16] and his associates focused on the development of tools for assisting antisubmarine warfare (ASW) analysts. The results suggest that some fairly simple cognitive aids could be developed that are capable of significantly improving the data fusion/analysis process.

21.3.3 Level 3: Threat Refinement

Level 3 processing involves interpreting a situation from a consequences point of view — assessing the meaning of the situation in terms of potential opportunities and threats. Alternative hypotheses are generated and projected into the future to determine the likely courses of action for engagements, and the consequences of those courses of action. The state of Level 3 processing is similar to that of Level 2. A number of prototype systems have been developed, but few have been deployed. The main focus of Level 3 processing has been the application of automated reasoning systems and techniques from the discipline of artificial intelligence. A special challenge for Level 3 processing is the determination of enemy intent. Conceptually, the determination of an enemy's intent involves a mind-reading exercise to determine what the enemy will do, under what circumstances, and with what motivation. When a well-known enemy doctrine exists, this problem can be modeled using a variety of techniques. However, in modern conflict situations, this doctrine is often unknown. Hence, automating the process of threat refinement is challenging. Another problem related to threat refinement is the role of adaptive intelligence opponents. How can engagements be modeled in which an opponent adapts to the actions of a protagonist? Much research has been performed in game theory to address this issue, but limited success has been realized in applying this work to realistic tactical situations.

21.3.4 Level 4: Process Refinement

The Level 4 process is a metaprocess — a process that monitors the overall data fusion process and seeks to optimize the data fusion within operational and physical constraints.[11,37] Functions within Level 4 processing include generation of sensor *look angles* (to indicate where to point the sensors to track targets), computation of measures of performance (MOP) and measures of effectiveness (MOE), determination of information needs and sources, and process optimization. Level 4 processing is relatively mature for single sensor environments. For one sensor or a few commensurate sensors, Level 4 processing is a routine problem of multiobjective optimization. This area has been researched extensively for applications such as industrial process control.

The Level 4 process becomes more challenging under a number of circumstances. These include use of a large number of sensors, use of codependent sensors, utilization of noncommensurate sensors (e.g., measuring very diverse physical phenomena on greatly different time scales), and use of sensors in a geographically distributed environment. Modern data fusion systems often involve geographically distributed collection and processing with adaptive systems that self-adjust for system failures and other problems.[36] Under these circumstances, developing global MOE and MOP models and optimizing overall system performance are difficult. Another challenge involves modeling sensor performance in realistic data collection environments. Finally, the most effective Level 4 process would link the information needs of a human decision maker to the sensor and source tasking in real time.

Much research remains to be performed in the Level 4 area. However, the improved intelligence and agility of modern sensors make this an area in which major improvements can be obtained with relatively modest effort. Current research by Nixon,[17] which uses economic theory to model resource utilization, is very intriguing.

21.3.5 Human-Computer Interface (HCI)

The human-computer interface (HCI) area in data fusion appears to be technology-rich and theory-poor. M. J. Hall, S. A. Hall, and Tate[9] point out the rapidly evolving HCI technology, which includes full-immersion, three-dimensional displays,[40] haptic interfaces,[41,42] three-dimensional sound, and other types of interfaces for accessing and analyzing data. However, they note that these interfaces reflect the *technology du jour* and have not been applied with a solid theoretical understanding of how humans access and respond to information displays. Many of the existing HCI for data fusion systems involve geographical information system (GIS) type displays and data access.[38,39] Although such displays are useful, no hard evidence exists to prove or disprove that they truly provide increased understanding of information from a data fusion system. Feran has argued that the HCI for intelligence systems can actually act as a bottleneck that limits the ability of a user to access and analyze data. Other studies have investigated the issue of trust in decision support systems (Llinas et al.[19]), and how the HCI affects the extent to which a user believes and trusts the results.

21.3.6   Database Management

Database management is important for data fusion systems for several reasons. First, the majority of software that must be developed for a data fusion system is database management software — even if a sophisticated commercial, off-the-shelf database management system (DBMS) package is used.[11] Data required for fusion systems range from sensor data (e.g., scalars, vectors, time series, and images) to information input by human users, environmental data, textual information, and knowledge such as doctrine. The database management for a data fusion system must simultaneously accept data at the rate provided by the contributing sensors and allow algorithms and users to rapidly retrieve large amounts of data using general Boolean queries. The combination of the complexity of the data sets and the need for real-time data storage and retrieval complicates database management for data fusion. In addition, the data associated with fusion systems often involve multiple levels of security — a challenge to do via a software approach. For all of these reasons, extensive special software must be implemented for data fusion systems.

21.4   Research Needs

A number of research areas could prove valuable to the data fusion community and improve the ability to develop robust systems. A summary of these research areas is shown in Figure 21.2 and described below.

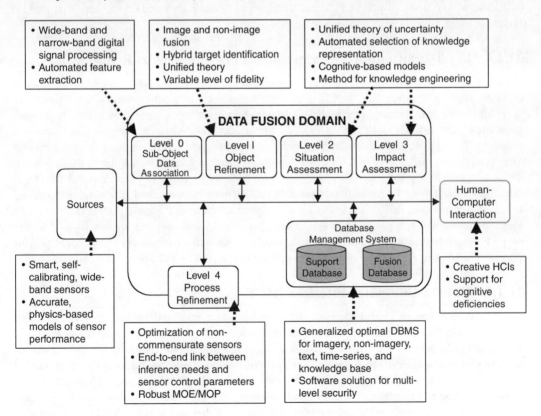

FIGURE 21.2   Technology needs in data fusion.

21.4.1 Data Sources

New sensor types and sources are always sought for data fusion applications. The rapid evolution of microprocessors and nanofabrication techniques provides a basis for rapid evolution of sensors. New smart, self-calibrating and wide-band sensors would benefit many DoD applications. In addition, accurate, physics-based models of sensor performance could be used to improve the downstream data fusion processing.

21.4.2 Source Preprocessing

Recent advances in digital signal processing and image processing have been based on new algorithms and improvements in computer processing speeds and data storage. Advances in source preprocessing likely will result from the application of new wide-band digital signal processing, which incorporates coherent processing (of multisensor data) and automated algorithm selection and utilization. For target classification and identification, the ability to perform automated feature extraction would be particularly useful.

21.4.3 Level 1: Object Refinement

Improvements in Level 1 processing are needed in several areas. These include data-level fusion of noncommensurate sensor data (e.g., fusion of image and nonimage data) using physics-based target and sensor models, and improved target identification using hybrid methods that incorporate target models, human-analyst information, and implicit information learned from the sensor data. Areas that are lacking include a better understanding of multiple methods of representing uncertainty and the means for selecting appropriate ways to represent information. One approach that could be fruitful involves investigating techniques that operate in a hierarchical manner at varying levels of fidelity (e.g., tracking of individual targets, target groups, and general target populations or classification methods that provide varying levels of target identity on demand).

21.4.4 Level 2: Situation Refinement and Level 3: Threat Refinement

Much work is needed in the Level 2 and Level 3 areas. Basic cognitive models for making inferences and decisions about a situation and threat are needed. A unified and practical theory (or calculus) of uncertainty is needed. Automated methods for selecting appropriate knowledge representation techniques are needed. New methods and tools are required to perform knowledge representation for automated reasoning. Techniques must be developed that are more robust and less fragile than current methods. Attempting both a *drill-down* approach and a *thin covering* approach could be helpful. In the drill-down method, investigators could select a very well-bounded problem in situation assessment and attempt to completely solve it using a combination of physical models, multiple automated reasoning methods, and ad hoc algorithms (i.e., drill down to obtain a complete solution to a narrow problem). In the thin covering approach, a broader problem would be selected and addressed, and a lesser level of fidelity would be possible than with the drill-down approach. The results of these approaches could provide valuable insight as to the optimal manner for approaching general Level 2 and Level 3 problems.

21.4.5 Human-Computer Interface (HCI)

The rapid evolution of HCI technologies should continue to be applied to data fusion systems. However, much more creativity is needed to improve the link between the fusion system and the human. The suggestions by M. J. Hall, S. A. Hall, and Tate[9] (e.g., deliberate synesthesia, time compression/expansion, negative reasoning enhancement, focus/de-focus, pattern morphing, and new uncertainty representation methods) provide an excellent starting point for new HCI research. In addition, more research is needed to understand human cognitive deficiencies and information access preferences. Based on this research, new tools should be developed to enhance the link between a data fusion system and effective human cognition. The focus of this research should be human-centered fusion.

21.4.6 Database Management

New DBMS models are needed for data fusion systems. The practice of combining existing techniques for representing images, signals, text, knowledge, and other data has been only mildly successful when applied to data fusion. New models should be developed that *begin with* the requirement for an integrated representation scheme. Software-based solutions are also required for multilevel security. Ongoing research in areas such as distributed data storage and retrieval, data compression, natural-language interfaces to DBMS, improved access and storage schemes, data mining, and related areas should be monitored and applied to the data fusion problem. This is an area in which the commercial market (e.g., for electronic commerce and business) will provide an impetus for significant improvement.

21.4.7 Level 4: Processing

Improvements in Level 4 processing could have a profound impact on the effectiveness of data fusion systems. The rapid advances in sensors and the ability to utilize hundreds or thousands of sensors provide both an opportunity and challenge for data fusion systems. New multiobjective, multiconstraint optimization methods are needed to effectively use these sensors. Special areas of research include the effective use of highly noncommensurate sensors (particularly those that operate on a greatly different time scale). The link between sensors and the human user must be strengthened to provide an information-based optimization. Research is needed to develop general measures of performance and measures of effectiveness.

21.4.8 Infrastructure Needs

To support the evolving research, a strong infrastructure is required for the data fusion community. The data fusion information access center (FUSIAC) could play a strong role in this infrastructure. Key elements that are needed include (1) a set of standard algorithms and software, (2) one or more test beds to provide a gold standard for algorithm evaluation, (3) warehouses of models for sensors and the environment, and (4) a communication forum. Of particular value would be a universal test case for evaluating algorithms. The image processing community, for example, has used a standard picture for evaluating and comparing algorithms. They have also made effective use of a visual programming toolkit (*Khoros*), funded by the Defense Advanced Research Projects Agency (DARPA), to perform rapid prototyping of image processing techniques. Such a toolkit would benefit the data fusion community.

21.5 Pitfalls in Data Fusion

After reviewing the state of data fusion technology and recommendations for future research, a practitioner might ask, "So what do I do tomorrow to implement a system? What problems and challenges must be addressed?" Several issues and resources are worth noting. First, Steinberg and Bowman[4] provide an overview of the general systems engineering approach for implementing data fusion systems. Engineering guidelines for selecting correlation algorithms are described by Llinas et al.[26] Several texts, such as those of Hall[11] and Waltz and Llinas,[35] provide detailed information on data fusion algorithms. Antony[20] describes issues in DBMS, and other texts focus on specific applications to target tracking (e.g., Blackman[22]) and signal processing techniques.[21]

Hall and Garga[37] have discussed the problem of implementing data fusion systems and identified a number of problems or pitfalls. These include the following dicta, which must be considered and addressed in order to implement an effective data fusion system.

- *There is no substitute for a good sensor.* No amount of data fusion can substitute for a single, accurate sensor that measures the phenomena that you want to observe.

- *Downstream processing cannot make up for errors (or failures) in upstream processing.* Data fusion processing cannot correct for errors in processing (or lack of preprocessing) of individual sensor data.

- *Sensor fusion can result in poor performance if incorrect information about sensor performance is used.* A common failure in data fusion is to characterize the sensor performance in an ad hoc or convenient way. Failure to accurately model sensor performance will result in corruption of the fused results.

- *There is no such thing as a magic or golden data fusion algorithm.* Despite claims to the contrary, there is no perfect algorithm that is optimal under all conditions. Often, real applications do not meet the underlying assumptions required by data fusion algorithms (e.g., available prior probabilities or statistically independent sources).

- *There will never be enough training data.* In general, there will never be sufficient training data to effectively support pattern recognition algorithms used for automatic target recognition or IFFN. Hence, hybrid methods must be used (e.g., model-based methods, syntax representations, or combinations of methods).

- *Quantifying the value of a data fusion system is difficult.* A challenge in data fusion systems is to quantify the utility of the system at a mission level. Although measures of performance can be obtained for sensors or processing algorithms, measures of mission effectiveness are difficult to define.[35]

- *Fusion is not a static process.* The data fusion process is not static; it is an iterative dynamic process that seeks to refine continually the estimates about an observed situation or threat environment.

21.6 Summary

The technology of multisensor data fusion has made major strides in the past two decades. Extensive research has been performed on data fusion algorithms, distributed architectures, automated reasoning techniques, and new resource allocation and optimization techniques. There is an emerging consensus in the data fusion community concerning basic terminology and engineering guidelines. Recent activities to initiate a data fusion information analysis center (FUSIAC) promise to accelerate the development of data fusion technology by increasing the communications among researchers and system implementers. However, despite these rapid advances, much research is needed to overcome current limitations and challenges in data fusion.

References

1. D. L. Hall, R. J. Linn, and J. Llinas, A survey of data fusion systems, *Proc. SPIE Conference on Data Structures and Target Classification*, 1991, 13–36.
2. O. Kessler, et al., *Functional Description of the Data Fusion Process*, Technical Report, Office of Naval Technology, Naval Air Development Center, Warminster, PA, January 1992.
3. F. E. White, *Data Fusion Lexicon*, Data Fusion Subpanel of the Joint Directors of Laboratories Technical Panel for C³ NOSC, San Diego, CA, 1991.
4. A. N. Steinberg and C. L. Bowman, Development and application of data fusion system engineering guidelines, *Proc. Nat. Symp. Sensor and Data Fusion (NSSDF)*, Lexington, MA, May 1998.
5. D. L. Hall and J. Llinas, A challenge for the data fusion community I: research imperatives for improved processing, *Proc. 7th Nat. Symp. Sensor Fusion*, 1994.
6. D. L. Hall and J. Llinas, An introduction to multisensor data fusion, *Proc. IEEE*, 85(1), 1997, 6–23.
7. D. L. Hall and A. K. Garga, Pitfalls in data fusion (and how to avoid them), *Proc. 2nd Internat. Conf. Information Fusion*, 1, 1999, 429–436, Sunnyvale, CA.
8. D. L. Hall and J. Llinas, From GI Joe to starship trooper: the evolution of information support for individual soldiers, *Proc. Internat. Conf. Circuits and Systems*, 1998.
9. M. J. Hall, S. A. Hall, and T. Tate, Removing the HCI bottleneck: how the human computer interface (HCI) affects the performance of data fusion systems, *Proc. 2000 Meeting of the MSS, Nat. Symp. Sensor and Data Fusion*, 2000.

10. A. Steinberg, Standardization in data fusion: new developments, *Proc. EuroFusion99: Internat. Conf. Data Fusion,* 1999, 269–278.

11. D. L. Hall, *Mathematical Techniques in Multisensor Data Fusion,* Artech House, Inc., Norwood, MA, 1992.

12. A. N. Steinberg and C. Bowman, Revisions to the JDL data fusion process model, *Proc. Nat. Symp. Sensor and Data Fusion (NSSDF),* Lexington, MA, May 1998.

13. C. Morefield, AlphaTech Corporation, private communication to D. L. Hall and A. N. Steinberg, May 18, 2000.

14. A. Poore, Multi-dimensional assignment formulation of data association problems arising from multi-target and multisensor tracking, *Computational Optimization Applications,* 3, 27–57, 1994.

15. D. L. Hall, R. J. Hansen, and D. C. Lang, The negative information problem in mechanical diagnostics, *Trans. ASME,* 119, 1997, 370–377.

16. J. G. Wohl, E. E. Entin, D. Serfaty, R. M. James, and J. C. Deckert, Human cognitive performance in ASW data fusion, *Proc. 1987 Tri-Service Data Fusion Symp.,* Johns Hopkins University, Applied Physics Laboratory, Laurel MD, 1987, 465–479.

17. M. Nixon, Application of economic auction methods to resource allocation and optimization, private communication to D.L. Hall, Washington, DC., April 17, 2000.

18. B. Feran, presentation to the National Reconnaissance Office (NRO), spring, 1999 (available on video from the NRO office in Chantilly, VA).

19. J. Llinas, C. Drury, W. Bialas, An-che Chen, *Studies and Analyses of Vulnerabilities in Aided Adversarial Decision-Making,* Technical Report, State University of New York at Buffalo, Dept. of Industrial Engineering, February 1997.

20. R. T. Antony, *Principles of Data Fusion Automation,* Artech House, Inc., Norwood , MA, 1995.

21. D. C. Swanson, *Signal Processing for Intelligent Sensing Systems,* Marcel Dekker, Inc., 2000.

22. S. S. Blackman, *Multiple-Target Tracking with Radar Applications,* Artech House, Inc., Norwood, MA, 1987.

23. Y. Bar-Shalom, Ed., *Multi-Target, Multi-Sensor Tracking: Applications and Advances,* Artech House, Norwood, MA. 1989.

24. K. Fukanaga, *Introduction to Statistical Pattern Recognition,* 2nd ed., Academic Press, NY, 1990.

25. D. Mush and B. Horne, Progress in supervised neural networks: what's new since Lippman? *IEEE Signal Processing Magazine,* 8–39, January 1993.

26. J. Llinas, B. Neuenfeldt, L. McConnell, D. Bohney, C. Bowman, D. Hall, J. Lochacki, and P. Applegate, Studies and analyses within project Correlation: an in-depth assessment of correlation problems and solution techiques, *Proc. 9th Nat. Symp. Sensor Fusion,* 1, 1996, 171–188.

27. R. Mahler, A unified foundation for data fusion, *Proc. 1994 Data Fusion Systems Conf.,* Applied Physics Laboratory, Johns Hopkins University, Laurel, MD, June 1987.

28. C. A. Barlow, L. D. Stone, and M. V. Hun, Unified data fusion, *Proc. 9th Nat. Symp. Sensor Fusion,* 1, 1996, 321–330.

29. J. Llinas and R. Antony, Blackboard concepts for data fusion and command and control applications, *Internat. J. Pattern Recog. and Artificial Intelligence,* 7(2), 1993.

30. D. L. Hall and R. J. Linn, Comments on the use of templating for multisensor data fusion, *Proc. 1989 Tri-Service Data Fusion Symp.,* 1, 345–354, 1989.

31. I. Watson and F. Marir, Case-based reasoning: a review, *The Knowledge Engineering Rev.,* 9(4), 1994 (http://www.surveying.salford.ac.uk...br-mirror/classroom/cbt-review.htm).

32. R. Gibson, D. L. Hall and J. Stover, An autonomous fuzzy logic architecture for multisensor data fusion, *Proc. 1994 Conf. Multi-Sensor Fusion and Integration for Intelligent Systems,* 1994, 143–150.

33. S. Russell and P. Norvig, *Artificial Intelligence: A Modern Approach,* Prentice Hall Series in Artificial Intelligence, Prentice Hall, NJ, 1995.

34. D. L. Hall, *Lectures in Multisensor Data Fusion,* Artech House, Inc., Norwood, MA, 2000.

35. E. Waltz and J. Llinas, *Multisensor Data Fusion,* Artech House, Inc., Norwood, MA, 1990.

36. D.L. Hall, P. Sexton, M. Warren, and J. Zmyslo, Dynamo: A tool for modeling integrated air defense systems, *Proc. 2000 Meeting of the MSS, Nat. Symp. Sensor and Data Fusion*, June 2000.

37. D. L. Hall and A. K. Garga, New perspectives on level four processing in data fusion systems, *Proc. SPIE AeroSense'99 Conf. Digitization of the Battlefield IV*, 1999.

38. B. E. Brendle, Jr., Crewman's associate: interfacing to the digitized battlefield, *Proc. SPIE: Digitization of the Battlefield II*, 1997, 195–202.

39. A. Steele, V. Marzen, and B. Corona, Army Research Laboratory advanced displays and interactive displays, Fedlab technology transitions, *Proc. SPIE: Digitization of the Battlespace IV*, 1999, 205–212.

40. The CAVE at NCSA, http://www.ncsa.uiuc.edu/VEG/ncsaCAVE.html.

41. R. E. Ellis, O. M. Ismaeil, and M. Lipsett, Design and evaluation of a high-performance haptic interface, *Robotica*, 14, 321–327, 1996.

42. http://haptic.mech.nwu.edu/

43. M. Piattelli-Palmarini, *Inevitable Illusions: How Mistakes of Reason Rule over Minds*, John Wiley & Sons, New York, 1994.

IV

Sample Applications

22 **A Survey of Multisensor Data Fusion Systems** *Mary L. Nichols* 22-1
Introduction • Recent Survey of Data Fusion Activities • Assessment of System Capabilities

23 **Data Fusion for Developing Predictive Diagnostics for Electromechanical
Systems** *Carl S. Byington and Amulya K. Garga* 23-1
Introduction • Aspects of a CBM System • The Diagnosis Problem • Multisensor Fusion
Toolkit • Application Examples • Concluding Remarks

24 **Information Technology for NASA in the 21st Century** *Robert J. Hansen,
Daniel Cooke, Kenneth Ford and Steven Zornetzer* 24-1
Introduction • NASA Applications • Critical Research Investment Areas for NASA • High-
Performance Computing and Networking • Conclusions

25 **Data Fusion for a Distributed Ground-Based Sensing System** *Richard R. Brooks* 25-1
Introduction • Problem Domain • Existing Systems • Prototype Sensors for
SenseIT • Software Architecture • Declarative Language Front-End • Subscriptions •
Mobile Code • Diffusion Network Routing • Collaborative Signal Processing •
Information Security • Summary

26 **An Evaluation Methodology for Fusion Processes Based on Information Needs**
Hans Keithley 26-1
Introduction • Information Needs • Key Concept • Evaluation Methodology

22

A Survey of Multisensor Data Fusion Systems

22.1 Introduction 22-1
22.2 Recent Survey of Data Fusion Activities 22-1
22.3 Assessment of System Capabilities.................................. 22-2
References 22-7

Mary L. Nichols
The Aerospace Corporation

22.1 Introduction

During the past two decades, extensive research and development on multisensor data fusion has been performed for the Department of Defense (DoD). By the early 1990s, an extensive set of fusion systems had been reported for a variety of applications ranging from automated target recognition (ATR) and identification-friend-foe-neutral (IFFN) systems to systems for battlefield surveillance. Hall, Linn, and Llinas[1] provided a description of 54 such systems and an analysis of the types of fusion processing, the applications, the algorithms, and the level of maturity of the reported systems. Subsequent to that survey, Llinas and Antony[2] described 13 data fusion systems that performed automated reasoning (e.g., for situation assessment) using the blackboard reasoning architecture. By the mid-1990s, extensive commercial off-the-shelf (COTS) software was becoming available for different data fusion techniques and for decision support. Hall and Linn[3] described a survey of COTS software for data fusion and Buede[4,5] performed surveys and analyses of COTS software for decision support.

This chapter presents a new survey of data fusion systems for DoD applications. The survey was part of an extensive effort to identify and assess DoD fusion systems and activities. This chapter summarizes 79 systems and provides an assessment of the types of fusion processing performed and their operational status.

22.2 Recent Survey of Data Fusion Activities

A survey of DoD operational, prototype, and planned data fusion activities was performed in 1999–2000. The data fusion activities that were surveyed had disparate missions and provided a broad range of fusion capabilities. They represented all military services. The survey emphasized the level of fusion provided (according to the JDL model described in Chapter 2 of this book) and the capability to fuse different types of intelligence data. A summary of the survey results is provided here.

In the survey, a data fusion system was considered to be more than a mathematical algorithm used to automatically achieve the levels of data fusion described in Chapter 2. In military applications, data fusion is frequently accomplished by a combination of the mathematical algorithms (or "fusion engines") and display capabilities with which a human interacts. Hence, the activities range from relatively small-scale algorithms to large-scale Command, Control, Intelligence, Surveillance, and Reconnaissance (C4I) systems, which use specific algorithms — such as trackers — in conjunction with a sophisticated display of data from multiple intelligence (Multi-INT) data types.

The objective in identifying the unique data fusion activities was to isolate the unique capabilities, both mathematical and display-related, of the activity. A master list was initiated, and the researcher applied expert judgment in eliminating activities for any of several reasons: (1) obsolete systems, (2) systems that were subsumed by other systems, (3) systems that did not provide unique fusion capabilities, (4) systems that were only data fusion enablers, and (5) systems that emphasized visualization.

Table 22.1 lists the resulting 79 unique DoD activities with their primary sponsoring service or organization. The list is intended to be a representative, rather than exhaustive, survey of all extant DoD fusion activities. The research and development activities, as well as the prototypical systems, are shown in bold type.

22.3   Assessment of System Capabilities

A primary goal of the survey was to understand the JDL fusion capabilities of the current operational fusion activities, as well as the emphasis of research and development activities. Recognizing that the activities often provided more than one level of fusion, all of the activities were assigned one or more levels. For instance, a typical operational fusion activity, such as the Global Command and Control System (GCCS), provides specialized algorithms for tracking to achieve Level 1 (object refinement) in addition to specific display capabilities aimed at providing the necessary information from which the analyst can draw Level 2 (situation refinement) inferences.

The capabilities were then counted for both the operational and non-operational activities, and a histogram was generated (see Figure 22.1). Note that Level 2 fusion was divided into two components, given that many operational military data fusion systems are said to facilitate Level 2 fusion through the display fusion of various INTs from which an analyst can draw Level 2 inferences.

The majority of operational data fusion activities provide Level 1 fusion. These activities include weapon systems, such as the Advanced Tomahawk Weapons Control System (ATWCS), and trackers, such as the Processor Independent Correlation and Exploitation System (PICES). A less common operational capability is algorithmic Level 2 fusion, which is provided by some operational systems such as the All Source Analysis System (ASAS).

The majority of the operational systems that are geared toward intelligence analysis have emerged from a basic algorithm to track entities of interest. In most cases, the trackers have operated from signals intelligence, or SIGINT. Gradually, these systems evolved by adding a display not only to resolve ambiguities in the tracking, but also to bring in additional INTs. As a result, most of the systems provide an underlying algorithmic fusion, in addition to a display that accommodates multi-INT data.

Algorithmic fusion to achieve Level 3 fusion is uncommon among the operational systems, and none of the operational systems provide Level 4 fusion. These capabilities, however, are being developed within the research and development (R&D) community and do exist in prototypical systems. In addition, R&D efforts are also focusing on Level 1 fusion, but generally with new intelligence data types or new combinations of intelligence data types.

The activities were also analyzed for their capability to fuse data algorithmically from multiple intelligence types. The use of more than one intelligence data type is becoming increasingly critical to solving difficult military problems. Multiple intelligence data types collected on a single entity can increase the dimensionality of that entity. Common intelligence data types that are in use today include SIGINT, infrared intelligence (IR), imagery intelligence (IMINT), moving target indicator (MTI), measurement and signatures intelligence (MASINT), and a combination of two or more of the above or multi-INT data types.

The pie charts in Figure 22.2 illustrate the capabilities to fuse data algorithmically from multi-INT data types for the surveyed activities. There are two pie charts (operational and non-operational) for each of four JDL levels of fusion. The pie chart in the upper left corner of the figure is interpreted as

TABLE 22.1 Recent Survey of DoD Data Fusion Activities

Activity Acronym	Data Fusion Activity	Primary Service
ABI	AWACS Broadcast Intelligence	USAF
ADSI	Air Defense Systems Integrator	Joint
AEGIS	AEGIS Weapon System	USN
AEPDS	Advanced Electronic Processing and Dissemination System	USA
AMSTE	Affordable Moving Surface Target Engagement	DARPA
ANSFP	Artificial Neural System Fusion Prototype	USAF
ARTDF	Automated Real-Time Data Fusion	USMC
ASAS	All Source Analysis System	USA
ATW	Advanced Tactical Workstation	USN
ATWCS	Automated Tomahawk Weapons Control System USN	USN
CAMDMUU	Connectionist Approach to Multi Attribute Decision Making under Uncertainty	USAF
CEC	Cooperative Engagement Capability	USN
CEE	Conditional Events and Entropy	USN
CV	Constant Vision	USAF
DADFA	Dynamic Adaptation of Data Fusion Algorithms	USAF
DADS	Deployable Autonomous Distributed System	USN
DDB	Dynamic Database	DARPA
E2C MCU	E2C Mission Computer Upgrade	USN
EASF	Enhanced All Source Fusion	USAF
EAT	Enhanced Analytical Tools	USAF
ECS	Shield Engagement Coordination System	USAF
ENT	Enhancements to NEAT Templates	USAF
ESAI	Expanded Situation Assessment & Insertion	USAF
FAST	Forward Area Support Terminal	USA
GALE-Lite	Generic Area Limitation Environment Lite	Joint
GCCS	Global Command and Control System	Joint
GCCS A	Global Command and Control System Army	USA
GCCS I3	Global Command and Control System Integrated Imagery and Intelligence	Joint
GCCS M	Global Command and Control System Maritime	USN
GDFS	Graphical Display Fusion System	USN
GISRC	Global Intelligence, Surveillance, and Reconnaissance Capability	USN
Hercules		Joint
IAS	Intelligence Analysis System	USMC
IDAS	Interactive Defense Avionics System	USAF
IFAMP	Intelligent Fusion and Asset Management Processor	USAF
ISA	Intelligence Situation Analyst	USAF
ISAT	Integrated Situation Awareness and Targeting	USA
IT	Information Trustworthiness	USA
ITAP	Intelligent Threat Assessment Processor	USAF
JIVA	Joint Intelligence Virtual Architecture	Joint
JSTARS CGS	Joint Surveillance Target Attack Radar Subsystem Common Ground Station	Joint
JTAGS	Joint Tactical Ground Station	Joint
KBS4TCT	Knowledge-Based Support for Time Critical Targets	USAF
LOCE	Linked Operational Intelligence Centers Europe	Coalition
LSS	Littoral Surveillance System	USN
MDBI&U	Multiple Database Integration & Update	USAF
MITT	Mobile Integrated Tactical Terminal	USA
Moonlight	Moonlight	Coalition
MSCS	Multiple Source Correlation System	USAF
MSFE	Multisource Fusion Engine	USAF
MSI	[E2C] Multisensor Integration	USN
MSTS	Multisource Tactical System	USAF
NCIF	Network Centric Information Fusion	USAF
NEAT	Nodal Exploitation and Analysis Tool	USAF
OBATS	Off-Board Augmented Theater Surveillance	USAF
OED	Ocean Surveillance Information System Evolutionary Development	USN

TABLE 22.1 (continued)   Recent Survey of DoD Data Fusion Activities

Activity Acronym	Data Fusion Activity	Primary Service
Patriot	Patriot Weapon System	Joint
PICES	Processor Independent Correlation Exploitation System	USN
QIFS	Quarterback Information Fusion System	USAF
SAFETI	Situation Awareness From Enhanced Threat Information	USAF
SCTT	SAR Contextual Target Tracking	USA
SMF	SIGINT/MTI Fusion	USA
Story Teller	EP3 Story Teller	USN
TADMUS	Tactical Decision Making under Stress	USN
TAS	Timeline Analysis System	USAF
TBMCS	Theater Battle Management Core Systems	USAF
TCAC	Technical Control and Analysis Center	USMC
TCR	Terrain Contextual Reasoning	USA
TDPS	Tactical Data Processing Suites	USAF
TEAMS	Tactical EA-6B Mission Support	USN
TERPES	Tactical Electronic Reconnaissance Processing Evaluation System	USMC
TES	Tactical Exploitation System	USA
TIPOFF	TIBS Integrated Processor and Online Fusion Function	Joint
TMBR	Truth Maintenance Belief Revision	USAF
TRAIT	Tactical Registration of Airborne Imagery for Targeting	USAF
TSA	Theater Situation Awareness	USAF
UGBADFT	Unified Generalized Bayesian Adaptive Data Fusion Techniques	USAF
VF	Visual Fusion	USA
WECM	Warfighter Electronic Collection and Mapping	USA

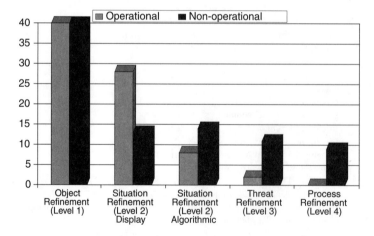

FIGURE 22.1 Comparison of fusion capabilities.

follows. All systems that provide algorithmic fusion to achieve Level 1 fusion (object refinement) were tallied according to the intelligence data type(s) used to achieve the fusion. The chart shows that the majority of operational systems use only SIGINT to achieve Level 1 fusion. By contrast, the pie chart in the upper right corner shows, for non-operational systems, a greater variety in the usage of other intelligence data types.

Other conclusions from this figure are that Level 2 fusion, which is achieved by operational systems, is primarily achieved using multi-INT data according to the pie chart in the lower left column, second row. All non-operational activities use multi-INT data to achieve Level 2 fusion. Few operational systems automatically integrate multi-INT data. Most data fusion systems display multi-INT data — sometimes on a common screen. Selected R&D systems are tackling the algorithmic integration of multi-INT data.

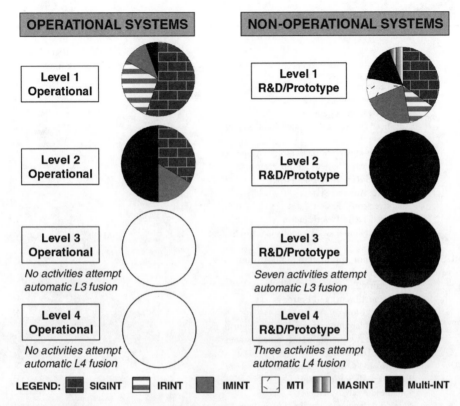

FIGURE 22.2 Algorithmic fusion by intelligence data type.

A common belief is that the realm of military data fusion is marked by numerous duplicative activities, which seems to imply a lack of organization and coordination. In reality, the 79 activities in this study reflect various relationships and migration plans. A close examination of the pedigree, current relationships, and progeny demonstrates some commonality and a greater degree of coordination than is often apparent. In addition, the DoD has established several migration systems to which existing fusion systems must evolve. Examples of these are the Global Command and Control System (GCCS) and the distributed common ground station (DCGS).

Specific operational intelligence systems have been identified as migration systems; others are viewed as legacy systems that will eventually be subsumed by a migration system. For nonintelligence systems, the fusion capabilities are frequently highly specialized and tailored to the overall mission of the system, such as weapon cueing. In addition, the non-operational systems are generally highly specialized in that they are developing specific technologies, but they also frequently leverage other operational and non-operational activities.

Table 22.2 shows the division of these activities into several categories:

1. Migration systems that are converging to a common baseline to facilitate interoperability with other systems
2. Legacy systems that will be subsumed by a migration system
3. Government-sponsored research and development and prototypes
4. Highly specialized capabilities that are not duplicated by any other system

The research and development activities, as well as the prototypical systems, are shown in boldface type.

In conclusion, the recent survey of DoD data fusion activities provides a snapshot of current and emerging fusion capabilities in terms of their level of fusion and their usage of various types of intelligence data. The survey also reflects a greater degree of coordination among military data fusion activities than was recognized by previous surveys.

TABLE 22.2   Status of Data Fusion Activities

Activity Acronym	Data Fusion Activity	Migration System	Legacy System	R&D / Prototype	Specialized Functions and Others
ABI	AWACS Broadcast Intelligence				X
ADSI	Air Defense Systems Integrator				X
AEGIS	AEGIS Weapon System	X			
AEPDS	Advanced Electronic Processing and Dissemination System		X		
AMSTE	Affordable Moving Surface Target Engagement			X	
ANSFP	Artificial Neural System Fusion Prototype			X	
ARTDF	Automated Real Time Data Fusion	X			
ASAS	All Source Analysis System	X			
ASF	Adaptive Sensor Fusion			X	
ATW	Advanced Tactical Workstation				X
ATWCS	Automated Tomahawk Weapons Control System USN	X			
CAMDMUU	Connectionist Approach to Multi Attribute Decision Making under Uncertainty			X	
CEC	Cooperative Engagement Capability			X	
CEE	Conditional Events and Entropy			X	
CV	Constant Vision				X
DADFA	Dynamic Adaptation of Data Fusion Algorithms			X	
DADS	Deployable Autonomous Distributed System			X	
DDB	Dynamic Database			X	
E2C MCU	E2C Mission Computer Upgrade				X
EASF	Enhanced All Source Fusion			X	
EAT	Enhanced Analytical Tools			X	
ECS	Shield Engagement Coordination System			X	
ENT	Enhancements to NEAT Templates			X	
ESAI	Expanded Situation Assessment and Insertion			X	
FAST	Forward Area Support Terminal		X		
GALE-Lite	Generic Area Limitation Environment Lite	X			
GCCS	Global Command and Control System	X			
GCCS A	Global Command and Control System Army	X			
GCCS I3	Global Command and Control System Integrated Imagery and Intelligence	X			
GCCS M	Global Command and Control System Maritime	X			
GDFS	Graphical Display Fusion System				X
GISRC	Global Intelligence, Surveillance, and Reconnaissance Capability			X	
Hercules				X	
IAS	Intelligence Analysis System	X			
IDAS	Interactive Defense Avionics System				X
IFAMP	Intelligent Fusion and Asset Management Processor			X	
ISA	Intelligence Situation Analyst				X
ISAT	Integrated Situation Awareness and Targeting			X	
IT	Information Trustworthiness			X	
ITAP	Intelligent Threat Assessment Processor			X	
JIVA	Joint Intelligence Virtual Architecture	X			
JSTARS CGS	Joint Surveillance Target Attack Radar Subsystem Common Ground Station				X
JTAGS	Joint Tactical Ground Station				X
KBS4TCT	Knowledge-Based Support for Time Critical Targets			X	
LOCE	Linked Operational Intelligence Centers Europe	X			
LSS	Littoral Surveillance System	X			
MDBI&U	Multiple Database Integration and Update			X	
MITT	Mobile Integrated Tactical Terminal		X		
Moonlight					X

TABLE 22.2 (continued)   Status of Data Fusion Activities

Activity Acronym	Data Fusion Activity	Migration System	Legacy System	R&D / Prototype	Specialized Functions and Others
MSCS	Multiple Source Correlation System				X
MSFE	Multisource Fusion Engine				X
MSI	[E2C] Multisensor Integration			X	
MSTS	Multisource Tactical System				X
NCIF	Network Centric Information Fusion			X	
NEAT	Nodal Exploitation and Analysis Tool			X	
OBATS	Off-Board Augmented Theater Surveillance			X	
OED	Ocean Surveillance Information System Evolutionary Development				X
Patriot	Patriot Weapon System				X
PICES	Processor Independent Correlation Exploitation System				X
QIFS	Quarterback Information Fusion System				X
SAFETI	Situation Awareness From Enhanced Threat Information			X	
SCTT	SAR Contextual Target Tracking			X	
SMF	SIGINT/MTI Fusion			X	
Story Teller	EP3 Story Teller				X
TADMUS	Tactical Decision Making under Stress			X	
TAS	Timeline Analysis System				
TBMCS	Theater Battle Management Core Systems	X			
TCAC	Technical Control and Analysis Center	X			
TCR	Terrain Contextual Reasoning			X	
TDPS	Tactical Data Processing Suites		X		
TEAMS	Tactical EA-6B Mission Support				X
TERPES	Tactical Electronic Reconnaissance Processing Evaluation System		X		
TES	Tactical Exploitation System	X			
TIPOFF	TIBS Integrated Processor and Online Fusion Function				X
TMBR	Truth Maintenance Belief Revision			X	
TRAIT	Tactical Registration of Airborne Imagery for Targeting			X	
TSA	Theater Situation Awareness				X
UGBADFT	Unified Generalized Bayesian Adaptive Data Fusion Techniques			X	
VF	Visual Fusion			X	
WECM	Warfighter Electronic Collection and Mapping			X	

References

1. Hall, D.L., Linn, R.J., and Llinas, J., Survey of data fusion systems, *Proc. SPIE Conf. Data Structure and Target Classification*, Orlando, FL, 1991, 147B, 13–36.
2. Llinas, J. and Antony, R. T., Blackboard concepts for data fusion applications, *Internat. J. Pattern Recognition and Artificial Intelligence*, 7(2), 1993, 285–308.
3. Hall, D. L. and Linn, R. J., Survey of commercial software for multi-sensor fusion, *Proc. SPIE Conf. Sensor Fusion: Aerospace Applications*, Orlando, FL, 1993.
4. Buede, D., Software review: Overview of the MCDA market, *J. Multi-Criteria Decision Analysis*, 1, 59–61, 1992.
5. Buede, D., Superior Design Features of Decision Analytic Software, *Computers Ops. Res.*, 19(1), 43–57, 1992.

23

Data Fusion for Developing Predictive Diagnostics for Electromechanical Systems

Carl S. Byington
The Pennsylvania State University

Amulya K. Garga
The Pennsylvania State University

23.1 Introduction 23-1
    Condition-Based Maintenance Motivation
23.2 Aspects of a CBM System 23-3
23.3 The Diagnosis Problem 23-4
    Feature-Level Fusion • Decision-Level Fusion • Model-Based Development
23.4 Multisensor Fusion Toolkit 23-7
23.5 Application Examples 23-8
    Mechanical Power Transmission • Fluid Systems • Electrochemical Systems
23.6 Concluding Remarks 23-29
Acknowledgments 23-30
References 23-30

23.1 Introduction

Condition-based maintenance (CBM) is a philosophy of performing maintenance on a machine or system only when there is objective evidence of need or impending failure. By contrast, time-based or use-based maintenance involves performing periodic maintenance after specified periods of time or hours of operation. CBM has the potential to decrease life-cycle maintenance costs (by reducing unnecessary maintenance actions), increase operational readiness, and improve safety.

Implementation of condition-based maintenance involves *predictive diagnostics* (i.e., diagnosing the current state or health of a machine and predicting time to failure based on an assumed model of anticipated use). CBM and predictive diagnostics depend on multisensor data — such as vibration, temperature, pressure, and presence of oil debris — which must be effectively fused to determine machinery health. Indeed, Hansen et al. suggested that predictive diagnostics involves many of the same functions and challenges demonstrated in more traditional Department of Defense (DoD) applications of data fusion (e.g., signal processing, pattern recognition, estimation, and automated reasoning).[1] This chapter demonstrates the potential for technology transfer from the study of CBM to DoD fusion applications.

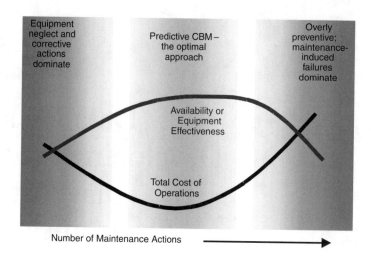

FIGURE 23.1　CBM provides the best range of operational availability or equipment effectiveness.

23.1.1　Condition-Based Maintenance Motivation

CBM is an emerging concept enabled by the evolution of key technologies, including improvements in sensors, microprocessors, digital signal processing, simulation modeling, multisensor data fusion, and automated reasoning. CBM involves monitoring the health or status of a component or system and performing maintenance based on that observed health and some predicted remaining useful life (RUL).[2-5] This *predictive maintenance* philosophy contrasts with earlier ideologies, such as *corrective maintenance* — in which action is taken after a component or system fails — and *preventive maintenance* — which is based on event or time milestones. Each involves a cost tradeoff. Corrective maintenance incurs low maintenance cost (minimal preventative actions), but high performance costs caused by operational failures. Conversely, preventative maintenance produces low operational costs, but greater maintenance department costs. Moreover, the application of statistical safe-life methods (which are common with preventative maintenance) usually leads to very conservative estimates of the probability of failure. The result is the additional hidden cost associated with disposing of components that still retain significant remaining useful life.

Another important consideration in most applications is the operational availability (a metric that is popular in military applications) or equipment effectiveness (more popular in industrial applications). Figure 23.1 illustrates regions of high total cost when overly corrective or overly preventive maintenance dominate. These regions also provide a lower total availability of the equipment. On the corrective side, equipment neglect typically leads to more operational failures during which time the equipment is unavailable. On the preventive side, the equipment is typically unavailable because it is being maintained much of the time. An additional concern that affects availability and cost in this region is the greater likelihood of maintenance-induced failures.

The development of better maintenance practices is driven by the desire to reduce the risk of catastrophic failures, minimize maintenance costs, maximize system availability, and increase platform reliability. These goals are desirable from the application arenas of aircraft, ships, and tanks to industrial manufacturing of all types. Moreover, given that maintenance is a key cost driver in military and commercial applications, it is an important area in which to focus research and development efforts. At nuclear power plants, for example, the operations and maintenance portion of the direct operating costs (DOC) grew by more than 120 percent between 1981 and 1991 — a level more than twice as great as the fuel cost component.[6]

A more explicit cost savings can be seen in Figure 23.2 derived from an Electric Power Research Institute study to estimate the costs associated with different maintenance practices in the utility industry. The first three columns were taken directly from the study and the fourth is estimated from some unpublished

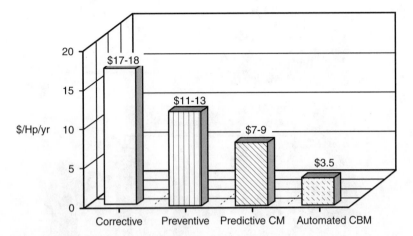

FIGURE 23.2   Moving toward condition monitoring and the optimal level of maintenance provided dramatic cost savings in the electric industry.

cost studies. Clearly, predictive practices provide cost benefit. The estimated 50 percent additional cost savings derived from predictive condition monitoring to automated CBM is manifested by the manpower cost focused on data collection/analysis efforts and cost avoidances associated with continuous monitoring and fault prediction.[7]

Such cost savings have motivated the development of CBM systems; furthermore, substantially more benefit can be realized by automating a number of the functions to achieve improved screening and robustness. This knowledge has driven CBM research and development efforts.

23.2   Aspects of a CBM System

CBM uses sensor systems to diagnose emerging equipment problems and to predict how long equipment can effectively serve its operational purpose. The sensors collect and evaluate real-time data using signal processing algorithms. These algorithms correlate the unique signals to their causes — for example, vibrational sideband energy created by developing gear-tooth wear. The system alerts maintenance personnel to the problem, enabling maintenance activities to be scheduled and performed before operational effectiveness is compromised.

The key to effectively implementing CBM is the ability to detect, classify, and predict the evolution of a failure mechanism with sufficient robustness — and at a low enough cost — to use that information as a basis to plan maintenance for mission- or safety-critical systems. "Mission critical" refers to those activities that, if interrupted, would prohibit the organization from meeting its primary objectives (e.g., completion and delivery of 2500 control panels to meet an OEM's assembly schedule). Safety critical functions must remain operational to ensure the safety of humans (e.g., airline passengers).

Thus, a CBM system must be capable of

- Detecting the start of a failure evolution,
- Classifying the failure evolution,
- Predicting remaining useful life with a high degree of certainty,
- Recommending a remedial action to the operator,
- Taking the indicated action through the control system,
- Aiding the technician in making the repair,
- Providing feedback for the design process.

These activities represent a closed-loop process with several levels of feedback, which differentiates CBM from preventive or time-directed maintenance. In a preventive maintenance system, time between overhaul

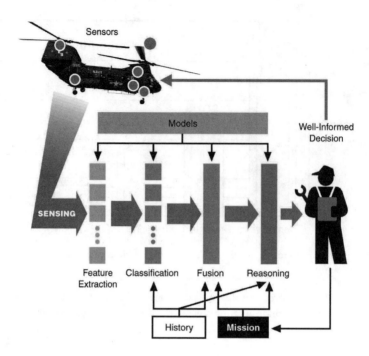

FIGURE 23.3 The success of CBM systems depends on: (1) The ability to design or use robust sensors for measuring relevant phenomena, (2) real-time processing of the sensor data to extract useful information (e.g., features or data characteristics) in a noisy environment, and to detect parametric changes that could indicate impending failure conditions, (3) fusion of multisensor data to obtain improved information (beyond what is available from a single sensor), (4) micro- and macro-level models that predict the temporal evolution of failure phenomena, and (5) automated approximate reasoning capable of interpreting the results of the sensor measurements, processed data, and model prediction in the context of an operational environment.[8]

(TBO) is set at design, based on failure mode effects, criticality analyses (FMECA), and experience with like machines' mortality statistics. The general concept of a CBM system is shown in Figure 23.3.

23.3   The Diagnosis Problem

Multisensor data fusion has been recognized as an enabling technology for both military and nonmilitary applications. However, improved diagnosis and increased performance do not result automatically from increased data collection. The data must be contextually filtered to extract information that is relevant to the task at hand. Another key requirement that justifies the use of data fusion is low false alarms. In general, there is a tradeoff between missed detections and false alarms, which is greatly influenced by the mission or operation profile. If a diagnostic system produces excessive false alarms, personnel will likely ignore it, resulting in an unacceptably high number of missed detections. However, presently data fusion is rarely employed in monitoring systems, and, when it is used, it is usually an afterthought. Data fusion can most readily be employed at the feature or decision levels.

23.3.1   Feature-Level Fusion

Diagnosis is most commonly performed as classification using feature-based techniques.[9] Machinery data are processed to extract features that can be used to identify specific failure modes. Discriminant transformations are often utilized to map the data characteristic of different failure mode effects into distinct regions in the feature subspace. Multisensor systems frequently use this approach because each sensor may contribute a unique set of features with varying degrees of correlation with the failure to be diagnosed. These features, when combined, provide a better estimate of the object's identity. Examples of this approach will be illustrated in the applications section.

23.3.2 Decision-Level Fusion

Following the classification stage, decision-level fusion can be used to fuse identity. Several decision-level fusion techniques exist, including voting, weighted decision, and Bayesian inference.[10] Other techniques, such as Dempster-Shafer's method[11,12] and generalized evidential processing theory,[13] are described in this text and in other publications.[14,15]

23.3.2.1 Voting

Voting, as a decision-level fusion method, is the simplest approach to fusing the outputs from multiple estimates or predictions by emulating the way humans reach some group agreement.[10] The fused output decision is based on the majority rule (i.e., maximum number of votes wins). Variations of voting techniques include weighted voting (in which sensors are given relative weights), plurality, consensus methods, and other techniques.

For implementation of this structure, each classification or prediction, i, outputs a binary vector, x_i, with D elements, where D is the number of hypothesized output decisions. The binary vectors are combined into a matrix X, with row i representing the input from sensor i. The voting fusion structure sums the elements in each column as described by Equation 23.1.

$$y(j)=\sum_{i=1}^{N} X(i,j) \qquad \forall j=1:D \qquad (23.1)$$

The output, $y(j)$, is a vector of length D, where each element indicates the total number of votes for output class j. At time k, the decision rule selects the output, $d(k)$, as the class that carries the majority vote, according to Equation 23.2 .

$$d(k)=\arg\max_{j} \; y(j) \qquad (23.2)$$

23.3.2.2 Weighted Decision Fusion

A weighted decision method for data fusion generates the fused decision by weighting and combining the outputs from multiple sensors. *A priori* assumptions of sensor reliability and confidence in the classifier performance contribute to determining the weights used in a given scenario. Expert knowledge or models regarding the sensor reliability can be used to implement this method. In the absence of such knowledge, an assumption of equal reliability for each sensor can be made. This assumption reduces the weighted decision method to voting. Note that at the other extreme, a weighted decision process could selectively weight sensors so that, at a particular time, only one sensor is deemed to be credible (i.e., weight = 1), while all other sensors are ignored (i.e., weight = 0).

Several methods can be used for implementing a weighted decision fusion structure. Essentially, each sensor, i, outputs a binary vector, x_i, with D elements, where D is the number of hypothesized output decisions. A binary *one*, in position j, indicates that the data was identified by the classifier as belonging to class j. The classification vector from sensor i becomes the ith row of an array, X, that is passed to the weighted decision fusion structure. Each row is weighted, using the *a priori* assumption of the sensor reliability. Subsequently, the elements of the array are summed along each column. Equation 23.3 describes this process mathematically.

$$y(j)=\sum_{i=1}^{N} w_i X(i,j) \qquad \forall j=1:D \qquad (23.3)$$

The output, $y(j)$, is a row vector of length D, where each element indicates the confidence that the input data from the multiple sensor set has membership in a particular class. At time k, the output decision, $d(k)$, is the class that satisfies the maximum confidence criteria of Equation 23.4.

$$\sum_{i=1}^{N} w_i = 1 \qquad (23.4)$$

This implementation of weighted decision fusion permits future extension in two ways. First, it provides a path to the use of confidence as an input from each sensor. This would allow the fusion process to utilize fuzzy logic within the structure. Second, it enables an adaptive mechanism to be incorporated that can modify the sensor weights as data are processed through the system.

23.3.2.3 Bayesian Inference

Bayes' theorem[16-18] serves as the basis for the Bayesian inference technique for identity fusion. This technique provides a method for computing the *a posteriori* probability of a particular outcome, based on previous estimates of the likelihood and additional evidence. Bayesian inference assumes that a set of D mutually exclusive (and exhaustive) hypotheses or outcomes exists to explain a given situation.

In the decision-level, multisensor fusion problem, Bayesian inference is implemented as follows. A system exists with N sensors that provide decisions on membership to one of D possible classes. The Bayesian fusion structure uses *a priori* information on the probability that a particular hypothesis exists and the likelihood that a particular sensor is able to classify the data to the correct hypothesis. The inputs to the structure are (1) $P(O_j)$, the *a priori* probabilities that object j exists (or equivalently that a fault condition exists), (2) $P(D_{k,i}|O_j)$, the likelihood that each sensor, k, will classify the data as belonging to any one of the D hypotheses, and (3) D_k, the input decisions from the K sensors. Equation 23.5 describes the Bayesian combination rule.

$$P(O_j|D_1,\ldots,D_K) = \frac{P(O_j)\prod_{k=1}^{K} P(D_k|O_j)}{\sum_{i=1}^{N} P(O_j)\prod_{k=1}^{K} P(D_k|O_j)} \qquad (23.5)$$

The output is a vector with element j representing the *a posteriori* probability that the data belong to hypothesis j. The fused decision is made based on the maximum *a posteriori* probability criteria given in Equation 23.6.

$$d(k) = \arg\max_{j} \left[P(O_j|D_1,\ldots,D_K) \right] \qquad (23.6)$$

A basic issue with the use of Bayesian inference techniques involves the selection of the *a priori* probabilities and the likelihood values. The choice of this information has a significant impact on performance. Expert knowledge can be used to determine these inputs. In the case where the *a priori* probabilities are unknown, the user can resort to the principle of indifference, where the prior probabilities are set to be equal, as in Equation 23.7.

$$P(O_j) = \frac{1}{N} \qquad (23.7)$$

The *a priori* probabilities are updated in the recursive implementation as described by Equation 23.8. This update sets the value for the *a priori* probability in iteration t equal to the value of the *a posteriori* probability from iteration $(t-1)$.

$$P_t(O_j) = P_{t-1}(O_j|D_1,\ldots,D_k) \qquad (23.8)$$

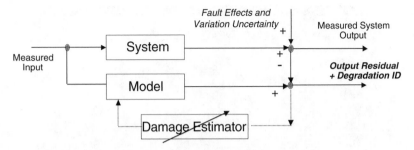

FIGURE 23.4 Adaptive concept for deterministic damage estimation.

23.3.3 Model-Based Development

Diagnostics model development can proceed down a purely data-driven, empirical path or model-based path that uses physical and causal models to drive the diagnosis. Model-based diagnostics can provide optimal damage detection and condition assessment because empirically verified mathematical models at many state conditions are the most appropriate knowledge bases.[19] The Pennsylvania State University Applied Research Laboratory (Penn State ARL) has taken a model-based diagnostic approach towards achieving CBM that has proven appropriate for fault detection, failure mode diagnosis, and, ultimately, prognosis.[20]

The key modeling area for CBM is to develop models that can capture the salient effects of faults and relate them to virtual or external observables. Some fundamental questions arise from this desired modeling. How can mathematical models of physical systems be adapted or augmented with separate damage models to capture symptoms? Moreover, how can model-based diagnostics approaches be used for design in CBM requirements such as sensor type, location, and processing requirements?

In the model-based approach, the physical system is captured mathematically in the form of empirically validated computational or functional models. The models possess or are augmented with damage association models that can simulate a failure mode of given severity to produce a symptom that can be compared to measured features. The failure mode symptoms are used to construct the appropriate classification algorithms for diagnosis. The sensitivity of the failure modes to specific sensor processing can be compared for various failure modes and evaluated over the entire system to aid in the determination of the most effective CBM approach.

23.3.3.1 Model-Based Identification and Damage Estimation

Figure 23.4 illustrates a conceptual method for identifying the type and amount of degradation using a validated system model. The actual system output response (event and performance variables) is the result of nominal system response plus fault effects and uncertainty. The model-based analysis and identification of faults can be viewed as an optimization problem that produces the minimum residual between the predicted and actual response.

23.4 Multisensor Fusion Toolkit

A multisensor data fusion toolkit was developed at the Penn State ARL to provide the user with a standardized visual programming environment for data fusion (see Figure 23.5).[21] With this toolkit, the user can develop and compare techniques that combine data from actual and virtual sensors. Detection performance and the number of false alarms are two of the metrics that can be used for such a comparison.

The outputs of one or more state/damage estimates can be combined with available usage information, based on feature vector classification. This type of a tool is an asset because it utilizes key information from multiple sensors for robustness and presents the results of the fusion assessment, rather than just a data stream. Furthermore, the tool is very useful for rapid prototyping and evaluation of data analysis and data fusion algorithms. The toolkit was written in Visual C++ using an object-oriented design approach.

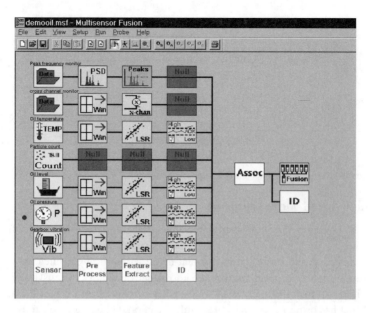

FIGURE 23.5 The Penn State ARL multisensor fusion toolkit is used to combine data from multiple sensors, improving the ability to characterize the current state of a system.

23.5 Application Examples

This section presents several examples to illustrate the development of a data fusion approach and its application to condition-based maintenance of real-world engineered systems. The topics chosen represent a range of machinery with different fundamental mechanisms and potential CBM strategies. The first example is a mechanical (gear/shaft/bearing) power transmission that has been tested extensively at Penn State ARL. The second example uses fluid systems (fuel/lubrication/hydraulic), and the third example focuses on energy storage devices (battery/fuel cells). All are critical subsystems that address fundamental CBM needs in the DoD and industry. Developers of data fusion solutions must carefully select among the options that are applicable to the problem. Several pitfalls in using data fusion were identified recently and suggestions were provided about how to avoid them.[22]

23.5.1 Mechanical Power Transmission

Individual components and systems, where a few critical components are coupled together in rotor power generation and transmission machinery, are relatively well understood as a result of extensive research that has been conducted over the past few decades. Many notable contributions have been made in the analysis and design, in increasing the performance of rotor systems, and in the fundamental understanding of different aspects of rotor system dynamics. More recently, many commercial and defense efforts have focused on vibration/noise analysis and prediction for fault diagnostics. Many employ improved modeling methods to understand the transmission phenomena more thoroughly,[23-26] while others have focused on detection techniques and experimental analysis of fault conditions.[27-33]

23.5.1.1 Industrial Gearbox Example

Well-documented transitional failure data from rotating machinery is critical for developing machinery prognostics. However, such data is not readily or widely available to researchers and developers. Consequently, a mechanical diagnostics test bed (MDTB) was constructed at the Penn State ARL for detecting faults and tracking damage on an industrial gearbox (see Figure 23.6).

FIGURE 23.6   The Penn State ARL mechanical diagnostics testbed is used to collect transitional failure data, study sensor optimization for fault detection, and to evaluate failure models.

23.5.1.1.1   *System and Data Description*

The MDTB is a motor-drivetrain-generator test stand. (A complete description of the MDTB can be found in Byington and Kozlowski.[34]) The gearbox is driven at a set input speed using a 30 Hp, 1750 rpm AC (drive) motor, and the torque is applied by a 75 Hp, 1750 rpm AC (absorption) motor. The MDTB can test single and double reduction industrial gearboxes with ratios from about 1.2:1 to 6:1. The gearboxes are nominally in the 5 to 20 Hp range. The motors provide about 2 to 5 times the rated torque of the selected gearboxes; thus, the system can provide good overload capability for accelerated failure testing.

The gearbox is instrumented with accelerometers, acoustic emission sensors, thermocouples, and oil quality sensors. Torque and speed (load inputs) are measured within 1% on the rig. Borescope images are taken during the failure process to correlate degree of damage with measured sensor data. Given a low contamination level in the oil, drive speed and load torque are the two major factors in gear failure. Different values of torque and speed will cause different types of wear and faults. Figure 23.7 illustrates potential regions of failures.

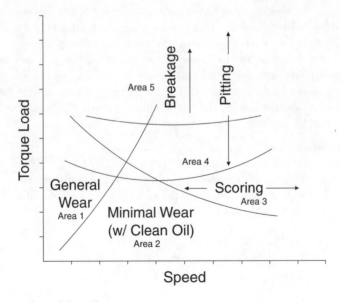

FIGURE 23.7   Regions of gear failures.[35]

TABLE 23.1   Wear Particle Morphology – Ferrography Descriptors

Particle	Description
Rubbing	Particles, 20 μm chord dimension and approx. 1 μm thick. Results from flaking of pieces from mixed shear layer-mainly benign.
Cutting	Swarf-like chops of fine wire coils, caused by abrasive cutting action.
Laminar	Thin, bright, free-metal particles, typically 1 μm thick, 20–50 μm chord width. Holes in surface and uneven edge profile. Gear-rolling element bearing wear.
Fatigue	Chunky, several microns thick from, e.g., gear wear, 20–50 μm chord width.
Spheres	Typically ferrous, 1 to 2 μm diameter, generated when micro-cracks occur under rolling contact fatigue condition.
Severe Sliding	Large/50 μm chord width, several microns thick. Surfaces heavily striated with long straight edges. Typically found in gear wear.

23.5.1.1.2   Gearbox Failure Conditions

In Area 1, the gear is operating too slowly to develop an oil film, so adhesive wear occurs. In Area 2, the speed is sufficiently fast to develop an oil film. The gears should be able to run with minimal wear. In Area 3, scoring is likely, because the load and speed are high enough to break down the existing oil film. Area 4 illustrates the dominance of pitting caused by high surface stresses that result from higher torque loads. As the torque is increased further, tooth breakage will result from overload and stress fatigue, as shown in Area 5.

Based on the above discussion, the MDTB test plan includes test runs that set the operating drive speed and load torque deep into each area to generate transitional data for each of the faults. These limits, of course, are not known exactly *a priori*. Areas 4 and 5 are the primary focal points because they contain critical and difficult-to-predict faults. Being able to control (to a degree) the conditions that affect the type of failure that occurs allows some control over the amount of data for each fault, while still allowing the fault to develop naturally (i.e., the fault is not seeded). If a particular type of fault requires more transitional data for analysis, adjustment of the operating conditions can increase the likelihood of producing the desired fault.

23.5.1.1.3   Oil Debris Analysis

Roylance and Raadnui[36] examined the morphology of wear particles in circulating oil and correlated their occurrences with wear characteristics and failure modes of gears and other components of rotating machinery. Wear particles build up over time even under normal operating conditions. However, the particles generated by *benign* wear differ markedly from those generated by the *active* wear associated with pitting, abrasion, scuffing, fracturing, and other abnormal conditions that lead to failure. Roylance and Raadnui[36] correlated particle features (quantity, size, composition, and morphology) with wear characteristics (severity, rate, type, and source).

Particle composition can be an important clue to the source of abnormal wear particles when components are made of different materials. The relationship of particle type to size and morphology has been well characterized by Roylance,[36] and is summarized in Table 23.1.

23.5.1.1.4   Vibration Analysis

Vibration analysis is extremely useful for gearbox analysis and gear failures because the unsteady component of relative angular motion of the meshing gears provides the major source of vibratory excitation.[37] This effect is largely caused by a change in compliance of the gear teeth and deviation from perfect shape. Such a modulated gear meshing vibration, $y(t)$, is given by:

$$y(t) = \sum_{n=0}^{N} X_n \big[1 + a_n(t)\big] \cos\big[2\pi f_m t + \phi_n + b_n(t)\big] \tag{23.9}$$

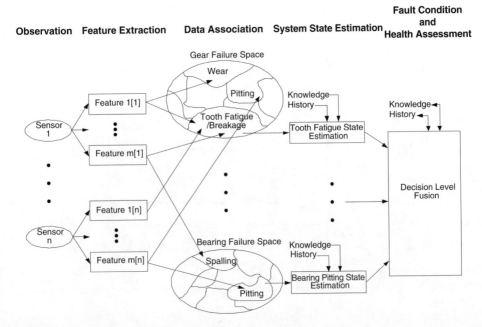

FIGURE 23.8 Oil/vibration data fusion process.[40]

where $a_n(t)$ and $b_n(t)$ are the amplitude and phase modulation functions. Amplitude modulation produces sidebands around the carrier (gear-meshing and harmonics) frequencies and is often associated with eccentricity, uneven wear, or profile errors. As can be seen from the equation, frequency modulation will produce a family of sidebands. These will both typically occur in gear systems, and the sidebands may either combine or subtract to produce an asymmetrical family of sidebands.

Much of the analysis has focused on the use of the appropriate statistical processing and transform to capture these effects. A number of figures of merit or features have been used to correlate mechanical faults. Moreover, short-time Fourier, Hilbert, and wavelet transforms have also been used to develop vibration features.[38,39]

23.5.1.1.5 Description of Features

Various signal and spectral modeling techniques have been used to characterize machinery data and develop features indicative of various faults in the machinery. Such techniques include statistical modeling (e.g., mean, rms, kurtosis), spectral modeling (e.g., Fourier transform, cepstral transform, autoregressive modeling), and time frequency modeling (e.g., short-time Fourier transform, wavelet transform, wideband ambiguity functions). Several oil and vibration features are now well described. These can be fused and integrated with knowledge of the system and history to provide indication of gearbox condition. In addition to the obvious corroboration and increased confidence that can be gained, this approach to using multiple sensors also aids in establishing the existence of sensor faults.

Features tend to organize into subspaces in feature space, as shown in Figure 23.8. Such subspaces can be used to classify the failure mode. Multiple estimates of a specific failure mode can be produced through the classification of each feature subspace. Other failure mode estimates can be processed at the same time as well. Note that a gearbox may deteriorate into more than one failure mode with several critical faults competing.

During 20+ run-to-failure transitional tests conducted on the MDTB, data were collected from accelerometer, temperature, torque, speed, and oil quality/debris measurements. This discussion pertains only to Test 14. Borescope imaging was performed at periodic intervals to provide damage estimates as ground truth for the collected data. Small oil samples of approximately 25 ml were taken from the gearbox during

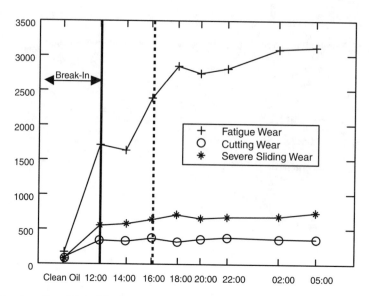

FIGURE 23.9   Number of particles/ml for fatigue, cutting, and sliding wear modes (larger than 15 microns) collected at various times.

the inspection periods. Post-run oil samples were also sent to the DoD Joint Oil Analysis Program (JOAP) to determine particle composition, oil viscosity, and particle wear type.

The break-in and design load time for Test 14 ran for 96 hours. The 3.33 ratio gearbox was then loaded at three times the design load to accelerate its failure, which occurred approximately 20 hours later (including inspection times). This load transition time was at approximately 1200 (noon), and no visible signs of deterioration were noted. The run was stopped every two hours for internal inspection and oil sampling. At 0200 (2:00 a.m.), the inspection indicated no visible signs of wear or cracks. After 0300, accelerometer data and a noticeable change in the sound of the gearbox were noted. Upon inspection, one of the teeth on the follower gear had separated from the gear. The tooth had failed at the root on the input side of the gear with the crack rising to the top of the gear on the load side (refer to Figure 23.9). The gearbox was stopped again at 0330, and an inspection showed no observable increase in damage. At 0500, the tooth from the downstream broken tooth had suffered surface pitting, and there were small cracks a millimeter in from the front and rear face of the tooth, parallel to the faces. The 0700 inspection showed that two teeth had broken and the pitting had increased, but not excessively, even at three times design load. Neighboring teeth now had small pits at the top-motor side corners.

On shutdown at 0815, with a significant increase in vibration (over 150% RMS), the test was concluded, and eight teeth had suffered damage. The damaged teeth were dispersed in clusters around the gear. There appeared to be independent clusters of failure processes. Within each cluster a tooth that had failed as a result of root cracking was surrounded by teeth that had failed due to pitting. On both clusters, the upstream tooth had failed by cracking at the root, and the follower tooth had experienced pitting.

Figure 23.9 shows the time sequence of three types of particle concentrations observed during this test run: fatigue wear, cutting wear, and severe sliding wear. Initial increases in particle counts observed at 1200 reflect debris accumulations during break-in. Fatigue particles manifested the most dramatic change in concentration of the three detectable wear particle types, nearly doubling between 1400 and 1800, suggesting the onset of some fault that would give rise to this type of debris. This data is consistent with the inspections that indicated pitting was occurring throughout this time period. No significant sliding or cutting wear was found after break-in.

Figure 23.10 illustrates the breakdown of these fatigue particle concentrations by three different micron size ranges. Between 1400 and 1800, particle concentrations increased for all three ranges with onset occurring later for each larger size category. The smallest size range rose to over 1400 particles/ml by

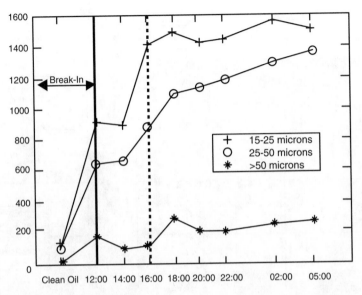

FIGURE 23.10   Number of fatigue-wear particles/ml by bin size collected at various times.

1800, while the particles in the midrange began to increase consistently after 1400 until the end of the run. The largest particle size shows a gradual upward trend starting at 1600, though the concentration variation is affected by sampling/measurement error. The observed trends could be explained by hypothesizing the onset of a surface fatigue fault condition sometime before 1600, followed by steadily generated fatigue wear debris.

Figures 23.11 and 23.12 show different features of the accelerometer data developed at Penn State ARL. The results with a Penn State ARL enveloping technique, Figure 23.11, clearly show evidence of some activity around 0200. The dashed line represents the approximate time the feature showed a notable change. This corresponds to the time when tooth cracking is believed to have initiated/propagated. The wavelet transform[40] is shown in Figure 23.12. It is believed to be sensitive to the impact events during breakage, and shows evidence of this type of failure after 0300. The processed indicators seem to indicate activity well before RMS levels provided any indication.

During each stop, the internal components appeared normal until after 0300, when the borescope verified a broken gear tooth. This information clearly supports the RMS and wavelet changes. The changes in the interstitial enveloping that occurred around 0200 (almost one hour earlier) could be considered as an early indicator of the witnessed tooth crack. Note that the indication is sensitive to threshold setting, and the MDTB online wavelet detection threshold triggered about an hour (around the same time as the interstitial) before that shown in Figure 23.12.

23.5.1.1.6   *Feature Fusion*[41]
Although the primary failure modes on MDTB gearboxes have been gear tooth and shaft breakage, pitting has also been witnessed. Based on the previous vibration and oil debris figures in this section, a good overlap of candidate features appears for both commensurate and noncommensurate data fusion. The data from the vibration features in Figure 23.13 show potential clustering as the gearbox progresses towards failure. Note from the borescope images that the damage progresses in clusters, which increase on both scales. The features in this subspace were obtained from the same type of sensor (i.e., they are commensurate). Often two noncommensurate features — such as oil debris and vibration — are more desirable.

Figure 23.14 shows a subspace example using a vibration feature and an oil debris (fatigue particle count) feature. There are fewer data points than in the previous example because the MDTB had to be shut down to extract an oil sample as opposed to using on-demand, DAQ collection of accelerometer data. During the progression of the run, the features seemed to cluster into regions that are discernible

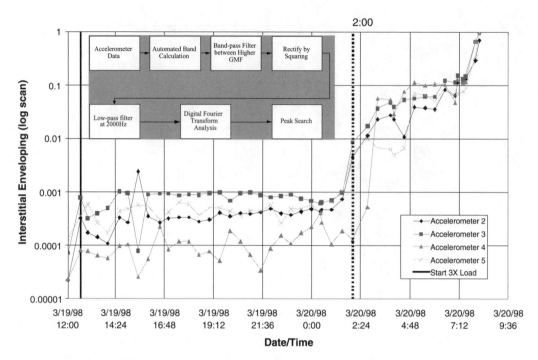

FIGURE 23.11   Interstitial enveloping of accelerometer.

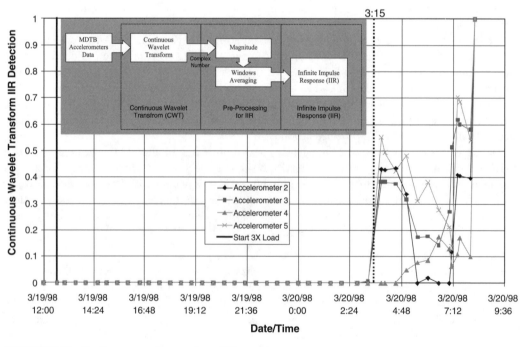

FIGURE 23.12   Continuous wavelet transform (IIR count).

and trackable. Subspaces using multiple features from both commensurate and noncommensurate sources would provide better information for classification, as would the inclusion of running conditions, system knowledge, and history. This type of association of data is a necessary step toward accomplishing more robust state estimation and higher levels of data fusion.

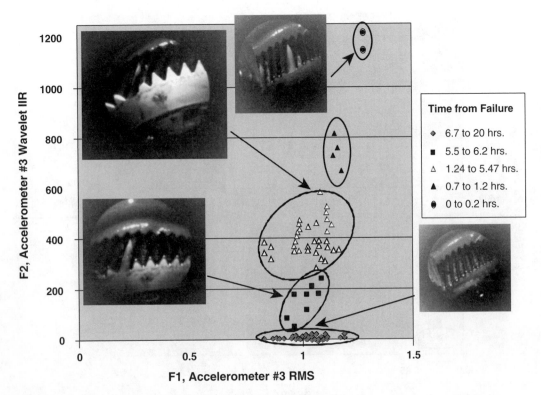

FIGURE 23.13   Feature subspace classification example.

23.5.1.1.7   *Decision Fusion Analysis*

Decision fusion is often performed as a part of *reasoning* in CBM systems. Automated reasoning and data fusion are important for CBM. Because the monitored systems exhibit complex behavior, there is generally no simple relationship between observable conditions and system health. Furthermore, sensor data can be very unreliable, producing a high false alarm rate. Hence, data fusion and automated reasoning must be used to contextually interpret the sensor data and model predictions. In this section, three automated reasoning techniques, neural networks, fuzzy logic, and expert/rule-based systems, are compared and evaluated for their ability to predict system failure.[42] In addition, these system outputs are compared to the output of a hybrid system that combines all three systems to realize the advantages of each. Such a quantitative comparison is essential in producing high quality, reliable solutions for CBM problems; however, it is rarely performed in practice.

Although expert systems, fuzzy logic systems, and neural networks are used in machinery diagnostics, they are rarely used simultaneously or in combination. A comparison of these techniques and decision fusion of their outputs was performed using the MDTB data. In particular, three systems were developed (expert system, fuzzy logic, and neural network) to estimate the remaining useful life of the gearbox during accelerated failure runs (see Figure 23.15). The inputs to the systems consisted of speed, torque, temperature, and vibration RMS in several frequency bands.

A graphical tool was developed to provide a quick visual comparison of the outputs of the different types of systems (see Figure 23.16). In this tool, colors are used to represent the relative levels of the inputs and outputs and a confidence value is provided with each output. The time to failure curves for the three systems and the hybrid system are shown in Figure 23.17. In this example, the fuzzy logic system provided the earliest warning, but the hybrid system gave the best combination of early warning and robustness.

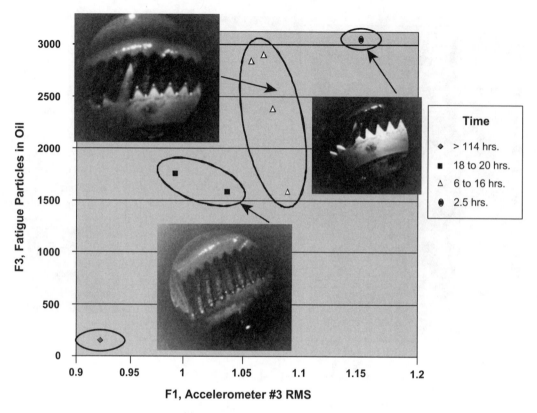

FIGURE 23.14   Non-commensurate feature subspace.

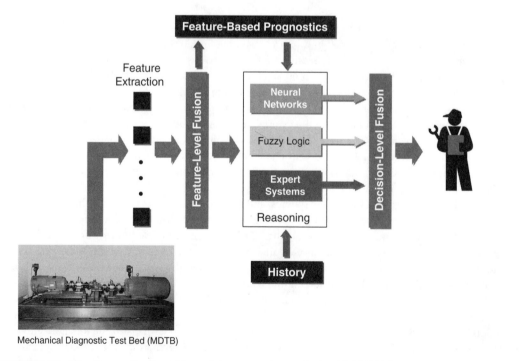

FIGURE 23.15   Flow diagram for comparison of three reasoning methods with MDTB data.

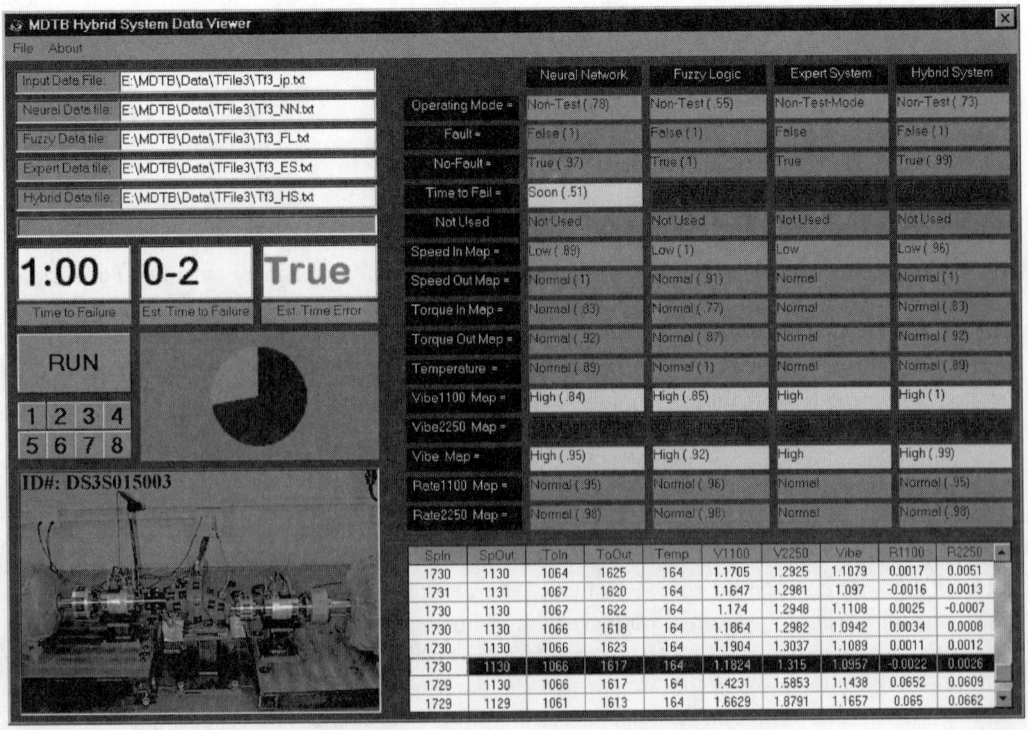

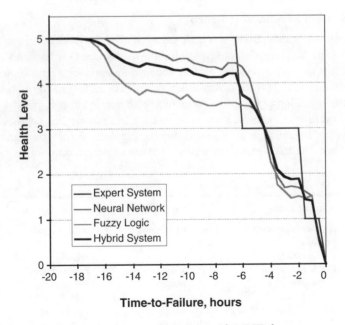

FIGURE 23.16 Graphical viewer for comparing the outputs of the reasoning systems.

FIGURE 23.17 Time-to-failure curves for the reasoning systems with MDTB data.

23.5.2 Fluid Systems

Fluid systems comprise lubrication,[43] fuel,[44] and hydraulic power application examples. Some efforts at evaluating a model-based, data fusion approach for fluid systems are discussed in the following sections. Such fluid systems are critical to many Navy engine and power systems and clearly must be part of the CBM solution.

23.5.2.1 Lubrication System Function

A pressure-fed lubrication system is designed to deliver a specific amount of lubricant to critical, oil-wetted components in engines, transmissions, and like equipment. The primary function of a lubricant is to reduce friction through the formation of film coatings on loaded surfaces. It also transports heat from the load site and prevents corrosion. The lubricating oil in mechanical systems, however, can be contaminated by wear particles, internal and external debris, foreign fluids, and even internal component (additive) breakdown. All of these contaminants affect the ability of the fluid to accomplish its mission of producing a lubricious (hydrodynamic, elastohydrodynamic, boundary, or mixed) layer between mechanical parts with relative motion.[45,46]

Lubricant contamination can be caused by many mechanisms. Water ingestion through seals (common in marine environments) or condensation will cause significant viscosity effects and corrosion. Fuel leakage through the (turbine fuel-lube oil) heat exchanger will also adversely affect lubricity. Moreover, fuel soot, dirt, and dust can increase viscosity and decrease the oil penetration into the loaded surface of the gears or bearings.[47] An often overlooked, but sometimes significant, source of contamination is the addition of incorrect or old oil to the system. Table 23.2 provides a list of relevant faults that can occur in oil lubrication systems and some wetted components' faults.

Many offline, spectroscopic and ferrographic techniques exist to analyze lubricant condition and wear-metal debris.[48-52] These methods, while time-proven for their effectiveness at detecting many types of evolving failures, are performed at specified time intervals through offline sampling.[53] The sampling interval is driven by the cost to perform the preventive maintenance versus the perceived degradation window over an operational time scale. The use of intermittent condition assessment will miss some lubricant failures. Moreover, the use of such offline methods is inconvenient and increases the preventive maintenance cost and workload associated with operating the platform.

23.5.2.2 Lubrication System Test Bench

A lubrication system test bench (LSTB) was designed to emulate the lubrication system of a typical gas turbine engine.[54,55] The flow rate relates to typical turbine speeds, and flow resistance can be changed in each of the three legs to simulate bearing heating and/or differences between various turbine systems. To simplify operation, the LSTB uses facility water, rather than jet fuel in the oil heat exchanger. The LSTB is also capable of adding a measured amount of foreign matter through a fixed-volume, dispensing pump, which is used to inject known amounts of metallic and nonmetallic debris, dirty oil, fuel, and water into the system. Contaminants are injected into a mixing block and pass through the debris sensors, upstream of the filter. The LSTB provides a way to correlate known amounts of contaminants with the

TABLE 23.2 Lubricant and Wetted Component Faults

Lubricant Faults	Gear Faults	Bearing Faults
Viscosity breakdown	Plastic deformation	Surface wiping
Oxidation	Pitting	Fatigue
Emulsification	Heavy scuffing	Fretting
Additive depletion	Chipping and tooth crack	Foreign debris
Sludge formation	Tooth breakage	Spalling
Fluid contamination	Case cracking	Inadequate oil film
External debris contam.	Surface fatigue	Overheating
Internal debris contam.	Abrasive wear	Corrosion
System leakage	Chemical wear	Cavitation erosion

system parameters and, thus, establishes a relationship between machinery wear levels, percentage of filter clogged, and viscosity of the lubricant.

The failure effects are in the areas of lubricant degradation, contamination, debris generated internally or externally, flow blockage, and leakage. These effects can be simulated or directly produced on the LSTB. Both degradation and contamination will result in changes in the oil transport properties. Water, incorrect oil, and sludge can be introduced in known amounts. Debris can be focused on metallic particles of 100-micron mean diameter, as would be produced by bearing or gear wear. Flow blockage can be emulated by restricting the flow through the control valves. Similarly, leakage effects can be produced by actual leaks or by opening the leg valves. Alternatively, seal leakage effects can cause air to flow into the lube system. This dramatically affects performance and is measurable. In addition, the LSTB can be used to seed mechanical faults in the pump, relief valve, and instrumentation. In the case of mechanical component failure, vibration sensors could be added.

23.5.2.3 TELSS Model and Metasensors

Note that association of failure modes to sensor and fused data signatures remains a hurdle in such CBM work. Evaluation of operational data gathered on the gas turbine engine provided some association to believed faults, but insufficient data on key parameters prevented the implementation of a fault tree or even an implicit association. Given the lack of failure test data and the limited data available on the actual engine, a simulation model was developed. The turbine engine lubrication system simulation (TELSS) output was used to generate virtual or metasensor outputs. This data was evaluated in the data fusion and automated reasoning modules.

The TELSS consists of a procedural program and a display interface. The procedural program is written in C code and uses the analog of electrical impedances to model the oil flow circuit. The model contains analytical expressions of mass, momentum, and energy equations, as well as empirical relationships. The interface displays state parameters using an object-oriented development environment. Both scripted and real system data can be run through the simulation. A great deal of effort was expended to properly characterize the Reynolds number and temperature-dependent properties and characteristics in the model. TELSS requires the geometry of the network, the gas generator speed, and a bulk oil temperature to estimate the pressures and flows throughout.[56]

23.5.2.4 Data Fusion Construct

The initial approach for lubrication system data fusion is summarized in Figure 23.18. This example follows the previous methodology of reviewing the data fusion steps within the context of the application. There are five levels in the data fusion process:

1. **Observation:** This level involves the collection of measured signals from the lubrication system being monitored (e.g., pressures, flow rates, pump speed, temperatures, debris sensors, and oil quality measurements).
2. **Feature extraction:** At this level, modeling and signal processing begins to play a role. From the models and signal processing, features (e.g., virtual sensor signals) are extracted; features are more informative than the raw sensor data. The modeling provides additional physical and historical information.
3. **Data association:** In this level, the extracted features are mapped into commensurate and non-commensurate failure mode spaces. In other words, the feature data is associated with other feature data based on how they reveal the development of different faults.
4. **System state estimation:** In this level, classification of feature subspaces is performed to estimate specific failure modes of the lubricant or oil-wetted components in the form of a state estimate vector. The vector represents a confidence level that the system is in a particular failure mode; the classification also includes information about the system and history of the lubrication system.
5. **Fault condition and health assessment:** For this level, system health decisions are made based on the agreement, disagreement, or lack of information that the failure mode estimates indicate. The decision processing should consider which estimates come from commensurate feature spaces and which features map to other failure mode feature subspaces, as well as the historical trend of the failure mode state estimate.

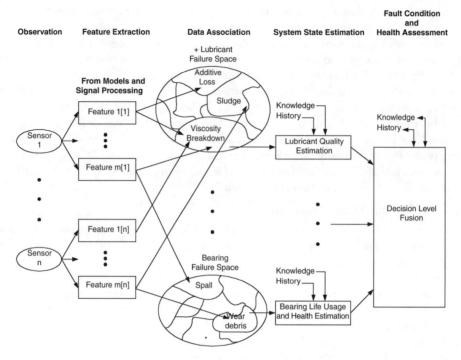

FIGURE 23.18   Lubrication system diagnostics/prognostics data fusion structure.

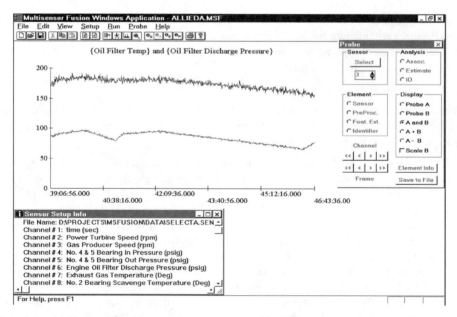

FIGURE 23.19   Processing of test stand data and correlation in the ARL data fusion toolkit.

23.5.2.5   Data Analysis Results

23.5.2.5.1   *Engine Test Cell Correlation*

Engine test cell data was collected to verify the performance of the system. The lubrication system measurements were processed using the Data Fusion Toolkit to produce continuous data through interpolation. Typical data is seen in Figure 23.19. This data provided the opportunity to trend variables against

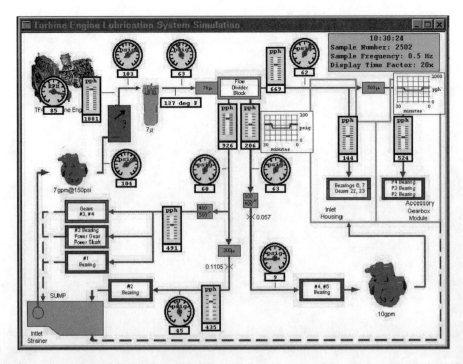

FIGURE 23.20 TELSS processing of operational data to produce metasensors. Massflow in pounds per hour is shown in bar scaled objects. Pressure throughout the circuit in pounds per square inch is illustrated by pressure gauge objects.

the fuel flow rate to the engine, gas generator speed and torque, and the power turbine speed and torque. Ultimately, through various correlation analysis methods, the gas generator was deemed the most suitable regression (independent) variable for the other parameters. It was used to develop three-dimensional maps and regressions with a measured temperature to provide guidelines for normal operation.

23.5.2.5.2   *Operational Data with Metasensor Processing*

Operational data was made available by a Navy unit that uses the gas turbine engines. The operational data was limited to the production engine variables, which consisted of one pressure and temperature. The TELSS model was embedded within the Multisensor Fusion Toolkit. The TELSS interface for an LPAS run is shown in Figure 23.20. Because the condition of the oil and filter was unknown for these runs, the type of oil and a specified amount of clogging was assumed. The variation of oil and types of filters can vary the results significantly. Different MIL-L-23699E oils, for many of which the model possesses regressions, can vary the flow rate predictions by up to 5 percent. Similar variation is seen when trying to apply the filter clogging to different vendors' filter products.

The TELSS simulation model can be used to simulate different fault conditions to allow data association and system state estimation. Figure 23.21 illustrates the output of the metasensor under simulated fault conditions of filter clogging with debris. Filter clogging is typically monitored through a differential pressure measurement or switch. This method does not account for other variables that affect the differential pressure. The other variables are the viscosity, or the fluid resistance to flow, which is dependent on temperature and the flow rate through the filter.

With this additional knowledge, the T-P-mdot relationship can be exploited in a predictive fashion. Toward the mid to latter portion of the curves, the pressure increases slightly, but steadily, as the flow rate remains constant or decreases. Meanwhile, the temperature increases around 1000 seconds and then decreases steadily from 1500 until the engine is shut off. Let us investigate these effects more closely. The increasing pressure drop from about 1200 to 1500 seconds occurs while the temperature and flow are approximately constant. This is one indication of a clogging effect. From 1500 through 2100, the flow

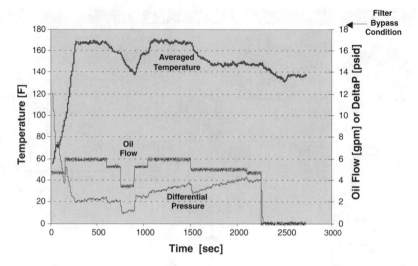

FIGURE 23.21   TELSS run illustrating the relationship between the system variables that can be fused to produce filter clogging association and estimates.

rate is at a lower level, but the pressure drop rises above its previous level at the higher oil flow rate. Looking only at these two variables could suffice; however, deeper analysis reveals that during the same timeframe, the temperature decreases, which means the viscosity (or resistance to flow) of the oil increases. This lower temperature would indicate that higher pressure drops could be expected for the same flow rate. This effect (increased viscosity due to lower temperature) is the reason why the pressure drop is so high at the beginning of the run. Consequently, this consideration actually adds some ambiguity to an otherwise crisp indication. The model analysis indicates, though, that the additional pressure drop caused by higher viscosity does not comprise the entire difference. Thus, the diagnosis of filter clogging is confirmed in light of all of the knowledge about the effects.

23.5.2.6   Health Assessment Example

The output from the TELSS model and Multisensor Toolkit was processed using an automated reasoning shell tool. The output of a shell that could be used to detect filter-clogging fractions is shown in the figures below. An expert system (ES), a fuzzy logic (FL) association, and a neural network (NN) perform the evaluations of filter clogging. The flow, temperature, and differential pressure were divided into three operational ranges. The ES was provided set values for fraction clogged. The FL was modeled with trapezoidal membership functions. The NN was trained using the fuzzy logic outputs.[56,57] For the first case shown, the combination of 4.6 gpm, 175°F, and 12 psid, the reasoning techniques all predict relatively low clogging. In the next case, the flow is slightly less, whereas the pressure is slightly higher at 12.5 psid. The NN evaluation quickly leans toward a clogged filter, but the other techniques lag in fraction clogged. The expert system is not sensitive enough to the relationships between the variables and the significance of the pressure differential increasing while the flow decreases markedly. This study and others conducted at ARL indicated that a hybrid approach based on decision fusion methods would allow the greatest flexibility in such assessments.

23.5.2.7   Summary

The objective of this fluid systems research was to demonstrate an improved method of diagnosing anomalies and maintaining oil lubrication systems for gas turbine engines. Virtual metasensors from the TELSS program and operational engine data sets were used in a hybrid reasoning shell. A simple module for the current-limited sensor suite on the test engine was proposed and recommendations for enhanced sensor suites and modules were provided. The results and tools, while developed for the test engine, are

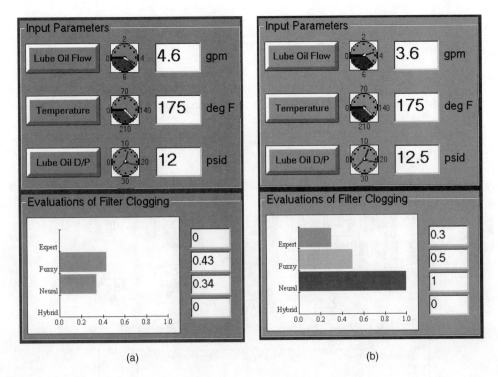

FIGURE 23.22 Hybrid reasoning shell evaluation (Cases 1 and 2).

applicable to all gas turbine engines and mechanical transmissions with similar pressure-fed lubrication systems.

As mentioned in a previous section, the ability to associate faulted conditions with measurable parameters is tantamount for developing predictive diagnostics. In the current example, metasensors were generated using model knowledge and measured inputs that could be associated to estimate condition. Development of diagnostic models results from the fusion of the system measurements as they are correlated to an assessed damage state.

23.5.3 Electrochemical Systems

Batteries are an integral part of many operational environments and are critical backup systems for many power and computer networks. Failure of the battery can lead to loss of operation, reduced capability, and downtime. A method to accurately assess the condition (state of charge), capacity (amp-hr), and remaining charge cycles (remaining useful life) of primary and secondary batteries could provide significant benefit. Accurate modeling characterization requires electrochemical and thermal elements. Data from virtual (parametric system information) and available sensors can be combined using data fusion. In particular, information from the data fusion feature vectors can be processed to achieve inferences about the state of the system.

This section describes the process of computing battery state of charge (SOC) — a process that involves model identification, feature extraction, and data fusion of the measured and virtual sensor data. In addition to modeling the primary electrochemical and thermal processes, it incorporates the identification of competing failure mechanisms. These mechanisms dictate the remaining useful life of the battery, and their proper identification is a critical step for predictive diagnostics.

Figure 23.23 illustrates the model-based prognostics and control approach that the battery predictive diagnostics project addresses. The modeling approach to prognostics requires the development of electrical, chemical, and thermal model modules that are linked with coupled parameters. The output of the

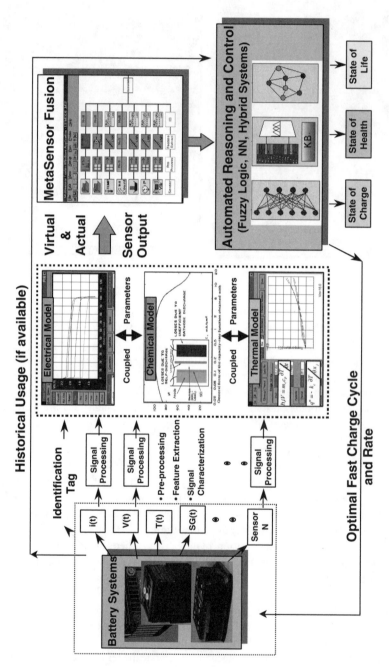

FIGURE 23.23 Model-based prognostics and control approach.

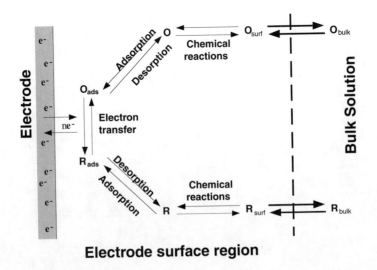

Electrode surface region

FIGURE 23.24 Electrode reaction process.[58,59]

models is then combined in a data fusion architecture that derives observational synergy, while reducing the false alarm rate. The reasoning system provides the outputs shown at the bottom right of the figure. Developments will be applicable to the eventual migration of the diagnosis and health monitoring to an electronic chip embedded into the battery (i.e., intelligent battery health monitor).

23.5.3.1 The Battery As a System

A battery is an arrangement of electrochemical cells configured to produce a certain terminal voltage and discharge capacity. Each cell in the battery is comprised of two electrodes where charge transfer reactions occur. The anode is the electrode at which an oxidation (O) reaction occurs. The cathode is the electrode at which a reduction (R) reaction occurs. The electrolyte provides a supply of chemical species required to complete the charge transfer reactions and a medium through which the species (ions) can move between the electrodes. Figure 23.24 illustrates the pathway ion transfer that takes place during the reaction of the cell. A separator is generally placed between the electrodes to maintain proper electrode separation despite deposition of corrosion products.[58] The electrochemical reactions that occur at the electrodes can generally be reversed by applying a higher potential that reverses the current through the cell. In situations where the reverse reaction occurs at a lower potential than any collateral reaction, a rechargeable or secondary cell can potentially be produced. A cell that cannot be recharged because of an undesired reaction or an undesirable physical effect of cycling on the electrodes is called a primary cell.[58]

Changes in the electrode surface, diffusion layer, and solution are not directly observable without disassembling the battery cell. Other variables such as potential, current, and temperature are observable and can be used to indirectly determine the performance of physical processes. For overall performance, the capacity and voltage of a cell are the primary specifications required for an application. The capacity is defined as the time integral of current delivered to a specified load before the terminal voltage drops below a predetermined cut-off voltage. The present condition of a cell is described nominally with the state-of-charge (SOC), which is defined as the ratio of the remaining capacity and the capacity. Secondary cells are observed to have a capacity that deteriorates over the service life of the cell. The term state-of-health (SOH) is used to describe the physical condition of the battery, which can rang from external behavior, such as loss of rate capacity, to internal behavior, such as severe corrosion. The remaining life of the battery (i.e., how many cycles remain or the usable charge) is termed the state-of-life (SOL).

23.5.3.2 Mathematical Model

An impedance model called the Randles circuit, shown in Figure 23.25, is useful in assessing battery condition. Impedance data can be collected online during discharge and charge to capture the full change

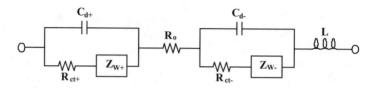

FIGURE 23.25   Two-electrode randles circuit model with wiring inductance.

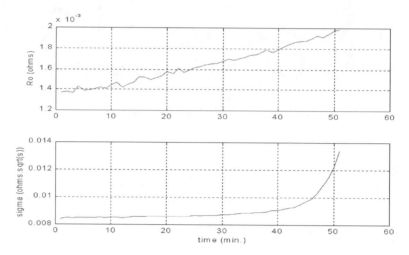

FIGURE 23.26   Nickel cadmium model parameters over discharge cycle.

of battery impedance and identify the model parameters at various stages of SOC, as well as over multiple cycles of the battery for SOH identification. Some of the identified model parameters of the nickel cadmium battery are shown in Figure 23.26 as the batteries proceed from a fully charged to a discharged state. Identification of these model-based parameters provides insight and observation into the physical processes occurring in the electrochemical cell.[60,61]

23.5.3.3   Data Fusion of Sensor and Virtual Sensor Data

The approach for battery feature data fusion is summarized in Figure 23.27. There are five levels in the data fusion processes: *observation* (data collection), *feature extraction* (computation of model parameters and virtual sensor features), *data association* (mapping features into commensurate and noncommensurate feature spaces), *system state estimation* (estimation of failure modes and confidence level), and *fault condition and health assessment* (making system health decisions).

Figure 23.28 illustrates the sensor and virtual sensor input to the data fusion processing. The outputs of the processing are the SOC, SOH, and SOL estimates that are fed into the automated reasoning processing. After the data association processing, an estimate of the failure mechanism is determined.

Two approaches for SOC prediction are described in the following sections. Each performs a kind of data fusion to utilize physically meaningful parameters to predict SOC. The definition of SOC is the amount of useful capacity remaining in a battery during a discharge. Thus, 100% SOC indicates full capacity and 0% SOC indicates that no useful capacity remains.

23.5.3.3.1   *ARMA Prediction of SOC*[62]

An effective way to predict SOC of a battery has been developed using ARMA model methodology. This model has performed well on batteries of various size and chemistry, as well as at different temperatures and loading conditions. ARMA models are a very common system identification technique because they are linear and easy to implement. The model used in this application is represented by the equation

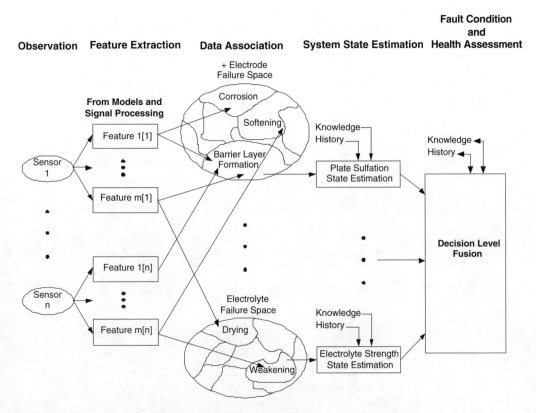

FIGURE 23.27 Battery diagnostics/prognostics data fusion structure.

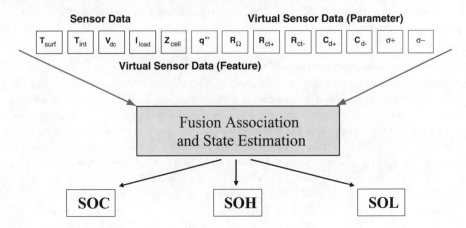

FIGURE 23.28 Generalized feature vector for battery predictive diagnostics.

$$y(t) = aX(t) + bX(t-1) + c_0 y(t-1) \qquad (23.10)$$

where y represents SOC, X represents a matrix of model inputs, and a, b, and c_0 represent the ARMA coefficients. Careful consideration was exercised in determining the inputs. These were determined to be V_D, I_D, R_Ω, θ, C_{DL}, and T_s, and output is SOC. The inputs were smoothed and normalized to reduce the dependence on noise and to prevent domination of the model by parameters with the largest magnitudes.

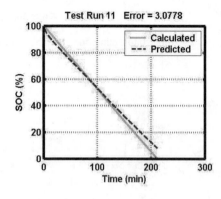 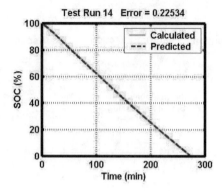

FIGURE 23.29   ARMA SOC prediction results for size-C lithium (CF)x batteries.

TABLE 23.3   Results of ARMA Model SOC Predictions

Chemistry	Size	# Cells	Type	Prediction Error (%)
Lithium[1]	C	1	Primary	2.18
Lithium[1]	2/3 A	1	Primary	2.87
NiCad[2]	D	1	Secondary	3.17
NiCad[2]	C	1	Secondary	4.50
Lead-Acid	12-Volt	6	Secondary	9.13

[1] Poly carbonmonofluoride lithium (spiral-type).
[2] Nickel-Cadmium.

The model was determined by using one of the runs and then tested with the remaining runs. Figure 23.29 shows the results for all eight size-C batteries. The average prediction error for lithium batteries was less than 3%, for NiCad less than 5%, and for lead acid less than 10%. These results are summarized in Table 23.3 below.

23.5.3.3.2   *Neural Networks Prediction of SOC*[62]

Artificial neural networks have been used successfully in both classification and function approximation tasks. One type of function approximation task is *system identification*. Although neural networks very effectively model linear systems, their main strength is the ability to model nonlinear systems using examples of input-output pairs. This was the basis for choosing neural networks for SOC estimation. Neural network SOC estimators were trained for lithium batteries of sizes C and 2/3 A under different loading conditions. For each type of battery, a subset (typically 3 to 6) of the available parameter vectors was chosen as the model input. Networks were trained to produce either a direct prediction of battery SOC or, alternatively, an estimation of initial battery capacity during the first few minutes of the run.

All networks used for battery SOC estimation had one hidden layer. The back propagation, gradient-descent learning algorithm is used, which utilizes the error signal to optimize the weights and biases of both network layers. The inputs to the network were a subset of I_D, V_D, R_Ω θ, and C_{DL}. As for the ARMA case, the inputs were smoothed and normalized. This led to smaller networks, which tend to be better at generalization. For time-delay neural networks, the selection of the number of delays and the length of the delays is crucial to the performance of the networks. Both short and long delays were tried during different training runs. The short delays gave better performance, which indicates that the battery SOC does not involve long time constants. This is also evident from the ARMA examples. Several types of neural networks were trained with battery data and extracted impedance parameters to directly predict the battery state-of-charge. Among the several training methods that were used, the Levenburg-Marquadt (L-M) provided the best results. The size of the network was also important in training. An excessively

TABLE 23.4   Error Rates for SOC Prediction Based on Initial Capacity Estimation with RBF Neural Networks

Battery Size [# Hidden Neurons]	Average Training Error [Training Set]	Average Testing Error	Maximum Testing Error [Run #]
Size C[6]	0.6%[13,14,16]	2.9%	6.8%[15]
Size 2/3 A[12]	0.8%[17,18,20]	2.9%	8.2%[23]

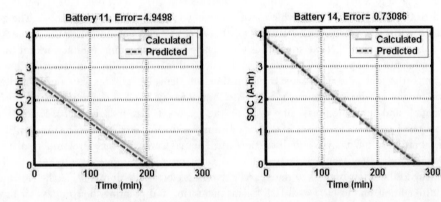

FIGURE 23.30   SOC prediction for size-C Lithium batteries using initial SOC estimation with RBF neural networks.

small network size results in inadequate training and a larger than necessary network size leads to over-training and poorer generalization.

Neural networks trained to directly provide the battery SOC provided consistently good performance. However, the performance was much better when the battery's initial capacity was first estimated. These networks were also trained to estimate the initial capacity of the battery during the first few minutes of the test. Then, the measured load current could easily be used to predict SOC because the current is the rate of change of charge. This is even more useful for SOH and SOL prediction because as the secondary batteries are reused they start at different initial capacity each time.

This method can also be used as a powerful tool in *mission planning*. Hypothetical load profiles can be used to predict whether a battery would survive or fail a given mission, thus preventing the high cost and risk of batteries failing in the field. The networks tend to slightly underestimate the battery SOC. This is a very important practical feature, since it results in a conservative estimate and avoids unscheduled downtime.

The results using radial basis function (RBF) neural networks were the best and are summarized in Table 23.4. The SOC plots are shown for size-C lithium batteries in Figure 23.30. The results are quite remarkable, considering that very little training data are required to produce the predictors. As more data are collected and several runs of each of level of initial battery SOC become available, the robustness of the predictors will likely improve. In addition, the neural network predictors have smaller error on outliers and provide a conservative prediction (i.e., they do not over-predict the SOC). Both of these advantages are very important in practical systems where certification and low false alarms are not just requirements, but can make the difference between a system that is actually used or shelved.

23.6   Concluding Remarks

The application of data fusion in the field of CBM and predictive diagnostics for engineered systems is rich with opportunity. The authors fully acknowledge that only a small amount of what is possible to accomplish with data fusion was presented in this chapter. The predictive diagnostics application domain is relatively new for data fusion, but the future is bright with the many analogies that can be drawn between more mature data fusion applications and the current one. The authors anticipate that continuing developments in actual and virtual feature fusion, as well as hybrid decision fusion of condition assessments, will tend to dominate the research field for some time.

Acknowledgments

The authors gratefully acknowledge support from Dr. Thomas McKenna and Dr. Phillip Abraham from the Office of Naval Research on Grant Numbers N00014-95-1-0461, N00014-96-1-0271, N00014-97-1-1008, and N00014-98-1-0795, under which most of this work was originally performed. The authors would also like to acknowledge the contributions of several past and present colleagues at the Applied Research Laboratory: Mr. Bill Nickerson for his vision, help in understanding a framework for the CBM problem, and contributions to the definitions and systems thinking sections; Mr. James Kozlowski for his work on gearbox and battery diagnostics data fusion; Mr. Derek Lang for his work on the multisensor toolkit and insight into decision fusion; Mr. Dan McGonigal for his help in programming the lubrication system example and automated reasoning work on the gearbox failure tests; Mr. Jerry Kasmala for his development and assistance in the use of the data fusion toolkit; Ms. Terri Merdes for her contributions to the oil-vibration fusion analysis and discussion; Mr. Thomas Cawley for his development of a battery mathematical model; Mr. Matthew Watson for his work on ARMA modeling for SOC prediction; Mr. Todd Hay for his contributions to the neural network prediction section; Ms. Natalie Nodianos for her insightful editing, and to Dr. David Hall for his mentoring and pointing the compass for the CBM data fusion effort at Penn State ARL.

References

1. Hansen, R. J., Hall, D. L., and Kurtz, S. K., A new approach to machinery prognostics, *Int'l. Gas Turbine and Aeroengine Congress and Exposition*, 94-GT-3, June 1994.
2. Kotanchek, M. E. and Hall, D. L., *CBM/IPD Process Model*, ARL TM 95-113, November 1995.
3. Byington, C. and Nickerson, G. W., Technology issues for CBM, *P/PM Technology Magazine*, Vol. 9, Issue 3, June 1996.
4. Byington, C., Monitoring the health of mechanical transmissions, *NCADT Quarterly*, No. 2, 1996.
5. Chamberlain, M., *U.S. Navy Pursues Air Vehicle Diagnostics Research*, Vertiflite, March/April 1994.
6. Douglas, J., The Maintenance Revolution, *EPRI J.*, 6–15, May/June 1995.
7. Condition-Directed Maintenance, Electric Power Research Institute, *Compressed Air Magazine*, August 1990.
8. Hall, D. L., Hansen, R. J., and Lang, D. C., The negative information problem of mechanical diagnostics, in *Proc. of the Int'l. Gas Turbine Institute Conf.*, Birmingham, England, ASME, June 1996.
9. Elverson, Bryan, Machinery fault diagnosis and prognosis, MS Thesis, The Pennsylvania State University, 1997.
10. Erdley, J., Data Fusion for Improved Machinery Fault Classification, M.S. Thesis in Electrical Engineering, The Pennsylvania State University, May 1997.
11. Dempster, A. P., *Generalization of Bayesian Inference*, in *J. Royal Statistical Soc.*, 30, 1968, 205–247.
12. Shafer, G., *A Mathematical Theory of Evidence*. Princeton University Press, Princeton, NJ, 1976.
13. Thomopoulos, S. C. A., Theories in distributed decision fusion: comparison and generalization, in *Proc. SPIE 1990 on Sensor Fusion III: 3-D Perception and Recognition*, Boston, MA, November 5–9, 1990, pp. 623–634.
14. Hall, D., *Mathematical Techniques in Multisensor Data Fusion*, Artech House Inc., 1992.
15. Llinas, J. and Waltz, E., *Multisensor Data Fusion*, Artech House, Inc., Boston MA, 1990.

16. Bayes, T., An essay towards solving a problem in the doctrine of chances, *Philosophical Trans. Royal Soc. London*, 53, 370–418, 1763.

17. Freund, J. E. and Walpole, R. E., *Mathematical Statistics*, 3rd. Ed., Prentice-Hall, Englewood Cliffs, New Jersey, 1980.

18. Pearl, J., *Probabilistic Reasoning in Intelligent Systems*, Morgan Kaufmann, San Mateo, California, 1988.

19. Natke, H. G. and Cempel, C., Model-based diagnosis — methods and experience, *51st MFPT Proc.*, April 1997.

20. Byington, C. S., Model-based diagnostics for turbine engine lubrication systems, *ARL* TM98-016, February 1998.

21. Hall, D. L. and Kasmala, G., A visual programming tool kit multisensor data fusion, in *Proceedings of the SPIE AeroSense 1996 Symposium*, Orlando, FL, Vol. 2764, 181–187, April 1996.

22. Hall, D. L. and Garga, A. K., 1999, Pitfalls in data fusion (and how to avoid them), *Proc. 2nd Internat. Conf. Information Fusion (FUSION 99)*, I, 429–436, 1999.

23. Begg, Merdes, Byington, Maynard, Dynamics Modeling for Mechanical Fault Diagnostics and Prognostics, *Maintenance and Reliability Conf.*, 1999.

24. Mitchell, L. D. and Daws, J. W., A basic approach to gearbox noise prediction, SAE Paper 821065, Society of Automotive Engineers, Warrendale, PA, 1982.

25. Ozguven, H. N. and Houser, D. R., Dynamic analysis of high speed gears by using loaded static transmission error, *J. Sound and Vibration*, 125(1), 71–83, 1988.

26. Randall, R. B., A new method of modeling gear faults, *Trans. ASME, J. Mechanical Design*, 104, 259–267, 1982.

27. Nicks, J. E. and Krishnappa, G., Evaluation of vibration analysis for fault detection using a gear fatigue test rig, C404/022, *IMEC E Proc.*.

28. Zakrajsek, J. J., An investigation of gear mesh failure prediction techniques, NASA Technical Memorandum 102340 (89-C-005), November 1989.

29. Tandon, N. and Nakra, B. C., Vibration and acoustic monitoring techniques for the detection of defects in rolling element bearings — a review, *Shaft & Vibration*, March 1992.

30. Li, C. J. and Wu, S. M., On-line detection of localized defects in bearings by pattern recognition analysis, *J. Engineering for Industry*, 111, 331–336, November 1989.

31. Rose, J. H., Vibration signature and fatigue crack growth analysis of a gear tooth bending fatigue failure, Presented at 44th Meeting of the Mechanical Failure Prevention Group (MFPG), April 2–5, 1990.

32. Mathew, J. and Alfredson, R. J., The condition monitoring of rolling element bearings using vibration analysis, *J. Vibration, Acoustics, Stress, and Reliability in Design*, 106, 447–453, July 1984.

33. Lebold, M., McClintic, K., Campbell, R., Byington, C., and Maynard, K., Review of vibration analysis methods for gearbox diagnostics and prognostics, *Proc. 53rd Meeting of the Society for Machinery Failure Prevention Technology*, Virginia Beach, VA, 623–634.

34. Byington, C. S. and Kozlowski, J. D., Transitional Data for Estimation of Gearbox Remaining Useful Life, *Proc. 51st Meeting of the Society for Machinery Failure Prevention Technology (MFPT)*, Virginia Beach, VA, Vol. 51, 649–657, April 1997.

35. Townsend, D. P., *Dudley's Gear Handbook*, 2nd Edition, McGraw Hill, New York, 1992.

36. Roylance, B. J. and Raadnui, S., The morphological attributes of wear particles — their role in identifying wear mechanisms, *Wear*, 175, 1994.

37. Brennan, M. J., Chen, M. H., and Reynolds, A. G., Use of vibration measurements to detect local tooth defects in gears, *Sound and Vibration*, November 1997.

38. Wang, W. J. and McFadden, P. D., Application of the wavelet transform to gearbox vibration analysis, American Society of Mechanical Engineers, *Structural Dynamics and Vibrations Proceedings 16th Annual Energy Sources Technology Conference and Exposition*, Jan. 31–Feb. 4, 1993, Vol. 52, pp. 13–20.

39. Ferlez, R. J. and Lang, D. C., Gear tooth fault detection and tracking using the wavelet transform, *Proc. 52nd Meeting of the MFPT,* 1998.

40. Byington, C. S. et al., Vibration and oil debris feature fusion in gearbox failures, *53rd Meeting of the Society for MFPT,* 1999.

41. Byington, C. S. et al., Fusion techniques for vibration and oil debris/quality in gearbox failure testing, *Internat. Conf. Condition Monitoring at Swansea,* 1999.

42. McGonigal, D. L., Comparison of automated reasoning techniques for CBM, M.S. Thesis in Electrical Engineering, The Pennsylvania State University, August 1997.

43. Byington, C. S., Model-based diagnostics for turbine engine lubrication systems, ARL TM98-016, February 1998.

44. Campbell, R. L and Byington, C. S., Gas turbine fuel system modeling and diagnostics, To be published in COMADEM Proceedings, December 2000.

45. Toms, L. A., *Machinery Oil Analysis: Methods, Automation and Benefits,* 1995.

46. Lubrication Fundamentals: Fundamental Concepts and Practices, *Lubrication Engineering Magazine,* 26–34, September 1997.

47. Eleftherakis, J. G. and Fogel, G., Contamination control through oil analysis, *P/PM Technology Magazine,* 62–65, October 1994.

48. Toms, A. M. and Powell, J. R., Molecular analysis of lubricants by FT-IR spectroscopy, *P/PM Technology Magazine,* 58–64, August 1997.

49. Richards, C., Oil analysis techniques advance with new information technology, *P/PM Technology Magazine,* 60–62, June 1995.

50. Neale, M. J., *Component Failures, Maintenance, and Repair,* Society of Automotive Engineers, Inc., Warrendale, PA, 1995.

51. Anderson, D. P., Rotrode filter spectroscopy — A method for the multi-elemental determination of the concentration of large particles in used lubricating oil, *P/PM Technology Magazine,* 88–89, September/October 1992.

52. Stecki, J. S. and Anderson, M. L. S., Machine condition monitoring using filtergram and ferrographic techniques, *Research Bulletin — Machine Condition Monitoring,* 3(1), September 1991.

53. Troyer, D. and Fitch, J. C., An introduction to fluid contamination analysis, *P/PM Technology Magazine,* 54–59, June 1995.

54. Byington, C. S., Model-based diagnostics of gas turbine engine lubrication systems, *Joint Oil Analysis Program International Condition Monitoring Program,* April 20–24, 1998.

55. Byington, C. S., Intelligent monitoring of gas turbine engine lubrication systems, ARL TN 98-016, February 1998.

56. Garga, A. K., A hybrid implicit/explicit automated reasoning approach for condition-based maintenance. *Proc. ASME Intelligent Ships Symp. II,* 1996.

57. Stover, J. A., Hall, D. L., and Gibson, R. E., A fuzzy-logic architecture for autonomous multisensor data fusion, *IEEE Trans. Industrial Electronics,* 403–410, June 1996.

58. Berndt, D., *Maintenance-Free Batteries,* 2nd Ed., John Wiley & Sons, New York, NY, 1997.

59. Bard, A. J. and Faulkner, L. R., *Electrochemical Methods: Fundamentals and Applications,* John Wiley & Sons, New York, 1980.

60. Boukamp, A., A nonlinear least squares fit procedure for analysis of immittance data of electrochemical systems, *Solid State Ionics,* 20(31), 1986.

61. Kozlowski, J. D. and Byington, C. S., Model-based predictive diagnostics for primary and secondary batteries, *Proc. of the 54th Meeting of the MFPT,* Virginia Beach, VA, May 1–4, 2000.

62. Kozlowski, J. D. et al., Model-based predictive diagnostics for electrochemical energy sources, *Proc. of IEEE Aerospace Conference,* 0-7803-6599-2/01, March 2001.

24

Information Technology for NASA in the 21st Century

Robert J. Hansen
University of West Florida

Daniel Cooke
NASA Ames Research Center

Kenneth Ford
Institute for Human and Machine Cognition

Steven Zornetzer
NASA Ames Research Center

24.1 Introduction 24-1
24.2 NASA Applications.. 24-2
        Unmanned Space Exploration • Manned Space
        Exploration • Earth Observation • Air Traffic
        Management • Next Generation System Design
24.3 Critical Research Investment Areas for NASA 24-3
        Automated Reasoning • Intelligent Data
        Understanding • Human-Centered Computing
24.4 High-Performance Computing and Networking........... 24-5
24.5 Conclusions 24-6

24.1 Introduction

The future of NASA is critically dependent on the development and implementation of new tools and methods from the information technology research community. A few examples are worth noting. The sophisticated unmanned exploration of Mars and other parts of the solar system, which will be aimed at answering fundamental science questions, such as the existence of early life forms in these environments, will require a new generation of automated reasoning tools. In addition, NASA's role in the development of new air traffic management tools and methods to be evaluated and deployed by the Federal Aviation Administration must involve new approaches to optimizing the combined performance of experts on the ground (air traffic controllers) and in the air (pilots) and the supporting information systems. Ongoing safe operation of the Space Shuttle depends on new capabilities for early identification of the precursors to failure of safety-critical system components from maintenance data and sensors distributed throughout the system. Use of the mountains of data generated by the Earth-observing satellites and next-generation space telescopes fielded by NASA demands fundamentally new methods of data interpretation and understanding. In addition, new aircraft and spacecraft designs depend on new high performance computing capabilities.

The complexity and diversity of such critical needs for the future has motivated NASA to develop an expanded information technology (IT) research and development (R&D) portfolio. The first step toward this end was an extensive strategic planning process for Computer Science/Information Technology R&D for the Agency. Beginning in 1996, NASA's Ames Research Center, located in the heart of Silicon Valley, assembled teams from the research community and the user communities served by the Agency to address two fundamental questions. First, what are those NASA applications domains for which revolutionary advances in information technology are the critical enabler for the future? Second, in light of the investments being made by the private sector and other parts of the public sector, how must NASA invest its IT research dollars to address these key applications domains?

Five critical applications domains were identified in which advances in IT are the critical "force multiplier" for NASA's future: (1) unmanned space exploration, (2) manned space exploration, (3) Earth observation, (4) air traffic management, and (5) design methodologies for complex vehicles and systems. On this basis NASA, working with its own computer scientists as well as those in academia and industry, determined that the critical areas for NASA investment in advanced computer science/IT research included: (1) automated reasoning, both for system autonomy and intelligent data understanding; (2) human-centered computing; and (3) high performance computing and networking. The following sections describe in detail both the challenges of the five applications domains and the resulting research agenda in the three investment areas.

24.2 NASA Applications

This section describes the five application areas introduced earlier: unmanned space exploration, manned space exploration, earth observation, air traffic management, and next generation system design.

24.2.1 Unmanned Space Exploration

The future of space science is tied to the ability to autonomously explore the solar system and beyond, and such unmanned missions are the necessary precursor to subsequent manned missions. Future unmanned space explorers must be capable of navigation, onboard interpretation of data collected by sensors on the spacecraft or its rovers, and real-time reconfiguration in response to failures of system components during the mission. The capability to perform this suite of tasks simultaneously and reliably on an extended deep space exploration mission is well beyond current capabilities in autonomous operation. In fact, NASA conducted its first in-space experiment in spacecraft autonomy only in 1999. This experiment demonstrated the capability of model-based reasoning to perform real-time system reconfiguration in response to simulated faults in spacecraft operation. This was a tremendous accomplishment, but it represents only the first step in the development and validation of the capabilities required for 21st century space exploration.

24.2.2 Manned Space Exploration

As currently envisioned, manned space missions will be a complex combination of human and robotic capabilities. The design and functioning of this unique human-system "team" presents challenging questions about how to optimize the team performance. Moreover, the duration of such missions imposes stringent requirements on system reliability well beyond what is presently achievable in complex hardware/software/human systems. Finally, today's concepts of manned exploration depend on the in-situ manufacture of fuel for the return trip from elements available on the planet by unmanned platforms prior to the launch of the manned mission. This will require a similar level of sophisticated autonomous capability to that associated with unmanned exploration.

24.2.3 Earth Observation

Earth-observing satellites deployed by NASA have successfully identified such important phenomena as ozone depletion in the vicinity of the South Pole and the influence of El Nino-La Nina phenomena on global weather patterns. The Agency is now in the midst of deploying a more sophisticated network of satellites with hyperspectral imaging capabilities. This new capability is expected to greatly enrich our understanding of Earth processes and lay the groundwork for applications such as improved weather forecasting, precision agricultural practices, and detection of conditions that are likely to lead to outbreaks of devastating diseases such as cholera.

Beyond the current era of satellite deployment, a complex sensor web made up of a large number of smaller satellites operating in a coordinated fashion is envisioned to provide more sophisticated predictive capabilities. This vision for enhanced Earth observation embodies the need for revolutionary IT advances of several types. For example, data understanding methodologies capable of operating on geographically distributed data sets will be required to identify previously unknown critical features and cause-effect relationships. A sophisticated form of onboard automated reasoning must coordinate the actions of the satellites in the sensor web and make real-time decisions concerning which data are critical to download for archival purposes. Very significant advances in high performance computing will be required to execute models of regional and global phenomena to merge with the data being collected to provide high-value, high-fidelity predictions of patterns and events.

24.2.4 Air Traffic Management

Behind the headlines about long flight delays at major airports is the sobering reality that air traffic is projected to grow more rapidly than the ability of the current air traffic management system, despite the new capabilities being deployed by the FAA as a part of the Free Flight initiative. In response to these projections and in partnership with the FAA, NASA has set a ten-year goal of developing the technology base to reduce accident rates per passenger mile by a factor of five, while increasing the capacity of the U.S. air transportation system by a factor of three. Realization of these national goals will depend on the development of fundamentally new models of air traffic management, which are rooted in IT advances such as (1) new simulation methods that provide the ability to evaluate the relative merit of candidate air traffic management architectures, accounting for the attendant complexities of interactions among humans, software, and hardware; (2) new approaches to achieving optimal "system" performance, where the system is understood as the combination of human experts and advanced intelligent agents embedded in the supporting information system; and (3) new human-computer interfaces that communicate complex information about the system in an intuitive, readily understood way.

24.2.5 Next Generation System Design

In recent years a broad realization has developed that fundamentally new system design methodologies are required. This realization resulted from the need to determine — *a priori* and in a virtual environment — optimal design/cost tradeoffs, the impact of design decisions on total lifecycle cost or performance, and the risks of component-level failures on system performance. This represents an extremely difficult set of modeling, simulation, and data management challenges. For example, methods are needed to assess performance at the system level, based on component-level simulations of varying fidelity ranging from fundamental physics simulations to simple statements of functionality. Another challenge of system design is the need to capture the underlying rationale employed by the best designers and its reuse as appropriate in all phases of the design of a new system. Similarly, means to identify and effectively reuse relevant data from past experiments, located in geographically distributed data sets of various formats, is a critical element.

24.3 Critical Research Investment Areas for NASA

Detailed analyses of these needs and the investments made within the private sector and other parts of the public sector, including the Departments of Defense, Energy, and Transportation, and the National Science Foundation made clear that NASA should emphasize research in three primary areas:

1. Automated reasoning, both in support of spacecraft autonomy and intelligent data understanding.
2. Human-centered computing, a relatively new interdisciplinary field associated with performance enhancement of individuals and teams through the synergistic application of emerging computer science, cognitive science, and perceptual science concepts.
3. High-performance computing and networking, focusing on both evolutionary and revolutionary advances in single site and distributed computing and data analysis.

Major NASA programs (such as the High-Performance Computing and Communication Program and the Aeronautics and Space Transportation IT Base Program) were redirected to best serve these goals, and a new Intelligent Systems Program was initiated to address shortfalls in longer term research in automated reasoning, intelligent data understanding, human-centered computing, and revolutionary computing concepts for the space environment. The following sections describe the three areas required to address NASA's needs. (Note that in this discussion a distinction is made between automated reasoning for system autonomy and for intelligent data understanding.)

24.3.1   Automated Reasoning

As previously noted, NASA must increasingly depend on the autonomous operation of highly capable spacecraft for deep space exploration. Moreover, the software embodiment of these capabilities requires high reliability software generation and test methods. Practical operational considerations dictate the need for increased system autonomy provided by automated reasoning. For example, communication delays will impact longer distance human and robotic missions. These delays will force future mission planners to face the reality that their missions will not be able to rely upon the ground-based remote control that has benefited past missions — whether the missions involve humans or robots. Greater autonomy for both types of missions must be a major characteristic of missions to Mars and beyond.

Given the lack of ground-based input, future autonomous systems for missions must accommodate uncertainty to a much greater extent than in the past. Whereas uncertain situations previously could default to humans on earth, the distances involved with future missions will reduce the reliance on earth-based remote control. Therefore, greater autonomy will be required in human and robotic missions.

The autonomous nature of past systems has been based on a systematic approach where scientists attempted to predict every eventuality that could occur and program a system response or action for each. If an event occurred that the system was not programmed to handle, earth-based controllers took over. These types of systems will not be adequate for future missions. Given the uncertainties involved in exploration, condition-response systems become fragile without the ground-based safety nets.

For example, the programming of past systems could be viewed in terms of a parent's instructions to a child. The child could be asked to react in the following ways:

- If someone hits you, do not hit back.
- Under no circumstance should you start a fight.
- Under no circumstance should you throw rocks at another, etc.

Raising a child in this manner would be difficult. Not every event or situation the child would encounter could be planned for in advance. Given the uncertainty of future events, the child is likely to encounter a situation where the system fails him or her. A better approach is to raise the child by instilling guiding principles such as: "Treat others the way you wish to be treated." This type of principle will better prepare the child to handle unforeseen situations. Likewise, autonomous systems of the future must accommodate guiding principles. A number of promising, model-based reasoning approaches — not to be confused with expert systems (e.g., Bayesian decision models, extended logic programming, neural network, and hybrid approaches) — produce more robust autonomous systems that can deal much more effectively with uncertainty.

24.3.2   Intelligent Data Understanding

NASA's present and future missions place a premium on the intelligent use of very large data sets. For example, satellites have recently been deployed for Earth science measurements (e.g., Landsat 7 and Terra.) When these elements of the Earth Observing System are fully deployed, they will generate more data per month than currently resides in all NASA databases on all topics. Another example is the Stratospheric Observatory for Infrared Astronomy (SOFIA.) This observatory will be deployed aboard a Boeing 747 in approximately two years to collect large data sets over the subsequent decade. A third example involves the large databases containing maintenance information on critical and complex systems as the Space Shuttle.

A new generation of data-understanding methodologies is required to allow scientists and engineers to make full use of these rich data sets. One can view these vast data sets as experimental or observed data. Empirical observations are typically reduced to concise theories, which explain the observations. In addition to data mining and knowledge discovery research, new data understanding methodologies are required to provide approaches that will better enable scientists to construct theories based on the data acquired. Beyond these efforts, a major result would be the reduction of these data sets into much smaller algorithmic units. These algorithms could be viewed as concise statements of the data — which provide more manageable, more easily understood representations of the data and potential reproduction of the data sets — thus resulting in much more significant data compression.

24.3.3 Human-Centered Computing

At the heart of research to improve air traffic management methodologies is the need to optimize the performance of geographically distributed teams of experts (e.g., pilots and air traffic controllers) and the supporting information system. Likewise, in the emerging reality of complex system design by teams of experts and in supporting data distributed around the globe there is a critical challenge to take a "system" view that is distinguished from current views of design. Specifically, formal approaches are called for — approaches that provide a more holistic view which includes the cognitive and perceptual abilities of humans, computational assets, and other intelligent agents and knowledge sources. In both of these cases, initiative resides both with the human experts and with the information system. Understanding and improving the performance of such "mixed-initiative" systems is a primary research focus of Human-Centered Computing within NASA. Enhancing the performance of a single expert rather than a team is of primary importance in some applications domains, and NASA is addressing this need as well as a part of its Human-Centered Computing thrust, where "system" refers to the individual and the supporting information system.

To better understand this revolutionary approach to systems design, consider past epochs of human experience. In the agrarian society of the past, humans equaled physical labor. Because humans spent the majority of their time performing labor, they had very little time to perform advanced problem solving, theory formulation, and the creative activities required for invention and discovery. In the industrial society, machines began performing physical labor and the humans equaled the brains of these machines. Humans had more freedom to perform advanced cognitive activities during this epoch, and science made great strides. In the information age, the brain of the human is extended and enhanced by a machine — the computer. Even trivial applications significantly extend human capabilities. Knowledge and the application of knowledge is embodied in software. For example, many people now prepare their taxes with the help of software tools that embody much of the skill and knowledge possessed by tax preparers. As the intelligence of systems increases, the more mundane and lower level reasoning activities can be delegated to computers, freeing the humans to perform the more advanced and creative cognitive activities. Results in this area will affect not only human exploration of distant planets, but also the abilities of humans to perform mission operations, air traffic control, and other high-level cognitive activities.

24.4 High-Performance Computing and Networking

Within NASA, three complementary approaches to high-performance computing are being pursued in parallel. First, working with industry partners, NASA is prototyping progressively larger, single system image machines based on commodity processors, and a new programming paradigm for such machines called multilevel parallelism. This combination of architecture and programming model has already shown extraordinary promise for complex numerical simulations. Typical results on production fluid dynamics and molecular dynamics codes have shown linear increases in sustained throughput with increasing processor count. These results extend to 512 processors, the largest machine tested to date. Moreover, the alterations in the vector version of the code to achieve these results with multilevel parallelism are a small fraction of that required to implement the Message Passing Interface (MPI). Evaluation of this combined architecture and programming paradigm is now underway for other types of simulations.

Second, in partnership with the National Science Foundation, NASA is performing the computer science research and demonstrations required to realize distributed, heterogeneous computing. Computations of the future are expected to require more computational capability than resides at any single site, necessitating the use of geographically distributed assets (e.g., computers and databases) connected over high-speed networks.

Third, research on new approaches to in-space computing are being studied. Size, weight, energy consumption, and space hazards are all problematical in this unique environment. Quantum and molecular computing may have the potential to overcome some of NASA's concerns about computing in space. The radiation and solar effects on computing could be offset by the massive parallelisms these approaches may offer. The size, weight, and power consumption concerns are also positively affected by these newer computing architectures. Perhaps the most important benefit is the new computational models and computer languages that may be implied by these approaches. Revolutionary computing approaches are radically different from the traditional von Neumann and the more conventional non-von Neumann approaches to architecture. As such, the computational models implied may provide radically new insights into problem solving — possibly helping scientists to find tractable solutions to problems for which only intractable algorithms are currently known. These algorithms may allow for feasible algorithms within the constraints of current technologies. More straightforward solutions to problems may result. (Currently, solutions to these problems are approximate solutions — given the intractability of the problems — making them much more complex to develop.) This space computing element is focused not on building quantum or molecular computers, but on the computational models and languages implied by these approaches, as well as in the development of specific NASA-relevant algorithms that would allow for the immediate exploitation of these technologies when they become available.

24.5   Conclusions

NASA's understanding of the importance of advanced IT to its future has grown dramatically in recent years, and this realization has resulted in careful strategic planning of IT R&D. To address shortfalls and opportunities identified, programs have been redirected to support new approaches to high performance computing, and a new Intelligent Systems Program has been initiated which focuses on automated reasoning for autonomy and data understanding, human-centered computing, and revolutionary computing concepts. These redirections and new activities are expected to position NASA to carry out the variety of difficult space exploration and earth science missions of the future, as well as serve as a critical contributor to enhanced air traffic management.

25

Data Fusion for a Distributed Ground-Based Sensing System

25.1	Introduction	25-1
25.2	Problem Domain	25-2
25.3	Existing Systems	25-3
25.4	Prototype Sensors for SenseIT	25-4
25.5	Software Architecture	25-5
25.6	Declarative Language Front-End	25-6
25.7	Subscriptions	25-6
25.8	Mobile Code	25-7
25.9	Diffusion Network Routing	25-7
25.10	Collaborative Signal Processing	25-7
25.11	Information Security	25-8
25.12	Summary	25-8
	Acknowledgments and Disclaimers	25-8
	References	25-8

Richard R. Brooks
The Pennsylvania State University

25.1 Introduction

Sensor Information Technology (SenseIT) is a program of the U. S. Department of Defense Advanced Research Projects Agency (DARPA) that began in fiscal year 1999. A number of research groups are exploring networking and organizational problems posed by large networks of unmanned, intelligent, battery-powered, wireless sensors.

The use of distributed sensor networks is of great interest to the Department of Defense (DoD). The work performed by this project is applicable to a number of heterogeneous unmanned sensing platforms. The deployment of large numbers of sensing nodes poses several new technical problems. Sensor networks resemble other wireless networks (such as cellular telephones) but have unique aspects, which are introduced in this chapter.

Several applications exist for this technology within the DoD, the most obvious being in the realm of intelligence, surveillance, and reconnaissance (ISR): automating deployment of heterogeneous sensor networks and interpretation of their readings. SenseIT could protect troops in the field as effectively as land mines, without many undesirable side effects. Civilian applications include law enforcement, agriculture, traffic monitoring, security, and environmental monitoring.

SenseIT provides a new perspective to programming embedded systems. It considers revolutionary approaches to coordination of multiple computers with power, time, bandwidth, and storage constraints. A chaotic network of small, inexpensive, unreliable systems form a large, powerful, dependable system.

An extreme example of this idea is a network made up of "smart dust."[1] SenseIT is defining the interface between cyberspace and the real world.

In the SenseIT framework, sensor fusion is a type of collaborative signal processing. SenseIT is both an application of sensor fusion, and a context for collaborative processing in evolving information infrastructures.

25.2   Problem Domain

Sensors return information about their environment. This requires physical interaction with the real world. Sensors may be *passive* or *active*. *Passive* sensors simply record emissions already present in the environment. *Active* sensors emit a signal and measure how the environment modifies the signal. Several sources describe sensor design in depth.[2-4]

The quality of sensor readings often depends on the proximity of the sensor to the event being observed. For events in hazardous environments, this places sensors where they can be easily detected and/or destroyed. This has a number of immediate consequences:

- Remotely operated, unmanned sensors are attractive.
- Sensor nodes should be autonomous with internal power supplies.
- Multiple nodes are needed to tolerate component failures.
- Sensor nodes should be inexpensive.
- Wireless communications increase deployment flexibility.

Hazardous sensing applications require networks of inexpensive, battery powered, autonomous, wireless, intelligent sensors.

For large numbers of electronic devices in hazardous terrain to be a reality, the devices must be very inexpensive. To keep expenses down, individual components will have severe constraints. The most obvious constraints are

- Power
- Communications bandwidth
- Sensor precision
- Local data storage
- CPU bandwidth

None of these constraints exist in isolation. For example, computation and communication use power. Additional operational constraints exist, such as

- Information latency
- Accuracy of results obtained
- Ability of the system to be reprogrammed
- System longevity

All of these constraints are tradeoffs. Increased system bandwidth reduces response latency. It also increases the power drain on the system, reducing system longevity. Increased accuracy may be achieved through massive redundancy, which increases system cost.

These tradeoffs are context dependent. Sometimes information is required promptly and accuracy can be sacrificed. Other times, the most important factor will be power conservation to increase system longevity. The tradeoffs need to be quantified. They will change as technology advances. Moore's law[5] promises that CPU bandwidth will become increasingly inexpensive and require less power. The power required to transmit data will likely remain almost constant. This implies that, in the future, as much processing as possible should take place local to the sensor in order to save battery power. One goal of SenseIT is to develop technology for working within equipment constraints. The sensor network must be flexible enough to adapt automatically to the dynamic environment it monitors.

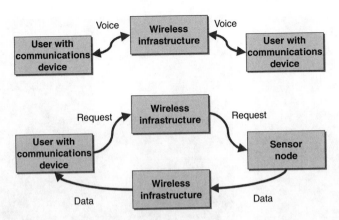

FIGURE 25.1 Cellular telephone wireless networks are symmetrical with voice traveling in both directions. Sensor network communication is asymmetric.

Another issue considered by SenseIT is system configuration and tasking. Configuring a system containing thousands of nodes is not trivial. Similarly, assigning duties to thousands of individual machines at run-time is an onerous task. Networks should configure themselves and adapt to changing conditions. Operators should be able to program the system by declaring the types of information required. They should not be required to assign lower-level tasks to individual nodes.

Many advances have been made in wireless networking. Up to now, wireless infrastructures were designed as extensions of the telephone network. Cellular telephone network traffic is symmetrical, supporting conversations between two individual communications devices. Connections are maintained for relatively long periods. Routing seeks a network path that can be maintained. Contrary to the telephone network, wireless devices are usually mobile. Communications devices are tracked as they move through the infrastructure. In most applications, routing tables are dynamically updated, or modified on the fly. The wireless infrastructure strives to keep an uninterrupted dialog between two or more parties.

Sensor networks require a radically different approach. Many, if not all, sensors have fixed positions. Tracking motion is not of primary importance. Communication is not between two individuals. In general, a user poses a query about a region. Sensors in the region return data. Figure 25.1 shows this asymmetry. A request is sent over the network. Data is returned. Connections are not maintained for long periods. Asymptotically optimal routing is irrelevant. Transient effects are very important. The network address of a sensor node will be less important than its physical location. Routing tables and network addresses may not be required at all.

Effective sensor networks will require advances in ad hoc wireless networking. The network must be able to configure itself and distribute tasks. Configuration and tasking will vary with the operational context. The network will consist of a large number of inexpensive components, which are individually unreliable and severely constrained. By coordinating activities, a reliable adaptive infrastructure will emerge. This represents a new application area and revolutionary approach to embedded systems programming.

25.3 Existing Systems

SenseIT takes place in the context of an evolving technology infrastructure. This section reviews systems that are precursors to SenseIT and their role in the current DoD infrastructure. An essential building block of SenseIT is wireless communications infrastructure. Since 1990, the military has had a Single Channel Ground and Airborne Radio System (SINCGARS), which provides connectivity to the armed forces via a common receiver and transmitter platform. It supplies large-scale networking connectivity but does not have the autonomy and adaptability required by sensor networks.

FIGURE 25.2 Prototype sensor used for SenseIT proof-of-concept demonstrations.

Similarly, unattended ground sensors (UGS) have been used by the military since the Vietnam War. The current generation of UGSs is the Improved Remotely Monitored Battlefield Sensor System (I-REMBASS), which remotely detects and classifies movements of enemy vehicles and personnel. The system supports magnetic, seismic, acoustic, infrared, and meteorological sensors. Unfortunately, these UGSs have limited processing capabilities and do not work cooperatively. I-REMBASS devices must be configured and monitored by a trained operator. One operator station can simultaneously display information from a number of sensors.

Other ISR equipment includes Unmanned Aerial Vehicles (UAVs) and Remotely Piloted Vehicles (RPVs) that perform aerial surveillance. Networked cooperation is very limited. Enemy forces easily detect these vehicles. UAVs require sophisticated onboard processing. This intelligence is currently devoted almost entirely to navigation.

Remote ISR can be provided either by satellite surveillance or the Joint Surveillance Target Attack Radar System (JSTARS). JSTARS is an airborne platform with advanced multi-mode radar. It is capable of viewing wide swaths of a battlefield. Ground reception of JSTARS information generally requires a large and expensive ground station. Satellite surveillance and JSTARS both provide invaluable ISR information from a distance, making them less vulnerable to attack. On the other hand, they are expensive centralized systems with potential single points of failure. Atmospheric effects can also obscure the battlefield from surveillance.

SenseIT has the potential to unite these existing technologies into a single framework. Furthermore, adding inexpensive sensor nodes to these systems will greatly extend their functionality with little additional cost.

25.4   Prototype Sensors for SenseIT

The current SenseIT program uses prototype intelligent sensor nodes developed by Sensoria. The nodes provide the capabilities necessary for this research, in an adaptable framework. Wireless communications, multiple sensing modalities, and modern computing facilities are combined in a compact battery-powered system.

Two processors provide computation support. The primary processor is a commercially available personal data assistant running Microsoft Windows CE. Most processing is done on this node. It has four megabytes of memory, which serves as main memory and file storage. Windows CE can be programmed using Microsoft visual programming languages. A second processor is used for real-time processing. It is the interface between the main processor and the wireless network, the sensors and the battery power supply. This processor uses a real-time executive and can also be reprogrammed.

In order to conserve power, the system has a power-saving mode capable of powering off the Windows CE processor. The preprocessor can shut down and reboot the Windows CE processor as necessary.

Three sensing modalities are supported by the prototype. Commercial microphones are used for acoustic sensing. Geophones measure seismic vibrations. Two-pixel infrared imagers detect motion. Seven different sensing configurations are possible, with any combination of the three. Fielded nodes would usually have a single sensing modality per node.

Two types of communications are available. IP wireline communications use an ethernet adapter attached to the Windows CE device for development support and system metrology. Documented pathologies of TCP/IP for wireless communications have prompted the developments of new protocols for wireless communications in the field.[6] Fielded systems will have a small number of gateway nodes connected to the sensor network (via a low-power radio) and the Internet (via a higher-power radio). As a result, sensor readings could be accessible from anywhere in the world.

The primary mode of communications in the field is an internal wireless transceiver, which uses time division multiplexing (TDMA) and frequency hopping ("blue tooth") to control communications between nodes. A master node establishes the frequencies used by nodes in its neighborhood for communications. In most situations, all nodes will be placed on the ground. The communications range of the nodes will be very limited, primarily due to reflections of communication signals from the ground known as "fading." This requires long-range communications to pass through multiple nodes on the way to their final destination.

Each node is equipped with a Global Positioning System (GPS). When nodes locate three satellites, they can accurately determine their position and the current time through triangulation. Position information is needed for data routing and querying. Just as important is the ability of GPS to provide a universal time stamp, which automatically solves many difficult coordination problems. Unfortunately, GPS's utility can be limited in urban environments, where obtaining a clear line-of-sight to satellites is difficult.

A large network of these nodes will be scalable and flexible. It will need to partition itself into a set of neighboring clusters. A number of multihop paths will exist between any two sensor nodes. New ad hoc networking methods are needed for organizing and routing data packets in this type of network.

25.5 Software Architecture

Figure 25.3 shows a functional decomposition of the system. In the figure, each component is labeled with the section that discusses it in this chapter. This section provides a brief overview of the components and how they interact.

Note that the majority of the system is autonomous. Users interact only with the declarative front end (described in Section 25.6). Network tasking is driven by the user expressing the type of information desired. This is translated into processes to be performed either by the network as a whole or by individual nodes.

A system of services and subscriptions has been developed to distribute tasks among nodes. Nodes are aware of their position and capabilities. They advertise the services they can provide to the network. Interested parties may subscribe to the services. This simple concept is a flexible way of matching needs to abilities. It is described in Section 25.7.

Mobile code (Section 25.8) is used to describe two approaches that cooperate to provide a flexible infrastructure for network operations. Mobile code can be downloaded and executed on any node when required. This capability allows any node to be dynamically reconfigured, reprogrammed, and retasked. Mobile code interacts with an *active network* language, which allows network routing protocols to be modified at run-time.

Section 25.9 describes some of the network routing innovations. Requests are made for data, rather than specific nodes. In addition, multihop routing provides a large number of routes through the network. Centralized coordination is inappropriate; therefore, determining how to make globally desirable routing decisions locally is very important.

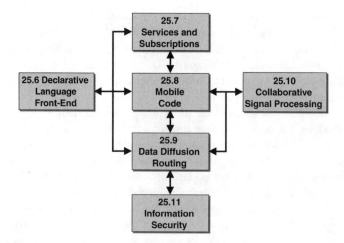

FIGURE 25.3 Functional decomposition of the SenseIT software architecture. Numbers refer to chapter sections describing the functionality.

Collaborative signal processing is described in Section 25.10. Collaboration is integrated into the routing protocols effectively combining data routing and sensor fusion. Constructing a large network of individually unreliable components adds many new constraints. Any approach must be an asynchronous algorithm. No assumptions can be made as to which inputs will be available and when they will arrive. Similarly, algorithms that can provide estimates quickly are desirable. Estimates should quickly converge to precise answers as more time is available for processing.

The final function shown in the diagram is information security. Section 25.11 discusses information security concerns. Sensor networks have mobile code and data. Transmission must be reliable. Corrupted programs may have catastrophic effects. The source of code and data must be verified. These problems are universal for networked systems. They are even more difficult for a system with limited bandwidth, power, storage, and computation.

25.6 Declarative Language Front-End

The user does not interface directly with sensor nodes or the network. Network configuration and tasking is done autonomously in response to user requests. Users interface with the network through a declarative front end located on a workstation on the Internet. The front-end communicates with the SenseIT network through a small number of gateway nodes with long-range radios. These nodes transfer tasks from the front-end to sensor nodes, and data from sensor-nodes to the user.

Front-end design is divided into three subtasks: graphical user interface, task generation, and execution. The graphical user interface is being developed cooperatively by USC ISI East and the Virginia Institute of Technology. Java applets present a graphical view of the region. Users can indicate regions of interest and the types of data desired. Data types include specific target types, movement direction, etc.

These declarations are passed to an agent-based system developed at the University of Maryland, which decomposes requests into specific tasks.[7] The tasks are passed to a distributed device database, Cougar, developed at Cornell.[8] The subscription method discussed in Section 25.7 maps information requirements to information providers. This meta-information is absolutely necessary. In general, transmitting housekeeping information needs to be avoided since it consumes system resources. Parts of Cougar are resident on the user workstation; other parts are query proxies that are resident on every SenseIT node.[8]

25.7 Subscriptions

Lincoln Laboratory proposed the use of a subscription service for data routing. The concept is simple and elegant. A similar concept for this type of application can be found in Marzullo.[9] Every node is aware of its physical configuration and position. As part of the network initialization process, each node advertises its position and capabilities. These advertisements are propagated through the network to the user workstation, where they are stored as metadata.

When users pose questions, the front-end uses the metadata to determine those nodes that are capable of providing the information desired. Among other things, the system can inform the user immediately when no relevant information sources exist.

25.8 Mobile Code

The SenseIT network is a chaotic system. Individual elements are unreliable. Each node has limited storage and power available. In some cases, deploying a network and waiting years to use it is desirable. During that time, system software and targets of interest can change. Two research groups are looking at how code mobility can create a more adaptable infrastructure.

The Pennsylvania State University Applied Research Laboratory is implementing mobile code software that allows legacy code to be downloaded and executed on demand. This has a number of advantages. Nodes can be implicitly reprogrammed. If a new target type is found in the field, the relevant classification programs can be downloaded for use. If local storage is not sufficient for all the tasks required by a node, tasks could be downloaded as needed. Programs not in use can be deleted. Halls[10] gives more information on applications of mobile code.

In addition, an active network system is being built on the mobile code infrastructure. The Network Flow Language (NFL), developed at Columbia University, allows network routing protocols to be modified at run-time. Extended markup language (XML) packets contain code defining new routing protocols. As they pass through the network, global network behavior can be reprogrammed.

25.9 Diffusion Network Routing

The SenseIT network was not designed for interpersonal communication. It was built for information retrieval. Because of this, addressing is not based on the recipient of the call; it is based on the geographical position of nodes. In addition, the recipient may not be unique. A query may concern a region simultaneously surveyed by multiple nodes.

Subsequently, researchers at the University of Southern California Information Sciences Institute[11] and UCLA have proposed routing algorithms based on information gradients and data diffusion. Requests and data flow through the network in a manner reminiscent of fluid flowing through a solid. An important part of these routing protocols is collaborative signal processing and data fusion. As data flows through the network, it can also be compressed. By combining individual readings and extracting salient observations, data integrity is increased. In this manner, the network becomes a data fusion and interpretation processing system.

25.10 Collaborative Signal Processing

The use of multiple sensors can increase data dependability. Just as repeating a single experiment a number of times provides statistics allowing more accurate determination of a parameter value.[12] When multiple sensors have readings of an area of interest, their data can be combined to derive a more accurate value. Of particular interest is the use of multiple readings for increased accuracy in target recognition.

Current work focuses on integrating information fusion into the diffusion routing algorithms. Readings from sensors in a neighborhood are combined to provide a more accurate reading. The initial approach uses a weighted voting algorithm[13] to provide a decision-in, decision-out level of fusion.[14] This will be a baseline for later work.

The distributed nature of the problem and unreliable nature of individual nodes renders many common methods, such as the Kalman filter, inappropriate. These approaches need to be extended. They need to be asynchronous algorithms,[15] accepting data as it arrives and tolerating failures of individual nodes. They should also be anytime algorithms[16] that provide accurate approximations at any time. The quality of the answer should quickly converge to the correct value.

25.11   Information Security

SenseIT considers several data security problems in developing a unique networking infrastructure. Adaptability and reconfiguration occurs at many different levels in the system. These capabilities are needed to guarantee that the system is immune to both accidental and intentional corruption.

Data and programs passing through the network must be reliable, but some information redundancy is required to guarantee against accidental corruption. On the other hand, compressing information reduces power consumption. These requirements are at odds. In addition, the system needs to be insensitive to jamming and interference.

The mobile code used in this program depends on a *trusted code* model. No attempt is made to verify that code does what it is intended to do. The source of the code is verified. If the code comes from a known provider, it will be executed. This is appropriate for SenseIT because the mode of delivery of programs does not add a new danger. The execution of Applets over the Internet is different, and a *sandbox* is required to protect systems using Internet browsers.

In addition to these factors, only authorized users should be able to interpret data from the SenseIT network. This implies that some type of data encryption is necessary. Data encryption can also be used to verify the source of programs and data on the network. An open problem is how to best distribute keys to the network before deployment. Another problem is the power required for data encryption.

In addition, automated target recognition (ATR) should be performed as close to the sensor as possible. This reduces the amount of data to a few bits, which reduces the power required for data transmission. Securely encrypting messages that consist of a very small number of bits is challenging.

25.12   Summary

This chapter provided a brief overview of the DARPA Sensor Information Technology (SenseIT) program, which is developing technologies required for large networks of intelligent sensors. These networks can be cost-effective for a number of military and civilian applications. To be effective, new approaches are needed for distributed systems and cooperative processing.

Acknowledgments and Disclaimers

Efforts sponsored by the Defense Advance Research Projects Agency (DARPA) and Air Force Research Laboratory, Air Force Materiel Command, USAF, under agreement number F30602-99-2-0520 (Reactive Sensor Network). The U.S. Government is authorized to reproduce and distribute reprints for Governmental purposes notwithstanding any copyright annotation thereon. The views and conclusions contained herein are those of the authors and should not be interpreted as necessarily representing the official policies or endorsements, either expressed or implied, of the Defense Advanced Research Projects Agency (DARPA), the Air Force Research Laboratory, or the U.S. Government

References

1. Wylie, M., Scientists develop a new breed of dust, http://www.msnbc.com/news/321983.asp?cp1 = 1, October 13, 1999.
2. Brooks, R.R. and Iyengar, S.S., *Multi-Sensor Fusion: Fundamentals and Applications with Software*. Prentice-Hall PTR, Upper Saddle River, NJ, 1998.
3. Hovanessian, S.A., *Introduction to Sensor Systems*, Artech House, Norwood, MA, 1988.
4. Fraden, J., *AIP Handbook of Modern Sensors: Physics, Designs, and Applications*, American Institute of Physics, New York, 1993.

5. Hwang, K., *Advanced Computer Architecture*, McGraw Hill, New York, NY, 1993.
6. Chaskar, H.M., Lakshman, T.V., and Madhow, U., TCP over wireless with link level error control: analysis and design methodology, *IEEE/ACM Trans. Networking*, Vol. 7, No. 5, 605–615, October 1999.
7. Subrahamian, V.A. et al., *Heterogeneous Agent Systems*, MIT Press, Cambridge, MA, 2000.
8. Bonnet, P. et al., *Query Processing in a Device Database System*, Ncstrl.cornell/TR99-1775, *http://www.ncstrl.org*, 1999.
9. Marzullo, K. and Wood, M.D., *Tools for Constructing Distributed Reactive Systems*, Ncstrl.cornell/TR-91-1193, http://www.ncstrl.org, 1991.
10. Halls, D.A., *Applying Mobile Code to Distributed Systems*, Ph.D. Dissertation in Computer Science, University of Cambridge, UK, June 1997.
11. Estrind, D. et al., Next century challenges: scalable coordination in sensor networks, *ACM MobiCom 99*, Seattle, WA, August 1999.
12. Montgomery, D.C., *Design and Analysis of Experiments*, Wiley, New York, 1984.
13. Saari, D.G., Geometry of voting: a unifying perspective, *Proc. Workshop on Foundations of Information/Decision Fusion with Applications to Engineering Problems*, DOE/ONR/NSF, Washington, DC, August 1996.
14. Dasarathy, B.V., *Decision Fusion*, IEEE Computer Society Press, Los Alamitos, CA, 1993.
15. Bertsekas, D.P. and Tsitsiklis, J.N., *Parallel and Distributed Computation: Numerical Methods*, Prentice Hall, Englewood Cliffs, NJ, 1989.
16. Zilberstein, S., *Operational Rationality Through Compilation of Anytime Algorithms*, Ph.D. Dissertation, Dept. of Computer Science, University of California at Berkeley, 1993.

26

An Evaluation Methodology for Fusion Processes Based on Information Needs

26.1 Introduction **26**-1
26.2 Information Needs **26**-2
    Database Analysis
26.3 Key Concept **26**-6
26.4 Evaluation Methodology **26**-6
References **26**-9

Hans Keithley
Office of the Secretary of Defense
Decision Support Center

26.1 Introduction

Fusion is a part of a larger Department of Defense (DoD) context — command, control, communication, computers, intelligence, surveillance and reconnaissance (C4ISR). C4ISR capabilities are enablers for the even larger context of information superiority. A key question that is asked at decision-making levels of the DoD is how C4ISR supports the military commander in the efficient execution of military operations. The question is germane at budget levels where, for example, C4ISR competes with weapons platforms for funding.

The Joint C4ISR Decision Support Center (DSC)* in DoD has performed numerous studies to determine the value of C4ISR in general and for fusion in particular. The DSC view is that value does not refer to measures of the technical merits of alternative ISR approaches; it refers instead to the value of C4ISR to support military command and control (C2). Increasingly, the C2 process has become a near-real-time decision based on perceived information. Obviously better ISR improves the data and information, but can it do so in a timely manner with high confidence? And what is the value of that information?

Fusion of information across several intelligence disciplines in the DoD context plays an essential role in producing the knowledge provided by C4ISR systems. The problem addressed by the DSC is to attempt to quantify this statement by directly evaluating the value of fusion in the satisfaction of "information needs," as defined by the DoD community. This chapter gives an overview of the methodology used to perform this type of evaluation.[1]

---

*The DSC is a part of the Office of the Secretary of Defense, C3I Directorate and has performed for 4 years studies and analyses for both OSD/C3I and the Joint Staff.

TABLE 26.1 Sources of DoD Information Needs

Sources of Information Needs
Army Tactical Needs Database
Customer Information Needs Assessment
Community Information Needs Forecast
Assured Support to Operational Commanders
Ground Information Requirements Handbook
U.S. Forces Korea Indication and Warning Requirements

TABLE 26.2 Description of Actions

Action	Description
Location	Where is it?
Track	How is it moving?
Identity	Who is it?
Status/Activity	What is it doing?
Capability	What can it do?
Intent	Who, what, when, where, how, why?

TABLE 26.3 Description of Objects

Object Class	Description
Order of Battle (OB)	Military (specifically mobile elements)
Infrastructure	Stationary (e.g., bridges)
Geospatial	Terrain, weather
Networks	Integrated air defense systems, computer
Political	Leaders, terrorists, etc.

26.2 Information Needs

The initial question is, "What are DoD Information Needs?" The approach used was to gather a number of published sources from various segments of the DoD community. The sources are shown in Table 26.1.

These sources contain information needs (i.e., user questions) that are not C4ISR oriented. In other words, the questions are asked without considering the mechanism used to provide the requested data or information. They, therefore, form a basis to evaluate multi-intelligence (INT) ISR systems. In general, different people with different terminology and objectives wrote each information need. To provide rigor in evaluation, these sources must be structured to rationalize the terminology describing requested information needs and associated quality requirements. They are, therefore, "mapped" to a canonical structure that fits the great majority of the information needs. The objective of this process is to facilitate evaluation of the satisfaction of these needs by ISR and fusion alternatives. The canonical information-needs structure contains the following four classes:

1. **Actions** — specifically requested data/information. Note that "detect" is implicitly included in location. Table 26.2 summarizes the action category.
2. **Objects** — information needs are expressed in terms of objects on the battlefield. These objects have different sizes, ranging from Corps to individual tank companies. Each object can have a different movement policy and different "signature" characteristics. The DSC has classified each object class in Table 26.3 into five size categories. For example, the OB class categories are: Corps, Unit, Battalion, Brigade, and Company objects. Each information need is associated with a specific object category.

Data ———————————————————————→ Knowledge

QR	Location	Track	Identity	Activity/State	Capability	Intent
5	5 meters	Vectors and prediction	Specify object and hierarchy	Many actions, states, and linkages	Many factors and influence	Desired end state and intent for future ops *known*
4	10 meters	Vectors	Specify object	Many actions and states; several linkages	Several factors and influence	Desired end state *known* and intent for future ops *determined*
3	20 meters	General speed and direction	Type object	Several actions and states; one linkage	Few factors and influence	Desired end state and intent for future ops *determined*
2	100 meters	Toward or away	Distinguish object	Few actions and states; no linkages	Few factors and no influence	Desired end state *determined* and intent for future ops *inferable*
1	1 Km	Stationary or not	Discriminate object	Single action or state	One factor and no influence	Desired end state *inferable* and intent for future ops *inferable*
0	10 Km	Detect	Detect	Detect	Detect	Desired end state *inferable* and intent for future ops *unknown*

FIGURE 26.1   Knowledge matrix.

3. **Functions** — the standard intelligence functions that support military operations; definitions for each function can be found in DoD joint publications.
4. **Missions** — intelligence needs are usually expressed in the context of specific military missions. Three important missions used for evaluation are
   - Precision engagement — (e.g., air strikes against fixed and moving targets, including time critical targets)
   - Dominant maneuver — ground maneuver (ground maneuver and halt phase)
   - SEAD — suppression of enemy air defenses.

The information needs (i.e., questions) surveyed usually imply a level of "quality" questions. For example "locate a tank to within 1000 meters to an implied confidence level." Timeliness requirements are occasionally included, but a scenario context usually provides this data. The canonical structure that includes *quality requirements* (QR) with actions is given in Figure 26.1. This is a "knowledge matrix" that allows the mapping of diverse information needs onto a single structure, which will be used for evaluation.

Each cell in Figure 26.1 has specific definitions that define the technical meaning of the amount of knowledge implied. In the analysis procedures, numbers are associated with each cell that reflect the *probability* that knowledge at that level can be achieved. The advantage of this structure is that all aspects of the C4ISR problem can be expressed on this matrix. *Each object on the battlefield has an associated knowledge matrix.* Both intelligence preparation of battlespace (IPB) knowledge and the capability of ISR sensors can be mapped onto this structure. The knowledge gain of fusion can also be shown in this same structure. In general, fusion will improve the accuracy and confidence of information (Level 1 of the Joint Directors of Laboratory (JDL) data fusion process model) and will also allow inference to capability and intent (Levels 2 and 3). The significance of this construct is that the information needs and the performance of C4ISR + fusion systems are expressed in the same matrix, which facilitates calculation of metrics that show the satisfaction of information needs.

A more detailed description of each cell in the knowledge matrix is available at www.dsc.osd.mil.

26.2.1 Database Analysis

The DSC database on information needs contains approximately 3500 specific questions or needs for the three missions evaluated. An analysis of this data provides quantification of the exact military information needs. Two types of analysis are presented here.

Figure 26.2 gives the distribution of the intelligence functions, see Table 26.4 for an explanation of the acronyms with respect to each of the missions. IPB is more important for dominant maneuver, while targeting and identification are strongly associated with precision engagement.

The distribution of actions and associated quality requirements are given in Figure 26.3. The figure shows the distribution of the information needs actions with respect to each mission. Note that *intent* is a question associated primarily with dominant maneuver, while *track* is primarily a precision engagement question.

TABLE 26.4 Description of Intelligence Functions

Function	Description
I&W	Indications and warning
IPB	Intelligence preparation of the battlespace
SD	Situation development
TGT	Targeting
FP	Force protection
BDA	Battle damage assessment

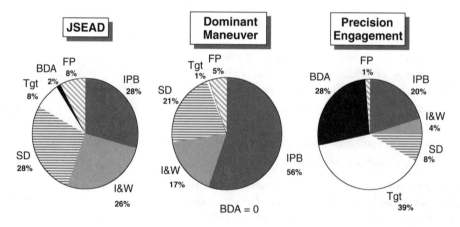

FIGURE 26.2 Distribution of information needs functions with respect to missions.

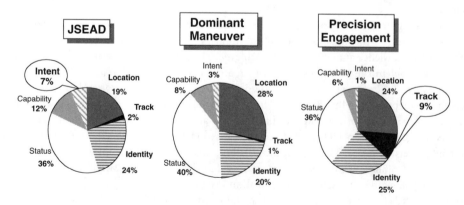

FIGURE 26.3 Distribution of information needs actions with respect to missions.

TABLE 26.5 Distribution of Actions to Support Operational Information Needs

Quality Requirement	Locate (%)	Track (%)	Identify (%)	Status/Act (%)	Capability (%)	Intent (%)
5	2	2	16	12	29	5
4	2	18	80	10	40	15
3	6	42	3	14	28	7
2	28	27	1	28	1	19
1	47	11	0	36	2	27
0	15	0	0	0	0	27
Totals	100	100	100	100	100	100

TABLE 26.6 Distribution of 3229 Information Needs (Actions vs. Functions)

Functions vs. Actions	Location	Track	Identity	Activity	Capability	Intent	Total
Bomb damage assessment	0	0	9	144	13	0	166
Force protection	17	1	12	59	7	1	97
Indications and warning	83	1	68	177	21	20	370
Intel prep. of the battlefield	218	0	324	391	237	100	1270
Situation development	228	58	210	211	30	27	764
Targeting	150	69	161	141	41	0	562
Totals	696	129	784	1123	349	148	3229

A second type of analysis involves the generation of statistics for each cell of the knowledge matrix. The total number of times an information need was stated for a particular action at a particular quality requirement was aggregated for comparison with other actions and quality requirements. Table 26.5 represents a summary of the actions and associated quality requirements of information needs for a composite of the three combat missions mentioned above. Table entries are in percentages. See Figure 26.1 for a description of the quality requirements. Note that Quality Level 0 does not mean "no information," it indicates "very poor information." The action "detect" is associated with a location QR of 0.

Examination indicates that the majority of the location information needs were for 1 km and some needs were for 5 meters. Remember that these numbers represent only the operational questions and do not reflect the total number of times in a warfighting scenario that the question was asked. Thus, although the number of 5-meter questions is small, it may have been asked many times and may be for very important targets. These dynamic effects are reflected in the evaluation phase by a detailed scenario that contains this data. With this proviso in mind, other observations can be stated: the most common tracking requirements are for general speed and direction. The numbers for identification (ID) indicate that operational information needs require a high level of ID (specify an object, such as an a T-72 tank). The highest level of ID in the knowledge matrix requires not only good physical ID, but also knowledge on the military hierarchy of the object. Requirements for status and activity do not reflect high quality requirements; the most frequently asked single question is for activity information to support Battle Damage Assessment (BDA). Capability information needs reflect higher quality requirements to know several to many factors that make up capability. Finally, intent requirements tend to be at the low end of the quality scale. This implies that intent be inferable from the perceived knowledge — users are not asking for a definitive statement on intent. This certainly does not mean that knowledge about intent is less important but that either (1) users prefer to infer this themselves (i.e., in the human mind), or (2) users may not be aware of the capability of current technology to provide automated support to intent estimation.

The analysis of the database is directed toward the generation of a metric to measure information needs satisfaction by ISR + fusion. The metric (see Section 26.4) is oriented around the intelligence functions and actions. Table 26.6 shows the distribution of 3229 individual information needs in the form to evaluate the metric. Individual needs for each function shown are aggregated over all quality levels and all three of the study missions. Analysis indicates the following:

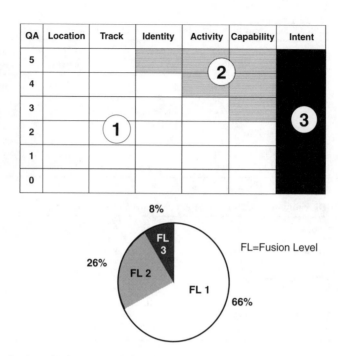

QA	Location	Track	Identity	Activity	Capability	Intent
5						
4				**2**		
3						**3**
2		**1**				
1						
0						

8%

26% FL 2 FL 3 FL 1 66%

FL=Fusion Level

FIGURE 26.4   Distribution of information needs with respect to fusion levels.

1. Targeting is driven primarily by location, tracking, and identification, but not intent. Level 1 fusion to support these actions must be near real time.
2. IPB is strongly driven by the actions' capabilities and intents. These are Levels 2-3 fusion products and need not be near real time, since IPB is a planning function.
3. BDA satisfaction is driven primarily by determining activity. This is a Levels 1-2 fusion function and supports a planning function.

The final analytic thread given here is the DSC mapping between the knowledge matrix structure and the three fusion Levels 1-3 in Figure 26.4. This mapping is subjective, but can be used to show the percentage of user information needs that are satisfied by the various fusion levels. The summary data indicates that about half of combat information needs are met with Level 1 fusion and that 8% require Level 3 fusion for satisfaction.

26.3   Key Concept

The knowledge matrix structure is essential to the process that measures the satisfaction of information needs. As shown in Figure 26.5, all parts of the problem can be expressed on the same matrix. The information needs are mapped to a specific set of cells in the matrix using the process suggested in Section 26.1. Mapped information needs are assumed to imply a 90% confidence requirement on the part of the user. Similarly, the performance of IPB and ISR can also be shown (also at 90% confidence) on the same matrix. Finally, the information gain resulting from fusion can be expressed on the knowledge matrix. Mathematically, all computations compute (i.e., estimate) the *probability of knowledge* at the implied action and quality level.

26.4   Evaluation Methodology

The evaluation process seeks to answer the question of the *value* of C4ISR in general and fusion in particular to the operational commander. This is accomplished by calculating the commander's information

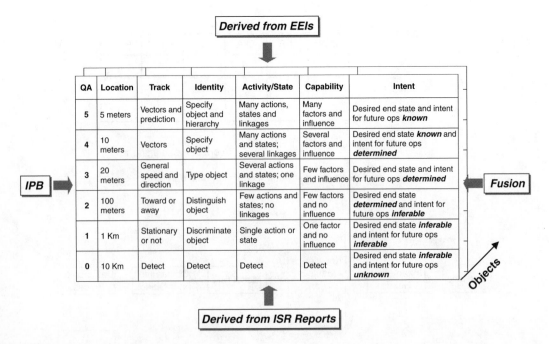

FIGURE 26.5 Key concept for evaluation.

needs satisfaction and then determining the value of this satisfaction to the military outcome. Only the first part of this process (needs satisfaction) is covered in this chapter. Steps in the process are

1. Choose an appropriate operational scenario. This should have hundreds to thousands of objects, each with one or multiple "signatures" that can be detected by ISR. This scenario is the basis for a Force-on-Force model that simulates the specified operational missions. The DSC uses a number of Force-on-Force models that have been validated on previous studies.
2. Map the appropriate operational information need to a set of knowledge matrices associated with that scenario object. Table 26.5 gives a cumulative count of this mapping for all objects included in a DSC scenario for operational missions.
3. Model the appropriate ISR suite. Prefusion ISR performance can be represented as a cell in the knowledge matrix at a 90% confidence level. Many ISR platforms perform multiple actions (e.g., a JSTARS will perform both tracking and some level of ID). In this case, entries are made in both action columns of the knowledge matrix.
4. Using ISR-generated observations, model fusion at all levels by predicting the performance of each fusion function. The DSC has created a fusion "process" model that shows the function of fusion to convert data to knowledge. This model calculates the *probability* that fusion will achieve knowledge at the appropriate quality level.[2] In the process model, Level 1 fusion acts to improve the accuracy or confidence for location, tracking, and identification actions. Levels 2 and 3 allow inference to activity, capability, and intent. See Figure 26.4 for the estimated mapping of fusion levels to knowledge.
5. Calculate an information needs "satisfaction" score for that object. The DSC fusion calculations in item 4 above represent probabilities of knowledge. The scoring uses the fact that the information need and the probability of satisfaction are expressed on the same knowledge matrix. This allows a linear summation calculation to obtain the needs satisfaction score.
6. Transform aggregate scores over the operational scenario into a metric.

The DSC metric was at a high enough level to provide senior decision makers information to decide on investment decisions. The format of the metric is shown in Figure 26.6 (numbers are representative

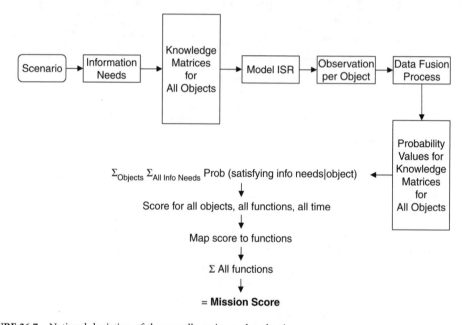

FIGURE 26.6 Form of the information needs satisfaction metric (notional numbers).

Mission	Function					
	I&W	IPB	SitA	FP	Tgt	BDA
JSEAD	.1	.2	.3	.4	.5	.6
Precision Engagement	.6	.5	.4	.3	.2	.1
Dominant Maneuver						

Score between 0 and 1

FIGURE 26.7 Notional depiction of the overall scoring and evaluation process.

only). Each score is a number between 0 and 1 that is proportional to a quantity measure. It is calculated as double sum of information needs satisfaction for a single object (in the scenario) and then over all the appropriate objects in the scenario. The summation over objects was carried out for each of the intelligence functions. Timeliness considerations are addressed implicitly in the scenario and warfighting calculations. Because the DSC scenarios have thousands of objects, averaging with Monte Carlo techniques is not necessary. Monte Carlo methods, however, would be appropriate for smaller scenarios having fewer objects.

The final metric represents the operational information needs satisfaction for a particular ISR configuration and a particular fusion configuration and processing topology. Other configurations and options yield a different score in the metric form. Comparison of the two (or more) metrics provides the required investment decision information.

Figure 26.7 shows a schematic of the overall process, in terms of the major steps leading to the computation of the Mission Score. The process involves objective, quantitative input and evaluations, and also a degree of subjective judgment. Part of the core of this process involves the knowledge matrices (see Figure 26.1) and, in turn, the probabilities assigned to the various cells of those matrices according to the capability of any given fusion process.

The total evaluation method outlined here is complex. It is intrinsically multi-INT and fusion itself is only a part of the process. In general, most aspects of the problem must be addressed simultaneously before an overview can be obtained. However, focusing on individual cells of the metric to show particular fusion aspects is also possible. For instance, one could focus on tracking to support precision engagement. The scenario could be abstracted to a smaller target set. Performing the evaluation across all INTs and all platforms is, however, essential. To do otherwise would generate a metric with too narrow of a focus.

The results of this process can be used to measure the value of fusion and more generally ISR. This type of analysis is required in the DoD to evaluate investment decisions between fusion processes and other aspects of military systems. Future developments involve using these techniques to directly evaluate information superiority in the context of realistic operational scenarios.

References

1. Multi-INT Fusion Performance Study, Joint C4ISR Decision Support Center study DSC-02-00, January 2001.
2. DSC Web site: www.dsc.osd.mil.

V

Resources

Web Sites and News Groups Related to Data Fusion..A-1
Data Fusion Web Sites • News Groups • Other World Wide Web Information •
Government Laboratories and Agencies

Web Sites and
News Groups Related
to Data Fusion

Data Fusion Web Sites...A-1
News Groups...A-3
Other World Wide Web Information...A-4
Government Laboratories and Agencies.....................................A-4

Data Fusion Web Sites

http://aeroweb.aero.csir.co.za/Iandmine/Sort%2Oby%2Osubject_1.html
http://asrt.cad.gatech edu/Mission-control/simon/paper1/intro.html
http://bacon.gmu.edu/c3i/resumes/buede.html
http://cma.cma.fr/PersonneI/Monget/cours/page34.htm
http://cmp.felk.cvut.cz/~kraus/Mirrors/iris.usc.edu/Vision-Notes/bibliography/compute456.html
http://cmp.felk.cvut.cz/~kraus/Mirrors/iris.usc.eduNision-Notes/bibliography/kwic/fus.html
http://cns-web.bu.edu/pub/dorman/sensor_fusion
http://concord.cscdc.be/meetings/transport_5th/presentations/ttcm5gen/tsldO23.htm
http://deis58.cineca.it/ernet/ernetbook/node111.html
http://dibemail.dibe.unige.it/department/imm/ISIP/arg4.html
http://ece-www.colorado.edu/~pao/bio.html
http://gregorysbooks.com/artech/electdef.htm
http://ic-www.arc.nasa.gov/ic/projects/condition-based-maintenance/bibliography.html
http://imagets10.univ.trieste.it/ipl/research.html
http://ispl.korea.ac.kr/~koppie/research/research.html
http://www.ll.mit.edu/Links/radsys.html
http://mphywww.tamu.edu/Gstudents/TongCV.html
http://roger.ucsd.edu/search/dMultisensor+data+fusion+%2D%2D+Congresses/-5,- 1/browse
http://sabre.afit.af.mil/MARION?S=MULTISENSOR+DATA+FUSION
http://schof.colorado.edu/~pao/research/fusion.html
http://scis.nova.edu/~dixonw/dcis791/deliva1.htm
http://sparky.mcmaster.ca/Technical_Reports/959493.html
http://srilanka.spie.org/web/abstracts/1600/1699.html
http://systeng.gmu.edu/InSERT/credits/PACHO~1. HTM
http://thorung.eeng.dcu.ie/~calcon/paul/Ann/lntro/intro7.html
http://uhavax.hartford.edu/disk$userdata/group/engineer/www/eefac.html
http://users.bart.nl/~lawick/projects.htm#fusion
http://vision.arc.nasa.gov/~al/hsr/fusion/97refs.html
http://watfast.uwaterloo.ca/~karray/newsl2.html
http://wisdom.arl.psu.edu/sigproc.html
http://www.aai.com/AAI/ARL/ARL.html
http://www.accurate-automation.com/team/team.htm

http://www.adelaide.edu.au/Calendar/maths/gcms_deg.htm
http://www.aptronix.com/
http://www.at.infowin.org/ACTS/lENM/CONCERTATION/MULTIMEDlA/events/spscal.htm
http://www.biggest.com/books/radar.htm
http://www.cdsp.neu.edu/info/faculty/shafai/publications.htm
http://www.clients.globalweb.co.uk/seo/Company/852673170963/-/
http://www.colorado.edu/CATECS/s98ecen.html
http://www.computer.org/conferen/proceed/8005abs.htm#E37E6
http://www.corptech.com/CompanyPages/S/101Z4K.cfm
http://www.crad.dnd.ca/program/inventory/cisys_e.html
http://www.crl.mcmaster.ca/cgi-bin/makerabs.pl?322
http://www.cs.kuleuven.ac.be/~scme/VOL_97/vol97nr27
http://www.cssip.edu.au/~mehmet/Tracking/Tracking.html
http://www.datafusion.cma.fr/conf/first/earsel_report.html
http://www.dc.teknowledge.com/teknowlogies.html
http://www.designlab.ukans.edu/res-oma.html
http://www.di.uoa.gr/~eusipco/call.html
http://www.dtic.mil/dtic/sbir/971af.html
http://www.ece.nwu.edu/~cdc/acc97/lgprog.txt
http://www.ee.psu.edu/faculty/hall/hall1.html
http://www.eecs.nwu.edu/research_programs/section-5.html
http://www.ei.org/spotlite/orderfrm/spot0087.htm
http://www.eleceng.adelaide.edu.au/Brochure/Theses.html
http://www.erim.org/lRIA/user-guide.html#sub45
http://www.fas.org/spp/military/docops/defense/97_dtos/jw_dtos/jw_dto_d.htm
http://www.gg.caltech.edu/~cs286ta/abstracts.html
http://www.ifp.uni-stuttgart.de/comm4/wg4_32/call.html
http://www.intellect-net.com/ablex/cs.htm
http://www.jhu-oei.com/r407.htm
http://www.king.ac.uk/geog/rgs-ibg/qmrgrss.htm
http://www.llnl.gov/sensor_technology/STR25.html
http://www.mitre.org/resources/centers/advanced_info/maybury/dckb/
http://www.nttc.edu/env/CMST/CMST_ch3.1.html
http://www.orincon.com/datasheet/multitrkdatdp.html
http://www.riam.kyushu-u.ac.jp/fracture/xavl 6.htm
http://www.site.gmu.edu/~plehner/rugs/DrChang/
http://www.spie.org/web/abstracts/2200/2232.html
http://147.163.15.8/workshop/Prog_Com.htm
http://bacon.gmu.edu/c3i/resumes/laskey.html
http://bert.cs.pitt.edu/Research/SSI.projects.html
http://bunny.cs.uiuc.edu/sigmod/jobs/mitre2-96.txt
http://cacfs.army.mil/bcs.html
http://decsai.ugr.es/ccma workshop/
http://diwww.epfl.ch/w3lami/team/lopez/cves.html
http://drogo.cselt.stet.it/fipa/yorktown/nyws006.htm
http://ece.wpi.edu/~nicolett/WECE/bu_fellow.html
http://esdis.gsfc.nasa.gov/esdis_proto/ETTW_FEB/CAMPBELL/INDEX.HTM
http://ftp.esf.org/pn/pn_002b.htm
http://gauss.nmsu.edu:8000/research.html
http://hermes.intelgate.com/fusafis.htm

http://hpdma2.math.unipa.it/Digesu/MacVis.html
http://iridium.nttc.edu/law/shows/fusion.txt
http://isse.gmu.edu/faculty/kersch/KRG/KR_DARPA_folder/KLAD.html

News Groups

comp.ai.neural.nets
comp.ai
sci.stat.math
sci.math.stat
comp.soft-sys.matlab
sci.image.processing
comp.ai.fuzzy
comp.ai.genetic
comp.ai.vision

IEEE Neural Networks Council
    http://www.ieee.org/ncc/index.html
Center for Neural Basis Cognition
    http://www.cs.cmu.edu/afs/cs/project/cnbc/CNBC.html
CMU Al Repository
    http://www.cs.cmu.edu/Web/Broupd/html/repository.html
MIT Al Lab
    http://www.ai.edu/
SPIE-International Society for Optical Engineering
    http://www.spie.org/
UCl Machine Learning Group
    http://www.icsi.uci.edu/AI/ML/Machine-Learning.html
UCSD Neuroscience
    http://salk.edu/NeuroWeb/
Genetic Programming
    http://www.salford.ac.uk/docs/eee/genetic.html
Artificial Intelligence Subject Index
    http://ai.iit.nrc.ca/misc.html
Classification Society of North America
    http://www.pitt.edu/~csnal
StatLib
    http://lib.stat.cmu.edu/
Statistical Software and Datasets
    http://www.nova.edu/lnter-Links/software/stats.html
Decision/Risk Analysis
    http://www.lumina.com/DA/
Decision Theory Group
    http://www.research.microsoft.com/research/dtg/
Pattern Recognition Bibliographies
    http://www.ph.tn.tudelft.nl/PRlnfo/biblio.html
Genetic Algorithms and Neural Networks
    http://www.cs.iastate.edu/~gannadm/homepage.html
TNO Physics and Electronics Laboratory
    http://www.tno.nI

Wavelet Signal Processing
   http://www.amara.com/current/wavelet.html
   http://www.math.sc.edu/~wavelet
Penn State ARL Toolkit
   http://www.wisdom.arl.psu.edu
Command and Control Discussion
   http://sm.ups.navy.mil/JW1D95/

Other World Wide Web Information

http://www.1d.com
http://www.afji.com/Mags/1996/Aug/Option.html
http://yorktown.dc.isx.com/
http://www.erg.sri.com/projects/badd
http://www.itsi.disa.mil/jta/
http://spider.ims.disa.mil/dii/index.html
http://jtfweb3.nosc.mil/
http://www.isi.edu/asd/gumps

Government Laboratories and Agencies

Rome Laboratory
   http://www.rl.af.mil
   http://www.rl.af.mil:800/index.html
Army CECOM
   http:llwww. monmouth.army.mil/cecom/cecom. html/
ARPA
   http://www.arpa.mil/
AFOSR
   http://web.fie.com/wed/fed/afr/afrindex.htm
Army Corps of Engineers
   http://www.usace.mil/usace-labs.html
Army Research Laboratory
   http://infro.arl.army.mil/
CRISP (Computer Retrieval of Information on Scientific Projects)
   gopher://gopher.nih.gov:70/11/res/crisp
CRIS (Current Research Information System)
   http://cristel.nal.usda.gov:8080
NASA
   http://www.gsfc.nasa.gov/NASA_homepage.html
Naval Research Laboratory
   http://www.nrl.navy.mil/
NSF
   http://www.nsf.gov/
ONR
   http://www.onr.navy.mil/
   http://www.fie.com
Locators (white and yellow pages; e-mail, etc.)
   http://www.excite.com/Reference/
Community of Science Funding (Opportunities Search Service)
   http://best.gdb.org/repos/fund/

Index

A

Adaptive ML-PDA, **8**-32
Advanced Sensor Exploitation Testbed (ASET), **20**-10
Advanced Tomahawk Weapons Control System (STWCS), **22**-2
Air Force Office of Scientific Research (AFOSR), **11**-16
Air Force Research Laboratory, **5**-10
Air Force Tactical Exploitation Capabilities (AFTENCAP) program, **17**-2
Air-to-air defense, **1**-3
Air-traffic control, **3**-1
Algorithms
  classic, **5**-6
  cluster, **3**-8
  complexity, **11**-14
  covariance intersection, **12**-5
  data association, **9**-18
  elitist, **5**-6
  genetic, **5**-4, **5**-5
  hill-climbing, **5**-4
  Lagrangian relaxation, **11**-12
  simulated annealing, **5**-4, **5**-5
  tabu search, **5**-5
All Source Analysis System (ASAS), **22**-2
Almost-Parallel Worlds Principle (APWOP), **14**-9
Antisubmarine Warfare (ASW), **19**-2
Architecture, **2**-4
  open layered, **16**-4
Artificial neural systems (ANS), **6**-19
Artificial neural networks (ANN), **21**-5
Assignment problem, **17**-11
  multidimensional, **11**-3
  N-dimensional, **11**-7
Association matrix, **9**-6
Association process, **2**-3
Automated reasoning, **24**-4
Automatic target
  cueing (ATC), **4**-3

  recognition (ATR), **4**-3, **4**-5, **4**-7, **4**-9, **22**-1, **25**-8
  ATR architectures, **4**-9
Autonomous vehicles, **12**-21
Axiomatic definition of probability, **7**-31

B

Battle Damage Assessment (BDA), **26**-5
Battlefield intelligence, **1**-4
Battlefield surveillance, **22**-1
Bayes
  filtering, **10**-3, **14**-10
  optimal data fusion, **7**-7, **14**-3
  theory/theory, **4**-6, **6**-5, **7**-6, **7**-11, **10**-2, **12**-4, **14**-4, **23**-6
Beamforming, **8**-15
Bearing measurement noise, **8**-10
Best fit, **7**-11
Biological fusion metaphor, **3**-7, **3**-9, **3**-10, **3**-13
Bit mapping, simple, **6**-1
Boundaries, **4**-6
Boyd's Observe, Orient, Decide and Act. *See* OODA.

C

Canonical node, **12**-9
Canonical problem-solving, **6**-18
CAPSOL, **19**-5
Cartesian coordinate frame, **13**-2, **13**-8
CECOM, **6**-25, **18**-25
Chi-square criterion, **17**-9
CI filter, **12**-13
Color normalization (CN), **4**-10
Command and Control (C2), **6**-2
Commensurate data, **1**-5, **6**-7
Communication intelligence (COMINT), **6**-13, **6**-14
Computer-Based Training (CBT), **19**-4, **19**-5
Concept of Operations (CONOPS), **16**-17
Condition-based maintenance, **1**-5, **23**-1
Coordinate selection, **8**-18

Covariance
    ellipse, **12**-6
    intersection (CI), **12**-4, **12**-5, **13**-1
Cramer-Rao lower bound, **8**-13, **14**-23

D

Data alignment, **16**-6, **16**-30
Data association, **16**-31, **23**-19
    algorithm, **9**-18
    hypothesis, **10**-13
    large-scale, **9**-15
Database design criteria, **18**-6
Database Management Systems (DBMS), **18**-1, **21**-7, **21**-9
Data flow, **2**-9
Data fusion
    applications, **1**-3
    application taxonomy, **4**-4
    Bayes-optimal, **7**-7
    defined, **2**-3
    engineering methodology, **16**-7
    image, **4**-1
    multilevel, **13**-19
    network, **12**-9
    node paradigm, **16**-6
    spatial, **4**-1, **4**-11, **4**-12, **4**-13
    system engineering, **16**-12
    track-to-track, **12**-21
Data Fusion Information Access Center (FUSIAC), **21**-9, **21**-10
Data registration, **5**-1
DBMS. *See* Database management systems
Decentralized data fusion, **12**-2
Decision-level fusion, **4**-8
Declarative knowledge class
    short-term, **6**-16
    medium-term, **6**-16
    long-term, **6**-16
    procedural, **6**-16
Defense Advanced Research Projects Agency (DARPA), **4**-2, **4**-12, **5**-10, **20**-11, **21**-9, **25**-1, **25**-8
Deliberate synesthesia, **19**-9
Dempster-Shafer method, **6**-5, **7**-1
Dempster-Shafer theory of evidence, **14**-2, **14**-10, **14**-27
Dempster's rules of combination, **4**-6
Diffusion network routing, **25**-7
Digital elevation models (DEMs), **4**-10, **4**-11
Distributed Common Ground Station (DCGS), **22**-5
Distributed data fusion network, **12**-2
Dummy report, **11**-5
Dynamic Multi-User Information Fusion (DMIF), **16**-14

E

Earth observing systems and information system (EOS-IS), **4**-13
Efficiency
    association, **18**-9
    complex query, **18**-9
    implementation, **18**-9
    overhead, **18**-9
    search, **18**-9
Electrochemical systems, **23**-23
Electronic countermeasures (ECMs), **8**-17, **8**-27
Electronic intelligence (ELINT), **1**-3, **6**-7, **6**-8, **6**-14
Electronic support measures (ESMs), **1**-3
Electro-optical (EO) imaging sensors, **4**-3, **8**-28
Enterprise Architecture Approach, **15**-5
    operational architecture (OA), **15**-5
    systems architecture (SA), **15**-5
    technical architecture (TA), **15**-5
Environmental monitoring, **1**-5
Environments, Open System (OSE), **16**-3
Equations
    Fokker-Planck (FPE), **14**-3, **14**-6
    prediction, **8**-5
    Ryser's, **9**-10
Error
    boxes, **3**-11
    ellipsoids, **3**-11, **3**-12, **3**-15
Estimation
    identity, **2**-4
    problems, **2**-3
    state, **2**-2, **2**-14, **8**-4, **23**-19
Euclidean distance, **18**-10
Evaluation methodology for fusion processes, **26**-1
Evaluation, Performance, **16**-18
Evidence combination, **6**-6
Expert systems, **13**-20, **16**-36, **23**-22

F

Failure Mode Effects, Criticality Analyses (FMECA), **23**-4
Feature extraction, **23**-19
Feature-level fusion, **4**-7, **23**-4, **23**-13
Fiddle factor, **9**-7
Filter (equations)
    nearest neighbor (NNF), **8**-1
    strongest neighbor (SNF), **8**-1
Filtering, multiple model, **12**-21
Finite-set statistics (FISST), **14**-2, **14**-7
Fitness function, **5**-6
FLIR. *See* Forward-looking infrared data
FOM. *See* Figures of merit
Fokker-Planck equation (FPE), **14**-3, **14**-6
Forward-looking infrared data, **1**-2, **4**-3, **4**-7, **6**-8, **6**-14
Frames of data, **11**-4
Function
    belief-mass, **14**-15
    credibility, **7**-12
    likelihood, **10**-4, **10**-6, **10**-7, **10**-9, **14**-11
    plausibility, **7**-12
Functional model, **2**-15
Fusion
    classes, **6**-16
    tree design, **16**-21, **16**-24, **16**-27
    post-detection, **4**-8
    process models, **6**-9
Fuzzy logic, **7**-1, **14**-2, **14**-27, **23**-22
    membership functions, **7**-11, **18**-23

sets, 7-10
set theory, 4-6
variables, 7-1

G

Gating
 criteria, 3-6
 fine, 11-10
 strategies/techniques, 3-7, 17-6
Gauss filter, second order, 13-7
Gaussian random process, 6-9
Geodesy, 4-2
Geographic information systems (GIS), 4-12
Global Command and Control System (GCCS), 22-5
Global Positioning System (GPS), 25-5
Global state estimation problems. *See* estimation.
Goal-driven problem solving, 16-5
Greedy Randomized Adaptive Local Search Procedure
 (GRASP), 11-10

H

Hard-decision
 M-of-N, 4-6
 weighted sum score, 4-6
HCI. *See* Human/computer interface
HEARSAY, 6-20
Heuristic,
 meta-, 5-5
High performance computing/networking, 24-5
High Range Resolution Radar (HRRR), 14-4
Human-centered computing, 24-5
Human-Computer Interface (HCI), 19-1, 21-6, 21-8
Hypothesis
 data association, 10-13
 evaluation, 16-6, 16-32, 17-4, 17-8
 generation, 16-6, 16-32, 17-4
 scan association, 10-14
 selection, 16-6, 16-32, 16-33, 17-4

I

Identification-Friend-Foe-Neutral (IFFN), 19-1, 22-1
Image intelligence (IMINT), 4-14, 7-19, 7-20, 22-2
Image registration, 5-3
Imagery
 dynamic, 4-10
 multiresolution, 4-10
 three-dimensional, 4-10
Information integration, 2-4
Information reduction factor, 8-14
Information security, 25-8
Informational states, 2-12
Infrared identification-friend-foe (IFF), 1-3
Infrared sensors (IR), 4-8, 22-2
Intelligence cycle, 2-17
Intelligence preparation of the battlefield (IPB), 4-3, 7-19,
 26-3

Interacting Multiple Model (IMM) estimator, 8-2, 8-18
IR. *See* Infrared sensors

J

Joint Assignment Matrix (JAM), 9-2, 9-6
 approximation methods, 9-14
 efficient computation, 9-8
 permanent approximations, 9-14
Joint Directors of Laboratories (JDL), 1-6
 data fusion working group, 1-6
JDL data fusion lexicon, 2-3
JDL process model, 1-6, 1-8, 2-1, 2-4, 2-5, 2-17, 4-5, 15-2,
 19-2, 21-2
 Level 1, 1-8, 11-1, 19-2, 20-5, 21-3, 21-8, 22-4, 26-3
 Level 2, 1-8, 1-9, 20-5, 21-5, 21-8, 22-2, 22-4, 26-3
 Level 3, 1-8, 1-9, 20-5, 21-6, 22-2, 26-3
 Level 4, 1-8, 1-9, 21-6, 21-9, 22-2
JDL revised process model/Hall, 19-10
 Level 5, 19-10
JDL revised process model Steinberg, 2-6, 2-13
 Level 0, 2-6, 2-7, 13-20, 26-5
 Level 1, 2-6, 2-7, 13-20
 Level 2, 2-6, 2-7, 13-20
 Level 3, 2-6, 2-7
 Level 4, 2-6, 2-7
Joint national intelligence development staff (JNIDS), 4-2
Joint probabilistic data association (JPDA), 8-2, 11-2, 21-4
Joint Surveillance Target Attack Radar System (JSTARS),
 25-4, 26-7
Jurkat-Ryser
 inequality, 9-13
 upper bound, 9-11

K

Kalman filter, 1-5, 3-4, 3-6, 3-14, 6-11, 8-23, 10-2, 10-6,
 12-2, 12-3, 12-4, 12-13, 13-2, 13-11, 16-36
 extended, 5-4, 8-18, 13-1, 16-36
KD-trees, 3-7, 3-9, 3-10
 layered box tree, 3-14
 priority, 3-13, 3-14
Knowledge
 behavioral, 6-11
 context, 6-5, 6-10
 non-sensor-derived, 6-5
Knowledge-based methods, 17-6

L

Lagrangian relaxation methods, 11-10, 11-12, 11-15
LANDSAT satellite, 1-4, 4-2, 4-14
Light amplification for detection and ranging (LADAR),
 4-3, 4-7, 4-8
Likelihood ratio detection, 10-23, 10-25, 10-27
Line of bearing, 10-7
Log-likelihood measure, 3-6
Low Observable (LO) targets, 8-27
Lubrication system test bench, 23-18

M

Magnetic resonance imaging (MRI), **5**-2
Mahalanobis distance, **3**-6
Mapping, charting and geodesy (MC&G), **4**-12
Markov densities, **14**-20
MATLAB source code, **12**-23
Maximum likelihood (ML) estimator, **8**-11, **8**-13, **8**-29, **8**-31, **8**-32, **17**-9
Measure of effectiveness (MOE), **4**-8, **16**-27, **20**-2, **20**-12, **21**-6
Measure of force effectiveness (MOFE), **4**-7, **20**-2, **20**-12
Measure of performance (MOP), **20**-2, **20**-12, **21**-6
Measurement and signature intelligence (MASINT), **6**-2, **7**-19, **22**-2
Measures of Detection Processing Performance (MODPP), **20**-12
Mechanical diagnostic test bed (MDTB), **1**-6, **23**-9
Medical
  diagnosis, **1**-5
  imaging, **4**-2
Memory
  long-term, **6**-14
  medium-term, **6**-14
  short-term, **6**-14
Metaphor
  biological fusion, **6**-1
  command and control (C2), **6**-2
  puzzle-solving, **6**-3
Metasensor processing, **23**-21
Millimeter wave (MMW), **4**-3, **4**-7, **4**-8, **4**-9
Minimum Mean Squared Error (MMSE), **13**-2
Mobile code, **25**-7
Model
  benign, **8**-25
  diffuse prior, **8**-6
  maneuver, **8**-25
  maneuver detection, **8**-25
  motion, **14**-12
  multitarget sensor, **14**-12
  Poisson, **8**-6
  syntax-based, **17**-6
MOE. *See* Measure of effectiveness
MOFE. *See* Measure of force effectiveness
Monte Carlo simulation, **8**-16, **12**-11, **13**-5, **13**-8, **13**-11, **13**-15, **20**-7
MOP. *See* Measure of performance
Moving target indication (MTI), **20**-14, **22**-2
Multiple frame assignments, **11**-16
Multiple hypothesis, **3**-5, **9**-4
Multiple hypothesis tracker (MHT), **8**-18, **10**-11, **10**-12, **10**-16, **10**-17, **10**-19, **10**-22, **11**-2, **21**-4
Multiple pane window, **11**-9
Multisensor fusion toolkit, **23**-7
Multiresolution methods, **5**-4
Multisource image processing system (MuSIP), **4**-13
Multitarget
  likelihood functions, **10**-9
  motion model, **10**-8
  process, **10**-9

N

Narrowband passive sonar tracking, **8**-8
Narrowband sonar, **8**-14
National aeronautics and space administration (NASA), **4**-2, **24**-1
NASA applications
  air traffic management, **24**-3
  earth observation, **24**-2
  manned space exploration, **24**-2
  unmanned space exploration, **24**-2
National imagery and mapping agency (NIMA), **4**-3
National spatial data infrastructure (NSDI), **4**-12
NCSA CAVE, 3-D full immersion imaging, **19**-3
Nearest neighbor
  global, **11**-2
  rule, **3**-4, **3**-5
Negative reasoning, **19**-10
Network Centric Warfare (NCW), **12**-1
Neural networks (NN), **13**-6, **23**-22, **23**-28, **23**-29
Noncommensurate data, **1**-5, **2**-2, **6**-7
Normalized Estimation Error Squared (NEES), **8**-17

O

Object-Oriented Database Management Systems (OODMS), **18**-3
Observation processes, **7**-5
Ocean surveillance, **1**-3, **1**-4
Office of Naval Research (ONR), **12**-21
Order of battle, **1**-3
OODA, **2**-5, **2**-17
  control loops, **16**-16
  fractal loops, **2**-18
Open System Environment (OSE), **16**-3

P

Paradigm, dual, **16**-10
Paradox
  Bertrand's, **7**-3
  Hughes', **7**-3, **7**-4
Pattern morphing methods, **19**-10
Pattern recognition, **17**-6
PDA. *See* Probabilistic data association
Perceptual states, **2**-12, **2**-13
Phased array radar, **11**-4
Processor Independent Correlation and Exploitation System (PICES), **22**-2
Pixel level fusion, **4**-6
Polar coordinates, **13**-8
Possibility
  measure, **7**-9
  theory, **7**-1, **7**-8
Predict, extract, match and search (PEMS), **2**-5
Probabilistic data association, **8**-2, **8**-3, **8**-6, **8**-21, **8**-29, **8**-31
  nonparametric, **8**-7
  parametric, **8**-7
Probability
  combination, **7**-15
    belief, **7**-16

possibility, 7-16
detection, 8-8
false alarm, 8-8
theory, 7-2
Problem-solving approaches, 16-5
goal-driven, 16-5
Processing
feature-level fusion, 1-6
decision-level fusion, 1-6
Processing approach
architectures, 1-5
Project correlation, 16-1, 17-1
Puzzle-solving metaphor, 6-3

R

Radar, phased array, 11-4
Random variables, 7-5
finite set, 14-9
variables, 7-5
vector, 14-9
Ratio
likelihood, 11-7
log-likelihood, 10-27
RCS. *See* Radar cross-section
Reasoning, 4-4, 16-36
Recursion
general MHT, 10-17
likelihood ration, 10-25
simplified, 10-26
Registration, 4-4, 5-2, 5-7, 5-9
Relational Database Management System (RDBMS), 18-2
Relational data model, 18-2
Relational states, 2-11
Remote sensing, 1-4
Requirements
analysis, 15-2
derivation, 15-1
Resampling, 4-11
Resource management, 16-2
problem, 16-11
trees, 16-4
Response
belief, 7-14
possibility, 7-14
probability, 7-13
Robotics, 1-5, 4-2, 5-2
Role of data fusion, 2-1
Root-Mean-Square Errors (RMSE), 8-26
Rough sets (Rgh), 17-9
Ryser's equations, 9-10

S

SA-6 Battery, 2-10
Scene understanding, 4-13
Scripts/frames, 16-36
SDI Surveillance Test Bed, 20-9

Semantic distance, 16-36
Sensor Information Technology (SenseIT), 25-1, 25-3
Sensor models, 14-11
Sensor resource management, 6-4
Sensors
acoustic, 6-14
ELINT, 6-10
optical, 6-14
tactile, 4-3
Set derivative, 14-16
Set integral, 14-16, 14-17
Signal intelligence (SIGINT), 4-14, 6-7, 7-19, 7-20, 22-2
Signal-to-Noise Ratio (SNR), 8-8, 8-16
Single Channel Ground and Airborne Radio System (SINCGARS), 25-3
Single Integrated Air Picture (SIAP), 11-16
Situation assessment, 14-2, 19-1
Situation awareness, 2-8, 2-9, 6-2, 6-4
Situation understanding, 6-24
Sliding windows, 11-15
Soft decision, 4-6
Spatial data structure, 18-11
SPOT system, 1-4, 4-2, 4-10
Stand-off Jammer (SOJ), 8-18
State and covariance update, 8-5
State-of-the-art, 1-8
Strategic defense initiative (SDI), 3-1, 3-15
national testbed, 3-15
scenario, 3-16
Suppression of Enemy Air Defenses (SEAD), 26-3
Surface-to-air defense, 1-3
SYMBOLICS computer, 1-1
Synthetic aperture radar (SAR), 1-3, 4-10, 4-14, 4-15, 6-13, 6-14, 20-11

T

Tactical erectable missile launcher (TEL), 6-12, 6-13
Target state, 10-3
TEC (U.S. Army topographic engineering center), 4-3
Templating, logical, 17-6
Ternary trees, 3-10, 3-12
Test and Evaluation (T&E), 20-3, 20-4, 20-5
Threat assessment, 2-4, 19-1
Track Motion Analysis (TMA), 8-2, 8-8
Track
assignment, 6-9
association, 6-9
formation, 8-19
initiation, 9-4, 11-8
maintenance, 8-20, 11-8
splitting, 3-5, 9-4
termination, 8-16
Track-before-detect, 10-30
Tracking
multihypothesis (MHT), 8-2, 8-18, 10-11, 10-12, 10-6, 10-17, 10-19, 10-22, 11-2
multitarget, 3-1
single target, 3-2, 3-3

Transformation
    affine, **5**-4, **5**-5
    nonaffine rubber sheet, **5**-4
    rubber sheet, **5**-4, **5**-5
    spherical-to-Cartesian, **8**-19
    wavelet, **5**-4, **5**-7
TRUST, **5**-5, **5**-7

U

Ultraviolet (UV), **4**-9
Unattended Ground Sensors (UGS), **25**-4
Uncertainty, **2**-2
Unified tracking recursion, **10**-11, **10**-12
Unmanned Aerial Vehicles (UAV), **25**-4

Unscented
    transformation (UT), **13**-1, **13**-2, **13**-5, **13**-7
    filter (UF), **13**-12
U.S. Army topographic engineering center (TEC), **4**-3

V

Virutal sensor data, **23**-26
Voting, **23**-5

W

Wavelet-based registration, **5**-7
Waterfall model, **2**-17
Weighted decision fusion, **23**-5